Often-Used Tables and Figures (Abridged Titles) (Except for those cited under Often-used Formulas)

Spectrochemical Series (page 71):

$I^- < Br^- < Cl^- < SCN^- < NO_3^- < F^- < OH^- < C_2O_4^{2-} < H_2O < NCS^-$
$< gly < C_5H_5N < NH_3 < en < NO_2^- < PPh_3 < CN^- < CO$

INTRODUCTION TO COORDINATION, SOLID STATE, AND DESCRIPTIVE INORGANIC CHEMISTRY

INTRODUCTION TO COORDINATION, SOLID STATE, AND DESCRIPTIVE INORGANIC CHEMISTRY

Glen E. Rodgers

Allegheny College

Boston, Massachusetts Burr Ridge, Illinois Dubuque, Iowa
Madison, Wisconsin New York, New York San Francisco, California St. Louis, Missouri

WCB/McGraw-Hill

A Division of The ***McGraw-Hill*** *Companies*

This book was set in Times Roman by Science Typographers, Inc.
The editors were Jennifer B. Speer and James W. Bradley;
the production supervisor was Kathryn Porzio.
The cover was designed by Glen E. Rodgers and Rafael Hernandez.
R. R. Donnelley & Sons Company was printer and binder.

INTRODUCTION TO COORDINATION, SOLID STATE, AND DESCRIPTIVE INORGANIC CHEMISTRY

This book is printed on acid-free paper.

3 4 5 6 7 8 9 0 DOW DOW 9 0 9 8 7

ISBN 0-07-053384-9

Library of Congress Cataloging-in-Publication Data

Rodgers, Glen E., (date).
Introduction to coordination, solid state, and descriptive inorganic chemistry/Glen E. Rodgers.
p. cm.
Includes index.
ISBN 0-07-053384-9
1. Coordination compounds. 2. Solid state chemistry.
3. Chemistry, Inorganic. I. Title.
QD474.R63 1994
546–dc20 93-15952

ABOUT THE AUTHOR

Glen E. Rodgers is Professor of Chemistry at Allegheny College in Meadville, Pennsylvania. Educated at Tufts University (B.S., 1966) and Cornell University (Ph.D., 1971), he taught for five years at Muskingum College before moving to his present position in 1975. He has taught introductory chemistry on several levels, chemistry for nurses, chemistry for nonscience majors, inorganic chemistry (on both the sophomore and advanced undergraduate levels), and numerous interdisciplinary courses with colleagues in history, education, english, philosophy, and psychology. He is the recipient of the 1993 Julian Ross Award, presented by Allegheny College "for singular accomplishments and contributions through excellence in teaching." His research interests include the synthesis and characterization of complexes of ferrocenyl ligands and the development of inorganic laboratory modules for use in integrated introductory laboratories. He lives with his wife Kathleen and their daughters Jennifer, Emily, and Rebecca in Meadville, Pennsylvania.

DEDICATED TO

Alger S. Bourn

MY HIGH SCHOOL CHEMISTRY TEACHER
WHO IGNITED A CONTINUING CURIOSITY

CONTENTS

Part B Solid State Chemistry

Part C Descriptive Chemistry of the Representative Elements

PREFACE

This brief and streamlined introduction to coordination, solid state, and descriptive main group inorganic chemistry is designed for the student who has completed a standard introductory course. It actively relies on the material mastered through the great efforts of teachers and students alike in the lectures, recitations, review sessions, late-night study groups, in- and out-of-office-hours discussions, and textbooks typically encountered in such courses.

The primary goal of this book is not to lay out the latest results in inorganic chemistry but rather to present a significant portion of this subdiscipline to new students in new ways. Designed and written *for students* rather than faculty (who, after all, already know most of this material), this book includes detailed physical and chemical explanations tailored especially for the sophomore student audience of the type that I have been enjoying for more than 20 years. I hope that these readers, like most of my own students, will find its conversational prose easy and enjoyable to read and understand. I further hope that these readers will soon come to realize that behind these paragraphs and chapters is a live human being who loves the challenge, lore, and relevance of this discipline called inorganic chemistry and who has tried, with some degree of success, to demonstrate to his readers why it is that he finds his chosen discipline so fascinating.

While making a significant portion of inorganic chemistry accessible, the book's second goal is to encourage students to organize material in their minds and not merely memorize a group of facts or trends. To the extent that this goal is reached, students will start to integrate the ideas presented here with the concepts established in their introductory courses in chemistry. One way to foster this integration is to ask questions within the body of the text. In this way, students are encouraged to become involved in thinking their way through the material. Another way to foster this integration and mastery is to present a number of end-of-chapter problems of varying difficulty. Many of the more than 900 problems included in this text are modeled after the questions I ask on my own quizzes and examinations. As such, they attempt to challenge students to apply what they have learned to new situations and thereby build a better and deeper understanding of the material. In addition, a fair number of the problems ask for short explanatory paragraphs, thereby emphasizing the importance of writing in the discipline of chemistry. Problems preceded by an asterisk are the ones that, in my experience seem to pose the greater challenges to students.

A third goal of the book is to bring an appropriate historical perspective to the material. American students, more so than those of other nations, seem to lack an appreciation of the centuries of human endeavor that have led to our present understanding of the chemical world around us. This book tries, in a modest way, to give an appreciation of how inorganic chemistry developed and who developed it. Moreover, the historical content is not presented just as an appendix but is quite often used as a foundation for content later in a chapter and in the problems.

A fourth goal of the text is to describe appropriate and attractive applications of inorganic chemistry. This practical perspective to the material leads to discussions of such diverse subjects as heavy metal poisoning and antidotes, antitumor chelating agents, the hydrogen economy, nuclear fusion, radiochemical chronometric techniques, the greenhouse effect, the threat to the stratospheric ozone layer, hard water, fireworks, ion-exchange materials, battery technology, fluoridation, and radon as a carcinogen.

As implied earlier, the material presented here does not rely on any knowledge obtained in a standard organic or physical chemistry course and therefore can be used by those students who are taking their first inorganic course *before* taking these traditional courses. (It is hoped that students who *have* taken these courses will find this book even more enlightening as they relate the material to that which they have already mastered.)

The special challenge of inorganic chemistry is its staggering diversity. The challenge of an author of an inorganic chemistry text is to decide what to include in a book that introduces this wide-ranging subdiscipline to students. I have, in almost all cases, decided that when it comes to the material to be included in these pages, "less is better than more." It is hoped that this philosophy will not discourage the adoption of this book but rather will be seen as an opportunity to build and present concepts and applications which the instructor finds particularly important and fascinating. Indeed, students reading this book should probably get used to the idea that many instructors will want to amplify and build upon the material presented here.

Consistent with the philosophy that less is better than more, this book does not contain the traditional chapters reviewing or expanding on atomic and molecular structure and other topics. Rather, it takes the view that these topics are developed adequately in most introductory courses and textbooks and need not be dwelt upon here. Also consistent with a *second-year* text, this book does not contain all the common pedagogical devices found in introductory books. (Chapter objectives, however, are set out in the accompanying Instructor's Manual in a format in which they can readily be made available to students. End-of-chapter summaries are provided, but in a narrative rather than an outline format. These summaries are also designed to serve as comprehensive chapter introductions and/or overviews.)

An important organizational feature to note is that the three main topic areas (coordination, solid state, and main-group descriptive inorganic chemistry) are presented in *self-standing* sections (Chapters 2 to 6, 7 and 8, and 9 to 19, respectively), written specifically so that a student can open the book to the beginning of any one of them and read without becoming frustrated at references to prior material. This feature also gives the instructor the opportunity to determine in what order the subjects will be presented or to cover only one or two of the topics rather than all three.

The five-chapter section on coordination chemistry covers history and nomenclature, structure, bonding theories (but not molecular orbital), rates and mechanisms, and applications. The two-chapter section on solid state chemistry is divided into structures and energetics.

The 11-chapter main-group descriptive section attempts to systematically construct a "network of interconnected ideas" that hopefully builds reader confidence and forges a deeper understanding of the periodic table. The network is introduced in Chapter 9 and in parts of Chapters 11 and 12. Hydrogen and oxygen chemistry are covered in Chapters 10 and 11, respectively. The eight chapters on the representative groups each include sections on (1) the history and discovery of the elements, (2) their fundamental properties as they relate to the network (including an overview of the hydrides, oxides, hydroxides and/or oxoacids, and halides of the group), (3) reactions and compounds of practical importance, and (4) topics of particular interest in a given group. Each of these eight chapters ends with a Selected Topics in Depth section that provides an opportunity to look at one topic in somewhat greater depth than normal.

ACKNOWLEDGMENTS AND KUDOS

Writing even a modest textbook of this type is the culmination of years of thinking, talking, criticizing, arguing, approaching problems in, and writing about chemistry. Through this process, every chemist develops his or her own unique interconnected network of ideas with which we approach the world of chemistry. Every chemist has special colleagues, mentors, and students who have had a larger influence on the construction of his or her individual networks than others. While not all these people can be mentioned here by name, I would be remiss if I did not cite the following for special recognition:

Alger Bourn, my high school chemistry teacher, who first excited me about chemistry;

Muriel Kendrick, who somehow browbeat into me the basics of grammar and sentence structure;

Robert Eddy at Tufts College, who infected me with the enthusiasm for chemistry as one of the liberal arts;

Mike Sienko and Bob Plane, at Cornell University, who showed their enthusiasm not only for excellence in research but in teaching as well;

Buddies in Plane's research group (particularly Dennis Strommen and Jol Sprowles) for constant conversations about the intricacies of chemistry and other far-ranging topics;

Rudy Gerlach at Muskingum College, my first teaching position, for many wonderful and detailed class postmortems and numerous other discussions about teaching, student viewpoints, and approaches to explanations of difficult concepts;

At Allegheny College: Richard Bivens for his continuing support when others would have discouraged the pursuit of such a project; Professors Bivens and Ann Sheffield, who read and commented freely on selected sections; in the English department, Paul Zolbrod, my colleague in a wonderful interdisciplinary experience, who always showed enthusiasm and was encouraging even in the darkest of days;

At the University of British Columbia: Brian James, Bill Cullen, and other members of the faculty, who gave this visiting professor of chemistry the opportunity to renew his knowledge of and enthusiasm for inorganic chemistry.

A textbook written specifically for students requires constant student input. I cannot adequately express my appreciation to the many students who carried around the mammoth early versions of the book and put up with the errors, unclear points, and other numerous vagaries inherent in using a textbook "in progress." Many of these students systematically kept notes and responded patiently and thoroughly to the numerous requests for feedback on the manuscript in its various forms over the last 6 years. Their contributions have been noted and are absolutely central to any success this book might enjoy.

Some students among the many deserve to be recognized by name: Liesl Rall for her careful and thorough reading and commentary on many chapters; Becky Spresser for her diligence, completeness, and humor in creating the data base of applications and history; Martin McDermot, Evan Ho, and Heather Dossat for continuing this process; and Doug Semian (who on his own time with no reimbursement from the author) for "translating" the complete text to an easier-on-the-eye, more compact version for student use.

Within the world of publishing, I am certainly indebted to all the editors and production people at McGraw-Hill for their patience and willingness to interact with a first-time author. And in this capacity I would like to thank Ray Chang for his concern and encouragement when the process of getting a book into print became particularly puzzling and even discouraging. I also would like to thank the following reviewers who commented freely on the various selected chapters or the entire text, particularly those who went beyond the basic requirements of reviewing to painstakingly note small points that make for clarity and precision in the writing: James P. Birk, Arizona State University; Donald L. Campbell, The University of Wisconsin–Eau Claire; John E. Frey, Northern Michigan University; Frank J. Gomba, U.S. Naval Academy; Timothy P. Hanusa, Vanderbilt University; Robert H. Harris, University of Nebraska; Ronald A. Krause, The University of Connecticut; Edward A. Mottel, Rose Hulman Institute of Technology; Philip H. Rieger, Brown University; Charles Scaife, Union College; and Steven H. Strauss, Colorado State University.

I am also indebted in advance to those who I hope will comment freely about the book and make suggestions for its improvement. Student and faculty readers alike are encouraged to send their comments to me at the Department of Chemistry, Allegheny College, Meadville, PA 16335. Please feel absolutely free to do so. Perhaps your efforts will pay off in much-improved future editions.

Families are always special, but mine *must* be the best. Daughters Jennifer, Emily, and Rebecca, in addition to being a constant source of joy and pride, have contributed immensely by protecting their father by answering the phone, being quiet around the house, solving some problems by themselves, and putting up with seemingly endless hours of "Dad up in his study." To my wife, Kathleen, I am indebted for the majority of good things that have happened to me during my adult life. In this endeavor, she has always had time for seemingly countless conversations about the uncertainties, frustrations, challenges, and joys of academic life in general, and book writing in particular. I could not have even attempted such a project without her love and support.

Glen E. Rodgers

INTRODUCTION TO COORDINATION, SOLID STATE, AND DESCRIPTIVE INORGANIC CHEMISTRY

CHAPTER 1

THE EVOLVING REALM OF INORGANIC CHEMISTRY

To the layperson interested in science, chemistry appears to present a united and somewhat forbidding front. In fact, to the uninitiated, all chemists seem to be alike. To paraphrase Gertrude Stein, a chemist is a chemist is a chemist. Yet shortly into a study of this wide-ranging and diverse science, it quickly becomes apparent that people who call themselves chemists are engaged in a mind-boggling variety of activities, many of which do not fit the usual stereotypes held by the public. Some chemists might spend their careers isolating, identifying, and characterizing the huge macromolecules of life, while others construct, test, and fine-tune intricate mathematical theories that describe the movements and energies of the smallest atoms and molecules. Some chemists *do* work in laboratories with the traditional scientific glassware and equipment; they are, in fact, surrounded by the sights, sounds, and smells often associated with chemistry. Others, however, work in quite different laboratories equipped with complicated instruments often interfaced with powerful computers. Still others, also classified well within the realm of chemistry, hardly ever step into a laboratory of any kind, except perhaps as a place to meet and engage their colleagues.

In this process of trying to make some sense of this proud and storied science, one quickly learns that it is commonly divided into subdisciplines such as biological, physical, analytical, organic, and inorganic. These divisions, although recognized to be arbitrary and still evolving, nevertheless significantly shape the way informed laypersons, students, and even professional chemists think about the science of chemistry. Indeed, chemists are often classified and labeled by the subdiscipline of chemistry they practice and the corresponding relatively small subsets of (1) courses they have taken and perhaps teach, (2) academic journals they read, referee, and publish in, and (3) books they read, edit, write, and teach and/or learn from.

Although serving some useful functions, the division of chemistry into five major subdisciplines is not as simplifying as it might appear. Principally this is because each of the major subdivisions still encompasses huge areas of knowledge. Even organic chemistry, which essentially restricts itself to the study of the compounds of *one* element, carbon, is difficult to describe and takes many years to master. Imagine then the challenge of describing and mastering *inorganic chemistry*, which is the study of the structures, properties, and reactions of all the elements and their compounds with the exception of the hydrocarbons and their immediate derivatives. All the *other* elements, all 109 of them, come under the realm of inorganic chemistry! This makes inorganic chemistry a huge umbrella label covering a vast and diverse field of study. Yet it may be the very diversity of inorganic chemistry that makes it so fascinating and attractive. Think of an aspect of why and how the world works the way it does and it quite likely can be better understood with a knowledge of inorganic chemistry.

To appreciate the present and future realm of inorganic chemistry better, it may be instructive to briefly investigate its past. Who were the first inorganic chemists? What did they study? How did the realm of inorganic chemistry evolve into what it is today? How would a knowledge of inorganic chemistry better inform everyone about how the world works the way it does?

Before chemistry was a separate discipline, indeed before the scientific revolution when the sciences began to be recognizable as separate areas of study, humankind was investigating chemical phenomena. The use of fire and the art of cooking, the smelting of ores into metals, the production of alloys such as bronze and brass, the preparation of glasses, cements, and explosives all were areas of chemical investigation before chemistry was recognized as a separate discipline.

During what period of time can we say that chemistry became a separate academic discipline? While opinions about establishing such a period of time and, indeed, what constitutes an independent academic discipline would certainly vary, one convenient time span might be the lifetime of Antoine Lavoisier (1743–1794). In 1743, the year of Lavoisier's birth in Paris, there were only 13 known elements. During his lifetime, that number roughly doubled (to about 28). More importantly, in great part through Lavoisier's efforts, accurate and reliable balances were developed that allowed for reproducible mass measurements. With the aid of this advance in technology, the laws of conservation of mass and definite proportion were devised. The definite composition of chemical compounds could then be accurately determined. The foundation of the science of chemistry had been laid. By 1794, the year Lavoisier was beheaded as a result of the French Revolution, it could be argued that chemistry had been established as a separate academic discipline. An individual chemist of that day, like Lavoisier himself, could describe and master most, if not all, of the chemical knowledge available to humankind. It was unnecessary to divide this knowledge into separate subdisciplines.

When was inorganic chemistry established? What was its realm at the outset? What types of problems did the first inorganic chemists address? Again, establishing such dates is arbitrary, but a convenient benchmark would be the year 1860, when the first International Chemical Congress took place in Karlsruhe, Germany. Given the availability of accurate mass measurements, the quantitative analysis of various minerals and ores had yielded 11 new elements during the period of time from Lavoisier's death to the convening of the Congress in Karlsruhe. The electrolysis of various salts had accounted for six more new elements and reactions with reducing agents (such as carbon and potassium) and acids for another 11. By the time chemists sat down together at Karlsruhe, there were about 60 known elements. Over the 66 years since Lavoisier's death, the number had doubled again and now more than half the elements were known.

As further background to the importance of the Congress, recall that the English chemist John Dalton had, at the start of the nineteenth century, established the first well-accepted atomic theory. His great accomplishment was marred, however, by his refusal to accept the new ideas of Amedeo Avogadro about the existence of diatomic molecular gases. This Dalton-Avogadro controversy went unresolved for about a half century (1811–1860) and resulted in great confusion among chemists. Then there were the awkward "pictograph" symbols for the elements that were another source of inefficiency and confusion. Jöns Jakob Berzelius of Sweden solved the problem of representation by proposing the system of chemical symbols for the elements that we use today. Berzelius also went on to suggest that all compounds be divided into either organic or inorganic. The very alphabet of inorganic chemistry, the elements and their compounds, was being identified and analyzed. Yet, the lack of a reliable set of atomic weights stood in the way. It was this problem that led to the first International Congress. Here, the ideas of Avogadro were revived by his fellow Italian, Stanislao Cannizzaro, and the first truly accurate tables of atomic weights were established.

On the more empirical front, industrial inorganic chemistry flowered. Some examples include the development of Portland cement in 1824, the patenting in 1831 of the contact process that was to revolutionize the production of sulfuric acid, the origin of the phosphate fertilizer industry in England in 1843, and the development of the diaphragm cell for the electrolytic generation of chlorine in 1851.

By 1860, chemistry was certainly too large a body of knowledge for one person to master. Inorganic and organic chemistry had been established as separate subdisciplines, and, indeed, organic chemistry was flourishing. Those in the inorganic area were still occupied with expanding the list of elements and determining the composition and nature of their compounds. The new technique of spectroscopy, established by the Germans Robert Bunsen and Gustav Kirchhoff in 1859, yielded six new elements within the next 15 years. Mineral analysis established eight lanthanides between 1879 and 1886. Also in 1886, the extremely reactive fluorine was isolated by Ferdinand Moissan. In the 1890s, William Ramsay and his fellow workers isolated most of the inert gases and Pierre and Marie Curie started their seminal work in radioactivity (isolating polonium and radium). This growing list of elements numbered approximately 83 by the turn of the century.

While the number of known elements steadily increased during this time, there was little order to the list despite some early attempts by Johann Döbereiner and then John Newlands. First proposed in the early part of the nineteenth century, Döbereiner's "triads" were groups of three elements (calcium, strontium, and barium, for example) in which the middle element had an atomic mass very close to the average of the other two. John Newlands' "law of octaves," proposed in 1866, suggested that the elements might be arranged in groups of seven with the eighth being much like the first in a manner similar to musical octaves. In 1869, the Russian chemist Dmitri Mendeleev brought significant order out of chaos by establishing his first periodic table and predicting the existence and properties of several still undiscovered elements. Over the next 15 years, the subsequent discoveries of these elements (gallium by Paul Lecoq de Boisbaudran, scandium by Lars Nilson, the germanium by Clemens Winkler) established the periodic law as the great organizing principle for the rapidly expanding realm of inorganic chemistry.

The progress in inorganic industrial chemistry continued during the period from 1860 to the turn of the century. New advances included the Solvay process for producing sodium carbonate and new economical ways of (1) producing steel (the Bessemer and open-hearth processes) and aluminum (the Hall-Héroult process), (2) recovering sulfur (the Frasch process) for use in making sulfuric acid, and (3) producing nitric acid from ammonia (the Ostwald process). The latter was one of the first great examples of the use of catalysis in industrial chemistry.

The turn of the century marked the beginning of a series of great new developments in physical chemistry. Late in the year 1900, Planck proposed that energy was quantized ($E = h\nu$), and this was soon followed by the revolutionary work of Einstein, Thomson, Rutherford, de Broglie, Pauli, and Schrödinger, among others. Along with these advances in atomic quantum theory, new ways at looking at chemical bonding were developed. Lewis structures (1923), valence-bond theory (1931), molecular orbital theory (early 1930s), and crystal field theory (1933) all quickly followed in succession. These new bonding theories were just the impetus that inorganic chemistry needed to move forward. The new theories helped to organize and explain the myriad numbers of compounds displayed by the 90 elements that were known by the 1920s.

From 1900 to 1950, the realm of inorganic chemistry continued to expand. In one of the most productive avenues of research, Alfred Werner worked through the first two decades of the new century to bring order to the mysterious set of cobalt ammonates and related compounds that had been painstakingly synthesized during the nineteenth century. His coordination theory gave us new ways to think about the structures, properties, and reactions of this new class of what he called "coordination compounds." The study of this coordination chemistry remains one of the primary components of inorganic chemistry to this day. It is the subject of the first part (Chapters 2–6) of this book.

In 1912, Max von Laue found that the recently discovered x-rays were diffracted by the regularly spaced layers of atoms in crystals. Not only did these experiments verify that x-rays were a high-frequency part of the electromagnetic spectrum, but they placed in the hands of physicists and chemists a powerful tool with which to investigate the structure of a variety of solid state compounds. The father-and-son team of William Henry and William Lawrence Bragg worked out the details of the structures of simple crystals like sodium chloride and coined such terms as ionic radii. (Henry Moseley also showed that x-rays could be related to the atomic number of an element. In so doing, he verified Mendeleev's periodic table and set it on an unchallengeable footing.) Starting about 1915, Max Born worked out a general expression for the lattice energy of a crystalline substance. From here the study of the structures and energetics of solid state inorganic substances was pursued with great vigor. It is the subject of the second part (Chapters 7 and 8) of this book.

With the electronic basis of the periodic table firmly in place by about the end of the first third of the twentieth century, the mighty challenge to inorganic chemists became apparent. First, they had to continue to detail the properties, structures, and reactions of the growing list of elements and their compounds. Furthermore, they had to rationalize the existing known chemistry with the new atomic and bonding theories. Finally, they had to make the periodic table a predictive tool for the organization of the chemistry of the elements. The compilation of the periodic chemistry of the elements had become the major province in the realm of inorganic chemistry. The last part of this book (Chapters 9 to 19) is devoted to (1) a fairly detailed introduction to the chemistry of the representative, or main-group, elements and their compounds and (2) the ways in which this chemistry can be understood in the light of the modern periodic table.

Other progress in inorganic chemistry that took place in the period 1900–1950 was derived from the work of Frederick Soddy who, in 1913, formulated the idea of isotopes. Within the next decade, principally through the work of Francis Aston and his series of spectrographs, the isotopes of most of the elements were investigated and categorized. The neutron was discovered in 1932, and shortly after Harold Urey discovered the isotopes of hydrogen. In the late 1930s and continuing on through World War II, nuclear fission was discovered and thoroughly investigated. Radioisotopes (isotopes of the normally stable elements) were synthesized and began to find uses in medicine and research. All these advances posed great new opportunities for inorganic chemists. For example, in the 1940s, carbon 14 dating was perfected by Willard Libby. The developments in nuclear chemistry required new ways for separating elements as well as new materials able to withstand the ravages of corrosion and high temperatures. The addition of the new artificial elements brought the number of elements over the 100 mark.

Little did we know that other new areas of study uncovered in the first 50 years of the twentieth century would turn out to be of such great significance in the second 50 years. These areas include (1) Svante Arrhenius' warning that we should be concerned about the warming of the atmosphere through the greenhouse effect (1908), (2) Alfred Stock's defining research on the hydrides of silicon and boron (starting in 1912), (3) the construction of the first Haber-Bosch ammonia synthesis plant (1913), (4) the discovery of ozone in the atmosphere by Charles Fabry (1913), (5) the synthesis and testing by Thomas Midgley of chlorofluorocarbons (CFCs) as ideal refrigerant gases (1928), (6) the investigation by H. T. Dean of the benefits of fluoridating public water supplies (1930s), (7) the development of the Xerox process by C. F. Carlson (starting in 1934), and (8) the chance discovery of the germanium transistor at the Bell Telephone Laboratories (1947).

Some have called the last half of the twentieth century the "renaissance of inorganic chemistry," but, in fact, all of chemistry has greatly flourished during this time. Certainly much of this chemistry should be placed at least partially in the realm of inorganic, but the lines between the various subdisciplines have become so blurred that it is impossible and, indeed, probably counterproductive to assign many developments to a given subdiscipline. One of the first events to take place during this period is representative of this blurring and/or crossing of subdisciplinary lines. In 1951, ferrocene was synthesized. This compound has an iron atom sandwiched between and bound to two planar cyclopentadienyl rings (C_5H_5) and was one of the most significant and important examples of an *organometallic* compound. Organometallic chemistry, characterized by the presence of metal-carbon bonds, bridges the organic and inorganic subdisciplines and has gone on to become one of the new and still-growing subdisciplines of chemistry. In 1960, Max Perutz determined the structure of hemoglobin by x-ray diffraction. This development was a key to the development of the still-growing field of bioinorganic chemistry.

During the 1960s, some unique developments in inorganic chemistry included (1) Neil Bartlett's announcement that he had synthesized the first compound of Ramsay's "inert" gases, (2) William Lipscomb and his group's work on the novel multicenter bonding in Stock's borohydrides, and (3) the chance discovery, by Barnett Rosenberg, that one of the simpler coordination compounds, *cis*-diamminedichloroplatinum(II), or "cisplatin," had significant activity as an antitumor agent.

Starting in the late 1950s and growing rapidly in the succeeding decades, the environmental movement fueled research in inorganic chemistry closely allied with physical and analytical chemistry. In the 1950s, photochemical smog was first explained as a series of reactions driven by the action of sunlight. The action of phosphates and other nutrients were carefully studied in the 1960s and early 1970s. Also during this time, the concerns over heavy-metal pollution mounted. In the 1970s, the energy crisis came to a head and with it the search for alternative fuels and means of transportation. As if all that were not enough, Rowland and Molina announced in 1974 that they had evidence that Midgeley's chlorofluorocarbons were most likely threatening the ozone layer of the stratosphere. The 1980s brought concerns over the origin and action of acid rain, and, in 1985, the ozone hole over Antarctica was discovered. These are all problems that will involve a great deal of inorganic chemistry to fully define and to solve. Most of them are treated at appropriate junctures in the last part of this book.

SUMMARY

The realm of inorganic chemistry has steadily evolved over the years. During the middle of the nineteenth century, inorganic chemistry was defined as one of the major subdisciplines of chemistry. In the latter part of the nineteenth century, inorganic chemists were primarily occupied organizing and filling in the blanks in the periodic table. Others concerned themselves primarily with developing a steady stream of improvements in inorganic industrial processes.

The first half of the twentieth century saw the realm of inorganic chemistry expand to include coordination compounds, radioisotopes, solid state structures and energetics, as well as research related to the advent and development of nuclear fission. Initial and defining work on the greenhouse effect, borohydrides, ammonia synthesis, atmospheric ozone, chlorofluorocarbons, fluoridation, the Xerox process, and transistors were reported. Also in the first half of the twentieth century, the principal challenge to inorganic chemists seemed to be defined: to fully investigate, understand, and predict the chemistry of the elements and their compounds in the light of the electronic basis of the periodic table.

In the second half of the twentieth century, new avenues of research have led to the new fields of organometallic and bioinorganic chemistry. In addition, the new compounds of the "inert" (now called noble) gases, the expansion of bonding theories to account for the structure and reactions of the borohydrides and related compounds, as well as the flowering of the environmental movement and the energy crisis have expanded the realm of inorganic chemistry still further. As we stand at the edge of the twenty-first century, the century of today's generation of students, inorganic chemistry promises to continue to be diverse, challenging, important, and, above all, truly fascinating.

PART A

COORDINATION CHEMISTRY

In this self-standing section, coordination chemistry is introduced over the course of five chapters. The titles of these chapters are listed below.

CHAPTER 2

AN INTRODUCTION TO COORDINATION CHEMISTRY

As noted in Chapter 1, one of the most productive areas of research in the twentieth century was Alfred Werner's development of coordination chemistry. It is a measure of Werner's impact on the realm of inorganic chemistry that the number, variety, and complexity of coordination compounds continues to grow even as we pass the centennial anniversary of his original work. Before we launch into a historical perspective on the development of this vital subsection of inorganic chemistry, we need to set the stage with a few important definitions.

Coordination chemistry concerns compounds in which a small number of molecules or ions called *ligands* surround a central metal atom or ion. Each ligand (from the Latin *ligare*, meaning "to bind") donates a pair of electrons to the metal. The metal-ligand bond, often represented as $M \leftarrow :L$, is an example of a *coordinate-covalent bond* in which both the electrons come from one atom. The *coordination number* is the number of ligands around a given metal atom or ion. Four or six (or occasionally other small integers) are typical values for these numbers. Collectively, the ligands are often referred to as the *coordination sphere* and, with the metal, are enclosed in square brackets when writing molecular formulas. For example, a typical formula might be $[ML_6]X_n$ or $M'_n[ML_4]$, where the M' is a simple metal cation and X might be a variety of anions. Note that in the first formula, the coordination sphere and the metal M constitute a cation while in the second they make an anion. Such coordinated metal ions are sometimes referred to as *complex* cations or anions.

Typically, coordination compounds are characterized by a wide range of bright colors. Varying the number and types of the ligands often significantly changes the color and also the magnetic characteristics of the compound. Some examples of coordinated (or complex) ions you may have encountered in earlier courses include the colorless $Ag(NH_3)_2^+$ cation (often discussed in connection with the group I qualitative analysis scheme), the dark-blue $Cu(NH_3)_4^{2+}$ ion (a good test for the presence of copper ions in solution), the deep-red $FeSCN^{2+}$ [a sensitive test for the presence of iron(III) ions], and typical aqueous cations, for example,

$Ca(H_2O)_6^{2+}$ and $Fe(H_2O)_6^{3+}$, which are more often abbreviated as $Ca^{2+}(aq)$ and $Fe^{3+}(aq)$, respectively.

Perhaps you have previously encountered coordination compounds (sometimes referred to as *transition metal complexes*) as part of a general chemistry course. Due to time considerations, this subject is usually covered only briefly, if at all, in such courses. In the first part (Chapters 2 to 6) of this book, however, coordination chemistry will be the sole focus of our attention. Accordingly, we will be able to systematically discuss the history, nomenclature, structures, bonding theories, reactions, and applications of such compounds. (After a physical chemistry course, more of the mathematical and abstract theoretical details are usually developed.) In this chapter, we will cover the historical perspective regarding such compounds, introduce some typical ligands, and start to develop a system of nomenclature.

2.1 THE HISTORICAL PERSPECTIVE

In earlier courses, the basics of atomic structure, the periodic table, and chemical bonding are investigated. The first two columns of Figure 2.1 are a chronological display of some of the concepts quite often discussed.

Starting at the top of the first column, you should recall how some early laws had firmly established that chemical compounds are always made up of the same definite composition by mass (Proust) and that this mass is always conserved in various reactions (Lavoisier). These empirical (from experiment) facts led to the first concrete atomic theory, developed at the outset of the nineteenth century by the English chemist John Dalton. Dalton assumed that atoms were hard, impenetrable spheres much like miniature billiard balls. He had no occasion (at least in writing) to speculate about their inner structures.

More than 40 elements were discovered and characterized during the nineteenth century. With the number of known elements increasing decade by decade (see Figure 9.2, for example), there was a variety of attempts to organize them in some coherent way. Dmitri Mendeleev, building on the work of others, noted that the properties of this growing list of elements seemed to vary in a periodic way and published his first periodic table in 1869. (For more on Mendeleev and his table, see Chapter 9, p. 214). Although more was becoming known about the properties of these elements, the inner structure of the constituent atoms remained a mystery.

Toward the end of the nineteenth century, a number of discoveries were made that began to reveal what might lie within the atom. Balmer devised a formula, based on a series of allowed integers, that organized (but did not explain) the visible spectrum of hydrogen. Roentgen and Becquerel discovered x-rays and radioactivity, respectively. J. J. Thomson discovered that electrons are a fundamental component of all matter. Ernest Rutherford, after seeing the results when alpha particles (emitted from radioactive atoms) were allowed to slam into thin gold foils, proposed that atoms were made up of very tiny, massive nuclei surrounded by electrons. (It did not take long for Neils Bohr to suggest that these electrons might be pictured as orbiting the nucleus like planets orbit the sun.) Rutherford's concept of the nucleus caught on rapidly, but the picture of orbiting electrons was soon replaced by electron "clouds" confined to an area about the nucleus by electrostatic forces. Atomic orbitals are a mental image or model still employed by chemists in the modern Schrödinger (quantum-mechanical) model of the atom. This model nicely accounts for the necessity of using integers (quantum numbers) to describe line spectra and the periodic table.

Atomic structure and the periodic table	Molecular structure and bonding	Coordination chemistry
1750: 1788: Law of Definite Composition, Proust 1790: Law of Conservation of Matter, Lavoisier		1798: First cobalt ammonates observed, Tassaert
1800: 1808: Dalton's Atomic Theory published in *New System of Chemical Philosophy* 1859: Spectroscope developed, Bunsen and Kirchhoff 1869: Mendeleev's First Periodic Table organizes 63 known elements 1885: Balmer formula for visible H spectrum 1894: First "Inert Gas" discovered 1895: X-rays discovered, Roentgen 1896: Radioactivity discovered, Becquerel	1830: The Radical Theory of Structure Liebig, Wohler, Berzelius, Dumas (organic compounds composed of methyl, ethyl, etc., radicals) 1852: Concept of Valence, Frankland (all atoms have a fixed valence) 1854: Tetravalent carbon atom, Kekule 1874: Tetrahedral carbon atom, Le Bel and Van't Hoff 1884: Dissociation theory of electrolytes, Arrhenius	1822: Cobalt ammonate oxalates prepared, Gmelin 1851: $CoCl_3 \cdot 6NH_3$, $CoCl_3 \cdot 5NH_3$, and other cobalt ammonates prepared Genth, Claudet, Fremy 1869: Chain Theory of Ammonates, Blomstrand 1884: Amendments to Chain Theory, Jørgensen 1892: Werner's dream about coordination compounds
1900: 1902: Discovery of the electron, Thomson 1905: Wave-Particle Duality of light, Einstein 1911: α-particle/gold foil experiment; Nuclear Model of the Atom, Rutherford 1913: Bohr Model of the Atom (electrons orbit nucleus) 1923: Wave-Particle Duality of electrons, De Broglie 1926: Schrödinger quantum mechanical atom (electrons in orbitals about nucleus; electron spectroscopy explained as transitions among orbitals)	1923: Electron Dot Diagrams, Lewis 1931: Valence Bond Theory, Pauling Heitler, London, Slater Early 1930s: Molecular Orbital Theory, Hund, Bloch, Mulliken, Hückel 1940: Valence Shell Electron Pair Repulsion (VSEPR) Theory, Sidgwick	1902: Three Postulates of Coordination Theory proposed, Werner 1911: Optical isomers of *cis*-$[CoCl(NH_3)(en)_2]X_2$ resolved, Werner 1914: Non-carbon-containing optical isomers resolved, Werner 1927: Lewis ideas applied to Coordination Compounds, Sidgwick 1933: Crystal Field theory, Bethe and Van Vleck
Modern Periodic Table including trends in periodic properties	Modern Concepts of Chemical Bonding	Modern Coordination Theory

FIGURE 2.1
The historical setting of coordination chemistry.

The second column of Figure 2.1 is a time line of some of the ideas concerning molecular structure and bonding. In Dalton's day, not all chemists would admit that atoms existed. Those that did (and undoubtedly a few of those who did not) could only speculate on how these fundamental particles might associate with or bond to each other. (It was even suggested at one point that each atom might have a characteristic number of embedded hooks that somehow held them tightly to other atoms.) As noted in the figure, the organic chemists led the way in conceiving new ideas concerning the basic structural units of carbon-based compounds. There seemed to be groups of atoms (for example, the methyl group, $CH_3—$, or the ethyl group, $CH_3CH_2—$), sometimes called radicals, that were present in a large number of compounds and stayed intact throughout various chemical reactions. By the middle of the nineteenth century, the concept of a fixed valence associated with each atom had been adopted in the effort to account for the nature of organic compounds and their constituent fragments. As Sir Edward Frankland stated it, "no matter what the character of the uniting atoms may be, the combining power of the attracting element... is always satisfied by the same number of atoms." So it was thought that carbon always had a fixed valence of 4, oxygen 2, hydrogen 1, and so forth.

By the end of the nineteenth century, experiments with electricity had indicated that it likely played a significant role in molecular bonding. After the discovery of the electron, G. N. Lewis proposed that these small negative particles might be the glue that holds atoms together. The number of electrons in the recently discovered "inert gases" seemed to be especially stable. The *octet rule* became the guide to chemical bonding. Various more sophisticated theories followed in the 1930s. Nevil Sidgwick proposed that electron pairs might repel each other and play a significant role in determining the shape of a molecule. Linus Pauling and others proposed that the overlap of atomic orbitals or special *hybrid orbitals* would bond one atom to another. Also developed during this time was the theory that molecules might be a group of nuclei held together by electronic confined waves appropriately called molecular (as opposed to atomic) orbitals. All of these ideas, from electron-dot diagrams to valence-shell electron-pair repulsion (VSEPR) theory to valence-bond (VB) and molecular orbital (MO) theories, still aid modern chemists in picturing the structure and bonding of compounds.

It is assumed that the above ideas are more or less familiar to you. (Some brief review will be provided as appropriate, but you might want to consult your general chemistry notes and book as needed.) Coordination compounds, however, outlined in the third column of Figure 2.1, are likely to be less familiar. How and by whom were these compounds discovered? What was the chain theory? Why did it lose out to Werner's coordination theory? Could these comparatively new and different compounds be accounted for using the ideas which had worked so well for the organic chemists? How did the ideas concerning atomic and molecular structure ultimately contribute to an understanding of these compounds? We take up the answers to such questions in the next section.

2.2 THE HISTORY OF COORDINATION COMPOUNDS

Early Compounds

At the very end of the eighteenth century Tassaert, a French chemist so obscure in the history of chemistry that his first name remains unknown, observed that ammonia combined with a cobalt ore to yield a reddish-brown mahogany-colored product. This was most likely the first known coordination compound. Throughout the first half of the nineteenth century many other, often beautifully crystalline examples of various cobalt *ammonates* were prepared. These compounds were strikingly colored, and the names given to them, for example, roseo-, luteo- (from the Latin *luteus*, "deep yellow"), and purpureocobaltic chlorides, reflected these colors. In the second half of the century, other ammonates, particularly those of chromium and platinum, were prepared. Despite various attempts, however, no theoretical basis was developed to satisfactorily account for these wondrous compounds.

Given the success of the organic chemists in describing the structural units and fixed atomic valences found in carbon-based compounds, it was natural that these ideas should be applied to the ammonates. The results, however, were disappointing. Consider the typical data for the cobalt ammonate chlorides listed in Table 2.1.

The formulas used in the last few decades of the nineteenth century indicated the ammonia-to-cobalt mole ratio but left the nature of the bonding between them to the imagination. (The compound with a 3:1 ammonia-to-cobalt ratio proved difficult to prepare. The corresponding iridium compound was used instead.) Conductivities measured when these compounds were dissolved in water are given qualitatively. Conductivity was then just starting to be taken as a measure of the number of ions produced in solution. The number of chloride ions precipitated by the addition of aqueous silver nitrate refers to the reaction represented in Equation (2.1):

$$AgNO_3(aq) + Cl^-(aq) \longrightarrow AgCl(s) + NO_3^-(aq) \tag{2.1}$$

Now how might you explain such data? More importantly from a historical point of view, how did the chemists of the late 1860s, who had been schooled in the relatively new but extraordinarily successful ideas of organic chemistry, explain such data? As shown in Figure 2.1, it seemed to have been fairly well established by then that all elements have a valence, sometimes called a *combining capacity*, which is a single fixed value. Furthermore, many workers had found that organic compounds could be pictured as vast chains of carbon atoms composed of radicals and groups of various types which also appeared to have fixed valences. For example, hexane, $CH_3{-}CH_2{-}CH_2{-}CH_2{-}CH_2{-}CH_3$, with its chain of six carbon atoms, could be pictured as containing monovalent methyl ($CH_3{-}$) groups on the ends with four divalent methylene (${-}CH_2{-}$) groups in the middle. Ordinary grain alcohol, of overall composition C_2H_6O, was composed of ethyl

TABLE 2.1
The cobalt ammonate chlorides (data available to Blomstrand, Jørgensen, and Werner)

Formula	Conductivity	No. of Cl^- ions precipitated
$CoCl_3 \cdot 6NH_3$	High	3
$CoCl_3 \cdot 5NH_3$	Medium	2
$CoCl_3 \cdot 4NH_3$	Low	1
$IrCl_3 \cdot 3NH_3$	Zero	0

(C_2H_5—) and hydroxyl (—OH) groups to yield a structural formula of C_2H_5—OH. Wood alcohol, CH_4O, was similarly displayed as CH_3—OH, composed of methyl and hydroxyl groups.

The Blomstrand-Jørgensen Chain Theory

In 1869, Christian Wilhelm Blomstrand first formulated his chain theory to account for the cobalt ammonate chlorides and other series of ammonates. Blomstrand, knowing that the fixed valence of cobalt was established at 3, chained together cobalt atoms, divalent ammonia groups, and monovalent chlorides to produce a picture of $CoCl_3 \cdot 6NH_3$, something like that shown in Figure 2.2*a*. (Actually, on the basis of some vapor density measurements, Blomstrand originally pictured the compound as dimeric.) Based on the prevailing ideas of the time, this was a perfectly reasonable structure. The divalent ammonia he proposed was consistent with a view of ammonium chloride as H—NH_3—Cl, the valence of 3 for cobalt was satisfied, nitrogen atoms were chained together much like carbon was in organic compounds, and the three monovalent chlorides were far enough removed from the cobalt atom to be available to be precipitated by aqueous silver chloride.

In 1884, Sophus Mads Jørgensen, a student of Blomstrand, proposed some amendments to his mentor's picture. First, he had new evidence that correctly indicated that these compounds were monomeric. Second, he adjusted the distance

(*a*) $CoCl_3 \cdot 6NH_3$

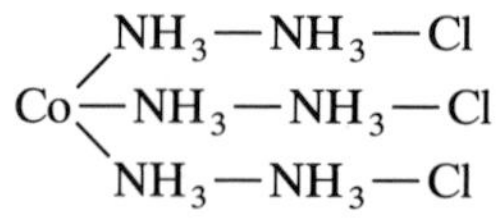

$$\mathrm{Co}\begin{array}{l}\diagup NH_3—NH_3—Cl\\ —NH_3—NH_3—Cl\\ \diagdown NH_3—NH_3—Cl\end{array}$$

(*b*)

(1) $CoCl_3 \cdot 6NH_3$

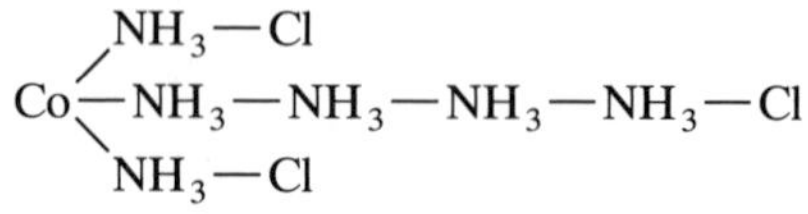

$$\mathrm{Co}\begin{array}{l}\diagup NH_3—Cl\\ —NH_3—NH_3—NH_3—NH_3—Cl\\ \diagdown NH_3—Cl\end{array}$$

(2) $CoCl_3 \cdot 5NH_3$

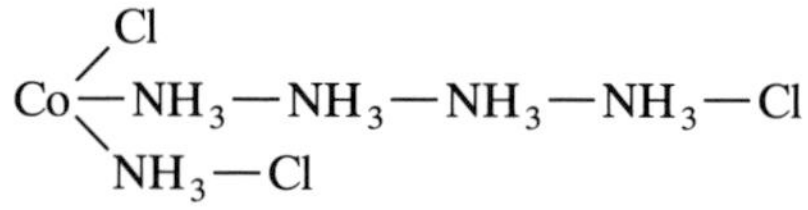

$$\mathrm{Co}\begin{array}{l}\diagup Cl\\ —NH_3—NH_3—NH_3—NH_3—Cl\\ \diagdown NH_3—Cl\end{array}$$

(3) $CoCl_3 \cdot 4NH_3$

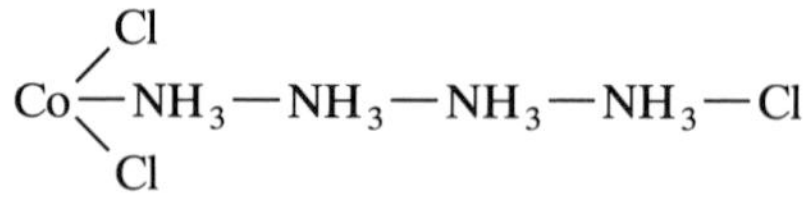

$$\mathrm{Co}\begin{array}{l}\diagup Cl\\ —NH_3—NH_3—NH_3—NH_3—Cl\\ \diagdown Cl\end{array}$$

(4) $IrCl_3 \cdot 3NH_3$

$$\mathrm{Ir}\begin{array}{l}\diagup Cl\\ —NH_3—NH_3—NH_3—Cl\\ \diagdown Cl\end{array}$$

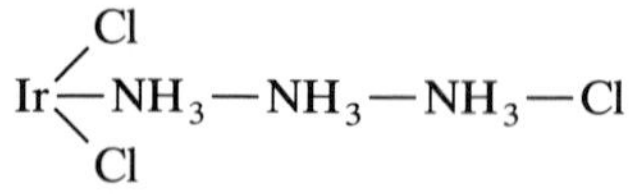

FIGURE 2.2
Representations of the cobalt ammonate chlorides by Blomstrand and Jørgensen: (*a*) Blomstrand's representation of $CoCl_3 \cdot 6NH_3$; (*b*) Jørgensen's representations of four members of the series with the iridium substituted for the intended cobalt in compound (4). [Adapted from Ref. 1, p. 6.]

of the chloride groups from the cobalt to account for the rates at which various chlorides were precipitated. The first chloride is precipitated much more rapidly than the others and so was put farther away and therefore less under the influence of the cobalt atom. His diagrams for the first three cobalt ammonate chlorides are shown in Figure 2.2*b*. Note that in the second compound, one chloride is now directly attached to the cobalt and therefore, Jørgensen assumed, unavailable to be precipitated by silver nitrate. In the third compound, two chlorides are similarly pictured. These amendments significantly improved the chain theory, but there remained a number of unanswered questions. For example, why are there only six ammonia molecules? Why not eight or ten? Why do we not see ammonia molecules that are chemically different depending on their positions in the chain? On balance, however, it appeared that the Blomstrand-Jørgensen theory of the cobalt ammonates was on the right track.

But was there a compound with only three ammonias? As shown in Figure 2.2*b*(4), the chain theory predicted that it should exist and, furthermore, should have one ionizable chloride. But this critical compound was not available. Jørgensen set out to prepare it to test his version of the chain theory. Try as he might, this excellent synthetic chemist could not come up with the desired cobalt compound. He did, however, manage to prepare, after considerable time and effort, the analogous iridium ammonate chloride. Alas, it was found to be a neutral compound with no ionizable chlorides. With no small amount of irony, the chain theory was in trouble thanks to the considerable efforts of one of its principal proponents.

The Werner Coordination Theory

Alfred Werner, a German-Swiss chemist, was torn between organic and inorganic chemistry. His first contributions (the stereochemistry, or spatial arrangements, of atoms in nitrogen compounds) were in the organic field, but there were so many intriguing inorganic questions being raised in those days that he decided that this was the area in which he would work. He observed the difficulties inorganic chemists were having in explaining coordination compounds, and he was aware that the established ideas of organic chemistry seemed to lead only into blind alleys and dead ends. In 1892, when Werner was only 26, his coordination theory came to him in a dream. He woke up and started to write it down, and by five o'clock in the morning it was essentially complete. But his new theory broke with the earlier traditions, and he had essentially no experimental proof to support his ideas. Jørgensen, Blomstrand, and others considered Werner to be an impulsive young man and his theory to be audacious fiction. Werner spent the rest of his life directing a systematic and thorough research program to prove that his intuition was correct.

Werner decided that the idea of a single fixed valence could not apply to cobalt and other similar metals. Working with the cobalt ammonates and other related series involving chromium and platinum, he proposed instead that these metals have two types of valence, a primary valence (*hauptvalenz*) and a secondary valence (*nebenvalenz*). The primary or ionizable valence corresponded to what we call today the *oxidation state*. For cobalt it is the +3 state. The secondary valence is more commonly called the *coordination number*. For cobalt it is 6. Werner maintained that this secondary valence was directed toward fixed geometric positions in space.

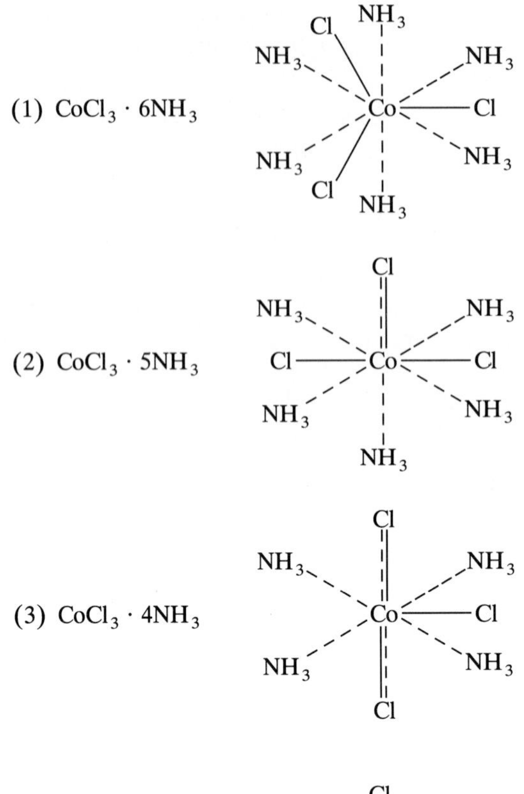

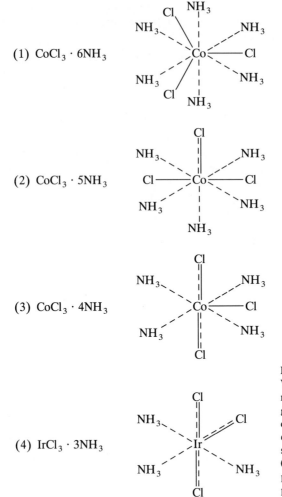

FIGURE 2.3
Werner's representations of the cobalt ammonate chlorides. The solid lines represent groups that satisfy the primary valence or oxidation state (+3) of cobalt, while the dashed lines represent those that satisfy the secondary valence, or coordination number (6). The secondary valence occupies a fixed position in space. [Adapted from Ref. 1, p. 7.]

Figure 2.3 shows Werner's early proposals for the bonding in the cobalt ammonates. He said that the cobalt must simultaneously satisfy both its primary and secondary valences. The solid lines show the groups which satisfy the primary valence, while the dashed lines, always directed toward the same fixed positions in space, show how the secondary valence was satisfied. In the first (1) compound, all the chlorides satisfy only the primary valence while the ammonias satisfy the secondary. In the second (2) compound, one chloride must do double-duty and help to satisfy both valences. The chloride which satisfies the secondary valence (and is directly bound to the Co^{3+} ion) was concluded to be unavailable for precipitation by silver nitrate. The third (3) compound has two chlorides doing double-duty and only one available for precipitation. The fourth (4) compound, according to Werner, should be a neutral compound with no ionizable chlorides. This was exactly what Jørgensen had found with the iridium compound.

Werner next turned to the geometry of the secondary valence (or coordination number). As shown in Table 2.2, six ammonias about a central metal atom or ion might assume one of several different common geometries including hexagonal planar, trigonal prismatic, and octahedral. The table compares some information

TABLE 2.2
The number of actual versus predicted isomers for three different geometries of coordination number 6. [Ref. 2.]

Formula	Hexagonal planar 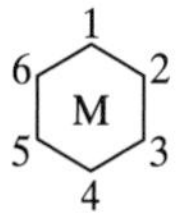	Trigonal prism 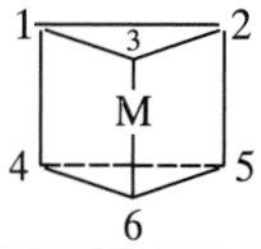	Octahedral 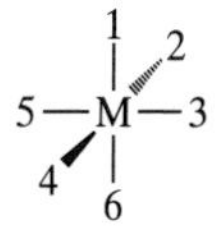	No. of actual isomers
	No. of predicted isomers (numbers in parentheses indicate position of the B ligands)			
MA_5B	One	One	One	One
MA_4B_2	Three	Three	Two	Two
	(1, 2)	(1, 2)	(1, 2)	
	(1, 3)	(1, 4)	(1, 6)	
	(1, 4)	(1, 6)		
MA_3B_3	Three	Three	Two	Two
	(1, 2, 3)	(1, 2, 3)	(1, 2, 3)	
	(1, 2, 4)	(1, 2, 4)	(1, 2, 6)	
	(1, 3, 5)	(1, 2, 6)		

about the predicted and actual number of isomers for a variety of substituted coordination compounds.

There are a few comments that need to be made about the information in this table before we can go on to discuss the significance of the data. Note first that the symbols for the compounds use M for the central metal and A's and B's for the various ligands. The numbers in parentheses for each isomer refer to the relative positions of the B ligands. *Isomers* are defined here as compounds which have the same numbers and types of chemical bonds but which differ in the spatial arrangements of those bonds. (A more detailed discussion of isomers is presented in Chapter 3.) The number of predicted isomers refers to the number of theoretically possible geometrical arrangements in space. For example, for the octahedral MA_5B case, there is only one possible geometry even though there are numerous ways to draw it. Figure 2.4*a* shows three equivalent ways to draw such an isomer. In each case the same configuration has simply been oriented differently in space so that the one B ligand is either up in the axial position or in a different equatorial position. In other words, all six octahedral positions are equivalent and it does not matter which position is occupied by the one B ligand. Figure 2.4*b* shows three equivalent configurations possible for the first MA_4B_2 isomer, while Figure 2.4*c* shows a like number for the second.

Now we can analyze the data in Table 2.2. For the MA_5B case only one isomer could actually be prepared experimentally, a result consistent with all three of the proposed geometries. For the MA_4B_2 case, however, Werner could prepare only two isomers. For the octahedral case, this actual number matched the possible number, but for the hexagonal planar and trigonal prism cases, there were three possible isomers. Assuming that Werner had not missed an isomer someplace, the data indicated that the "positions in fixed space" for six ligands is octahedral. The same type of analysis for the MA_3B_3 case gives a similar result. Only the

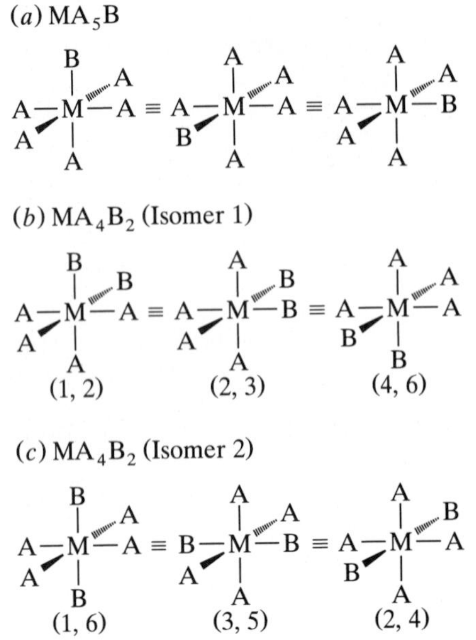

(*a*) MA_5B

(*b*) MA_4B_2 (Isomer 1)

(1, 2) (2, 3) (4, 6)

(*c*) MA_4B_2 (Isomer 2)

(1, 6) (3, 5) (2, 4)

FIGURE 2.4
Equivalent configurations for some octahedral isomers.

octahedral configuration gives the same number of isomers as could actually be prepared.

Given these results (obtained by analyzing a large number of series of coordination compounds), Werner was able to predict that there would be two isomers found for the $CoCl_3 \cdot 4NH_3$ case. These proved somewhat difficult to prepare, but in 1907 Werner was finally successful. He found two isomers, one a bright green and the other a vivid violet color. Now although all this would be considered "negative" evidence (as opposed to conclusive proof) by a philosopher of science (it was the *absence* of an isomer which constituted the evidence), the case for the coordination theory was growing stronger. Werner's positive proof will be discussed in the next chapter (pp. 41–42) when we consider the optical activity of coordination compounds. The "negative" proof, however, was enough for Jørgensen. In 1907, he dropped his opposition to Werner's "audacious" coordination theory.

All of this goes to demonstrate, as so often is the case in science, that sometimes we need to take risks. We must occasionally follow our intuitions or, in Werner's case, his dream, and advocate a new and sometimes poorly supported way of thinking about a phenomenon in order to make a truly revolutionary advance. Blomstrand and Jørgensen tried to extend the established ideas of organic chemistry to account for the newer coordination compounds. In doing so, one could argue, they actually impaired progress in the understanding of this branch of chemistry. The trick, of course, is to know when to stick to the established ideas and when to break away from them. Werner chose the latter course and, 20 years later in 1913, received the Nobel Prize in chemistry.

2.3 THE MODERN VIEW OF COORDINATION COMPOUNDS

Today, the molecular formulas of coordination compounds are represented in a manner that makes it clearer which groups are part of the coordination sphere and which are not. As indicated in the introduction to this chapter, the metal atom or ion and the ligands coordinated to it are enclosed in square brackets. It follows that the cobalt ammonate chlorides can be represented as:

(1) $CoCl_3 \cdot 6NH_3$	$[Co(NH_3)_6]Cl_3$
(2) $CoCl_3 \cdot 5NH_3$	$[Co(NH_3)_5Cl]Cl_2$
(3) $CoCl_3 \cdot 4NH_3$	$[Co(NH_3)_4]Cl_2]Cl$
(4) $CoCl_3 \cdot 3NH_3$	$[Co(NH_3)_3Cl_3]$

The ammonia molecules and chloride ions inside the brackets satisfy the coordination number of cobalt. The chlorides in the coordination sphere do double-duty, also helping to satisfy the +3 oxidation state of the cobalt. The chlorides outside the brackets, sometimes called *counterions*, help satisfy only the oxidation state. They are the only ionic chlorides available to be precipitated by silver nitrate. For example, if compound (2) is placed in water and treated with aqueous silver ions, the resulting reaction would be that represented by Equation (2.2):

$$[Co(NH_3)_5Cl]Cl_2(s) + 2Ag^+(aq) \longrightarrow 2AgCl(s) + [Co(NH_3)_5Cl]^{+2}(aq) \tag{2.2}$$

Although cobalt compounds were the most prevalent subject of his research program, Werner and his collaborators worked with other metals as well. As an example, consider the following series of platinum ammonates presented in their modern format. Note that in this case the series is extended to include anionic complex ions. The counterions in these latter cases, compounds (6) and (7), are potassium cations:

(1) $[Pt(NH_3)_6]Cl_4$

(2) $[Pt(NH_3)_5Cl]Cl_3$

(3) $[Pt(NH_3)_4Cl_2]Cl_2$

(4) $[Pt(NH_3)_3Cl_3]Cl$

(5) $[Pt(NH_3)_2Cl_4]$

(6) $K[Pt(NH_3)Cl_5]$

(7) $K_2[PtCl_6]$

Chromium complexes were also investigated. In 1901, Werner used the results of molecular weight determinations and conductivities to propose that the two known compounds of formula $CrCl_3 \cdot 6H_2O$ should be represented as the violet $[Cr(H_2O)_6]Cl_3$ and the emerald-green $[Cr(H_2O)_4Cl_2]Cl \cdot 2H_2O$.

Ammonia was certainly one of the most famous ligands to be investigated by Werner. It is referred to as a *monodentate* ligand, defined as one that donates only a single pair of electrons to a metal atom or ion. The word *monodentate* comes from the Greek *monos* and the Latin *dentis* and, not unexpectedly, literally means "one tooth." A monodentate ligand, then, has only one pair of electrons with which to "bite" the metal. Some other common monodentate ligands are shown in Table 2.3. (The nomenclature for these ligands will be discussed in the next section.) Not surprisingly, there are bidentate, tridentate, and, in general, multidentate ligands as well. A few of the other common multidentate ligands are also

TABLE 2.3
Common monodentate, multidentate, bridging, and ambidentate ligands

	Usually monodentate ligands		
F^-	fluoro		
Br^-	bromo		
I^-	iodo		
CO_3^{2-}	carbonato		
NO_3^-	nitrato		
SO_3^{2-}	sulfito		
$S_2O_3^{2-}$	thiosulfato		
SO_4^{2-}	sulfato	Common bridging ligands	
CO	carbonyl	Common bridging ligands	
Cl^-	chloro	Common bridging ligands	
O^{2-}	oxo	Common bridging ligands	
O_2^{2-}	peroxo	Common bridging ligands	
OH^-	hydroxo	Common bridging ligands	
NH_2^-	amido	Common bridging ligands	
CN^-	cyano	Common bridging ligands	Ambidentate ligands
SCN^-	thiocyanato	Common bridging ligands	Ambidentate ligands
NO_2^-	nitro	Common bridging ligands	Ambidentate ligands
H_2O	aquo		
NH_3	ammine		
CH_3NH_2	methylamine		
$P(C_6H_5)_3$	triphenylphosphine		
$As(C_6H_5)_3$	triphenylarsine		
N_2	dinitrogen		
O_2	dioxygen		
NO	nitrosyl		
C_2H_4	ethylene		
C_5H_5N	pyridine		

	Multidentate ligands	
$NH_2CH_2CH_2H_2$	ethylenediamine (en)	(2)
$CH_3\underset{O}{\underset{\|}{C}}\overset{(-)}{C}H\underset{O}{\underset{\|}{C}}CH_3$	acetylacetonato (acac)	(2)
$C_2O_4^{2-}$	oxalato	(2)
$NH_2CH_2COO^-$	glycinato (gly)	(2)
$NH_2CH_2CH_2NHCH_2CH_2NH_2$	diethylenetriamine (dien)	(3)
$N(CH_2COO)_3^{3-}$	nitrilotriacetato (NTA)	(4)
$(OOCCH_2)_2NCH_2CH_2N(CH_2COO)_2^{4-}$	ethylenediamine-tetraacetato (EDTA)	(6)

given in Table 2.3. The denticities of these ligands are given in parentheses. For example, the denticity of ethylenediamine is 2.

Ethylenediamine, shown in Figure 2.5, was a bidentate ligand of particular importance in the work of both Werner and Jørgensen. Notice that both of the nitrogen atoms in this compound have a lone pair of electrons that can be donated to a metal. Notice also that when both electron pairs interact with the same metal, the resulting configuration rather resembles a crab clutching at its prey. Multidentate ligands that form one or more rings with a metal atom in this manner are called *chelates* or *chelating agents*, terms derived from the Greek *chele*, meaning "claw." Incidentally, if you have had any exposure to organic chemistry, you may be a little surprised at the name ethylenediamine. Modern organic chemists would call this compound 1,2-diaminoethane, but the older term using ethylene as the $—C_2H_4—$ radical seems to be a permanent fixture in the nomenclature of coordination chemistry.

Two other general types of ligands are represented in Table 2.3 and should be briefly mentioned here. The first are the common *bridging ligands*, defined as those containing two pairs of electrons donated to two metal atoms simultaneously. The interaction of such ligands with metal atoms can be represented as $M \leftarrow :L: \rightarrow M$. Ligands of the bridging type include amide (NH_2^-), carbonyl (CO), chloride (Cl^-), cyanide (CN^-), hydroxide (OH^-), nitrite (NO_2^-), oxide (O^{2-}), peroxide (O_2^{2-}), sulfate (SO_4^{2-}), and thiocyanate (SCN^-). Werner was very active in preparing a number of cobalt ammonia or cobalt ethylenediamine compounds containing ligands of this type. The second type of ligand to include at this point is the *ambidentate ligand*. These are ligands that, depending on the experimental conditions and the metals involved, can use one of two different atoms to donate a pair of electrons to a metal. If we represent this type of ligand as $:AB:$, then it can form one of two possible coordinate-covalent bonds, either $M \leftarrow :AB:$ or $:AB: \rightarrow M$, with a metal atom. Common ambidentate ligands include cyanide, thiocyanate, and nitrite.

2.4 AN INTRODUCTION TO THE NOMENCLATURE OF COORDINATION COMPOUNDS

The nomenclature of coordination compounds will be introduced in two sections. Here we consider the basics of naming ligands (including multidentate, ambidentate, and bridging) which occur in simple neutral as well as ionic coordination compounds. In Chapter 3 we will concentrate on the nomenclature for compounds for which a variety of isomers are possible.

Table 2.4 gives some rules for naming ligands and simple coordination compounds. Note that the name of anionic ligands is modified by removing the *-ide* suffix of halides, oxides, hydroxides, and so forth, or the last *e* of an *-ate* or *-ite* ending and replacing these with *-o*. Accordingly, fluoride becomes fluoro, nitrate becomes nitrato, sulfite becomes sulfito, and so forth. (As usual in nomenclature, there are a few exceptions; for example, amide becomes amido and the N-bonding form of the ambidentate nitrite becomes nitro.) The very few positive ligands are modified by adding an *-ium* suffix to the root name. The names of neutral ligands are usually not modified, but there are a few common neutral ligands which have special names. For example, water becomes aquo, ammonia is called ammine, carbon monoxide is carbonyl, and nitrogen(II) oxide is nitrosyl. Molecular oxygen and nitrogen are referred to as dioxygen and dinitrogen, respectively.

In naming a coordination compound, the cation is named first and then the anion (just as for ordinary salts, for example, sodium chloride or ammonium nitrate). For a given complex, the ligands are always named first in alphabetical order followed by the name of the metal. (Note that in writing the formulas for

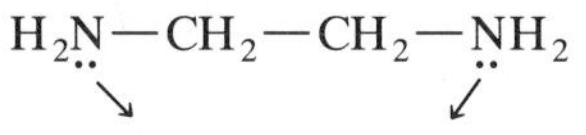

$H_2\ddot{N}—CH_2—CH_2—\ddot{N}H_2$

(*a*)

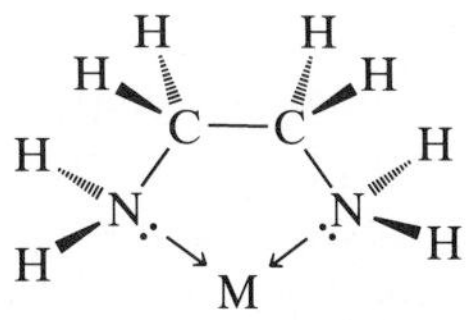

(*b*)

FIGURE 2.5
The bidentate ethylenediamine ligand.

TABLE 2.4
Nomenclature rules for simple coordination compounds

Ligands

1. Anionic ligands end in *-o*.

F^-	fluoro	NO_2^-	nitro	SO_3^{2-}	sulfito	OH^-	hydroxo
Cl^-	chloro	ONO^-	nitrito	SO_4^{2-}	sulfato	CN^-	cyano
Br^-	bromo	NO_3^-	nitrato	$S_2O_3^{2-}$	thiosulfato	NC^-	isocyano
I^-	iodo	CO_3^{2-}	carbonato	ClO_3^-	chlorato	SCN^-	thiocyanato
O^{2-}	oxo	$C_2O_4^{2-}$	oxalato	CH_3COO^-	acetato	NCS^-	isothiocyanato

2. Neutral ligands are named as the neutral molecule.

C_2H_4	ethylene	$(C_6H_5)_3P$	triphenylphosphine
$NH_2CH_2CH_2NH_2$	ethylenediamine	CH_3NH_2	methylamine

3. There are special names for four neutral ligands:

   H_2O aquo NH_3 ammine CO carbonyl NO nitrosyl

4. Cationic ligands end in *-ium*.

   $NH_2NH_3^+$ hydrazinium

5. Ambidentate ligands are indicated by:
   a. Using special names for the two forms, for example, nitro and nitrito for $—NO_2^-$ and $—ONO^-$
   b. Placing the symbol of the coordinating ligand in front of the name of the ligand, for example, *S*-thiocyanato and *N*-thiocyanato for —SCN and —NCS

6. Bridging ligands are indicated by placing a μ- before the name of the ligand.

Simple coordination compounds

1. Name the cation first, then the anion.
2. List the ligands alphabetically.
3. Indicate the number (2, 3, 4, 5, 6) of each type of ligand by:
   a. The prefixes *di-*, *tri-*, *tetra-*, *penta-*, *hexa-* for:
      (1) All monoatomic ligands
      (2) Polyatomic ligands with short names
      (3) Neutral ligands with special names
   b. The prefixes *bis-*, *tris-*, *tetrakis-*, *pentakis-*, *hexakis-* for:
      (1) Ligands whose names contain a prefix of the first type (*di-*, *tri-*, etc.)
      (2) Neutral ligands without special names
      (3) Ionic ligands with particularly long names
4. If the anion is complex, add the suffix *-ate* to the name of the metal. (Sometimes the *-ium* or other suffix of the normal name is removed before adding the *-ate* suffix. Some metals, such as copper, iron, gold, and silver, use the Latin stem for the metal and become cuprate, ferrate, aurate, and argentate, respectively.)
5. Put the oxidation state in roman numerals in parentheses after the name of the central metal.

coordination compounds, the opposite order is followed, with the symbol for the metal preceding the formulas for the ligands.) The oxidation state of the metal is indicated by Roman numerals in parentheses after the name. If the complex is an anion, the *-ate* suffix is added to the name of the metal. Occasionally, the *-ium* or other suffix has to be removed from the name of the metal before the *-ate* is added. For example, chromium becomes chromate, manganese becomes manganate, and molybdenum becomes molybdenate. Some metals, such as copper, iron, gold, and silver, retain the Latin stem for the metal and become cuprate, ferrate, aurate, and argentate in an anionic setting.

The number of ligands is indicated by the appropriate prefix given in the table. Note that there are two sets of prefixes, one (*di-*, *tri-*, *tetra-*, etc.) for monoatomic ions, polyatomic ions with short names, or the special neutral ligands noted above, and a second (*bis-*, *tris-*, *tetrakis-*, etc.) for ligands which already contain a prefix from the first list, for example, ethylenediamine or triphenylphosphine, or for ligands whose names commonly appear in parentheses. The use of parentheses is not as systematic in practice as might be expected. Generally, neutral ligands without special names and ionic ligands with particularly long names are enclosed in parentheses. So, for example, acetylacetonato is generally enclosed in parentheses while oxalato is not.

There are two ways to handle ambidentate ligands. One is to use a slightly different form of the name depending on the atom that is donating the electron pair to the metal. The second is to actually put the symbol of the donating atom before the name of the ligand. So —SCN might be called thiocyanato or *S*-thiocyanato, while —NCS would be isothiocyanato or *N*-thiocyanato. $—NO_2^-$ and —ONO, however, are most always referred to as nitro and nitrito, respectively.

Bridging ligands are designated by placing a Greek μ before the name of the ligand. So a bridging hydroxide (OH^-), amide (NH_2^-), or peroxide (O_2^{2-}) ligand becomes μ-hydroxo, μ-amido, or μ-peroxo, respectively. If there is more than one of a given bridging ligand, the prefix indicating the number of ligands is placed after the μ. For example, if there are two bridging chloride ligands, they are indicated as μ-dichloro. If there are more than one different bridging ligands, they are given in alphabetical order.

The best way to understand Table 2.4 and the above explanation is by a series of examples. First we consider naming compounds for which we are given a formula.

Example 1. Name the compound $[Co(NH_3)_4Cl_2]Cl$.

We start by naming the complex cation. The ligands are named alphabetically with ammine first and then chloro. There are four ammonias and two chlorides, so the prefixes *tetra-* and *di-* are used. The cobalt oxidation state is determined by tracing the charges back as follows. The net charge on the complex cation must be +1 to balance the one −1 chloride anion. Since there are two −1 chlorides in the coordination sphere, the cobalt must be +3 in order for the net charge on the cation to come out as +1. With all this in mind, the full name of the compound is

tetraamminedichlorocobalt(III) chloride

Example 2. Name the compound $(NH_4)_2[Pt(NCS)_6]$

Here we have a platinum-containing complex anion and the common ammonium ion, NH_4^+, as the cation. Given that the ligand is written with the N symbol first, we know that it is the isothiocyanato (or, alternatively, *N*-thiocyanato) form of the ambidentate ligand. There are six of these ligands, so we use the *hexa-* prefix. The anion must have a net charge of −2 to balance the two +1 ammonium cations. Since the thiocyanate ion is also −1, the platinum oxidation state must be +4 to give a net −2 charge on the anion. Because the platinum is contained in a complex anion, its *-um* suffix is removed and replaced with *-ate*. Accordingly, the full name of the compound is

ammonium hexaisothiocyanatoplatinate(IV)

Example 3. Name the compound $[Cu(NH_2CH_2CH_2NH_2)_2]SO_4$.

As in Example 1, we again have a complex cation. The ligand is ethylenediamine, which is often abbreviated as en so that the formula for this compound is usually shortened to $[Cu(en)_2]SO_4$. There are two ethylenediamine ligands, but since it is a neutral ligand with *di-* within its name, the prefix *bis-* is used. The copper oxidation state will be the same as the net charge on the complex cation (since the ligands are neutral). That charge must be $+2$ to balance the -2 of the sulfate ion. The full name of this compound is

bis(ethylenediamine)copper(II) sulfate

Example 4. Name the compound $[Ag(CH_3NH_2)_2][Mn(H_2O)_2(C_2O_4)_2]$.

In this case both the cation and the anion are complex. Starting with the cation, there are two methylamine ligands for which we will use the *bis-* prefix. In the anion, the two aquo (water) ligands come alphabetically before the two oxalato ligands. The name manganese is amended to manganate because this metal is in a complex anion. The oxidation states here must be such that a 1:1 ratio of the cation and anion results. This could be Ag(I)/Mn(III), Ag(II)/Mn(II), or similar values in which, in this case, the sum of the oxidation states is 4. Considering the chemistry of silver and manganese, the first case is the most appropriate. (In time, you will become familiar enough with common transition metal oxidation states that you will not need to debate such matters.) The full name of this compound is

bis(methylamine)silver(I) diaquodioxalatomanganate(III)

Example 5. Name the compound $[(NH_3)_4Co \begin{smallmatrix} OH \\ OH \\ OH \end{smallmatrix} Co(NH_3)_4](NO_3)_3$.

This is our first example of a bridged compound. The three hydroxides bridge between the two cobalt ions. We name such compounds from left to right and remember to put a μ in front of the bridging ligands. The oxidation states of the metals could be (III) and (III) or (II) and (IV) or (I) and (V) or any other combination adding up to 6, but even from our brief exposure to cobalt chemistry, you would probably (and correctly) chose the first alternative. The full name of the compound is

tetraamminecobalt(III)-μ-trihydroxotetraamminecobalt(III) nitrate

Now we turn to a few examples in which we are given the names of some coordination compounds and are asked to supply the correct formulas. Incidentally, the rules for writing these formulas, as for all chemical compounds, are determined by the International Union of Pure and Applied Chemistry (IUPAC). The IUPAC rules concerning the order in which the formulas of the ligands in a coordination compound should be written are unexpectedly complicated and generally not treated in a textbook at this level. Instead, we will follow the common and simplifying (but not officially correct) practice of writing the ligand formulas of a coordination compound in the same order they are named, that is, in alphabetical order by the first letter of the name of the ligand.

Example 6. Write the formula for the compound (acetylacetonato)tetraaquocobalt(II) chloride.

The formula for the bidentate acetylacetonate ligand is given in Table 2.3, but this -1 anion is usually abbreviated as acac. The acac and four waters constitute the coordination sphere that with the cobalt(II) cation are set apart in brackets. The net charge on the complex cation is $+1$ (because the acac is -1), so one chloride counteranion is needed. The formula of the compound would be

$$[Co(acac)(H_2O)_4]Cl$$

Example 7. Write the formula for the compound triamminechloro (ethylene)nitroplatinum(IV) phosphate.

This compound has four different types of ligands in the coordination sphere: NH_3, Cl^-, C_2H_4, and NO_2^- (bonded through the nitrogen). The only real difficulty in constructing this formula is figuring out how many cations and anions there must be. The cation has a net charge of $+2$, and the anion is -3. Therefore, there must be three cations and two anions to ensure electrical neutrality. The formula for this compound is

$$[Pt(NH_3)_3Cl(C_2H_4)NO_2]_3(PO_4)_2$$

Example 8. Write the formula for tetraamminechromium(III)-μ-amido-μ-hydroxobis(ethylenediamine)iron(III) sulfate.

The OH^- and NH_2^- are bridging ligands between the cobalt and iron cations. The overall charge on this huge cation is $+4$ ($+6$ from the two $+3$ cations and -2 from the two -1 anions). Therefore there must be two -2 sulfates in the formula which is given below.

$$[(NH_3)_4Cr\begin{smallmatrix} H_2 \\ N \\ \diagup\diagdown \\ \diagdown\diagup \\ O \\ H \end{smallmatrix}Fe(en)_2](SO_4)_2$$

SUMMARY

Coordination compounds are typically characterized by four or six ligands in a coordination sphere surrounding a metal atom or ion. This chapter starts a systematic investigation of coordination chemistry by putting its history into perspective, introducing some typical ligands, and setting down the basics of a scheme of nomenclature.

The discovery and explanation of coordination compounds should be viewed against the larger picture of progress in understanding atomic structure, the periodic table, and molecular bonding. The contributions of Proust and Lavoisier, among others, led Dalton to formulate the first concrete atomic theory in 1808. Mendeleev published his first periodic table in 1869. With the discoveries of x-rays, radioactivity, electrons, and the nucleus at the turn of the last century, the modern quantum-mechanical picture of the atom started to emerge in the 1920s. This model gives a theoretical explanation for atomic line spectra and the modern periodic table.

Organic chemists led the way in picturing molecular bonding. They relied on concepts such as radicals (which kept their identity through various reactions) and atoms with a fixed valence or combining power. Once the electron was discovered in the early part of this century, Lewis was able to explain some aspects of bonding on the basis of his electron-dot formulas and the octet rule. The valence-shell electron-pair repulsion (VSEPR), valence-bond (VB), and molecular orbital (MO) theories followed in the 1930s.

Coordination compounds were first prepared in the late 1700s. Over the next century, many compounds were synthesized and characterized but little progress was made in formulating and accounting for their molecular structures. Attempts to apply the concepts of radicals, chains of self-linked atoms, and a constant fixed valence (all ideas which had been so successful in organizing organic compounds) did not work well for coordination compounds.

The Blomstrand-Jørgensen chain theory was the most successful of the early theories which attempted to explain the known series of cobalt ammonates. This theory combined trivalent cobalt atoms, divalent ammonia radicals, and monovalent chlorides to produce structures that accounted for some of the formulas, conductivities, and reactions of these compounds. However, when an analog of a critical compound was finally synthesized, the chain theory's prediction was wrong and it started to lose favor.

Alfred Werner literally dreamed up the modern theory of coordination compounds in 1892. He envisioned that metals had two types of valence, which we refer to today as oxidation state and coordination number. Some ligands satisfy only the coordination number, while others simultaneously satisfy the oxidation state. These ideas explain why some chlorides in the cobalt ammonate chlorides are ionizable and some are not. By comparing the actual number of known isomers with the number which should exist for various geometries, Werner was able to conclude that the six ligands in the cobalt ammonates were in an octahedral arrangement.

Ammonia is an example of a monodentate ligand, meaning that it can donate only one lone pair of electrons to a given metal atom. Ethylenediamine, on the other hand, is bidentate. When both nitrogen atoms donate a pair of electrons to a metal, a ring including the metal atom is formed. Ligands which form rings in this manner are called chelates or chelating agents. Various multidentate, bridging, and ambidentate ligands are given in a list of common ligands (Table 2.3).

The nomenclature of simple coordination compounds is developed in a set of rules for referring to ionic and neutral ligands, the number of each type of ligand, and the oxidation state of the metal. A number of examples of naming compounds and writing formulas are given.

PROBLEMS

***2.1.** Briefly state, in your own words, Dalton's atomic theory. How might the concept of embedded hooks have been used to explain the existence of molecules formed between the atoms that Dalton pictured?

2.2. The law of definite composition says that the elements in a given compound are always present in the same proportions by mass. How might this observation have led early chemists to the concept of fixed valence?

2.3. The law of multiple proportions says that two elements can combine in several different proportions by mass to form different compounds. For example, carbon forms both a monoxide and a dioxide. Is this law, as exemplified by these two oxides of carbon, consistent or inconsistent with the concept of fixed valence? Briefly explain your answer.

***2.4.** Briefly explain how the experiment in which alpha particles were shot at thin gold foils led Ernest Rutherford to propose that the atom contains a nucleus.

2.5. Write a concise paragraph explaining how a knowledge of the modern quantum-mechanical or Schrödinger atom can be applied to explain the emission line spectra of elements.

***2.6.** Write a brief, concise paragraph explaining how the rules governing the relationships of the four quantum numbers determine the shape of the modern periodic table.

2.7. Briefly explain how the concept of valence led to the famous blanks left in Mendeleev's early periodic tables.

2.8. Briefly summarize how the ideas of (*a*) chains of carbon atoms, (*b*) single fixed valences for all atoms, and (*c*) groups ("radicals") of atoms also of fixed combining capacity led to disappointing results in picturing the bonding in coordination compounds.

2.9. If you had been in Jørgensen's place in the late 1890s, how might you have attempted to explain the two isomers of $CoCl_3 \cdot 4NH_3$?

2.10. In the same way that the ammonia molecule was thought to have an overall valence of 2 and was represented as $—NH_3—$ in the Blomstrand-Jørgensen chain theory, water can be represented as $—H_2O—$. Several compounds containing chromium(III), water, and chloride are given below.

Formula	No. of Cl^- ions precipitated
(1) $CrCl_3 \cdot 6H_2O$	3
(2) $CrCl_3 \cdot 5H_2O$	2
(3) $CrCl_3 \cdot 4H_2O$	1

(*a*) Write a balanced equation for the reaction of compound (1) with an aqueous silver nitrate solution, $AgNO_3(aq)$.

(*b*) How might you suspect that the conductivity of aqueous solutions of these compounds might vary? Briefly rationalize your answer.

(*c*) Draw appropriate diagrams of compound (2) in the above series using
   (i) the Blomstrand-Jørgensen chain theory
   (ii) the Werner coordination theory
   (iii) the present method of representing coordination compounds.

2.11. The primary compounds considered by Blomstrand, Jørgensen, and Werner, as they struggled to provide a theory for what we now know as *coordination compounds*, were, as we have seen, the cobalt ammonates. Another series known in the late 1800s was the platinum ammonate chlorides for which the following data had been tabulated.

Compound	No. of Cl^- ions precipitated with $AgNO_3$	Conductivity	No. of isomers known
(1) $PtCl_2 \cdot 4NH_3$	2	|	1
(2) $PtCl_2 \cdot 3NH_3$	1	Decreases	1
(3) $PtCl_2 \cdot 2NH_3$	0	↓	2

(*a*) Write structural formulas for these three compounds as Blomstrand and Jørgensen might have written them.
(*b*) Write structural formulas for these compounds [including the two isomers of compound (3)] as Werner would have written them.
(*c*) Write structural formulas as they might be represented today. (*Hint*: The geometry about the platinum is square planar.)

***2.12.** In what way would you suspect that the Jørgensen idea of putting less-reactive chlorides next to the cobalt atom and those more reactive further away influenced Werner's thinking?

2.13. Coordination compounds of formula MA_4 might be square planar or tetrahedral. How many isomers would you predict to exist for compounds of formula MA_2B_2 for these two geometries? $Pt(NH_3)_2Cl_2$ has two known isomers, while $[CoBr_2I_2]^{2-}$ has only one. Speculate on the structures of these complexes.

2.14. Suppose that Werner's *nebenvalenz* (secondary valence) had turned out to be directed toward the corners of a trigonal prism. Draw all the possible isomers of tetraamminedichlorocobalt(III) chloride.

2.15. Suppose that Werner's *nebenvalenz* had turned out to be directed toward the corners of a hexagon. Draw all the possible isomers of the tetraamminedichloroplatinate(IV) cation.

2.16. Suppose that Alfred Werner's *nebenvalenz* had turned out to be hexagonal planar instead of octahedral. Draw and name *all* the possible isomers of $Cr(CO)_3Cl_3$.

2.17. Suppose that Alfred Werner's *nebenvalenz* had turned out to be trigonal prismatic instead of octahedral. Draw and name *all* the possible isomers of $Cr(CO)_3Cl_3$.

2.18. Given the compound $Cr(H_2O)_3Cl_3$.
(*a*) Draw a diagram showing how Werner would have represented this compound. Explain which ligands satisfy the primary and secondary valences.
(*b*) How many geometrical isomers of $Cr(H_2O)_3Cl_3$ would be possible if the coordination sphere of this complex was trigonal prismatic? How many if it was octahedral? Justify your answer using clearly drawn diagrams.

2.19. Compare and contrast the terms *ambidentate* and *ambidextrous*.

***2.20.** Draw diagrams similar to that of Figure 2.5 for the bidentate oxalate ligand. (*Hint:* This anion has resonance structures.)

2.21. Combinations of cobalt(III), ammonia, nitrite (NO_2^-) anions, and potassium (K^+) cations result in the formation of a series of seven coordination compounds.
(*a*) Write the modern formulas for the members of this series. (*Hint:* Some compounds do not contain all four of the above components.)
(*b*) How many ionic nitrites would there be in each of these compounds?
(*c*) How many isomers would each of these compounds have assuming they have octahedral coordination spheres?

2.22. Combinations of iron(II), H_2O, Cl^-, and NH_4^+ can result in the formation of a series of seven coordination compounds, one of which is $[Fe(H_2O)_6]Cl_2$.
(*a*) Write the modern formulas for the other members of the series. (*Hint:* Some compounds do not contain all four of the above components.)
(*b*) How many chlorides could be precipitated from each of these compounds by reacting them with aqueous silver nitrate?
(*c*) How many isomers would each of these compounds have assuming they have octahedral coordination spheres?

2.23. Name the following compounds:
(*a*) $[Pt(NH_3)_4Cl_2]SO_4$
(*b*) $K_3[Mo(CN)_6F_2]$
(*c*) K[Co(EDTA)]
(*d*) $[Co(NH_3)_3(NO_2)_3]$
(*e*) $[Fe(en)_3][IrCl_6]$

2.24. Name the following compounds:
(*a*) $[Pt(NH_3)_6]Cl_4$
(*b*) $[Ni(acac)\{P(C_6H_5)_3\}_4]NO_3$
(*c*) $(NH_4)_4[Fe(C_2O_4)_3]$
(*d*) $W(CO)_3(NO)_2$
(*e*) $[VCl_2(en)_2]_4[Fe(CN)_6]$

2.25. Name the following compounds:
(*a*) $[Pt\{P(C_6H_5)_3\}_4](CH_3COO)_4$
(*b*) $Ca_3[Ag(S_2O_3)_2]_2$
(*c*) $Ru\{As(C_6H_5)_3\}_3Br_2$
(*d*) $K[Cd(H_2O)_2(NTA)]$
(*e*) $[Ag(NH_3)_2]_2[Co(NH_3)ONO)_5]$

2.26. Many coordination compounds were initially named for their color or for the person who first synthesized them. Name the following compounds using modern nomenclature:
(*a*) a "roseo" salt: $[Co(NH_3)_5H_2O]Br_3$
(*b*) purpureocobaltic chloride: $[Co(NH_3)_5Cl]Cl_2$
(*c*) Zeise's salt: $K[PtCl_3(C_2H_4)]$
(*d*) Vauquelin's salt: $[Pd(NH_3)_4][PdCl_4]$

2.27. Name the following compounds:

(*a*) 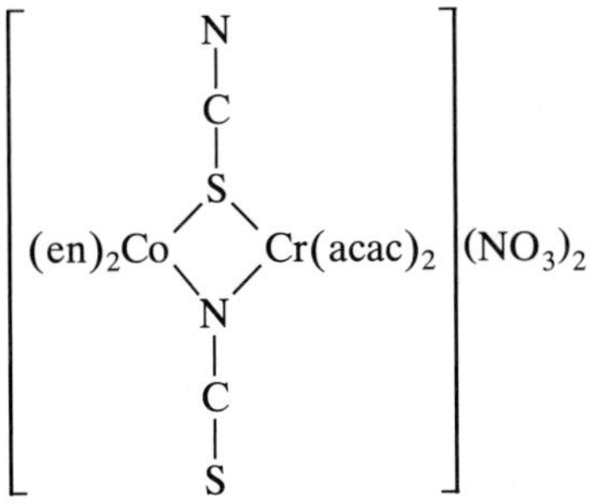
```
 ┌              N               ┐
 │              |               │
 │              C               │
 │              |               │
 │              S               │
 │            /   \             │
 │ (en)2Co            Cr(acac)2 │(NO3)2
 │            \   /             │
 │              N               │
 │              |               │
 │              C               │
 │              |               │
 │              S               │
 └                              ┘
```

(*b*) 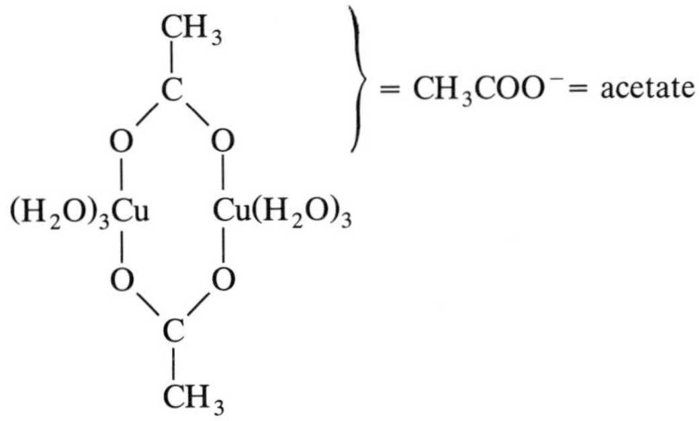
```
              CH3        }
               |         }
               C         }  = CH3COO⁻ = acetate
             /   \       }
            O     O
            |     |
     (H2O)3Cu     Cu(H2O)3
            |     |
            O     O
             \   /
               C
               |
              CH3
```

(*c*)
```
                               H
                               |
                               O
                             /   \
[(SCN)3(H2O)2Cr                    Co(NH3)5]SO4
```

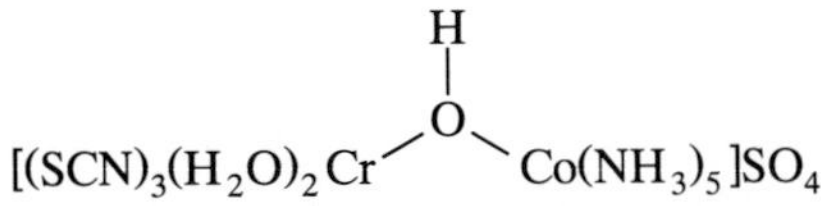

2.28. Write formulas for the following compounds:
(*a*) pentaammine(dinitrogen)ruthenium(II) chloride
(*b*) aquobis(ethylenediamine)thiocyanatocobalt(III) nitrate
(*c*) sodium hexaisocyanochromate(III)
(*d*) hexaamminecobalt(III) pentachlorocuprate(II)

2.29. Write formulas for the following compounds:
(*a*) bis(methylamine)silver(I) acetate
(*b*) barium dibromodioxalatocobaltate(III)

(*c*) carbonyltris(triphenylphosphine)nickel(0)
(*d*) tetrakis(pyridine)platinum(II) tetrachloroplatinate(II)

2.30. Write formulas for the following compounds:
(*a*) tetrakis(pyridine)bis(triphenylarsine)cobalt(III) chloride
(*b*) ammonium dicarbonylnitrosylcobaltate(-I).
(*c*) potassium octacyanomolybdenate(V)
(*d*) diamminedichloroplatinum(II) (Peyrone's salt)

2.31. Write formulas for the following compounds:
(*a*) pentamminechromium(III)-μ-hydroxopentamminechromium(III) chloride
(*b*) diammine(ethylenediamine)chromium(III)-μ-bis(dioxygen)tetraamminecobalt-(III) bromide.

CHAPTER 3

STRUCTURES OF COORDINATION COMPOUNDS

In Chapter 2 we started to investigate the nature of coordination compounds. We saw how Werner was able to account for the structures of the cobalt ammonates by assuming the metal to have two types of valences that today we call oxidation state and coordination number. After defining an isomer, we demonstrated that, assuming an octahedral configuration of the six ligands about the cobalt, there were only two possible isomers for a compound of formula MA_4B_2. Since Werner could prepare only two isomers, he and others were convinced that the octahedral configuration was correct.

Here in Chapter 3 we will conduct a systematic investigation of the structures of coordination compounds. We start by elaborating on the various types of isomers possible and then set out on a tour of the most common coordination numbers. We will find that coordination number 6 usually corresponds to an octahedral configuration, while if the coordination number is 4, both tetrahedral and square planar geometries are possible. The nomenclature necessary to describe the isomers encountered will be developed as we proceed.

3.1 STEREOISOMERS

Table 3.1 shows the relationship among the various types of isomers. Now the word isomer (from the Greek *isomers*) literally means having equal parts, so compounds are isomers if they have the same number and types of parts, in this case, *atoms*. As shown in Table 3.1 isomers can be subdivided into two major categories depending on whether they have the same number and types of *chemical bonds* or not. Those with the same numbers and types of bonds are called stereoisomers and are the subject of this section. Structural isomers, which have differing numbers and types of bonds, are taken up in Section 3.6.

Stereoisomers differ in the spatial arrangements of their bonds. If the spatial arrangements result in different geometries, they are known simply as *geometric isomers*. The nomenclature for these isomers most often distinguishes between the

TABLE 3.1
Types of Isomers

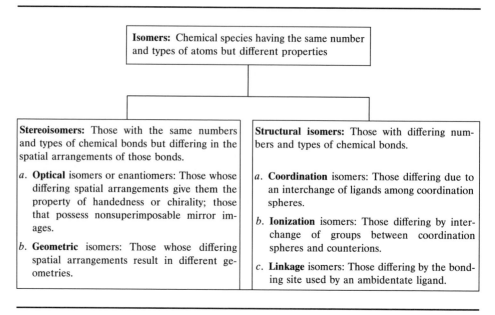

Isomers: Chemical species having the same number and types of atoms but different properties

Stereoisomers	Structural isomers
Stereoisomers: Those with the same numbers and types of chemical bonds but differing in the spatial arrangements of those bonds.	**Structural isomers:** Those with differing numbers and types of chemical bonds.
a. **Optical** isomers or enantiomers: Those whose differing spatial arrangements give them the property of handedness or chirality; those that possess nonsuperimposable mirror images.	*a.* **Coordination** isomers: Those differing due to an interchange of ligands among coordination spheres.
b. **Geometric** isomers: Those whose differing spatial arrangements result in different geometries.	*b.* **Ionization** isomers: Those differing by interchange of groups between coordination spheres and counterions.
	c. **Linkage** isomers: Those differing by the bonding site used by an ambidentate ligand.

possible geometrical arrangements by appending a descriptive prefix such as *cis-*, *trans-*, *mer-*, or *fac-* before the name of the compound. Specific examples will be encountered as we make our tour through the most common types of coordination spheres.

The second type of stereoisomer occurs in molecules that have the property of chirality, a word that comes from the Greek *cheir*, meaning "the hand." *Chirality* literally means handedness. Now hands, as we all know, come in two forms, left-handed and right-handed, that are nonsuperimposable mirror images of each other. Try a little experiment. Place your hands palm to palm with the fingers lined up thumb to thumb, index finger to index finger, and so forth. Note that your hands are mirror images of each other. Now try to maneuver your hands so that, when looking at the back of them, for example, each hand looks exactly the same. In other words, try to superimpose one hand on the other. Try as you might, the hands cannot be made to look the same; they are nonsuperimposable mirror images of each other. They are chiral. Other objects (gloves, scissors, fencing foils, various tools) are also chiral; that is, they come in either a left- or right-handed form. When you buy a baseball glove, for example, you have to specify whether you want a left-handed or a right-handed glove.

Molecules also come in handed, or chiral, forms that are nonsuperimposable mirror images of each other. Such molecules are known as optical isomers or enantiomers. The word *enantiomer* comes from the Greek words *enantios*, meaning "opposite," and *meros*, meaning "part," so that *enantiomers* are the matched left- and right-handed forms of a given chiral molecule. While these enantiomers always have identical melting points, boiling points, dipole moments, solvent capabilities, and so forth, one property that distinguishes them is the ability to rotate the plane of polarized light in opposite directions.

Light is commonly pictured as oscillating electric and magnetic fields occupying perpendicular planes both of which include the direction of travel, as shown in Figure 3.1*a*. Note that there are an infinite number of planes which fit the above description. In ordinary unpolarized light, no one plane is preferred over another. This situation is depicted in Figure 3.1*b*. If, however, ordinary light is passed through (or sometimes reflected off) an appropriate medium, it can be polarized, or made to act as if it is confined to just one given plane. Polarization is represented in Figure 3.1*c*. Polarized light has only one possible plane for its oscillating electric or magnetic field.

In the early 1880s, it was discovered that certain substances (for example, turpentine and other organic liquids, aqueous sugar solutions, quartz, and other minerals) are *optically active*; that is, they are able to rotate the plane of a polarized light beam of a particular wavelength. This situation is shown in Figure 3.2. What could cause such an effect? Pasteur, in the middle of the nineteenth century, suggested that the optical activity of these substances might arise from the handedness of the molecules themselves. Today, we recognize that Pasteur was absolutely right. One enantiomer of a chiral molecule will rotate polarized light in one direction, while on the other hand (pun intended), the other enantiomer will rotate the light an equal amount in the opposite direction. If there are equal numbers of molecules of both enantiomers present, what is known as a "racemic mixture," then there will be no net rotation of the plane of the polarized light.

How can we tell if a given molecule is chiral or not? One way is to actually construct its mirror image and see if it is superimposable on the original. If the mirror image is nonsuperimposable, the molecule is chiral. It will exist as enantiomers that, when separated, will be optically active. However, the actual construction of mirror images is often time-consuming and cumbersome. A second, quicker, and almost always reliable method of testing for chirality is to look for an *internal mirror plane*, that is, a plane of symmetry which passes through the molecule such that any given component atom either sits in the plane or can be reflected through it into another, exactly equivalent atom. A molecule that does not possess such an internal mirror plane will be chiral. (There are a few exceptions to this rule, but they are beyond the scope of this text.)

Figure 3.3 shows several molecules, some that have an internal mirror plane and some that do not. Figure 3.3*a* shows three (of the possible five) internal mirror planes in the square planar tetrachloroplatinate(II) ion. Consider, for example, the plane marked M_3. Note that it contains the platinum atom and two of the chlorine atoms, while the remaining two chlorine atoms are reflected through the plane into each other. Since this molecule contains an internal mirror plane, it is nonchiral. Figure 3.3*b* and *c* shows tetrahedral species. The bromodichloroiodozincate(II) ion, $[ZnBrCl_2I]^{2-}$, shown in Figure 3.3*b*, is nonchiral because it has an internal mirror plane containing the zinc, bromine, and iodine atoms and reflecting the chlorine atoms into each other. Try as we might, however, no internal mirror planes can be identified in the bromochlorofluoroiodomethane molecule of Figure 3.3*c*. This chiral molecule is optically active and possesses left- and right-handed enantiomers.

The nomenclature for optical isomers most often recognizes the existence of enantiomers by putting the symbol R/S- before the name of the compound. R stands for *rectus* (from the Greek word meaning "right-handed"), while S stands for *sinister* (meaning "left-handed"). The R/S-designation, then, indicates the existence of these enantiomers. The symbol d/l- before the name of a chiral

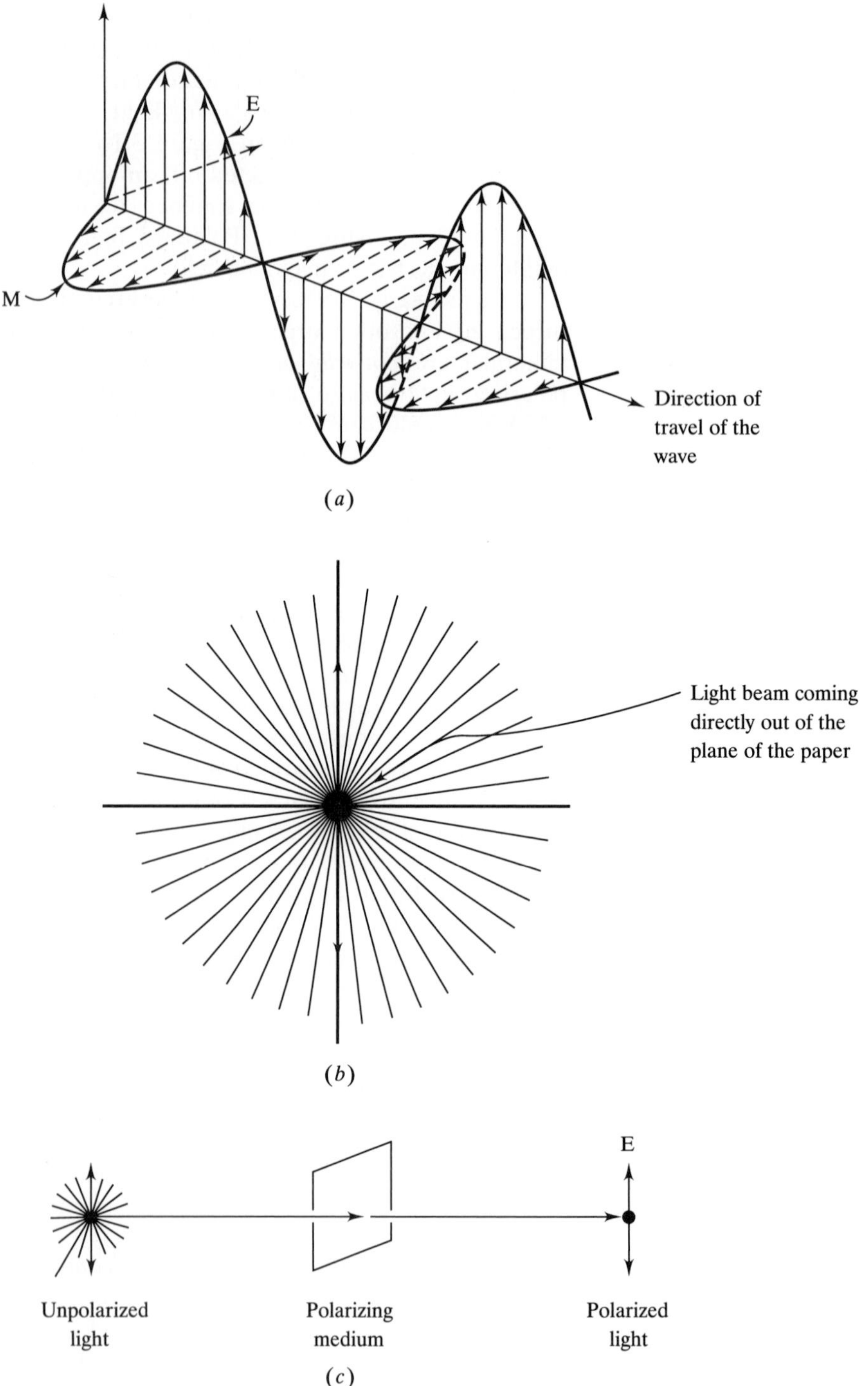

FIGURE 3.1
Light as electromagnetic radiation. (*a*) Light as perpendicular electric (E) and magnetic (M) fields. (*b*) Some of the possible planes of the electric field in unpolarized light. (*c*) Unpolarized light is passed through a polarizing medium that allows only one orientation (plane) of the electric field to pass.

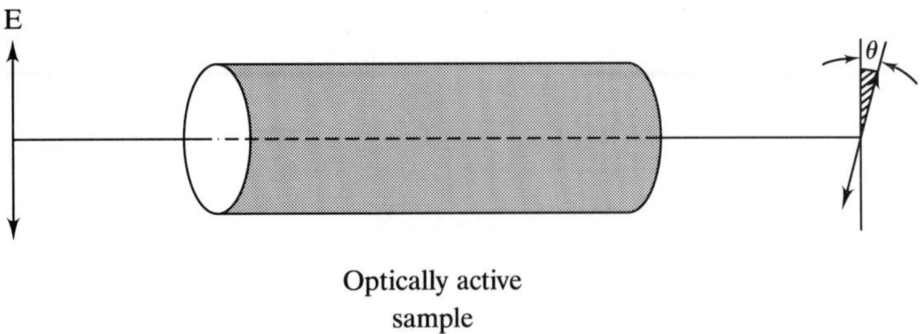

FIGURE 3.2
Passing polarized light through an optically active sample causes the plane of the oscillating electric field (and, not shown, the perpendicular plane of the oscillating magnetic field) to be rotated by an angle θ.

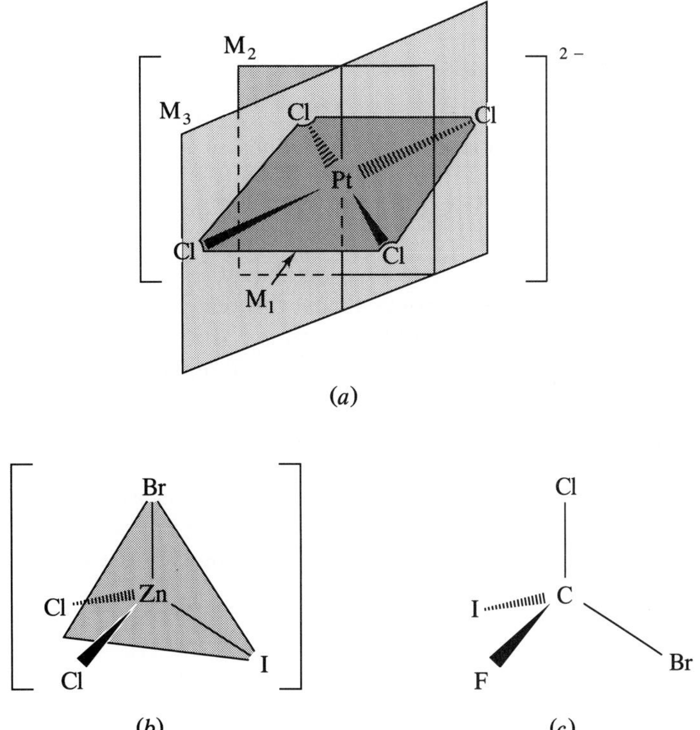

FIGURE 3.3
(*a*) The planar $PtCl_4^{2-}$ has five internal mirror planes, three of which are shown. (*b*) $[ZnCl_2BrI]^{2-}$ has one internal mirror plane that bisects the Cl-Zn-Cl angle. (*c*) CBrClFI has no internal mirror planes and is therefore chiral.

compound centers on the opposite abilities of a pair of enantiomers to rotate the plane of polarized light. The d stands for *dextrorotatory* (rotating light to the right), while the l is for levorotatory (rotating light to the left). (Incidentally, it should be pointed out that there is not a direct correspondence between the R/S- and d/l- terminology. That is, for example, rectus (R) molecules do not always rotate polarized light to the right.) Finally, the capital Greek letters Δ and Λ also occasionally serve to indicate the chirality of a compound. Specific examples of this nomenclature will follow in the tour of common coordination sphere geometries that follows.

3.2 OCTAHEDRAL COORDINATION SPHERES

We saw in Chapter 2 (pp. 16–18) how Werner determined that cobalt, chromium, and a variety of other metals have a secondary valence directed to the corners of an octahedron. Now we turn our attention to a detailed description of this, the most common geometry encountered for coordination compounds. We will see that a wide variety of both geometrical and optical stereoisomers are encountered. At the outset, we restrict our investigation to compounds involving only monodentate ligands and then move on to some more complicated cases involving multidentate, chelating ligands.

Compounds with Monodentate Ligands

Recall that all six positions in an octahedron are equivalent. Therefore, when one of the monodentate A ligands in a coordination compound of formula MA_6 is replaced with a different ligand B, only one possible configuration exists for the resulting MA_5B complex. (There may be different ways to draw them, but all such structures are equivalent. See Figure 2.4*a* for further details.) With only one possible configuration, octahedral MA_5B does not have geometrical isomers. You should satisfy yourself that this one possible structure contains at least one internal mirror plane and therefore is not chiral.

When a second B ligand replaces an A ligand to produce a complex of formula MA_4B_2, two geometrical isomers are possible. As an example, consider the cation $[Co(NH_3)_4Cl_2]^+$ that occurs in compounds such as tetraamminedichlorocobalt(III) chloride, $[Co(NH_3)_4Cl_2]Cl$. The isomer with the two chloride ligands across from each other, as shown in Figure 3.4*a*, is called *trans* (meaning "over" or "across"), while the one with two adjacent chlorides, shown in Figure 3.4*b* is called *cis* (meaning "on the same side of"). The full name of the

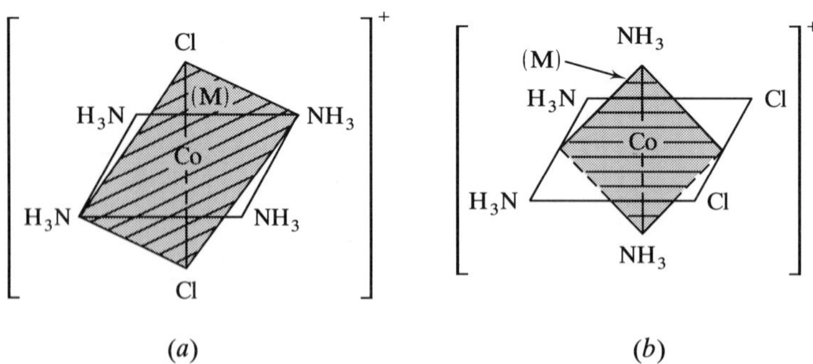

FIGURE 3.4
(*a*) The *trans*-tetraamminedichlorocobalt(III) cation has the two chloride ligands across from each other. (*b*) The *cis*-tetraamminedichlorocobalt(III) has the two chloride ligands that are adjacent to each other. Both these geometrical isomers possess an internal mirror plane (*M*) and are nonchiral.

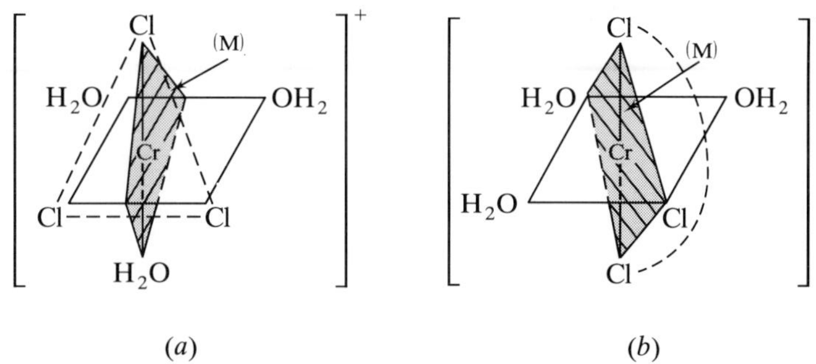

FIGURE 3.5
(*a*) *cis*- or *fac*-triaquotrichlorochromium(III) with the triangular face of chloride ligands outlined (dashed lines). (*b*) *trans*- or *mer*-triaquotrichlorochromium(III) with the chlorides along half the meridian outlined (dashed lines). Both geometrical isomers possess an internal mirror plane (*M*) and are nonchiral.

complete compound would be *cis*- or *trans*-tetraamminedichlorocobalt(III) chloride. It was his preparation of two and only two geometric isomers of compounds of this type that convinced Werner (and his rival, Jørgensen, for that matter) that cobalt complexes are octahedral. Note in Figure 3.4 that both of these geometrical isomers contain an internal mirror plane and are nonchiral.

The replacement of a third A ligand with a B produces a complex of formula MA_3B_3. Again, there are two possible geometrical isomers, occasionally also referred to using the *cis*- and *trans*- prefixes. An example is triaquotrichlorochromium(III) shown in Figure 3.5. Because the cis isomer has chlorides (and waters) at the corners of a triangular face of the octahedron, it is more common to refer to it as the facial isomer and give it the prefix *fac*-. The trans isomer has chlorides (and also waters) along half of a meridian of the octahedron and is called the meridional isomer and given the *mer*- prefix. (A meridian is a great circle of the earth passing through the geographical poles.) As shown in the figure, both isomers have an internal mirror plane and are not chiral.

Neither time nor space allows many more examples of octahedral coordination spheres containing only monodentate ligands, but we should consider one that has both geometrical and optical isomers. Consider the complex of general formula $MA_2B_2C_2$, a somewhat more complicated case than the ones previously encountered. It is best to approach the possibilities systematically. Start by putting the A ligands trans to each other and then placing the B and C ligands. Figure 3.6*a* and *b* shows the two possible placements of B's and C's. The first has both trans B's and C's, while the second has both cis B's and C's. Note that the case where, for example, the B's are trans and the C's are cis is not possible. (You should be able to convince yourself that this is true.)

Moving now to the cases that have cis A's, Figure 3.6*c* has trans B's and cis C's, Figure 3.6*d* trans C's and cis B's, and Figure 3.6*e* has both cis B's and C's. There are, therefore, five geometrical isomers in this case. Are any of these chiral? The figures show internal mirror planes for all the cases except the last. Accordingly, the cis-cis-cis isomer is chiral; it has optical isomers.

Now we must consider the nomenclature for the above case. Consider the compound diamminediaquodicyanocobalt(III) chloride that contains the $[Co(NH_3)_2(H_2O)_2(CN)_2]^+$ cation. The possible structures are as shown in Figure

trans A's:

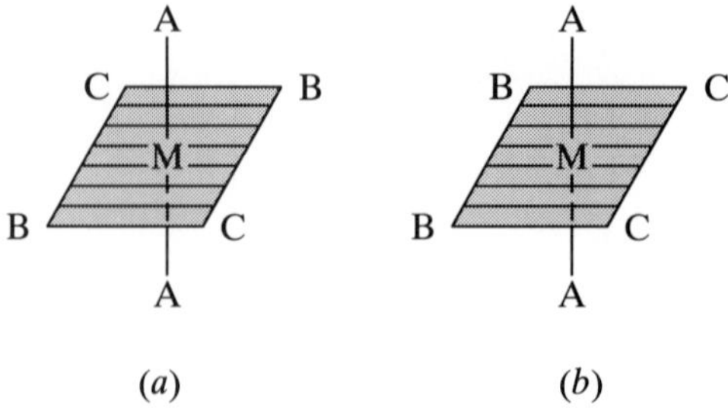

cis A's:

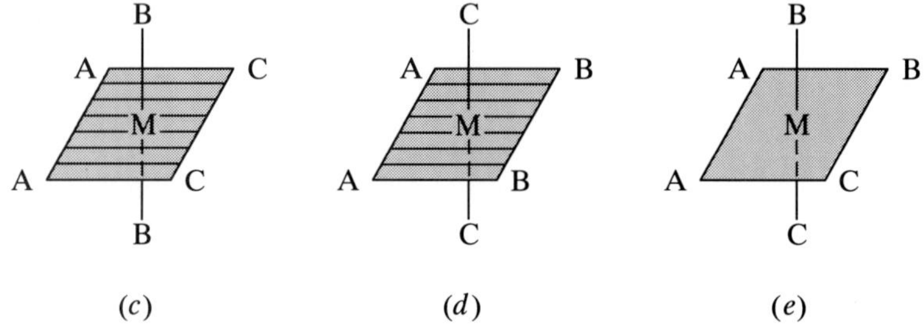

FIGURE 3.6
The five geometric isomers of $MA_2B_2C_2$. (*a*) and (*b*) have the two A ligands in the trans positions, while (*c*), (*d*), and (*e*) have them in the cis positions. (*a*) to (*d*) have internal mirror planes, but the all-cis isomer (*e*) has no mirror plane and is chiral.

3.6 with $M = Co^{3+}$, $A = NH_3$, $B = H_2O$, and $C = CN^-$. The set of names for the five geometric isomers, as given below, must clearly indicate the arrangements of the ligands in each case.

1. *trans*-diammine-*trans*-diaquodicyanocobalt(III)
2. *trans*-diammine-*cis*-diaquodicyanocobalt(III)
3. *cis*-diammine-*trans*-diaquodicyanocobalt(III)
4. *cis*-diammine-*cis*-diaquo-*trans*-dicyanocobalt(III)
5. *R/S*-*cis*-diammine-*cis*-diaquo-*cis*-dicyanocobalt(III)

Note that only in the last two cases must the cis/trans nature of the cyanide ligands be specified. In the first three cases, they have to be trans, cis, and cis, respectively. Why must this be so? It comes down to the fact alluded to earlier that once one pair of ligands is cis in this coordination sphere, there must always be another cis pair. To see why this is so, take the second case, shown in general in Figure 3.6*b*, as an example. Note that once the A's are specified to be trans and the B's to be cis, the C ligands (CN^- in our specific compound) are required to be cis. In the last two cases in which both the A's and B's are cis, however, the C's could be either trans as shown in case 4 or cis as in case 5. The last geometrical isomer, the cis-cis-cis case, is chiral, and so the *R/S*- prefix must be added to so indicate the existence of optical isomers. Compounds of this type are difficult to resolve into enantiomers. In fact, this latter compound was not resolved until 1979.

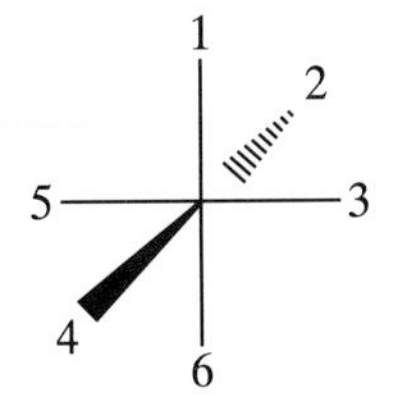

FIGURE 3.7
The numbering system for referring to ligand positions in more complicated octahedral cases.

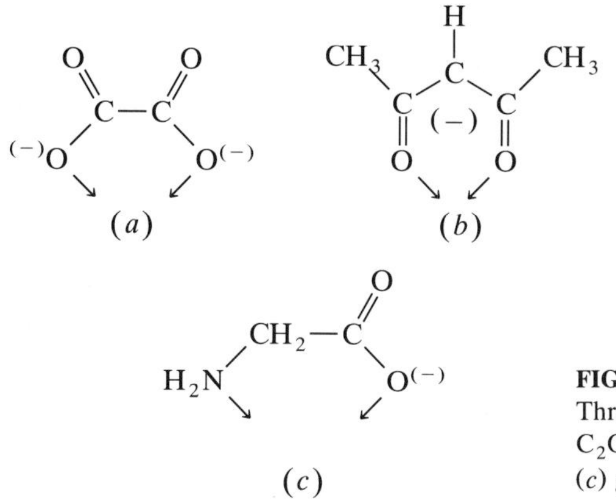

FIGURE 3.8
Three common chelating ligands: (*a*) oxalate, $C_2O_4^{2-}$; (*b*) acetylacetonate, $CH_3COCHCOCH_3$; (*c*) glycinate, $NH_2CH_2COO^-$.

A large number of coordination compounds contain only monodentate ligands; some of them have an impressive number of geometric and optical isomers. For example, a compound of formula MA_2BCDE has nine geometric isomers, six of which are chiral. For MABCDEF there are 15 geometric isomers, all of which are chiral. In these more involved cases, the use of *trans*- and *cis*- prefixes is awkward. Instead, it is more convenient to use a numbering system like the one shown in Figure 3.7. Using this system, the cis-cis-cis case above becomes *R*/*S*-1,2-diammine-3,4-diaquo-5,6-dicyanocobalt(III).

Compounds with Chelating Ligands

Multidentate ligands, often called chelating agents, were introduced in Chapter 2 (p. 20). In addition to ethylenediamine, $NH_2CH_2CH_2NH_2$, shown in Figure 2.5, three other common chelating agents are displayed in Figure 3.8. Ethylenediamine, oxalate, and acetylacetonate are symmetrical chelating ligands (both halves of the ligand are the same), while glycinate is unsymmetrical. Confining ourselves to a discussion of symmetrical bidentates (represented as A–A) for the moment, we start with octahedral complexes of general formula $M(A\text{–}A)_2B_2$.

It is important to realize that the coordinating sites of bidentate ligands generally can span only the cis positions of an octahedron. (They are not long enough to span across the trans positions.) With this in mind, we look at a specific example of a $M(A\text{–}A)_2B_2$ complex, the diaquodioxalatochromate(III) anion, $[Cr(H_2O)_2(C_2O_4)_2]^-$. The two possible geometric isomers are shown in Figure 3.9. The trans isomer possesses an internal plane of symmetry (that includes the metal and the oxalates) and is nonchiral. The cis isomer does not have an internal plane and is optically active.

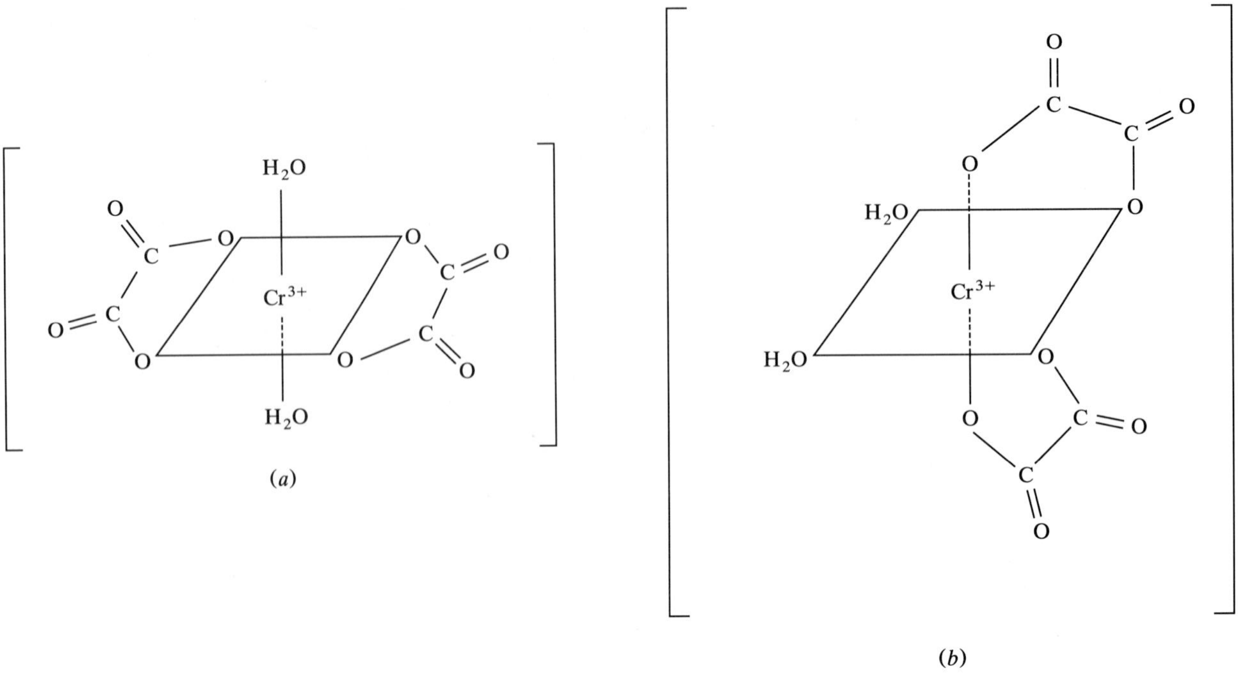

(a)

(b)

FIGURE 3.9
The two geometric isomers of the $[Cr(H_2O)_2(C_2O_4)_2]^-$ ion: (*a*) *trans*- and (*b*) *R/S-cis*-diaquodioxalatochromate(III).

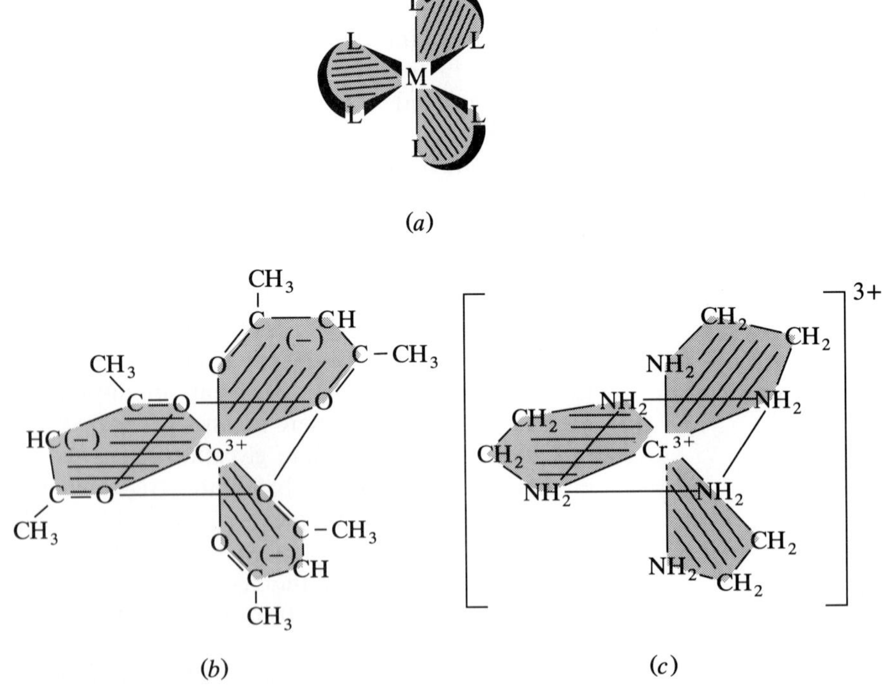

FIGURE 3.10
Propellerlike tris(A–A)M complexes: (*a*) view showing similarity to a propeller, (*b*) *R/S*-tris(acetylacetonato)cobalt(III), and (*c*) the *R/S*-tris(ethylenediamine)chromium(III) cation.

When three bidentate ligands surround a metal, propellerlike complexes result. Two examples of these, R/S-tris(acetylacetonato)cobalt(III) and the R/S-tris(ethylenediamine)chromium(III) cation, are shown in Figure 3.10. These are both chiral species.

Recall (Chapter 2, p. 18) that comparing the number of predicted *geometric* isomers with the actual number that could be synthesized constituted impressive but "negative" evidence for octahedral secondary valences. The positive proof that Werner so desperately needed was obtained by resolving the *optical* isomers of compounds containing chelating ligands. In 1911, Werner and Victor King, an American doctoral student, were able to report the synthesis and resolution of the optical isomers of *cis*-$[Co(NH_3)Cl(en)_2]X_2$, where X = Cl, Br, or I.

How might you carry out such a resolution? Werner and King used a right-handed form of the chiral anion 3-bromocamphor-9-sulfonate to replace the halide anion in the racemic mixture of the above salt. Now, by definition, the racemic mixture contains equal numbers of the right- and left-handed enantiomers of the cation. The right-handed enantiomer of the resolving anion will form salts with the two "hands" of the cation. These salts will have different spatial

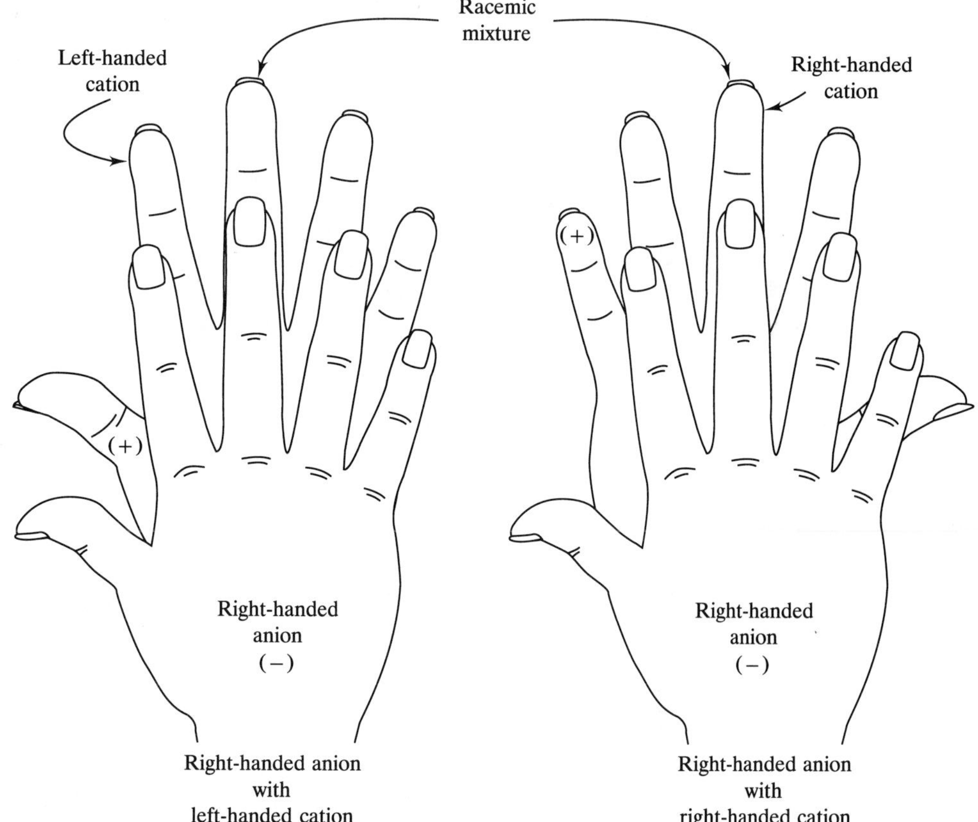

FIGURE 3.11
A representation of the resolution of a racemic mixture of the enantiomers of a cation (+) using a right-handed form of a chiral anion (−). The left- and right-handed forms of the cations have different positions relative to the right-handed form of the anion.

relationships among the component atoms and therefore different properties such as solubilities and melting points. The situation is represented using actual hands in Figure 3.11. Note that the distances between the positions of the fingers of the right-left pair is different from the positions in the right-right pair. Salts analogous to the right-left pair would have different properties than those analogous to the right-right pair and therefore could be successfully separated.

Werner and his students went on, in a remarkably short period of time, to resolve a large number of chiral chelate-containing coordination compounds. Indeed, this work led directly to Werner's receiving the Nobel Prize in chemistry in 1913. However, there was one nagging doubt left to be resolved. All of the chiral compounds synthesized up to that time (by far the majority by organic chemists) had contained carbon. Could it be, as some of Werner's critics argued, that this element had some mysterious, special ability to produce optical isomerism? Perhaps it was not the octahedral configuration of chelating ligands around a metal atom that produced the optical activity but rather merely this special ability of carbon atoms. To dispel this last doubt, Werner set out to resolve an optically active coordination compound that contained absolutely no carbon. In 1914, Werner and Sophie Matissen (Werner was ahead of his time in that he supervised the work of many women doctoral students) reported the resolution of the amazing compound shown in Figure 3.12. (Somewhat ironically, this compound had been synthesized some 16 years earlier by Jørgensen and his students.) In this compound, the chelating ligands are themselves complex cations with the hydroxide groups bridging between the peripheral cobalt(III) ions in the ligands and the one in the very center of the complex. Note that this chiral compound contains no carbon. Chirality could no longer be asserted to be the exclusive province of carbon chemistry.

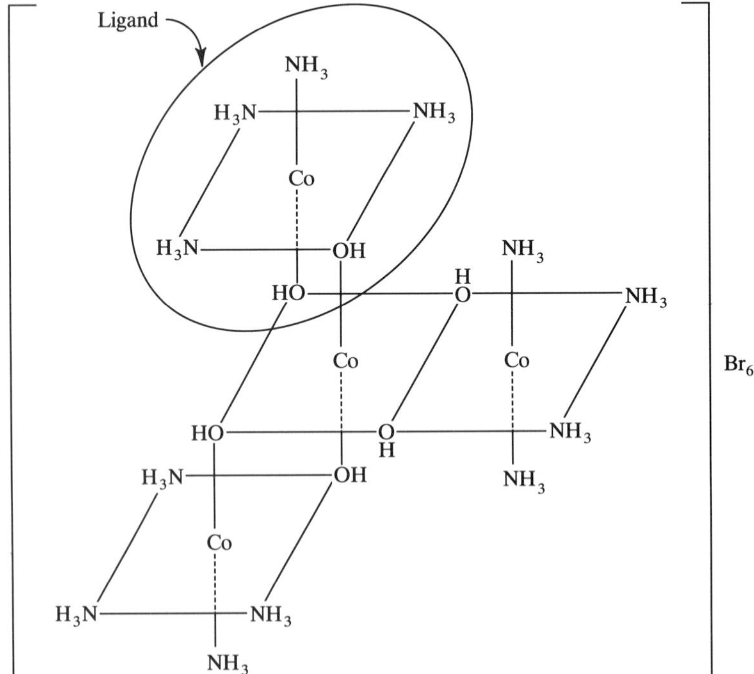

FIGURE 3.12
$\{Co[Co(OH)_2(NH_3)_4]_3\}Br_6$, a chiral coordination compound containing no carbon resolved by Werner and Matissen.

3.3 SQUARE PLANAR COORDINATION SPHERES

While coordination number 6 is certainly the most prevalent, 4 is also fairly common. The geometry associated with four ligands around a central metal is usually either tetrahedral or square planar. While it makes sense that the larger the size of the ligand, the fewer of them could fit around a small metal cation, a detailed explanation of the reasons that a given metal takes on a coordination number of 4 rather than 6 and square planar rather than tetrahedral, will have to await our discussion of bonding theories (Chapter 4) and, to some extent, solid state structures (Chapter 7). For now, it is sufficient to say that square planar complexes are the most common in d^8 metals such as Ni(II), Pd(II), Pt(II), and Au(III) and in the d^9 Cu(II).

Analogous to the MA_5B case discussed earlier, there is only one possible configuration for square planar MA_3B. For MA_2B_2 there are two geometric isomers, cis and trans. For example, diamminedichloroplatinum(II), $[Pt(NH_3)_2Cl_2]$, is shown in Figure 3.13. As is commonly the case in square planar complexes, the plane of the molecule is the internal mirror plane, and these compounds are rarely chiral. It follows that more complicated compounds, like those of the type MA_2BC —take, for example, the cation in amminechlorobis(pyridine)platinum(II) chloride, $[Pt(NH_3)Cl(py)_2]Cl$—also have two possible geometric isomers, but neither of these is chiral. As in the more complicated octahedral cases, the nomenclature for MABCD compounds can often be simplified using a numbering system. Consider amminebromochloro(pyridine)platinium(II) as an example. Keeping the ammine in place in the upper left corner, the other three ligands can be placed trans to it as shown in Figure 3.14*a* to *c*. The first isomer can be named fairly unambiguously as *trans*-amminebromochloro(pyridine)platinum(II), but the other two are both *cis*-amminebromo compounds. Accordingly, the numbering system shown in Figure 3.14*d* makes life much easier.

Recall that the introduction of chelating ligands into octahedral coordination compounds made for a greater incidence of chiral species. While this is not generally true for square planar cases, there are a few well-established examples where asymmetric chelating ligands produce chiral square planar complexes. One such example is given in Figure 3.15.

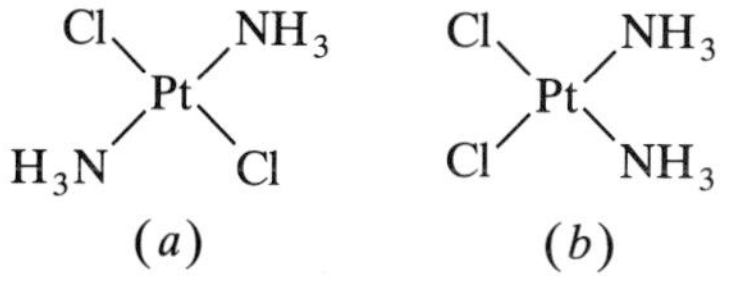

Cl NH$_3$ / Pt / H_3N Cl
(*a*)

Cl NH$_3$ / Pt / Cl NH$_3$
(*b*)

FIGURE 3.13
(*a*) *trans*- and (*b*) *cis*-diamminedichloroplatinum(II).

NH$_3$ Cl / Pt / py Br
(*a*)

NH$_3$ Br / Pt / py Cl
(*b*)

NH$_3$ Br / Pt / Cl py
(*c*)

1 2 / M / 4 3
(*d*)

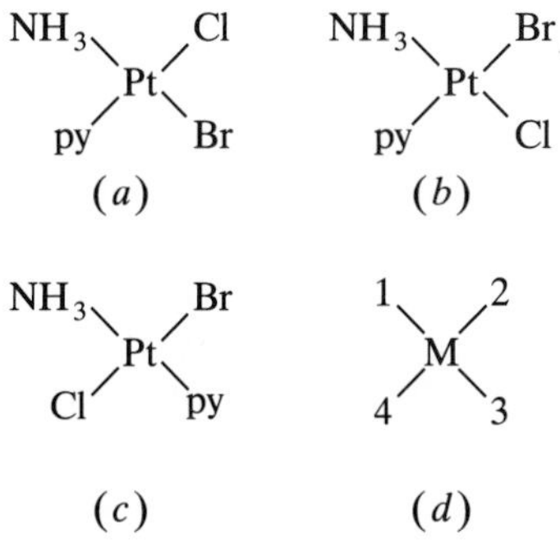

FIGURE 3.14
The geometric isomers of $Pt(NH_3)BrCl(py)$: (*a*) 1-ammine-3-bromochloro(pyridine)platinum(II), (*b*) 1-amminebromo-3-chloro(pyridine)platinum(II), and (*c*) 1-amminebromochloro-3-(pyridine)platinum(II). (*d*) General numbering scheme for square planar coordination compounds. [Ref. 2, p. 305.]

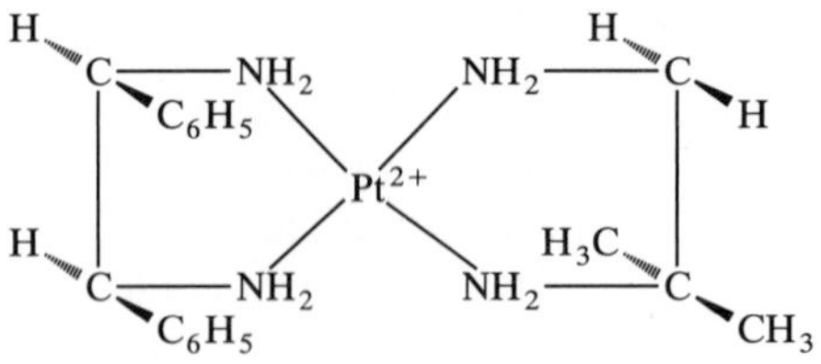

FIGURE 3.15
A square planar chiral complex of platinum(II) with an asymmetric derivative of ethylenediamine as the chelating agent.

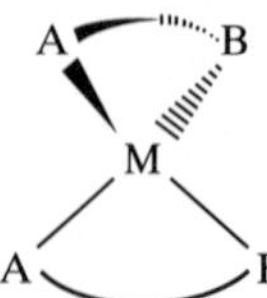

FIGURE 3.16
A tetrahedral chiral coordination compound results when an asymmetric ligand (A–B) is chelated to a metal (M).

3.4 TETRAHEDRAL COORDINATION SPHERES

Since all four positions of a tetrahedron are adjacent to each other, there can be no cis/trans geometric isomers. To convince yourself that this is true, start with an MA_4 tetrahedral configuration and replace one A ligand with a B. The resulting MA_3B configuration is the same no matter which A ligand was initially replaced. Now replace a second A ligand. Again, no matter which of the remaining A ligands is replaced, the resulting MA_2B_2 looks exactly the same. The B ligands cannot be across from each other in one structure and on the same side in another.

There are at least two ways to generate chiral tetrahedral structures. The first is a compound analogous to the bromochlorofluoroiodomethane pictured in Figure 3.3*c*. Compounds of this type containing a central transition metal atom are difficult to prepare but in principle are chiral. Chirality can also be achieved by using asymmetric chelating ligands. Figure 3.16 shows two general asymmetric ligands A–B chelated to a metal atom. Note that the structure contains no internal mirror plane and is therefore chiral.

3.5 OTHER COORDINATION SPHERES

While octahedral, square planar, and tetrahedral were certainly the most common geometries encountered by Werner and his contemporaries, there is an ever-growing list of others. Most of these have come to light with the introduction, in the 1960s, of increasingly routine structural determination methods such as x-ray diffraction techniques. Some representative coordination numbers and corresponding structures are given in Table 3.2 and Figure 3.17.

Coordination number 2 is uncommon among transition-metal compounds. As represented in Table 3.2, the rather limited number of these species that do exist are confined almost exclusively to complexes of silver(I), gold(I), mercury(II), and (not shown) copper(I).

Although some compounds would appear to have a coordination number 3, only a few, on close inspection, actually do. For example, cesium trichlorocuprate(II), $CsCuCl_3$, is actually made up of chloride-bridged chains of tetrahedral $CuCl_4^{2-}$ units. Genuine examples of coordination number 3 often involve large, bulky ligands. Two relatively simple examples are given in Table 3.2.

Coordination number 4, as we have seen, involves either tetrahedral or square planar geometries. The former is commonly found for various d^5 or d^{10} electronic configurations, while the latter, as previously noted, is most often found with d^8 and, occasionally, d^9 metals. Often, the energetic difference between these two configurations is very small. A few representative examples of both these geometries are given in Table 3.2. The number of coordination compounds with a coordination number of 4 is second only to those in which the coordination number is 6.

Coordination number 5 also has two predominate geometries, square pyramidal and trigonal bipyramidal, that differ only slightly in energy. In fact, such five-coordinate compounds are often examples of *fluxional compounds*, those which exist in two or more chemically equivalent configurations so rapidly interconverted that some physical measurements are unable to distinguish one from the other. Figure 3.18 shows what is known as the Berry mechanism for the interconversion of the two forms of iron pentacarbonyl, $Fe(CO)_5$, a typical fluxional compound. Notice that only very small, low-energy movements of the ligands are involved in converting one trigonal bipyramid into a square pyramidal transition

TABLE 3.2
Survey of coordination sphere geometries

Coordination number	Structure	Examples
2	Linear	$Ag(NH_3)_2^+$, $AuCl_2^-$, $Hg(CN)_2$
3	Trigonal	HgI_3^-, $Pt(PPh_3)_3$
4	Tetrahedral	MnO_4^-, $CoBr_4^{2-}$, ReO_4^-, $Ni(CO)_4$
	Square planar	$PdCl_4^{2-}$, $Pt(NH_3)_4^{2+}$, $Ni(CN)_4^{2-}$
5	Trigonal bipyramidal	$CuCl_5^{3-}$, $CdCl_5^{3-}$, $Ni(CN)_5^{3-}$
	Square pyramidal	$VO(acac)_2$, $Ni(CN)_5^{3-}$
6	Octahedral	$Co(NH_3)_6^{3+}$
	Trigonal prismatic	$Re(S_2C_2Ph_2)_3$
7	Pentagonal bipyramidal	ZrF_7^{3-}, HfF_7^{3-}, $V(CN)_7^{4-}$
	Trigonally capped octahedron	NbF_7^{2-}, TaF_7^{2-}
	Tetragonally capped trigonal prism	$NbOF_6^{3-}$
8	Square antiprism	ReF_8^{2-}, TaF_8^{3-}, $W(CN)_8^{4-}$
	Triangular-faced dodecahedron	$Mo(CN)_8^{4-}$, $Ti(NO_3)_4$
9	Tricapped trigonal prism	ReH_9^{2-}

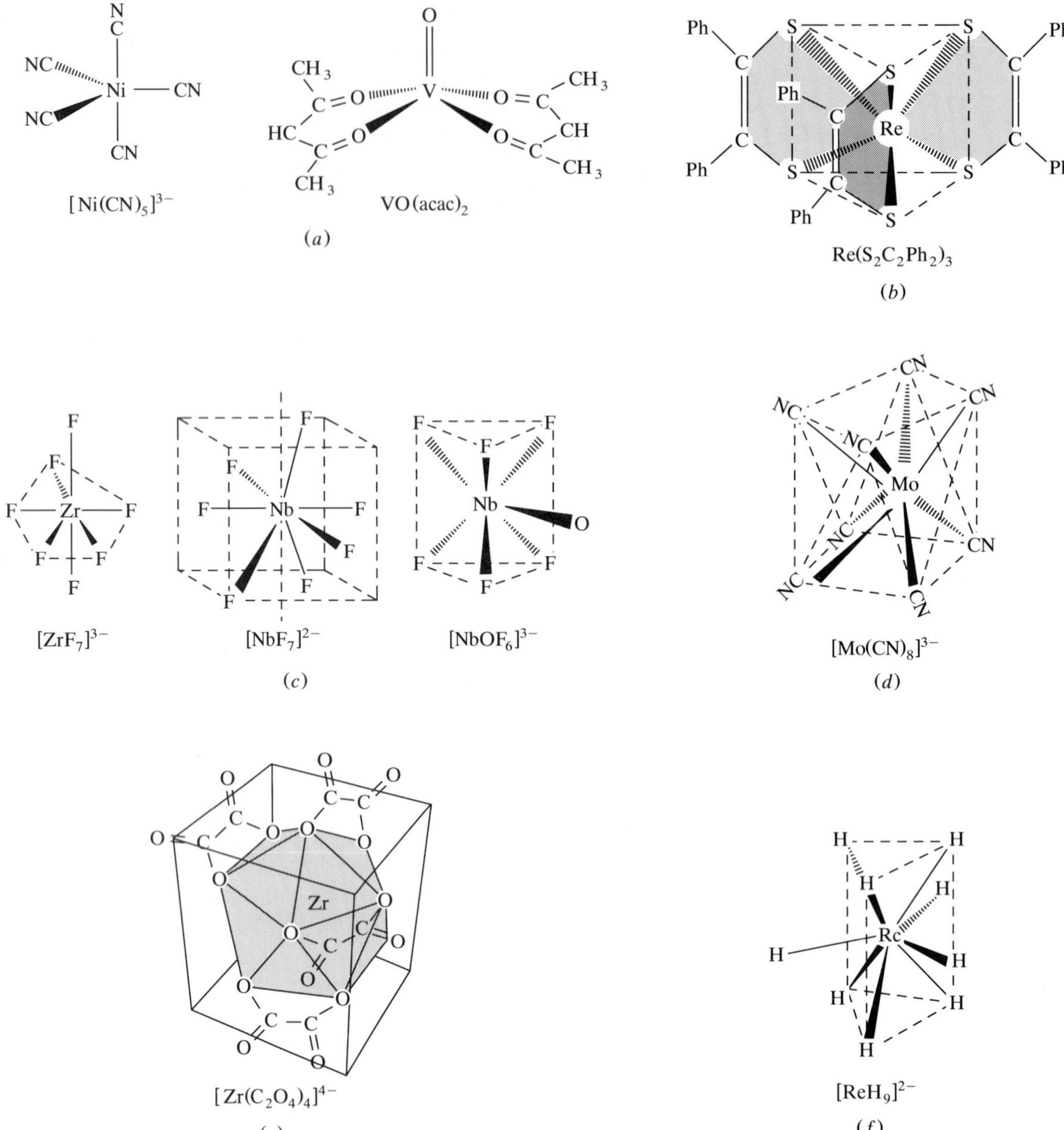

FIGURE 3.17
Some representative examples of coordination numbers 5 through 9: (*a*) trigonal bipyramidal $[Ni(CN)_5]^{3-}$ and square planar $VO(acac)_2$; (*b*) trigonal prismatic $Re(S_2C_2Ph_2)_3$; (*c*) pentagonal bipyramidal $[ZrF_7]^{3-}$, trigonally capped octahedral $[NbF_7]^{2-}$, and tetragonally capped trigonal prismatic $[NbOF_6]^{3-}$; (*d*) square antiprismatic $[Mo(CN)_8]^{3-}$; (*e*) triangular-faced dodecahedron $[Zr(C_2O_4)_4]^{4-}$; (*f*) tricapped trigonal prismatic $[ReH_9]^{2-}$.

state and then back into a second trigonal bipyramid. As the labeling of the carbonyl ligands shows, this mechanism makes it easily possible for a given ligand to switch back and forth among the axial and equatorial positions. This interconversion often occurs so rapidly that some physical methods, notably nuclear magnetic resonance (nmr) spectroscopy, cannot distinguish between the axial and equatorial ligands of the trigonal bipyramid. Sometimes, at lower temperatures,

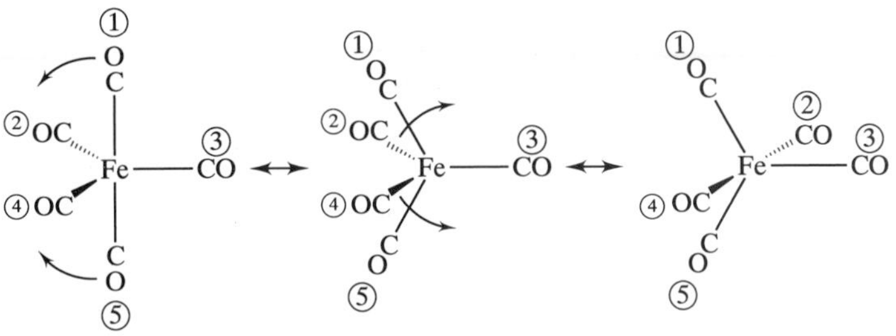

FIGURE 3.18
The Berry mechanism for the interconversion of trigonal bipyramidal and square pyramidal forms of $Fe(CO)_5$, a representative fluxional compound. A given ligand may be axial in one of the trigonal bipyramids but equatorial in another.

this rapid interconversion can be slowed down to the point that the axial and equatorial ligands can be identified.

Certainly not all five-coordinate coordination compounds are fluxional. Table 3.2 shows a few examples of species that adopt either a square pyramidal or trigonal bipyramidal configuration. Note, however, that the pentacyanonickelate(II) anion, $Ni(CN)_5^{3-}$, exists in both forms. In fact, both forms can be found in the same compound, $[Cr(en)_3][Ni(CN)_5]\cdot 1.5H_2O$.

We have seen that an octahedral configuration is by far the most prevalent geometry in coordination compounds. Table 3.2 and Figure 3.17 both show, however, a rare example of a trigonal prismatic geometry for coordination number 6. Most evidence points to an interaction among the sulfur atoms that stabilizes the trigonal prism over the expected octahedral configuration.

Coordination number 7 actually has at least three common geometries that, again, are very close in energy. In most cases, higher coordination numbers are only favored with large metals (usually second- and third-row transition metals) and/or small ligands like fluoride. A similar situation exists for coordination numbers 8 and 9. Table 3.2 and Figure 3.17 show representative examples of these higher coordination numbers.

3.6 STRUCTURAL ISOMERS

Structural isomers were defined in Table 3.1 as those with differing numbers and types of chemical bonds. There have been a number of names given to a variety of these isomers, but we will restrict ourselves to three: coordination, ionization, and linkage.

Coordination isomers are those characterized by an interchange of ligands among coordination spheres. One set of compounds is $[Pt^{II}(NH_3)_4][Pt^{IV}Cl_6]$ and $[Pt(NH_3)_3Cl][Pt(NH_3)Cl_5]$. Two compounds with the same number and types of atoms but different numbers and types of chemical bonds, they are further characterized by the exchange of an ammine and a chloride between the platinum(II) square planar and the platinum(IV) octahedral coordination spheres. A second set of coordination isomers, involving the exchange of ligands between Co(III) and Cr(III), is $[Co(NH_3)_6][Cr(CN)_6]$ and $[Co(NH_3)_5CN][Cr(NH_3)(CN)_5]$. A third example involving two coordinated metals spanned by bridging ligands is shown in Figure 3.19.

Ionization isomers are those characterized by an interchange of groups between the coordination sphere and the counterions. Starting with

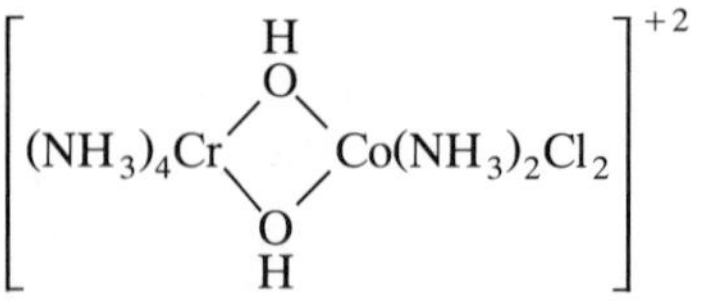

$$\left[(NH_3)_4Cr\begin{matrix}H\\O\\ \\O\\H\end{matrix}Co(NH_3)_2Cl_2\right]^{+2}$$

tetraamminechromium(III)-μ-dihydroxodiamminedichlorocobalt(III)

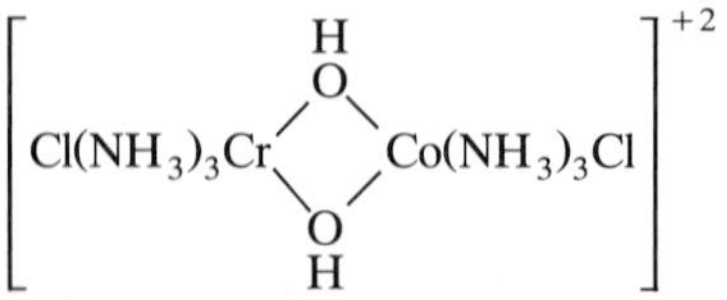

$$\left[Cl(NH_3)_3Cr\begin{matrix}H\\O\\ \\O\\H\end{matrix}Co(NH_3)_3Cl\right]^{+2}$$

triamminechlorochromium(III)-μ-dihydroxotriamminechlorocobalt(III)

FIGURE 3.19
Two coordination isomers of a bridged cation.

$[CoCl(en)_2(NO_2)]SCN$, for example, the thiocyanate counterion can be exchanged with either a chloride or a nitrite ligand in the coordination sphere to produce $[Co(en)_2(NO_2)SCN]Cl$ and $[CoCl(en)_2SCN)]NO_2$, respectively. $[Cr(H_2O)_6]Cl_3$, $[Cr(H_2O)_5Cl]Cl_2 \cdot H_2O$, and $[Cr(H_2O)_4Cl_2]Cl \cdot 2H_2O$ are ionization isomers in which the chloride counterions are exchanged with waters in the coordination sphere. Once outside the coordination sphere, the former aquo ligands become waters of hydration.

Linkage isomers are those that result when an ambidentate ligand (see Chapter 2, p. 21) switches its coordinating atom. The first identified instance of linkage isomerism involved the nitrite ion, NO_2^-. The Lewis structure of this anion involves resonance structures as shown in Figure 3.20*a*. The resonance hybrid, Figure 3.20*b*, involves π-electron density spread out among the three atoms of the anion. The structure of the hybrid shows that both the nitrogen and oxygen atoms have a pair of electrons which can be donated to a metal atom. If the nitrogen atom is the coordinating atom, then the ligand is referred to as *nitro*, while if the oxygen is bound, the ligand is called *nitrito*. Nitrite complexes were not only the first but they are still among the most common examples of linkage isomerism.

In 1894, Jørgensen prepared two forms of the compound now formulated as $[Co(NH_3)_5(NO_2)]Cl_2$. One form was yellow, the other red. These turned out to be the oldest known examples of linkage isomerism. One is the nitro (N-bonded), the other nitrito (O-bonded). But which is which? Jørgensen and Werner worked together on this problem, but it is important to realize that they had none of the fancy instruments used today to help distinguish between such isomers. Instead, they compared the colors of the two forms with those of other known complexes. Compounds that were known to contain six Co^{3+}–N interactions, such as those containing the $[Co(NH_3)_6]^{3+}$ and $[Co(en)_3]^{3+}$ cations, were uniformly yellow. On the other hand, compounds that contained five Co^{3+}—N bonds and one

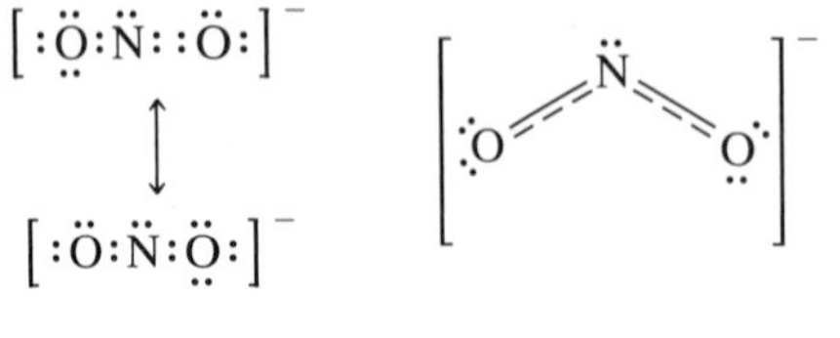

$[:\ddot{O}:\ddot{N}::\ddot{O}:]^-$
↕
$[:\ddot{O}::\ddot{N}:\ddot{O}:]^-$

(*a*)

(*b*)

FIGURE 3.20
(*a*) Two resonance structures and (*b*) the resonance hybrid for the nitrite ion, NO_2^-.

Co^{3+}—O bond, such as $[Co(NH_3)_5H_2O]^{3+}$ and $[Co(NH_3)_5NO_3]^{2+}$, were red. Therefore, it followed that the yellow isomer was N-bonded (nitro) and the red was O-bonded (nitrito). The correct formulas and corresponding names are

$[Co(NH_3)_5NO_2]Cl_2$ (yellow)
Pentaamminenitrocobalt(III) chloride

$[Co(NH_3)_5ONO]Cl_2$ (red)
Pentaamminenitritocobalt(III) chloride

The thiocyanate ion, SCN^-, is also known to form a variety of linkage isomers. Analogous to the above pentaamminecobalt(III) compounds, $[Co(NH_3)_5NCS]^{2+}$ is nitrogen-bonded. The corresponding pentacyanocobalt(III) compound, however, is S-bonded. The full formulas and corresponding names are

$[Co(NH_3)_5NCS]Cl_2$
Pentaammineisothiocyanatocobalt(III) chloride

$K_3[Co(CN)_5SCN]$
Potassium pentacyanothiocyanatocobaltate(III)

There are a few isolated examples of cyanide, CN^-, serving as an ambidentate ligand. One example occurs with the cobalt(III) complexes of the tetradentate ligand triethylenetetramine (trien), $NH_2CH_2CH_2NHCH_2CH_2NHCH_2CH_2NH_2$:

$[Co(CN)_2(trien)]^+$
Dicyano(triethylenetetramine)cobalt(III) cation

$[Co(NC)_2(trien)]^+$
Diisocyano(triethylenetetramine)cobalt(III) cation

SUMMARY

To discuss the structures of coordination compounds, we first defined various types of isomers, compounds that have the same number and types of atoms but differ in their chemical properties. Stereoisomers have the same number and types of chemical bonds, while structural isomers do not. Stereoisomers can be subdivided into geometric and optical isomers.

Optical isomers are chiral, that is, they are "handed" or come in left- and right-handed forms called enantiomers that rotate the plane of polarized light in opposite directions. A racemic mixture contains equal numbers of both enantiomers.

A quick test for chirality is to look for the presence of an internal mirror plane. A molecule that lacks such a plane is most always chiral or optically active and quite often is designated by the *R* (rectus, or right-handed)/*S* (sinister, or left-handed) scheme of nomenclature.

The most common geometry encountered in coordination chemistry is octahedral. Starting with a compound of formula MA_6 and successively replacing the monodentate A ligands with others designated B and C results in a large number of various isomers. MA_4B_2 has two geometric isomers, designated cis and trans. MA_3B_3 also has two geometric isomers that are usually designated facial (*fac-*) and meridonal (*mer-*). The general compound $MA_2B_2C_2$ has five geometric isomers, one of which is chiral.

Additional opportunities for optical activity arise in molecules containing chelating ligands (A–A). For example, a compound of general formula $M(A–A)_2B_2$ has two geometric isomers, one of which is chiral. The propellerlike tris-chelated

$M(A{-}A)_3$ compounds are always chiral. Werner and his collaborators took advantage of the ready optical activity of such complexes to provide additional proof that the coordination spheres of his many and varied cobalt compounds were octahedral. After 20 years of work, Werner's coordination theory was thoroughly established and he received the Nobel Prize in chemistry in 1913. In a final positive proof, Werner and co-workers forever laid to rest the claim that carbon somehow was responsible for chirality by preparing and resolving a chiral compound containing no carbon atoms at all.

A coordination number of 4 is second in prevalence to the octahedral coordination number 6. Two geometries are common for coordination number 4. Square planar compounds, while seldom chiral, often have geometric isomers. Tetrahedral compounds, since all the positions of a tetrahedron are adjacent to each other, cannot have geometric isomers. There are a limited number of examples of chiral tetrahedral compounds that arise when the ligand itself is asymmetric. Other coordination numbers, not known in Werner's time, have come to light in the last 30 years with the advent of modern structure-determining techniques. Fluxional compounds have two or more chemically equivalent configurations so rapidly interconverted that some physical measurements are unable to distinguish one from the other.

Structural isomers, those with differing numbers and types of chemical bonds, can be subdivided into coordination, ionization, and linkage isomers. Coordination isomers are characterized by an interchange of ligands among coordination spheres, while ionization isomers have an interchange between the coordination spheres and the counterions. Linkage isomers result when an ambidentate ligand switches its coordinating atom. Nitrite (NO_2^-), thiocyanate (SCN^-), and cyanide (CN^-) are commonly responsible for most linkage isomers.

PROBLEMS

3.1. Two compounds have the formula C_2H_6O. One, known as common grain alcohol or ethanol, is composed of the ethyl (C_2H_5—) radical bound to the hydroxyl (—OH) group, while the second, dimethyl ether, an extremely volatile organic solvent not suitable for human ingestion, has two methyl (CH_3—) groups bound to a central oxygen atom. Classify these compounds: are they (*a*) isomers, (*b*) structural isomers, (*c*) stereoisomers? More than one of these terms may apply.

3.2. Consider a common chaise lounge used to relax in the sun on warm summer days. It can be spread out into an approximately flat configuration like a cot or it can be made to look rather chairlike or it can be folded up for storage. Are these three forms more like structural or stereoisomers? Briefly rationalize your answer.

3.3. Consider the following two chain theory formulations for the structure of $CoCl_3 \cdot 6NH_3$. Would these be considered stereo- or structural isomers? Briefly explain your answer.

$$\mathrm{Co}\begin{cases}\mathrm{NH_3{-}Cl}\\ \mathrm{NH_3{-}NH_3{-}NH_3{-}NH_3{-}Cl}\\ \mathrm{NH_3{-}Cl}\end{cases} \qquad \mathrm{Co}\begin{cases}\mathrm{NH_3{-}NH_3{-}Cl}\\ \mathrm{NH_3{-}NH_3{-}Cl}\\ \mathrm{NH_3{-}NH_3{-}Cl}\end{cases}$$

3.4. Given the isomers listed for the hexagonal planar and trigonal prism coordination spheres shown in Table 2.2, are these structural isomers or stereoisomers? Briefly justify your answer.

3.5. Look carefully at a good pair of scissors. Does it have an internal mirror plane or not? On the basis of this observation, are these scissors handed or not?

3.6. Carefully explain in one well-composed paragraph, using your own words, what it means to say a molecule is optically active.

3.7. Given the isomers listed for the hexagonal planar coordination sphere shown in Table 2.2, are these geometric isomers or optical isomers? For any geometrical isomers given in the table for hexagonal planar, which, if any, of these are chiral? Briefly justify your answers.

3.8. Given the isomers listed for the trigonal prism coordination sphere shown in Table 2.2, are these geometric isomers or optical isomers? For any geometrical isomers given in the table for trigonal prism, which, if any, of these are chiral? Briefly justify your answers.

3.9. Redraw the structure of the tetrachloroplatinate(II) ion shown in Figure 3.3*a*. Beside it draw its mirror image. Can these mirror images be superimposed on each other? What conclusion regarding the chirality of this anion can you draw from this exercise? Is this the same conclusion arrived at by considering whether the molecule has an internal mirror plane?

3.10. Redraw the structure of the bromodichloroiodozincate(II) ion shown in Figure 3.3*b*. Beside it draw its mirror image. Can these mirror images be superimposed on each other? What conclusion regarding the chirality of this anion can you draw from this exercise? Is this the same conclusion arrived at by considering whether the molecule has an internal mirror plane?

3.11. Redraw the structure of the bromochlorofluoroiodomethane molecule shown in Figure 3.3*c*. Beside it draw its mirror image. Can these mirror images be superimposed on each other? What conclusion regarding the chirality of this molecule can you draw from this exercise? Is this the same conclusion arrived at by considering whether the molecule has an internal mirror plane?

3.12. Originally the four atoms bound to a tetravalent carbon atom were thought to occupy the corners of a square. However, when four different atoms were bound to a carbon atom, the resulting molecules were found to be optically active. Will (*a*) the square planar molecule and/or (*b*) a tetrahedral molecule adequately account for the chirality of this type of molecule? Briefly justify your answer.

3.13. Restate, in your own words, the conditions for chirality that depend on (*a*) the superimposability of mirror images and (*b*) the presence of an internal mirror plane.

3.14. Given AlClBrI and PClBrI, draw a diagram showing the molecular shape of these molecules and then draw their mirror images. Which of these species has superimposable mirror images? Are any of these species chiral?

3.15. For the molecules given in Problem 3.14, which have internal mirror planes? For those that do, draw the plane on your diagram of the molecular shape. Are any of these species chiral?

3.16. Would you expect any of the possible geometrical isomers of PF_2Cl_3 to be chiral? Why or why not? Draw diagrams to support your answer.

3.17. Draw structural formulas for the following compounds:
(*a*) *cis*-chlorocarbonylbis(triphenylphosphine)iridium(I)
(*b*) *trans*-chloronitrotetraamminechromium(III) nitrate
(*c*) *trans*-diammine-*trans*-bis(pyridine)dinitritocobalt(III) nitrate

3.18. How many additional internal mirror planes are there in the first four parts of Figure 3.6? Draw diagrams that illustrate these additional planes.

*__3.19.__ Draw and name all the possible stereoisomers of:
(*a*) $[Fe(NH_3)_3(H_2O)ClF]$
(*b*) $[Fe(NH_3)_2(H_2O)_2ClF]NO_2$

3.20. Draw structural formulas for the following compounds:
(*a*) potassium *trans*-dichlorobis(oxalato)cobaltate(III)
(*b*) (acetylacetonato)-*cis*-dichloro-*trans*-bis(triphenylarsine)rhenium(III)
Are either of these compounds chiral? Briefly support your answer.

3.21. Write formulas for the following molecules or ions:
(*a*) trioxalatochromate(III)
(*b*) *cis*-dichloro(ethylenediamine)platinum(II)
Are either of these compounds chiral? Briefly support your answer.

3.22. Write formulas for the following compounds:
(*a*) (ethylenediaminetetraacetato)cobaltate(III)
(*b*) dichloro(nitrilotriacetato)chromate(III)
Are either of these compounds chiral? Briefly support your answer.

3.23. Briefly explain how the synthesis and resolution of cobalt complexes containing chelating agents played a vital role in establishing the validity of Werner's coordination theory.

3.24. Draw diagrams showing the structures of and name all the stereoisomers of $[CoCl_2(en)_2]Cl$.

3.25. Jørgensen had synthesized $[CoCl_2(en)_2]Cl$ which was known to come in two forms named for their colors: violeo (cis) and praseo (trans). Werner cited the existence of these two (and only two) isomers as proof of an octahedral coordination sphere. Given that the bidentate ethylenediamine can only span cis positions, how many geometric isomers would this compound have if it assumed a trigonal prismatic coordination sphere? Of these, how many would be chiral?

3.26. In 1901, Edith Humphrey, a doctoral student of Alfred Werner, synthesized two isomers of bis(ethylenediamine)dinitrocobalt(III) nitrate. Draw diagrams showing the structures of all the stereoisomers possible for this compound.

3.27. The first published consideration by Werner of the possibility of chiral forms of coordination compounds came in 1899 when he described bis(ethylenediamine)oxalatocobalt(III) chloride. How many geometric and optical isomers does this compound have? Draw diagrams showing the molecular structures of these isomers.

***3.28.** If the compound of Figure 3.15 were actually tetrahedral rather than square planar as shown, would it still be chiral? Briefly justify your answer.

3.29. Nitrilotriacetate (NTA), $N(CH_2COO)_3^{3-}$, is a tetradentate ligand with the following structure:

$$:N\begin{cases} CH_2COO^- \\ CH_2COO^- \\ CH_2COO^- \end{cases}$$

Suppose you are given the following coordination compound:

$$K_2[CoCl(NTA)(SCN)]$$

(*a*) Name the compound.
(*b*) Determine the number of stereoisomers the above compound would have if it were octahedral. Justify the number of chiral species.

3.30. How many stereoisomers would the compound of Problem 3.29 have if it were trigonal prismatic? Justify the number of any chiral species.

3.31. Given the compound $Ni(gly)_2$. Would it be optically active if it were tetrahedral? Square planar? Briefly discuss your answer.

3.32. (*a*) Draw out the Lewis structure of the nitrate ion and discuss its potential to serve as an ambidentate ligand.
(*b*) Why did Jørgensen and Werner identify the complex cation $[Co(NH_3)_5NO_3]^{2+}$ as definitely containing five $Co^{3+}—N$ and one $Co^{3+}—O$ interactions?

3.33. Briefly explain why bis(ethylenediamine)dinitritocobalt(III) chloride, as opposed to bis(ethylenediamine)dinitricobalt(III) chloride, should be the same color as bis(ethylenediamine)dinitratocobalt(III) chloride.

3.34. When Werner's student Edith Humphrey first synthesized the complex salts of formula $[Co(en)_2(NO_2)_2]X$, they were taken to be O-bonded forms of nitrite. Only later were they identified as N-bonded. Write the formulas of and name both the linkage isomers of this compound where X is $AuCl_4^-$.

3.35. Draw a Lewis structure of the thiocyanate ion that contains a triple carbon-nitrogen bond. When thiocyanate binds to a metal through the sulfur atom, studies show that the M—S—C angle is approximately 108–109°, but when it binds through the nitrogen atom, the M—N—C bond angle is linear. On the basis of your Lewis structure for SCN^-, briefly rationalize this difference.

3.36. Draw a Lewis structure of the cyanide ion, CN^-. Would you expect the cyano and isocyano forms of this ambidentate ligand to have linear or bent M—C—N and M—N—C bonds? Briefly discuss your answer.

***3.37.** On the basis of your answers to Problems 3.35 and 3.36, combined with your knowledge of ammonia as a ligand, speculate why $[Co(NH_3)_5NCS]Cl_2$ is N-bonded but $K_3[Co(CN)_5SCN]$ is S-bonded. Consider the amount of space occupied by both S- and N-bonded forms of the thiocyanate ligand as part of your answer.

3.38. *Trans*-bis(ethylenediamine)dithiocyanatocopper(II) forms three linkage isomers. Write structural formulas and name each isomer.

3.39. How many coordination isomers could be formed starting with $[Cu(NH_3)_4][PtBr_4]$? Write the formula for and name each isomer.

3.40. Write the formula for and name one coordination *or* ionization isomer of the following compounds:
(*a*) $[Co(NH_3)_5(NO_3)]SO_4$
(*b*) $[Cr(en)_3][Cr(C_2O_4)_3]$
Indicate the type of isomerization that applies in each case.

3.41. Write the formula for and name one coordination *or* ionization isomer of the following compounds:
(*a*) $[Pt(NH_3)_4Cl_2]Br_2$
(*b*) $[Cu(NH_3)_4][PtCl_4]$.
Indicate the type of isomerization that applies in each case.

3.42. By writing formulas or drawing structures related to the compound of formula $[Pt(NH_3)_4(C_2O_4)]Cl_2$, give one example of as many of the following types of isomers as possible: geometric, optical, linkage, coordination, and ionization. Name each compound that you cite as an illustration.

3.43. By writing formulas or drawing structures related to the compound of formula $[Pd(NH_3)_2(NO_2)_2]$, give one example of as many of the following types of isomers as possible: geometric, optical, linkage, coordination, and ionization. Name each compound that you cite as an illustration. For simplicity, assume that any nitrite ions serving as ligands in a given compound are all either N- or O-bonded.

3.44. By writing formulas or drawing structures related to the compound of formula $[VCl_2(en)_2]NO_2$, give one example of as many of the following types of isomers as possible: geometric, optical, linkage, coordination, and ionization. Name each compound that you cite as an illustration. For simplicity, assume that any nitrite ions serving as ligands in a given compound are all either N- or O-bonded.

CHAPTER 4

BONDING THEORIES FOR COORDINATION COMPOUNDS

In the last two chapters we have been considering the history, nomenclature, and structures of coordination compounds. In these earlier discussions we introduced the metal-ligand (M—L) coordinate-covalent bond in which the ligand donates a pair of electrons to the metal atom or ion. Now we are in a position to consider the nature of the M—L bond in greater detail. Is it primarily an ionic interaction between ligand electrons and a positively charged metal cation? Or should the M—L bond be more properly described as predominantly covalent in character? Whatever the character of the bond, the description of M–L interactions must account for (1) the stability of transition metal complexes, (2) their electronic and magnetic characteristics, and (3) the variety of striking colors displayed by these compounds.

In this chapter we will investigate the rise of various bonding theories that have been applied to Werner's coordination compounds. We will see how Lewis electron-dot diagrams were applied to coordination compounds and how the familiar octet rule was translated into the effective atomic number rule for these compounds. Shortly thereafter, in the 1930s, the crystal field theory (CFT), valence bond theory (VBT), and molecular orbital theory (MOT) were conceived, and there followed an intense struggle to determine the most effective theoretical basis for the bonding in transition metal complexes. As it turns out, the valence bond theory has now fallen into relative disuse and will be treated but briefly in this chapter. Although of great promise, the molecular orbital theory is rather abstract and does not lend itself easily to quantification. This leaves the crystal field theory (in which the M—L bond is treated as purely electrostatic in nature) as the best of the three approaches to the bonding in coordination compounds, particularly at an introductory inorganic level.

4.1 EARLY BONDING THEORIES

Lewis Acid-Base Theory

As depicted in Figure 2.1 and outlined in the Summary of Chapter 2, viable bonding theories started to emerge from the quantum-mechanical model of the atom in the 1920s. G. N. Lewis proposed his now familiar electron-dot diagrams and octet rule for simple compounds in the early 1920s, and by the end of the decade, these ideas were applied to coordination compounds by Nevil Sidgwick. It was he who first proposed the idea of the coordinate-covalent bond referred to in earlier chapters.

To put the various descriptions of the M—L bond into proper historical perspective, we start with a brief review of various acid-base theories. In the latter part of the nineteenth century, Svante Arrhenius's ideas about the existence of ions in the aqueous solutions of salts, acids, and bases led to his classification of acids and bases as substances that liberated hydrogen (H^+) or hydroxide (OH^-) ions, respectively. Regarded today as rather limited in scope, the Arrhenius definitions still remain useful in classifying compounds and their resulting reactions. Later, in the early 1920's, J. N. Brønsted and Thomas Lowry independently proposed a more general theory that classified acids as proton donors and bases as proton acceptors. The Brønsted-Lowry theory was not restricted to aqueous solutions, did not require the presence of a hydroxide group in a base, and encompassed more compounds as acids and bases. It proved to be a broader and more generally useful acid-base theory than that of Arrhenius.

Lewis proposed his still broader and more useful theory of acids and bases in the late 1920s and early 1930s. Classifying acids as electron-pair acceptors and bases as electron-pair donors, he thereby liberated acid-base theory entirely from its former dependence on the presence of hydrogen. The advantage of the Lewis theory is that a larger number of reactions can be classified as acid-base than under either the Arrhenius or Brønsted-Lowry definitions. The classic example used to demonstration the more general nature of the Lewis theory is the gas-phase reaction between boron trifluoride and ammonia represented in Equation (4.1). Note that BF_3 neither provides hydrogen ions in solution nor is a proton donor as required by the Arrhenius and Brønsted-Lowry theories, respectively. Similarly, ammonia neither provides hydroxide ions in solution nor acts as a proton acceptor. Therefore, this reaction is not an acid-base reaction under these more restricted definitions. Boron trifluoride is, however, an electron-pair acceptor and ammonia an electron-pair donor, so the reaction can be classified as acid-base under the Lewis definitions. (In a sense, the product of a reaction between a Lewis acid and a Lewis base could be called a *Lewis salt*. However, the more technical term for such a product is a *Lewis adduct*.)

$$BF_3(g) + :NH_3(g) \longrightarrow F_3B \longleftarrow :NH_3(s) \tag{4.1}$$

$BF_3(g)$	$:NH_3(g)$	$F_3B \longleftarrow :NH_3(s)$
Electron-pair acceptor (Lewis acid)	Electron-pair donor (Lewis base)	A coordinate-covalent bond between the acid and the base (Lewis "salt" or adduct)

Sidgwick applied these ideas to coordination compounds. He noted that compounds such as the cobalt ammonates, described so ably by Alfred Werner's coordination theory, could also be classified as Lewis adducts. Equation (4.2) shows the formation of the hexaamminecobalt(III) cation from the Co^{3+} cation and six ammonia molecules. Note that the metal cation is an electron-pair acceptor (Lewis acid) and each ammonia molecule is an electron-pair donor (Lewis base).

$$\underset{\text{Lewis acid}}{Co^{3+}(aq)} \quad + \quad \underset{\text{Lewis bases}}{6:NH_3(aq)} \quad \longrightarrow \quad \underset{\substack{\text{Coordination compound}\\ = \text{Lewis salt or adduct}}}{\begin{array}{c} \ddot{N}H_3 \\ \downarrow \;\swarrow :NH_3 \\ H_3N: \rightarrow Co^{3+} \leftarrow :NH_3 \\ H_3N: \nearrow \uparrow \\ \ddot{N}H_3 \end{array}} \tag{4.2}$$

The resulting coordination compound, then, can be thought of as a Lewis salt or adduct.

Parallel to the octet rule that applies to the Lewis structures drawn for simple molecules, the *effective atomic number* (*EAN*) rule states that in a coordination compound, the sum of the electrons of the metal plus those donated by the ligands should be equal to the number of electrons associated with the next-higher noble gas. For example, in $[Co(NH_3)_6]^{3+}$, the cobalt +3 ion has (27 − 3 =) 24 electrons, while the six ammine ligands donate (6 × 2 =) 12 electrons for a total of 36 electrons. There are also 36 electrons associated with krypton, the next-higher noble gas found at the right end of the period containing cobalt. Like the octet rule, the EAN rule is often violated but provides a convenient and useful rule of thumb. A large variety of coordination compounds (for example, $Ni(CO)_4$, $[Pd(NH_3)_6]^{4+}$, and $PtCl_6^{2-}$), involving all three rows of transition metals, satisfy the EAN rule. Unfortunately, there are also a large number of perfectly stable compounds (for example, $[Cr(NH_3)_6]^{3+}$, $[PdCl_4]^{2-}$, and $[Pt(NH_3)_4]^{2+}$) in which the EAN rule is violated.

The Crystal Field, Valence-Bond, and Molecular Orbital Theories

As outlined in the introduction, the 1930s saw the emergence of three theories (crystal field, valence-bond, and molecular orbital) that today, with the benefit of perfect hindsight, we recognize held great promise for effective and useful ways of visualizing the bonding in coordination compounds. The crystal field theory (CFT), developed to a large extent by the physicists Hans Bethe and John Van Vleck, regarded the metal-ligand bond as exclusively electrostatic. CFT considered the effect of an octahedral, tetrahedral, or square planar "field" of ligands on the energies of the atomic orbitals of the metal. Mostly the province of physicists, CFT was largely ignored by inorganic chemists until the 1950s. The CFT and its amended version admitting some covalent character [called the adjusted CFT or the ligand field theory (LFT)], is now a principal bonding theory for coordination compounds. One of its primary attractions is its conceptual simplicity, making it of great value, particularly when one is approaching the problem of the bonding in coordination compounds for the first time.

The valence-bond theory (VBT), primarily the work of Linus Pauling, regarded bonding as characterized by the overlap of atomic or hybrid orbitals on individual atoms. (Atomic orbitals are sometimes referred to as "native" orbitals—those original or indigenous to a free atom—whereas linear combinations of these native atomic orbitals constitute hybrid orbitals.) VBT has been largely successful in accounting for the structure of many simple molecules, particularly those encountered in organic chemistry. For an octahedral coordination compound, the VBT theory envisions the overlap among an octahedrally shaped d^2sp^3 hybrid orbital of the metal with appropriate atomic or hybrid orbitals of the ligands. Both of the bonding electrons constituting the M—L bond are donated by the ligand. Although VBT was the principal way in which inorganic chemists visualized coordination compounds until the 1950s, it has now fallen into disfavor due to its inability to account for various magnetic, electronic, and spectroscopic properties of these compounds.

The molecular orbital theory (MOT), developed piecemeal by a number of chemists and physicists, pictures the electrons of a molecule setting up standing- or confined-wave patterns of electrons that are under the influence of two or more nuclei. These *molecular* orbitals, analogous to the familiar atomic orbitals, are spread out over all the atoms of the entire molecule. Molecular orbitals, like their atomic cousins, can be arranged in energy levels that account for the stability of various molecules. Molecular orbital theory would seem to be the best way, in the final analysis, to understand chemical bonding in most all molecules, including coordination compounds. Conceptually, it is simple to understand, but its quantitative application to relatively simple polyatomic molecules including coordination compounds is highly abstract and mathematical. It is not the place for a beginner to start to visualize bonding in these compounds.

Given the above background on the basic tenets and principal advantages and disadvantages of the CFT, VBT, and MOT, it is not surprising that we will concentrate on CFT for the visualization of the bonding in coordination compounds.

4.2 THE CRYSTAL FIELD THEORY

Crystal field theory assumes that all M–L interactions are purely electrostatic in nature. More specifically, it considers the electrostatic effect of a field of ligands on the energies of a metal's valence-shell orbitals. To discuss the CFT we need only be aware of two fundamental concepts: (1) the coulombic theory of electrostatic interactions and (2) the shapes of the valence orbitals of transition metals, that is, the *nd* orbitals ($n = 3$ for the first row of transition metals, and so forth). The first concept involves only the familiar ideas of the repulsion of like and the attraction of dislike electrical charges. Quantitatively, the potential energy of two charges Q_1 and Q_2 separated by a distance r is given by the formula shown in Equation (4.3). The second concept, the shapes of the d orbitals, requires a little more development.

$$\text{Potential energy} = \frac{Q_1 Q_2}{r} \tag{4.3}$$

The Shapes of the 3*d* Orbitals

Orbitals, as you recall from earlier courses, are but confined or standing electron waves. In other words, when a negative electron, treated as a wave as allowed by the wave-particle duality, is confined to the area about a positive nucleus, certain allowed wave patterns or *orbitals* are set up. (Much like various allowed wave patterns are set up in a guitar string or a kettledrum head.) The shapes of various orbitals represent the probabilities of finding an electron in the area surrounding the nucleus of an atom. These orbitals are described by a set of quantum numbers (n, l, m), and each has a characteristic energy. The familiar hydrogen-like (those

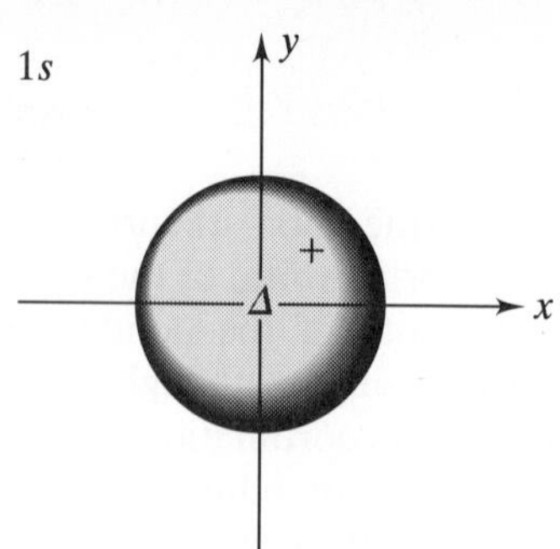

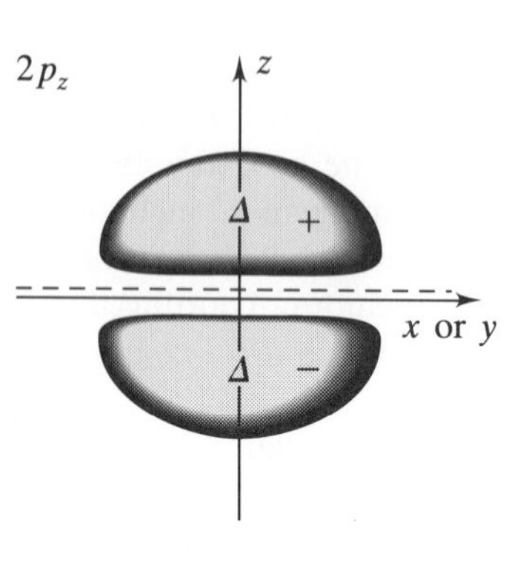

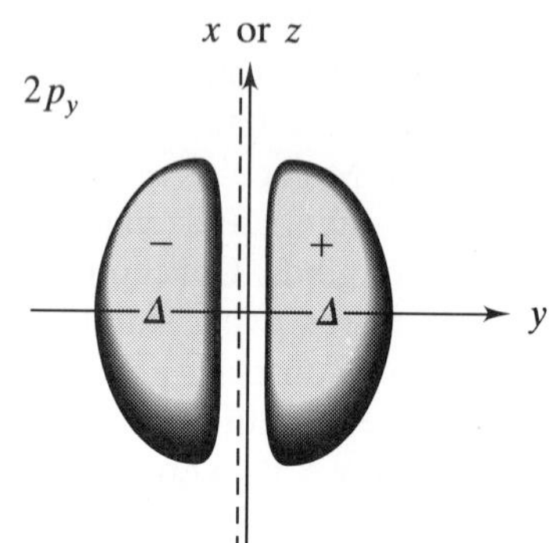

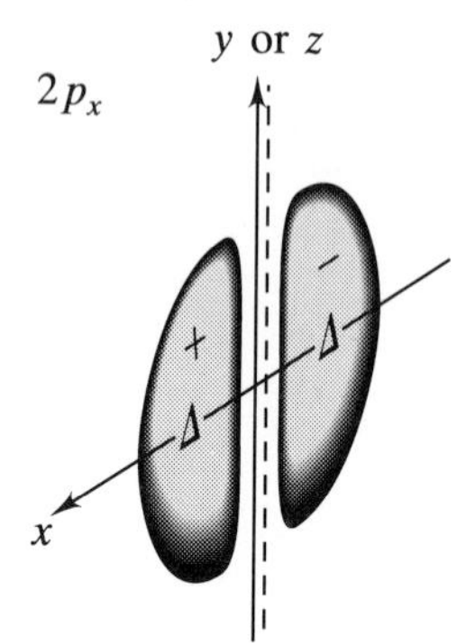

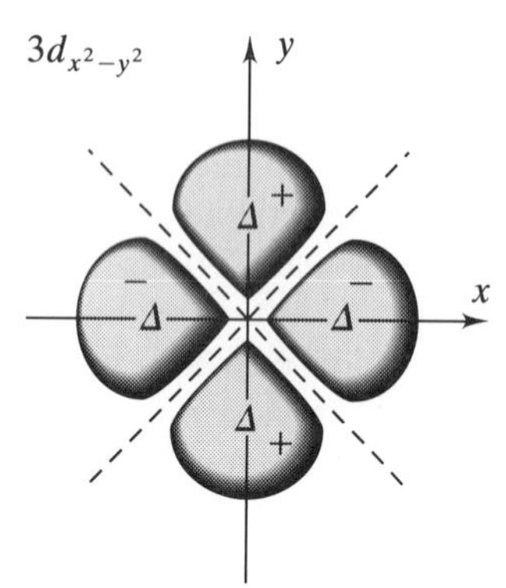

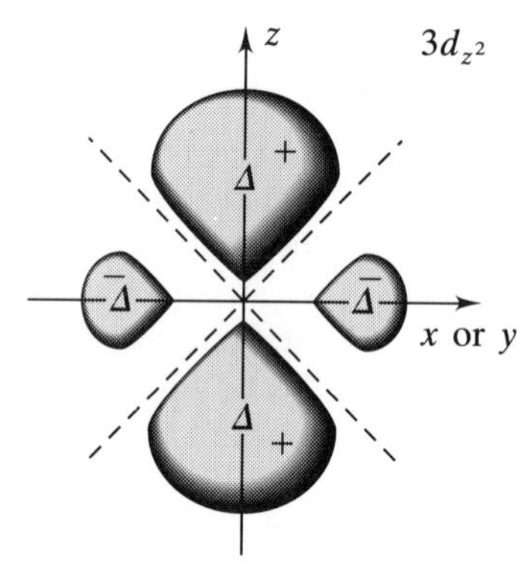

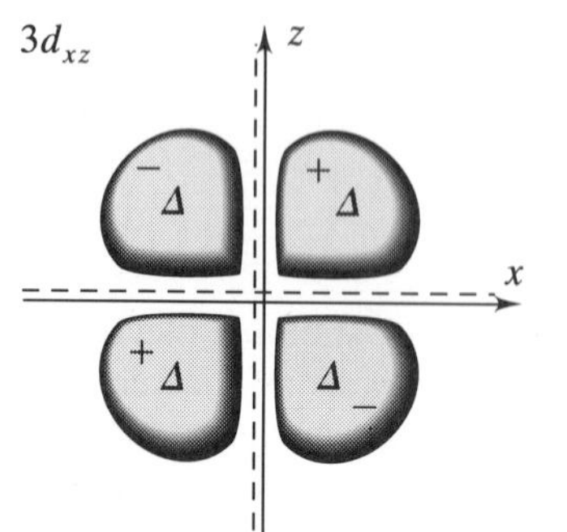

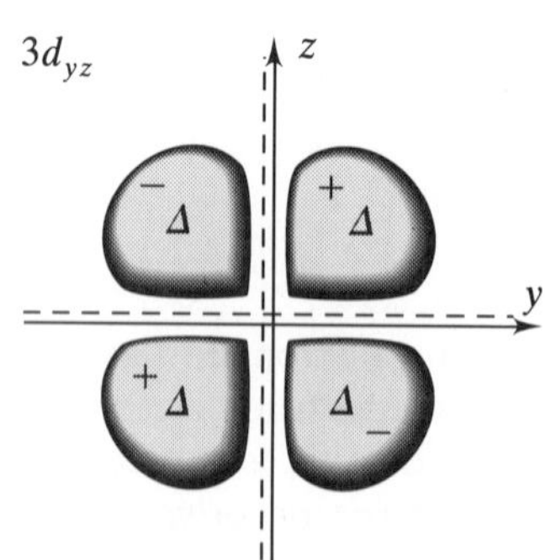

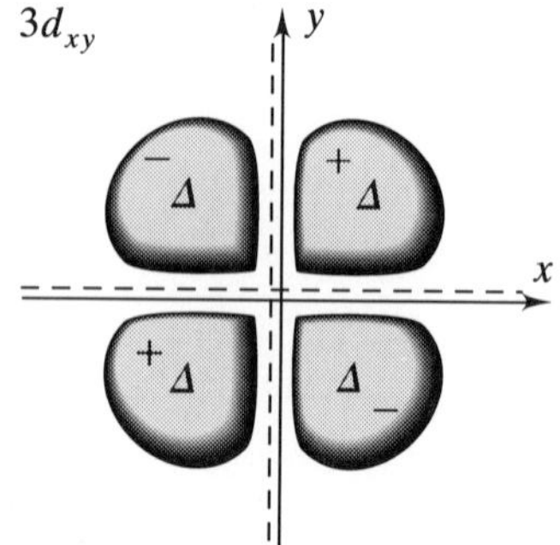

FIGURE 4.1
Cross-sectional sketches of the hydrogen-like 1*s*, 2*p*, and 3*d* orbitals. The orbitals are shown as 90 percent boundary surfaces with plus and minus signs representing phases and triangles representing the points of maximum electron probability. Nodes are given as dashed lines.

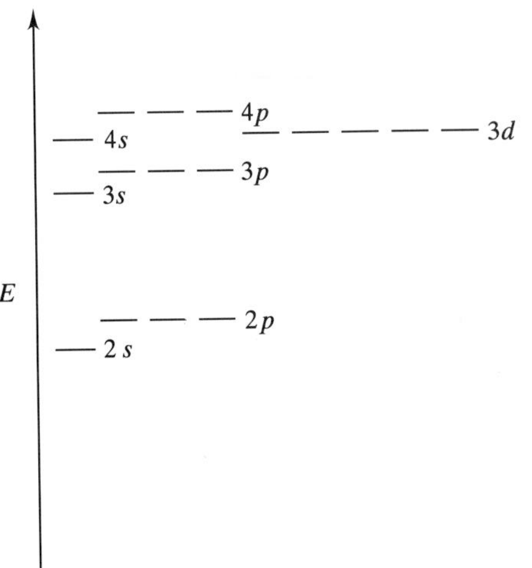

FIGURE 4.2
The relative energies of atomic orbitals in a many-electron atom such as a typical first-row transition metal.

generated in a one-electron case, typically taken to be the hydrogen atom) orbitals are shown in Figure 4.1. Figure 4.2 shows the relative energies of the 1*s*, 2*s*, 2*p*, 3*s*, 3*p*, 4*s*, and 3*d* orbitals in a typical first-row transition metal.

There are a few special notes to be made about these orbitals as depicted in the figures. First, you should recognize (in Figure 4.2) the familiar pattern of the energy-level diagram. Within a shell (orbitals with the same principal quantum number n) the orbitals are divided into subshells (those with the same orbital quantum number l). For example, the $n = 3$ orbitals are divided into the 3*s*, 3*p*, and 3*d* subshells that increase in energy as l increases from 0 to 1 to 2. You should also recall that the three 3*p* orbitals as well as the five 3*d* orbitals are degenerate; that is, they have the same energies. It is the effect of a field of ligands on the energies of the five *nd* orbitals that is a principal focus of the CFT.

In closely examining Figure 4.1, you may perhaps be a little surprised at the nearly spherical (or circular in two dimensions) shape of the *p* and *d* orbitals. In many introductory courses (and particularly in the models commonly shown in these courses) these orbitals look something like balloons tied together at the nucleus. In fact, a more accurate pictorial representation is given in Figure 4.1. Each of these orbitals, represented in two-dimensional cross-sectional diagrams, looks like it has been cut out of a circular piece of "cloth." The outer solid lines represent 90 percent boundary surfaces (inside of which the electron will be found 90 percent of the time), while the dashed lines represent nodes, areas where the electron has a zero probability of existing. The plus and minus signs indicate the phase of the orbital in a given area, while the small triangles show the points of maximum electron probability in a given "lobe" of the orbital. The actual three-dimensional orbitals can be visualized by rotating the lobes of the two-dimensional cross-sectional diagrams about the appropriate axis or axes. For example, the three-dimensional $3d_{z^2}$ orbital is generated by rotating the two-dimensional cross section about the z axis, as shown in Figure 4.3. Not too surprisingly, given the shapes indicated, the sum of the electron probabilities in a given subshell (3*p* or 3*d*, for example) is a sphere.

One of the principal differences between the three *p* orbitals and the five *d* orbitals is that while the former set has three identical orbitals oriented along the x, y, and z axes, respectively, the latter has four identical orbitals (d_{xy}, d_{yz}, d_{xz},

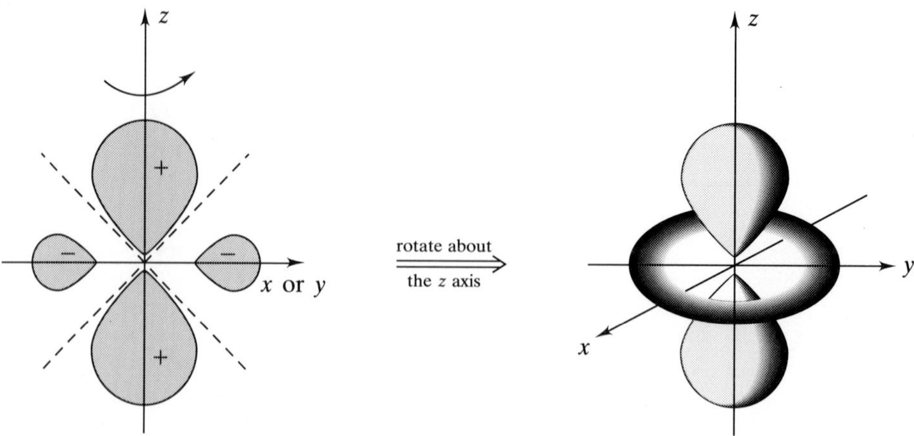

FIGURE 4.3
The rotation of the cross-sectional sketch of a $3d_{z^2}$ orbital about the z axis generates a three-dimensional representation of that orbital.

and $d_{x^2-y^2}$) and one (the d_{z^2}) that looks like it is special, that is, rather different from the other four. This distinction between the d_{z^2} orbital and the other four needs to be addressed in some detail because a thorough knowledge of the shapes of the d orbitals is essential to an understanding of CFT.

Actually, it turns out that the Schrödinger mathematics which yields the probability of finding electrons in various orbitals can be carried out a number of different ways yielding a variety of solutions. The five $3d$ orbitals depicted in Figure 4.1 are but one possible set of orbitals. (They turn out to be the most convenient set for reasons that are unnecessary to the following argument.) Another solution of the Schrödinger mathematics yields, in fact, *six dependent* orbitals. Dependent orbitals are so named because any one of them can be expressed as a linear combination of two others. These six dependent d orbitals are the d_{xy}, d_{yz}, d_{xz}, and $d_{x^2-y^2}$ orbitals shown in Figure 4.1 plus two more similar to $d_{x^2-y^2}$, labeled $d_{z^2-y^2}$ and $d_{z^2-x^2}$. These latter two dependent orbitals are shown in the top part of Figure 4.4. Note that they, like the $d_{x^2-y^2}$ orbital, have lobes pointing along the axes given in their designations. The six dependent d orbitals, like those in other subshells, are degenerate; that is, they have the same energies in a free metal atom or ion.

The five *independent* $3d$ orbitals that we normally use are generated from the six *dependent* orbitals by taking a linear combination of (i.e., adding) the $d_{z^2-y^2}$ and $d_{z^2-x^2}$ orbitals to generate the d_{z^2} orbital, as represented in Figure 4.4. Note here that both of these dependent orbitals have a probability of finding the electrons along the positive and negative z axis, but only the $d_{z^2-y^2}$ has probability along the y axis, while only the $d_{z^2-x^2}$ has probability along the x axis. Therefore, when these two orbitals are added together, the resulting d_{z^2} orbital has twice as great an electron probability along the z axis as it does along the other two axes. The points of maximum probability (given by the triangles), however, remain the same in all three orbitals. Now we are ready to consider the effect of a crystal field of ligands on the five traditional $3d$ orbitals. Keep in mind, however, that the d_{z^2} orbital is not as special as it looks. It is merely a linear combination of two dependent orbitals that look exactly like the other four orbitals in the subshell.

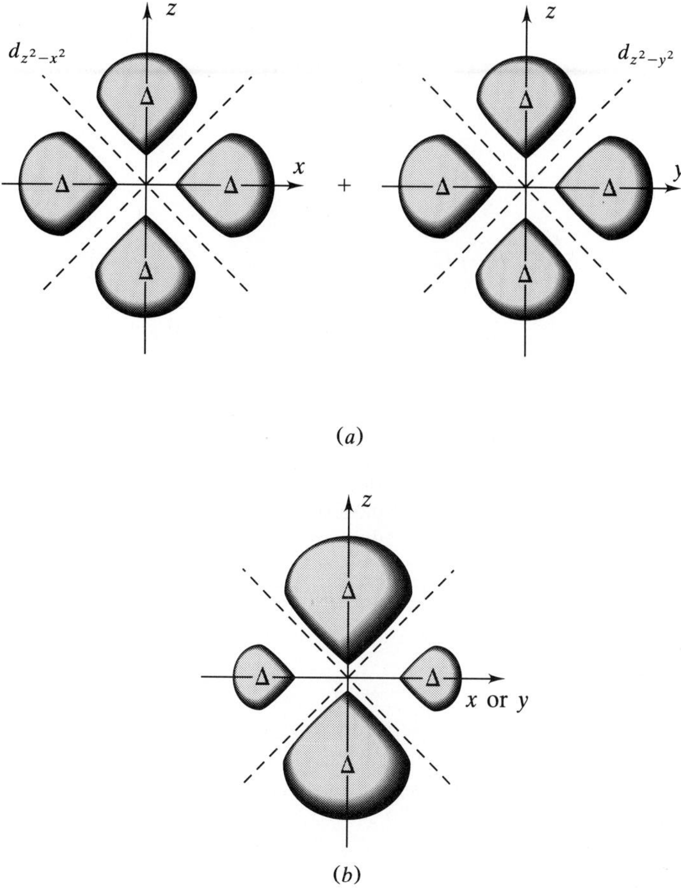

FIGURE 4.4
(a) The dependent $d_{z^2-x^2}$ and $d_{z^2-y^2}$ orbitals. (b) The d_{z^2} orbital that results from a linear combination of the above two dependent orbitals. Note that the $3d_{z^2}$ orbital is the same size as its constituent dependent orbitals but has double the probability of finding an electron along the $+z$ and $-z$ axes.

Octahedral Fields

Suppose we start with a first-row transition metal ion, M^{n+}, containing some unspecified number of $3d$ electrons. CFT considers what would happen when an octahedral field of ligands is constructed around this metal. Assuming that each of the six ligands has a pair of electrons to "donate" to the metal, there will be a total of 12 ligand electrons to consider. Figure 4.5 shows the physical construction of the octahedral field about the metal in four stages and the energy corresponding to each stage.

Stage I has the metal and the 12 electrons an infinite distance apart. The five $3d$ orbitals of this "free" metal ion will not be affected at all by the ligand electrons and will remain degenerate. We will take the energy of this first stage to be zero and to be the reference point for the other three. In stage II, the 12 ligand electrons are brought up around the metal to form a spherical shell at the appropriate M–L distance. Since the ligand electrons are smeared out into a spherically symmetric cloud, each of the metal $3d$ orbitals will be equally affected

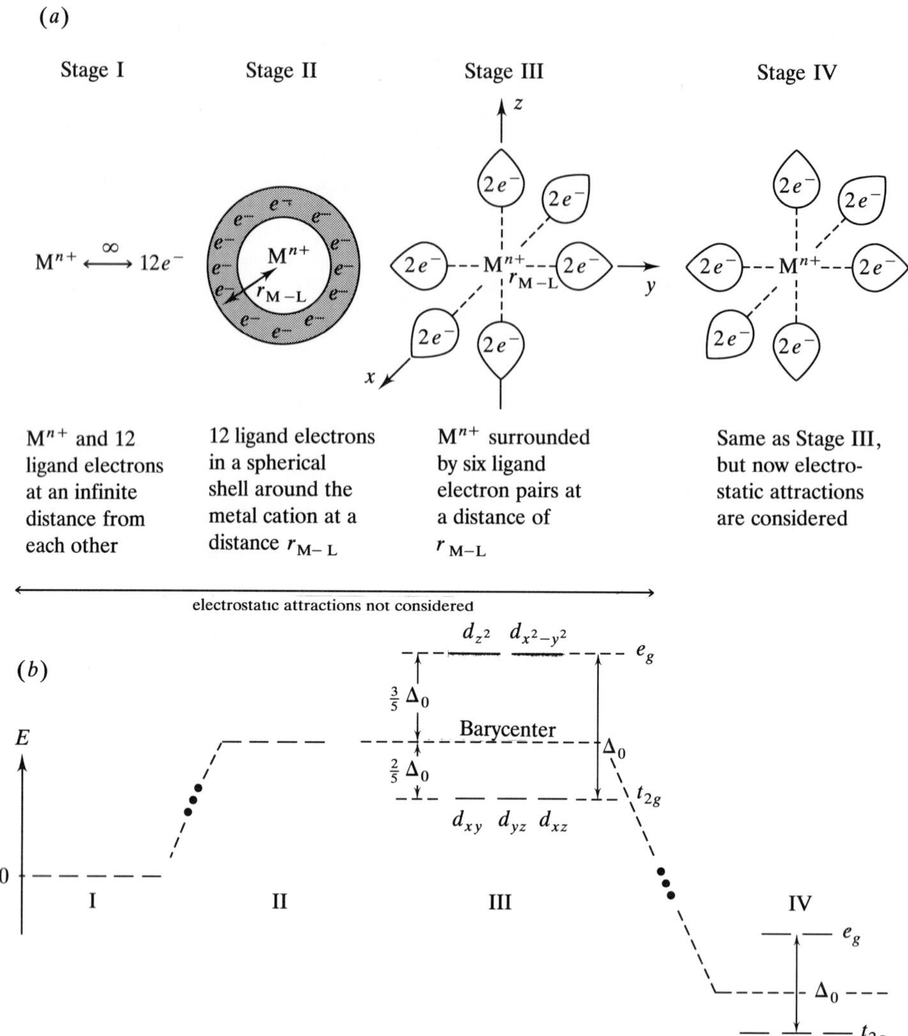

FIGURE 4.5
(a) Four hypothetical stages of the construction of an octahedral field about a metal cation and (b) the energies of the d orbitals corresponding to each stage. Note that the relative energy changes are not to scale. The split between the d orbitals is actually much smaller than the energy changes in going from stages II to III or III to IV. [Ref. 3, pp. 150–151.]

and their degeneracy will remain unbroken. However, since any electrons in these $3d$ orbitals will be repelled by the ligand electrons, the potential energy of the system will increase. [Both Q_1 and Q_2 in Equation (4.3) will be negative, so the potential energy of the system will be positive.] At this point, the effect of the electrostatic attractions between the positive metal cation and the 12 ligand electrons has not been considered.

In stage III the 12 electrons are actually arranged into an octahedral field at the same M–L distance as in stage II. (For convenience, the ligands are pictured as being along the three cartesian coordinates.) Since the distance between the metal and the ligand electrons has remained the same, the net potential energy of the system does not change. Another way to say this is that the *barycenter*, the average energy of a set of orbitals in the same subshell, remains constant in going from stage II to stage III.

The energy-level diagram for stage III shows three of the 3d orbitals decreasing in energy (relative to the barycenter) and two increasing. What are the factors involved in these changes in energy? Note that the two orbitals ($d_{x^2-y^2}$ and d_{z^2}) that increase in energy point directly along the cartesian axes and therefore directly at the ligands. Any electrons in these orbitals will be repelled by the ligand electrons; the energy of these orbitals is therefore higher than the barycenter. Does it bother you that the differently shaped d_{z^2} orbital has the same energy as the $d_{x^2-y^2}$ orbital? It should not. Recall that the d_{z^2} orbital is but a linear combination of the $d_{z^2-y^2}$ and $d_{z^2-x^2}$ dependent orbitals that look like and have the same energy as the $d_{x^2-y^2}$ orbital. (All three of these orbitals point directly at the ligands.) Therefore, it follows that the $d_{x^2-y^2}$ and d_{z^2} orbitals should be of the same energy. Note that these two orbitals are referred to as the e_g set. The basis of this symbol is derived from various symmetry arguments that we will not cover here.

The three orbitals (d_{xy}, d_{yz}, and d_{xz}) that decrease in energy relative to the barycenter point in between the ligands, and therefore the metal electrons that occupy them are farther away from the ligand electrons than they were in stage II. Therefore, there is a decrease in the energies of these orbitals. The collective symmetry symbol for these three orbitals in an octahedral field is t_{2g}.

Next we turn to a discussion of the quantitative aspects of the changes in the energies of the e_g and t_{2g} sets. First we define the *crystal field* (*CF*) *splitting energy* Δ as the difference between the energies of the d orbitals as a result of the application of a field of ligands. Since this is an octahedral field, we designate the CF splitting energy as Δ_o, where the o stands for octahedral. Next note that since the position of the barycenter remains unchanged, the total energy decrease of one set of orbitals must be equal to the total energy increase of the other. Therefore, since there are two e_g orbitals, they must increase by $\frac{3}{5}\Delta_o$ while the three t_{2g} orbitals must decrease by $\frac{2}{5}\Delta_o$. Another way to express this relationship is

$$\begin{gathered}\text{Energy decrease} = \text{Energy increase} \\ 3(\tfrac{2}{5}\Delta_o) = 2(\tfrac{3}{5}\Delta_o)\end{gathered} \tag{4.4}$$

So far we have discussed three of the four stages of the construction of an octahedral field of ligands about a metal ion shown in Figure 4.5. To this point the energy of the system has increased (which usually means the process will not occur spontaneously), but there is one additional factor to consider. We have yet to account for the effect of the electrostatic attractions of the 12 ligand electrons and the positively charged metal ion. This factor, represented in stage IV, results in a decrease in the barycenter of the system and a complex that is lower in energy than the free metal ion located an infinite distance from the 12 ligand electrons. [The potential energy of this last step adds a negative contribution to the overall energy of the system because one charge (on the metal ion) is positive and the other (the charges of the ligand electrons) is negative.]

This completes a description of the basic CFT as it applies to octahedral fields. We proceed now to briefer treatments of tetragonally distorted, square planar, and tetrahedral fields. Following these sections, we will be in a position to investigate the consequences and applications of the crystal field theory.

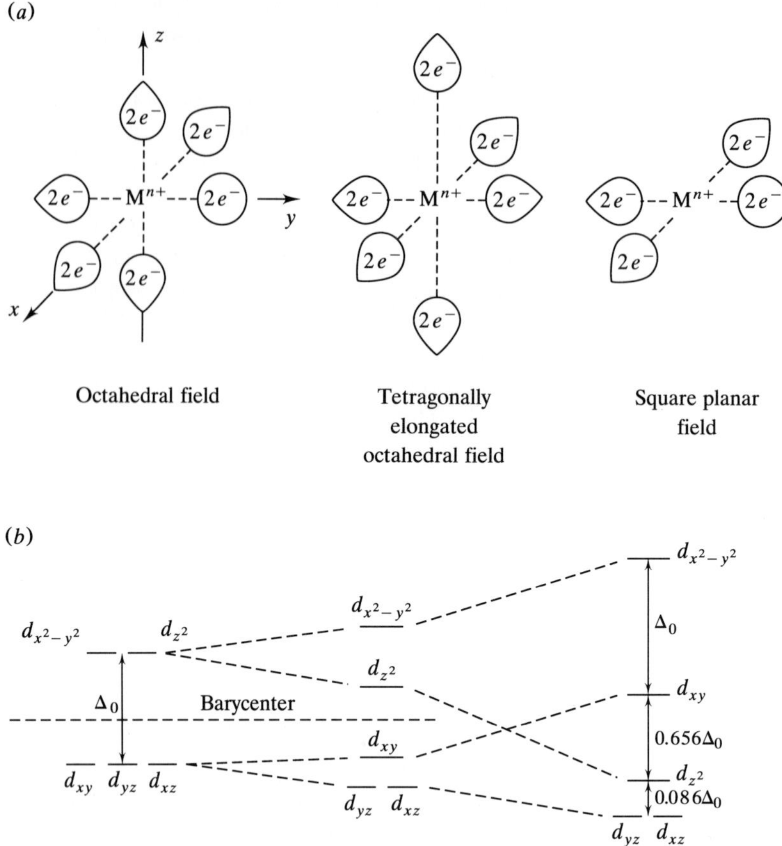

FIGURE 4.6
(*a*) The gradual removal of the *z*-axis ligands results in the progression from an octahedral to a tetragonally elongated octahedral to a square planar field of ligands. (*b*) The change in the energies of the *d* orbitals in a central metal atom or ion corresponding to three fields. [(*b*) from Ref. 4, p. 643.]

Tetragonally Distorted Octahedral and Square Planar Fields

A tetragonally distorted octahedral field is produced when the *z*-axis ligands are either pushed closer to or pulled away from the metal atom or ion. Moving the *z*-axis ligands closer produces a tetragonal compression, while pulling them away produces a tetragonal elongation of the original octahedral field. We will consider the elongation here. (The phrase "tetragonally distorted" is derived from the fact that when viewed down the *z* axis, an octahedron looks like a tetragon, a little-used word for a four-sided figure.) By convention (that is, by a procedure agreed upon by a large number of practitioners in a field for the sake of uniformity and convenience), we choose to change the distance from the metal to the ligands along the *z* axis rather than the *x* or the *y*. In most cases where there is any special or unique axis of any type (another example is the designation of the bonding axis), the *z* axis should be designated as that axis.

The left side of Figure 4.6 shows both the physical and energetic changes accompanying a tetragonal elongation. Carefully note that when the two *z*-axis ligands are pulled away from the metal, the *x*- and *y*-axis ligands are shown to simultaneously move somewhat closer. The slight movement of the *x*- and *y*-axis ligands is allowed because the removal of the *z*-axis ligands creates a small amount of empty space around the metal. The coulombic attraction between ligand electrons and the charge of the metal then pulls these ligands in closer to fill in that empty space.

Turning now to the energy changes accompanying a tetragonal elongation, we will assume that the simultaneous movement of all six ligands keeps the barycenter of the tetragonally elongated field the same as that in the original octahedral field. Note that all the orbitals with a z-axis component (d_{yz}, d_{xz}, d_{z^2}) become more stable; that is, their energies decrease. This is readily understood in terms of electrostatics. When the z-axis ligand electrons are pulled back, a metal electron will more readily occupy, for example, the d_{z^2} orbital because there is now less electron-electron repulsion and the potential energy will be lower. The change in the energy of the d_{z^2} orbital is significantly greater than that in the d_{xz} and d_{yz} orbitals because the former points directly along the z axis and its energy is more directly affected by movements of ligands along that axis. The metal orbitals in the xy plane increase in energy because their occupation results in more electron-electron repulsions and corresponding increases in their potential energies. The energy of the $d_{x^2-y^2}$ orbital is affected more than that of the d_{xy} orbital again because the former points directly at the x- and y-axis ligands.

What happens if we continue the movement of the z-axis ligands until they are completely removed from the metal? The result, as shown in Figure 4.6, is a square planar field of ligands. Note that the metal orbitals with a z component continue to become more stable (decrease in potential energy) to the point that the energy of the d_{z^2} orbital falls below that of the d_{xy}. The differences in energy between the various orbitals in the square planar case are indicated in the figure. Note that we cannot assume that the square planar and octahedral barycenters are at the same energy because there are no longer six ligands at the M–L distance but only four ligands at a somewhat smaller M–L distance. (See Problem 4.21 for an opportunity and a hint on how to calculate the position of the barycenter within the square planar crystal field.)

Tetrahedral Fields

The tetrahedral crystal field is more difficult to visualize than the octahedral and related square planar cases. One way to simplify the tetrahedral case is to consider its relationship to a cubic field. Figure 4.7 shows a cubic field, that is, a field of

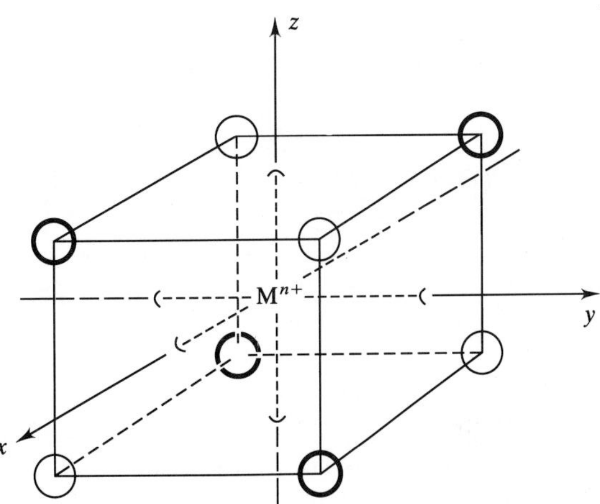

FIGURE 4.7
Eight ligands of a cubic field surrounding a central metal atom or ion. The ligands in bold print represent one of the two tetrahedra that together constitute the cubic field. The cartesian axes project from each face of the cube.

eight ligands at the corners of a cube containing the metal ion at the center. As illustrated, the cube can be thought of as the sum of two tetrahedra. (One tetrahedron of ligands is depicted in bold print for clarity.) By convention, the cartesian coordinates are shown coming out of the center of each of the six cube faces.

Each of the five d orbitals is schematically shown in a tetrahedral field in Figure 4.8. A quick inspection of this figure reveals why this case is not as simple as the others discussed to this point. *None* of the d orbitals points directly at any of the ligands. Instead, they *all* point, to some degree, in between the ligands. To make a distinction among sets of orbitals as we did in the octahedral case, the distance from the "tip" of any given orbital to a ligand is indicated in the figure. (The distance is given in terms of l, the length of the side of the cube.) Notice that the top three orbitals (d_{xy}, d_{yz}, and d_{xz}) are ($l/2 =$) 0.50 l from each ligand, while the bottom set ($d_{x^2-y^2}$ and d_{z^2}) is a little farther away at ($l\sqrt{2}/2 =$) 0.707 l. It follows that the top set of orbitals is relatively less stable than the bottom. (In other words, metal electrons that occupy the top set experience more electron-electron repulsions and are of higher potential energy.)

Figure 4.9 summarizes the crystal field splitting in the tetrahedral case. The difference in energy of the two sets is symbolized by Δ_t, where t stands for tetrahedral. Using the same reasoning as explained for the octahedral field, the top set of orbitals (now referred to as the t_2 set) is $\frac{2}{5}\Delta_t$ above the barycenter, while the bottom set (the e set) is $\frac{3}{5}\Delta_t$ below. Due to the less clear-cut distinction between these two sets of orbitals as compared to the octahedral case, it is not surprising that Δ_t is only about half that obtained for the octahedral case.

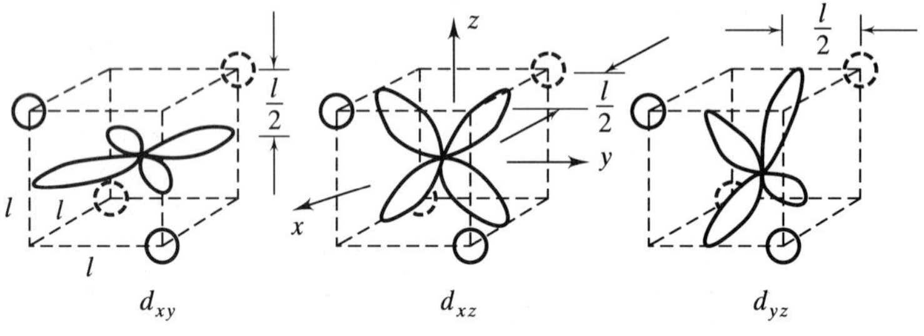

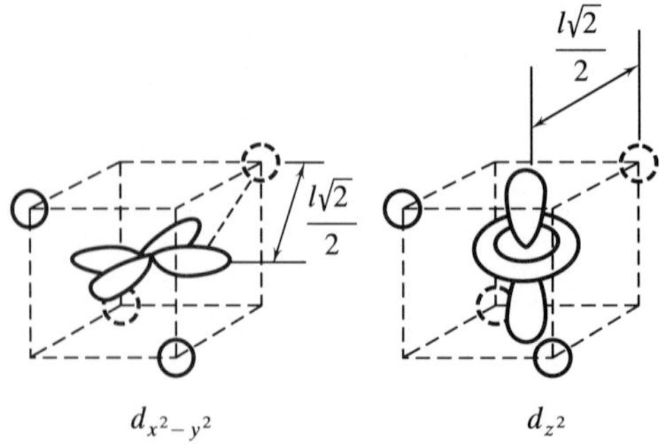

FIGURE 4.8

The five d orbitals of a metal atom or ion in a tetrahedral field. The top three orbtials (d_{xy}, d_{xz}, and d_{yz}) are a distance $l/2$ from the ligands, while the bottom two ($d_{x^2-y^2}$ and d_{z^2}) are farther away at $l\sqrt{2}/2$. l is defined as the length of cube edge.

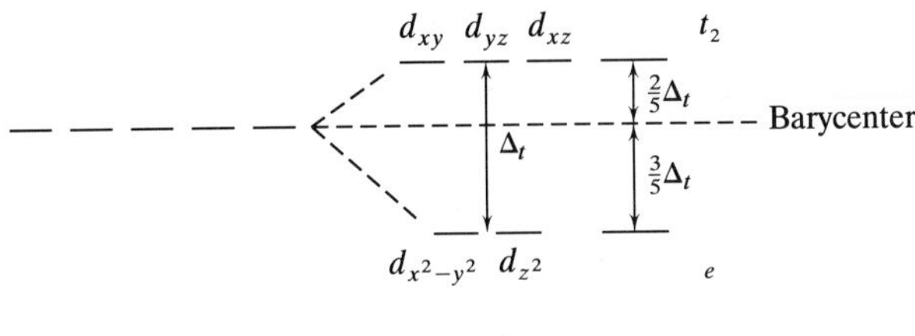

FIGURE 4.9
The crystal field splitting of the d orbitals by a tetrahedral field.

4.3 CONSEQUENCES AND APPLICATIONS OF CRYSTAL FIELD SPLITTING

As we have just seen, the essential stability of coordination compounds, according to CFT, is the direct result of a release of energy accompanying the electrostatic or coulombic interaction of the ligand electrons with a metal ion. Now we are in a position to quantify that stability and go on to see how CFT can also account for the colors as well as the electronic and corresponding magnetic properties of these compounds. Explanation of these properties will give us confidence that this whole picture of the bonding in coordination compounds is a useful model which should be continued. As with any model, particularly such a conceptually simple one as this, we must also consider its various deficiencies and how they can be addressed.

Crystal Field Splitting Energies versus Pairing Energies

We start with a simple case. Suppose we are given two nondegenerate valence electronic energy levels, called E_1 and E_2, separated by an energy difference labeled Δ. As shown in Figure 4.10, we wish to consider placing two electrons in the orbitals corresponding to these energy levels. The first electron, as expected, will occupy the lowest energy level, E_1. Where the second electron goes, however, needs some discussion.

In what energy level would you expect to find the second electron? Assuming, properly so, that a given energy level can hold two electrons and drawing on past experience with the electronic configurations of atoms and the Aufbau principle, it would not be surprising if you chose case I shown in the figure. Here the second electron pairs with the first and also occupies the lowest energy level. There is, of course, another possibility. The second electron might occupy the higher energy level, E_2, as shown in case II. Such a configuration would usually be designated as an *excited state*, but let us not jump (pun intended) to that conclusion. Instead, consider the relative energies of these two cases.

Case I has a net energy of $2E_1 + P$, where P represents the *pairing energy*, defined as the positive contribution to the potential energy of a system due to the pairing of two electrons in the same energy level. A detailed consideration of pairing energy, as it turns out, can be highly involved and not worth an extended discussion here. At first blush, however, why do you suppose it would cost energy to pair electrons? Certainly one of the most important factors would be electron-electron repulsions. After all, these energy levels are associated with orbitals, and putting two electrons in the same orbital or volume of space results in a repulsive force between them and a positive contribution to the potential energy of the system.

Case II, on the other hand, has a net energy of $E_1 + E_2$, but E_2 is equal to $E_1 + \Delta$, the difference or split between the two energy levels. So the net energy of

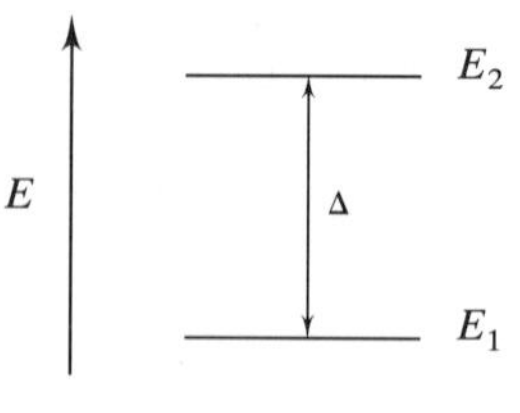

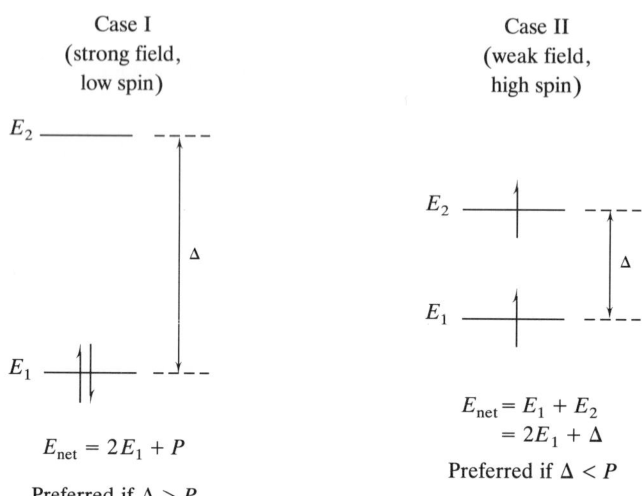

FIGURE 4.10
Strong-field, low-spin and weak-field, high-spin cases in two nondegenerate energy levels. The relative values of Δ and P determine the state of lowest energy. The low-spin case is preferred if $\Delta > P$, while the high-spin case is preferred if $\Delta < P$. [Adapted from Ref. 4, p. 644.]

case II is $2E_1 + \Delta$. Under what circumstances will a system adopt case I or case II? We know that it will adopt the configuration of lowest energy, which in turn will be determined by the relative magnitudes of Δ and P. If Δ is a large number, larger than P, case I, called the *strong-field case*, will be of lower energy and will be the more stable. If Δ is a relatively small number, then case II, called the *weak-field case*, will result. Also note that the strong-field case results in fewer (none in this particular example) unpaired electrons. The two electron spins would cancel each other for a net electron spin of zero. We call this the *low-spin case*. Conversely, the weak-field case has more unpaired electrons (two in this example) and is appropriately called the *high-spin case*.

Crystal Field Stabilization Energies

What happens when we extend these arguments to systems containing degenerate energy levels such as those produced in a metal by an octahedral crystal field? Table 4.1 shows the possible electronic configurations that result for varying numbers of d electrons. [Note that electrons are removed from a transition metal in the order ns electrons first, then $(n-1)d$ electrons. It follows, for example, that while the electronic configuration of titanium is $[Ar]4s^23d^2$, that of Ti^{3+} is $[Ar]3d^1$. Ti^{3+}, then, is called a d^1 case.]

TABLE 4.1
Electronic configurations and crystal field stabilization energies for metal ions in octahedral complexes

No. of *d* electrons	Unsplit case	High-spin case (CFSE)	Low-spin case (CFSE)
1	↑ _ _ _ _	↑ _ _ _ _ t_{2g}^1 ($\frac{2}{5}\Delta_o$)	
2	↑ ↑ _ _ _	↑ ↑ _ _ _ t_{2g}^2 ($\frac{4}{5}\Delta_o$)	
3	↑ ↑ ↑ _ _	↑ ↑ ↑ _ _ t_{2g}^3 ($\frac{6}{5}\Delta_o$)	
4	↑ ↑ ↑ ↑ _	$t_{2g}^3e_g^1$ ↑ ↑ ↑ ↑ _ ($\frac{3}{5}\Delta_o$)	↑↓ ↑ ↑ _ _ t_{2g}^4 ($\frac{8}{5}\Delta_o - P$)
5	↑ ↑ ↑ ↑ ↑	$t_{2g}^3e_g^2$ ↑ ↑ ↑ ↑ ↑ (0)	↑↓ ↑↓ ↑ _ _ t_{2g}^5 ($\frac{10}{5}\Delta_o - 2P$)
6	↑↓ ↑ ↑ ↑ ↑	$t_{2g}^4e_g^2$ ↑↓ ↑ ↑ ↑ ↑ ($\frac{2}{5}\Delta_o$)	↑↓ ↑↓ ↑↓ _ _ t_{2g}^6 ($\frac{12}{5}\Delta_o - 2P$)
7	↑↓ ↑↓ ↑ ↑ ↑	$t_{2g}^5e_g^2$ ↑↓ ↑↓ ↑ ↑ ↑ ($\frac{4}{5}\Delta_o$)	↑↓ ↑↓ ↑↓ ↑ _ $t_{2g}^6e_g^1$ ($\frac{9}{5}\Delta_o - P$)
8	↑↓ ↑↓ ↑↓ ↑ ↑	↑↓ ↑↓ ↑↓ ↑ ↑ $t_{2g}^6e_g^2$ ($\frac{6}{5}\Delta_o$)	
9	↑↓ ↑↓ ↑↓ ↑↓ ↑	↑↓ ↑↓ ↑↓ ↑↓ ↑ $t_{2g}^6e_g^3$ ($\frac{3}{5}\Delta_o$)	
10	↑↓ ↑↓ ↑↓ ↑↓ ↑↓	↑↓ ↑↓ ↑↓ ↑↓ ↑↓ $t_{2g}^6e_g^4$ (0)	

For d^1, d^2, and d^3 cases, there is no choice as to what orbitals will be occupied. The t_{2g} orbitals are of lowest energy, and the first three electrons can occupy them without having to pair up. For the d^4 case, however, there is a choice. The fourth electron can either pair up in one of the t_{2g} orbitals or go to the higher-energy e_g orbital. Which resulting electronic configuration, t_{2g}^4 or $t_{2g}^3e_g^1$, will be favored? To answer this question, we introduce the concept of *crystal field stabilization energy* (*CFSE*), defined as the decrease in energy, relative to the unsplit state, of a coordination compound caused by the splitting of the metal *d* orbitals by a field of ligands. CFSEs are given in parentheses for each electronic configuration in Table 4.1. For example, since the t_{2g} orbitals are $\frac{2}{5}\Delta_o$ lower in energy than the barycenter, the CFSE of a d^1 case is just $\frac{2}{5}\Delta_o$. For the d^2 case, the CFSE is twice $\frac{2}{5}\Delta_o$, or $\frac{4}{5}\Delta_o$, and so forth.

For the d^4 case, the CFSEs are different for the high- and low-spin cases. For the low-spin case, there are now four electrons in the t_{2g} set, but two of them must be paired. In the unsplit state, no pairing is necessary. Accordingly, the CFSE is $4(\frac{2}{5}\Delta_o)$ minus P, the pairing energy. That is, since it costs energy to pair two of the electrons, P must be subtracted from the stabilization energy that otherwise results from four electrons occupying the t_{2g} set. For the high-spin case, three electrons are $\frac{2}{5}\Delta_o$ lower in energy but one is $\frac{3}{5}\Delta_o$ higher. Therefore, the CFSE is $3(\frac{2}{5}\Delta_o) - 1(\frac{3}{5}\Delta_o)$, or just $\frac{3}{5}\Delta_o$. Having now calculated these CFSEs, can we decide which configuration is more stable?

As in the nondegenerate case described above, the relative stability of the two cases comes down to the difference between Δ_o and P. To see this, the expressions for the CFSEs in each case have been recast below:

Low spin	High spin
(t_{2g}^4)	$(t_{2g}^3 e_g^1)$
$\text{CFSE} = \frac{8}{5}\Delta_o - P$	$\text{CFSE} = \frac{3}{5}\Delta_o = \frac{8}{5}\Delta_o - \Delta_o$

Note that the only difference between these energies is the relative values of Δ_o versus P. If $\Delta_o > P$, the low-spin case will have the greater CFSE and will be favored. If, on the other hand, $\Delta_o < P$, the high-spin case has a greater CFSE and is favored. Similar results are obtained in the d^5 to d^7 cases where both high- and low-spin cases are possible. The d^8 to d^{10} cases have only one possible electronic configuration. So, in summary, while the degenerate energy levels resulting from an octahedral field look somewhat more complicated, the relative stability of the high- and low-spin cases is still decided by the relative magnitudes of Δ_o and the pairing energy P.

Factors Affecting the Magnitude of the Crystal Field Splitting Energies

Pairing energies, particularly for the first-row transition metals, are relatively constant. Therefore, the choice between strong- and weak-field, low- and high-spin cases comes down to the magnitude of the crystal field splitting energies Δ. The higher Δ is, the greater the chance of a strong-field–low-spin electronic configuration. The lower Δ is, the greater the chance of a weak-field–high-spin electronic configuration.

We have already discussed one factor affecting the magnitude of crystal field splitting energies, namely, the geometry of the field. For a given metal ion and set of ligands, Δ_o is about twice as large as Δ_t. In fact, as a rule of thumb, Δ_t is always smaller than P and tetrahedral complexes are always weak-field–high-spin. Square planar fields (see Figure 4.6) have splitting energies approximating those of octahedral fields and can be either high- or low-spin.

What other factors bear upon the degree to which d orbitals are split by a crystal field? Consider next the properties of the metal cations. For example, the greater the charge on a metal ion, the greater the magnitude of Δ. For a given metal ion, M^{n+}, for example, Δ is always larger for a +3 charge than for a +2. Table 4.2 shows a variety of metal ions and ligand systems. Note that in each case the splitting for the +3 ion is larger than that of the corresponding +2. What is the rationale for this observation? Simply put, using the electrostatic CFT: the greater the charge on the metal, the more the ligands are pulled in toward it and therefore the more the ligand electrons are able to affect or split the energies of the metal d orbitals.

The size of the metal also has an effect on the crystal field splitting. Compare similarly charged metal ions (for example, in Table 4.2, Cr^{3+} and Mo^{3+}; Co^{3+}, Rh^{3+}, and Ir^{3+}) and note that for the larger second- and third-row transition metal ions, the Δ_o are always larger. The rationale here is that the larger ion has more room about it so that a given set of ligands can approach closer to it without elbowing each other out. (It helps in this and other discussions to follow, to define the term *steric hindrance*, the impediment to the formation of a given configuration of a molecule due to the spatial interference or crowding out of various atoms or groups of atoms in the molecule. Therefore we can say that given field of ligands

TABLE 4.2
Octahedral crystal field splitting energies Δ_o, cm^{-1}

$(M')^{2+}$		$(M')^{3+}$		$(M'')^{3+}$		$(M''')^{3+}$	
			Cr^{2+}, Cr^{3+}, Mo^{3+}				
$[CrCl_6]^{4-}$	13,000	$[CrCl_6]^{3-}$	13,200	$[MoCl_6]^{3-}$	19,200		
$[Cr(H_2O)_6]^{2+}$	14,000	$[Cr(H_2O)_6]^{3+}$	17,400				
		$[Cr(NH_3)_6]^{3+}$	21,500				
$[Cr(en)_3]^{2+}$	18,000	$[Cr(en)_3]^{3+}$	21,900				
		$[Cr(CN)_6]^{3-}$	26,600				
			Co^{2+}, Co^{3+}, Rh^{3+}, Ir^{3+}				
$[Co(H_2O)_6]^{2+}$	9,300	$[Co(H_2O)_6]^{3+}$	18,200	$[RhCl_6]^{3-}$	20,000	$[IrCl_6]^{3-}$	25,000
$[Co(NH_3)_6]^{2+}$	10,100	$[Co(NH_3)_6]^{3+}$	22,900	$[Rh(H_2O)_6]^{3+}$	27,000	$[Ir(NH_3)_6]^{3+}$	41,000
$[Co(en)_3]^{2+}$	11,000	$[Co(en)_3]^{3+}$	23,200	$[Rh(NH_3)_6]^{3+}$	34,100	$[Ir(en)_3]^{3+}$	41,400
		$[Co(CN)_6]^{3-}$	33,500	$[Rh(en)_3]^{3+}$	34,600		
				$[Rh(CN)_6]^{3-}$	45,500		
			Mn^{2+}, Mn^{3+}				
$[MnCl_6]^{4-}$	7,500	$[MnCl_6]^{3-}$	20,000				
$[Mn(H_2O)_6]^{2+}$	8,500	$[Mn(H_2O)_6]^{3+}$	21,000				
$[Mn(en)_3]^{2+}$	10,100						
			Fe^{2+}, Fe^{3+}				
$[Fe(H_2O)_6]^{2+}$	8,500	$[FeCl_6]^{3-}$	11,000				
$[Fe(CN)_6]^{2+}$	32,800	$[Fe(H_2O)_6]^{3+}$	14,300				
		$[Fe(CN)_6]^{3-}$	35,000				

experiences less steric hindrance about a larger metal ion.) Again, the closer the ligands can approach, the more the ligand electrons are able to affect or split the energies of the metal *d* orbitals.

Note how well the crystal field theory has accounted for the degree of splitting of the *d* orbitals caused by the type of field and the charge and size of the central metal. We turn now, however, to a consideration of the relative ability of *ligands* to split *d* orbitals. Table 4.2 shows that for the 11 metal cations listed, the five ligands can be arranged in the following order of increasing ability to split metal *d* orbitals: Cl^-, H_2O, NH_3, en, and CN^-. An expanded list of this sort, called the *spectrochemical series*, is shown below:

$$I^- < Br^- < Cl^- < SCN^- < NO_3^- < F^- < OH^- < C_2O_4^{2-} < H_2O$$

$$< NCS^- < \text{gly} < C_5H_5N < NH_3 < \text{en} < NO_2^- < PPh_3 < CN^- < CO$$

Now, the question is, can we make any sense out of this series using CFT? This, as it turns out, is a somewhat difficult task. The effect of ligand size makes some sense in that the larger or bulkier ligands seem to be concentrated at the low end of the series. As the ligands decrease in size, for example, for the halides in the order I^-, Br^-, Cl^-, to F^-, their ability to split the *d* orbitals increases. This is consistent with earlier crystal field arguments. Due to increased steric hindrance, larger, bulkier ligands should not be able to approach very closely to a metal ion and, therefore, should not be able to greatly affect the relative energies of the *d* orbitals.

There are several aspects of the spectrochemical series, however, that are *not* readily explained in terms of CFT. For example, one would think that ligands which carry full negative charges would split *d* orbitals better than neutral ligands of approximately the same size. This is not necessarily the case. For example, water is higher in the series than is OH^-. Or, for a second example, one would think that the greater the dipole moment of a ligand (resulting in a higher concentration of electron density on the donating atom of the ligand), the higher in the series it would be. Again, this is not borne out by the data: ammonia has a smaller dipole moment than water, yet NH_3 is higher in the series.

Perhaps most surprising of all, look closely at the ligands on the high end of the spectrochemical series. For example, triphenylphosphine, PPh_3, is a very bulky, neutral ligand with a low dipole moment, yet it is very high in the series. Carbon monoxide or carbonyl, CO, is neutral and has a very modest dipole moment, yet it is listed as the highest in the series. Clearly the crystal field theory, with its assumption of completely electrostatic M–L interactions, does not appear to rationalize the spectrochemical series particularly well. Short of abandoning the CFT altogether, what must we do to modify it so that a degree of explanation of the series can be achieved? It seems logical that we need to investigate the possibility of admitting some covalent contributions to the M—L bond.

The last factor affecting the splitting of *d* orbitals to be discussed, then, is the degree and nature of covalent M–L interactions. Ligand electrons, after all, must be associated with various orbitals that may be capable of overlapping various metal orbitals. Some of the possible metal-ligand covalent interactions are shown in Figure 4.11.

The top half of the figure shows interactions of the sigma (σ) type. Recall that sigma bonds involve the head-on overlap of atomic orbitals. As shown, the most important metal orbitals capable of sigma bonding are of the *p* and *d* types. Specifically, in first-row transition metals, these would be the 4*p* and 3*d* orbitals. For head-on overlap to occur, the 3*d* orbitals must be those of the e_g set, namely, $d_{x^2-y^2}$ and d_{z^2}. Only these orbitals point directly at the ligands. Consistent with the nature of the coordinate-covalent bond proposed by Sidgwick, the e_g metal orbitals are generally empty and both of the bonding electrons come from the ligand. (In octahedral complexes, as previously discussed, the e_g orbitals are of higher energy and the ones most likely to be empty.) The corresponding ligand orbitals are of two major types: *p* orbitals and various hybrid orbitals (*sp* and sp^3 are shown using H_2O, CO, and Cl^- as representative ligands). Using these orbitals, a degree of M–L overlap and corresponding covalent character can be postulated. It turns out that since all ligands are capable of such interactions, these are not of great importance for explaining the above irregularities (at least as seen from the viewpoint of the electrostatic CFT) in the spectrochemical series.

The bottom half of Figure 4.11 shows interactions of the pi (π) type. Recall that these involve parallel overlap of the orbitals participating in the bond. The metal orbitals of primary importance here are the t_{2g}-type *d* orbitals. These orbitals point in between the ligands and are appropriately positioned (it is often said that they are *of the correct symmetry*) to form π bonds with ligand orbitals. Ligand orbitals capable of pi bonding are the *p*- and *d*-type orbitals, as well as the so-called π^* antibonding orbitals. (Antibonding molecular orbitals result from the out-of-phase overlap of *p* orbitals within a given ligand. Of the ligands we have considered, CO, CN^-, NO, and ethylene, C_2H_4, possess such orbitals. More on the general nature of such orbitals, derived from a qualitative consideration of molecular orbital theory, can be found in most general chemistry textbooks.)

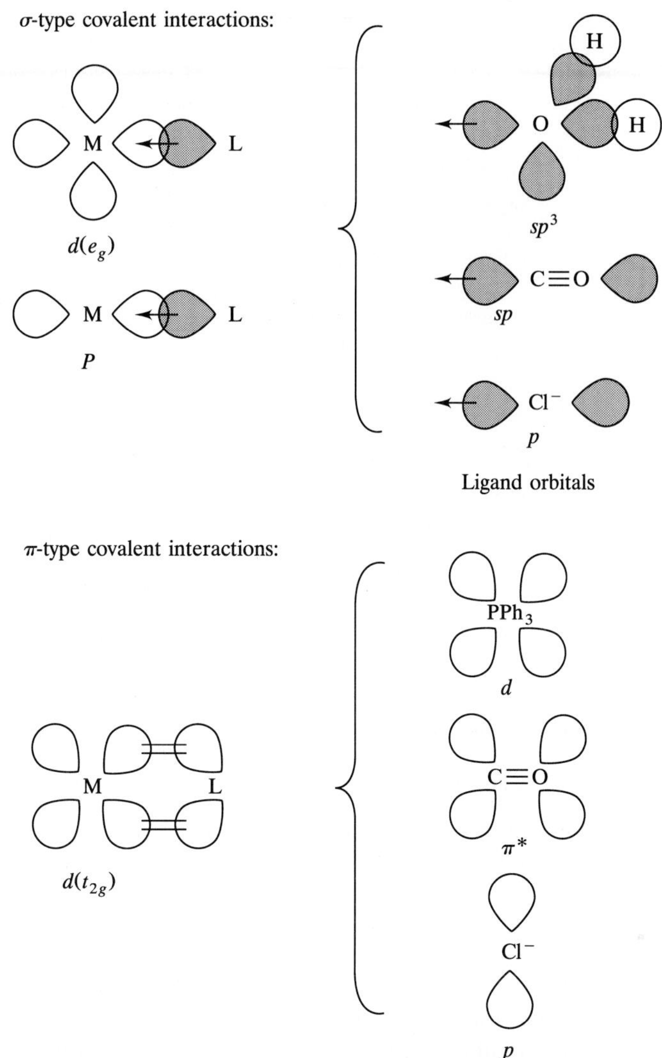

FIGURE 4.11
Some possible σ and π covalent metal-ligand interactions. Participating metal orbitals are shown on the left, while ligand orbitals (on four representative ligands, H_2O, CO, Cl^-, and PPh_3) are listed on the right. Sigma (σ) bonding electrons are donated by the ligand. Pi (π) bonding electrons can come from either the metal or the ligand.

Now what happens when M—L π bonds are formed? Furthermore, can these π interactions at least partially explain the above irregularities in the spectrochemical series?

Figure 4.12 shows two types of π bonding that can occur in coordination compounds. The first (Figure 4.12*a*) is between a filled metal orbital and an unfilled ligand orbital. (The term *backbonding* is often used to refer to this type of interaction. After a ligand forms a σ bond with a metal—and some electron density is transferred from the ligand to the metal as a result—the metal is said to be returning, or backbonding, some of this electron density to the ligand.) Backbonding, then, results in some electron density being transferred to the ligand and thereby produces a greater negative charge on the ligand and a greater positive

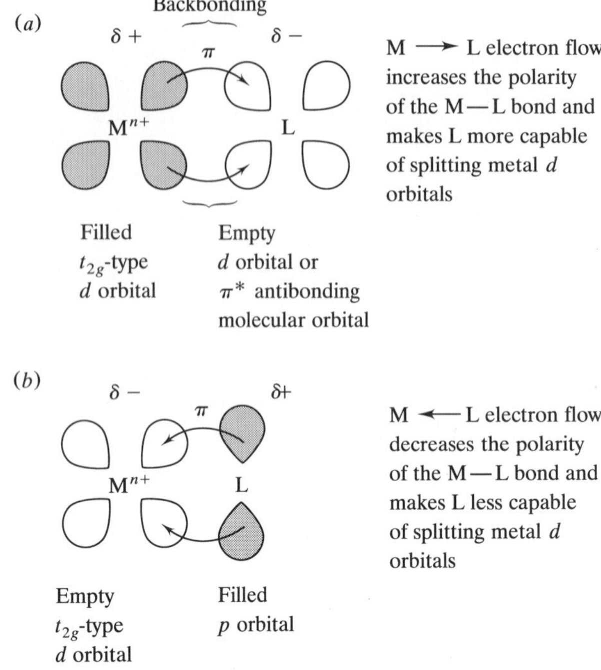

FIGURE 4.12
Two types of M — L π bonds. (*a*) Backbonding, a π interaction between a filled metal orbital and an unfilled ligand orbital, results in greater polarity of the M — L bond and an enhanced splitting of the metal *d* orbitals. L can be ligands such as the phosphines, PR_3; the arsines, AsR_3; cyanide, CN^-; and carbonyl, CO. (*b*) A π interaction between an empty metal orbital and a filled ligand orbital yields a less polar M — L bond and a diminished splitting. Here, L can be ligands such as hydroxide, OH^-; oxide, O^{2-}; or the halides, I^-, Br^-, Cl^-.

charge on the metal. That is to say, the M—L bond becomes more polar due to backbonding. This enhanced polarity results in a greater electrostatic interaction between the metal and the ligand and, in CFT terms, a greater split between the *d* orbitals of the metal. Some of the ligands capable of this type of interaction are the phosphines, carbonyl, and the isoelectronic cyanide ion. Note that these ligands appear high in the spectrochemical series.

The second type of M—L π bonding is shown in Figure 4.12*b*. Here the covalent overlap is between a filled ligand orbital and an unfilled metal orbital. This time the sharing results in a transfer of π electron density from the ligand to the metal making the M—L bond less polar. Therefore, the M–L electrostatic interaction results in a lesser split between the *d* orbitals of the metal. Ligands capable of this type of interaction are hydroxide, oxide, and the halides. Note that these ligands tend to be concentrated toward the low end of the spectrochemical series.

We have seen that the crystal field theory accounts well for the effect of metal charge and size on the magnitude of the crystal field splitting energy. To explain the relative ability of ligands in splitting *d* orbitals, however, a certain degree of covalent character in the M—L bond must be admitted. This modification to the CFT allows for the explanation of several of the major irregularities in the spectrochemical series.

Magnetic Properties

As discussed above, the magnitude of the crystal field splitting energy largely determines the number of unpaired electrons in a given compound. This in turn, as we will see in this section, has a direct bearing on the magnetic properties of coordination compounds.

Molar susceptibility χ_M can be defined as a measure of the degree to which a mole of a substance interacts with an applied magnetic field. It is can be measured in a special apparatus called a Guoy balance, shown schematically in Figure 4.13. This balance is set up so that the sample, typically contained in a small glass tube, is suspended halfway down into the gap between the poles of a strong magnet. Using a sensitive balance, the masses of the sample with the magnet on and off are measured. The difference between these masses leads to a value of the gram susceptibility χ_g, which is then converted to a molar susceptibility χ_M.

Diamagnetism is an induced property of all compounds that results in the substance being repelled by a magnetic field. *Paramagnetism* is a property of compounds with one or more unpaired electrons that results in a substance being drawn into an applied magnetic field. Paramagnetism is several orders of magnitude (powers of 10) stronger than diamagnetism so that a paramagnetic substance is drawn into the applied field and appears to weigh more when the magnet is on. Substances that are only diamagnetic appear to weigh slightly less when the magnet is on.

Molar susceptibility is a macroscopic property that reflects the magnetic moment μ, a microscopic property of electrons. The general relationship between

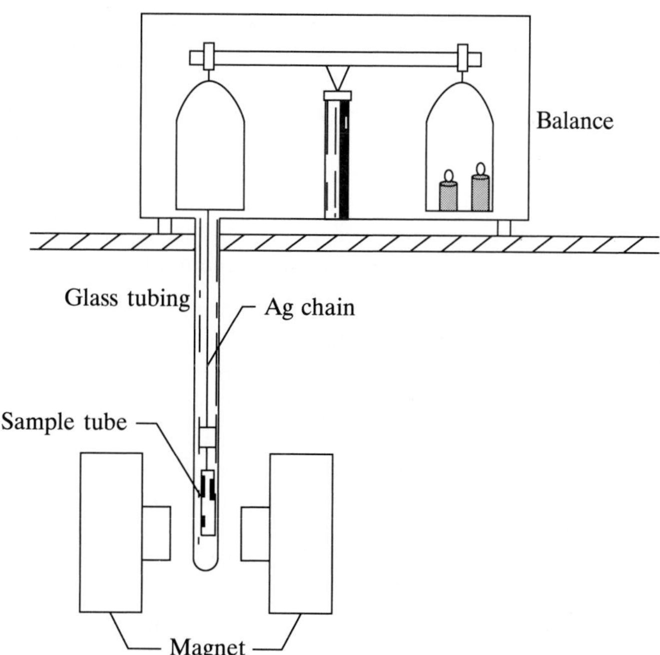

FIGURE 4.13
A schematic diagram of a Guoy balance for the measurement of magnetic susceptibilities. The sample is suspended between the poles of a powerful magnet and weighed with the magnet on and off. The difference in weights is related to the gram and molar susceptibilities. [Adapted from Ref. 5.]

TABLE 4.3
Spin-only magnetic moments for one to five unpaired electrons

No. of unpaired electrons n	Spin-only magnetic moment μ_S, BM
1	1.73
2	2.83
3	3.87
4	4.90
5	5.92

χ_M and μ is given in Equation (4.5):

$$\mu = 2.84\sqrt{\chi_M T} \tag{4.5}$$

where T = temperature, K
μ = magnetic moment, cgs units called Bohr magnetons (BM)
χ_M = molar susceptibility, $(\text{BM})^2\text{K}^{-1}$

Magnetic moments result when charged particles are put into motion. Classically, we can visualize two types of electron motion that give rise to magnetic moments. The first is the electron spinning about its own axis. The moment resulting from this "electron spin" is called the *spin-only magnetic moment* μ_S. The second is the electron orbiting about the nucleus resulting in an *orbital magnetic moment* μ_L. While the theoretical basis of magnetic behavior is significantly beyond the scope of this book, it turns out that μ_S contributes much more to observed magnetic moments (especially for the first-row transition metals) than does the orbital moment. μ_S can be related to the number of unpaired electrons as shown in Equation (4.6). Table 4.3 shows the spin-only magnetic moment for one to five unpaired electrons.

$$\mu_S = \sqrt{n(n+2)} \tag{4.6}$$

where n = number of unpaired electrons
μ_S = spin-only magnetic moment, units known as Bohr magnetons (BM)

The result of these relationships is that a measurement of the molar susceptibility of a paramagnetic substance can be converted into a magnetic moment. These moments can then be compared with spin-only moments to give a measure of the number of unpaired electrons in a compound.

For example, take two iron(III) compounds, potassium hexacyanoferrate(III), $K_3[Fe(CN)_6]$, and potassium hexafluoroferrate(III), $K_3[FeF_6]$. The cyanides are strong-field ligands, while the fluorides are weak. Furthermore, $Fe^{3+}(d^5)$ is capable of low- and high-spin states, so these compounds may have dissimilar electronic configurations and therefore different magnetic characteristics. If we measure the molar susceptibility of each at room temperature (25° C) (and correct for the various diamagnetic contributions involved in each complex), the results are 1.41×10^{-3} and 14.6×10^{-3} $(\text{BM})^2\text{K}^{-1}$, respectively. Using Equation (4.5), experimentally derived magnetic moments can be calculated as shown below:

$$K_3[Fe(CN)_6]: \quad \mu = 2.84\sqrt{(1.41 \times 10^{-3})(298)} = 1.84 \text{ BM}$$

$$K_3[FeF_6]: \quad \mu = 2.84\sqrt{(14.6 \times 10^{-3})(298)} = 5.92 \text{ BM}$$

From Table 4.3, we see that the experimental magnetic moment of the cyanide compound is consistent with one unpaired electron. (An experimental moment slightly different from the spin-only value is usually attributed to a small contribution from the orbital motion of the unpaired electrons.) Such a result is, in fact, consistent with a strong-field–low-spin t_{2g}^5 state having one unpaired electron. The fluoride compound, on the other hand, has an experimental moment consistent with a weak-field–high-spin $t_{2g}^3e_g^2$ state with five unpaired electrons.

We see, then, that the magnetic properties of coordination compounds are consistent with the crystal field theory. Furthermore, these properties can be used to substantiate the spectrochemical series of ligands.

Absorption Spectroscopy and the Colors of Coordination Compounds

One of the most striking properties observed from the very beginning of the almost two centuries' work with coordination compounds is their large variety of often bright and vivid colors. Color we know to be the result of the absorption of a portion of the visible spectrum. Those frequencies not absorbed are reflected off or transmitted through a substance to our eyes to produce the sensation we call color. But what is it about coordination compounds that makes them display such a wonderful variety of colors?

To answer this question, let us start with a hypothetical situation. Suppose, for example, a given coordination compound, ML_6, is orange while a different compound containing the same metal, ML'_n (where L and L′ are different ligands and n may or may not be 6), is purple. Can we propose reasons why these compounds are colored? Furthermore, do the different colors of the two compounds containing the same metal lead us to make some tentative qualitative conclusions about L and L′, or perhaps about n? Take ML_6 first. We have said that it transmits or reflects orange. Therefore it most likely absorbs visible light of the higher frequencies, as shown in Figure 4.14*a*. ML'_n, on the other hand, would appear to absorb lower-frequency visible light. Recall that the frequency absorbed is directly related by $E = h\nu$, the equation first proposed by Max Planck, to the energy absorbed. Therefore, as indicated in the figure, there seems to be a greater difference between the energy levels in ML_6 than there is in ML'_n. The key to the colors of these compounds is that these energy levels (responsible for the absorption of visible light and therefore the colors displayed) are assigned to the various degenerate sets of *d* orbitals attributed to coordination compounds by crystal field theory.

An explanation of the difference in the colors of ML_6 and ML'_n might involve L′ being lower in the spectrochemical series than L and therefore ML'_6 absorbing a lower frequency of light. Or, another possibility might be a change in the geometry of the crystal field, perhaps to a tetrahedral field, in which the energy difference between sets of *d* orbitals is characteristically smaller than in octahedral fields. Different combinations of ligands, coordination numbers, and metals (including their oxidation states), then, would seem to be responsible for the variety of colors displayed by coordination compounds.

One excellent example of the above changes in color is a type of "invisible ink" based on cobalt complexes. The form of the ink that one writes with is a very slightly pink aqueous solution that is almost colorless and therefore nearly invisible when it dries. In this form the message can be passed in front of an unsuspecting unobserver who sees, on cursory inspection, just an ordinary piece of blank paper. The pink color of the ink is attributed to the hexaaquocobalt(II) cation, as shown in Equation (4.7). When the receiver of the message exposes the paper to a source

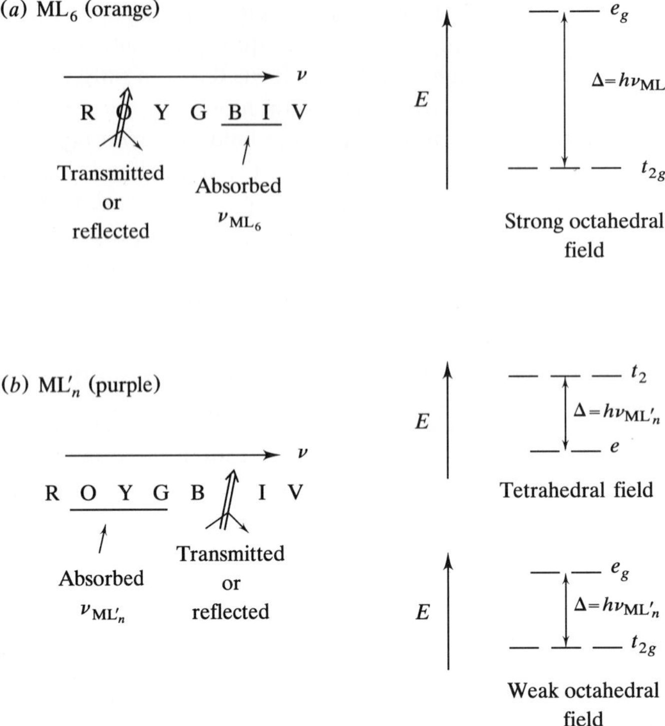

FIGURE 4.14
Two hypothetical colored coordination compounds. (*a*) ML_6 absorbs higher and transmits or reflects lower frequencies of visible light and appears orange. The absorbed frequency might correspond to the large energy difference between the degenerate sets of *d* orbitals in a strong octahedral field. (*b*) ML'_n absorbs lower and transmits or reflects higher frequencies and appears purple. The lower frequency absorbed might correspond to a tetrahedral or a weak octahedral field. The letters ROYGBIV correspond to the colors of visible light from lower to higher frequencies.

of heat (for example, a match or a hair dryer), the water is driven out of the original complex and the message "develops" into a rather dark blue form, attributed to the tetrachlorocobaltate(II) ion also shown in Equation (4.7):

$$\underset{\text{Pink}}{2[Co(H_2O)_6]Cl_2(s)} \xrightarrow{\text{heat}} \underset{\text{Blue}}{Co[CoCl_4](s)} + 12H_2O(g) \qquad (4.7)$$

The colors of these two cobalt(II) complexes are different because the six water ligands are driven off, and only the limited number of chloride ligands are left. The octahedral hexaaquo complex is characterized by a larger Δ because water is higher in the spectrochemical series than chloride and because the field is octahedral. The different crystal field splittings, then, are responsible for the different colors of the two complexes. The situation is illustrated below.

$[Co(H_2O)_6]^{2+}$: R O Y G B I V ⟶ ν (R O Y: Reflected; G B I V: Absorbed)

$[CoCl_4]^{2-}$: R O Y G B I V ⟶ ν (R O Y: Absorbed; G B I V: Reflected)

Now it should not come as a great surprise that the above is a rather oversimplified presentation of the reasons for different colors in coordination compounds. Some of the details and complications come to light (visible light of course!) when we start looking at actual uv–visible spectra of coordination compounds. Some of these, for example, the spectrum of $Ti(H_2O)_6^{3+}(aq)$ given in Figure 4.15, show a direct connection between the frequencies transmitted and the color displayed. Solutions of Ti^{3+} absorb in the green frequencies and transmit the red and blue. Therefore, they appear a red-violet to our eyes. The wavelength at the maximum absorption is about 520 nm. But now let us investigate how these spectra lead to values of Δ for transition metal complexes.

First, a word about the units of wavelength, frequency, and energy commonly employed in discussing the uv–visible spectra of coordination compounds. Wavelength is commonly given in nanometers (1 nanometer $= 10^{-9}$ meter). Ordinarily, we would expect frequencies to be given in hertz (or cycles per second), but these units, as shown in Equation (4.8) for the $Ti(H_2O)_6^{3+}$ case, result in inconvenient numbers:

$$\nu = \frac{c}{\lambda} = \frac{3.00 \times 10^{8} \text{ m/s}}{520 \times 10^{-9} \text{ m}} = 5.77 \times 10^{14} \text{ s}^{-1} \text{ (Hz)} \tag{4.8}$$

Instead, frequencies (and also energies as it turns out) are tabulated in terms of a special unit called a *wave number*, $\bar{\nu}$, that is just the reciprocal of the wavelength in centimeters. The frequency absorbed by $Ti(H_2O)_6^{3+}$, then, would be as calculated in Equation (4.9):

$$\bar{\nu}\ (\text{cm}^{-1}) = \frac{1}{\lambda} = \frac{1}{(520 \times 10^{-9} \text{ m})(100 \text{ cm/m})} = 19{,}200 \text{ cm}^{-1} \tag{4.9}$$

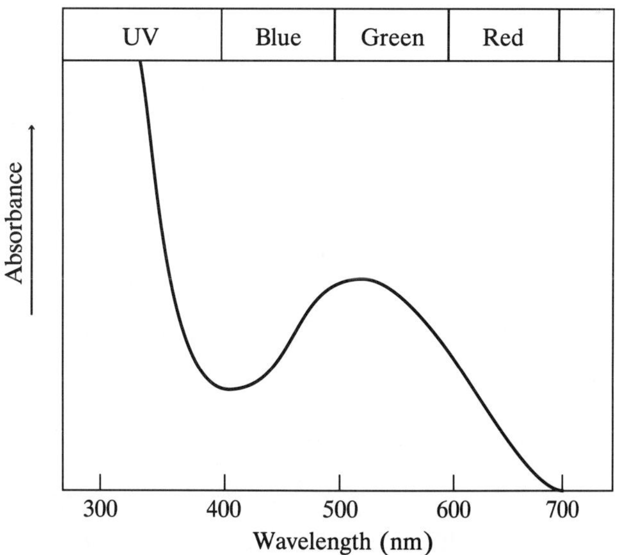

FIGURE 4.15
The visible absorption spectrum of $[Ti(H_2O)_6]^{3+}$. Solutions containing only this species absorb green light but transmit blue and red and thus appear red-violet to our eyes. The wavelength at maximum absorption is approximately 520 nm corresponding to a frequency (and crystal field splitting energy) of 19,200 cm^{-1}. [Adapted from Ref. 1, p. 36.]

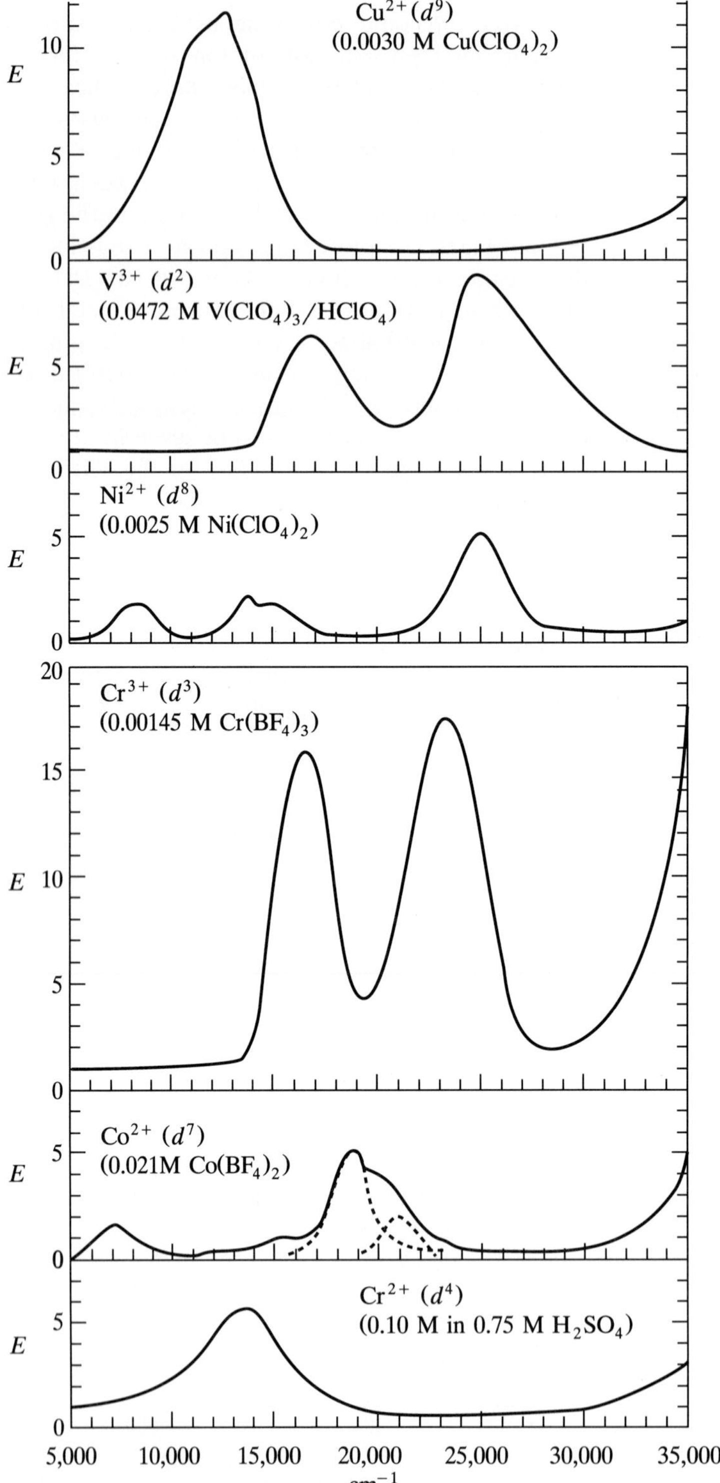

FIGURE 4.16
Some examples of the absorption spectra of first-row transition metal complexes. [Adapted from Ref. 6.]

For reasons of historical context and convenience, crystal field splitting energies (Δ's) are also displayed in these same units. For some transition metal ions (those with one, four, six, and nine *d* electrons, for example), there is a direct correspondence between the energy absorbed and the crystal field splitting energy. For $Ti(H_2O)_6^{3+}$, Δ_o is, in fact, tabulated as 19,200 cm^{-1}. A value of Δ_o in kilojoules per mole can be calculated as shown in Equation (4.10), but these units are not commonly used in displaying splitting energies. Rather, as we saw in Table 4.2, they are given in reciprocal centimeters (cm^{-1}), or wave numbers.

$$\Delta_o\ (\text{kJ/mol}) = \frac{hc}{\lambda}N \tag{4.10}$$

$$= \underbrace{\frac{(6.626 \times 10^{-34}\,\text{J}\cdot\text{s})(3.00 \times 10^{8}\ \text{m/s})}{520 \times 10^{-9}\ \text{m}}}_{\text{J/ion}} \times \underbrace{6.023 \times 10^{23}}_{\text{ions/mol}}$$

$$= 230{,}000\ \text{J/mol} = 230\ \text{kJ/mol}$$

When we change the ligand coordinated to the titanium(III), the wavelength at maximum absorption changes. For hexachlorotitanate(III), $TiCL_6^{3-}$, λ_{max} = 770 nm. This shift is attributed to a different ability of the chloride ligands to split the *d* orbitals of the Ti(III). In this case, Δ_o comes out to be 13,000 cm^{-1} (160 kJ/mol), consistent with chloride being lower in the spectrochemical series.

Before leaving this subject, it should be noted that for most coordination compounds, there is not a direct correspondence between the frequencies absorbed and the crystal field splitting energies. Some examples are shown in Figure 4.16. The reasons for this are complicated and considerably beyond the scope of this discussion, but they have to do with the fact that the orbital and spin electronic motions are not independent of each other. They, in fact, become "coupled" together to produce a more complicated situation (even when viewed with the classical electron-as-a-particle model). The resulting splitting energies are described by what is known as the *vector model of the atom* and the Russell-Saunders spin-orbit coupling scheme not covered here.

SUMMARY

The earliest modern (post-quantum-mechanical) picture of the bonding in coordination compounds was an extension of Lewis electron-dot ideas. Sidgwick proposed that the metal-ligand interaction was best thought of as a coordinate-covalent bond in which both bonding electrons are donated to the metal atom or ion by a ligand. This picture classifies ligands as Lewis bases (electron-pair donors) and metals as Lewis acids (electron-pair acceptors). The resulting complexes often follow the effective atomic number rule.

The crystal field, valence-bond, and molecular orbital theories were all viable explanations of the bonding in coordination compounds. CFT treats the M–L interaction as exclusively electrostatic, while VBT treats it as the overlap of appropriate native and hybrid atomic orbitals. MOT constructs multinuclear molecular orbitals analogous to the mononuclear atomic orbitals. CFT, particularly with an allowance for some small degree of covalent character, is the most valuable bonding theory available, particularly at the beginning level.

A consideration of crystal field theory starts with two fundamental concepts: the coulombic theory of electrostatic interactions and a detailed knowledge of the shapes of d orbitals. Two-dimensional cross-sectional diagrams of hydrogen-like orbitals show orbitals as being "cut out of circular pieces of cloth" by various nodes. The sum of the electron probabilities in a given subshell is a sphere. The five d orbitals appear to be composed of four similar and one, the d_{z^2}, special orbital. To see that the d_{z^2} is not unique in shape or energy, a set of six dependent d orbitals is first visualized. The d_{z^2} orbital turns out to be just a linear combination of two of these dependent orbitals that look exactly like the other four.

When an octahedral field is constructed around a transition metal atom or ion, the five independent metal d orbitals are split into two groups, the lower-energy t_{2g} orbitals that point in between the ligands and the higher-energy e_g orbitals that point directly at them. The energies of these two degenerate sets can be calculated relative to the barycenter, the average energy of a set of orbitals in the same subshell. The difference between these sets of orbitals is called the crystal field splitting energy Δ.

When an octahedral field is tetragonally elongated by gradually removing the z-axis ligands, the energies of the d orbitals are changed. Any ligand with a z-axis component becomes more stable. Completely removing the z-axis ligands results in a square planar field of ligands. The differences between the energies of the various orbitals in a square planar field are tabulated relative to that found in an octahedral field. A tetrahedral field can be visualized as half a cubic field. In both the tetrahedral and cubic cases, all the orbitals point in between the ligands to varying degrees. Three orbitals (the t_2 set) point less directly at the ligands than the other two (the e set). The tetrahedral crystal field splitting energy is about one-half that of an octahedral field.

The relative values of the pairing energy P and the crystal field splitting energy Δ determine whether a complex will be strong-field–low-spin or weak-field–high-spin. Both these possibilities exist for octahedral d^4, d^5, d^6, and d^7 cases. Crystal field stabilization energies (CFSEs) can be calculated for all cases.

The factors affecting the magnitude of the crystal field splitting energy include (1) the geometry of the field, (2) the charge and size of the metal, (3) the ability of the ligand to split d orbitals, and (4) the degree and nature of covalent contributions to the M—L bond. The wholly electrostatic crystal field theory accounts nicely for the first two factors. It does not, however, do well at rationalizing the spectrochemical series in which ligands are ordered by their increasing ability to split d orbitals. To start to explain this series, a degree of covalent M–L character must be introduced. Covalent interactions, particularly the π type, can be pictured as modifying the polarity of the M–L interaction and thereby partially accounting for the positions of some ligands in the series.

Confirmation of crystal field theory comes from a consideration of the magnetic properties of coordination compounds. Molar susceptibilities, derived from measurements on a Guoy balance, can be related to the magnetic moment of the complex. A comparison of this experimentally derived magnetic moment with "spin-only" moments yields a measure of the number of unpaired electrons in the compound. The results derived from a consideration of magnetic properties are consistent with the crystal field theory.

Compounds are colored because they absorb some wavelengths of visible light while reflecting or transmitting others. The differences in the energies of various sets of d orbitals, caused by the presence of a field of ligands, are such that light of visible frequencies is absorbed. Different combinations of ligands, coordination numbers, and metals result in the large variety of colors displayed by these substances.

The frequencies of light absorbed and the related crystal field splitting energies are usually given in reciprocal centimeters (cm^{-1}), or wave numbers. The frequencies derived from uv–visible spectra can, in some cases, be directly related to the crystal field splitting energy. In other cases, however, the coupling of orbital and spin electronic motions makes such a direct correspondence invalid. The spectrochemical series follows directly from such considerations of uv–visible spectra.

PROBLEMS

4.1. The classic acid-base reaction between hydrogen ions and hydroxide ions to produce water can be viewed from the Arrhenius, Brønsted-Lowry, and Lewis definitions. Assign each reactant as an acid or a base under each definition and briefly justify your assignments.

4.2. The reaction of boric acid, $B(OH)_3$, with hydroxide ions, OH^-, to produce tetrahydroxoborate, $B(OH)_4^-$, is better classified as a Lewis acid-base reaction than it is under the Arrhenius or Brønsted-Lowry definitions. Write an equation for this reaction and assign the two reactants as Lewis acids or bases.

4.3. Hemoglobin is the agent of oxygen transfer (respiration) in most animals. In this process, a diatomic oxygen molecule is bound to the central iron ion of the hemoglobin, and the resulting oxyhemoglobin is transported from the lungs to the tissue. In what sense can respiration be viewed as an acid-base reaction?

4.4. The reaction between the oxide ion and carbon dioxide to yield carbonate can be classified as an acid-base reaction. Under what acid-base definition(s) can this classification be made? Using which definition(s) is such a classification not possible? Briefly explain your answer.

4.5. Determine the effective atomic number (EAN) of the metal in each of the following coordination compounds or complex ions:
(*a*) $[IrCl_6]^{3-}$
(*b*) $Fe(CO)_5$
(*c*) $Cr(CO)_6$
(*d*) $[Co(NH_3)_2(NO_2)_4]^-$
(*e*) $[RuCl_2(PPh_3)_3]$
Which of these species follows the EAN rule?

4.6. Determine the effective atomic number (EAN) of the metal in each of the following coordination compounds or complex ions:
(*a*) $[Cu(NH_3)_4]^+$
(*b*) $[Ag(NH_3)_2]^+$
(*c*) $[Fe(CN)_6]^{4-}$
(*d*) $[Mo(CO)_6]$
(*e*) $[Fe(C_2O_4)_3]^{3-}$
Which of these species follows the EAN rule?

4.7. Using Coulomb's law [Equation (4.3)], briefly explain why we say that energy is released when a proton and an electron, starting an infinite distance from each other, are brought together to form a hydrogen atom.

4.8. Using Coulomb's law [Equation (4.3)], briefly explain why we say that the potential energy of a system consisting of two electrons increases when they are brought from an infinite distance apart to a position where they are side by side.

4.9. Sketch out well-labeled drawings of the $3d_{xy}$ and $3d_{x^2-y^2}$ orbitals.

4.10. Sketch out well-labeled drawings of all the $3d$ orbitals that point in between the cartesian axes.

4.11. Sketch out well-labeled drawings of all the $3d$ orbitals that point along the cartesian axes.

4.12. Sketch out well-labeled drawings of all the dependent $3d$ atomic orbitals that point along the cartesian axes.

4.13. In your own words, explain how the $3d_{z^2}$ orbital is related to the other four $3d$ orbitals.

4.14. In your own words, explain how the six dependent $3d$ orbitals are condensed into the five independent $3d$ orbitals we normally consider.

4.15. How would the $3p$ orbitals split in an octahedral field? Briefly justify your answer.

4.16. Using the six dependent d orbitals instead of the five we commonly use today:

(*a*) Draw a well-labeled energy-level diagram showing the splitting of the six d orbitals in an octahedral field. Label Δ_o and determine the position of the barycenter of the subshell.

(*b*) Using d_{xz} and $d_{z^2-x^2}$ as representative examples, briefly explain the relative positions of the d orbitals in the above energy-level diagram.

4.17. Would the change in entropy associated with the construction of an octahedral field of ligands about a metal ion as described in Section 4.2 be positive or negative? Briefly justify your answer. Does your result mean that the change in enthalpy of this process must necessarily be negative in order for the process to be spontaneous? Again, briefly justify your answer.

4.18. Briefly explain in your own words why the d_{z^2} and $d_{x^2-y^2}$ orbitals should be degenerate in an octahedral field even though they are of quite different shapes.

4.19. Suppose that an octahedral field is tetragonally compressed along the z axis. (That is, assume the z-axis ligands get closer to the metal while the x- and y-axis ligands move farther away.) Draw a well-labeled diagram showing the effect on the energy of the five d orbitals starting from their positions in an octahedral field. Briefly rationalize the change in the energy of each orbital.

4.20. Suppose that the tetragonal compression described in Problem 4.19 continues until a linear crystal field results. An unlabeled energy-level diagram for the resulting linear field is given below.

(*a*) Position each orbital in the diagram and briefly rationalize your placements.

(*b*) Given the energy differences in the diagram, calculate the position of the barycenter in the linear field. (*Hint:* Recall that the net increase in energy from the barycenter must equal the net decrease.)

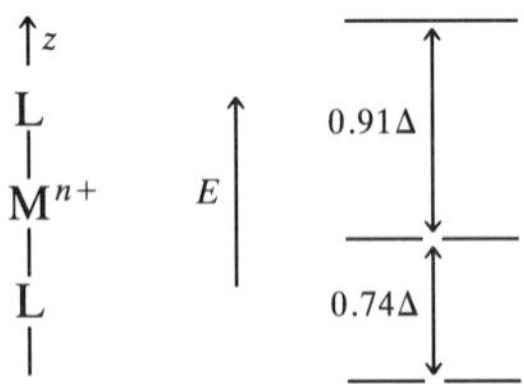

*** 4.21.** Use Figure 4.6 to determine the position of the barycenter for a square planar crystal field. (*Hint:* Recall that the net increase in energy from the barycenter must equal the net decrease.)

4.22. Sketch an energy-level diagram for the $3p$ orbitals in a square planar field lying in the xy plane. Label all orbitals and their relative energies.

4.23. Would a tetragonally elongated d^9 complex be more or less stable than an octahedral complex? Draw energy-level splitting diagrams to support your answer.

4.24. (*a*) Using the cartesian coordinates relative to eight ligands at the corners of a cube as shown in Figure 4.7, show the splitting of the d orbitals in a "cubic field" of ligands.

(*b*) Using the d_{xy} and $d_{x^2-y^2}$ orbitals as examples, explain why you placed these two orbitals as you did in the above crystal field splitting diagram.

4.25. The ion $[PaF_8]^{3-}$ has a cubic arrangement of ligands around the protactinium ion.

(*a*) What is the d-orbital splitting pattern in this ion?

(*b*) Label the principal energy split in the pattern Δ_c, establish a barycenter, and estimate the energy of each d orbital relative to that barycenter.

4.26. The crystal field splitting diagram for a *square pyramidal* crystal field is given below. (The z axis goes through the top of the pyramid, while the four equatorial ligands are located along the + and − x and y axes.)

(*a*) Assign each energy level to an appropriate $3d$ orbital. Briefly justify your assignments.

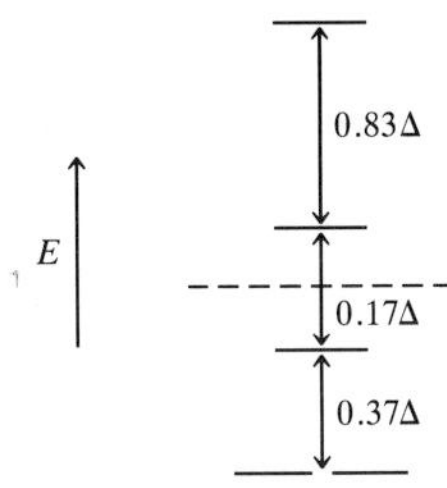

(*b*) Demonstrate that the dotted line in the above energy-level diagram represents the barycenter.

4.27. Draw an energy-level diagram for the six dependent d orbitals in a linear crystal field (assume that the two ligands lie along the positive and negative z axes). Place a barycenter in your diagram. Briefly rationalize your energy-level diagram.

4.28. Draw a well-labeled diagram showing how the triply degenerate p orbitals in a free atom or ion should split in a linear crystal field oriented along the z axis. Show the position of the barycenter and the energy of each orbital in terms of the Δ_L, the crystal field splitting energy for a linear field.

4.29. The crystal field splitting diagram for a trigonal bipyramidal crystal field is shown below. From the information given, *quantitatively* determine the position of the barycenter (indicated qualitatively as a dotted line).

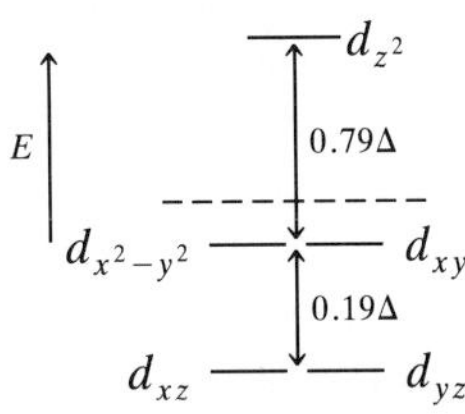

4.30. Tetrahedral complexes are most always high-spin, while octahedral complexes can be either high- or low-spin depending on the metal and/or ligand. What is the most likely explanation of this experimental result?

4.31. Suppose that both high- and low-spin cases are possible for tetrahedral complexes. What numbers of d electrons, d^n, would have both low- and high-spin possibilities? To support your answer, give the electronic configurations (in terms of t_2 and e) for these possibilities.

4.32. Given that only the split between the top two energy levels of a square planar field can give rise to high- and low-spin states, what numbers of d electrons, d^n, would have both possibilities? Briefly support your answer.

4.33. Given that the split between the top two energy levels in a trigonal bipyramidal field (see Problem 4.29) can give rise to high- and low-spin states, what numbers of d electrons, d^n, would have both possibilities? Briefly support your answer.

4.34. How many valence-shell d electrons are there in each of the following ions: Cr^{3+}, Co^{2+}, Pd^{4+}, Pt^{2+}, and Cu^{2+}?

4.35. How many unpaired electrons would there be in each of the following cases:
(*a*) d^4, octahedral, low-spin
(*b*) d^6, tetrahedral, high-spin
(*c*) d^9, square planar
(*d*) d^7, octahedral, high-spin
(*e*) d^2, cubic
(*f*) d^8, octahedral with tetragonal elongation

4.36. Calculate the CFSE in the octahedral d^3 and d^8 cases.

4.37. Verify the CFSEs found in Table 4.1 for the d^5, d^6, and d^7 cases (both high- and low-spin).

4.38. Determine the crystal field stabilization energy (CFSE) in terms of Δ and P for each of the following:
(*a*) $[Fe(CN)_6]^{4-}$
(*b*) $[Ru(NH_3)_6]^{3+}$
(*c*) $[Co(NH_3)_6]^{3+}$
Where there are high- and low-spin possibilities, briefly justify which you choose to calculate a CFSE for.

4.39. Determine the crystal field stabilization energy (CFSE) in terms of Δ and P for each of the following:
(*a*) $[Fe(H_2O)_6]^{3+}$
(*b*) $[PtCl_6]^{2-}$
(*c*) $[Cr(NH_3)_6]^{3+}$
Where there are high- and low-spin possibilities, briefly justify which you choose to calculate a CFSE for.

4.40. Complexes with eight d electrons (d^8) are sometimes square planar and sometimes octahedral. Consider the high-spin complexes of $[NiCl_4]^{2-}$ and $[NiCl_6]^{4-}$. Given the splitting diagrams for these two fields and the results of Problem 4.21, calculate the crystal field stabilization energies for the two d^8 cases and predict which configuration is favored under what conditions.

4.41. Using the results of Problem 4.29, determine the crystal field stabilization energy of a d^6 ion in a trigonal bipyramidal field. (If both high- and low-spin cases are possible, calculate the CFSE for both cases and state the conditions under which the low-spin case will be more stable.)

4.42. Given a metal ion with a d^7 configuration and the results of your calculation in Problem 4.20, calculate the crystal field stabilization energy for the linear field in a high-spin configuration.

4.43. The crystal field splitting diagram for a square pyramidal crystal field is given in Problem 4.26. When one ligand is substituted for another in an octahedral coordination sphere, the first step is often the loss of one ligand to form a square pyramidal intermediate. When this step is accompanied by a gain in crystal field stabilization energy (CFSE) relative to that of the octahedral complex, the substitution is often rapid, while if CFSE is lost, the substitution is slow. Calculate the CFSE for both the octahedral and square pyramidal forms for a strong-field Co^{3+} complex. Would you suspect that the substitution reactions of such complexes would be rapid or slow?

4.44. Apply the term *steric hindrance* in a short description of the movement of the *xy* ligands in a tetragonal compression.

4.45. The crystal field splitting energies Δ_o for PtF_6^{2-}, $PtCl_6^{2-}$, and $PtBr_6^{2-}$ are 33,000, 29,000, and 25,000 cm^{-1}, respectively. Are these small or large values? Briefly discuss these values with respect to others found in Table 4.2.

4.46. Would you expect a high- or low-spin state for a $[Ru(NH_3)_6]^{2+}$ complex? Carefully explain your answer.

4.47. Arrange the following ions in order of increasing split between the t_{2g} and e_g sets of d orbitals, that is, in order of increasing Δ_o.

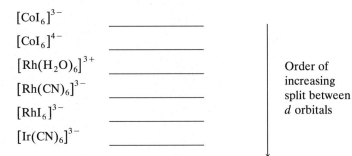

$[CoI_6]^{3-}$	______________	Order of increasing split between *d* orbitals ↓
$[CoI_6]^{4-}$	______________	
$[Rh(H_2O)_6]^{3+}$	______________	
$[Rh(CN)_6]^{3-}$	______________	
$[RhI_6]^{3-}$	______________	
$[Ir(CN)_6]^{3-}$	______________	

4.48. Briefly rationalize the placement of the *iodo complexes* in the series constructed in Problem 4.47.

4.49. How many unpaired electrons would $[Re(CN)_6]^{3-}$ and $[MnCl_6]^{4-}$ have? Give electronic configurations for each ion in terms of the t_{2g} and e_g sets. Carefully rationalize your answers.

4.50. Briefly explain why phosphines (PR_3) are generally stronger-field ligands than ammines (NR_3).

4.51. On the basis of crystal field theory, the fluoride ion should be a particularly strong-field ligand while the triphenylphosphine, PR_3 (where $R = C_6H_5$), molecule should be an extremely weak-field ligand. Briefly rationalize this difference on the basis of the purely electrostatic crystal field theory.

***4.52.** Sulfides, R_2S, can serve as ligands. Would you expect them to be relatively high, medium, or low in the spectrochemical series? (*Hint:* Consider the possibility that these ligands are capable of π bonding.)

4.53. Estimate the spin-only magnetic moment for a d^6 ion in octahedral and tetrahedral fields generated by weak- and strong-field ligands.

4.54. $Na_2[Ni(CN)_4]\cdot 3H_2O$ is essentially diamagnetic. Speculate on the basic geometry of the complex anion, $[Ni(CN)_4]^{2-}$. Briefly explain your answer.

4.55. The experimental magnetic moments of four manganese complexes are given below. State whether the complexes are high-spin or low-spin. Also write down the electronic configurations (within the t_{2g} and e_g sets of $3d$ orbitals) that are consistent with these observed magnetic moments.

Complex	μ_{exptl}, **BM**
$[Mn(CN)_6]^{4-}$	1.8
$[Mn(CN)_6]^{3-}$	3.2
$[Mn(NCS)_6]^{4-}$	6.1
$[Mn(acac)_3]$	5.0

4.56. A given square pyramidal RuL_5 complex containing Ru^{2+} (d^6) has a magnetic moment of 2.90 BM. Calculate the CFSE of this complex. (*Hint:* Start by placing the six electrons into the square pyramidal splitting diagram given in Problem 4.26.)

4.57. Which of the iodo complexes in the series constructed in Problem 4.47 do you suspect would have the highest molar susceptibility? Briefly explain. As part of your answer, calculate a value for the spin-only moment for that complex.

4.58. A complex of nickel(II), $[NiCl_2(PPh_3)_2]$, has an experimental magnetic moment of 2.96 BM. The analogous complex of palladium(II) is diamagnetic. Draw and name all the possible isomers that will exist for the palladium compound. Briefly justify your answer. (Assume that both compounds have the same geometry around the central atom.)

4.59. The magnetic susceptibilities of the following two ruthenium complexes, $[RuF_6]^{4-}$ and $[Ru(PR_3)_6]^{2+}$, are 1.01×10^{-2} and approximately zero, respectively. With how many unpaired electrons are these magnetic results consistent? (Assume $T = 298$ K.) Briefly justify your answer.

4.60. Briefly explain qualitatively why many cyano complexes of divalent transition metal ions are yellow, whereas many aquo complexes of these ions are blue or green.

4.61. Would you be surprised to learn that all the complexes in the series constructed in Problem 4.47 are various shades of magenta (purplish-red)? Briefly rationalize your answer.

4.62. Diamagnetic ($\mu = 0$) complexes of cobalt(III) such as $[Co(NH_3)_6]^{3+}$, $[Co(en)_3]^{3+}$, and $[Co(NO_2)_6]^{3-}$ are orange-yellow. In contrast, the paramagnetic complexes $[CoF_6]^{3-}$ and $[Co(H_2O)_3F_3]$ are blue. Explain qualitatively this difference in color and magnetic moment.

4.63. A solution of $[Ni(H_2O)_6]^{2+}$ is green and paramagnetic ($\mu = 2.90$ BM), while a solution of $[Ni(CN)_4]^{2-}$ is colorless and diamagnetic. Suggest a qualitative explanation for these observations. Include diagrams showing the molecular geometry and the *d*-orbital energy levels of these complex ions as part of your answer.

4.64. In one concise paragraph, indicate how one might go about constructing a part of the spectrochemical series using visible spectroscopy, that is, by measuring the wavelengths of visible light absorbed by compounds.

4.65. There is extensive experimental evidence to support crystal field theory. Name and briefly describe the nature of two pieces of such evidence.

CHAPTER 5

RATES AND MECHANISMS OF REACTIONS OF COORDINATION COMPOUNDS

Having discussed the history and nomenclature (Chapter 2), structures (Chapter 3), and the bonding (Chapter 4) of coordination compounds, we turn now to a treatment of the reactions of these compounds. We will start with a discussion of the common types of reactions usually encountered, but this simple categorization will lead quite quickly into the more important questions of the how these reactions actually take place. For example, we will start to consider whether reactions happen primarily because (1) two reactant molecules collide to produce an intermediate that subsequently falls apart into the product molecules or (2) a reactant molecule falls apart first and a resulting fragment then collides with another reactant to make a product molecule. These are questions concerning the pathways, or *mechanisms*, for reactions involving coordination compounds. We will then see that the favored mechanism, that is, the one which is the most apt to happen, is the one characterized by the smallest energy increase accompanying the formation of a reaction intermediate. An investigation of energy changes will involve our knowledge, albeit recently acquired in the last chapter, of the bonding in various reactant, product, and possible intermediate molecules.

5.1 A BRIEF SURVEY OF REACTION TYPES

Figure 5.1 classifies the more common reactions of coordination compounds. Starting in the center with a general compound $[ML_n]^{n+}$, substitution, dissociation, addition, redox or electron transfer reactions, and the reaction of a coordinated ligand are indicated. Note that there is one major oversimplification in Figure 5.1, namely, that all the ligands in the starting compound are shown to be identical.

(These are known as *homoleptic* compounds, those in which all the ligands attached to the central metal atom or ion are identical.) Most reactions of coordination compounds actually involve heteroleptic compounds (those with more than one type of ligand), but for reasons of simplicity, these are not specifically represented in the figure. Additionally, as we will see, there are a significant number of reactions that are a combination of more than one of the types shown in the figure. With these caveats (notes of warning), we can now turn to a brief discussion of each of these reaction types.

Substitution reactions, shown vertically at the top of the figure, are the most common. These involve the substitution of one ligand in a coordination sphere for another without any change in the coordination number or the oxidation state of the metal. Sometimes there is a wholesale replacement of all the ligands of a homoleptic compound with a different ligand as shown in Equation (5.1), but more often only a fraction of the original ligands are replaced as shown in Equations (5.2) and (5.3):

$$[Cu(H_2O)_4]^{2+} + 4NH_3(aq) \longrightarrow [Cu(NH_3)_4]^{2+} + 4H_2O \tag{5.1}$$

$$[Cr(CO)_6] + 3PPh_3 \longrightarrow [Cr(CO)_3(PPh_3)_3] + 3CO(g) \tag{5.2}$$

$$[PtCl_4]^{2-} + NH_3 \longrightarrow [Pt(NH_3)Cl_3]^- + Cl^-(aq) \tag{5.3}$$

The tendency of such substitution reactions to take place is often tabulated in terms of *stepwise* and *overall* equilibrium constants. For example, the reaction of the tetraaquocopper(II) ion with ammonia in aqueous solution, shown in Equation (5.1) above, can be broken down into the stepwise replacement of one water ligand at a time as given in Equations (5.4) to (5.7):

$$[Cu(H_2O)_4]^{2+} + NH_3 \longrightarrow [Cu(NH_3)(H_2O)_3]^{2+} + H_2O \tag{5.4}$$

$$K_1 = \frac{[\{Cu(NH_3)(H_2O)_3\}^{2+}]}{[\{Cu(H_2O)_4\}^{2+}][NH_3]}$$

$$[Cu(NH_3)(H_2O)_3]^{2+} + NH_3 \longrightarrow [Cu(NH_3)_2(H_2O)_2]^{2+} + H_2O \tag{5.5}$$

$$K_2 = \frac{[\{Cu(NH_3)_2(H_2O)_2\}^{2+}]}{[\{Cu(NH_3)(H_2O)_3\}^{2+}][NH_3]}$$

$$[Cu(NH_3)_2(H_2O)_2]^{2+} + NH_3 \longrightarrow [Cu(NH_3)_3(H_2O)]^{2+} + H_2O \tag{5.6}$$

$$K_3 = \frac{[\{Cu(NH_3)_3(H_2O)\}^{2+}]}{[\{Cu(NH_3)_2(H_2O)_2\}^{2+}][NH_3]}$$

$$[Cu(NH_3)_3(H_2O)]^{2+} + NH_3 \longrightarrow [Cu(NH_3)_4]^{2+} + H_2O \tag{5.7}$$

$$K_4 = \frac{[\{Cu(NH_3)_4\}^{2+}]}{[\{Cu(NH_3)_3(H_2O)\}^{2+}][NH_3]}$$

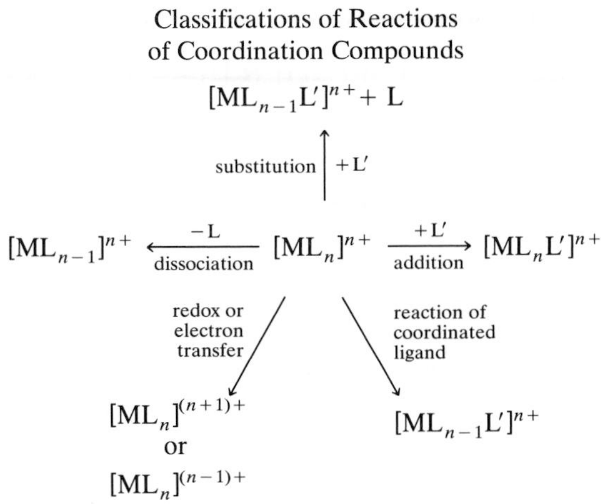

FIGURE 5.1
Five different types of reactions of coordination compounds: substitution, dissociation, addition, redox or electron transfer, and the reaction of a coordinated ligand.

The equilibrium constant K_n for each successive replacement of one ligand for another is called, logically enough, a *stepwise equilibrium constant*. Since these reactions are carried out in aqueous solutions where the concentration of water is very nearly constant, the concentration of water, $[H_2O]$, does not appear in the mass action expressions for the equilibrium constants but rather is incorporated into the value of each K. (You should be familiar with this tactic from your previous studies, for example, the expressions for K_a and K_b used in discussing aqueous acid-base equilibrium problems.) The overall replacement of the four water ligands with ammonia ligands [Equation (5.1)] can be described by an *overall equilibrium constant*, symbolized by a β. Since four ligands are replaced, this is designated β_4. The overall constant is just the product of the four stepwise constants as shown in Equation (5.8):

$$\beta_4 = \frac{\left[\{Cu(NH_3)_4\}^{2+}\right]}{\left[\{Cu(H_2O)_4\}^{2+}\right][NH_3]^4} = K_1K_2K_3K_4 \tag{5.8}$$

As we have seen in Chapter 3 on the structures of coordination compounds, there are many more examples of heteroleptic complexes than homoleptic. These, too, undergo substitution reactions. Equations (5.9) and (5.10) show two common examples. The first, the reaction of a complex with water, is an *aquation reaction*, while the second, the reaction with an anion, is called an *anation reaction*. We will return to a more detailed discussion of these various reactions in Section 5.3.

$$[Co(NH_3)_5NO_3]^{2+} + H_2O \longrightarrow [Co(NH_3)_5H_2O]^{3+} + NO_3^- \tag{5.9}$$

$$[Co(NH_3)_5(H_2O)]^{3+} + Cl^- \longrightarrow [Co(NH_3)_5Cl]^{2+} + H_2O \tag{5.10}$$

Dissociation reactions, shown in the center left of Figure 5.1, involve a decrease in the number of ligands and sometimes, but not always, a decrease in the coordination number. An example involving a decrease in coordination number is the "invisible ink" reaction discussed in the last chapter and repeated below in Equation (5.11). A dissociation reaction in which the coordination number does not decrease because a bidentate ligand is replaced by two monodentate ligands is given in Equation (5.12).

$$2[Co(H_2O)_6]Cl_2 \longrightarrow Co[CoCl_4] + 12H_2O \qquad (5.11)$$

Red — Blue

$$[Cr(en)_3]Cl_3 \xrightarrow{120°C} [CrCl_2(en)_2]Cl + en \qquad (5.12)$$

Addition reactions, shown in the center right of Figure 5.1, are accompanied by an increase in the coordination number of the metal. For steric reasons, most addition reactions occur with complexes in which the metal has a low coordination number initially. For example, Equation (5.13) shows a tetrahedral titanium compound accepting two additional chloride ligands to produce an octahedral hexachlorotitanate(IV) complex. Equation (5.14) shows a square planar bis(acetylacetonato)copper(II) molecule accepting a pyridine (py) ligand to form a square pyramidal product.

$$TiCl_4 + 2Cl^- \longrightarrow [TiCl_6]^{2-} \qquad (5.13)$$

$$Cu(acac)_2 + py \longrightarrow Cu(acac)_2py \qquad (5.14)$$

Oxidation-reduction or electron-transfer reactions, shown in the lower left of Figure 5.1, involve the oxidation or reduction of a coordinated transition metal atom or ion. The figure represents the simplest case of electron transfer, namely, one in which the coordination number remains constant. Equation (5.15) shows an example in which the hexaammineruthenium(III) ion is reduced by the addition of chromium(II) ion, while Equation (5.16) shows the oxidation of a hexacyanoferrate(II) complex by hexachloroiridate(IV). In neither of these complexes is the coordination number altered.

$$[Ru(NH_3)_6]^{3+} + Cr^{2+} \longrightarrow [Ru(NH_3)_6]^{2+} + Cr^{3+} \qquad (5.15)$$

$$[Fe(CN)_6]^{4-} + [IrCl_6]^{2-} \longrightarrow [Fe(CN)_6]^{3-} + [IrCl_6]^{3-} \qquad (5.16)$$

There are examples, not represented in Figure 5.1, in which the oxidation of a complexed metal atom or ion is accompanied by the addition of one or more ligands so as to increase the coordination number. These are common enough reactions that they are designated *oxidative-addition reactions*. One of the very first reactions of this type was first characterized by L. Vaska in the 1960s and is shown in Equation (5.17):

$$trans\text{-}[Ir(CO)Cl(PPh_3)_2] + HCl \longrightarrow [Ir(CO)Cl_2H(PPh_3)_2] \qquad (5.17)$$

Note that in the reactant, quite often referred to as *Vaska's compound*, the coordination number of the iridium(I) ion is 4, while in the product the iridium is formally of oxidation state +3 and has a coordination number of 6.

The opposite of oxidative-addition, logically enough, is called *reductive-elimination*. Equation (5.18) shows the reductive elimination reaction of carbonylchlorodihydridobis(triphenylphosphine)iridium(III) to form diatomic hydrogen and Vaska's compound:

$$[Ir(CO)ClH_2(PPh_3)_2] \longrightarrow [Ir(CO)Cl(PPh_3)_2] + H_2 \qquad (5.18)$$

Reactions of coordinated ligands, shown in the lower right of Figure 5.1, are reactions of a ligand that take place without the breaking of the metal-ligand bond. Equation (5.19) shows a water ligand in hexaaquochromium(III) reacting with a hydroxide ion to produce the corresponding hydroxo complex. Here the $Cr—OH_2$ bond has not been broken, but the water has lost a proton to produce the hydroxo ligand. Equation (5.20) shows the reaction of pentaamminecarbonatocobalt(III) with acid to form pentaammineaquocobalt(III). This reaction proceeds without the breaking of the $Co—OCO_2$ bond. Another common ligand that reacts while coordinated is acetylacetonate. Equation (5.21) shows the replacement of the central hydrogen atom of acac with a bromine atom.

$$[Cr(H_2O)_6]^{3+} + OH^- \longrightarrow [Cr(H_2O)_5(OH)]^{2+} + H_2O \tag{5.19}$$

$$[Co(NH_3)_5CO_3]^+ + 2H_3O^+ \longrightarrow [Co(NH_3)_5(H_2O)]^{2+} + 2H_2O + CO_2 \tag{5.20}$$

$$Cr\left(\begin{matrix} O—C\diagup^{CH_3} \\ \quad\quad \diagdown \\ \quad\quad\quad C—H \\ \quad\quad \diagup \\ O—C\diagdown_{CH_3} \end{matrix}\right)_3 + 3Br_2 \longrightarrow Cr\left(\begin{matrix} O—C\diagup^{CH_3} \\ \quad\quad \diagdown \\ \quad\quad\quad C—Br \\ \quad\quad \diagup \\ O—C\diagdown_{CH_3} \end{matrix}\right)_3 + 3HBr \tag{5.21}$$

Before leaving this brief survey of reaction types, a word is in order concerning how the course of these various reactions might be monitored. As we have already discussed in Chapter 4, different complexes have different uv–visible (or absorption) spectra. The substitution, dissociation, addition, or alteration of a ligand or ligands often changes the maximum wavelength of uv–visible light absorbed. The same can be said when the oxidation state of the metal changes. These changes in the uv–visible spectra (often accompanied by changes in the actual color of the compounds) can be used to monitor the rate of these reactions. Changes in the molar susceptibility can be used in a similar manner. Other physical techniques such as infrared (ir) and nuclear magnetic resonance (nmr) spectroscopy, not discussed in this text, can also be used to monitor the progress of a given reaction.

5.2 LABILE AND INERT COORDINATION COMPOUNDS

To classify the varying rates of reaction (most commonly with regard to substitution) of coordination compounds, Henry Taube, who received the 1983 Nobel Prize in chemistry for his work in the kinetics of coordination compounds, suggested the terms labile and inert. If we consider a 0.1 M aqueous solution, a *labile* coordination compound is one that under these circumstances, has a half-life of less than a minute. (Recall that *half-life* is the amount of time required for the concentration of the reactant to decrease to half its initial concentration.) An *inert* compound, on the other hand, is one with a half-life greater than a minute.

We must clearly keep in mind that the terms labile and inert are *kinetic* terms. They refer to the rate of a reaction that in turn is governed by its energy of activation E_a. (Here, you may want to review your knowledge of kinetics as introduced in previous courses.) These terms relate to how fast a compound reacts rather than to how stable it is. Stable and unstable are *thermodynamic* terms. They are related to the changes in free energy, enthalpy, and entropy of the compound. Reactions with a large negative change in free energy and a correspondingly large positive equilibrium constant go spontaneously from left to right; the products of such reactions are considered more stable than the reactants. The change in free energy can in turn be related to the changes in enthalpy and entropy (recall that $\Delta G = \Delta H - T\Delta S$).

To illustrate the difference between kinetic lability and thermodynamic stability, we consider some specific examples. Take the familiar Werner complex cation hexaamminecobalt(III), $[Co(NH_3)_6]^{3+}$. It reacts spontaneously with acid. In fact, the equilibrium constant for the reaction corresponding to Equation (5.22) is very large, in the vicinity of 10^{30}:

$$[Co(NH_3)_6]^{3+} + 6H_3O^+ \longrightarrow [Co(H_2O)_6]^{3+} + 6NH_4^+ \qquad (5.22)$$

Therefore, we would say that this cation is *unstable* toward reaction with acid. On the other hand, it takes several days at room temperature to get this reaction to go significantly from left to right even in 6 *M* HCl. Accordingly, the rate of this reaction is so slow that $[Co(NH_3)_6]^{3+}$ must be classified as *inert* under these circumstances. This complex cation, then, is unstable (thermodynamically) but inert (kinetically) toward reaction with acid.

In contrast, tetracyanonickelate(II), $[Ni(CN)_4]^{2-}$, is exceptionally *stable* (thermodynamically). The equilibrium constant for its formation, represented in Equation (5.23), is very large, also in the vicinity of 10^{30}:

$$Ni^{2+} + 4CN^- \longrightarrow [Ni(CN)_4]^{2-} \qquad K = 10^{30} \qquad (5.23)$$

At the same time, this complex anion is *labile*; that is, the cyanide ligands in the coordination sphere exchange rapidly with those found free in an aqueous solution. This exchange rate can be measured when carbon 14–labeled cyanide ions are placed in solution with the complex as represented in Equation (5.24):

$$[Ni(CN)_4]^{2-} + 4\,{}^{14}CN^- \longrightarrow [Ni({}^{14}CN)_4]^{2-} + 4CN^- \qquad (5.24)$$

The labeled cyanides exchange places with their unlabeled counterparts quite rapidly. In a matter of seconds, in fact, half the unlabeled cyanides are replaced with labeled ligands. The tetracyanonickelate(II) ion, then, is stable but labile.

In summary, some coordination compounds are kinetically inert, while others turn out to be labile. Furthermore, this lability seems to be unrelated to the thermodynamic stability of the compound. Now, being a veteran chemistry student trained to ask critical questions, you are about to ask how can we tell which complexes will be inert and which will be labile? As you might suspect, this is indeed a crucial question. It turns out that *complexes of the first-row transition metal ions, with the exception of Cr^{3+} and Co^{3+}, are generally labile, while most second- and third-row transition metal ions are inert*. But *how*, you ask, do we explain such a statement? *Why*, for example, should the rates of reactions involving Co^{3+} and Cr^{3+} be different from those involving other first-row transition metal atoms and cations? *What* is it about these particular cations that makes them so inert? To start to answer such queries, we turn now to a discussion of some of the most extensively studied reactions of coordination compounds, those involving the substitution of octahedral complexes.

5.3 SUBSTITUTION REACTIONS OF OCTAHEDRAL COMPLEXES

Possible Mechanisms

The overall substitution reaction of a homoleptic complex was represented in Figure 5.1. To discuss mechanisms, however, it will be helpful to consider a more specific situation. Take, for example, an octahedral coordination compound containing a metal bound to five inert ligands (L) and one labile ligand (X) that is about to be replaced by an incoming ligand (Y). The overall equation for this reaction is

$$ML_5X + Y \longrightarrow ML_5Y + X \qquad (5.25)$$

(1) $$ML_5X \xrightarrow[k_1]{\text{slow}} ML_5 + X$$

five-coordinate intermediate

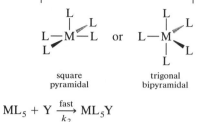

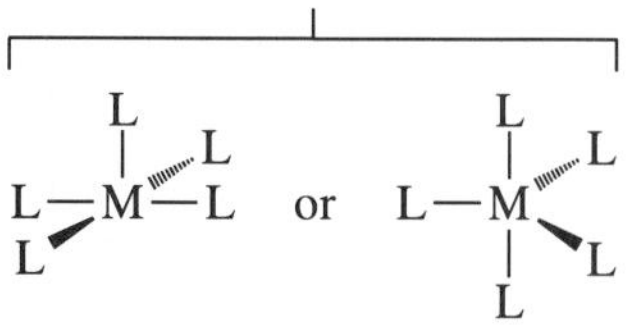

square pyramidal or trigonal bipyramidal

(2) $$ML_5 + Y \xrightarrow[k_2]{\text{fast}} ML_5Y$$

Overall rate = rate of rate-determining step (1)

$$= -\frac{\Delta[ML_5X]}{\Delta t} = k_1[ML_5X]$$

FIGURE 5.2
The dissociative (D) mechanism for the substitution of one ligand for another in an octahedral complex, ML_5X (L = inert ligands, X = labile ligand, and Y = incoming ligand). The mechanism assumes that the first step (1), the breaking of the M—X bond to form the five-coordinate ML_5 intermediate, is the rate-determining step. The rate of the reaction is found to be dependent only upon the concentration of the original complex, $[ML_5X]$.

How might this substitution actually take place on the molecular level? Or, to restate the question, what is the *reaction mechanism*, that is, the sequence of molecular-level steps involved in the reaction? There would, at first blush, seem to be at least two major possibilities. One is called the *dissociative* (D) mechanism, the other *associative* (A).

In the dissociative mechanism, we picture the X ligand breaking off from the reactant to produce the five-coordinate ML_5 intermediate. (This intermediate might well assume a square pyramidal or perhaps a trigonal bipyramidal geometry.) In a second, faster reaction the intermediate and the incoming ligand join together to yield the product. The complete mechanism is shown in Figure 5.2.

Note that we have assumed that the slower first step, the dissociation, is the rate-determining step. (At this point, you may want to review your knowledge of elementary reaction mechanisms as covered in previous courses. Recall that the rate-determining step, sometimes called the *bottleneck step*, is the slowest elementary reaction in a mechanism. The rate of this step determines the overall rate of the reaction.) Note that the dissociative mechanism predicts that the rate of the overall substitution reaction is dependent only on the concentration of the original complex, $[ML_5X]$, and is independent of the concentration of the incoming ligand, [Y].

The other logical pathway by which a substitution reaction might take place is the associative mechanism. In this case, the rate-determining step is the collision between the original complex, ML_5X, and the incoming ligand, Y, to produce a seven-coordinate intermediate, ML_5XY. (This intermediate might assume a "bicapped octahedral" structure, in which the X and Y ligands share one of the normal octahedral sites, or possibly a pentagonal bipyramidal geometry.) The second, faster step is the dissociation of the X ligand to produce the desired product. This mechanism is shown in Figure 5.3. Note that the associative mechanism predicts that the rate of the substitution reaction will be dependent upon the concentration of *both* ML_5X and Y.

So, you may be beginning to think, deciding whether a mechanism is associative or dissociative should not be particularly difficult. It looks like all we need do is determine the rate law for the reaction: if the rate depends upon

(1) $$ML_5X + Y \xrightarrow[k_1]{\text{slow}} ML_5XY \quad \text{(seven-coordinate intermediate)}$$

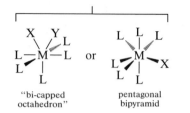

"bi-capped octahedron" or pentagonal bipyramid

(2) $$ML_5XY \xrightarrow[k_2]{\text{fast}} ML_5Y + X$$

$$\text{Overall rate} = \text{rate of rate-determining step (1)}$$
$$= -\frac{\Delta[ML_5X]}{\Delta t} = k_1[ML_5X][Y]$$

FIGURE 5.3
The associative (A) mechanism for the substitution of one ligand for another in an octahedral complex, ML_5X (L = inert ligands, X = labile ligand, and Y = incoming ligand). The mechanism assumes that the first step (1), the formation of an M—Y bond to form the seven-coordinate ML_5XY intermediate, is the rate-determining step. The rate of the reaction is found to be dependent upon both the concentration of the original complex, $[ML_5X]$, and that of the incoming ligand, [Y].

$[ML_5X]$ only, it is dissociative, but if the rate also depends on [Y], it is associative. Not surprisingly, coordination chemistry kinetics turns out not to be quite this simple. Two complications come immediately to mind. First, the actual mechanisms may be more complicated than the clearly differentiated D and A mechanisms outlined above. Second, there may be special experimental conditions that "mask" the dependence of a rate on the concentration of the incoming ligand.

First, consider the idea that the purely associative and dissociative mechanisms, while useful as starting points in a discussion of the kinetics of coordination compounds, are rather idealized and perhaps somewhat oversimplified possibilities. Is it not just a little unrealistic, for example, to suppose that the M—X bond fully breaks and the resulting five-coordinate ML_5 intermediate exists for a substantial period of time before it subsequently collides with Y to produce the final product? Or may we be expecting too much to suppose that the ML_5XY intermediate in the associative pathway has a significant measurable lifetime before the M—X bond breaks and ML_5Y results? After all, if the real mechanisms were this neat and simple, we should have been able to isolate a variety of ML_5 and/or ML_5XY intermediates and thereby add to the evidence that a given reaction is D or A. Unfortunately, isolation of such intermediates is rather rare. Given this discussion, we need to consider one other type of reaction pathway, called the *interchange* (I) mechanism.

Consider a situation where the entering ligand, Y, sits just outside the coordination sphere of the ML_5X complex and, as the M—X bond starts to break and X starts to move away, the M—Y bond simultaneously starts to form and Y moves into the coordination sphere. In this case, the attacking (Y) and leaving (X) ligands gradually interchange places in the coordination sphere of the metal and no distinct five- or seven-coordinate intermediate would be formed (or, therefore, be available to be isolated). Such a reaction pathway, sometimes called a *concerted* or an *interchange* mechanism, may well be more realistic than a purely D or A mechanism.

Having described the above I mechanism, it is important to note that we do not have to totally discard the idea of dissociative and associative mechanisms. For example, if the M—X bond breaks preferentially in the interchange, then the interchange is closer to a dissociative than to an associative mechanism. In this case, we could designate the mechanism I_d (*interchange-dissociative*), where the d subscript indicates the dissociative nature of the interchange. Similarly, when M—Y bond formation is favored somewhat, the mechanism would be designated as I_a (*interchange-associative*).

So, in summary, we have described five types of mechanisms, D, A, I, I_d, and I_a. Let us assume from here on out that while we will speak generally about dissociative and associative mechanisms, the terms D and A are reserved for situations in which five- and seven-coordinate intermediates have actually been isolated. If no intermediates have been isolated, the designations I_d and I_a would seem to be more appropriate.

Experimental Complications

In Figures 5.2 and 5.3 we considered the rate laws that would result from dissociative and associative mechanisms, respectively. In the end, you will recall, it appeared that these pathways could be readily differentiated by the dependence or lack of dependence of their rate laws on the concentration of the entering ligand, Y. As noted earlier, we need to discuss the possibility that this dependence may be masked by certain experimental conditions.

Common examples of this masking of a concentration dependence are substitution reactions carried out in aqueous solution. Again considering the same overall substitution of Y for X in the ML_5X reactant, two probable rate-determining steps are shown in Figure 5.4. Note that the dissociative mechanism still predicts a rate dependence only on the concentration of the initial complex, $[ML_5X]$. This result is shown in Equation (5.26).

For the associative mechanism, the reaction of the starting complex with a *water molecule* is assumed to be the rate-determining step. Such an assumption is certainly not unexpected because the concentration of water in a dilute aqueous solution is very large (approximately 55.6 M), much larger than the concentration of Y (perhaps about 0.1 M). The likelihood of the starting complex colliding with a water molecule is therefore much greater than a collision with Y. As shown in Equation (5.27), the rate of this step is dependent upon *both* the concentrations of ML_5X and H_2O. But now we have to recognize, as noted previously (p. 91), that $[H_2O]$ is so large that, in this situation, it is essentially a constant. Therefore, as shown, the two constants in Equation (5.27), k' and $[H_2O]$, are combined to yield Equation (5.28). Note that while the result appears to be first-order only in $[ML_5X]$, we know that it is also first-order in $[H_2O]$. This latter dependence, however, has been masked or hidden by the nature of the experimental situation. For this reason, Equation (5.28) is often said to represent a "pseudo" first-order reaction. In the cases discussed then, the rate laws for both the dissociative and associative mechanisms are, for practical purposes, identical. In this situation we cannot decide between the two mechanisms through an analysis of the observed rate laws.

(*a*) Dissociative:

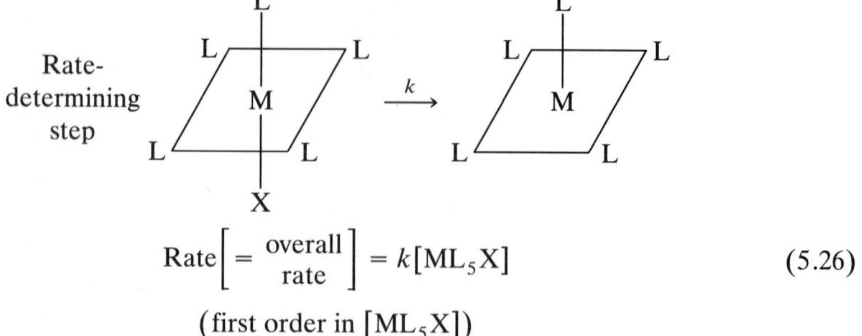

$$\text{Rate}\left[=\begin{matrix}\text{overall}\\ \text{rate}\end{matrix}\right]=k[ML_5X] \tag{5.26}$$

(first order in $[ML_5X]$)

(*b*) Associative:

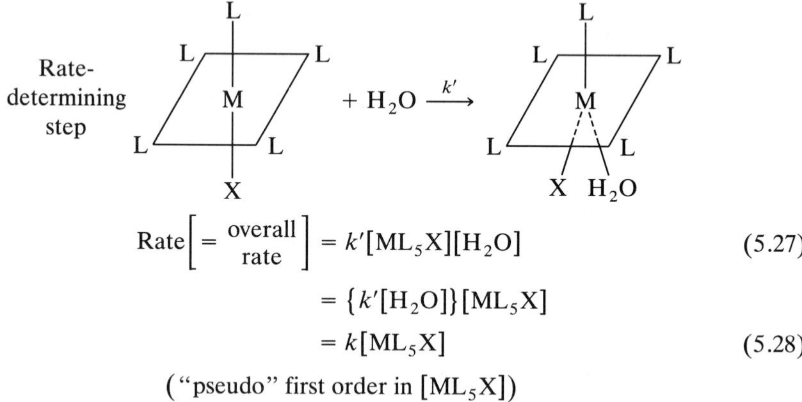

$$\text{Rate}\left[=\begin{matrix}\text{overall}\\ \text{rate}\end{matrix}\right]=k'[ML_5X][H_2O] \tag{5.27}$$

$$=\{k'[H_2O]\}[ML_5X]$$

$$=k[ML_5X] \tag{5.28}$$

("pseudo" first order in $[ML_5X]$)

FIGURE 5.4
The masking of concentration dependence in aqueous solution. (*a*) A dissociative mechanism is first-order in the concentration of the complex reactant. (*b*) An associative mechanism is first-order in both the complex reactant and the incoming water ligand. However, since the concentration of water in aqueous solution is so large and very nearly constant, $[H_2O]$ is combined with k' to produce an overall rate constant k. The resulting rate law is "pseudo" first-order. The dependence of the associative mechanism on the concentration of the incoming water ligand has been masked.

Evidence for Dissociative Mechanisms

Throughout the above discussion, we still have not answered the fundamental question: What is the preferred mechanism of substitution reactions of octahedral coordination compounds? The answer is that *dissociative mechanisms are preferred*. In this section we investigate some of the evidence that supports such a conclusion. We will cite three different types of reactions: (1) the rates of exchange of water molecules, (2) anation reactions, and (3) aquation reactions.

We start with a strange set of measurements. These involve the exchange between the ordinary water molecules in the hydration spheres of various metal ions and isotopically labeled *bulk* water (H_2O^{17}). (Section 11.2 presents a more detailed description of water structure.) A general equation representing such reactions is given in Equation (5.29):

$$M(H_2O)_6^{n+} + H_2O^{17} \xrightarrow{k} M(H_2O)_5(H_2O^{17})^{n+} + H_2O \tag{5.29}$$

The measurements of the rates of these reactions are carried out by a variety of methods, some very sophisticated, depending upon the rate of the exchange. Such methods demonstrate that these reactions are first-order in the concentration of the original hydrated cation, $M(H_2O)_6^{n+}$, that is, the rate law for these reactions is as given in Equation (5.30). As discussed earlier, the fact that these reactions are first-order is consistent with but certainly not definitive proof of a dissociative mechanism.

$$-\frac{\Delta[M(H_2O)_6^{n+}]}{\Delta t} = k_1[M(H_2O)_6^{n+}] \tag{5.30}$$

The relative rates for a large number of metal ions are given in Figure 5.5 in terms of the log of the rate constant for the reaction. Note that these rate constants (and therefore the rates themselves) vary over approximately 16 orders of magnitude, an amazing range of rates.

Close examination of this data reveals that there is a definite increase in the rate of exchange (reflected in an increase in the rate constant) going down various groups of the periodic table. Look, for example, at the results for Group 1A (Li^+, Na^+, K^+, Rb^+, and Cs^+) or for Group 2A (Be^{2+}, Mg^{2+}, Ca^{2+}, Sr^{2+}, and Ba^{2+}). Comparison of these two groups also reveals that the higher +2 charged ions of Group 2A have slower rates than the +1 charged Group 1A ions. Can we explain

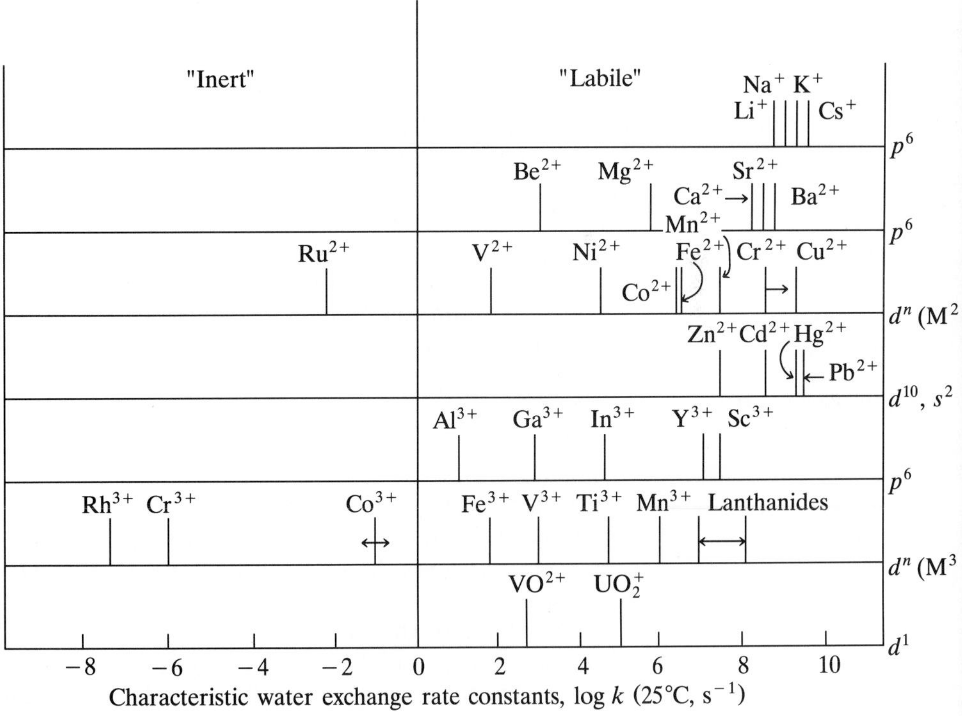

FIGURE 5.5
Rate constants for water exchange for various ions.

$$[M(H_2O)_n]^{m+} + H_2O^{17} \xrightarrow{k} [M(H_2O)_5(H_2O^{17})]^{m+} + H_2O$$

Data are tabulated as the log of the rate constant at 25°C. Inert hydrated ions, those that only slowly exchange water molecules between the hydration sphere and bulk water structure, are given on the left, while labile hydrated ions are on the right. [Ref. 7.]

such results and is this explanation consistent with a dissociative mechanism for these exchange reactions?

The explanation revolves around the effect of the charge and the size of a metal cation on the strength of the $M^{n+}—OH_2$ bond. It should make sense to you that the higher the charge, the greater the electrostatic attraction between the positively charged metal and the electron pair of the water ligand. Therefore the strength of the $M^{n+}—OH_2$ bond should increase with increasing charge. The larger the metal ion, on the other hand, the farther the center of its positive charge is from the ligand electron pair and, consequently, the weaker the bond. These two factors can be combined in what is known as the *charge density*, defined as the ratio of the charge on a metal cation over its ionic radius. By definition, the charge density of a metal cation increases with increasing charge and decreasing size. It follows from the above discussion that as charge density increases so also should the strength of the $M—OH_2$ bond. (Charge density is further discussed in Sections 8.1 and 9.2.)

Now we turn to a consideration of whether the data for these exchange reactions are consistent with a dissociative mechanism. Within a given group, all the cations are the same charge and only the size varies. So, in Group 1A, as we go down the group, the cations are getting larger and the charge density decreases. This in turn means the $M^+—OH_2$ bond is getting weaker and more easily broken. It is, of course, the breaking of the $M^{n+}—OH_2$ bond that is the rate-determining step in a dissociative mechanism. Therefore, as the bond gets weaker and easier to break, it makes sense that the rate of the water exchange reaction should increase.

When we turn to the doubly charged Group 2A ions, the charge density should be larger and therefore the strength of the $M^{2+}—OH_2$ bonds should be greater. Accordingly, it follows from a dissociative mechanism that the rate of exchange should be slower. Comparison of the data for Groups 1A and 2A shows that this is exactly the case. So, we see that these data are consistent with (but not necessarily proof of) a dissociative mechanism for these reactions. We will postpone a discussion of the remainder of this table, specifically, the trends within the transition metals, until the next section.

Anation reactions were defined earlier as those in which an anion replaces another ligand in a given complex. Equation (5.31) represents the anation of a hexaaquo metal cation with a -1 anion:

$$[M(H_2O)_6]^{n+} + X^- \longrightarrow [M(H_2O)_5X]^{(n-1)+} + H_2O \tag{5.31}$$

Such reactions presumably could proceed by either a dissociative or an associative mechanism, but most evidence seems, again, to favor the dissociative. For example, Table 5.1 shows the rate constants for three anations of $[Ni(H_2O)_6]^{2+}$ as well as the reactions with neutral ammonia and water ligands. Note that these constants and therefore the rates of reaction show little or no dependence (less than one order of magnitude) on the identity of the entering ligand. Furthermore, the values are all very similar to that of the rate of water exchange. (See the value of log k for $L = H_2O$ in Table 5.1; also note that this value is consistent with that obtained by reading Figure 5.5.) These data are consistent with a dissociative rate-determining step in which a water molecule breaks away from the nickel(II) and, in a succeeding fast step, is replaced by L.

Aquation reactions involve the replacement of a ligand (other than water itself) with a water molecule. A considerable amount of work has been done on reactions of this type involving a variety of inert cobalt(III) complexes. For

TABLE 5.1
Rate constants for substitution reactions of $[Ni(H_2O)_6]^{2+}$

$[Ni(H_2O)_6]^{2+} + L \xrightarrow{k} [Ni(H_2O)_5L]^{n+} + H_2O$

L	k, s^{-1}	log k
F^-	8×10^3	3.9
SCN^-	6×10^3	3.8
CH_3COO^-	30×10^3	4.3
NH_3	3×10^3	3.5
H_2O	25×10^3	4.4

Data taken from R. G. Wilkins, *Acc. Chem. Res.*, **3**: 408(1970).

example, Equation (5.32) represents the aquation reactions of various pentaammine(ligand)cobalt(III) cations:

$$[Co(NH_3)_5L]^{n+} + H_2O \xrightarrow{k} [Co(NH_3)_5(H_2O)]^{(n+1)+} + L^- \qquad (5.32)$$

Table 5.2 gives some kinetic and thermodynamic data related to this reaction. First note that the rate constants, unlike those for the anation reactions just discussed, now *do* seem to vary quite significantly with various ligands, L^-. Such a variation is consistent with a rate-determining step in which M—L bonds of varying strength are broken.

TABLE 5.2
Rate constants for the aquation of pentaammine(ligand)cobalt(III) complexes and equilibrium constants for the anation of pentaammineaquocobalt(III) with various anions

	L	k, s^{-1}	K_a, M^{-1}	
Slowest rate of reaction ↑	NCS^-	5.0×10^{-10}	470	Strongest M—L bonds ↑
	F^-	8.6×10^{-8}	20	
	$H_2PO_4^-$	2.6×10^{-7}	7.4	
	Cl^-	1.7×10^{-6}	1.25	
	Br^-	6.3×10^{-6}	0.37	
	I^-	8.3×10^{-6}	0.16	
↓ Fastest rate of reaction	NO_3^-	2.7×10^{-5}	0.077	↓ Weakest M—L bonds

The rate constants k refer to the following aquation reactions:

$$[Co(NH_3)_5L]^{2+} + H_2O \xrightarrow{k} [Co(NH_3)_5(H_2O)]^{3+} + L^-$$

The equilibrium constants K_a refer to the following anation reactions:

$$[Co(NH_3)_5(H_2O)]^{3+} + L^- \overset{K_a}{\rightleftharpoons} [Co(NH_3)_5L]^{2+} + H_2O$$

The slowest rates of aquation correspond to the largest equilibrium constants for anation.

Data taken from F. Basolo and R. G. Pearson, *Mechanisms of Inorganic Reactions, A Study of Metal Complexes in Solution*, 2d ed., Wiley, New York, 1968, pp. 164–166.

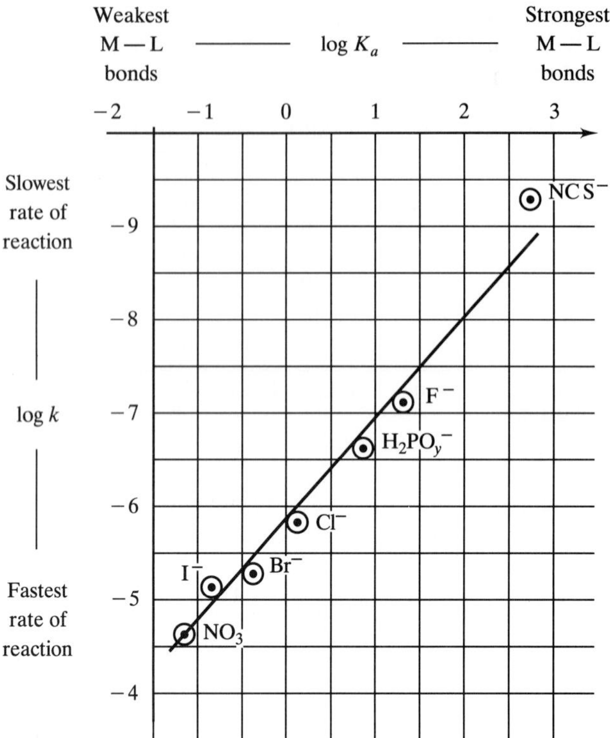

FIGURE 5.6
A plot of log K_a versus log k for a variety of pentaammine(ligand)cobalt(III) complex cations. The rate constants k refer to the following aquation reactions:

$$[Co(NH_3)_5L]^{2+} + H_2O \xrightarrow{k} [Co(NH_3)_5(H_2O)]^{3+} + L^-$$

The equilibrium constants K_a refer to the following anation reactions:

$$[Co(NH_3)_5(H_2O)]^{3+} + L^- \overset{K_a}{\rightleftharpoons} [Co(NH_3)_5L]^{2+} + H_2O$$

The value of K_a is a measure of the M—L bond strength with the strongest bonds having the largest values of K_a. The value of k is a measure of the rate of the aquation. The smallest values of log k have the slowest rates of reaction. The plot shows that the stronger the M—L bond, the slower the rate of aquation.

Can we correlate the above aquation rate constants with a measure of the M—L bond strength? The third column of Table 5.2 shows the equilibrium constants for the reactions in which a water ligand is replaced by the L^- anion in each case. Notice that the only major difference in these various reactions would seem to be the strength of the M—L bond. In fact, it can be shown that log K_a for these reactions is directly proportional to the M—L bond strength. (See Problem 5.30 for an opportunity to demonstrate this relationship.) The relationship between log K_a and k is in turn plotted in Figure 5.6. It demonstrates a straight-line relationship between these two parameters and therefore a strong correlation between the rate constant and the M—L bond strength. That is, as the M—L bond strength increases, it becomes more difficult to remove the L, and the rate decreases. The evidence for dissociative pathways for octahedral substitution reactions continues to mount.

TABLE 5.3
Rate constants for the aquation of *trans*-bis(substituted-ethylenediamine)-dichlorocobalt(III) complexes:

$$trans\text{-}[CoCl_2(H_2N{-}CR_1R_2{-}CR_3R_4{-}NH_2)_2]^+ + H_2O \xrightarrow{k} trans\text{-}[(H_2O)ClCo(H_2N{-}CR_1R_2{-}CR_3R_4{-}NH_2)_2]^{2+} + Cl^-$$

	Groups on bidentateamine					
	R_1	R_2	R_3	R_4	k, s^{-1}	
Increasing ↓	H	H	H	H	3.2×10^{-5}	↓
bulk of	CH_3	H	H	H	6.2×10^{-5}	Increasing
bidentate	CH_3	H	CH_3	H	4.2×10^{-4}	rate of
amine	CH_3	CH_3	H	H	2.2×10^{-4}	reaction
ligand	CH_3	CH_3	CH_3	CH_3	3.2×10^{-2}	

Data taken from F. Basolo and R. G. Pearson, *Mechanisms of Inorganic Reactions, A Study of Metal Complexes in Solution*, 2d ed., Wiley, New York, 1968, p. 162.

Other aquation reactions give further evidence concerning the most favored mechanism of octahedral substitution reactions. Consider the data for the aquation of various bidentate amine complexes of cobalt(III) as shown in Equation (5.33) and Table 5.3:

$$[CoCl_2(H_2N{-}CR_1R_2{-}CR_3R_4{-}NH_2)_2]^+ + H_2O \longrightarrow [Co(H_2O)Cl(H_2N{-}CR_1R_2{-}CR_3R_4{-}NH_2)_2]^{2+} + Cl^- \tag{5.33}$$

Notice that the more methyl groups (and therefore the bulkier the bidentate amine), the faster the rate of aquation. To see if this is consistent with either a dissociative or associate mechanism, consider the rate-determining step in each case as shown in Figure 5.7.

In Figure 5.7*a*, the associative rate-determining step is shown. What would happen to the rate of such a reaction as the bulk of the bidentate amine increased? Simply put, the incoming water ligand would find it more difficult to make its way into the metal to form the seven-coordinate intermediate, and therefore the rate would decrease. This is not what the data in Table 5.3 indicate happens, and therefore such data are not consistent with an associative mechanism.

In Figure 5.7*b*, on the other hand, the dissociative rate-determining step is shown. Here the increased bulk of the bidentate amine would increase the steric hindrance around the metal and help to force the chloride ligand out of the coordination sphere. The dissociation of the chloride ligand would be more favorable as the bulk of the amine increased, and therefore the rate of the reaction would increase. This increase with the bulk of the ligand is exactly what Table 5.3 indicates happens. Therefore, the data are consistent with a dissociative mechanism.

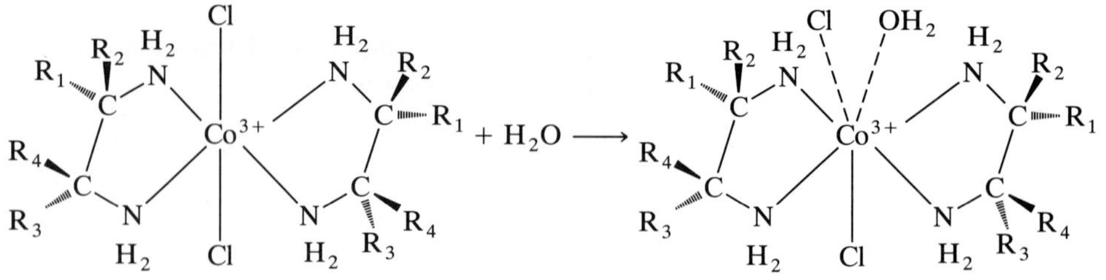

(*a*) Associative rate-determining step

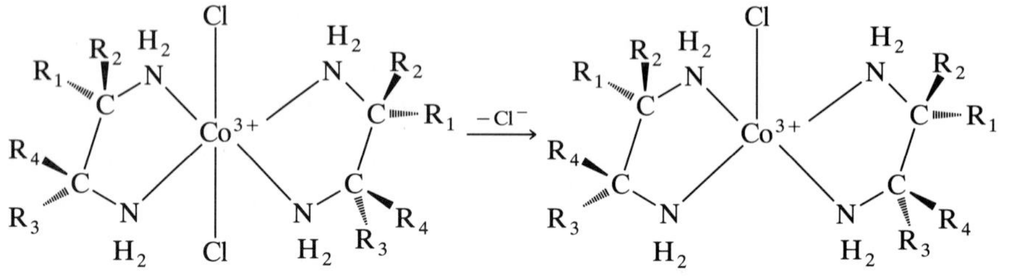

(*b*) Dissociative rate-determining step

FIGURE 5.7
(*a*) Associative rate-determining step. The entering water ligand attacks the Co^{3+} to form the seven-coordinate intermediate of the associative mechanism. As the bulk of the bidentate amine increases, the water ligand will find it more difficult to get to the metal cation. Therefore, the rate of the rate-determining step should decrease with the increasing bulk of the amine. (*b*) Dissociative rate-determining step. The chloride ligand dissociates to form the five-coordinate intermediate of the dissociative mechanism. As the bulk of the bidentate amine increases, the chloride ligand will experience more steric hindrance and will be pushed out of the coordination sphere more easily. Therefore, the rate of the rate-determining step should increase with the increasing bulk of the amine.

Finally, consider the effect of changing the overall charge on a complex. Again, much work has been done with cobalt(III) compounds. For the two reactions shown in Equations (5.34) and (5.35), note that the faster rate (larger rate constant) goes with the smaller net charge:

$$[Co(NH_3)_5Cl]^{2+} + H_2O \longrightarrow [Co(NH_3)_5(H_2O)]^{3+} + Cl^- \qquad (5.34)$$

$$k = 6.7 \times 10^{-6}\ s^{-1}$$

$$trans\text{-}[Co(NH_3)_4Cl_2]^{+} + H_2O \longrightarrow [Co(NH_3)_4(H_2O)Cl]^{2+} + Cl^- \qquad (5.35)$$

$$k = 1.8 \times 10^{-3}\ s^{-1}$$

Is this consistent with a dissociative or associative mechanism? In the dissociative step, the greater the charge on the complex cation, the more difficult it would be to remove a chloride anion from it and therefore the slower rate of dissociation. This is consistent with the rates that are given. In the associative mechanism, on the other hand, the greater the net charge, the more attracted the incoming water ligand would be and the faster the rate-determining step. Such a mechanism is inconsistent with the data.

To conclude this section, we have seen that various pieces of data from (1) rates of water exchange, (2) anations, and (3) various aquation reactions all seem to generally favor a dissociative mechanism for substitution reactions of octahedral compounds.

Explanation of Inert versus Labile Complexes

Now that we have fairly well established that the dissociative mechanism generally applies for the substitution reactions of octahedral complexes, we are in a good position to begin to answer some of our earlier (p. 94) critical questions about inert versus labile complexes. As defined earlier, labile and inert are kinetic terms describing the rates of reactions of coordination compounds. Rates, in turn, you should recall from earlier courses, are dependent on the magnitude of the energy of activation E_a of the rate-determining step.

Figure 5.8 shows a typical reaction profile, that is, a plot of potential energy versus *reaction pathway*. Recall that the reactants must be converted to a *transition state* or *activated complex* before they can be transformed into the products. The energy difference between the reactants and the transition state is called the *energy of activation* and must be attained before a reaction can take place. In general, the rate-determining steps of slow reactions are characterized by high activation energies and, conversely, those of fast reactions by low E_a's. The Arrhenius equation, $k = Ae^{-E_a/RT}$, gives the exact dependence of the rate constant (and therefore the rate itself) on the energy of activation. (R is the gas constant, T is the temperature in degrees kelvin, and A is a constant often called the *collision frequency*.)

Now for a substitution reaction in which the rate-determining step is the dissociation of a ligand, what are the various factors that contribute to the activation energy? Some of these were discussed earlier but not in the context of E_a's. For example, we said earlier that the size and charge of the metal cation influence the strength of the M—L bond. Now we can say that, since this bond is broken during the rate-determining step, it directly affects the magnitude of E_a. Metals with larger charge densities [or charge-to-radius (Z/r) ratios] have stronger M—L bonds, higher energies of activation, and therefore slower rates of substitution. We also discussed steric hindrance about the metal and the overall charge on a complex as having a bearing on the rate of substitution reactions. Complexes with little steric hindrance or a high overall charge will have larger M—L bond strengths, higher energies of activation, and slower rates of reaction.

Given these factors, can we start to explain why, as noted earlier, the first-row transition metals (with the exception of Co^{3+} and Cr^{3+}) are generally

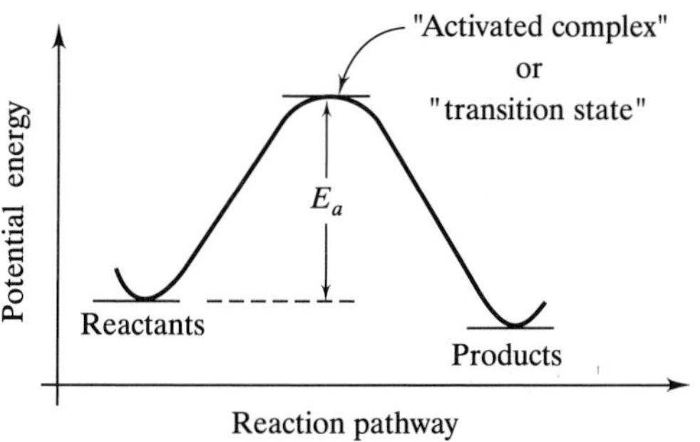

FIGURE 5.8
A reaction profile: a plot of potential energy versus reaction pathway. Reactants must acquire the energy of activation in order to achieve the transition state or activated complex before they can be transformed into products. The higher the energy of activation, the slower the reaction.

labile while the second- and third-row metals are inert? Based on the above factors, we can appreciate, in part, why the larger metals are more apt to be inert. Because they are significantly larger than the first-row elements, there is less steric hindrance among the ligands about them and therefore less tendency for a given ligand to be forced from the coordination sphere. In addition, these metals are often more highly charged than their lighter congeners, leading to stronger M—L bonds that must be broken during the rate-determining step.[72]

Assuming a significant covalent contribution to the M—L bond interaction (see the discussion in Chapter 4, p. 72, for further details), an additional reason for the inertness of the second- and third-row metals can be offered. These larger metals use $4d$ and $5d$ orbitals in their sigma bond interactions with a ligand. These larger $4d$ and $5d$ orbitals extend farther out toward the ligand and overlap better with its orbitals. Therefore, this additional covalent contribution to the M—L bond strength makes these M—L (where M = second- or third-row transition metals) bonds stronger and more difficult to break in the rate-determining step of a dissociation reaction.

So now that we have a fair understanding of why the second- and third-row metals are inert, we can turn to a discussion of the relative labilities of the first-row ions. Recall that most of these are labile except for Co^{3+} and Cr^{3+}, which are inert. Why should this be? The key to this mystery, it turns out, lies in the change in the crystal field stabilization energy on going from the octahedral reactant to the five-coordinate activated complex. Table 5.4 shows CFSEs for octahedral and square pyramidal fields. Most importantly to the argument presented here is the

TABLE 5.4
Changes in crystal field stabilization energies

The CFSEs (in units of Δ_o) for octahedral (oct) and square pyramidal (sp) fields are shown followed by the change in CFSE for the process ML_6 (octahedral) → ML_5 (square pyramid). A (+) indicates a gain in CFSE during the process and a (−) a loss in CFSE

d^n	CFSE ML_6 (oct)	CFSE ML_5 (sp)	ΔCFSE			
d^1	0.40	0.46	+0.06			
d^2	0.80	0.91	+0.11			
d^3	1.20	1.00	−0.20			
	Low-spin–strong field			**High-spin–weak field**		
	CFSE ML_6 (oct)	**CFSE ML_5 (sp)**	**ΔCFSE**	**CFSE ML_6 (oct)**	**CFSE ML_5 (sp)**	**ΔCFSE**
d^4	$1.60 - P$	$1.46 - P$	−0.14	0.60	0.91	+0.31
d^5	$2.00 - 2P$	$1.91 - 2P$	−0.09	0	0	0
d^6	$2.40 - 2P$	$2.00 - 2P$	−0.40	0.40	0.46	+0.06
d^7	$1.80 - P$	$1.91 - P$	+0.11	0.80	0.91	+0.11
	CFSE ML_6 (oct)	**CFSE ML_5 (sp)**	**ΔCFSE**			
d^8	1.20	1.00	−0.20			
d^9	0.60	0.91	+0.31			
d^{10}	0	0	0			

change in CFSE shown in each case. A plus sign means there is a gain in CFSE on going from the octahedral reactant to the square pyramidal intermediate, while a negative sign represents a loss in CFSE.

What is the consequence of a gain or loss of CFSE on going from the octahedral reactant to the square pyramidal transition state? It makes sense that if there is additional crystal field stabilization energy in the transition state, then its formation is favored and the rate-determining step is faster. On the other hand, if there is less CFSE in the transition state than in the reactants, this would make it less stable (higher in energy) and more difficult to achieve. Therefore, the reaction would be slower.

To see the effect of the change of crystal field stabilization energy (ΔCFSE) more clearly, consider the general substitution reaction shown in Equation (5.36):

$$[CrL_5X]^{3+} + Y \longrightarrow [CrL_5Y]^{3+} + X \qquad (5.36)$$

The reaction profile for the rate-determining step, the dissociation of the X ligand, is shown schematically in Figure 5.9. The top curve (dashed line) shows the energy

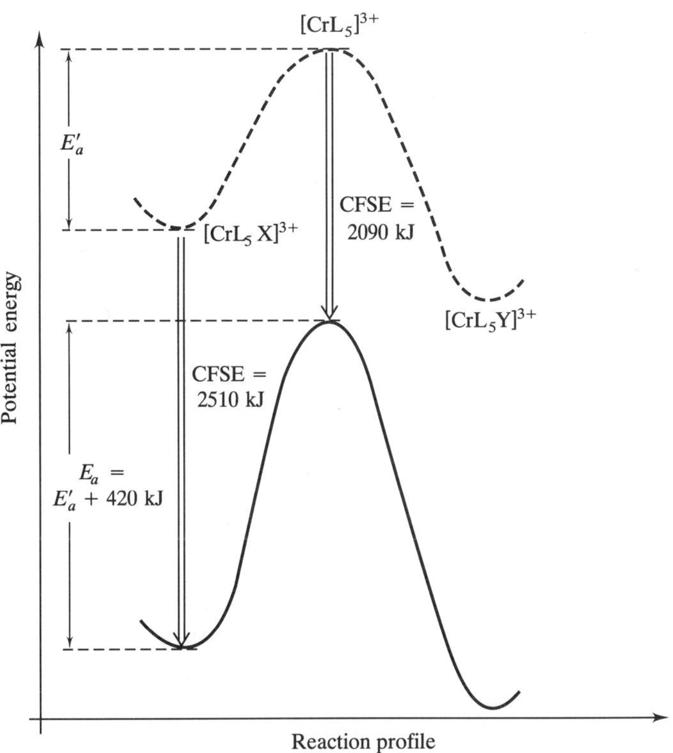

FIGURE 5.9
The reaction profile for the rate-determining step of a substitution reaction of $[CrL_5X]^{3+}$ assuming a dissociative mechanism. The top curve (dashed line) shows an energy of activation of E'_a without considering the effect of crystal field stabilization energy (CFSE). The lower curve (solid line) is the result of subtracting the CFSEs for the complex reactant, product, and transition state. If X = L = H_2O, the energy of activation E_a is 420 kJ/mol greater in the lower curve as the result of the loss of CFSE on going from the octahedral reactant to the square pyramidal transition state.

changes without CFSEs. Now consider the effect of subtracting the CFSEs for both the starting and transition state complexes. If Δ_o of the octahedral reactant is about 2090 kJ (the value for X = L = H_2O), then the CFSE = $1.20\Delta_o$ = 2510 kJ/mol. (See the d^3 entry in Table 5.4 for CFSEs in terms of Δ_o for Cr^{3+} complexes.) Therefore, the starting complex is 2510 kJ/mol more stable due to the effect of its octahedral crystal field. For the $[Cr(H_2O)_5]^{3+}$ transition state, CFSE = Δ_o = 2090 kJ/mol, meaning that the transition state is 2090 kJ more stable due to the effect of its square pyramidal crystal field. The resulting change in CFSE (ΔCFSE = $-0.20\Delta_o$ = -420 kJ) on going from the reactant to the transition state is a loss of 420 kJ. The lower curve (solid line) is the resulting reaction profile. Carefully note that the energy of activation of the lower curve is now 420 kJ/mol higher than that shown in the top curve. The result is that *the loss of CFSE has been added directly onto the energy of activation* for the reaction.

Can we use the results given in Table 5.4 to explain why Cr^{3+} and Co^{3+} complexes are inert? The table indicates that d^3, d^6 (low-spin–strong-field), and d^8 have the most negative ΔCFSEs. Of these cases, two also occur in +3 charged metals, Cr^{3+} (d^3) and Co^{3+} (d^6). (Due to the +3 charge, Co^{3+} complexes are generally low-spin–strong-field.) These two cases, then, involve strong $M^{3+}—L$ bonds and a loss of CFSE on going to the transition state. Accordingly, the dissociative rate-determining steps involving Cr^{3+} and low-spin Co^{3+} have high energies of activation and are comparatively slow. Here we have a rationale for the fact that Cr^{3+} and Co^{3+} complexes are "inert."

What about other metal cations that have a charge of only +2 but involve a loss of CFSE? Figure 5.5 shows that V^{2+} (d^3) and Ni^{2+} (d^8), while not considered inert, have some of the lowest rate constants for water exchange among the labile metals. Also note that Cu^{2+} (d^9) and Cr^{2+} (d^4), with only +2 charges and significant *gains* in CFSE on going to the transition state, have among the fastest rates of water exchange. So we see that data from studies of the rate of water exchange of various metals correlate well with the results of Table 5.4.

5.4 REDOX OR ELECTRON-TRANSFER REACTIONS

In the above substitution reactions, the metal oxidation states remain constant; no electrons are transferred to or from the metal ions. Now we turn our attention to reactions in which electrons are, in fact, transferred from one metal to another. Recall that if an atom or ion, usually a complexed metal cation in the cases we are about to consider, loses electrons, its oxidation state increases and it is said to be oxidized. If the metal gains electrons, its oxidation state decreases and it is reduced. (Some students remember these definitions by the mnemonic LEO goes GER that stands for *L*oses *E*lectrons *O*xidized, *G*ains *E*lectrons *R*educed. The theoretical basis of redox reactions is more fully developed in Chapter 12, where we consider the strength of various oxidizing and reducing agents using the concept of standard reduction potentials.)

What is the specific sequence of molecular-level steps that results in an electron being transferred from one coordinated metal ion to another? That is, what are the possible mechanisms for these redox reactions? There appear to be two possibilities, first delineated by Taube in the early 1950s. In the outer-sphere mechanism the coordination spheres of the metals stay intact, while in the inner-sphere mechanism they are altered in some way.

TABLE 5.5
Some outer-sphere electron-transfer reactions

Self-exchange reactions	k, M^{-1} s^{-1}
$[Mn(CN)_6]^{3-} + [Mn(CN)_6]^{4-} \longrightarrow [Mn(CN)_6]^{4-} + [Mn(CN)_6]^{3-}$	$\geq 10^4$
$[IrCl_6]^{3-} + [IrCl_6]^{2-} \longrightarrow [IrCl_6]^{2-} + [IrCl_6]^{3-}$	$\approx 10^3$
$[Ru(NH_3)_6]^{3+} + [Ru(NH_3)_6]^{2+} \longrightarrow [Ru(NH_3)_6]^{2+} + [Ru(NH_3)_6]^{3+}$	$\approx 8 \times 10^2$
$[Fe(CN)_6]^{4-} + [Fe(CN)_6]^{3-} \longrightarrow [Fe(CN)_6]^{3-} + [Fe(CN)_6]^{4-}$	$\approx 7 \times 10^2$
$[Ru(H_2O)_6]^{2+} + [Ru(H_2O)_6]^{3+} \longrightarrow [Ru(H_2O)_6]^{3+} + [Ru(H_2O)_6]^{2+}$	≈ 44
$[Co(H_2O)_6]^{3+} + [Co(H_2O)_6]^{2+} \longrightarrow [Co(H_2O)_6]^{2+} + [Co(H_2O)_6]^{3+}$	≈ 5
$[Co(en)_3]^{3+} + [Co(en)_3]^{2+} \longrightarrow [Co(en)_3]^{2+} + [Co(en)_3]^{3+}$	$\approx 1 \times 10^{-4}$
$[Co(C_2O_4)_3]^{3-} + [Co(C_2O_4)_3]^{4-} \longrightarrow [Co(C_2O_4)_3]^{4-} + [Co(C_2O_4)_3]^{3-}$	$\approx 1 \times 10^{-4}$
$[Cr(H_2O)_6]^{3+} + [Cr(H_2O)_6]^{2+} \longrightarrow [Cr(H_2O)_6]^{2+} + [Cr(H_2O)_6]^{3+}$	$\approx 2 \times 10^{-5}$
$[Co(NH_3)_6]^{3+} + [Co(NH_3)_6]^{2+} \longrightarrow [Co(NH_3)_6]^{2+} + [Co(NH_3)_6]^{3+}$	$\approx 1 \times 10^{-6}$
Cross reactions	
$[Fe(CN)_6]^{4-} + [IrCl_6]^{2-} \longrightarrow [Fe(CN)_6]^{3-} + [IrCl_6]^{3-}$	$\approx 4 \times 10^5$
$[Cr(H_2O)_6]^{2+} + [Fe(H_2O)_6]^{3+} \longrightarrow [Cr(H_2O)_6]^{3+} + [Fe(H_2O)_6]^{2+}$	$\approx 2 \times 10^3$
$[Cr(H_2O)_6]^{2+} + [Ru(H_2O)_6]^{3+} \longrightarrow [Cr(H_2O)_6]^{3+} + [Ru(H_2O)_6]^{2+}$	$\approx 2 \times 10^2$
$[Ru(NH_3)_6]^{3+} + [V(H_2O)_6]^{2+} \longrightarrow [Ru(NH_3)_6]^{2+} + [V(H_2O)_6]^{3+}$	$\approx 8 \times 10^1$
$[Co(NH_3)_6]^{3+} + [Ru(NH_3)_6]^{2+} \longrightarrow [Co(NH_3)_6]^{2+} + [Ru(NH_3)_6]^{3+}$	$\approx 1 \times 10^{-2}$
$[Co(NH_3)_6]^{3+} + [V(H_2O)_6]^{2+} \longrightarrow [Co(NH_3)_6]^{2+} + [V(H_2O)_6]^{3+}$	$\approx 4 \times 10^{-3}$
$[Co(en)_3]^{3+} + [V(H_2O)_6]^{2+} \longrightarrow [Co(en)_3]^{2+} + [V(H_2O)_6]^{3+}$	$\approx 2 \times 10^{-4}$
$[Co(NH_3)_6]^{3+} + [Cr(H_2O)_6]^{2+} \longrightarrow [Co(NH_3)_6]^{2+} + [Cr(H_2O)_6]^{3+}$	$\approx 9 \times 10^{-5}$
$[Co(en)_3]^{3+} + [Cr(H_2O)_6]^{2+} \longrightarrow [Co(en)_3]^{2+} + [Cr(H_2O)_6]^{3+}$	$\approx 2 \times 10^{-5}$

Outer-Sphere Mechanisms

To make it possible for an electron to move from one metal ion to another, it makes sense that they should be as close to each other as possible. Assuming, for the moment, that their two coordination spheres stay intact, the distance between two metal ions will be at a minimum when the two coordination spheres are in contact. When an electron is transferred among metal ions whose intact coordination spheres are in contact at their outer edges, this is referred to as an *outer-sphere* mechanism.

A variety of reactions are thought to occur via such a mechanism. Some examples with their corresponding (second-order) rate constants are shown in Table 5.5. The entries at the top of table are *self-exchange* reactions in which two coordinated ions, identical in every way except for the oxidation state of the metal ion, simply exchange an electron. One metal ion is oxidized, the other reduced, but no net reaction actually takes place because the products are indistinguishable from the reactants. The lower set of reactions, labeled *cross reactions*, involve a transfer (or a "crossing over") of an electron between different coordinated metal ions. These examples, as shown, do result in net reactions.

Consider as an example for discussion the self-exchange reaction between the hexaaquoruthenium(III) and the hexaaquoruthenium(II) ions shown in Equation

(5.37):

$$[Ru(H_2O^{17})_6]^{3+} + [Ru(H_2O)_6]^{2+} \rightarrow [Ru(H_2O^{17})_6]^{2+} + [Ru(H_2O)_6]^{3+} \quad (5.37)$$

Reactions like this one are followed by labeling one of the complexes with a radioactive isotopic tracer. In this case the water molecules in the reactant Ru^{3+} complex contain the O^{17} isotope instead of the normal O^{16}. This reaction has been determined to be first-order in the concentrations of *both* reactants (second-order overall) and to have a rate constant (see Table 5.5) of about 44 $M^{-1}s^{-1}$. On the other hand, the rate of water exchange in $[Ru(H_2O)_6]^{2+}$ is considerably slower (with a rate constant between 10^{-2} and 10^{-3} s^{-1} according to Figure 5.5). The rate of water exchange in $[Ru(H_2O)_6]^{3+}$ would be still slower because of the +3 charge of the central metal ion. Given the relative slowness of these water exchange reactions, it follows that the above reaction cannot proceed by a mechanism involving the breaking of the $Ru^{n+}—OH_2$ bonds. [Also note that if the water molecules did dissociate from their respective cations, the H_2O^{17} molecules would be randomly distributed among the products and not concentrated in the hexaaquoruthenium(II) product complex.] We are left with an electron transfer between intact coordination spheres as the most plausible mechanism for this reaction.

So what is the mechanism of this outer-sphere electron transfer? The first step, represented in Equation (5.38), would be the collision of the two reactant coordinated ions to form an *outer-sphere complex*, or what is sometimes called an *ion pair*. Next the electron transfer, represented in Equation (5.39), takes place instantaneously within this ion pair. Finally, as shown in Equation (5.40), the two product ions separate.

$$[Ru(H_2O^{17})_6]^{3+} + [Ru(H_2O)_6]^{2+} \rightleftharpoons [Ru(H_2O^{17})_6]^{3+}/[Ru(H_2O)_6]^{2+} \quad (5.38)$$

$$[Ru(H_2O^{17})_6]^{3+}/[Ru(H_2O)_6]^{2+} \longrightarrow [Ru(H_2O^{17})_6]^{2+}/[Ru(H_2O)_6]^{3+} \quad (5.39)$$

e^-

$$[Ru(H_2O^{17})_6]^{2+}/[Ru(H_2O)_6]^{3+} \longrightarrow [Ru(H_2O^{17})_6]^{2+} + [Ru(H_2O)_6]^{3+} \quad (5.40)$$

There is one additional point to be made concerning this mechanism. The electron transfer in the outer-sphere complex will be extremely fast to the point that we may view it as occurring instantaneously. (Very light electrons move much faster than the heavier and therefore more lumbering and cumbersome complex ions.) But there is a problem here. If the M–L distances (Ru^{n+}–OH_2 distances in the above example) are very different in the two complexes, there will be a large energy barrier, that is to say, energy of activation involved in bringing about the electron transfer. Consequently, it makes sense that the greater the difference in the M–L distances in the reactants, the slower the reaction. Incidentally, the best mental image of this process is to think of the two M–L distances coming to an intermediate point at which the electron transfer takes place. The additional energy needed to adjust the M–L distances to this intermediate value contributes directly to the energy of activation. Now the $Ru^{2+}—OH_2$ and $Ru^{3+}—OH_2$ bond lengths are fairly similar, 2.03 and 2.12 Å, respectively, so this self-exchange reaction, as we have noted, takes place fairly rapidly.

What would happen if the M—L bond lengths in the two complex ions were more radically different? Examples of this situation are the self-exchange reactions involving Co^{3+} and Co^{2+} ions. (Note how these reactions are concentrated toward the lower ends of both parts of Table 5.5.) Specifically, the self-exchange reaction involving the hexaammine complexes, as shown in Equation (5.41), is a case where the M—L bond lengths are different enough that this reaction is very slow:

$$\underset{\substack{Co^{3+}—NH_3 \\ = 1.94\ \text{Å}}}{[Co(NH_3)_6]^{3+}} + \underset{\substack{Co^{2+}—NH_3 \\ = 2.11\ \text{Å}}}{[Co(NH_3)_6]^{2+}} \longrightarrow [Co(NH_3)_6]^{2+} + [Co(NH_3)_6]^{3+} \qquad k \approx 10^{-6}\ M^{-1}\ s^{-1} \tag{5.41}$$

A word is in order as to why the Co^{n+}–NH_3 distances are so dissimilar. Octahedral Co^{3+} (low-spin) has an electron configuration of t_{2g}^6 with all the metal electrons pointing in between the ligands. Octahedral Co^{2+} has an electron configuration of $t_{2g}^6e_g^1$ (if low-spin) or $t_{2g}^5e_g^2$ (if high-spin). In either case, not only is the charge acting on the water ligands smaller in Co^{2+} but the seventh electron (and the eighth if high-spin) points directly at the ligands. Therefore, M–L distances are considerably greater in Co(II) complexes.

An example of a rapid self-exchange reaction in which the change in M–L distances is particularly small is that involving hexacyanoferrate complexes as shown in Equation (5.42):

$$\underset{\substack{Fe^{2+}—CN^- \\ = 1.92\ \text{Å}}}{[Fe(CN)_6]^{4-}} + \underset{\substack{Fe^{3+}—CN^- \\ = 1.95\ \text{Å}}}{} \longrightarrow [Fe(CN)_6]^{3-}\ [Fe(CN)_6]^{3-} + [Fe(CN)_6]^{4-} \qquad k \approx 700\ M^{-1}\ s^{-1} \tag{5.42}$$

Note that in these two cations [Fe^{2+} (low-spin): t_{2g}^6; Fe^{3+} (low-spin): t_{2g}^5] an electron is merely transferred among the t_{2g} sets of orbitals (that point between the ligands), and therefore the M–L distances do not change appreciably.

Inner-Sphere Mechanisms

As we have just discussed, the coordination spheres of both reactants remain intact during an outer-sphere electron-transfer reaction. The same is not true, however, of the inner-sphere reactions we now start to consider. *Inner-sphere* electron-transfer reactions involve the formation of a bridged complex in which the two metal ions are connected by a bridging ligand that helps to promote the electron transfer. Often, but not always, the bridging ligand itself is transferred from one metal center to the other. Ligand transfer, then, is a good sign (but not absolute proof) that an inner-sphere mechanism is in operation. Of course, if there is not an available bridging ligand, then the inner-sphere mechanism cannot be correct. If a bridging ligand is available but not transferred, either an inner- or an outer-sphere mechanism may be possible.

The first and now classic set of reactions involving inner-sphere electron-transfer mechanisms was reported in 1953 by Taube and his group. The overall reaction is given in Equation (5.43):

$$[Co^{III}(NH_3)_5X]^{2+} + [Cr^{II}(H_2O)_6]^{2+} + 5H^+ \longrightarrow$$

e^-

$$[Cr^{III}(H_2O)_5X]^{2+} + [Co^{II}(H_2O)_6]^{2+} + 5NH_4^+ \quad (5.43)$$

$$X^- = F^-, Cl^-, Br^-, I^-, NCS^-, NO_3^-, CN^-, \ldots$$

Note that Co^{3+} is reduced to Co^{2+}, while Cr^{2+} is oxidized to Cr^{3+}. A bridging ligand (X^-) is transferred from the cobalt coordination sphere to that of the chromium.

As we did for the outer-sphere case, we start by showing the sequence of the steps of the mechanism. First we note that of the two complex reactants, one, the Co(III) complex, is inert while the other is labile. This leads us to postulate that the first step [shown in Equation (5.44)] is the dissociation of a water molecule from the labile complex. The second step [Equation (5.45)] is the formation of the bridged complex connecting the two metal ions. The third and rate-determining step [Equation (5.46)] is the actual electron transfer along the bridge set up by the X^- ligand. Fourth, as shown in Equation (5.47), the Co^{2+} is now the labile metal, and it dissociates the bridging ligand forming two separate complexes again. Lastly [Equation (5.48)], the ammine ligands in the Co^{2+} complex are protonated and replaced by waters.

$$\underset{\text{Labile}}{[Cr^{II}(H_2O)_6]^{2+}} \longrightarrow [Cr^{II}(H_2O)_5]^{2+} + H_2O \quad (5.44)$$

$$\underset{\text{Inert}}{[Co^{III}(NH_3)_5X]^{2+}} + [Cr^{II}(H_2O)_5]^{2+} \longrightarrow \left[(NH_3)_5Co^{III}{-}X^-{-}Cr^{II}(H_2O)_5\right]^{4+} \quad (5.45)$$

$$[(NH_3)_5Co^{III}{-}X^-{-}Cr^{II}(H_2O)_5]^{4+} \longrightarrow \left[(NH_3)_5Co^{II}{-}X^-{-}Cr^{III}(H_2O)_5\right]^{4+} \quad (5.46)$$

e^-

$$[(\underset{\text{Labile}}{NH_3)_5Co^{II}}{-}X^-{-}\underset{\text{Inert}}{Cr^{III}(H_2O)_5}]^{4+} \longrightarrow \left[Cr^{III}(H_2O)_5X\right]^{2+} + \left[Co^{II}(NH_3)_5\right]^{2+} \quad (5.47)$$

$$\underset{\text{Labile}}{[Co^{III}(NH_3)_5]^{2+}} + 5H^+ \xrightarrow{H_2O} [Co^{II}(H_2O)_6]^{2+} + 5NH_4^+ \quad (5.48)$$

There are a number of variations on this basic scheme that, taken together, lend a great deal of support to the inner-sphere or *bridging-ligand* mechanism. For example, if X^- is the chloride ion and it is isotopically labeled with Cl^{36}, the label is always transferred to the chromium coordination sphere. Conversely, if the reaction is carried out in a solution containing free $^{36}Cl^-$ ions, none of these labeled ions are found in the products. If X^- is *S*-thiocyanate, SCN^-, the product

TABLE 5.6
Comparative rate constants for the inner-sphere electron-transfer reaction
$[Co(NH_3)_5X]^{2+} + [Cr(H_2O)_6]^{2+}$

X^-	k, $M^{-1}\,s^{-1}$
F^-	2.5×10^5
Cl^-	6.0×10^5
Br^-	1.4×10^6
I^-	3.0×10^6

Data taken from F. Basolo and R. G. Pearson, *Mechanisms of Inorganic Reactions, A Study of Metal Complexes in Solution*, 2d ed., Wiley, New York, 1968, p. 481.

contains predominantly *N*-thiocyanate, NCS^-, as one would predict if the SCN^- acts as a bridging ligand. (See Problem 5.50 for an opportunity to work through this mechanism.)

Table 5.6 shows the variation of the rates of the above reaction as the X^- bridging ligand varies among the four halides. Can we make any conclusions about the relative abilities of these ligands in facilitating the rate-determining electron-transfer step? It appears that the larger the halide, the faster the reaction. Why would this be? The reason seems to be connected to the polarizability of the halide. (*Polarizability* is the ease with which the electron cloud of an atom, molecule, or ion can be distorted so as to set up a dipole moment. Generally, large and diffuse species whose electron clouds are held relatively weakly by their nuclear charges are more polarizable than small and compact ones. This property is discussed in somewhat greater detail in Chapter 9, pp. 225–226.) Once the bridged intermediate is formed as shown in Figure 5.10, the halide can be polarized by the more highly charged Co^{3+} cation. The induced dipole in the halide then attracts the electron from the Cr^{2+} and facilitates the electron transfer. It follows that the larger and more polarizable the halide is, the greater the dipole moment that can be induced in it, and the easier the electron transfer.

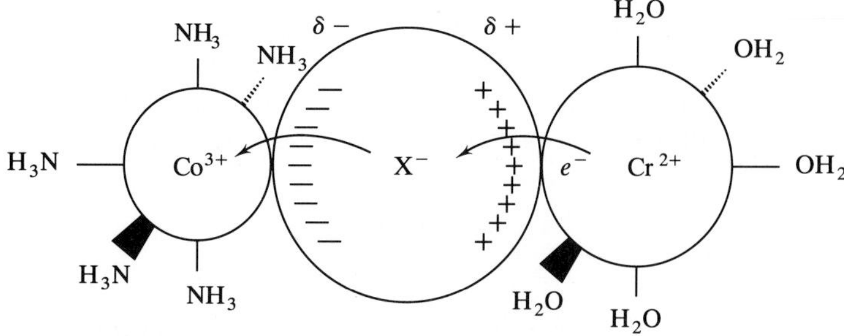

FIGURE 5.10
The electron-transfer step in the inner-sphere complex, $[(NH_3)_5Co^{III}—X—Cr^{II}(H_2O)_5]$. The more highly charged Co^{3+} polarizes the bridging ligand X^-. The resulting induced dipole of X^- helps to facilitate the transfer of an electron from the Cr^{2+} to the Co^{3+}.

5.5 SUBSTITUTION REACTIONS IN SQUARE PLANAR COMPLEXES: THE KINETIC TRANS EFFECT

A general substitution reaction of square planar complexes is shown in Equation (5.49):

$$\mathrm{L_2ML_2} + \mathrm{X} \longrightarrow \mathrm{L_3MX} + \mathrm{L} \tag{5.49}$$

Whereas, as we discussed in Section 5.3, the overwhelming bulk of evidence for substitution reactions of octahedral complexes indicates that they most often proceed via a dissociative mechanism, most of the evidence gathered in working with square planar complexes indicates that they are substituted via an associative pathway. This difference is not particularly surprising. After all, a square planar complex has two places (above and below the plane of the molecule) where an incoming ligand can readily attack the metal to form a five-coordinate intermediate.

As we have seen, square planar complexes usually occur with the d^8 metals such as Pt(II), Pd(II), Ni(II), and Au(III). The platinum(II) complexes are particularly inert and, because these rather slow reactions can be followed by fairly traditional and straightforward methods, have been extensively studied and analyzed.

A particular feature of these substitution reactions is the *kinetic trans effect*, defined as the relationship between the rate of substitution of square planar complexes and the nature of the species trans to the ligand being displaced. To understand this effect more clearly, consider the general substitution reactions given in Figure 5.11. The nonlabile ligands A and B can be ranked in order of their abilities to displace the ligands trans to them. For example if the nonlabile A ligand is a better "trans director" than B, then reaction I occurs. Conversely, if B is the better trans director, then the ligand trans to B is preferentially displaced and reaction II occurs.

After a large number of comparative experiments have been carried out, a *trans-directing series*, such as the one shown below, can be constructed:

$$CN^- \geq CO > NO_2^- > I^-, SCN^- > Br^- > Cl^- > py \geq NH_3 > H_2O$$

Such a series, although empirically derived, can be most useful in preparing high-purity isomers of square planar complexes. To demonstrate this, suppose we start with the tetrachloroplatinate(II) ion, $PtCl_4^{2-}$, and examine some of its typical substitution reactions. First, consider two additional empirical facts. (1) It is generally easier to replace a chloride ligand bound to Pt(II) than it is to replace other ligands. That is, other ligands will most always displace chlorides, and the only way to carry out the reverse is to swamp the system with a large excess of Cl^-

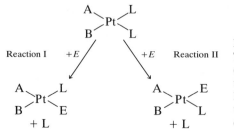

$$\mathrm{(A)(B)Pt(L)(L)} \xrightarrow[\text{Reaction I}]{+E} \mathrm{(A)(B)Pt(L)(E)} + \mathrm{L} \qquad \mathrm{(A)(B)Pt(L)(L)} \xrightarrow[\text{Reaction II}]{+E} \mathrm{(A)(B)Pt(E)(L)} + \mathrm{L}$$

FIGURE 5.11
The kinetic trans effect. An entering ligand (E) can displace either of two leaving ligands (L) depending upon the trans-directing ability of the nonlabile ligands (A and B). If A is a better trans director than B, reaction I occurs. If B is the better trans director, reaction II occurs.

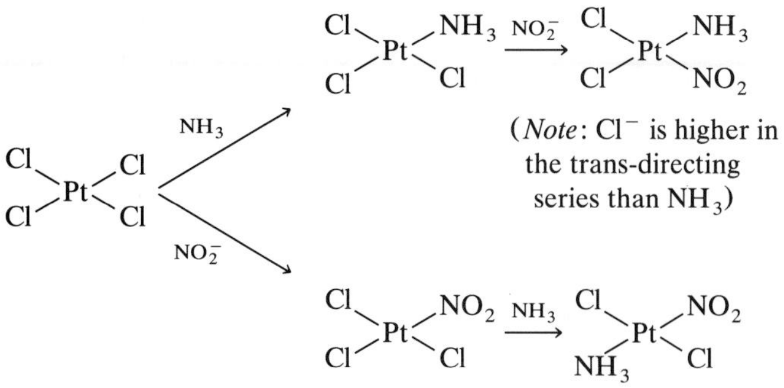

$$[PtCl_4] \xrightarrow{NH_3} cis\text{-}[PtCl_3(NH_3)] \xrightarrow{NO_2^-} [PtCl_2(NH_3)(NO_2)]$$

(*Note*: Cl^- is higher in the trans-directing series than NH_3)

$$[PtCl_4] \xrightarrow{NO_2^-} [PtCl_3(NO_2)] \xrightarrow{NH_3} [PtCl_2(NO_2)(NH_3)]$$

(*Note*: NO_2^- is higher in the trans-directing series than Cl^-)

FIGURE 5.12
Cis- and *trans*-amminedichloronitroplatinium(II) can be prepared in high purity by varying the order in which the ammine and nitrite ligands are added to tetrachloroplatinate(II). Charges on the various complexes are omitted for clarity.

(an example of Le Chatelier's principle) as indicated by the unequal arrows in Equation (5.50):

$$[PtCl_4]^{2-} + L \rightleftharpoons [PtCl_3L]^- + Cl^- \tag{5.50}$$

(2) In a substitution reaction in which there is more than one possibility as to which chloride will be replaced, the trans-directing series is used to predict which possibility will be realized.

For example, suppose we wish to prepare *cis*- and *trans*-amminedichloronitroplatinate(II). How would we proceed? The procedure is outlined in Figure 5.12. If we start with the tetrachloroplatinate(II) and react it first with ammonia and then with nitrite, the cis isomer is prepared in high purity. Note that when amminetrichloroplatinate(II) is treated with nitrite, a chloride trans to another chloride is replaced preferentially to the chloride trans to ammine. This takes advantage of the fact that chloride is higher in the trans-directing series than is ammonia. On the other hand, when trichloronitroplatinate(II) is treated with ammonia, the chloride trans to the nitro ligand is replaced preferentially. Here, the nitrite ion is higher in the series than is chloride.

How can we rationalize the kinetic trans effect? One still popular but only partial explanation was devised by A. Grinberg in the 1930s. It maintains that the trans-directing ability of a ligand is related to its polarizability. As noted above in the section on inner-sphere electron-transfer reactions, polarizability can be thought of as the ease with which a dipole moment can be induced in a species. Figure 5.13 illustrates the polarization theory. The first step is the induction of an instantaneous dipole in the trans-directing ligand by the platinum cation. Second, this latter dipole in turn induces a dipole in the large, polarizable platinum cation. The $Pt^{2+}—Cl^-$ bond is somewhat weakened by the repulsion between the negatively charged ligand and the negative end of the dipole induced in the cation. Therefore, the chloride trans to A is preferentially replaced. Support for this theory is demonstrated by looking at the trans-directing series. The more polarizable ligands, such as $—SCN^-$ and I^- and the ligands containing π clouds (for example, CO and CN^-), are high in the series, while the less polarizable ligands

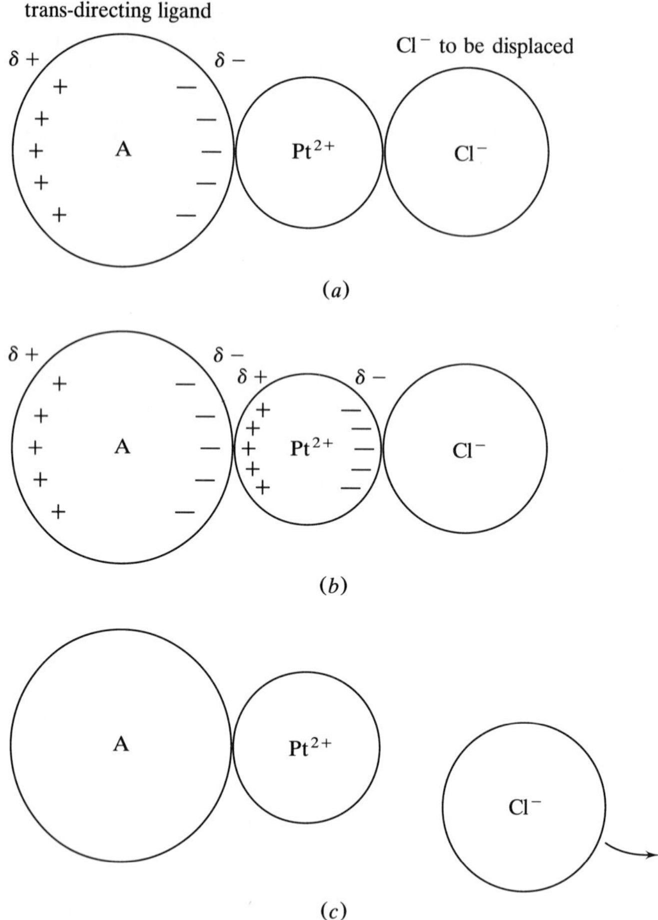

FIGURE 5.13
The polarization theory for explaining the kinetic trans effect in square planar platinum(II) complexes. (*a*) The platinum(II) cation induces a dipole in the polarizable trans-directing ligand A. (*b*) The induced dipole in ligand A induces a dipole in the polarizable Pt^{2+} cation. (*c*) The chloride anion trans to A is more easily released due to the extra repulsive forces between its negative charge and the induced dipole of the platinum(II) cation.

such as ammonia and water are lower in the series. Additional support comes from the observation that platinum complexes demonstrate a more pronounced trans effect than those of the less polarizable palladium(II) and nickel(II) cations.

SUMMARY

This chapter starts with a brief survey of five types of reactions of coordination compounds: (1) substitution, (2) dissociation, (3) addition, (4) redox or electron transfer, and (5) reactions of coordinated ligands. The tendency for a coordination compound to undergo a substitution reaction is tabulated by means of stepwise and overall equilibrium constants. Aquation and anation reactions are special examples of substitution reactions. Addition and dissociation reactions involve adding or removing ligands from the coordination sphere, respectively, while reduction and oxidation reactions involve adding or removing electrons from the metal. Oxidative-addition and reductive-elimination reactions, each combining two of the above reaction types, are logical opposites of each other. Reactions of coordinated ligands occur without breaking or forming any metal-ligand bonds.

Labile and *inert* are kinetic terms that classify coordination compounds by how fast they react. Labile compounds react quickly and inert compounds slowly. These kinetic terms should not be confused with the thermodynamic terms *stable* and *unstable*. Compounds can be thermodynamically unstable but kinetically inert or, conversely, stable but labile. Complexes of the first-row transition metals, with the exception of Cr^{3+} and Co^{3+}, are generally labile, while coordination compounds of second- and third-row transition metal ions are inert.

There are two possible starting points for the study of the mechanisms of octahedral substitution reactions. The rate-determining step of the dissociative (D) mechanism involves the breaking of an M—L bond to form a five-coordinate intermediate which then rapidly adds a new ligand. In the associative (A) mechanism, the rate-determining step is the collision of the new ligand with the octahedral reactant to produce a seven-coordinate intermediate from which an original ligand is then rapidly expelled. The D mechanism, which predicts the rate of the reaction to be first-order only in the original coordination compound, should be clearly distinguishable from an A mechanism that is consistent with a rate law which is first-order in *both* the reactants.

There are, however, two experimental complications that blur the above distinctions. First, the D and A mechanisms turn out to be idealized and oversimplified extremes. Reactions more likely proceed by an interchange (I) mechanism in which bond breaking and forming happen nearly simultaneously. If bond breaking is favored (even if only slightly), the mechanism is termed I_d (interchange-dissociative), while if bond making is favored, an I_a (interchange-associative) is indicated. These last two terms are particularly favored if no five- or seven-coordinate intermediates, respectively, can be isolated. The second experimental complication involves the masking of the rate dependence on the concentration of the incoming ligand. This masking effect is common, particularly in reactions carried out in aqueous solution.

Despite these complications, many pieces of evidence (drawn from the analysis of water exchange, anation, and aquation reactions) favor a dissociative mechanism. First, rates of water exchange are consistent with a dissociative mechanism in which the rate-determining step is the breaking of the $M—OH_2$ bond. Among other factors, this bond-breaking step is dependent upon the charge density of the metal ion. The larger the charge density, the stronger the $M—OH_2$ bond and the slower the bond-severing step. The rates of anation reactions of hydrated complexes are consistent with the breaking of the $M—OH_2$ bond as the initial and rate-determining step. Finally, the rates of aquation reactions are dependent upon the strength of the bond between the metal and ligand being replaced. The rates of aquation reactions vary considerably with the identity of the displaced ligand, the bulk of the inert ligands remaining in the coordination sphere, and the overall charge of the complex.

With the dissociation mechanism firmly established for octahedral substitution reactions, an explanation of the lability of complexes can be formulated in terms of the energy of activation of the bond-breaking step. The size and charge of a metal ion as well as any changes in the crystal field stabilization energy (CFSE) upon bond dissociation influence the energy of activation. Using these parameters, the trends in the lability of coordination compounds can be rationalized. Specifically, complexes of Cr^{3+} (d^3) and Co^{3+} (d^6), are inert because of their high charge and significant loss of CFSE upon bond dissociation.

Henry Taube formulated two principal mechanisms for electron transfer reactions: outer-sphere and inner-sphere. A variety of self-exchange reactions and cross reactions occur by an outer-sphere mechanism. After the formation of an ion-pair composed of the intact coordination spheres of the two reacting metals, an electron is transferred and the ion pair dissociated. The faster outer-sphere redox reactions are those whose M–L distances need only be adjusted slightly in the course of the reaction. An inner-sphere mechanism involves the formation of a bridged complex in which the bridging ligand facilitates the electron transfer and is often transferred from one metal to the other. The more polarizable the potential bridging ligand, the faster the reaction.

Substitution reactions in square planar complexes take place via associative mechanisms. In the kinetic trans effect, a series can be constructed in which species are put in order of their ability to labilize ligands trans to themselves. Using this series, high-purity square planar isomers can be readily synthesized. The trans-directing ability of a ligand is also directly related to its polarizability.

PROBLEMS

5.1. Demonstrate that the product of K_1 to K_4 for the stepwise replacement of waters in $[Cu(H_2O)_4]^{2+}$ by ammonias [given in Equations (5.4) to (5.7)] results in the expression for β_4 given in Equation (5.8).

5.2. The water ligands in the hexaaquonickel(II) cation can be replaced by ammonia.

(*a*) Show chemical equations and write mass action expressions for the stepwise replacement of each water with one ammonia.

(*b*) Write an equation for the overall replacement of the six water ligands by six ammonia ligands and then demonstrate that β_6, the overall equilibrium constant, is given by the product of the six stepwise constants.

5.3. Give the stepwise and overall equilibrium constants for the reaction

$$Cr(CO)_6 + 3PPh_3 \longrightarrow Cr(CO)_3(PPh_3)_3 + 3CO$$

5.4. Give the stepwise and overall equilibrium constants for the reaction

$$[Fe(H_2O)_6]^{3+} + 3en \longrightarrow [Fe(en)_3]^{3+} + 6H_2O$$

5.5. If you were to write the mass action expression to represent the reaction of dilute aqueous acetic acid, $HC_2H_3O_2(aq)$, with water to yield aqueous acetate and hydronium ions, would water generally be included? Why or why not? Where would the value of $[H_2O]$ be found?

5.6. Write the mass action expression for the equilibrium constant of the weak base ammonia, NH_3, in aqueous solution corresponding to the following reaction:

$$NH_3 + H_2O \rightleftharpoons NH_4^+ + OH^-$$

5.7. Classify each of the following reactions. More than one of the five basic classifications may apply.

(*a*) $W(CO)_6 \rightarrow W(CO)_5 + CO$

(*b*) $[Co(CN)_5]^{2-} + I^- \rightarrow [Co(CN)_5I]^{3-}$

(*c*) $[PdCl_4]^{2-} + C_2H_4 \rightarrow [PdCl_3(C_2H_4)]^- + Cl^-$

(*d*) $IrBr(CO)(PPh_3)_2 + HBr \rightarrow IrBr_2(CO)H(PPh_3)_2$

(*e*) $[Co(NH_3)_5Cl]^{2+} + [Cr(H_2O)_6]^{2+} \rightarrow [Co(NH_3)_5(H_2O)]^{2+} + [Cr(H_2O)_5Cl]^{2+}$

5.8. Classify each of the following reactions. More than one of the five basic classifications may apply.

(*a*) $[Co(NH_3)_5NO_3]^{2+} + [Ru(NH_3)_6]^{2+} \rightarrow [Co(NH_3)_5NO_3]^{+} + [Ru(NH_3)_6]^{3+}$

(*b*) $[PtCl_6]^{2-} \rightarrow [PtCl_4]^{2-} + Cl_2$

(*c*) $[Cr(H_2O)_6]^{3+} + H_2O \rightarrow [Cr(H_2O)_5OH]^{2+} + H_3O^+$
(*d*) $Cr(CO)_6 + 3PPh_3 \rightarrow Cr(CO)_3(PPh_3)_3 + 3CO$
(*e*) $Zn(NH_3)_4^{2+} + 2NH_3 \rightarrow Zn(NH_3)_6^{2+}$

5.9. In Equation (5.19) (repeated below), suppose that the hydroxide reactant were to be labeled with O^{17}, where in the products would the label end up? Briefly explain your answer.

$$[Cr(H_2O)_6]^{3+} + OH^- \longrightarrow [Cr(H_2O)_5(OH)]^{2+} + H_2O$$

5.10. In Equation (5.20) (repeated below), suppose that the oxygen atoms of the carbonate ligand were labeled with O^{17}, where in the products would the labels end up? Briefly explain your answer.

$$[Co(NH_3)_5CO_3]^+ + 2H_3O^+ \longrightarrow [Co(NH_3)_5(H_2O)]^{2+} + 2H_2O + CO_2$$

5.11. Briefly speculate on why many cobalt(III) compounds are prepared by oxidizing Co(II) salts rather than by substituting ligands in cobalt(III) complexes.

5.12. Briefly speculate on why many chromium(III) compounds are prepared by reducing chromates and dichromates rather than by substituting ligands in chromium(III) complexes.

5.13. Classify the following complex ions as inert or labile:
(*a*) $[Co(NH_3)_6]^{2+}$
(*b*) $[Co(NH_3)_5NO_2]^{2+}$
(*c*) $[CoI_6]^{3-}$
(*d*) $[Fe(H_2O)_5(NCS)]^{2+}$
(*e*) $[Ni(en)_3]^{2+}$
(*f*) $[IrCl_6]^{3-}$

5.14. Classify the following complex ions as inert or labile:
(*a*) $[Ru(NH_3)_5py]^{3+}$
(*b*) $[Cr(acac)_3]$
(*c*) $[Cr(en)_3]Cl_2$
(*d*) $[Mo(CN)_6]^{3-}$
(*e*) $[V(H_2O)_6]^{3+}$
(*f*) $[MnF_6]^{3-}$

5.15. Of the following cyanide complexes, $[Ni(CN)_4]^{2-}$, $[Mn(CN)_6]^{3-}$, and $[Cr(CN)_6]^{3-}$, which one would you most expect to be (*a*) labile, (*b*) inert? Briefly justify your answer.

5.16. In terms of Figure 5.2, would you expect the concentration of ML_5 to be relatively low or high (compared to the concentrations of the reactants and products) throughout the reaction? Briefly explain your answer in terms of its relative production and consumption during the reaction.

***5.17.** Suppose that the rate-determining step in Figure 5.2 was the second step in which the intermediate ML_5 and Y come together to form the product. In this case, what rate law is predicted?

***5.18.** Suppose that the rate-determining step in Figure 5.3 was the second step in which the seven-coordinate intermediate breaks apart to form ML_5Y and X. In this case, what rate law is predicted?

***5.19.** Suppose that the rate-determining step in Figure 5.3 was the second step in which the seven-coordinate intermediate breaks apart to ML_5Y and X. Would you expect the concentration of ML_5XY to be relative low or high (compared to the concentrations of the reactants and products)? Briefly explain your answer in terms of its relative production and consumption during the reaction.

5.20. Draw a diagram showing an interchange (*I*) mechanism in the following three stages:
(*a*) The resulting situation after an incoming Y ligand has taken its place just outside the coordination sphere of the starting ML_5X complex

(*b*) The midpoint of the interchange between the attacking Y and the leaving X ligand when the M—X and M—Y bonds are approximately at half strength

(*c*) The situation after the M—X bond has been fully replaced by the M—Y bond but the X ligand now resides just outside the coordination sphere of the product ML_5Y complex.

5.21. Assuming no experimental complications, explain, in your own words, some ways in which one might expect to differentiate between an associative (*A*) and dissociative (*D*) mechanism for the substitution of octahedral complexes.

5.22. Suppose the exchange of water reactions of Figure 5.5 were carried out by an associative mechanism rather than a dissociative mechanism. Would you expect, in this case, the rates to increase or decrease with (*a*) increasing charge of the metal or (*b*) increasing size of the metal? Briefly explain your answer.

5.23. In one well-written paragraph, summarize the evidence that favors a dissociative or interchange-dissociative mechanism for the substitution of octahedral coordination compounds.

5.24. The anation of $[Co(H_2O)(CN)_5]^{2-}$ in which the water ligand is replaced by a variety of anions (X^-) has been shown to proceed by a dissociative (*D*) mechanism. (In this case the five-coordinate intermediate has been identified.) Write the steps for this mechanism. Do you expect the rates of these reactions to depend upon the identity of the X^- ligand? Why or why not?

5.25. The rate constants for some anations of $[Cr(NH_3)_5H_2O]^{3+}$ are shown below. Are these data consistent with a dissociative or associative mechanism? Write an equation for the rate-determining step of these reactions and briefly justify your answer.

$$[Cr(NH_3)_5H_2O]^{3+} + L^- \xrightarrow{k} [Cr(NH_3)_5L]^{2+} + H_2O$$

L^-	k, $M^{-1}\,s^{-1}$	log k
NCS^-	4.2×10^{-4}	−3.4
Cl^-	0.7×10^{-4}	−4.1
Br^-	3.7×10^{-4}	−3.4
CF_3COO^-	1.4×10^{-4}	−3.9

Data from the work of D. Thusius, *Inorg. Chem.*, **10**:1106 (1971).

5.26. The rate constants (at 45°C) for the anation of $[Co(NH_3)_5H_2O]^{3+}$ with three anions are shown below. Are these data consistent with a dissociative or associative mechanism? Write an equation for the rate-determining step of these reactions and briefly justify your answer.

$$[Co(NH_3)_5H_2O]^{3+} + L^{n-} \xrightarrow{k} [Co(NH_3)_5L]^{(3-n)+} + H_2O$$

L^{n-}	k, s^{-1}	log k
NCS^-	1.6×10^{-5}	−4.8
Cl^-	2.1×10^{-5}	−4.7
SO_4^{2-}	2.4×10^{-5}	−4.6

From R. G. Wilkins, *The Study of Kinetics and Mechanism of Reactions of Transition Metal Complexes*, Allyn and Bacon, Boston, 1974, p. 188.

5.27. The rate constant for the exchange of the water ligand in $[Co(NH_3)_5H_2O]^{3+}$ is 1.0×10^{-4} s^{-1} at 45°C. Compare this value with the data shown in Problem 5.26. What conclusions can be drawn from the comparison?

∗**5.28.** The rate constants at the same temperature for the anation of $[Cr(H_2O)_6]^{3+}$ with a series of four anions are shown below.

(*a*) Are these data consistent with a dissociative or associative mechanism?

(*b*) Write an equation for the rate-determining step of these reactions and briefly justify your answer.

(*c*) Rationalize the trends in the rate constant data.

$$[Cr(H_2O)_6]^{3+} + L^- \xrightarrow{k} [Cr(H_2O)_5L]^{2+} + H_2O$$

L^-	k, M^{-1} s^{-1}	log k
NCS^-	1.8×10^{-6}	−5.7
NO_3^-	7.3×10^{-7}	−6.1
Cl^-	2.9×10^{-8}	−7.5
I^-	8.0×10^{-10}	−9.1

From the work of S. T. D. Lo and D. W. Watte, *Aust. J. Chem.*, **28**:491, 501(1975).

∗**5.29.** The rate of exchange of the water ligand in $[Cr(H_2O)_6]^{3+}$ is 3×10^{-6} s^{-1} at 25°C. Compare this value with the data shown in Problem 5.28. What conclusions can be drawn from the comparison?

∗**5.30.** For reactions of the general type

$$[Co(NH_3)_5H_2O]^{3+} + L^- \overset{K_a}{\rightleftharpoons} [Co(NH_3)_5L]^{2+} + H_2O$$

demonstrate that log K_a is proportional to D_{M-L}. Recall that $\Delta G° = \Delta H° - T\Delta S°$ and that $\Delta G° = = -RT \ln K = -2.3RT \log K$. First determine the relationship between $\Delta H°$ and the M—L bond strength (D_{M-L}) by analyzing what bonds are broken and formed in this reaction. Assume that $\Delta S°$ is approximately constant for various reactions of this type.

5.31. The rate constants at 25°C for the aquation of complexes of general formula $[CoCl(L)(en)_2]^{n+}$ are given in the following table. Are these data consistent with an associative or dissociative mechanism? Briefly rationalize your answer.

$$cis\text{-}[Co(L)Cl(en)_2]^{n+} + H_2O \xrightarrow{k} cis\text{-}[Co(H_2O)Cl(en)_2]^{2+} + L$$

L	k, s^{-1}	log k
OH^-	1.2×10^{-2}	−1.9
Cl^-	2.4×10^{-4}	−3.6
NH_3	5×10^{-7}	−6.3

From the work of M. L. Tobe et al.; for example, see *Sci. Prog.*, **48**:484 (1960).

5.32. Compare and rationalize the rate constants for the aquation reactions of the following *trans*-cobalt(III) complexes.

$$[Co(NH_3)Br(en)_2]^{2+} + H_2O \xrightarrow{k=1.2\times10^{-6}\,s^{-1}} [Co(NH_3)(H_2O)(en)_2]^{3+} + Br^-$$

$$[CoBr_2(en)_2]^{+} + H_2O \xrightarrow{k=1.4\times10^{-4}\,s^{-1}} [Co(H_2O)Br(en)_2]^{2+} + Br^-$$

5.33. In your own words, write a concise paragraph explaining the relationship between the rates of substitution reactions and changes in the crystal field stabilization energy. Assume that the dissociation of a ligand from an octahedral reactant to form a five-coordinate intermediate is the rate-determining step.

5.34. The splitting diagram for a square pyramidal crystal field is given below. Verify the change in crystal field stabilization energy given in Table 5.4 for a metal ion with a d^3 electronic configuration.

—	$+0.914\Delta$
—	$+0.086\Delta$
------	Barycenter
—	-0.086Δ
— —	-0.457Δ

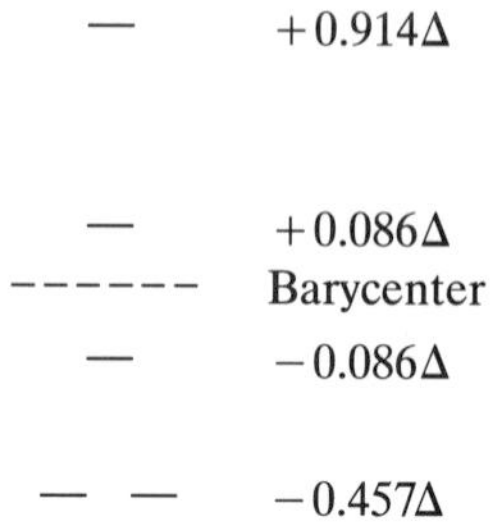

5.35. Using the square pyramidal crystal field splitting diagram given in Problem 5.34, verify the change in the CFSE given in Table 5.4 for a metal ion with a low-spin d^6 electronic configuration.

5.36. The splitting diagram for a trigonal bipyramidal crystal field is given below. Calculate the change in CFSE for a d^3 metal ion. Compare your result with that given in Table 5.4. Given your result, would you conclude that these ions are kinetically inert?

$-d_{z^2}$	0.707Δ
$E\uparrow$ ------	Barycenter
$d_{x^2-y^2}$— $-d_{xy}$	-0.082Δ
d_{xz}— $-d_{yz}$	-0.272Δ

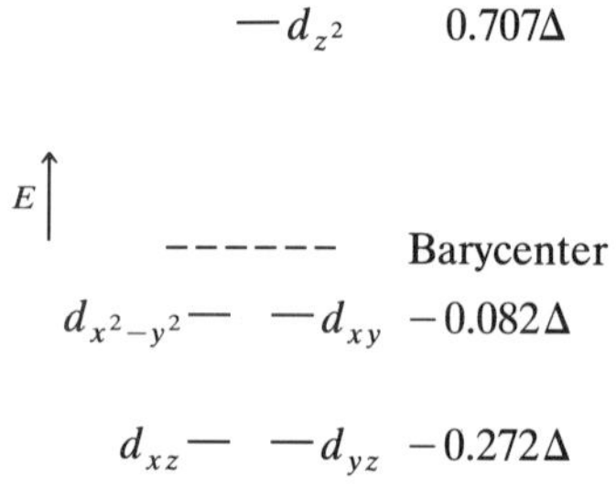

5.37. Using the splitting diagram for a trigonal bipyramidal crystal field given in Problem 5.36, calculate the change in CFSE for an M^{2+} d^8 ion when the field changes from octahedral to trigonal bipyramidal. (Assume high-spin states.)

(*a*) Would you conclude that these ions are kinetically inert?

(*b*) Would you also conclude that a trigonal bipyramidal transition state is favored over one that is square pyramidal? Why or why not?

5.38. The Fe^{3+} and V^{3+} ions, while +3 charged like Cr^{3+} and Co^{3+}, are not classified as inert. Briefly explain why not.

5.39. Octahedral Ni^{2+} complex ions, while experiencing reductions in crystal field stabilization energies upon losing ligands to form square pyramidal transition states, are not classified as inert. Briefly explain why not.

***5.40.** Although octahedral Cu^{3+} complexes would most likely be inert, not many have been observed.

(*a*) Explain why such compounds would be expected to be inert.

(*b*) Explain why not many such compounds are known.

5.41. Both $[Fe(CN)_6]^{4-}$ and $[IrCl_6]^{2-}$ exchange their ligands rather slowly, yet the cross reaction between them (shown below) occurs very rapidly ($k = 4 \times 10^5\ M^{-1}\ s^{-1}$). Propose a mechanism to account for this observation.

$$[Fe(CN)_6]^{4-} + [IrCl_6]^{2-} \rightarrow [Fe(CN)_6]^{3-} + [IrCl_6]^{3-}$$

5.42. Discuss the probable mechanism of the following reaction:

$$[Co(NH_3)_6]^{3+} + [Ru(NH_3)_6]^{2+} \rightarrow [Co(NH_3)_6]^{2+} + [Ru(NH_3)_6]^{3+}$$

Also speculate why this reaction is rather slow ($k = 1 \times 10^{-2}\ M^{-1}\ s^{-1}$). (*Hint:* Examine the *d*-orbital occupancy during the reaction.)

5.43. The self-exchange electron-transfer reaction between $[Co(en)_3]^{3+}$ and $[Co(en)_3]^{2+}$ is rather slow ($k \approx 10^{-4}$). Explain this observation on the basis of *d*-orbital occupations.

5.44. Are the units of the rate constants in Table 5.5 consistent with an overall second-order reaction? Demonstrate your answer clearly.

5.45. Given the second-order nature of outer-sphere electron-transfer reactions, speculate on a rate-determining step for these reactions.

5.46. The M–L distances in tris(ethylenediamine) complexes of Ru(II) and Ru(III) are 2.12 and 2.10 Å, respectively. Speculate on the mechanism and rate of the self-exchange electron-transfer reaction between these two complexes. As part of your answer write an equation representing the reaction.

***5.47.** In addition to the energy involved in readjusting the M—L bond lengths in the coordinated metal ions that serve as reactants, there are several other contributions to the energy of activation of an outer-sphere redox reaction. Speculate on what these contributions might involve.

5.48. For each of the following electron-transfer reactions, speculate whether the mechanism is outer-sphere or inner-sphere:
(*a*) $[IrCl_6]^{2-} + [W(CN)_8]^{4-} \rightarrow [IrCl_6]^{3-} + [W(CN)_8]^{3-}$
(*b*) $[Co(NH_3)_5CN]^{2+} + [Cr(H_2O)_6]^{2+} \rightarrow [Cr(H_2O)_5NC]^{2+} + [Co(NH_3)_5(H_2O)]^{2+}$
(*c*) $[{*}Cr(H_2O)_6]^{2+} + [Cr(H_2O)_5F]^{2+} \rightarrow [{*}Cr(H_2O)_5F]^{2+} + [Cr(H_2O)_6]^{2+}$

5.49. Write out the mechanism for the inner-sphere electron-transfer reaction between $[Co(NH_3)_5SCN]^{2+}$ and $[Cr(H_2O)_6]^{2+}$ that produces the pentaaquo-*N*-thiocyanatochromium(III) cation as one of its products.

5.50. While the *N*-thiocyanato form of the $[Cr(H_2O)_5NCS]^{2+}$ is the principal product (approximately 70 percent) of the above inner-sphere electron transfer (see Problem 5.49), the *S*-thiocyanato linkage isomer also forms. Write a mechanism that shows how the *S*-linked isomer might be formed.

5.51. For reactions of the general type

$$[{*}Cr(H_2O)_6]^{2+} + [Cr(H_2O)_5X]^{2+} \longrightarrow [{*}Cr(H_2O)_5X]^{2+} + [Cr(H_2O)_6]^{2+}$$

rate constants increase in the order $X^- = F^-, Cl^-, Br^-$. Write out the mechanism for the above reaction and discuss the trend in the rate constant data in the light of your proposed mechanism.

5.52. Given that the substitution of square planar complexes occurs by an associative mechanism, speculate (*a*) on the order of the reaction with respect to the various reactants and (*b*) on the dependence of the rate of these reactions on the steric bulk of the nonreacting ligands of the complex, the bulk and charge of the entering ligands, and the overall charge carried by the reactant complex.

5.53. Using the kinetic trans effect, show how the three possible geometric isomers of amminebromochloro(pyridine)platinum(II) can be prepared.

5.54. "Cisplatin," *cis*-diamminedichloroplatinum(II), is an extremely potent antitumor agent. Show how it can be prepared in high purity to the exclusion of the trans geometric isomer by employing the kinetic trans effect.

5.55. Given the polarization theory for explaining the kinetic trans effect, where would you speculate that the hydroxide ion, OH^-, might show up in the trans-directing series? Justify your answer.

5.56. Given the polarization theory for explaining the kinetic trans effect, where would you speculate that the organic sulfides, R_2S, might show up in the trans-directing series? Justify your answer.

5.57. Predict the products of the following reactions:

(*a*) $[Pt(CO)Cl_3]^- + py \rightarrow$

(*b*) $\left[\begin{matrix} Cl & & Cl \\ & Pt & \\ H_3N & & NO_2 \end{matrix}\right]^- + NH_3 \rightarrow$

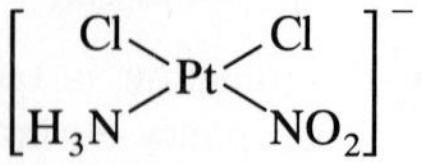

(*c*) $[PtCl_3SCN]^{2-} + H_2O \rightarrow$

(*d*) $[PtCl_3CN]^{2-} + NH_3 \rightarrow$

CHAPTER 6

APPLICATIONS OF COORDINATION COMPOUNDS

What do winning silver and gold from their ores, blueprints, nickel purification, photographic hypo, food preservation, detergent builders, the color of blood, carbon monoxide poisoning, laetrile, antidotes for lead poisoning, British anti-lewisite, blue cross, and antitumor agents all have in common? The answer is that each involves an application, in one way or another, of coordination compounds. In this chapter we will systematically investigate the above applications and a number of others. We start with complexes involving monodentate ligands and then move on to those involving various chelating multidentates.

6.1 APPLICATIONS OF MONODENTATE COMPLEXES

Many students have prepared coordination compounds in prior chemistry courses but often have not fully understood or even identified them as such. For example, in the Group I qualitative analysis scheme, lead(II), mercury(I), and silver(I) are isolated as the white precipitates $PbCl_2$, Hg_2Cl_2, and AgCl, respectively. To separate the silver from the other two cations, aqueous ammonia is added to form the linear diamminesilver(I) complex, $Ag(NH_3)_2^+$, as shown in Equation (6.1). Since neither lead(II) chloride nor mercury(I) chloride react similarly, the silver is successfully separated from the other two metals.

$$AgCl(s) + 2NH_3(aq) \rightarrow Ag(NH_3)_2^+(aq) + Cl^-(aq) \qquad (6.1)$$

Copper(II) is sometimes included in the Group I qualitative analysis scheme. Its chloride is soluble, but it does form a tetraammine complex as shown in Equation (6.2). This square planar complex absorbs visible light to produce a characteristic deep-blue color that can be used to indicate the presence of copper ions.

$$Cu^{2+}(aq) + 4NH_3(aq) \rightarrow \underset{\text{Deep blue}}{Cu(NH_3)_4^{2+}(aq)} \qquad (6.2)$$

Similarly, the presence of iron(III) in a water supply can be detected by adding a small amount of potassium thiocyanate, KSCN, to produce the characteristic deep-red thiocyanatoiron(III) complex as shown in Equation (6.3). To be more specific, the product of this reaction is actually pentaaquothiocyanatoiron(III), but it is often abbreviated as shown.

$$Fe^{3+}(aq) + SCN^-(aq) \longrightarrow \underset{\text{Red}}{FeSCN^{2+}(aq)} \qquad (6.3)$$

These latter two coordinated cations, $FeSCN^{2+}$ and $Cu(NH_3)_4^{2+}$, are deeply colored due to their strong absorption of visible light. We know from Chapter 4 that the energy differences among *d* orbitals caused by various crystal fields most often correspond to visible frequencies of light. Therefore, transition metal complexes are almost invariably colored. In fact, this property has made various transition metal coordination compounds useful as the colored components of pigments, dyes, inks, and paints. For example, prussian blue, first discovered in the early 1700s, is an iron cyanide complex produced by adding potassium hexacyanoferrate(II) to any iron(III) salt as shown in Equation (6.4). Turnbull's blue, long thought to be a different compound but now known to be identical to prussian blue, is produced analogously by adding potassium hexacyanoferrate(III) to any iron(II) salt:

$$\left.\begin{array}{l} Fe^{III}(aq) + K_4[Fe^{II}(CN)_6](aq) \\ Fe^{II}(aq) + K_3[Fe^{III}(CN)_6](aq) \end{array}\right\} \longrightarrow \underset{\text{Prussian or Turnbull's blue}}{Fe_4^{III}[Fe^{II}(CN)_6]_3 \cdot 4H_2O} \qquad (6.4)$$

(In the mid-1800s it was found that both of these pigments could be infused into special papers called *blueprints* that found a special use for the quick and cheap preparation of multiple copies of large-scale architectural and engineering drawings.) The structure of these hexacyanoferrate pigments eluded characterization for many years, but in the 1970s it was demonstrated to be as shown in Figure 6.1. Here, the cyanide ligands bridge between the iron(II) and iron(III) cations as shown.

Cyanide complexes are also used to separate gold and silver from their ores. The crushed ore, containing minute quantities of the free metal, is subjected to a dilute solution of a cyanide salt while simultaneously being oxidized by blowing air through it. As a result, soluble dicyanoargentate(I) and dicyanoaurate(I) complexes are formed as shown in Equation (6.5). The silver or gold (or sometimes a Ag–Au alloy) is then isolated by reaction with a good reducing metal such as zinc as shown in Equation (6.6). As detailed in Section 6.5, cyanide is an extremely toxic material and must be handled with great care.

$$4M(s) + 8CN^-(aq) + O_2(g) + 2H_2O(l) \longrightarrow$$
$$4[M(CN)_2]^-(aq) + 4OH^-(aq) \qquad M = Au, Ag \quad (6.5)$$

$$Zn(s) + 2[M(CN)_2]^-(aq) \longrightarrow$$
$$[Zn(CN)_4]^{2-}(aq) + 2M(s) \qquad M = Au, Ag \quad (6.6)$$

Another valuable transition metal, nickel, is purified by a process first formulated by Ludwig Mond in the 1890s. In the *Mond process*, the impure nickel metal is subjected to a warm (approximately 75°C) stream of carbon monoxide gas.

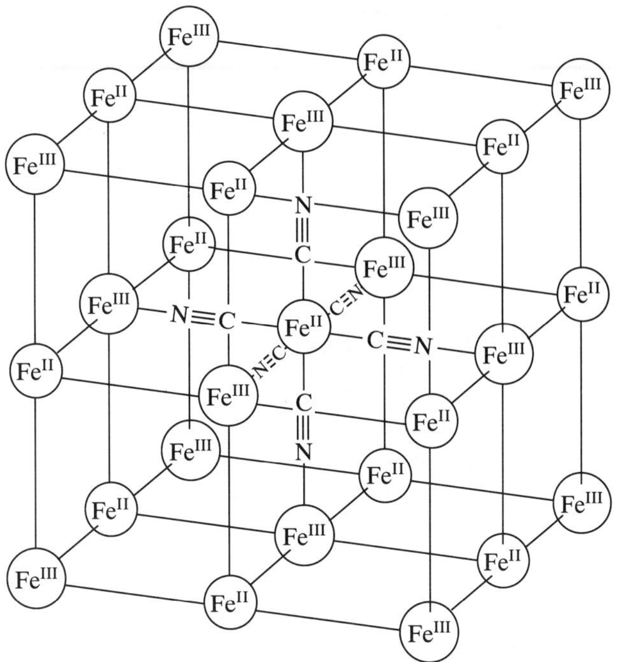

FIGURE 6.1
A portion of the structure of prussian or Turnbull's blue, iron(III) hexacyanoferrate(II) tetrahydrate. Waters of hydration have been omitted for clarity. The cyanide ligands bridge between the iron(II) and iron(III) cations. Some of the cyanide ions are occasionally replaced with water molecules bound to the Fe(III) cations. Occasional vacancies (not shown) are also in the structure. [Adapted from Ref. 8.]

The gaseous, tetrahedral tetracarbonylnickel(0), $Ni(CO)_4$ (often commonly called nickel tetracarbonyl), is immediately formed and allowed to pass into another chamber at about 225°C. (Other transition metals present as impurities are not similarly complexed.) At the higher temperature, the equilibrium between solid nickel, carbon monoxide, and the nickel carbonyl [as shown in Equation (6.7)] is reversed and pure nickel is deposited. The highly toxic carbon monoxide gas is recycled continuously throughout the process. Nickel tetracarbonyl, even more toxic than CO and certainly the most hazardous nickel compound known, is a volatile liquid at room temperature. Extreme care must be taken in its handling.

$$Ni(s) + 4CO(g) \underset{225°C}{\overset{75°C}{\rightleftharpoons}} Ni(CO)_4(g) \tag{6.7}$$

A final example of the application of complexes involving monodentate ligands is the formation of bis(thiosulfato)argentate(I) ions during the use of photographic "fixers." Black-and-white photography depends upon the sensitivity of various silver halides to visible light. These salts, most often silver bromide, but occasionally the iodide for particularly "fast" film, are evenly embedded into the gelatinous surface of film. When exposed to visible light, the halide loses its electron which is in turn taken on by the silver cations to produce atomic silver and a dark, or "exposed," spot on the film. After developing (the enhancing of these dark spots), the excess (and unexposed) silver halide must be removed from the film. This "fixing" of the image is accomplished with sodium thiosulfate "*hypo*" or ammonium thiosulfate "*ammonium hypo*" as illustrated in Equation (6.8). Both the ionic products of this reaction are water-soluble and easily washed away from the surface of the film.

$$AgBr(s) + Na_2S_2O_3(aq) \longrightarrow Na_3[Ag(S_2O_3)_2](aq) + NaBr(aq) \quad (6.8)$$

$$\left[O_3S{-}S \rightarrow Ag \leftarrow S{-}SO_3 \right]^{3-}$$

6.2 TWO KEYS TO THE STABILITY OF TRANSITION METAL COMPLEXES

To more fully understand the above applications as well as those to come, it is useful to introduce two ideas at this point. The first is the theory of hard and soft acids and bases, and the second is the chelate effect.

Hard and Soft Acids and Bases

As discussed in Chapter 4, coordination compounds are Lewis salts or adducts composed of a Lewis acid (the metal atom or ion) and various numbers of Lewis bases (the ligands). Lewis acids accept electron pairs, while Lewis bases donate them. In the early 1960s, Ralph Pearson introduced the idea of hard and soft acids and bases. *Hard acids* are defined as small, compact, highly charged electron-pair acceptors, while *hard bases* are small, highly electronegative electron-pair donors. *Soft acids and bases*, on the other hand, are large, diffuse, and polarizable species. Under these definitions, metal cations such as Al^{3+} and Cr^{3+} are hard acids, while ligands such as F^-, NH_3, and H_2O are hard bases. Conversely, metal cations like Hg^{2+}, Ag^+, and Au^+ and ligands containing phosphorus, sulfur, and diffuse π bonds (for example, CN^- and CO) are soft. A more complete listing of hard and soft acids and bases is given in Table 6.1.

TABLE 6.1
Partial list of hard and soft acids and bases

	Acids	Bases*
Hard	H^+, Li^+, Na^+, K^+ Be^{2+}, Mg^{2+}, Ca^{2+} Al^{3+} Cr^{3+}, Co^{3+}, Fe^{3+} Ti^{4+}, Zr^{4+}, Hf^{4+}, Th^{4+} Cr^{6+}	NH_3, RNH_2 H_2O, OH^-, O^{2-} F^-, Cl^- NO_3^-, ClO_4^- SO_4^{2-} PO_4^{3-} CH_3COO^-
Borderline	Fe^{2+}, Co^{2+}, Ni^{2+}, Cu^{2+} Zn^{2+}, Sn^{2+}, Pb^{2+} Ru^{2+}, Rh^{2+}, Ir^{3+}	N_2, py NO_2^-, SO_3^{2-} Br^-
Soft	Cd^{2+}, Hg_2^{2+}, Hg^{2+}, Cu^+, Ag^+, Au^+, Tl^+ Pd^{2+}, Pt^{2+}, Pt^{4+} M^0 (metal atoms)	H^- CN^-, C_2H_4, CO PR_3, AsR_3, R_2S, RSH SCN^-, $S_2O_3^{2-}$ I^-

*R = CH_3—, CH_3CH_2—, C_6H_5—, etc.

The special utility of the hard-soft–acid-base (HSAB) idea is that *soft acids preferentially bind with soft bases and hard acids with hard bases*. This rule, largely based on observations, is useful for rationalizing and predicting the relative stabilities of transition-metal complexes and other compounds. For example, take the last three applications of coordination compounds covered in the preceding section. The bis(thiosulfato)argentate(I) complex formed in fixing black-and-white photographic images is characterized by a soft-soft $Ag^+—S$ coordinate-covalent bond rather than a soft-hard $Ag^+—O$ one. Nickel(0), a soft metal atom, forms stable complexes with soft carbonyl molecules that have large, diffuse π-electron clouds of electrons. Or, finally, note that silver(I) and gold(I), large, diffuse second- and third-row $+1$ transition metal cations, form strong bonds with the soft cyanide ligand.

We can see from these examples that the HSAB idea is certainly a useful one. However, some caution is in order. While many complexes follow the HSAB rules, a number of perfectly stable ones do not. For example, the diamminesilver(I) complex mentioned above involves a soft acid and a hard base, while the thiocyanatoiron(III) complex has the converse. So, while the HSAB rule is a handy way to organize our ideas about the stability of transition metal complexes, it certainly is not foolproof and must be regarded only as a useful rule of thumb.

There have been various attempts to rationalize the HSAB idea, but, to date, its theoretical basis is incompletely understood. Most of these rationales center around the idea that hard-hard interactions tend to be stabilized by strong ionic forces, while soft-soft interactions are stabilized by covalent bonds and/or London dispersion forces.

The Chelate Effect

The *chelate effect* may be defined as the unusual stability of a coordination compound involving a chelating, multidentate ligand as compared to equivalent compounds involving monodentate ligands. To see the magnitude of this effect, consider the following two reactions represented in Equations (6.9) and (6.10). Note that the overall stability constant β of the tris(ethylenediamine) complex is about 10 orders of magnitude greater than that of the equivalent hexaammine complex.

$$Ni^{2+}(aq) + 6NH_3(aq) \longrightarrow [Ni(NH_3)_6]^{2+} \qquad \beta = 4.0 \times 10^8 \qquad (6.9)$$

$$Ni^{2+}(aq) + 3en(aq) \longrightarrow [Ni(en)_3]^{2+} \qquad \beta = 2.0 \times 10^{18} \qquad (6.10)$$

What could be the cause of such a large difference in thermodynamic stability? After all, the number of $Ni^{2+}—N$ coordinate-covalent bonds is 6 in both the products of these two reactions, so their enthalpy changes (ΔH) should be fairly similar. That seems to leave entropy as the major explanation for the effect. Indeed, the rationale for the chelate effect can be understood in two ways, both related to the relative probabilities that the two reactions will occur. First, consider the number of reactants and products in the two cases. As written more explicitly in Equations (6.11) and (6.12), it is apparent that the number of ions and molecules scattered throughout the water structure in the first reaction stays the same (seven in both the reactants and the products). In the second reaction, however, three ethylenediamine molecules replace six water molecules in the coordination sphere, and the number of particles scattered at random throughout the aqueous solution increases from four to seven. The larger number of particles distributed randomly in the solution represents a state of higher probability or higher entropy for the products of the second reaction. Therefore, the second reaction is favored over the first due to this entropy effect.

$$\underbrace{[Ni(H_2O)_6]^{2+} + 6NH_3}_{7 \text{ "particles"}} \longrightarrow \underbrace{[Ni(NH_3)_6]^{2+} + 6H_2O}_{7 \text{ "particles"}} \tag{6.11}$$

$$\underbrace{[Ni(H_2O)_6]^{2+} + 3en}_{4 \text{ "particles"}} \longrightarrow \underbrace{[Ni(en)_3]^{2+} + 6H_2O}_{7 \text{ "particles"}} \tag{6.12}$$

Recall that the relative importance of the enthalpy and entropy changes in a reaction is given by the expression for the free-energy change shown in Equation (6.13). A large increase in the entropy of a reaction is reflected in a more negative value for ΔG°. The equilibrium constant, in turn, is related to the change in free energy by the expression given in Equation (6.14). It should be clear that a more negative ΔG° will be reflected in a larger positive value for the equilibrium constant. (You may find it useful to review these concepts from your previous chemistry experiences.) So we see that the significant increase in the entropy in the second reaction is what causes, in large part, the higher value of β, the overall stability constant for that reaction.

$$\Delta G^\circ = \Delta H^\circ - T\Delta S^\circ \tag{6.13}$$

$$\Delta G^\circ = -RT \ln K \tag{6.14}$$

A second way to explain the chelate effect is to look at what happens once one end of the bidentate ethylenediamine ligand attaches to the metal. The situation is shown in Figure 6.2. Note that the concentration of the second nitrogen Lewis base (the other amine group of the ethylenediamine) is now greater in the vicinity of the Ni^{2+} ion than it would be if the second nitrogen Lewis base (a second monodentate NH_3 ligand, for example) were free to randomly roam about any place in the solution. Often this situation is expressed in terms of the "local" or "effective concentration" of the second ligand. In other words, the concentration of the second Lewis base in the local vicinity of the metal ion is effectively larger when the second group is attached to the first by a relatively short two-carbon chain. Therefore, the reaction involving the bidentate is more probable and has a correspondingly higher equilibrium constant.

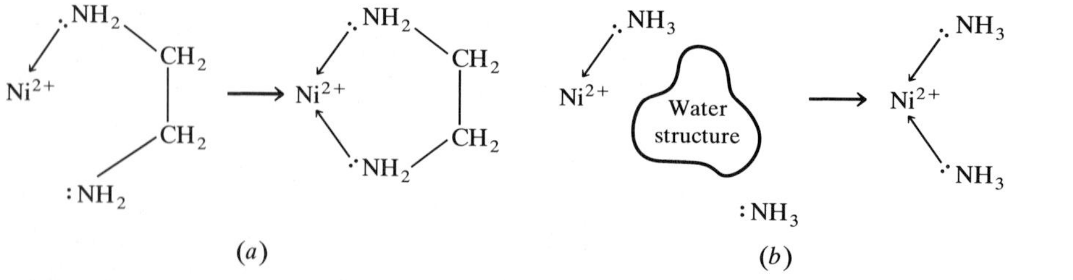

FIGURE 6.2
The chelate effect. (*a*) The second amine of an ethylenediamine molecule complexes to a nickel(II) cation to form a five-membered ring. (*b*) A second ammonia molecule binds to the same cation. The probability of the first reaction occurring is greater because the second ligand is tethered to the nickel by the chelate chain. The "local concentration" of the second ligand in the vicinity of the metal ion is greater in the first case.

Given the above comments about local concentration, it is not surprising that the length of the chelate chain has a direct bearing on the probability of the second Lewis base forming a coordinate-covalent bond with the metal. An ethylenediamine ligand has two carbons between the Lewis bases and forms a five-membered ring (including the metal) when both ends are coordinated. Such a ring structure is very stable. Six-membered rings are also stable, but when the chains and rings become much larger than this, the enhancement of the local concentration is diminished, and the resulting complexes become less stable.

6.3 APPLICATIONS OF MULTIDENTATE COMPLEXES

Not surprisingly, a large number of applications take advantage of the chelate effect. In this section we start by discussing *complexometric quantitative analytical methods*, which are those involving the formation of a complex as the key to the measurement of the amount of a material in a sample.

Gravimetric methods are those that involve the production, isolation, and weighing of a solid to determine the amount of a material in a sample. Perhaps the example of this method most often cited, although not complexometric, is the precipitation of AgCl(*s*) in the determination of the percentage of silver and/or chloride in a substance. One of the most common *complexometric* gravimetric procedures involves the use of a bidentate chelating agent, dimethylglyoximate ($dmgH^-$), to determine nickel. The structure of $dmgH_2$ is shown in Figure 6.3 as well as the $Ni(dmgH)_2$ complex. Note that two five-membered rings are formed in the complex and also, incidentally, that the two rings interact with each other via hydrogen bonding.

Another common gravimetric procedure involves the determination of aluminium with a chelating agent known as 8-hydroxyquinoline (sometimes called 8-quinolinol, or oxime). This complexing agent, also shown in Figure 6.3, is an example of what is called a *heterocycle*. One of the delocalized rings in the compound contains more than one type of atom, in this case, carbon and nitrogen. Again, the bidentate ligand forms stable five-membered rings with the metal ion.

Titrimetric methods involve the formation of a metal complex as the endpoint of a titration. In these procedures a solution of the metal ion is usually titrated against a known concentration of a chelating agent. Before the titration begins, a minute amount of an indicator (also a ligand) is added. The indicator forms a colored coordination compound with a small amount of the metal ion. The remainder of the metal, the so-called free metal, is uncomplexed. As the titration progresses, the free-metal ions are first complexed by the chelating agent. However, when all the free metal has been reacted, the chelating agent then starts to remove the metal from the M-indicator complex. The free indicator is of a different color than the M-indicator complex, and the solution changes color to indicate that the stoichiometric endpoint has been reached.

Examples of complexometric titrations include the use of triethylenetetramine (trien) to titrate copper(II), 1,10-phenanthroline (phen) to titrate iron(III), and ethylenediaminetetraacetic acid (H_4EDTA) to titrate a large variety of $+2$ and $+3$ metal ions. These chelating agents and their representative metal complexes are shown in Figure 6.4. Note that H_4EDTA can be thought of as a derivative of ethlyenediamine in which the four amine hydrogens are all replaced with acetic acid moieties ($-CH_2COOH$). When EDTA chelates a metal cation, it can occupy all six octahedral sites and forms five stable rings with each metal.

(a) $dmgH_2$ $Ni(dmgH)_2$

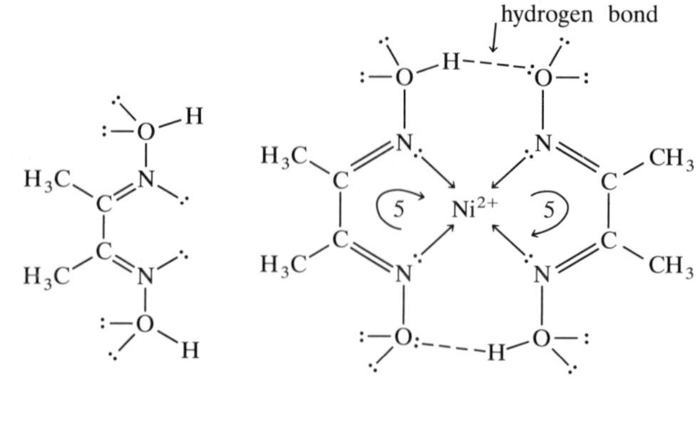

(b) oximeH $Al(oxime)_3$

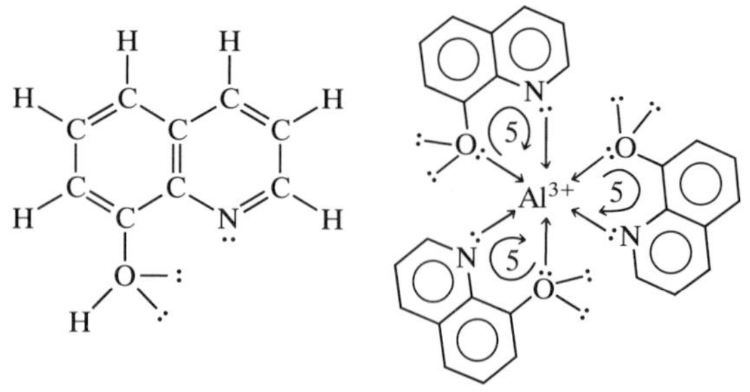

FIGURE 6.3
The structures of the uncoordinated ligands and the resulting coordination compounds involved in the gravimetric determinations of nickel(II) and aluminum(III): (*a*) the free dimethylglyoxime ligand, $dmgH_2$, and the bis(dimethylglyoximato)nickel(II), $Ni(dmgH)_2$, complex and (*b*) the free 8-hydroxyquinoline or oxime ligand, oximeH, and the tris(8-hydroxyquinolate) aluminum(III), $Al(oxime)_3$, complex.

Due to the especially stable complexes formed by EDTA, it is put to a large number of uses. In one of the most common, EDTA titrations are used to determine the concentrations of hard water ions, Ca^{2+} and Mg^{2+}, in a natural water supply. (More on hard water can be found in Section 6.4 as well as in Section 13.3, pp. 336–337.) Because EDTA ties up so many $+2$ and $+3$ cations in complex form, these metals are not chemically available to participate in ways they normally would. For this reason, EDTA is sometimes referred to as a *sequestering agent*, one that isolates and chemically sets aside metal ions. Left free, these trace metals often catalyze various reactions, many of which are undesirable. For example, trace metals catalyze the decomposition of food and various other consumer products, making them rancid and discolored. To retard these metal-catalyzed decomposition reactions, small amounts of EDTA are added to meats, mayonnaise, and soap (among other products) to increase their shelf life. Moreover, EDTA is used to control the levels of trace metal ions in a number of manufacturing situations including the textile, paper, dairy, and rubber industries.

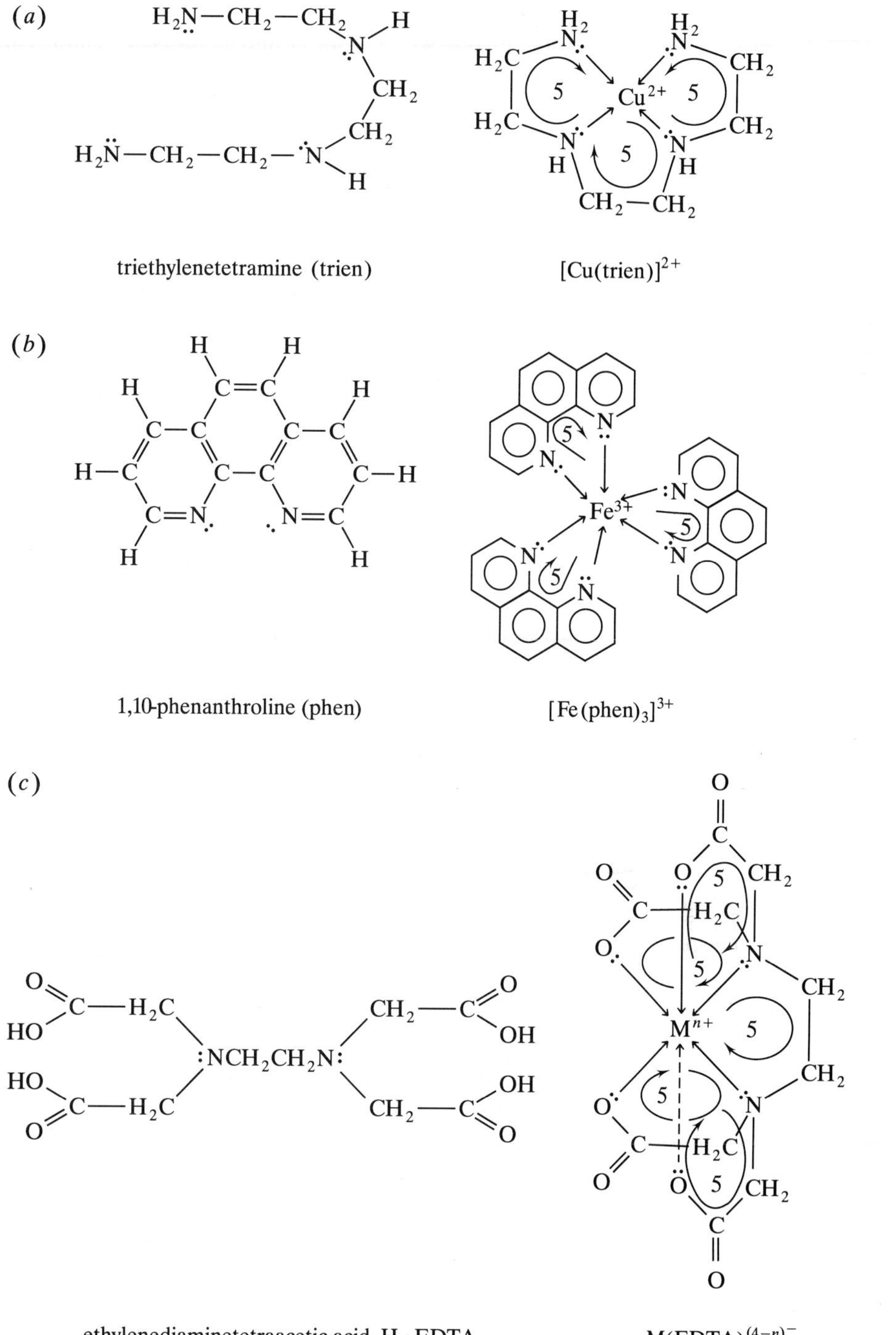

(*a*)

triethylenetetramine (trien) $[Cu(trien)]^{2+}$

(*b*)

1,10-phenanthroline (phen) $[Fe(phen)_3]^{3+}$

(*c*)

ethylenediaminetetraacetic acid, H_4EDTA $M(EDTA)^{(4-n)-}$

FIGURE 6.4
Three different chelating agents: (*a*) triethylenetetramine (trien), (*b*) 1,10-phenonthroline (phen), and (*c*) ethylenediaminetetraacetic acid (H_4EDTA) and the resulting coordinated ions employed in various complexometric titrations.

EDTA is also used to remove the calcium carbonate and sulfate scales that form in hot water boilers and heaters. Calcium carbonate precipitates out when hard water is heated because the bicarbonate present in any naturally aerated water supply decomposes under the action of heat into carbonate, water, and carbon dioxide as shown in Equation (6.15). (This reaction is the source of the very small bubbles that form on the sides of a beaker before water heated in it comes to a boil.) The carbonate is then available to be precipitated by calcium, magnesium, or iron cations that are often present in a natural water supply. This precipitation reaction is shown for calcium in Equation (6.16):

$$2HCO_3^-(aq) \xrightarrow{\text{heat}} H_2O + CO_2(g) + CO_3^{2-} \tag{6.15}$$

$$Ca^{2+}(aq) + CO_3^{2-}(aq) \longrightarrow CaCO_3(s) \tag{6.16}$$

Calcium sulfate forms in hot water boilers and heaters because it is less soluble in hot water than cold. EDTA is also used as an antidote for heavy-metal poisoning, a topic we will take up in Section 6.5.

6.4 CHELATING AGENTS AS DETERGENT BUILDERS

In the late 1940s, when automatic clothes washers first came on the market, a large problem was soon encountered. When ordinary laundry soap was used in these appliances, a white (more often gray when used to clean something dirty), gelatinous, gummy precipitate, the product of various reactions between soap and hard water ions such as Ca^{2+}, Mg^{2+}, and Fe^{3+} often formed. This precipitate was not only unpleasant because it was often deposited on the wash but, more importantly to the washer manufacturer, it literally gummed up the works of the washer. The tiny holes that allow the wash water to be drained from the tub of an applicance were soon plugged by this gummy precipitate. The solution to this problem was the formulation of new synthetic detergents (*syndets*). Tide, manufactured in 1947 by Proctor & Gamble, was the first syndet to appear.

Synthetic detergents are still made up of two principal components: (1) the surfactant, the soaplike molecule that does the actual cleaning, and (2) the builder, an agent added to sequester, or set aside, hard water ions such as Ca^{2+}, Mg^{2+}, and $Fe^{2+/3+}$ that would precipitate or otherwise interfere with the action of the surfactants. The original builders were chelating agents.

The most common builders in the 1950s and early 1960s were the phosphates, represented in Figure 6.5*a* by tripolyphosphate (TPP), $P_3O_{10}^{5-}$. As shown, TPP is a tridentate chelating agent. The most common form of TPP, the simple sodium salt, is inexpensive to manufacture, able to establish and maintain the

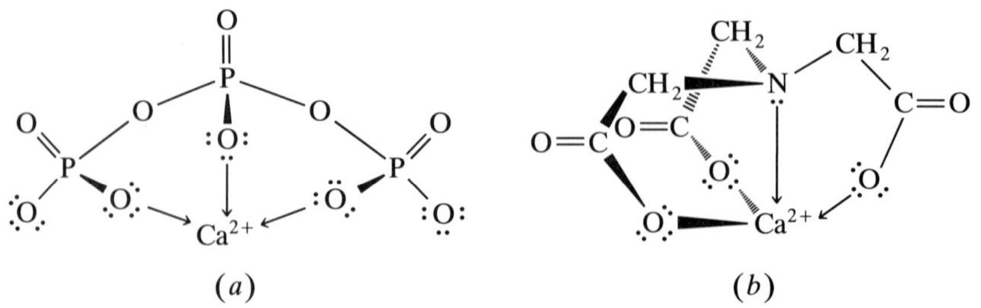

(*a*) (*b*)

FIGURE 6.5
Two chelating detergent builders; (*a*) tripolyphosphate (TPP) and (*b*) nitrilotriacetate (NTA). Each is shown complexed to Ca^{2+}, a representative hard water ion.

proper pH for the most effective action of surfactants, and is particularly efficient at chelating and therefore sequestering hard water ions.

There really appeared to be only one problem with phosphates. As many gardeners know, phosphates (as well as carbonates, nitrates, potassium and magnesium salts, etc.) are nutrients. As the popularity of phosphate detergents grew, tons and tons of them, largely untreated by wastewater treatment plants, were dumped into our streams, rivers, ponds, and lakes. There, it was found, these phosphates (along with a large and complicated variety of nutrients from other sources like phosphates, nitrates, nitrites, and ammonia from nondetergent domestic and industrial sources, as well as phosphates from agricultural and feedlot runoffs) were responsible for vast blooms of algae and other water plants.

By the late 1960s, detergent-related phosphates had become identified as one of the primary, controllable causes of advanced eutrophication. *Eutrophication* itself is a natural process by which a body of water gradually ages by filling up with aquatic plant life to eventually become first a swamp and ultimately a meadow. Normally, such a process takes thousands of years. *Advanced eutrophication* is the acceleration of the natural aging process of a body of water caused by overfertilization with nutrients, resulting in a rapid and excessive growth of aquatic plant life. As a result of their association with advanced eutrophication, detergent-related phosphates were severely limited in some areas and even banned in others. Many states passed laws limiting the phosphorus content in detergents to 8.7%, which corresponds to 34% sodium tripolyphosphate.

The search was on for a replacement for phosphates as the primary detergent builder. One candidate, developed at considerable cost by Procter & Gamble, was the quadridentate chelating agent nitrilotriacetic acid (H_3NTA). In a thought process similar to what we went through with H_4EDTA, H_3NTA can be thought of as a derivative of ammonia in which all three hydrogen atoms are replaced with acetic acid moieties ($-CH_2COOH$). In Figure 6.5*b*, the resulting anion, nitrilotriacetate, is shown chelating a representative hard water ion. It is an excellent chelating agent and syndet builder. Unfortunately, some early reports (1970) suggested that NTA could also complex mercury, cadmium, and lead and, in these complexed forms, carry them across various body barriers (the placental barrier and the blood-brain barrier, to name two) and cause birth defects and brain damage. In the mid-1970s, NTA was also implicated as a cause of kidney damage and cancer. These early reports now appear to be largely unsubstantiated. In 1980, in fact, the Environmental Protection Agency reported that there was no reason to regulate NTA in detergents. The damage was done, however, and NTA is still not widely used as a syndet builder although it has found significant use in other countries such as Canada and Germany.

So what has replaced phosphates in syndets? A quick survey of laundry products shows that many of them use "suspended sodium carbonate," Na_2CO_3, as a builder. Carbonates precipitate hard water ions as a fine, granular precipitate, a process represented in Equation (6.17). Silicates also precipitate out hard water ions [see Equation (6.18)] and have found limited use as phosphate replacements. Phosphates themselves are still used (although limited to 8.7 percent in laundry detergents in many states), particularly in industrial-strength cleaners and in dishwasher detergents.

$$Ca^{2+}(aq) + CO_3^{2-}(aq) \longrightarrow CaCO_3(s) \tag{6.17}$$

$$Ca^{2+}(aq) + SiO_3^{2-}(aq) \longrightarrow CaSiO_3(s) \tag{6.18}$$

6.5 BIOINORGANIC APPLICATIONS OF COORDINATION CHEMISTRY

Biochemistry, the chemistry of living systems, is substantially informed by a knowledge of inorganic chemistry in general and coordination chemistry in particular. In fact, bioinorganic chemistry is a relatively new and still-growing interdisciplinary field of chemistry. One of the most productive areas investigated by bioinorganic chemists over the last 30 years is the nature and action of hemoglobin in respiration. Others include therapeutic chelating agents and platinum antitumor agents.

Respiration

Respiration, the process by which a living organism takes in oxygen and gives off carbon dioxide in return, is accomplished in all but a few animals by the action of hemoglobin, a tetrameric protein with an astounding molecular weight (at least to an inorganic chemist) of about 64,500 u. Two views of hemoglobin are given in Figure 6.6. The first is an overall representation of the entire molecule, not easy to do in any great detail for a molecule of this size! Most of this four-part molecule is protein and not dealt with here, but embedded in these protein chains are four planar disks called *heme* groups. Figure 6.6*b* shows a more detailed view of one of them. In each heme, four nitrogen atoms are coordinated to an Fe^{2+} cation to form a square planar array. Directly beneath the iron cation, a histidine group, derived from the protein infrastructure of the molecule, occupies a fifth site. The sixth position around the iron is available to transport an oxygen molecule from the lungs, gills, or just the skin of an organism to the various cells that need it.

In human beings and most animals, hemoglobin is found in the blood and, in fact, is what gives blood its characteristic color. (In those species without hemoglobin, the blood is either a different color or colorless.) A vastly oversimplified representation of respiration is given in Equation (6.19):

$$\text{Hb} + \text{O}_2(g) \rightleftharpoons \text{HbO}_2 \qquad (6.19)$$

Here Hb represents one of the heme sites in hemoglobin. Oxygen and the hemoglobin (a dark purplish-red color) are in equilibrium with HbO_2, oxyhemoglobin (a bright scarlet). When the free heme sites in blood encounter a high concentration or partial pressure of oxygen gas in the alveoli (small air sacs) of the lungs, the equilibrium shifts to the right. When the oxyhemoglobin encounters a cell in which the partial pressure of oxygen is low, the above equilibrium shifts back to the left and the oxygen is released.

There has been a great deal of discussion among bioinorganic chemists over the years about the mode of bonding of the diatomic oxygen molecule to the Fe^{2+} in heme. As shown in Figure 6.6*b*, most of the evidence seems to favor a bent structure consistent with the oxygen donating one of its lone pairs to the iron. The ability of each heme group to take on an oxygen molecule seems to be dependent on how many of the other three groups are currently coordinated to other oxygen molecules. That is, the four heme groups seem to bind oxygen molecules cooperatively. When one heme group takes on an oxygen, the other groups become even more receptive to the binding of a second oxygen. Evidently, the binding of one oxygen, through a series of not particularly well understood steps, opens up channels, making it easier for succeeding oxygen molecules to make their way to other heme groups. Rather surprisingly, when one thinks about it in some detail, the oxygen does not oxidize the Fe^{2+} ion to Fe^{3+}. Evidently, the protein structure

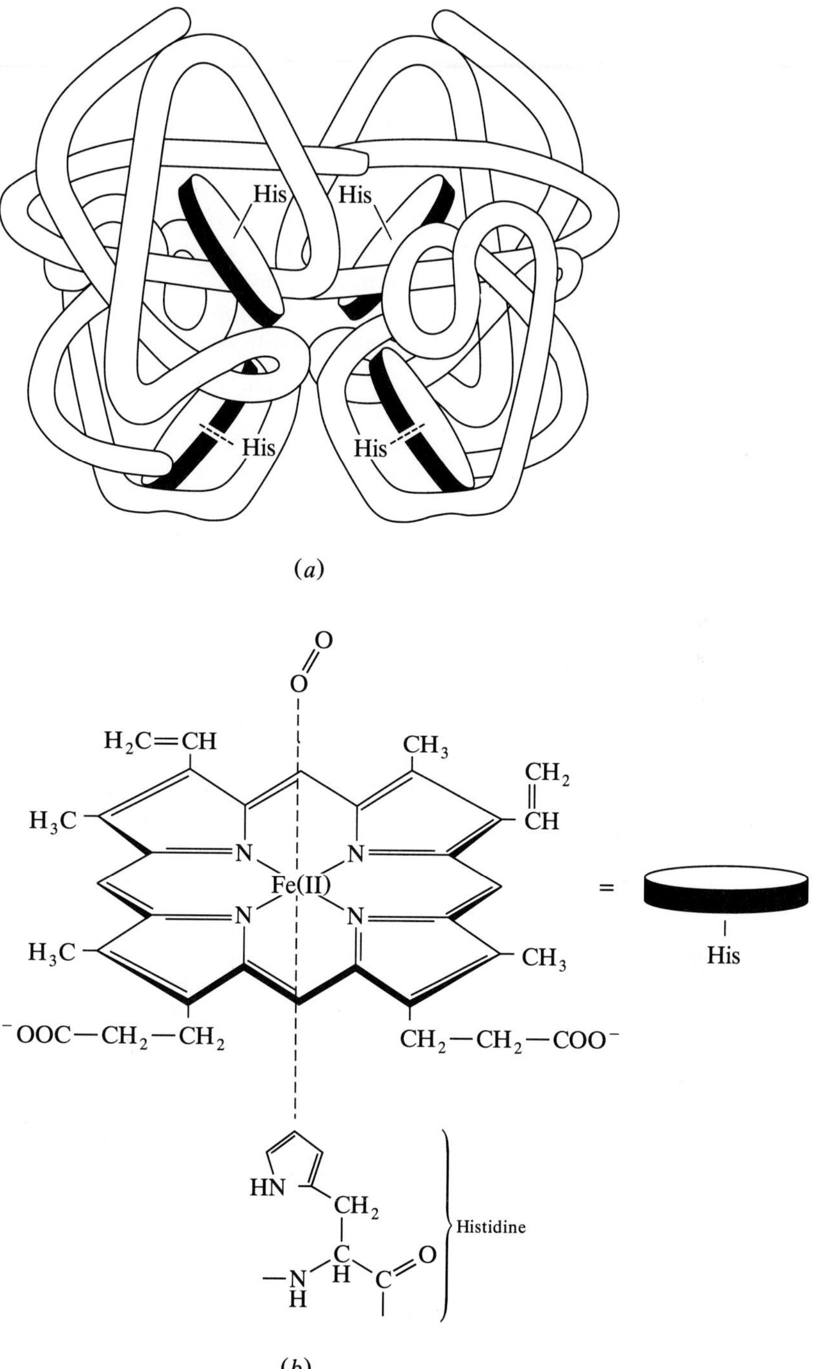

FIGURE 6.6
Hemoglobin and heme. (*a*) Hemoglobin is a tetrameric protein containing four heme groups, shown as planar disks. The tubular sections are protein structure, while histidine groups are identified as His. [Ref. 9.] (*b*) Heme is the iron-containing complex of hemoglobin that actually transports oxygen molecules in the process of respiration. Four nitrogen atoms in heterocyclic rings form a square planar array around the Fe(II), while a histidine group (also attached to the protein part of the hemoglobin) and an oxygen molecule are bound in the axial positions. [Ref. 10.]

provides protection against this oxidation. In the absence of this protective sheathing, the iron is oxidized and no longer binds oxygen. Instead, a water molecule occupies the sixth site, and the compound is a brown color. (This oxidation also occurs in hemoglobin outside the body of the organism and is responsible for the color of dried blood and old meat.)

When hemoglobin is exposed to both oxygen and carbon monoxide gas, the carbonyl complex, called *carboxyhemoglobin*, is formed preferentially. The resulting reaction is represented in Equation (6.20). The equilibrium constant for the formation of carboxyhemoglobin is about 250 times larger than that for oxyhemoglobin. The result is that breathing carbon monoxide deprives the cells of oxygen, and the victim is asphyxiated.

$$Hb + CO(g) \rightleftharpoons HbCO \qquad (6.20)$$

The cyanide ion, CN^-, isoelectronic with carbon monoxide, also binds to the iron cation in heme groups. Instead of primarily attacking hemoglobin, however, cyanide interferes with the function of a group of compounds called the *cytochromes*. Cytochromes of various types (labeled c, a, a_3, etc.) are involved in the reduction of molecular oxygen to water as part of the cellular respiration process. When the iron cations in the heme groups of cytochromes are bound preferentially by the cyanide ion instead of molecular oxygen, O_2, cellular respiration (the conversion of O_2 to water) essentially stops.

Cyanide poisoning can come about in a variety of ways. In the chemical laboratory and workplace, one must always be on guard when employing cyanide salts (such as potassium cyanide, KCN) or gaseous hydrogen cyanide, HCN. In addition to being used to extract gold and silver from their ores and to make various dyes (as previously mentioned), cyanides are also used in metal refining and plating and to extract silver from exposed photographic and x-ray films. Hydrogen cyanide gas is used as a fumigant to kill various pests found in houses, warehouses, and ship's holds.

Other sources of cyanides are less obvious. For example, the seeds of various fruits such as apples, cherries, peaches, apricots, and plums contain compounds called *cyanogenetic glycosides* that release cyanide on digestion. The fatal dose for a small child is only 5 to 25 seeds. They are only dangerous, however, if the seed capsules are broken. Laetrile, claimed to be a cancer cure, is reported to be made from apricot kernels and contains a cyanide-releasing substance. It has caused fatal cyanide poisoning. Finally, sodium pentacyanonitrosylferrate(III), $Na_2[Fe(CN)_5NO]$, sometimes called sodium nitroprusside, is useful for controlling high blood pressure, but it, too, releases cyanide ions, and an overdose may lead to cyanide poisoning.

Therapeutic Chelating Agents for Heavy Metals

What is to be done when a young child is brought into a hospital emergency room with lead poisoning caused by eating lead-containing paint chips or drinking orange juice from improperly treated earthenware pottery? These and other similar emergencies occur daily all over the world. Indeed, lead is sometimes called the "everywhere poison" because it is distributed so pervasively, some would say perversely, in and among the places we live and work. (Lead is found in paint, pottery, solder, crystal glass, water pipes, and the soil and vegetation of roadside areas, just to name a few places.) The average levels of lead found in modern adults is about 10 times that found in Egyptian mummies. (See Section 15.3 for more information on lead compounds and toxicology.) Children are particularly susceptible to the many effects of lead poisoning: fatigue, tremors, empairment of motor ability, diminished IQ, slower reaction times, poorer eye-to-hand coordination. Lead is also able to cross over the placental and blood-brain barriers and causes brain damage and birth defects. In severe cases, it produces coma and death.

The mode of action of lead and other heavy metals is to bind to various amino acids that make up proteins. Proteins, huge biological macromolecules that control a wide variety of functions in the body, often contain sulfur groups that can function as Lewis bases. These soft sulfur atoms are easily coordinated to various heavy-metal cations. The resulting stable, soft-soft complexes render the protein incapable of functioning normally.

When a child suffering from lead poisoning is brought in for treatment, the first priority is to reduce the concentration of lead in the bloodstream as quickly as possible. The first line of treatment is most often EDTA therapy. As we saw in Section 6.3, ethylenediaminetetraacetic acid is an excellent hexadentate chelating agent. It effectively complexes and sequesters almost all the +2 and +3 metals including lead. It is for this very reason, however, that EDTA cannot be used for long-term treatment; it is an indiscriminate chelator and, left to its own, would chelate vital metal ions including Mg^{2+}, Fe^{2+}, Cu^{2+}, and, most importantly, Ca^{2+}. (To help minimize calcium loss, EDTA is commonly administered as calcium salts such as CaH_2EDTA and $CaNa_2EDTA$.) Moreover, EDTA cannot be administered orally because lead in the gastrointestinal tract may be chelated and spread to the rest of the body. Instead, it is given intravenously or intramuscularly. EDTA therapy is also used for iron, manganese, zinc, copper, beryllium, and cobalt detoxification.

The second line of treatment against lead poisoning often involves a chelating agent called *penicillamine*. Its structure is shown in Figure 6.7*a*. As shown, it is a tridentate chelating agent but, more importantly, contains a sulfur atom in its $CH_3-\overset{|}{\underset{CH_3}{\underset{|}{C}}}-S-H$ group that subsequently loses a proton to yield an additional lone pair attached to the sulfur. The soft-soft interaction between a sulfur lone

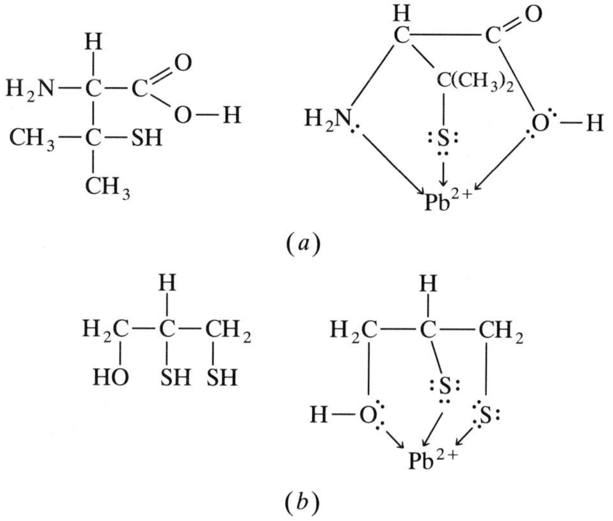

(*a*)

(*b*)

FIGURE 6.7
The free ligands and lead complexes of (*a*) penicillamine and (*b*) British anti-lewisite.

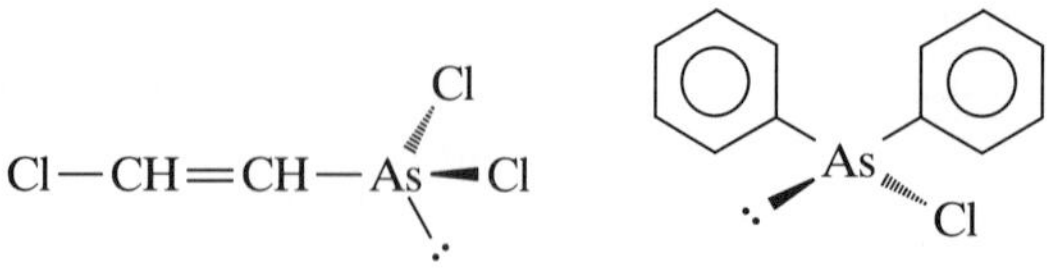

$Cl-CH{=}CH-AsCl_2$ (*a*)

$(C_6H_5)_2AsCl$ (*b*)

FIGURE 6.8
Two arsenical blistering agents: (*a*) lewisite (2-chloroethenylarsenic dichloride) and (*b*) blue cross (diphenylarsenic chloride).

pair and a Pb^{2+} cation makes penicillamine a more specific antidote for lead and other heavy, soft metals such as gold, bismuth, and mercury. It is also effective against the soft nickel tetracarbonyl. Penicillamine is water-soluble and can be administered orally.

Another lead antidote goes under the unique name of British anti-lewisite (BAL). Its structure is shown in Figure 6.7*b*. More properly, it should be called 2,3-dimercaptopropanol, and its two soft mercapto (— SH) groups make it particularly effective against soft metals. In fact, it was originally developed during World War I as an antidote against arsenical (arsenic-containing) blistering agents like *lewisite* (named after Winford Lee Lewis, an American chemist), and *blue cross*. The structures of these compounds are shown in Figure 6.8. (Some further information on arsenicals can be found in Section 16.1.) Given its origin, it is not surprising that BAL is also effective against arsenic poisoning even if a lethal dose has been taken. It also can be used for gold detoxification but should not be used against mercury or bismuth because the chelated metals may be spread throughout the body. BAL is not water-soluble and cannot be administered orally.

Both penicillamine and BAL have been used for treatment of Wilson's disease, a metabolic disorder characterized by the buildup of copper in the body. One striking symptom of this disease is *Kayser-Fleischer rings of the cornea*, copper- or bronze-colored rings that develop in the eyes of people afflicted with this malady. This symptomatic buildup of copper is treated with penicillamine. Since there is no known cure for Wilson's disease, these treatments must be continued throughout the lifetime of the patient. Triethylenetetramine (trien) has also been used against Wilson's disease.

Lead poisoning has a long and deadly history. Nicander, the Greek physician and poet, reported incidents of plumbism (lead poisoning) more than 20 centuries ago. The first incident in the modern era to bring mercury and its hazards to the public eye occurred at Minamata Bay, Japan, in 1953. Here many fishermen and their families were stricken with mercury poisoning when they ate fish and shellfish that contained high amounts of mercury, ultimately traced to the effluent of a nearby poly(vinylchloride) factory. Mercury in fish, particularly those such as tuna, marlin, and swordfish at the top of aquatic food chains, soon became a newsworthy issue. Another significant source of mercury is seed grain that is often treated with a mercurial fungicide. Unfortunately, this seed grain is not always used as it was intended. When it is fed directly to humans or to farm animals used for human consumption, it often causes tragic results. Other sources of mercury in the environment include mining, fuel combustion, the chlor-alkali process for making chlorine (see Section 18.2), paints, thermometers, explosives, electrical devices, and batteries.

The symptoms of mercury poisoning include tremors, dizziness, lack of coordination, thirst, vomiting, diarrhea, and ultimately, at high-enough exposures, coma, brain damage, and death. Mercury poisoning often results in irreversible damage, but under some circumstances it can be treated with penicillamine and

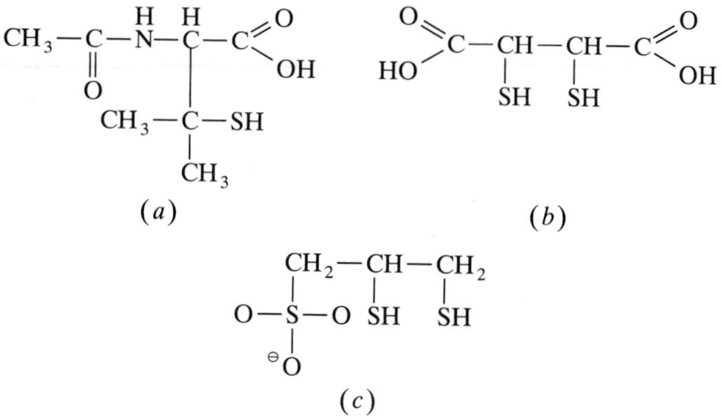

(*a*) (*b*) (*c*)

FIGURE 6.9
Three additional mono- and dimercapto chelating agents used in treating mercury and other heavy-metal poisoning: (*a*) *N*-acetyl-penicillamine (NAPA), (*b*) 2,3-dimercaptosuccinic acid (DMSA), and (*c*) 2,3-dimercaptopropane-1-sulfonate (DMPS).

other mono- and dimercapto chelating agents that contain soft Lewis bases. Three additional mono- and dimercapto chelating agents used in treating mercury and other heavy metals are shown in Figure 6.9.

Platinum Antitumor Agents

In 1964, Barnett Rosenberg and his associates were trying to assess the effect of electric fields on the rate of growth of *Escherichia coli* (or *E. coli*) cells. They found, quite by accident, that a small amount of the platinum from the electrodes they were using was transformed into *cis*-diamminedichloroplatinum(II), $PtCl_2(NH_3)_2$, and was responsible for a radical slowing down of the rate of cell division. By the early 1970s, direct clinical studies had shown that this coordination compound, which quickly became known as *cisplatin*, was able to completely stop the growth of various solid tumors, particularly those associated with testicular and ovarian cancers. Additional animal studies showed that it had activity against a wide variety of carcinomas. Like most antitumor drugs, cisplatin was found to have some serious side effects, the most notable being severe kidney damage.

These discoveries have led to a concerted effort to determine how cisplatin works against these tumors. Such information, it is hoped, will be useful in leading to modifications that will be more effective and/or have fewer and less severe side effects. Cisplatin and some related antitumor compounds are shown in Figure 6.10. It is significant to note that all of these are cis complexes. The corresponding trans compounds show no antitumor activity whatsoever.

The mechanism of these compounds against tumors seems to be related to their ability to form a chelated complex with some of the nitrogen bases of deoxyribonucleic acid (DNA), one of the molecular building blocks of life. Specifically, when DNA replicates, it splits apart into two strands each of which must be faithfully copied. (A single strand of DNA is composed of a series of nucleotides, each of which in turn is composed of three parts: a phosphate, a sugar, and a nitrogen base.) The relative arrangements of these are represented in Figure 6.11.

One mechanism that has been proposed for the action of cisplatin and its derivatives is also represented in Figure 6.11. As we noted in Section 5.5, square

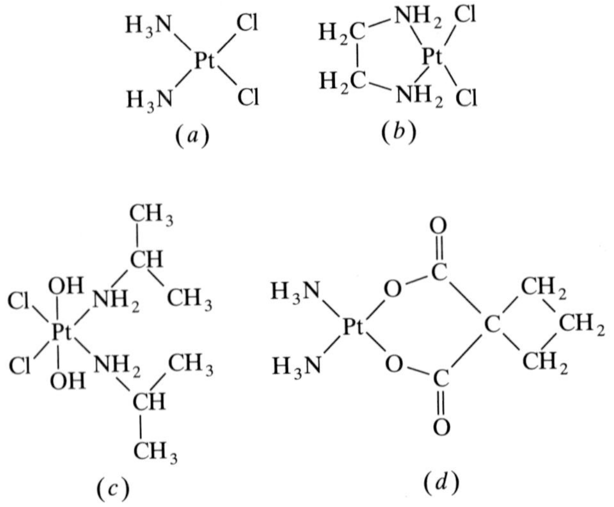

FIGURE 6.10
Cisplatin and some related antitumor compounds: (*a*) *cis*-diamminedichloroplatinum(II), cisplatin; (*b*) *cis*-dichloro(ethylenediamine)platinum(II); (*c*) *cis*-dichloro-*trans*-dihydroxobis(isopropylamine)platinum(IV); (*d*) *cis*-diammine-1,1-cyclobutanedicarboxylateplatinum(II).

planar platinum complexes have exceptionally labile chloride ligands. These chlorides may be replaced with nitrogen and oxygen atoms in the heterocyclic nitrogen bases of DNA. (See p. 131 for a definition of a heterocycle.) Specifically shown is the base guanine. Once the oxygen and one of the nitrogens of this nitrogen base form a complex with the platinum cation, the DNA can no longer be replicated faithfully. Instead, the formation of this complex leads to a mistake (or mutation) in the DNA replication and the destruction of the cancerous cell. For reasons that are not clear, cancerous cells are unable to repair such mistakes quickly enough but normal, healthy cells often are.

Current work centers on synthesizing cisplatin derivatives that show significant antitumor activity but have fewer side effects. The compounds shown in Figure 6.10*c* and *d* are but two of the possibilities currently under investigation.

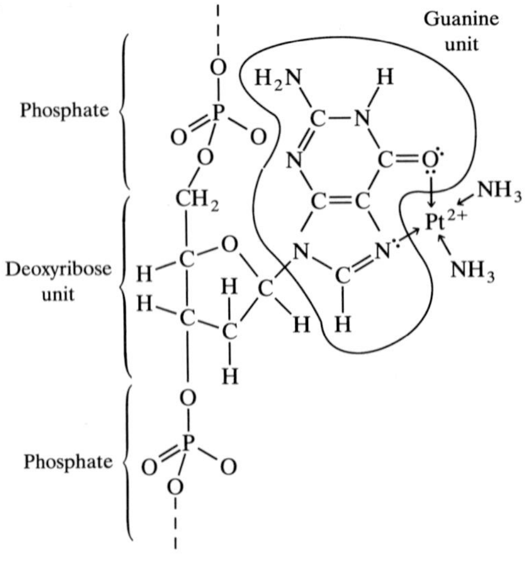

FIGURE 6.11
One possible mechanism for the action of cisplatin against tumors. Deoxyribonucleic acid (DNA) is a polymer made up of a series of nucleotides, each in turn composed of a phosphate, sugar, and nitrogen base. The nitrogen base shown here is guanine. Cisplatin may, after losing its two labile chloride ligands, complex the heterocyclic nitrogen base and make the DNA less capable of accurate replication.

SUMMARY

Coordination compounds have a wide variety of applications. Complexes of monodentate ligands are used in qualitative analysis, identification of copper(II) and iron(II), dyes [prussian (Turnbull's) blue], the separation of gold and silver from their ores, the purification of nickel, and as fixers in black-and-white photography.

Two ideas that aid in a fuller understanding of the applications of coordination compounds are the hard-soft–acid-base (HSAB) theory and the chelate effect. Metals (Lewis acids) and ligands (Lewis bases) can be categorized as hard or soft. *Hard* refers to substances with small, compact, difficult-to-distort electron clouds, while *soft* substances are large, diffuse, and easily distorted. Hard-hard and soft-soft interactions are often found to be more favorable than hard-soft. The chelate effect refers to the unusual stability of a coordination compound involving a chelating multidentate ligand. This effect can be explained in two different ways, both of which are related to changes in entropy upon complexation.

Complexometric analytical methods are based on an understanding of coordination chemistry. Gravimetric complexometric procedures include the production, isolation, and weighing of a coordination compound, often involving multidentate chelating agents. The analysis of nickel using dimethylglyoxime and iron with 8-quinolinol are two examples. The best example of a complexometric titration involves the use of ethylenediaminetetraacetic acid (EDTA) to titrate a wide variety of +2 and +3 cations in the presence of suitable complexometric indicators. The special stability of M-EDTA complexes leads to a number of applications of this hexadentate chelating effect.

Synthetic detergents are composed of a surfactant and a builder. The latter is often a chelating agent such as the tripolyphosphate ion. It sequesters hard water cations, rendering them incapable of interfering with the surfactant. Phosphates, however, are nutrients and have been implicated as a major cause of advanced eutrophication. A variety of phosphate replacements have been considered including the controversial nitrilotriacetic acid (NTA). Suspended sodium carbonate is currently a popular detergent builder, but phosphates are still used in a large number of products.

Bioinorganic applications of coordination compounds include hemoglobin, therapeutic chelating agents, and antitumor compounds. Hemoglobin is responsible for respiration or oxygen transport in most animals. The working part of hemoglobin is a heme group containing an Fe^{2+} cation coordinated to four nitrogen atoms and a histidine group. The sixth octahedral site is available to bind oxygen molecules. The cooperative interaction between oxygen molecules and the four heme groups of hemoglobin involve bent $Fe—O=O$ bonds but no oxidation of the iron(II) cations. Carbon monoxide and cyanide bind more strongly to the iron cations in the heme groups of hemoglobin and various cytochromes, respectively, than does oxygen, making these extremely dangerous reagents.

Therapeutic chelating agents are used as antidotes for heavy metal poisoning. Lead poisoning renders proteins incapable of performing their normal functions but can be treated with chelating agents such as EDTA, penicillamine, and British anti-lewisite (BAL). EDTA is a nonselective chelating agent and must be quickly replaced with penicillamine and/or BAL that contain soft sulfur-containing Lewis bases which are more selective for heavy metal cations. Wilson's disease and mercury poisoning are also treated with a variety of therapeutic chelating agents.

Almost 30 years ago, cisplatin, a square planar coordination compound of platinum(II), was found to be an effective antitumor agent. It evidently forms complexes with some of the nitrogen bases of DNA, producing mutations during the replication of cancer cells. Derivatives that have less severe side effects are currently in the development stages.

PROBLEMS

6.1. In terms of electronic configurations and crystal field theory, why is diamminesilver(I) colorless whereas tetraamminecopper(II) is colored? Be specific.

6.2. Speculate on the colors of aqueous solutions containing the $Au(CN)_2^-$ and $Ag(CN)_2^-$ ions. Briefly justify your answer.

***6.3.** The Liebig method for the determination of cyanide involves a titration against a reagent such as aqueous silver nitrate. Explain how this might work and include an equation as part of your answer.

6.4. Identify the substances being oxidized and reduced when cyanide complexes are used to separate gold from their ores.

6.5. Given what you know about the temperature dependence of the Mond process, would you suspect the formation of nickel tetracarbonyl from nickel and carbon monoxide to be endothermic or exothermic? Briefly explain your answer.

6.6. In a short paragraph, rationalize why ionic forces should be more important in hard-hard–acid-base interactions than in soft-soft.

6.7. In a short paragraph, rationalize why London dispersion forces should be more important in soft-soft–acid-base interactions than in hard-hard.

6.8. Briefly rationalize why *S*-thiocyanato complexes are found with metals such as mercury but *N*-thiocyanato are found with cobalt.

6.9. The overall equilibrium constant for the reaction of hexaaquoiron(II) with 2 moles of acetate to form diacetatotetraaquoiron(II) is 120, whereas that for the reaction with 1 mole of malonate, $^-OOCCH_2COO^-$, is 630. Write equations representing these reactions and briefly rationalize the differences in the magnitudes of the equilibrium constants.

6.10. The overall equilibrium constant for the reaction of hexaaquocobalt(II) with 2 moles of acetate to form diacetatotetraaquocobalt(II) is 80, whereas that for the reaction with 1 mole of oxalate, $^-OOCCOO^-$, is on the order of 5.0×10^4. Write equations representing these reactions and briefly rationalize the differences in the magnitudes of the equilibrium constants.

6.11. Briefly rationalize the fact that complexes of acetone, $CH_3\underset{\underset{O}{\|}}{C}CH_3$, are not particularly stable but those of acetylacetonate are.

***6.12.** The equilibrium constant for the reaction of hexaaquocobalt(II) with 2 moles of malonate, $^-OOCCH_2COO^-$, is approximately 5.0×10^3. Use this and the information given in Problem 6.10 to decide whether malonate or oxalate complexes of Co^{2+} are more stable. Draw the possible structures of the resulting chelate complexes. What does your answer lead you to conclude about the relative stability of five- and six-membered chelate rings in this case?

6.13. Using the information given in this chapter, determine an equilibrium constant for the following reaction:

$$[Ni(NH_3)_6]^{2+}(aq) + 3en(aq) \longrightarrow [Ni(en)_3]^{2+}(aq) + 6NH_3(aq)$$

Is this reaction thermodynamically favorable? Briefly rationalize your answer.

6.14. Using the data given in Problem 6.10, determine an equilibrium constant for the following reaction:

$$[Co(CH_3COO)_2(H_2O)_4] + C_2O_4^{2-} \longrightarrow [Co(H_2O)_4(C_2O_4)] + 2CH_3COO^-$$

Is this reaction thermodynamically favorable? Briefly rationalize your answer.

6.15. A 0.3456-g sample of a nickel ore yields 0.7815 g of $Ni(dmgH)_2$ in a gravimetric determination. Determine the percentage of nickel in the sample.

6.16. An aluminum ore contains approximately 24% aluminum. About what mass, in grams, of ore should be treated with 8-quinolinol to produce about 0.50 g of precipitate?

6.17. A 0.2000-g sample containing calcium is titrated against a 0.04672 *M* EDTA solution. If 23.94 mL of the solution is required to reach an endpoint using a complexometric indicator, determine the percentage of calcium in the sample.

6.18. An aqueous solution containing Pb^{2+} and Cd^{2+} can be titrated using EDTA. If an excess of cyanide is added to the solution, the cadmium is said to be masked by the formation of a cyanide complex. Suppose that 20.00 mL of a solution containing only lead and cadmium is first titrated against 45.94 mL of a 0.02000 *M* solution of EDTA. After adding an excess of sodium cyanide, only 34.87 mL of the same EDTA solution produces an endpoint. Calculate the molarities of both the lead and the cadmium in the original solution.

6.19. A 0.2005-g sample of copper ore is dissolved and diluted to the mark in a 100-mL volumetric flask. Then 10.00 mL of the solution is removed and titrated against a 0.01000 *M* solution of trien. If 22.75 mL of the trien solution is required, calculate the percentage of copper in the original sample.

6.20. Suppose that 43.28 mL of a 0.1000 *M* solution of 1,10-phenanthroline is required to titrate a $Fe^{3+}(aq)$ solution. How many grams of iron were present in the solution?

6.21. Verify that the limit of 8.7% by weight of phosphorus in a detergent corresponds to 34% sodium tripolyphosphate, $Na_5P_3O_{10}$.

6.22. Sodium acid pyrophosphate, $Na_4[O_3POPO_3]$, acts as a sequestering agent to help keep potatoes from darkening after heating. The darkening is due to the formation of an iron complex. Draw a diagram showing the nature of this complex. Assume an Fe-pyrophosphate ratio of 1 : 1.

6.23. Explain to an economics major who paid reasonably close attention in his or her high school chemistry class why phosphates have been an important component of synthetic detergents. Extra points may be awarded for presenting evidence that this was done in person.

6.24. The pentasodium salt of diethylenetriaminepentaacetic acid can be used for chelating water hardness ions and a variety of heavy metal ions. Draw the structural formula of this multidentate ligand. Would you suspect that the full denticity of this ligand would be employed in its interaction with the above metals?

6.25. Explain to a biology major who struggled through his or her introductory college chemistry class why human blood is of two shades of red, a dark red when returning to the heart and a bright red when leaving the lungs. Extra points may be awarded for presenting evidence that this was done in person.

6.26. Sodium nitroprusside, $Na_2[Fe(CN)_5NO]$, is used to control high blood pressure. It liberates CN^- after its administration. Explain to a physician who was forgotten all but his or her most elementary chemistry why this might be a problem?

*** 6.27.** When an oxygen molecule binds to hemoglobin, there is substantial evidence that the iron(II) changes from a low-spin d^6 state to a high-spin state. How might this affect the size of the iron cation and its ability to fit in the square planar site in heme? There is evidence that this size change is what initiates the cooperative interaction among the four heme sites in the protein.

6.28. Diethylenetriaminepentaacetate has been administered orally to mice and found to significantly increase the urinary excretion of manganese. Draw a structure for this chelating agent and discuss its mode of action.

6.29. Lead is more often found in nature as its sulfide than its oxide. Provide a brief rationale for this fact.

6.30. The word *mercaptan*, from which the phrase "mercapto group" (—SH) is derived, comes from the Latin words for "mercury capture." Mercury is often found associated with sulfur-containing groups. Provide a brief rationale for this fact.

6.31. Of EDTA, penicillamine, and BAL, which, if any, would be optically active? Briefly discuss your answer.

6.32. Histidine and cysteine are used for cobalt detoxification. Based on the structures given below, rationalize this fact.

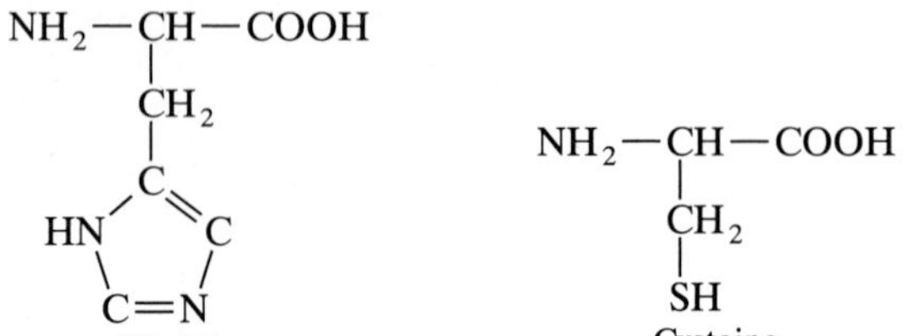

$NH_2—CH—COOH$
$|$
CH_2
$|$
C
$HN \quad C$
$C=N$

Histidine

$NH_2—CH—COOH$
$|$
CH_2
$|$
SH

Cysteine

6.33. Of histidine and cysteine (see Problem 6.32 for structures), which would be a better antidote against heavy metals?

6.34. The structures of serine and cysteine are shown below. Which of these amino acids would be the better antidote against heavy metals? Briefly rationalize your answer.

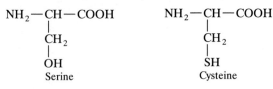

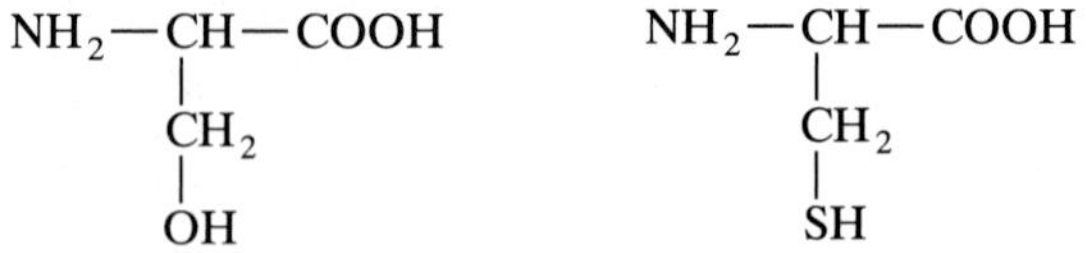

$NH_2—CH—COOH$
$|$
CH_2
$|$
OH

Serine

$NH_2—CH—COOH$
$|$
CH_2
$|$
SH

Cysteine

***6.35.** The structures of methionine and cysteine are shown below. Which of these amino acids would be the better antidote against heavy metals?

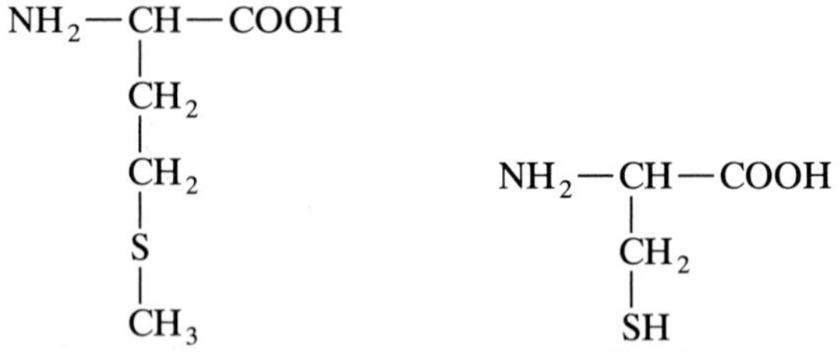

$NH_2—CH—COOH$
$|$
CH_2
$|$
CH_2
$|$
S
$|$
CH_3

Methionine

$NH_2—CH—COOH$
$|$
CH_2
$|$
SH

Cysteine

6.36. Explain to an English major who paid reasonably close attention in his or her AP chemistry high school class why cisplatin works against tumors but the corresponding trans compound does not. Extra points may be awarded for presenting evidence that this was done in person.

PART B

SOLID STATE CHEMISTRY

In this self-standing section, solid state chemistry is introduced over the course of two chapters. The titles of these chapters are listed below.

CHAPTER 7

SOLID STATE STRUCTURES

As outlined in Chapter 1, the realm of inorganic chemistry was considerably expanded in the early twentieth century when it was discovered that x-ray diffraction could provide detailed information on the structures of solids. At the same time that the intricacies of the crystalline state gradually came to light, the theoretical basis describing the forces which hold these structures together was conceived, tested, and refined. These next two chapters explore some of what is now known about both the structures and the energetics of the inorganic solid state.

Solids are usually composed of atoms, molecules, or ions arranged in a rigid, repeating geometrical pattern of particles known as a *crystal lattice*. Before looking at a variety of crystal lattices, we start this chapter with a survey of crystal types based on the nature of the forces among particles.

7.1 TYPES OF CRYSTALS

Crystals are usually categorized by the type of interactions operating among the atoms, molecules, or ions of the substance. These interactions include ionic, metallic, and covalent bonds as well as intermolecular forces such as hydrogen bonds, dipole-dipole forces, and van der Waals forces.

Ionic Crystals

As you know from general chemistry, when two elements, one a metal with a low ionization energy and the other a nonmetal with a high exothermic electron affinity, are combined, electrons are transferred to produce cations and anions. These ions are held together by nondirectional, electrostatic forces known as *ionic bonds*. Figure 7.1 illustrates a hypothetical view of the formation of sodium chloride, NaCl, perhaps the most common example of an ionic crystal, from its constituent atoms. Picking any sodium cation at random, note that a chloride anion could approach from any direction; that is, there is no particular direction in which the cation-anion interaction will be stronger. Stated another way, the anion could

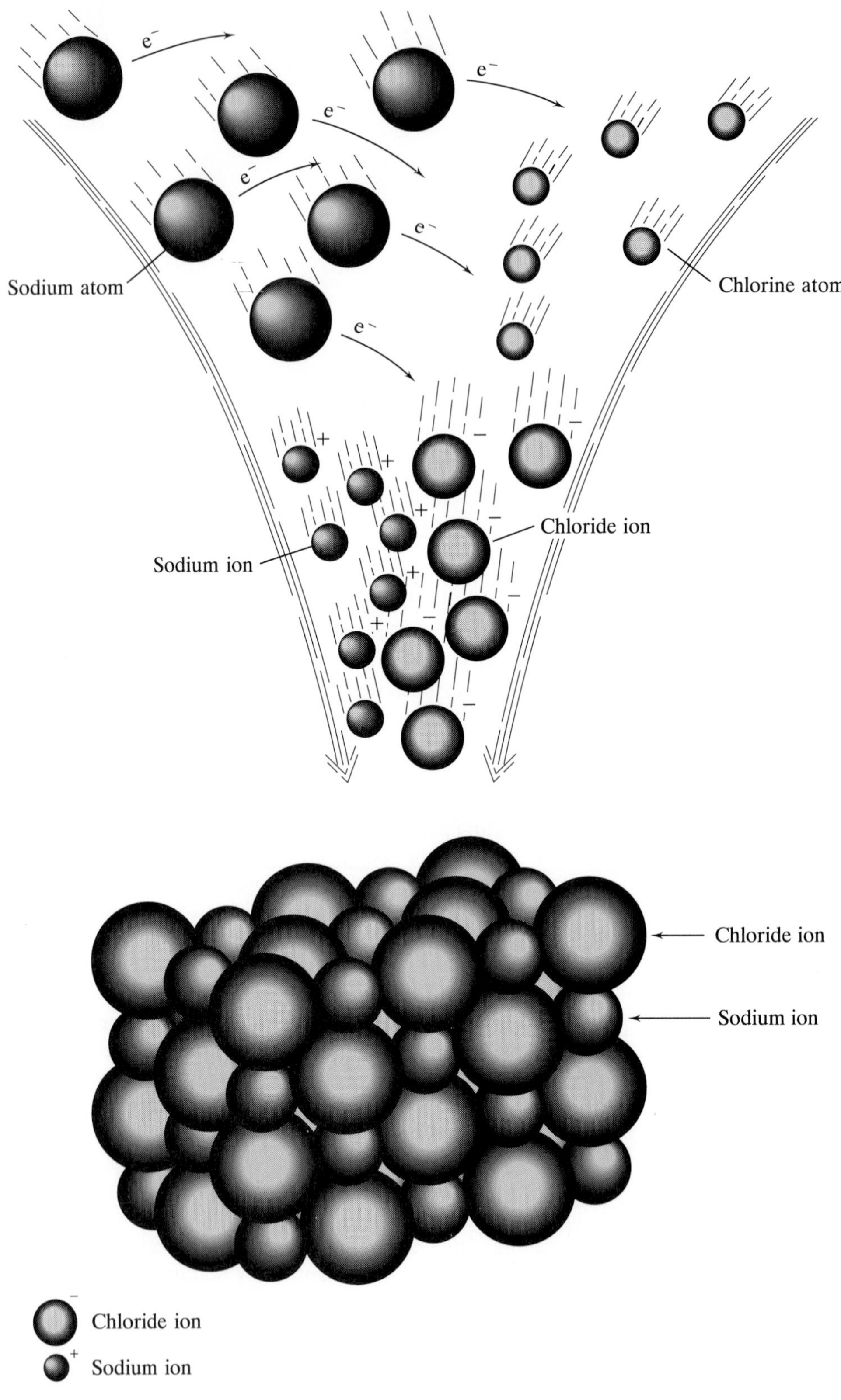

FIGURE 7.1
In a hypothetical view of the formation of sodium chloride, the constituent elements are combined, electrons are transferred, and ionic bonds among the sodium and chloride ions are formed. [Ref. 11.]

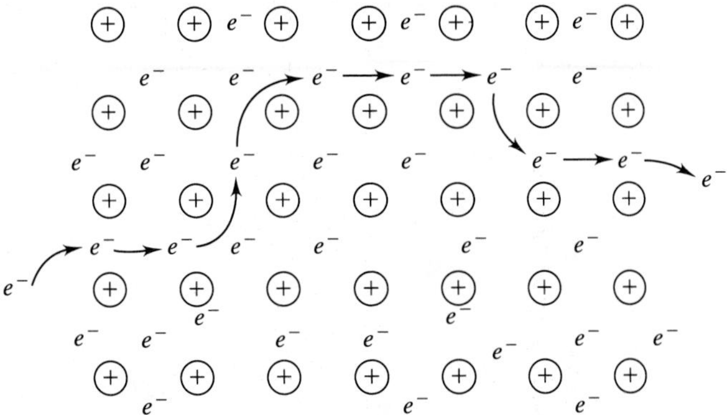

FIGURE 7.2
Electrical conductivity in a metallic lattice. A lattice of cations is held together by a "sea" of electrons. An electron enters from the left and bumps other electrons down the lattice until an electron emerges from the right side. [Adapted from Ref. 12.]

come toward the cation from any direction or point in space and still experience the same electrostatic force. Thus we say that ionic bonding is *nondirectional*. As we will see later, the arrangement of the anions around the cation is not determined by a preferred direction necessary for maximization of the ionic interaction but rather by the relative sizes and charges of the cations and anions. Some examples of compounds which form ionic crystals are cesium chloride (CsCl), calcium fluoride (CaF_2), potassium nitrate (KNO_3), and ammonium chloride (NH_4Cl).

Metallic Crystals

Metals, those elements generally from the left side of the periodic table, form crystals in which each atom has been ionized to form a cation (of charge dependent on its electronic configuration) and a corresponding number of electrons. The cations are pictured to form a crystal lattice which is held together by a "sea of electrons"—sometimes called a *Fermi sea*. The electrons of the sea are no longer associated with any particular cation but are free to wander about the lattice of cations. Given the above description, we can define a *metallic crystal* as a lattice of cations held together by a sea of free electrons. A general two-dimensional representation of such a crystal is shown in Figure 7.2. Note that the sea analogy allows us to picture electrons flowing from one place in the lattice to another. That is, suppose we shape the metal (copper is a good example) into a wire. If we put electrons in one side of the wire, electrons will be bumped along the lattice of cations until some electrons will be pushed out the other end. The result of this electron flow is electric conductivity, one of the most characteristic properties of metals.

Covalent Crystals

A *covalent crystal* is composed of atoms or groups of atoms arranged into a crystal lattice which is held together by an interlocking network of covalent bonds.

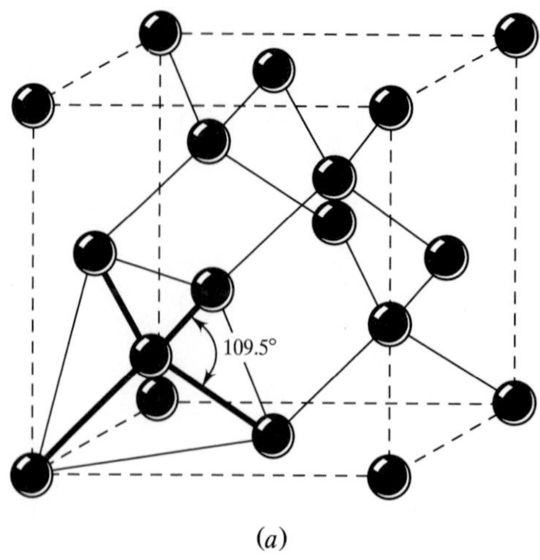

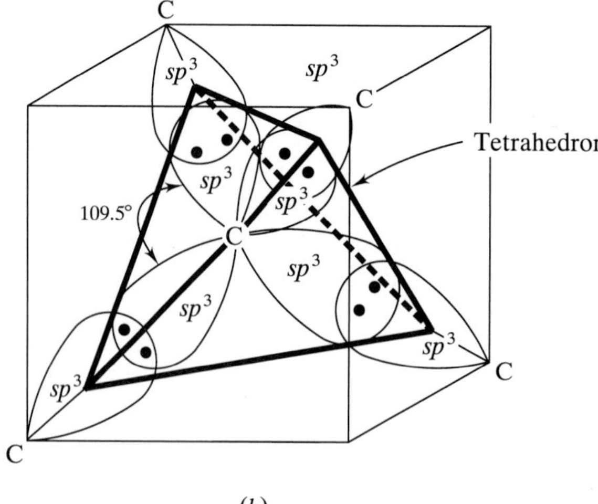

FIGURE 7.3
The diamond structure as an example of a covalent crystal in which directional covalent bonds are formed among the individual atoms. (*a*) A segment of the lattice showing each carbon atom surrounded by a tetrahedron of four other carbons. [Ref. 13.] (*b*) Each carbon is sp^3-hybridized, and the overlap of these hybrid orbitals is maximized by maintaining C—C—C bond angles of 109.5°. [Adapted from Ref. 12.]

Covalent bonds (the result of the sharing of one or more pairs of electrons in a region of orbital overlap between two or perhaps more atoms) are directional interactions as opposed to ionic and metallic bonds which are nondirectional. A good example of a covalent crystal is diamond, shown in Figure 7.3. Note that each carbon atom is best thought of as being sp^3-hybridized and that to maximize the overlap of these hybrid orbitals, a C—C—C bond angle of 109.5° is necessary. These interactions are therefore directional in nature. Other compounds which form covalent crystals are silicon dioxide (SiO_2), graphite, elemental silicon (Si), and boron nitride (BN).

Atomic-Molecular Crystals

When the atoms or molecules of a crystal lattice are held together by relatively weak intermolecular forces such as (in increasing order of strength) London

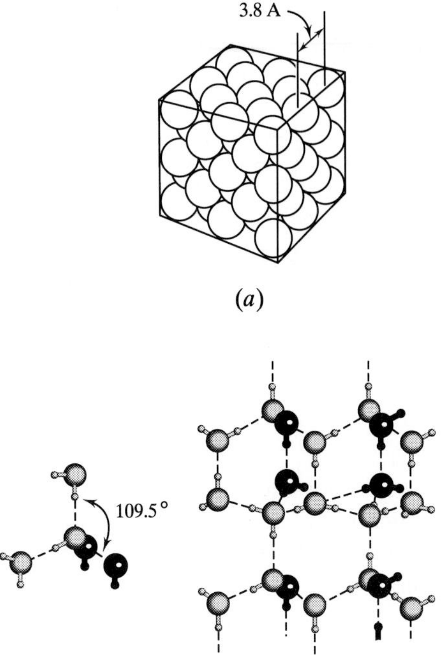

FIGURE 7.4
Two examples of atomic-molecular crystals. (*a*) Argon atoms held together by nondirectional London dispersion intermolecular forces. [Ref. 12.] (*b*) Water molecules held together by directional hydrogen bonds. Due to the structure of an individual water molecule, the most effective H—O --- H bond angle is 109.5°. [Ref. 14.]

dispersion forces, dipole-dipole forces, or hydrogen bonds, an *atomic or molecular crystal* results. You may want to go back and review these forces from your general chemistry course. Hydrogen bonds, particularly those in water, are covered in Chapter 11. Intermolecular forces can be either nondirectional, as in the case of crystals of argon (Figure 7.4*a*), or directional, as in the case of ice (Figure 7.4*b*). In the latter case, the H—O---H angle is 109.5°, an angle determined by the geometry of individual water molecules. Other compounds which form atomic-molecular crystals are dry ice (CO_2) and the solid forms of methane (CH_4), hydrogen chloride (HCl), and phosphorus (P_4).

7.2 A-TYPE CRYSTAL LATTICES

Space Lattices and Unit Cells

Independent of the nature of the forces involved, a crystal lattice can be described using the following concepts. A *space lattice* is the pattern of points which describes the arrangements of ions, atoms, or molecules in a crystal lattice. A *unit cell* is the smallest, convenient microscopic fraction of a space lattice which (1) when moved a distance equal to its own dimensions in various directions generates the entire space lattice and (2) reflects as closely as possible the geometrical shape or symmetry of the macroscopic crystal. Before looking at a variety of three-dimensional examples of space lattices and unit cells, we consider the simpler two-dimensional portion of a space lattice shown in Figure 7.5.

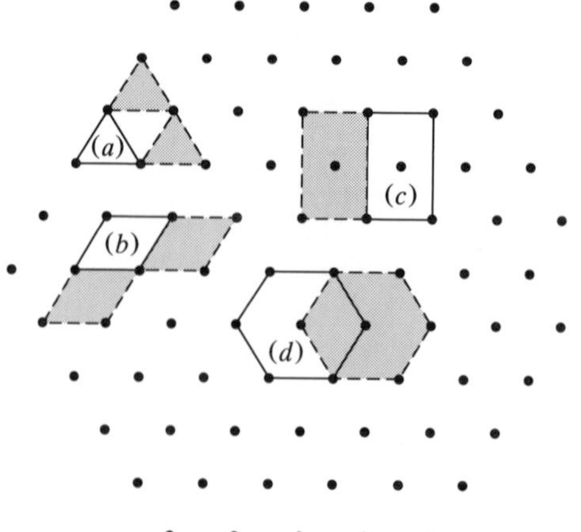

FIGURE 7.5
A portion of a hexagonal, two-dimensional space lattice showing four possible unit cells: (*a*) an equilateral triangle which does not generate the entire lattice; (*b*) a rhombus and (*c*) a rectangle which do generate the lattice but do not reflect the overall symmetry of the macroscopic crystal; and (*d*) a hexagon which is the unit cell of choice.

Note that we do not know what the points in this lattice represent or the type of interaction which exists among them. Furthermore, we must assume that these points represent a crystal which contains, for the sake of argument, 1 mole or Avogadro's number (6.02×10^{23}) of atoms, molecules, or ions. In other words, these points stretch out in two dimensions as far as the mind can imagine. Our task is to describe this entire space lattice in terms of a single unit cell. What geometric shapes should we consider for this cell? We start with the equilateral triangle marked (*a*) in the figure. Is it a unit cell? When moved a distance equal to its own dimensions in a variety of directions, it generates other equilateral triangles which are shaded in the figure. Can we generate the entire space lattice this way? No, we cannot, because half of the space lattice (that in the upside-down equilateral triangles) is unaccounted for. This triangle is not a unit cell.

Consider the rhombus marked (*b*). Would it be a unit cell? It would be a better candidate than the triangle because all the space can be accounted for when it is moved. How about rectangle (*c*)? Again, it does generate the entire lattice. Note, however, that neither the rhombus nor the rectangle reflects the overall shape or symmetry of the space lattice, or at least that section of lattice represented in the figure. Finally, look at the hexagon marked (*d*). Note that in addition to generating the entire space lattice, its shape is the most reflective of the overall shape of the lattice. The hexagon, therefore, is the best unit cell for this space lattice. (Note that it does not matter that the hexagonal unit cells overlap each other.)

A-Type Lattices

A-type lattices are those in which all the atoms, ions, or molecules of the crystal are the same size and type, "type A." We will explore many of the various lattices that are possible in this, the simplest of the lattice types. First, however, we should briefly consider an important simplification. It turns out to be convenient to represent the particles of a crystal as hard spheres. Of course, we know that atoms, ions, and molecules are not hard spheres; atoms and simple ions are electron clouds with a nucleus in the center. Molecules are overlapping electron clouds which hold a given number of nuclei together. Nevertheless, the model employing spheres to represent such entities works very well indeed, although later we will need to point out where this model starts to break down. With the hard-sphere simplification in mind, we can proceed to look at the various possibilities for

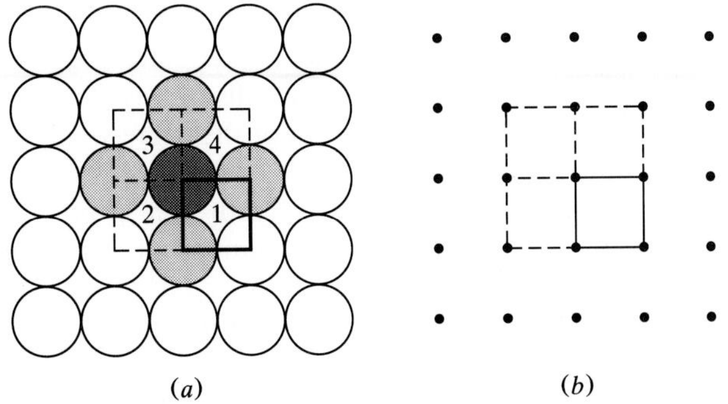

FIGURE 7.6
One layer of a simple cubic lattice. (*a*) A layer of spheres in which a given sphere (shaded) has four nearest neighbors (lightly shaded spheres) and a coordination number of 4. The square unit cell (solid lines connecting the centers of four spheres) contains one-fourth of each of the four spheres and therefore a total of 1 ($= 4 \times \frac{1}{4}$) sphere. The given sphere is in four different unit cells (numbered). (*b*) A portion of the space lattice for the layer and a corresponding unit cell. Three other unit cells are shown with dashed lines.

A-type lattices. We will start with the easiest case to visualize, the simple cubic lattice, and work our way up in complexity.

A visualization of the simple cubic lattice starts with the layer of hard spheres shown in Figure 7.6*a*. Notice that all the spheres touch at right angles to each other and that each sphere has four *nearest neighbors*, atoms which touch a given sphere. We define the *coordination number* as the number of nearest neighbors; it follows that the coordination number is 4 in this case. Figure 7.6*b* shows a portion of the two-dimensional space lattice corresponding to the same arrangement. Both representations illustrate a unit cell (a square) for this layer. Note that the unit cell connects the centers of four spheres and that only one-fourth of each sphere is actually in the unit cell. It follows that the number of complete spheres in the unit cell is 1 ($= 4 \times \frac{1}{4}$). Another way to visualize this situation is to note that the shaded sphere is in four different unit cells (three others are shown with dashed lines) and therefore can only be one-fourth in the original unit cell.

If we stack another layer of spheres directly on top of the first, we have a *simple cubic lattice*. Figure 7.7*a* shows a portion of this lattice which illustrates that

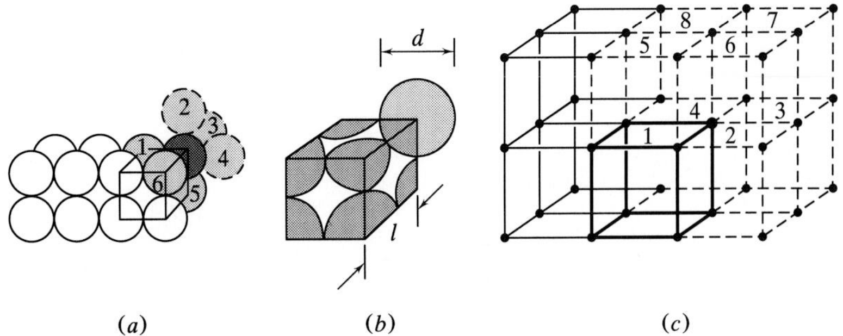

FIGURE 7.7
The simple cubic lattice. (*a*) The unit cell of the lattice is shown in solid lines. The shaded sphere has a coordination number of 6. (*b*) A close-up showing that the shaded sphere is one-eighth in the unit cell. (*c*) A portion of the space lattice showing that the highlighted point is in eight different unit cells (numbered on top faces of the cubes). The original unit cell is shown in bold solid lines; the remaining seven unit cells are shown in dashed lines.

the coordination number of each sphere is now 6. The unit cell is now a cube connecting the centers of the appropriate eight spheres. Figure 7.7*b* shows a close-up of a given sphere which illustrates that it is only one-eighth in the unit cell. Figure 7.7*c* shows a portion of the corresponding space lattice. Notice that the highlighted point is in eight unit cells (seven others are shown with dashed lines) and therefore can only be one-eighth in a given unit cell. It follows from both of the above arguments that the number of complete spheres in the unit cell is $1\ (= 8 \times \frac{1}{8})$.

Three other aspects of the simple cubic lattice are appropriate to discuss at this point: the dimensions of the unit cell, the fraction of space occupied by the spheres, and the density. First, note in Figure 7.7*b* that the spheres touch each other along the unit cell edge. If we let l be the cell edge and d be the diameter of a sphere, it follows that $l = d$. Second, we can calculate the fraction of the unit cell actually occupied by the spheres. Since the unit cell reflects the entire lattice, this fraction will also indicate what percentage of the entire lattice is occupied by the spheres and will be a measure of the effectiveness with which the spheres are packed. Since there is only one sphere per unit cell and the volume of a cubic unit cell is given by l^3 (and therefore d^3), it follows that

$$\begin{array}{l}\text{Fraction of space}\\ \text{occupied by spheres}\end{array} = \frac{\left[(4\pi/3)(d/2)^3\right] \times 1}{d^3} = 0.52 \qquad (7.1)$$

Note that the d^3 terms cancel and that the fraction of space occupied is 0.52. That is, about half of the space is occupied by the spheres.

TABLE 7.1
A-type lattices

Name	Coord. no.	No. of spheres per unit cell	Spheres touching along	Fraction of space occupied by spheres	Density expression
Simple cubic	6	$8 \times \frac{1}{8} = 1$	Cell edge $= l = d$	$\frac{\left[\frac{4}{3}\pi(d/2)^3\right] \times 1}{d^3}$ $= 0.52$	$\frac{1\ \text{atom}(\text{AW}/6.02 \times 10^{23})(\text{g/atom})}{d^3\ \text{cm}^3}$
Body-centered cubic (bcc)	8	$1 + (8 \times \frac{1}{8})$ $= 2$	Body diagonal $= 2d = l\sqrt{3}$	$\frac{\left[\frac{4}{3}\pi(d/2)^3\right] \times 2}{(2d/\sqrt{3})^3}$ $= 0.68$	$\frac{2\ \text{atoms}(\text{AW}/6.02 \times 10^{23})(\text{g/atom})}{(2d/\sqrt{3})^3\ \text{cm}^3}$
Cubic close-packed (ccp) or Face-centered cubic (fcc)	12	$6(\frac{1}{2}) + 8(\frac{1}{8})$ $= 4$	Face diagonal $= 2d = l\sqrt{2}$	$\frac{\left[\frac{4}{3}\pi(d/2)^3\right] \times 4}{(d\sqrt{2})^3}$ $= 0.74$	$\frac{4\ \text{atoms}(\text{AW}/6.02 \times 10^{23})(\text{g/atom})}{(d\sqrt{2})^3\ \text{cm}^3}$
Hexagonal close-packed (hcp)	12	$2(\frac{1}{2}) + 3$ $+ 12(\frac{1}{6})$ $= 6$	See problem set solutions for Problem 7.11	$\frac{\left[\frac{4}{3}\pi(d/2)^3\right] \times 6}{\left[24\sqrt{2}\,(d/2)^3\right]}$ $= 0.74$	$\frac{6\ \text{atoms}(\text{AW}/6.02 \times 10^{23})(\text{g/atom})}{24\sqrt{2}\,(d/2)^3\ \text{cm}^3}$

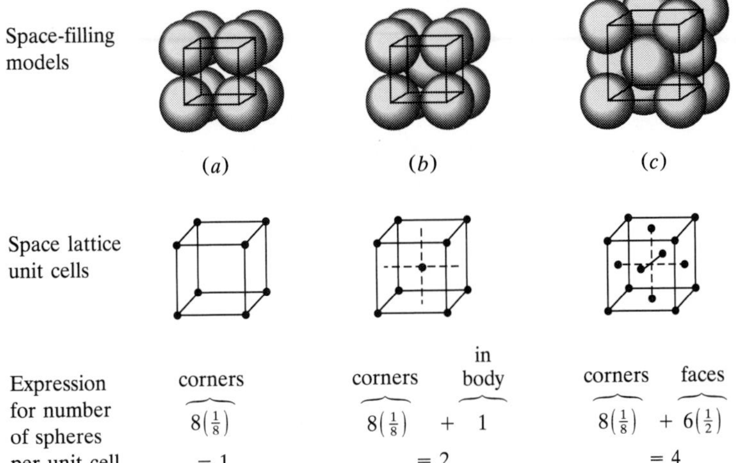

FIGURE 7.8
Unit cells of (*a*) simple, (*b*) body-centered, and (*c*) face-centered cubic lattices. The top drawings represent the actual "space-filling" unit cells. The bottom drawings show only the centers of each sphere in the space lattice. Note that the unit cell includes only the volume of the cube connecting the centers of the spheres at the corners of the cube. The number of spheres per unit cell is shown under each unit cell.

Third, we can calculate the density of such a configuration. Remember that each sphere represents an atom, ion, or molecule for which we can calculate a mass. If we assume that the sphere is an atom and that its mass, as commonly calculated in general chemistry, is atomic weight (AW) divided by Avogadro's number (in units of grams/mole divided by atoms/mole, which equals grams/atom), then the expression for the density in a simple cubic lattice is

$$\text{Density}\ (\text{g/cm}^3) = \frac{1\ \text{atom}(\text{AW}/6.02 \times 10^{23})(\text{g/atom})}{d^3\ \text{cm}^3} \tag{7.2}$$

The above discussion for the simple cubic lattice is summarized in Table 7.1.

A second A-type lattice is called *body-centered cubic* (bcc) and, as the name implies, differs from the simple cubic lattice in that a second sphere is placed in the center of the cubic cell. A unit cell is shown in Figure 7.8*b*. Whereas the eight spheres at the corners are only one-eighth in the unit cell, the center sphere is completely incorporated in the body of the cell and therefore has a coordination number of 8. The number of spheres per unit cell is accordingly 2 [$= 1 + 8(\frac{1}{8})$]. Since the corner spheres no longer touch each other (and therefore $l \neq d$), we must derive a new relationship between the cell edge and the sphere diameter. Note from Figure 7.9*a* that the spheres *do* touch each other along the body diagonal of the cube and that the body diagonal $= 2d$. The relationship between the body diagonal and the cell edge is shown in Figure 7.9*b*. Therefore, it follows that

$$l = 2d\sqrt{3} \tag{7.3}$$

With this result we can rationalize the expressions for the fraction of space occupied by the spheres and the density as given in Table 7.1. Note that the body-centered cubic lattice is a more efficient way to pack spheres (68 percent compared to 52 percent) than the simple cubic.

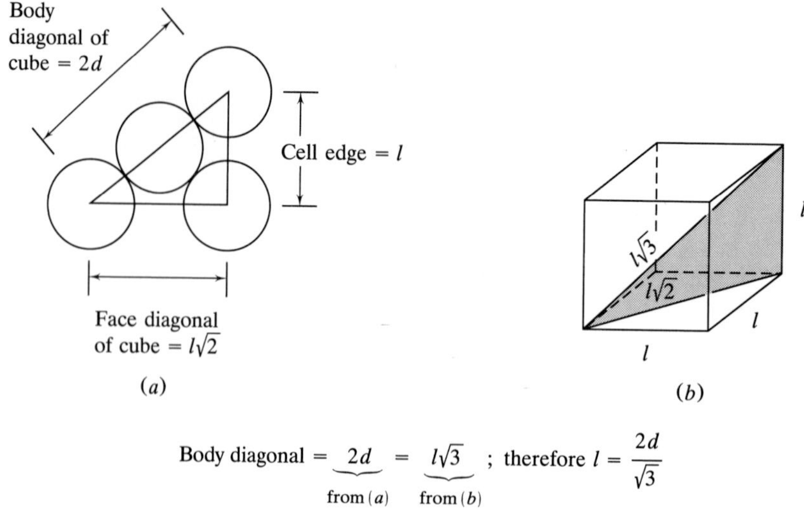

FIGURE 7.9
The relationship between sphere diameter and cell edge in a body-centered unit cell. (*a*) The spheres touch along the body diagonal but not along the face diagonal or the cell edge. (*b*) If the cell edge is l, the face diagonal is $l\sqrt{2}$ and the body diagonal is $l\sqrt{3}$. Solving for the cell edge yields $l = 2d/\sqrt{3}$.

You may already have noticed that Figure 7.6*a* is not the best (the most efficient) way to pack spheres in a layer. To increase the efficiency of packing we can fit a given sphere in the crevice or depression between two others as shown in Figure 7.10*a*. Consider this to be layer A of the cubic close-packed (ccp) and hexagonal close-packed (hcp) structures. The second layer (layer B) is laid down such that a given sphere fits in the triangular depression left by three spheres in layer A. These two layers together are shown in Figure 7.10*b*. Note that there are now two types of triangular depressions in layer B: those marked *a* which have no spheres below them and those marked *b* which have layer A spheres directly below them.

If the third layer is placed in the *a* depressions, they create a new layer C. The resulting ABCABC packing scheme (shown in Figure 7.10*c*) is known as the *cubic close-packed* (*ccp*) *structure* and is our third A-type lattice. Although it is not apparent from the view shown in Figure 7.10*c*, the ccp structure, as its name indicates, has a cubic unit cell. The relationship between the ABC layers and the unit cell is more clearly shown in Figures 7.11 and 7.12. Note that when the three layers of Figure 7.11*a* are reoriented to the views shown in Figure 7.11*b* and *c*, the *face-centered unit cell* (a cubic unit cell with additional points in the center of each of the six faces of the cube) becomes apparent. Figure 7.13 shows clearly that the coordination number of a given sphere is 12. It is most important that you convince yourself that a cubic close-packed structure has a face-centered cubic (fcc) unit cell. A good summary is the "equation"

$$\text{ccp} = \text{ABCABC} = \text{fcc}$$

The face-centered cubic unit cell, which is also shown in Figure 7.8 along with the simple and body-centered cubic cells, contains a total of four spheres. As in the simple cubic, the eight corner spheres are one-eighth in the cell. The six spheres in the cube faces are one-half in a given unit cell. Accordingly, as shown in Figure 7.13*a*, there are 4 $[= 8(\frac{1}{8}) + 6(\frac{1}{2})]$ spheres per cell. Also note the spheres are now in contact along a face diagonal which leads, as shown in Figure 7.13*b*, to

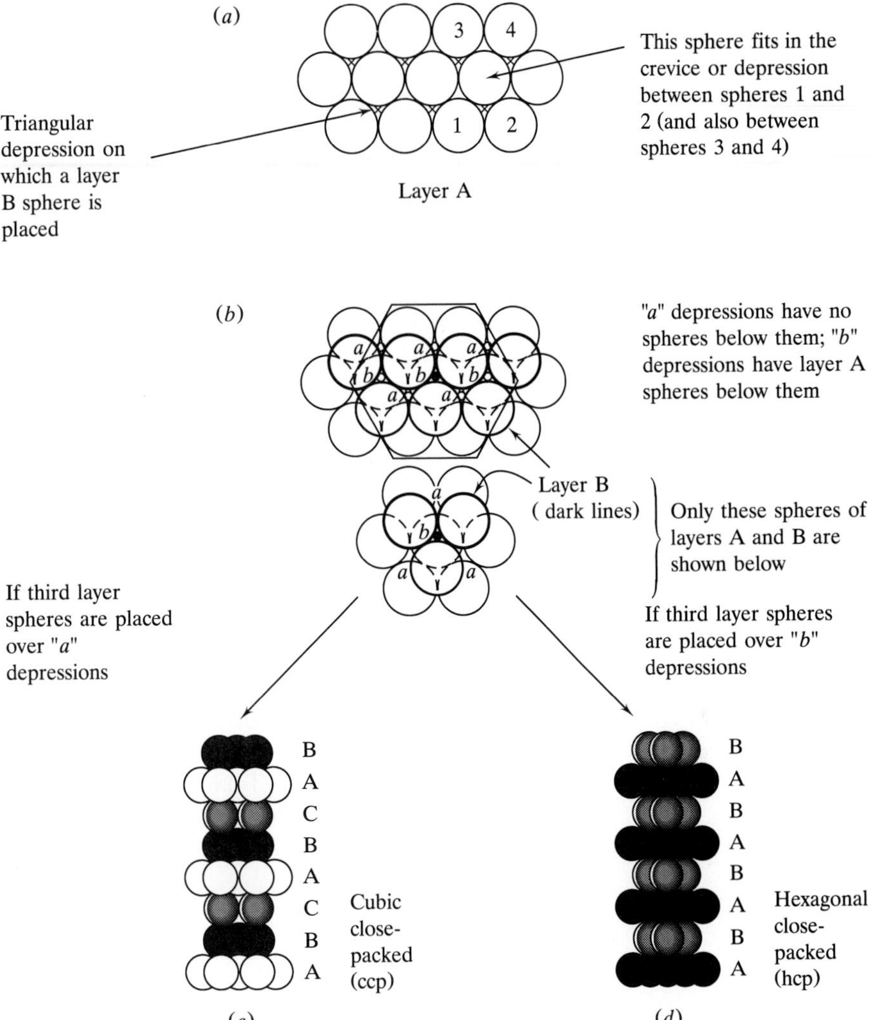

FIGURE 7.10
Cubic and hexagonal close-packed structures. (*a*) Layer A of both structures. (*b*) Layers A and B of both structures showing two types of depressions. (*c*) ABCABC layer scheme produces the cubic close-packed structure. (*d*) ABABAB layer scheme produces the hexagonal close-packed structure.

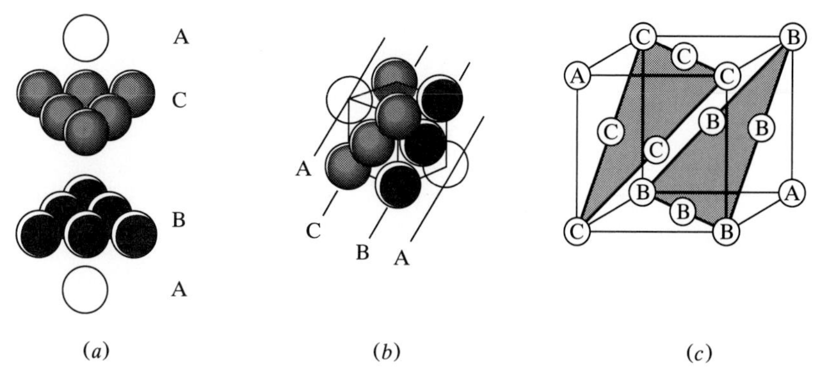

FIGURE 7.11
Three views of the cubic close-packed (ccp) structure which has a face-centered cubic (fcc) unit cell. (*a*) The three ABCABC layers. (*b*) A space-filling view of the fcc unit cell showing the orientation of the ABCABC layers. (*c*) A space lattice view of the fcc unit cell with the ABCABC layers.

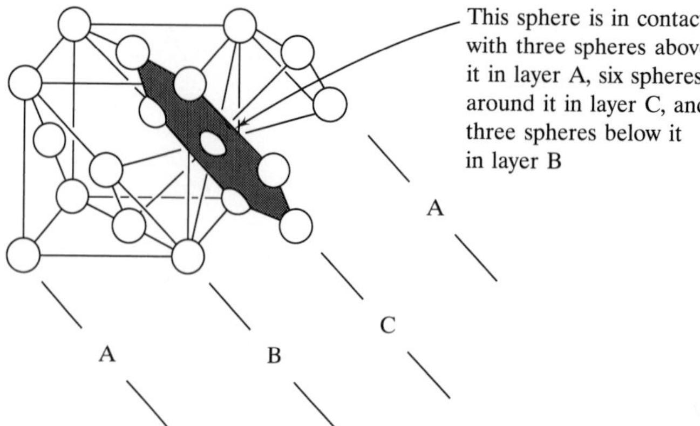

FIGURE 7.12
The face-centered cubic unit cell with extra spheres to show both the ABCABC layering scheme and that a given sphere (center of hexagon) has a coordination number of 12. [Ref. 15.]

the relationship $l = d\sqrt{2}$. As we did previously with the simple cubic and body-centered cubic cases, we can use this result to rationalize the expressions for the fraction of space occupied by the spheres and the density of a face-centered cubic (= ccp) unit cell as given in Table 7.1. Note that the face-centered cubic unit cell (along with the hexagonal close-packed) is the most efficient way to pack spheres in three-dimensional space (74 percent).

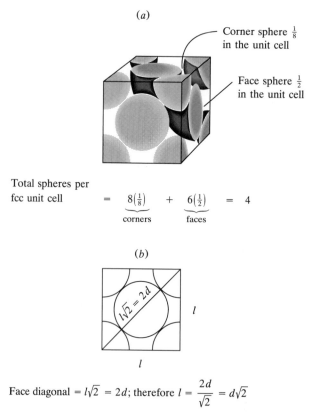

(*a*)

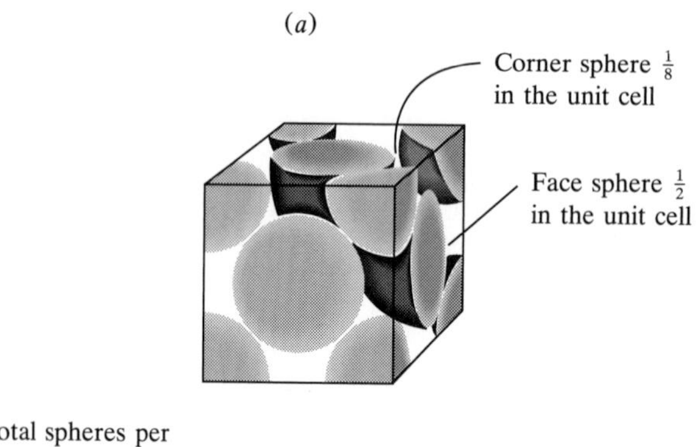

Total spheres per fcc unit cell $= \underbrace{8(\tfrac{1}{8})}_{\text{corners}} + \underbrace{6(\tfrac{1}{2})}_{\text{faces}} = 4$

(*b*)

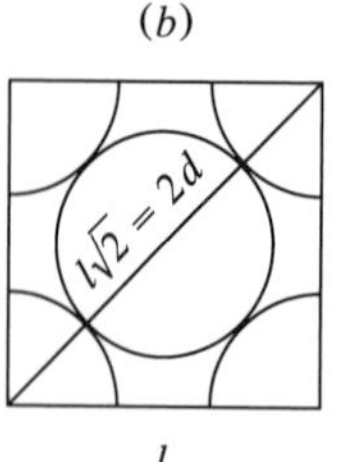

Face diagonal $= l\sqrt{2} = 2d$; therefore $l = \dfrac{2d}{\sqrt{2}} = d\sqrt{2}$

FIGURE 7.13
(*a*) One face-centered unit cell showing that the corner spheres are one-eighth and the face spheres one-half in the unit cell for a total of four spheres. (*b*) One face of the fcc unit cell showing the relationship between the sphere diameter and the unit cell edge.

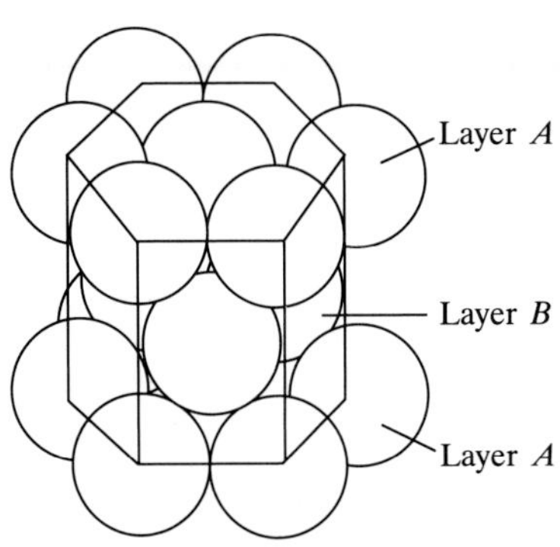

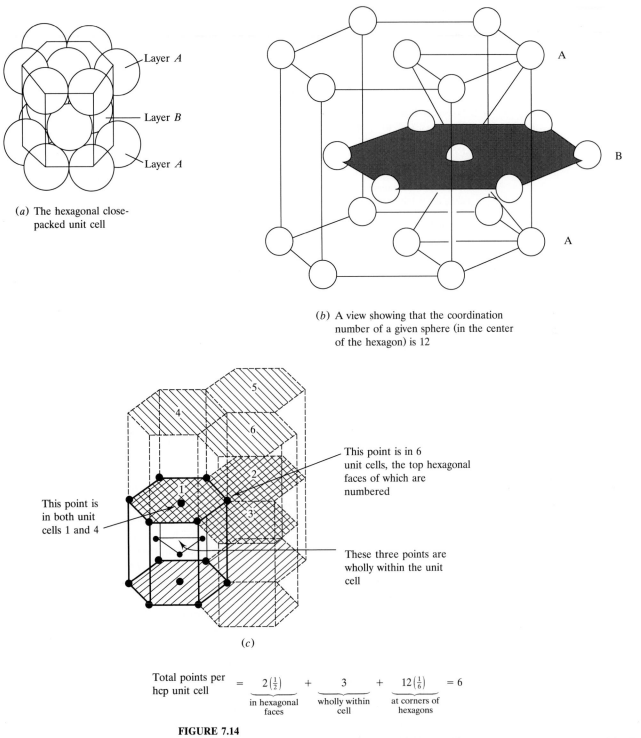

(*a*) The hexagonal close-packed unit cell

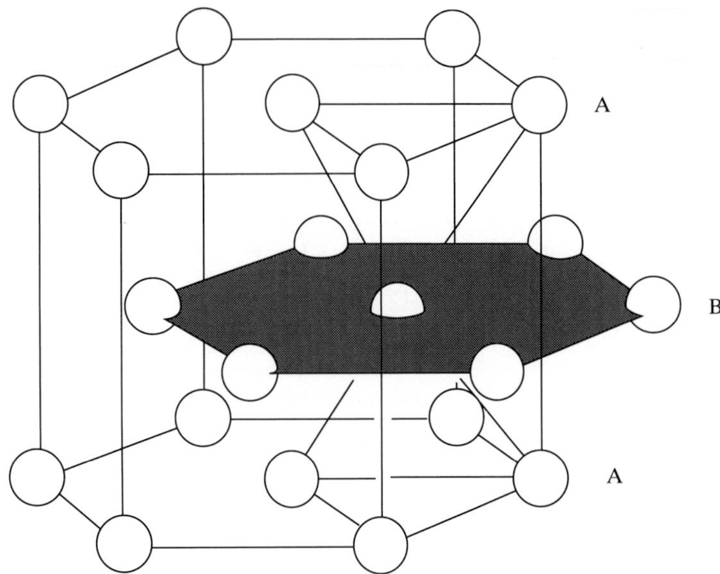

(*b*) A view showing that the coordination number of a given sphere (in the center of the hexagon) is 12

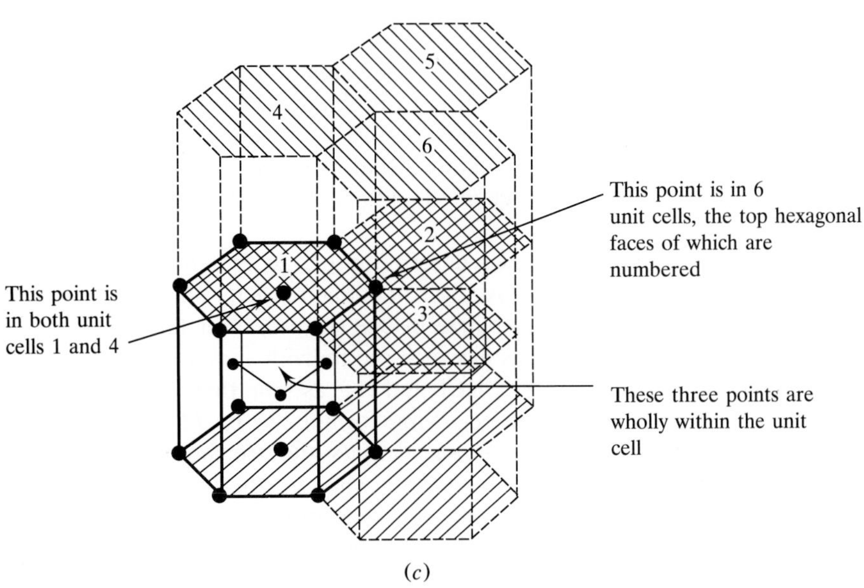

$$\text{Total points per hcp unit cell} = \underbrace{2\left(\tfrac{1}{2}\right)}_{\text{in hexagonal faces}} + \underbrace{3}_{\text{wholly within cell}} + \underbrace{12\left(\tfrac{1}{6}\right)}_{\text{at corners of hexagons}} = 6$$

FIGURE 7.14
Three views of the hexagonal close-packed unit cell showing (*a*) the ABABAB layering scheme, (*b*) that the coordination number of a given sphere (center of hexagon) is 12, and (*c*) that the number of spheres per unit cell is 6. [(*a*), Ref. 16; (*b*), Ref. 15.]

																	He
																	hcp
Li	Be											B	C	N	O	F	Ne
bcc	hcp											rh	d	hcp	sc	—	ccp
Na	Mg											Al	Si	P	S	Cl	Ar
bcc	hcp											ccp	d	sc	ortho	tet	ccp
K	Ca	Sc	Ti	V	Cr	Mn	Fe	Co	Ni	Cu	Zn	Ga	Ge	As	Se	Br	Kr
bcc	ccp	hcp	hcp	bcc	bcc	bcc	bcc	hcp	ccp	ccp	hcp	sc	d	rh	hcp	ortho	ccp
Rb	Sr	Y	Zr	Nb	Mo	Tc	Ru	Rh	Pd	Ag	Cd	In	Sn	Sb	Te	I	Xe
bcc	ccp	hcp	hcp	bcc	bcc	hcp	hcp	ccp	ccp	ccp	hcp	tet	tet	rh	hcp	ortho	ccp
Cs	Ba	La	Hf	Ta	W	Re	Os	Ir	Pt	Au	Hg	Tl	Pb	Bi	Po	At	Rn
bcc	bcc	hcp	hcp	bcc	bcc	hcp	hcp	ccp	ccp	ccp	rh	hcp	ccp	rh	mono	—	ccp

Note: hcp = hexagonal close-packed, ccp = cubic close-packed, bcc = body-centered cubic, sc = simple cubic, tet = tetragonal, rh = rhombohedral, d = diamond, ortho = orthorhombic, mono = monoclinic.

FIGURE 7.15
The most stable crystal structures assumed by the elements in their solid phase.

Return now to Figure 7.10*b*. If the spheres of the third layer are placed in the *b* depressions, they generate another layer A, and the ABABAB packing scheme as shown in Figure 7.10*d* results. This is the *hexagonal close-packed* (*hcp* = *ABABAB*) *lattice* and is the last of our A-type structures. Figure 7.14*a* shows the hexagonal prismatic unit cell of the hcp structure. Figure 7.14*b* emphasizes that the coordination number of a given sphere is 12. Note from Figure 7.14*c* that there are six spheres per unit cell. The volume of the unit cell turns out to be as given in Table 7.1. See the solution to Problem 7.11 for the details of how this expression is derived. With the above result, the fraction of space occupied and density expressions follow. Again note that the hcp and ccp lattices result in 74 percent of the space being occupied and are the most efficient packing schemes possible.

This completes our detailed description of the four most important A-type lattices and their unit cells. Analysis of Figure 7.15 shows that more than 80 percent of the elements crystallize in one of these four lattices. In addition, a number of molecular substances in which the individual molecules closely approximate spheres (for example, CH_4, HCl, and H_2) assume these structures. There are actually 14 possible A-type lattices. First formulated by M. A. Bravais in 1850, these are still referred to as the *Bravais lattices* and are shown in Figure 7.16. We will have greater occasion to refer to these when we discuss AB_n-type lattices.

The interatomic distances between two atoms in contact in any of these lattices can be determined by x-ray diffraction. While the radius of an isolated atom is essentially infinite, the radius of an atom in contact with a neighboring atom in a lattice is just half the interatomic distance between them. If the elements are metals, a compilation of *metallic* radii can be obtained; if the crystal consists of nonbonded atoms held together by van der Waals intermolecular forces, then *van der Waals* radii are the result. In covalent crystals and, of course, in discrete molecules, atoms are covalently bonded, and therefore *covalent* radii can be tabulated. These three types of radii are given in Table 7.2.

How good is this model of solid state chemistry? To demonstrate, we will calculate the density of an element listed in Figure 7.15 and compare our result with the known density. Take copper as an example. It assumes a cubic close-packed structure which means it has a face-centered cubic unit cell. The accepted value for the metallic radius of copper as found in Table 7.2 is 1.28 Å. We know from

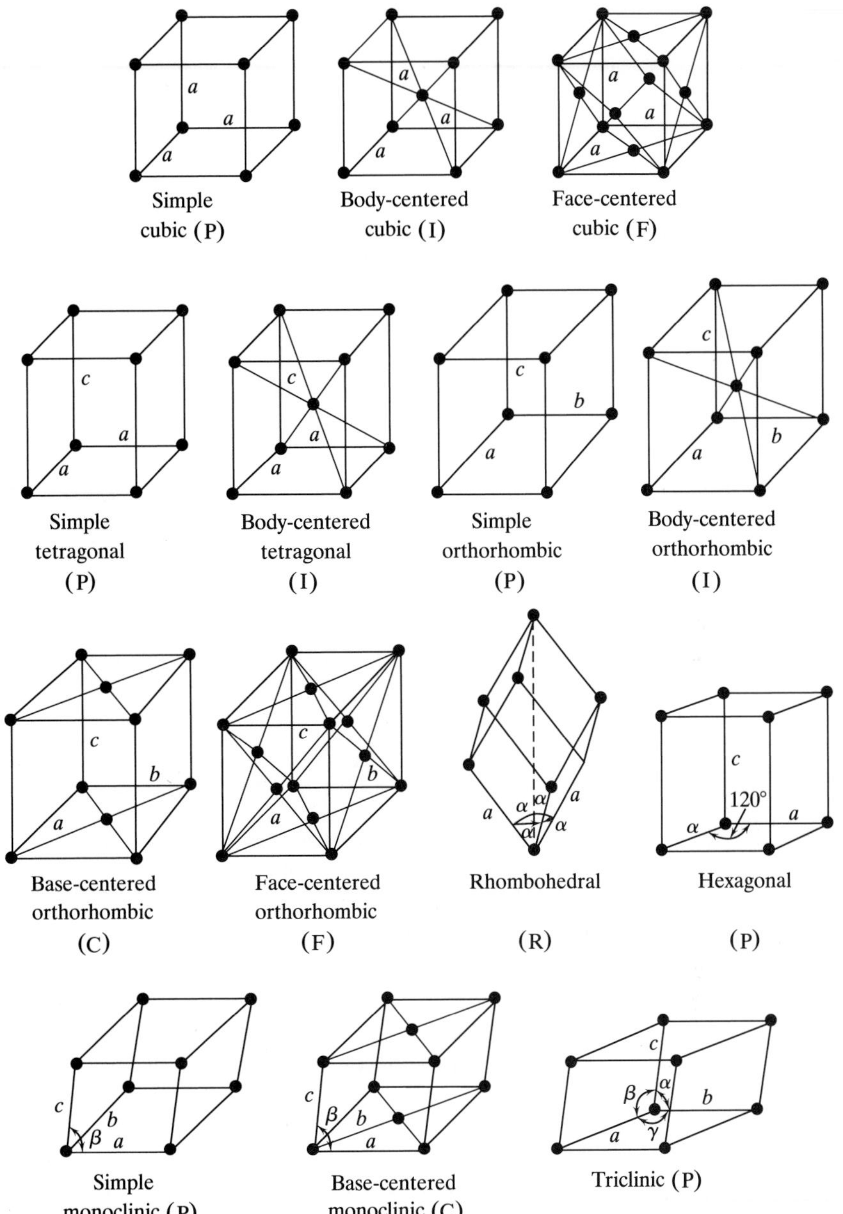

FIGURE 7.16
The 14 Bravais lattices which are composed of seven essential unit cells: cubic, tetragonal, orthorhombic, rhombohedral, hexagonal, monoclinic, and triclinic. (P = primitive or simple; I = body-centered; F = face-centered; C = base-centered.)

our previous discussions that there will be four copper atoms per unit cell and that the atoms touch along the face diagonal. Using the expression for the density found in Table 7.1, we can calculate the density as follows:

$$\text{Density} = \frac{4 \text{ atoms}\left(\dfrac{63.54 \text{ g/mol}}{6.02 \times 10^{23} \text{ atoms/mol}}\right)}{\left[2(1.28 \times 10^{-8})\sqrt{2}\right]^3 \text{ cm}^3} = 8.90 \text{ g/cm}^3 \qquad (7.4)$$

TABLE 7.2
Selected metallic, van der Waals, and covalent radii

KEY: Symbol
Metallic radius*
van der Waals radius
Covalent radius
(all values in Å units)

H																	He
...																	...
1.20																	1.40
0.37																	...
Li	Be											B	C	N	O	F	Ne
1.57	1.12											...	....	....	....	....	...
1.82	....											...	1.70	1.55	1.52	1.47	1.54
1.34	0.90											0.82	0.77	0.75	0.73	0.72	...
Na	Mg											Al	Si	P	S	Cl	Ar
1.91	1.60											1.43	....	....	....	....	...
2.27	1.73											...	2.10	1.80	1.80	1.75	1.88
1.54	1.30											1.18	1.17	1.06	1.02	0.99	...
K	Ca	Sc	Ti	V	Cr	Mn	Fe	Co	Ni	Cu	Zn	Ga	Ge	As	Se	Br	Kr
2.35	1.97	1.64	1.47	1.35	1.29	1.37	1.26	1.25	1.25	1.28	1.37	1.53	1.39	....	....	....	...
2.75	....	....	....	....	....	....	....	....	1.63	1.43	1.39	1.87	....	1.85	1.90	1.85	2.02
1.96	1.74	1.44	1.32	1.25	1.27	1.46	1.20	1.26	1.20	1.38	1.31	1.26	1.22	1.20	1.16	1.14	1.15
Rb	Sr	Y	Zr	Nb	Mo	Tc	Ru	Rh	Pd	Ag	Cd	In	Sn	Sb	Te	I	Xe
2.50	2.15	1.82	1.60	1.47	1.41	1.35	1.34	1.34	1.37	1.44	1.52	1.67	1.58	1.61			
....	....	....	....	....	....	....	....	....	1.63	1.72	1.58	1.93	2.17	....	2.06	1.96	2.16
2.11	1.92	1.62	1.48	1.37	1.45	1.56	1.26	1.35	1.31	1.53	1.48	1.44	1.41	1.40	1.36	1.33	1.26
Cs	Ba	La	Hf	Ta	W	Re	Os	Ir	Pt	Au	Hg	Tl	Pb	Bi			
2.72	2.24	1.88	1.59	1.47	1.41	1.37	1.35	1.36	1.39	1.44	1.55	1.71	1.75	1.82			
....	....	....	....	....	....	....	....	....	1.72	1.66	1.55	1.96	2.02	....			
2.25	1.98	1.69	1.49	1.38	1.46	1.59	1.28	1.37	1.28	1.43	1.51	1.52	1.47	1.46			

*Metallic radii vary somewhat with coordination numbers. For simplicity these values are for atoms with a coordination number = 12.

The experimental density is 8.96 g/cm^3. The error between the calculated and actual densities is about 0.7 percent. Not bad for a model that represents copper atoms as hard spheres like billiard balls.

7.3 AB$_n$-TYPE CRYSTAL LATTICES

AB$_n$-type lattices are those in which the spheres representing the atoms, ions, or molecules are of two different sizes. The most common example of these lattices are ionic crystals in which the anion is larger than the cation. In this case it is best to picture the anions forming an A-type lattice and the cations fitting into the "holes" in that lattice. To the extent that the crystal is purely ionic, the packing assumed by the anions will be in large measure determined by the relative sizes of the two species. That is, the holes in the anionic lattice must be of the proper size to adequately accommodate the cations. The first topic we need to cover, then, is the number and type of holes present in A-type lattices.

Cubic, Octahedral, and Tetrahedral Holes

The hole left in the center of a simple cubic unit cell is referred to as a *cubic hole*. Figure 7.17 shows a space-filling model of such a unit cell with one of the corner

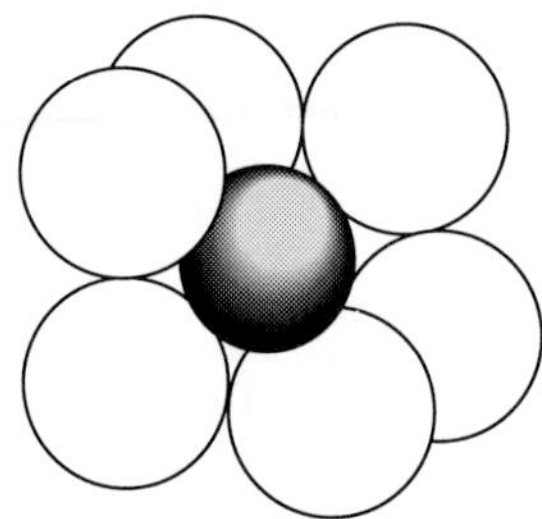

FIGURE 7.17
A space-filling representation of a cubic hole (shaded) within a simple cubic unit cell. (A corner sphere has been removed for clarity.) [Ref. 17.]

spheres removed and the hole outlined for clarity. The radius of this hole is about three-quarters that of the spheres which form it (see Problem 7.18). We decided earlier that there is one sphere per simple cubic unit cell. The cubic hole is completely within the unit cell so that there is also one hole per unit cell. It follows that there is one cubic hole per sphere in this lattice.

The holes in cubic and hexagonal close-packed lattices are either tetrahedral or octahedral. Ultimately we will want to know not only where these holes are located but also how many of them there are per sphere. Figure 7.18*a* illustrates a *tetrahedral hole* found within a tetrahedron formed by the larger spheres. Often when looking at a drawing or a model of a ccp or hcp lattice, the tetrahedron is conveniently identified either as one sphere sitting on the depression formed by three others (Figure 7.18*b*) or as a triangle of three perched atop a single sphere (Figure 7.18*c*). An *octahedral hole* is found within an octahedron of larger spheres, which can be viewed either as a square of four spheres with a fifth and sixth sitting

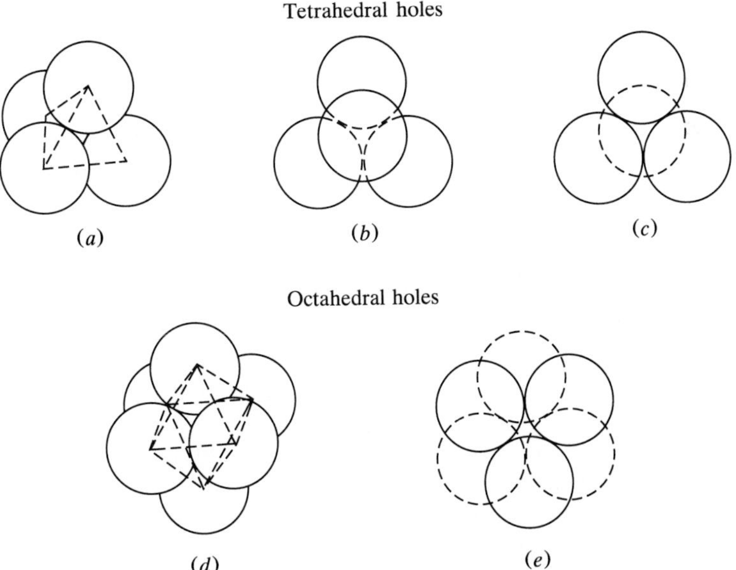

FIGURE 7.18
Various ways to visualize tetrahedral and octahedral holes. A tetrahedral hole in the center of (*a*) a tetrahedron viewed from the side, (*b*) a triangle of spheres with a fourth sphere sitting on top, and (*c*) a triangle of spheres placed atop a single sphere. An octahedral hole in the center of (*d*) an octahedron viewed as four spheres in a square with one on top and one on the bottom and (*e*) an octahedron viewed as a triangle of spheres placed atop another triangle of spheres rotated 60° from the first.

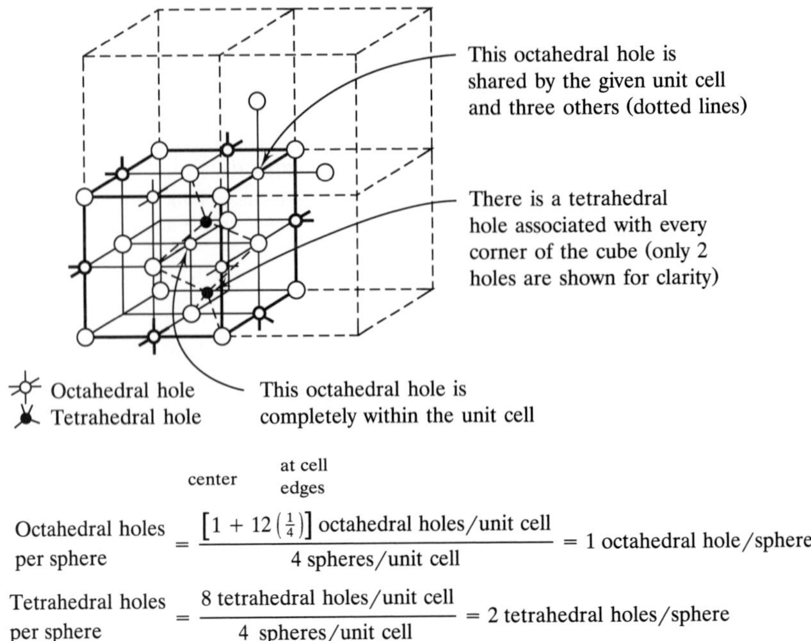
This octahedral hole is shared by the given unit cell and three others (dotted lines)
There is a tetrahedral hole associated with every corner of the cube (only 2 holes are shown for clarity)
Octahedral hole
Tetrahedral hole
This octahedral hole is completely within the unit cell

$$\text{Octahedral holes per sphere} = \frac{[\overset{\text{center}}{1} + \overset{\text{at cell edges}}{12\left(\frac{1}{4}\right)}]\text{ octahedral holes/unit cell}}{4\text{ spheres/unit cell}} = 1\text{ octahedral hole/sphere}$$

$$\text{Tetrahedral holes per sphere} = \frac{8\text{ tetrahedral holes/unit cell}}{4\text{ spheres/unit cell}} = 2\text{ tetrahedral holes/sphere}$$

FIGURE 7.19
A face-centered cubic unit cell showing the positions and numbers of octahedral and tetrahedral holes per unit cell. There are octahedral holes in the center and in the middle of the 12 edges of the unit cell and a tetrahedral hole associated with each corner of the unit cell. Since there are four spheres per unit cell, there are one octahedral hole and two tetrahedral holes per sphere.

above and below it (Figure 7.18*d*) or often more easily as a triangle of three spheres sitting above a second triangle of three rotated 60° relative to the first (Figure 7.18*e*).

Where are these holes located within a close-packed lattice? Looking back at Figure 7.10*b*, you should be able to confirm that the octahedral holes are the *a* depressions while the tetrahedral holes are the *b* depressions. Now how many holes are there per sphere? Consider the face-centered unit cell of a cubic close-packed lattice as shown in Figure 7.19. Remember that we have determined (see Figure 7.13) that there are four spheres per fcc unit cell. Note that there is one octahedral hole in the center of the cell and 12 others located on each edge of the cube. Each of these 12 holes is shared among four unit cells as shown, therefore each is one-fourth in the given unit cell. As indicated in the figure, there is a total of four octahedral holes per unit cell. This leads to the conclusion that there is one octahedral hole per sphere. There is one tetrahedral hole associated with each of the eight corners of the unit cell. Similarly, this leads to the conclusion that there are two tetrahedral holes per sphere. These ratios of holes per sphere are the same for a hexagonal close-packed structure.

Radius Ratios

You may have noticed that a tetrahedral hole is quite small while an octahedral hole is a little larger. (It turns out that the former is about one-quarter the radius

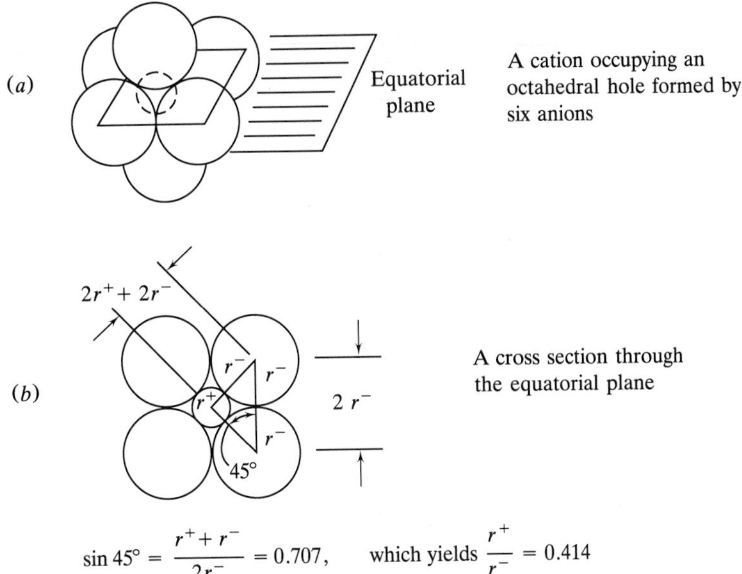

$$\sin 45° = \frac{r^+ + r^-}{2r^-} = 0.707, \quad \text{which yields } \frac{r^+}{r^-} = 0.414$$

FIGURE 7.20
(*a*) A three-dimensional and (*b*) cross-sectional view of a cation occupying an octahedral hole in the A-type lattice of anions. The radius-ratio r^+/r^- characteristic of an octahedral hole is 0.414.

of the spheres which form it, while the latter is about four-tenths the radius.) So the relative sizes of the holes are as shown below.

Size of hole	cubic	>	octahedral	>	tetrahedral
Coord. no.	8		6		4

Now if a smaller species, such as a cation, were of just the right size to fit snugly into one of these holes, it would have the coordination number as indicated above. Note that the coordination number is directly proportional to the size of the hole.

How can we calculate the relative size of a hole and therefore of the cation that would just fit within it? Consider Figure 7.20, which shows a cross section of an octahedral hole taken through the equatorial plane. A cation of just the right size is shown occupying the hole. Some analytical geometry and trigonometry yields the result that the ratio of the radius of the cation to the radius of the anion, assumed to be the larger species, is 0.414. Similar calculations for trigonal, tetrahedral, and cubic holes yield the following radius rations. (A trigonal hole, one in the center of a triangle of spheres, is so very small that it is seldom occupied.)

Type of hole	r^+/r^-
Trigonal	0.155
Tetrahedral	0.225
Octahedral	0.414
Cubic	0.732

TABLE 7.3
Correlation among radius ratios, maximum coordination numbers, and types of holes occupied

r^+/r^-	0.155	to	0.225	to	0.414	to	0.732	to	higher values
Maximum C.N. possible		3		4		6		8	
Type of hole		Trigonal		Tetrahedral		Octahedral		Cubic	

What information do these radius ratios provide? Table 7.3 shows how coordination number (C.N.) and type of hole correspond to the ratios. This table provides a guide to what type of hole will be occupied in a given situation. Take NaCl as an example. The radius of Na^+ is 1.16 Å, while the radius of Cl^- is 1.67 Å; r^+/r^- comes out to be 0.695, which falls within the 0.414 to 0.732 octahedral range. Referring to the table, we predict that the Cl^- will form an A-type lattice, cubic close-packed it turns out, and that the Na^+ cations will fit into the octahedral holes in that lattice. The maximum C.N. of the cation is predicted and in fact turns out to be 6.

Note that 0.414, the radius ratio we calculated when the cation just fits into the octahedral hole in the anionic A-type lattice, is the *lower limit* to the range where octahedral holes will be occupied. In other words, the radius ratio can be greater than but not less than the ideal value. Why should this be? What must happen to cause the radius ratio to deviate from the ideal value of 0.414? If the cation increases in size, causing an increase in the ratio, it will force the anions to lose contact with each other. Energetically, this is favorable because separating the negatively charged anions will reduce the repulsive forces among them and result in a more stable crystal lattice. Thus, even if the cation is larger than the ideal octahedral hole, it will force the anions apart and continue to occupy this site.

Again starting at a radius ratio of 0.414 such that the cation just fits the octahedral hole in the lattice of the anions, consider what must happen if the cation decreases in size causing the radius ratio to fall to values less than 0.414. The anions would still be attracted to the cation which is now "rattling around" in a hole bigger than it is. The anions would crowd around the cation and therefore be bumping into one another. This would lead to large repulsions between the electron clouds of the anions which would be an energetically unfavorable situation. Accordingly, if the cation is smaller than the ideal octahedral hole, the most favorable response is for the cation to switch to a smaller hole. For example, in beryllium sulfide, BeS, the radius ratio is 0.35. If the beryllium cations were to occupy octahedral holes in the close-packed lattice of sulfide anions, there would be too much repulsion among the anions, and the beryllium cations would switch to occupying a tetrahedral hole to produce a more energetically favorable situation.

Finally, you should also note that Table 7.3 indicates the *maximum* coordination number for a given radius ratio. For example, in ZnS, the radius ratio comes out to be 0.52, indicating that the maximum coordination number of the zinc cations is 6. In accordance with the above discussion, the coordination number cannot be 8 because the small zinc cation would rattle around in the bigger cubic hole. The coordination number could, however, be 4 as this would result in the doubly negative sulfide (S^{2-}) ions arranged in a tetrahedron around the zinc cation being forced apart. In fact, the zinc cations turns out to occupy tetrahedral holes and have a coordination number of 4. The crystal structures for ZnS will be discussed in more detail in the next section.

The above discussion illustrates the general utility of Table 7.3, but it is clear that we must approach these matters with great caution. In a later section (p. 175) we will see that the correlation between the actual coordination number and that predicted by radius ratios varies from 33 to 100 percent depending on the structure. Overall, the correlation is about two in three. One of several reasons why Table 7.3 can be considered only a guide is the uncertainty in determining ionic radii.

Ionic Radii

As noted earlier (p. 162) when discussing metallic, van der Waals, and covalent radii, x-ray diffraction methods traditionally yield accurate information on interatomic distances. In ionic crystals, the analogous *interionic distances* have been available from these measurements. What could not be determined, however, was where one ion stopped and another started. That is, it was known how to apportion the interionic distance between the cation and anion. More recently, it

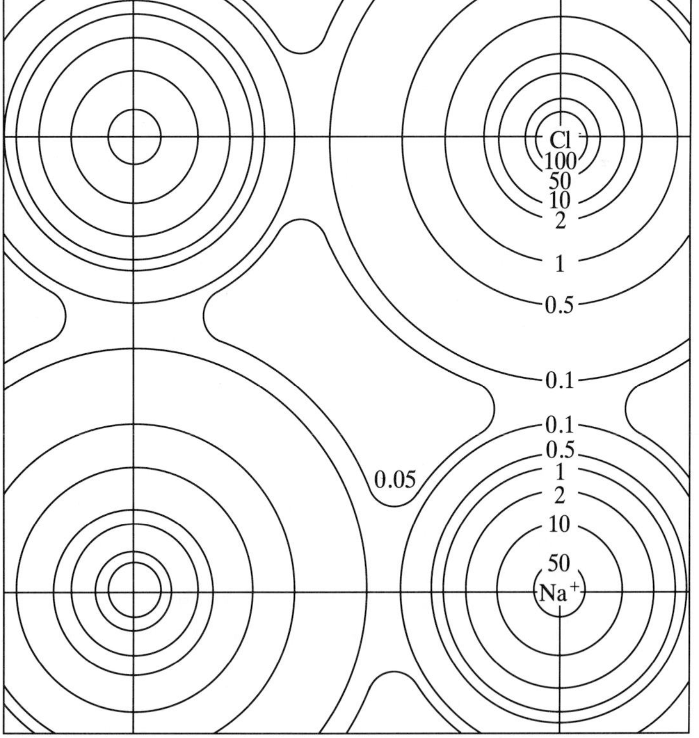

FIGURE 7.21
A high-resolution x-ray diffraction map of the electron-density contours in sodium chloride. Numbers indicate the electron density (electrons/Å^3) along each contour line. The "boundary" of each ion is defined as the minimum in electron density between the ions.

TABLE 7.4

Shannon-Prewitt ionic radii in Ångstrom units (Å) of cations.* [Ref. 2.]

Charge on cation is the same as the group number except as noted in actinides.

C.N.	1A	2A	3B	4B	5B	6B	7B	1B	2B	3A	4A	5A	6A	7A
	Li	Be								B	C	N		
3	—	0.30								0.15	0.06	0.044		
4	0.73	0.41								0.25	0.29	—		
6	0.90	0.59								0.41	0.30	0.27		
	Na	Mg								Al	Si	P	S	Cl
4	1.13	0.71								0.53	0.40	0.31	0.26	0.22
6	1.16	0.86								0.675	0.540	0.52	0.43	0.41
8	1.32	1.03								—	—	—	—	—
	K	Ca	Sc	Ti	V	Cr	Mn	Cu	Zn	Ga	Ge	As	Se	Br
4	1.51	—	—	0.56	0.495	0.40	0.39	0.74	0.74	0.61	0.530	0.475	0.42	0.39
6	1.52	1.14	0.885	0.745	0.68	0.58	0.60	0.91	0.880	0.760	0.670	0.60	0.56	0.53
8	1.65	1.26	1.010	0.88	—	—	—	—	1.040	—	—	—	—	—
	Rb	Sr	Y	Zr	Nb	Mo	Tc	Ag	Cd	In	Sn	Sb	Te	I
2	—	—	—	—	—	—	—	0.81	—	—	—	—	—	—
4	—	—	—	0.73	0.62	0.55	0.51	1.14	0.92	0.76	0.69	—	0.57	0.56
								1.16						
6	1.66	1.32	1.040	0.86	0.78	0.73	0.70	1.29	1.09	0.940	0.830	0.74	0.70	0.67
8	1.75	1.40	1.159	0.98	0.88	—	—	1.42	1.24	1.06	0.95	—	—	—
	Cs	Ba	La	Hf	Ta	W	Re	Au	Hg	Tl	Pb	Bi	Po	At
4	—	—	—	0.72	—	0.56	0.52	—	1.10	0.89	—	—	—	—
6	1.81	1.49	1.172	0.85	0.78	0.74	0.67	1.51	1.16	1.025	0.79	0.90	0.81	0.76
8	1.88	1.56	1.300	0.97	0.88	—	—	—	1.28	1.12	0.915	—	—	—

C.N.															
	La	Ce	Pr	Nd	Pm	Sm	Eu	Gd	Tb	Dy	Ho	Er	Tm	Yb	Lu
6	1.172	1.15	1.13	1.123	1.11	1.098	1.087	1.078	1.063	1.052	1.041	1.030	1.020	1.008	1.00
7	1.24	1.21	—	—	—	1.16	1.15	1.14	1.12	1.11	—	1.085	—	1.065	—
8	1.300	1.283	1.266	1.249	1.233	1.219	1.209	1.193	1.180	1.167	1.155	1.144	1.134	1.125	1.11
	Ac^{III}	Th^{IV}	Pa^{V}	U^{VI}	Np^{IV}	Pu^{IV}	Am^{IV}	Cm^{IV}	Bk^{IV}	Cf^{IV}				No^{II}	
6	1.26	1.08	0.92	0.87	1.01	1.00	0.99	0.99	0.97	0.961				1.24	
8	—	1.19	1.05	1.00	1.12	1.10	1.09	1.09	1.07	1.06					

*Values from R. D. Shannon, *Acta Crystallogr.*, **A32**:751 (1976). The reference includes other oxidation states and coordination numbers.

has been possible to produce high-resolution x-ray studies such as that shown in Figure 7.21 for NaCl. By locating the minimum of electron density along the interionic distance, accurate values for the radius of both the cation and the anion could be determined. These studies are the basis of the Shannon-Prewitt ionic radii given in Tables 7.4 to 7.6.

Note that these tables show that ionic radius varies somewhat with coordination number. This variation for the sodium cation is shown below.

C.N.	Na^+ radius, Å
4	1.13
6	1.16
8	1.32

TABLE 7.5

Shannon-Prewitt ionic radii in Ångstrom units (Å) of cations with variable oxidation states.* [Ref. 2.]

C.N. = 6 unless otherwise noted

Oxidation number													
	Ti	V	Cr	Mn	Fe	Co	Ni	Cu	Zn	Ga	Ge	As	Se
2	1.00	0.93	0.94 0.87^{LS}	0.970 0.81^{LS}	0.77(4) 0.92 0.75^{LS}	0.72(4) 0.88.5 0.79^{LS}	0.69(4) 0.63(4)^{Sq} 0.83	0.71(4) 0.71(4)^{Sq} 0.87	0.74(4) 0.88				
3	0.810	0.780	0.755	0.785 0.72^{LS}	0.785 0.69^{LS} 0.63(4)	0.75 0.68.5^{LS}	0.74 0.70^{LS}	0.68^{LS}		0.76		0.72	
4	0.745	0.72	0.69	0.67	0.725	0.67	0.62^{LS}				0.67		0.64
	Zr	Nb	Mo	Tc	Ru	Rh	Pd	Ag	Cd	In	Sn	Sb	Te
2	—	—	—	—	—	—	0.78(4)^{Sq}	0.93(4)^{Sq}	1.09	—	1.22^{+}(8)	—	—
3	—	0.86	0.83	—	0.82	0.805	0.90	0.81(4)^{Sq}	—	0.94	—	0.94(5)	—
4	0.86	0.82	0.79	0.785	0.76	0.74	0.755	—	—	—	0.83	—	1.11
	Hf	Ta	W	Re	Os	Ir	Pt	Au	Hg	Tl	Pb	Bi	Po
1	—	—	—	—	—	—	—	1.51	1.33 1.11(3)	1.64	—	—	—
2	—	—	—	—	—	—	0.74(4)^{Sq}	—	1.16	—	1.33	—	—
3	—	0.86	—	—	—	0.82	—	0.82(4)^{Sq}	—	1.025	—	1.17	—
4	0.85	0.82	0.80	0.77	0.770	0.765	0.765	—	—	—	0.915	—	1.08
	Ce^{IV}	Pr^{IV}	Sm^{II}	Eu^{II}	Tb^{IV}	Tm^{II}	Yb^{II}						
	1.01	0.99	1.41(8)	1.31	0.90	1.17	1.16						
	Pa^{IV}	U^{IV}	Np^{VI}	Pu^{VI}	Am^{III}	Cm^{III}	Bk^{III}	Cf^{III}					
	1.04	1.03	0.86	0.85	1.115	1.11	1.10	1.09					

*Values from R. D. Shannon, *Acta Crystallogr.*, **A32**:751 (1976). Low-spin values (LS) and values for square planar (Sq) coordination are designated by superscripts.

†Value for C.N. 8 from R. D. Shannon and C. T. Prewitt, *Acta Crystallogr.*, **B25**:925 (1969). The value is probably doubtful, since it was not included in the revised tabulation (footnote a).

TABLE 7.6

Shannon-Prewitt ionic radii in Ångstrom units (Å) for common anions.* [Ref. 2.]

C.N. = 6 unless otherwise noted

	OH^-	H^-
	1.23	1.53†
N^{3-}	O^{2-}	F^-
1.32	1.26	1.19
(C.N. 4)	S^{2-}	Cl^-
	1.70	1.67
	Se^{2-}	Br
	1.84	1.82
	Te^{2-}	I^-
	2.07	2.06

*R. D. Shannon, *Acta Crystallogr.*, **A32**:751 (1976).

†D. F. C. Morris and G. L. Reed, *J. Inorg. Nucl. Chem.*, **27**:1715 (1965).

TABLE 7.7
AB-type lattices: radius ratios, structures, and coordination numbers

		Anions		Cations	
Structure	r^+/r^{-}*	**Structure**	**C.N.**	**Structure**	**C.N.**
Sodium chloride (NaCl)	$\frac{1.16}{1.67} = 0.69$	ccp(fcc)	6	All oct. holes	6
Zinc blende (ZnS)	$\frac{0.88}{1.70} = 0.52$	ccp (fcc)	4	Half tetra. holes	4
Wurtzite (ZnS)	$\frac{0.88}{1.70} = 0.52$	hcp	4	Half tetra. holes	4
Cesium chloride (CsCl)	$\frac{1.81}{1.67} = 1.08$†	Simple cubic	8	All cubic holes	8

*Shannon-Prewitt ionic radii for C.N. = 6.
†$r^-/r^+ = 0.93$.

These results make sense because as the number of anions around the cation increases, they will crowd each other out to some extent and make the apparent cationic radius increase. For purposes of calculating radius ratios in order to use Table 7.2 to determine the hole occupied and the coordination number, we usually use C.N. = 6 values.

AB Structures

Table 7.7 presents information about radius ratios, coordination numbers, and the structure of both the cations and the anions for four representative AB-type ionic lattices. Figure 7.22 gives a drawing of each of these lattices. In each case, you should compare the information given in these two sources and, if possible, study a model of the lattice as you read the following short descriptions.

Sodium chloride, or "rock salt," is one of the most commonly encountered AB structures. As previously noted, its radius ratio is consistent with sodium cations occupying octahedral holes in the cubic close-packed lattice of chloride anions. Figure 7.22*a* shows the ABCABC layer structure of chlorides which is consistent with a face-centered cubic unit cell. The sodium ions are shown in the octahedral holes in both the center of the unit cell and at the cell edges. Figure 7.22*b* shows a larger portion of the space lattice, again indicating the octahedral environment around each cation. It also shows that the crystal could be represented by a unit cell with fcc sodium ions and chloride ions in the octahedral holes. You should confirm that the coordination number of both the cations and anions is 6 and that there are a total of four chloride ions $[8(\frac{1}{8}) + 6(\frac{1}{2})]$ and four sodium ions $[1 + 12(\frac{1}{4})]$ per unit cell, consistent with a 1:1 stoichiometry for this compound.

Using the above description, we should be able to calculate a value for the density of sodium chloride. Taking the unit cell as the basis for the determination (as we did for A-type structures), we can calculate the mass of the four NaCl formula units and divide that by the volume of the unit cell. Remember that the

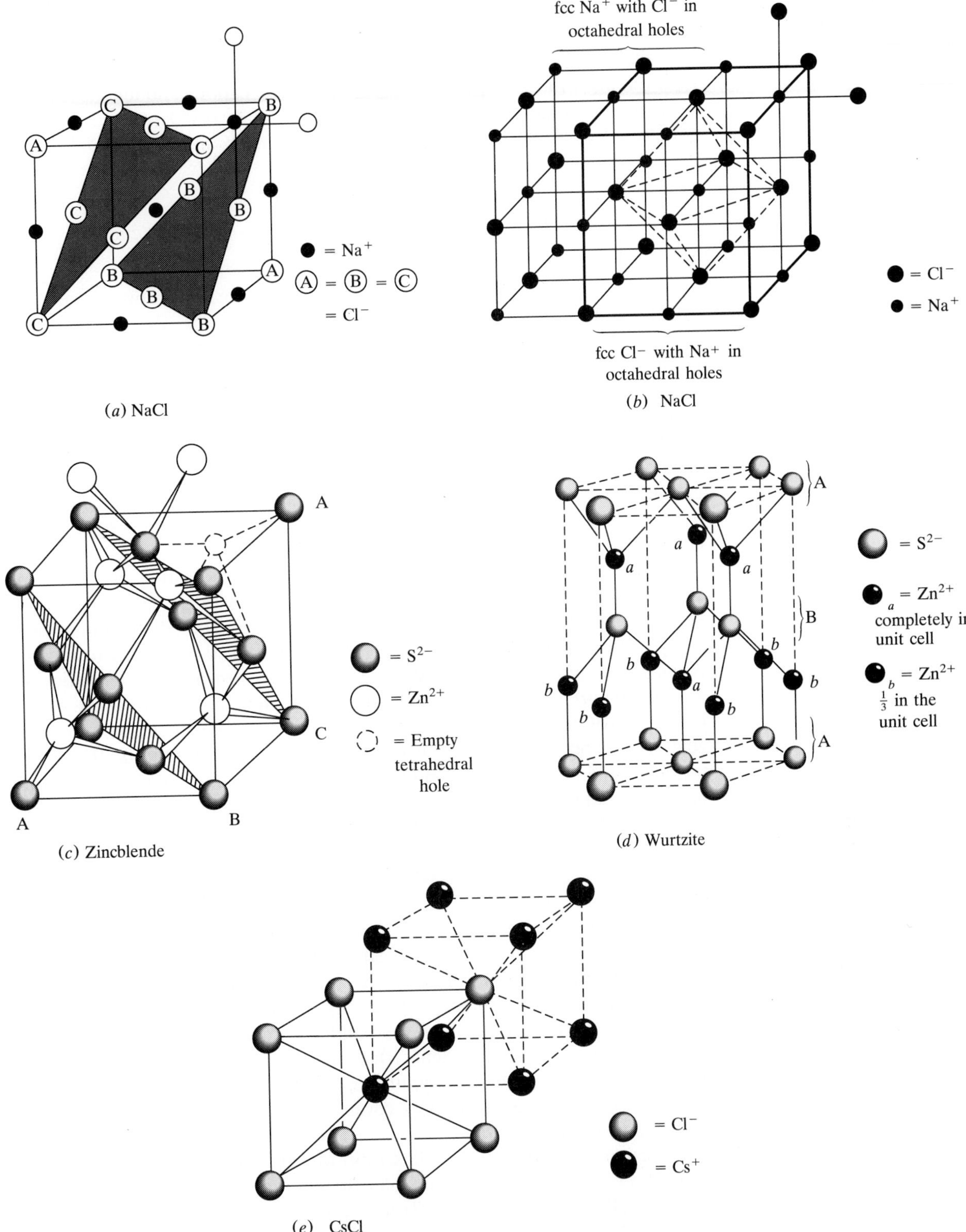

(*a*) NaCl

(*b*) NaCl

(*c*) Zincblende

(*d*) Wurtzite

(*e*) CsCl

FIGURE 7.22
AB-type structures. (*a*) NaCl emphasizing the ABCABC layers of the chloride anions with sodium cations in octahedral holes. (*b*) NaCl viewed as fcc Cl^- with Na^+ in octahedral holes, or vice versa. (*c*) Zinc blende with Zn^{2+} occupying half the tetrahedral holes in fcc S^{2-}. (*d*) Wurtzite with Zn^{2+} occupying half the tetrahedral holes in hcp S^{2-}. (*e*) CsCl with Cs^+ in the cubic holes of Cl^-, or vice versa. [Ref. 18.]

chloride anions are not in contact because the sodium cations are somewhat larger than the ideal radius for an octahedral hole. The cations and anions are, however, in contact along the cell edge so that $l = 2r(Na^+) + 2r(Cl^-) = 2(1.16) + 2(1.67) = 5.66$ Å. The density is calculated as shown in Equation (7.5):

$$\text{Density} = \frac{\dfrac{4(22.99) + 4(35.45)}{6.02 \times 10^{23}}\,\text{g}}{(5.66 \times 10^{-8})^3\ \text{cm}^3} = 2.14\ \text{g/cm}^3 \tag{7.5}$$

The actual density of NaCl is 2.165 g/cm^3. The difference between the actual and calculated densities is about 1 percent.

While the coordination number of the cation in NaCl comes out as predicted by the radius-ratio guideline, and the density calculated for NaCl is very close to the actual value, it is instructive to know how good the correlation between the known crystal structure and the structure predicted by radius-ratio calculations is in each of AB_n cases we will discuss. Table 7.8 shows that 58 percent of the compounds which assume a rock salt structure were consistent with the radius-ratio calculation. Again we see that while radius ratios are a useful guideline, we must use caution when drawing conclusions from a purely ionic, hard-sphere model for crystal structures.

The radius ratio of zinc sulfide, ZnS, is 0.52, indicating that the maximum coordination number for the zinc cations should be 6. As previously described, however, the zinc cations occupy tetrahedral holes not octahedral holes in the sulfide lattice. If the sulfides form a cubic close-packed array, the resulting lattice is called *zinc blende*. If the sulfides are hexagonal close-packed, the lattice is called *wurtzite*. Figure 7.22*c* shows a view of the zinc blende structure which emphasizes both the ABCABC arrangement and the fcc unit cell of the anions. Note that the zinc cations occupy four of the eight tetrahedral holes. One of the four empty tetrahedral holes is indicated in the figure. Also note that the occupied tetrahedral holes themselves form a tetrahedron. Consistent with a 1 : 1 stoichiometry, there are four $[8(\frac{1}{8}) + 6(\frac{1}{2})]$ sulfide ions to match the four zinc cations found completely within the unit cell.

Figure 7.22*d* shows the wurtzite structure. The sulfides form an ABABAB (hcp) layer structure, and the zinc cations occupy half the tetrahedral holes in the lattice. Consistent with a 1 : 1 stoichiometry, there are six $[12(\frac{1}{6}) + 2(\frac{1}{2}) + 3]$ sulfide anions as well as six $[4 + 6(\frac{1}{3})]$ zinc cations. Table 7.8 shows a very low correlation (33 percent) between known structure and calculated radius ratios for the zinc blende and wurtzite structures. One important reason for the low correlation seems to be the degree to which the bonding in crystals of this type is of covalent character. There is good evidence that in zinc sulfide, the zinc atoms and sulfur atoms, to a significant degree, assume sp^3 hybrid orbitals and that the compound should be described as an infinite covalent lattice like diamond rather than as a purely ionic lattice.

Cesium chloride has a radius ratio of 1.08 because, using Shannon-Prewitt radii, the cesium cation is larger than the chloride anion. In this case we should actually calculate r^-/r^+ (= 0.93) and assume that the cations form the A-type lattice and the chlorides fill the appropriate holes. Note that 0.93 falls in the cubic hole/C.N. = 8 range of Table 7.3. As shown in Figure 7.22*e*, the cesiums form a simple cubic lattice and the chloride anions occupy the cubic holes. Alternatively, the chloride anions can be pictured as forming the A-type lattice with the cesium

TABLE 7.8
Compounds having common AB structures

Structure	Compounds	Correlation percent*
Rock salt	Alkali halides† and hydrides	58
	AgF, AgCl, AgBr	
	Monoxides of Mg, Ca, Sr, Ba, Mn, Fe, Co, Ni, Cd	
	Monosulfides of Mg, Ca, Sr, Ba, Mn, Pb	
	TiC, VC, InP, InAs, SnP, SnAs	
	NH_4I, KOH, KSH, KCN	
Zinc blende	BeX, ZnX, CdX, HgX (X = S, Se, Te)	33‡
	Diamond, silicon, germanium, gray tin	
	SiO_2 (cristobalite)	
	BN, BP, SiC, CuX (X = F, Cl, Br, I)	
	XY (X = Al, Ga, In; Y = P, As, Sb)	
Wurtzite	ZnX (X = O, S, Te)	33‡
	CdX (X = S, Se)	
	BeO, MgTe, NH_4F	
	MN (M = Al, Ga, In)	
	AgI, NH_4F	
Cesium chloride	CsX (X = Cl, Br, I, SH, CN)	100
	TlX (X = Cl, Br, I, CN)	
	NH_4X (X = Cl, Br, CN)	

*Percentage of compounds for which the radius ratio correlates with (i.e., is consistent with) the known structure. Taken from L. C. Nathan, *J. Chem. Educ.*, **62**(3):215 (1985). Shannon-Prewitt radii for C.N. = 6 used for all radii. Results are only slightly improved when using radii where the correct coordination number of each ion is taken into account.

†Some, such as KF, RbF, and CsF, have cations which are larger than the anions.

‡This result is for zinc blende and wurtzite structures taken together.

cations in the cubic holes. Using the solid lines as the unit cell, note that the coordination number of both the cation and anion is 8. Note also that there is a total of one $[8(\frac{1}{8})]$ chloride per unit cell and, of course, one cesium cation in the body consistent with a 1 : 1 stoichiometry. Table 7.8 shows that the greatest correlation (100 percent) between known structure and calculated radius ratios occurs for the CsCl structure.

Table 7.8 also shows a number of compounds that assume the above AB structures. The diamond structure is listed as showing a zinc blende structure where, of course, all the spheres are the same size. Refer to Figure 7.3*a* and see if you can confirm this comparison. This situation is also consistent with the covalent character of both diamond and zinc sulfide.

AB_2 Structures

Table 7.9 and Figure 7.23 present information and drawings of AB_2 structures. Before considering each of these in turn, let's look more closely at the relationship between the coordination numbers of the cation and anion. In AB compounds the

TABLE 7.9
AB_2-type lattices: Radius ratios, structures, and coordination numbers

		Anions		Cations	
Structure	r^+/r^-*	**Structure**	**C.N.**	**Structure**	**C.N.**
Fluorite (CaF_2)	$\frac{1.14}{1.19} = 0.96$	1. Simple cubic 2. All tetra.	4	1. Half of cubic holes 2. fcc	8
Cadmium iodide (CdI_2)	$\frac{1.09}{2.06} = 0.53$	hcp	3	Half of oct. holes	6
Rutile (TiO_2)	$\frac{0.745}{1.26} = 0.59$	Simple tetra.	3	Oct. holes	6

*Shannon-Prewitt ionic radii for C.N. = 6.

coordination numbers of each ion were identical. This turns out to be an application of the relationship found in Equation (7.6).

$$(\text{C.N. of A}) \times (\text{no. of A in formula}) = (\text{C.N. of B}) \times (\text{no. of B in formula}) \tag{7.6}$$

Since the number of A spheres equals the number of B spheres in AB compounds, the coordination numbers are also equal.

Fluorite, CaF_2, shows a radius ratio of 0.96, which predicts that the calcium ions will occupy cubic holes formed by the fluoride anions. You should note in Figure 7.23*a* that the calcium ions do indeed occupy such sites. However, as required by stoichiometry (see Problem 7.28), half of the cubic holes go unoccupied. The unit cell of the lattice, therefore, cannot be the simple cubic of fluorides with one calcium in the body. Rather, a larger unit cell of fcc calcium ions with fluorides filling the tetrahedral holes is the more appropriate description. Note that the coordination number of the fluorides is 4, which is consistent with Equation (7.6). Table 7.10 indicates that there is a 90 percent correlation between the known crystal structure and the calculated radius ratio for compounds which assume the fluorite structure.

Cadmium iodide, CdI_2, shows a radius ratio of 0.53, which predicts that the cadmium cations will occupy octahedral holes. Figure 7.23*b* shows the resulting structure. Again, for stoichiometric reasons (see Problem 7.28), only half the octahedral holes can be occupied. Note that the iodides are hexagonal close-packed and that the occupied octahedral holes occur in layers. A closer inspection of this lattice shows that the iodide layers which have cadmium cations in octahedral holes between them are closer together than those where the octahedral holes are empty. This certainly makes sense from an electrostatic point of view. The correlation between known structure and radius-ratio calculations is 74 percent for the CdI_2 structure.

The rutile structure of titanium dioxide, TiO_2, is not close-packed. The radius ratio of 0.59 falls in the octahedral hole/C.N. = 6 region, and the drawing of the structure in Figure 7.23*c* shows this to be the case. Note that this is not a cubic unit cell but rather is tetragonal (see Figure 7.16). The coordination number

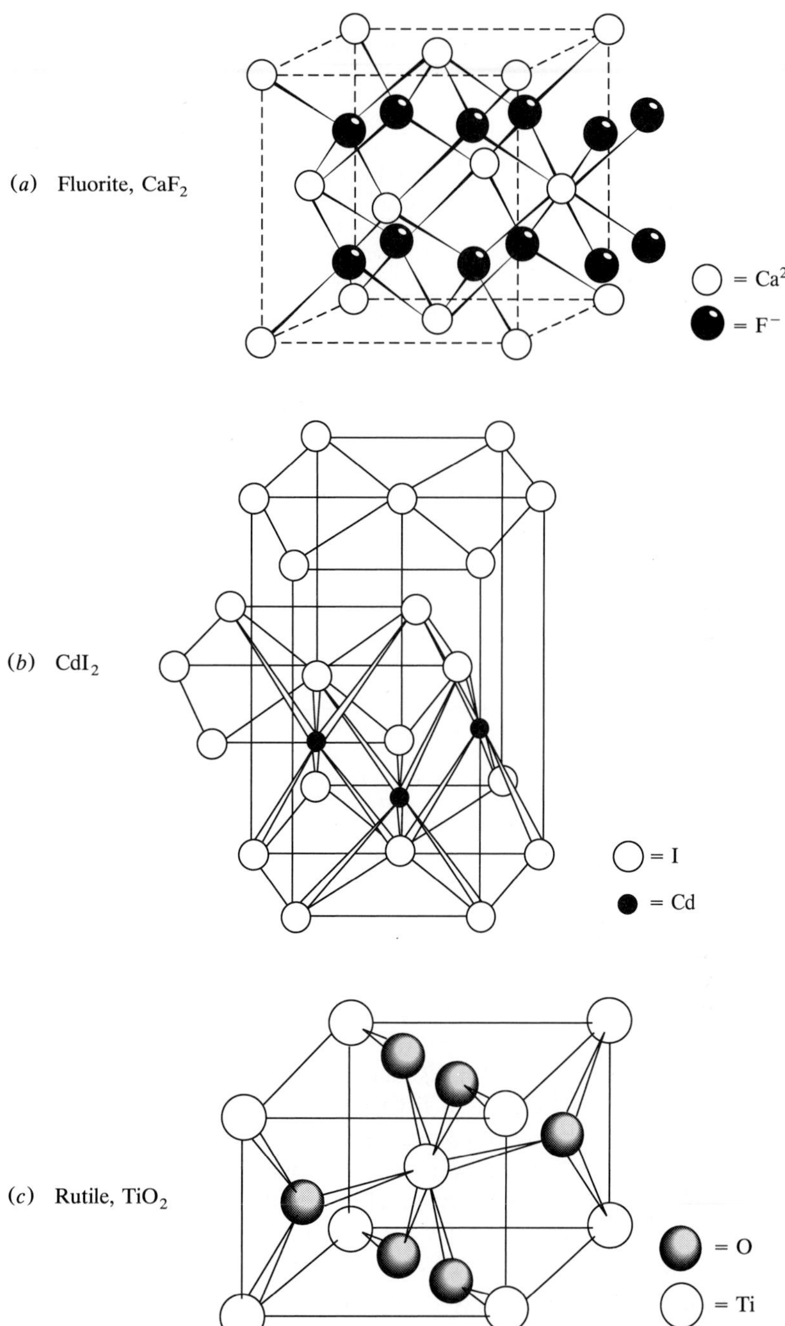

FIGURE 7.23
AB_2-type structures. (*a*) Fluorite, CaF_2, with F^- occupying all the tetrahedral holes in fcc Ca^{2+} or with Ca^{2+} occupying half the cubic holes in simple cubic F^-. (*b*) Cadmium iodide, CdI_2, with Cd^{2+} occupying half the octahedral holes of hcp I^-. (*c*) Rutile, TiO_2, with Ti^{4+} occupying octahedral holes in a non-close-packed array of oxide ions.

TABLE 7.10
Compounds having common AB_2 structures

Structure	Compounds	Correlation percent*
Fluorite	MF_2 (M = Ca, Cd, Hg, Sn) MCl_2 (M = Sr and Ba) MO_2 (M = Po, Zr, Hf, Ce, Pr, and various actinides) Transition metal dihydrides	90
Antifluorite†	M^I_2X (M^I = Li, Na, K; X = 0, S, Se, Te) Rb_2X (X = O, S)	50
Cadmium iodide	MBr_2 (M = Mg, Fe, Co, Ni, Cd) MCl_2 (M = Ti, V) MI_2 (M = Mg, Ca, Ti, Mn, Fe, Co, Ni, Zn, Cd, Pb) $M(OH)_2$ (M = Mg, Ca, Mn, Fe, Co, Ni, Cd) MS_2 (M = Ti, Zr, Pt, Sn) MSe_2 (M = Ti, V) MTe_2 (M = Ti, V, Co, Ni)	74
Rutile	MF_2 (M = Mg, Cr, Mn, Fe, Co, Ni, Cu, Zn) MO_2 (M = Si, Ge, Sn, Pb, Te, Ti, V, Cr, Mn, Zr, Mo, Os, Ir, Ce)	75

*Percentage of compounds for which the radius ratio correlates with (i.e., is consistent with) the known structure. Taken from L. C. Nathan, *J. Chem. Educ.* **62**(3):215 (1985). Shannon-Prewitt radii for C.N. = 6 used for all radii. Results are only slightly improved when using radii where the correct coordination number of each ion is taken into account.

†In the antifluorite structure the positions of the cations and anions are reversed from that in fluorite.

of the oxides is 3, which is consistent with the rule stated earlier. The correlation between known structure and radius-ratio calculations is 75 percent for the rutile structure.

Table 7.10 shows some compounds that assume the above AB_2 structures. Included is a listing for the antifluorite structure in which the positions of the cations and anions are reversed from fluorite. The primary example of antifluorite is lithium oxide, Li_2O, in which the oxides are cubic close-packed (face-centered cubic) with the smaller lithium cations occupying all the tetrahedral holes.

7.4 STRUCTURES INVOLVING POLYATOMIC MOLECULES AND IONS

We have discussed some of the more straightforward and common AB_n-type structures in which the lattice points represent the positions of atoms or ions. When monoatomic species are replaced by polyatomic species, the situation can become significantly more complicated and generally beyond the scope of this text. There are, however, a number of examples where these polyatomic entities simply assume arrays directly analgous to those we have just discussed. For example, solid carbon dioxide, or dry ice, shown in Figure 7.24*a*, is just a face-centered array of

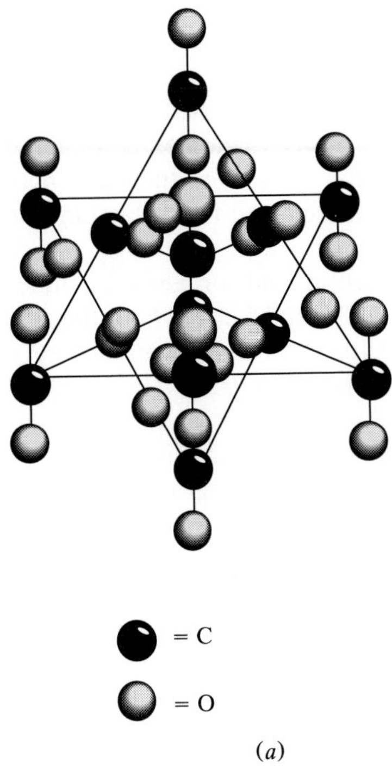

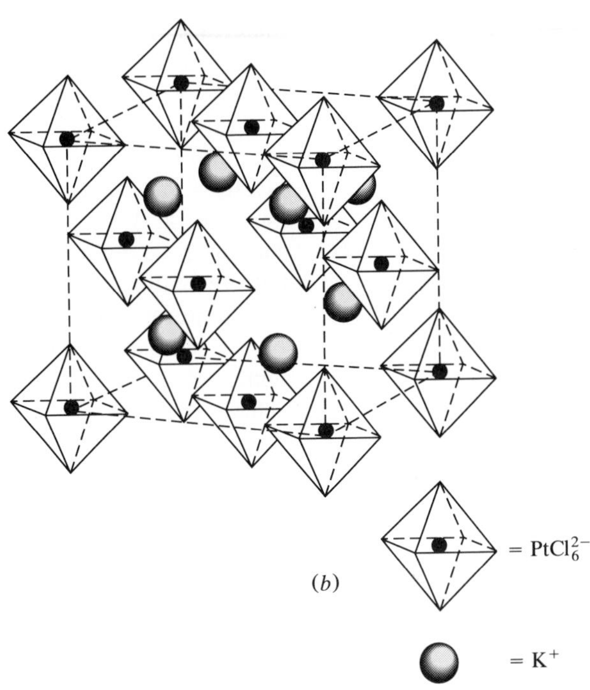

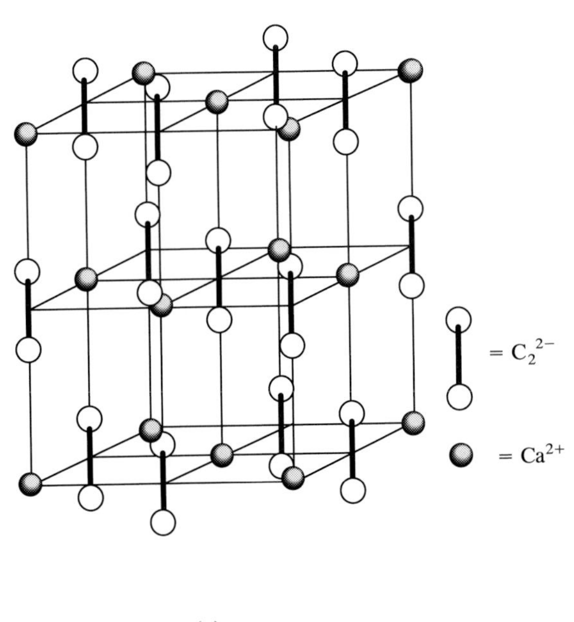

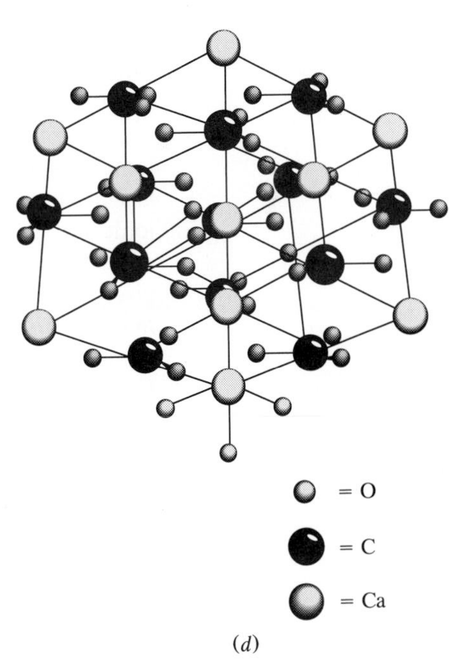

FIGURE 7.24
Structures with polyatomic molecules or ions. (*a*) Dry ice, CO_2(s), with CO_2 molecules in fcc array. (*b*) Potassium hexachloroplatinate(IV), K_2PtCl_6, in antifluorite structure with K^+ occupying all the tetrahedral holes in fcc $PtCl_6^{2-}$. (*c*) Calcium carbide, CaC_2, with C_2^{2-} occupying octahedral holes in elongated fcc Ca^{2+}. (*d*) Calcium carbonate, $CaCO_3$, with Ca^{2+} in distorted octahedral holes of face-centered rhombohedral CO_3^{2-}. [(*a*) *and* (*d*), *Klinger Educational Products*.]

CO_2 molecules. Figure 7.24*b* shows a unit cell of potassium hexachloroplatinate(IV), K_2PtCl_6, which has an octahedral $PtCl_6^{2-}$ complex anion of the type discussed in Chapter 3. The compound crystallizes in an antifluorite structure with the $PtCl_6^{2-}$ anions in a face-centered cubic unit cell and the potassium ions in all the tetrahedral holes.

Many compounds with a polyatomic cation and/or anion assume a rock salt–type structure. These range from KSH, KCN, and FeS_2, where the polyatomic ions are fairly simple (see Tables 7.8 and 7.10), to a more complicated example such as hexaamminecobalt(III) hexachlorothallate(III), $[Co(NH_3)_6][TlCl_6]$, where both the cation and anion are complex ions. In all these cases the anions assume a face-centered cubic array and the cations occupy the octahedral holes.

Examples where the cubic symmetry has been degraded are simply too numerous to be treated in a systematic way here. Two examples, however, are calcium carbide, CaC_2 (which contains the C_2^{2-} anion), and calcium carbonate, shown in Figure 7.24*c* and *d*. The former is a tetragonal unit cell, while the latter is a rhombohedral. See Figure 7.16 for the details of these unit cells.

7.5 DEFECT STRUCTURES

To this point we have assumed that solids always form perfect crystal lattices. Given what you learned about the spontaneity of reactions in general chemistry, you could probably predict that structures in which various imperfections are randomly distributed throughout the lattice are of high statistical probability, i.e., high entropy, and can be formed fairly readily. Examples of imperfections include (1) simple vacancies, (2) unexpected occupation of interstitial sites, (3) the incorporation of impurities, that is, atoms or ions other than those of the parent crystal, and (4) various lattice imperfections.

A missing atom in a metallic or covalent crystal or a missing pair of ions in an ionic crystal, known as a *Schottky defect*, is shown in Figure 7.25*a*. In a *Frenkel defect*, shown in Figure 7.25*b*, a cation is displaced from its normal position in an ionic lattice and occupies another hole, or *interstitial site*. In both of the above defects, the stoichiometry of the crystal is maintained. If a cation is missing from a given site, its loss may be electrically balanced by another cation losing one or more extra electrons. A good example of this is iron(II) oxide, FeO. If some number of Fe^{2+} ions are missing, twice that number of others may be oxidized to Fe^{3+}, which would keep the sample neutral but result in an oxygen-to-iron ratio a little greater than 1. For this reason, the actual iron (II) oxide stoichiometry is commonly about $Fe_{0.95}O$. This is a good example of a *nonstoichiometric compound* (one in which there is a nonintegral ratio of atoms) and is shown for a part of one layer of the FeO crystal lattice in Figure 7.25*c*.

When a given ion, atom, or molecule is replaced with a similar-sized species of a different element or compound, the sample is impure. This is a common occurrence in nature and often results in some precious minerals such as rubies or emeralds. (See Chapter 14, p. 364.) Occasionally, samples are purposely "doped" with an impurity to achieve a given property. We will investigate this procedure in detail when we discuss semiconductors in Chapter 15. There are a variety of lattice imperfections which are dependent on the conditions present when crystallization occurs. One such lattice defect, an *edge dislocation* in a metallic crystal, is shown in Figure 7.25*d*.

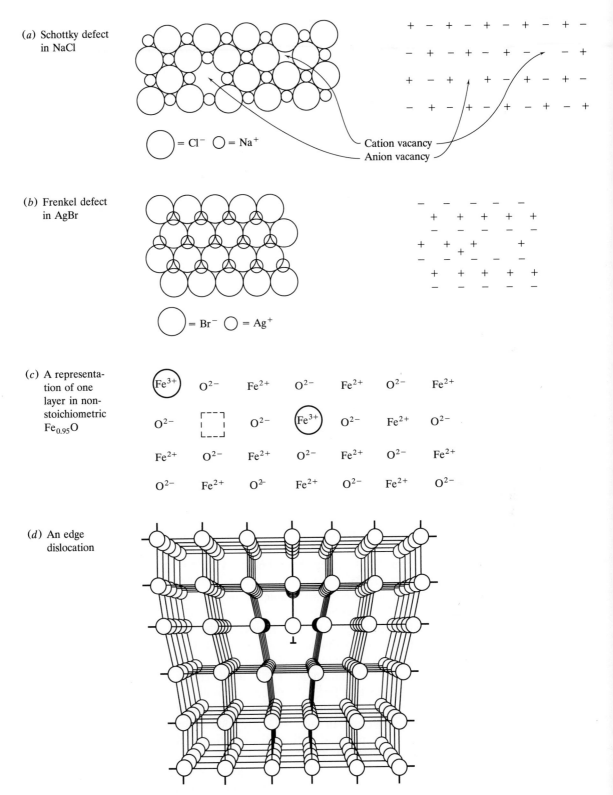

(*a*) Schottky defect in NaCl

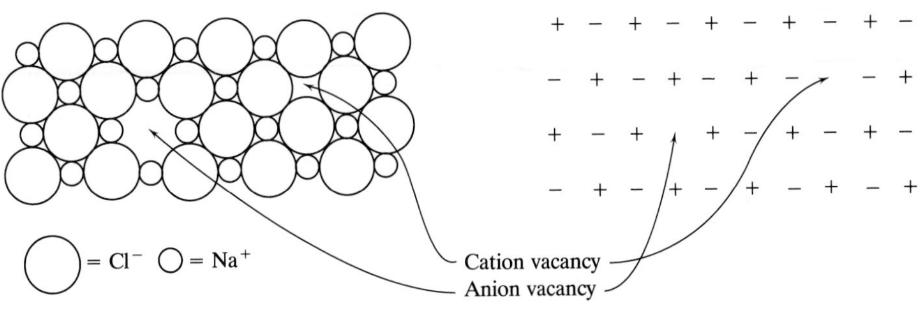

(*b*) Frenkel defect in AgBr

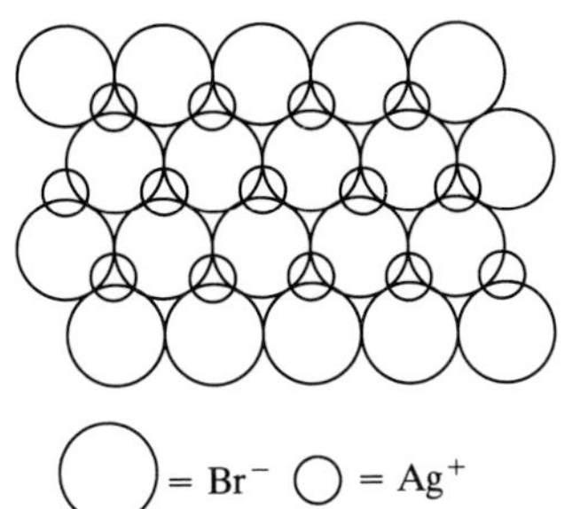

(*c*) A representation of one layer in nonstoichiometric $Fe_{0.95}O$

Fe^{3+} O^{2-} Fe^{2+} O^{2-} Fe^{2+} O^{2-} Fe^{2+}

O^{2-} ☐ O^{2-} Fe^{3+} O^{2-} Fe^{2+} O^{2-}

Fe^{2+} O^{2-} Fe^{2+} O^{2-} Fe^{2+} O^{2-} Fe^{2+}

O^{2-} Fe^{2+} O^{2-} Fe^{2+} O^{2-} Fe^{2+} O^{2-}

(*d*) An edge dislocation

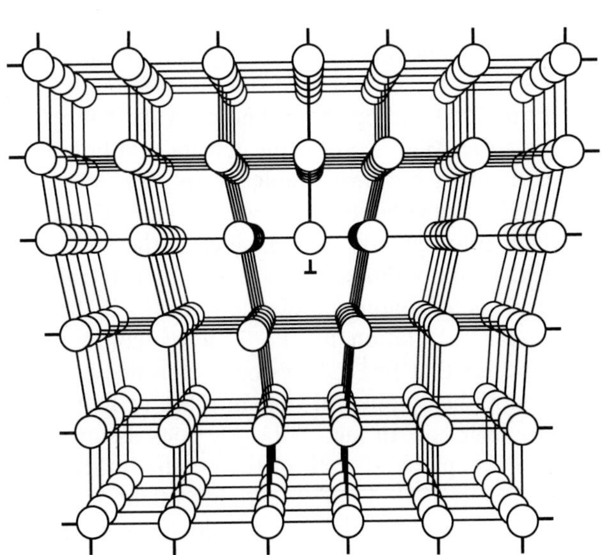

FIGURE 7.25
Solid state defect structures including (*a*) Schottky defect, (*b*) Frenkel defect, (*c*) nonstoichiometric $Fe_{0.95}O$, and (*d*) an edge dislocation. [(*a*), (*b*), Ref. 2, pp. 230, 231; (*d*), Ref. 18.]

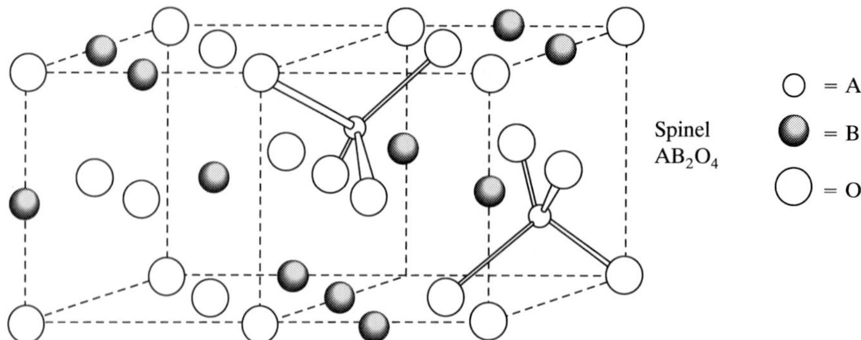

FIGURE 7.26
A portion of the space lattice of the spinel structure ($A^{II}B_2^{III}O_4$) showing the A(II) cations (small open circles) occupying 2 of the possible 16 tetrahedral sites (or one-eighth of the tetrahedral holes) and the B(III) cations (small solid circles) occupying four of the possible eight octahedral sites (or one-half of the octahedral holes). [Ref. 19.]

7.6 SPINEL STRUCTURES: CONNECTING CRYSTAL FIELD EFFECTS WITH SOLID STATE STRUCTURES

If you have already covered crystal field theory (Chapter 4), you are now in a position to bring both crystal field theory and solid state structures together to better understand a class of compounds called the *spinels*. $MgAl_2O_4$ is representative of these compounds that are of general formula $A^{II}B_2^{III}O_4$. The oxide anions are found to be cubic close-packed (face-centered cubic) and, in a *normal* spinel, the A(II) cations occupy one-eighth of the tetrahedral holes, while the B(III) cations occupy one-half of the octahedral holes. A section of the lattice is shown in Figure 7.26. Be sure you can verify that this figure and the above fractions of holes occupied result in the correct stoichiometry (one A for four oxides; two B's for four oxides) (see Problem 7.45). While most of the over 100 compounds classified as spinels are normal, a significant minority are *inverse* in which half the B(III) cations have exchanged places with all the A(II) cations. Why should this be so?

One important factor in determining whether a spinel will be normal or inverse is the crystal field stabilization energy (CFSE) of the cations occupying the tetrahedral and octahedral holes. Take $NiFe_2O_4$ as an example. The Ni^{2+} has a d^8 structure. Will it prefer the tetrahedral hole it would occupy in a normal situation or might it be more stable in an octahedral hole? As developed in Section 4.3, CFSE is calculated as shown in Figure 7.27*a*. Note that the oxides are weak-field (high-spin) ligands, and, as always, the splitting of the *d* orbitals in an octahedral field is approximately twice that of a tetrahedral field of the same ligands. The result is a greater stabilization energy if the Ni^{2+} occupies an octahedral site. A similar calculation for Fe^{3+}, a d^5 configuration, is given in Figure 7.27*b* and indicates that Fe^{3+} shows no preference based on CFSE. Accordingly, if the Ni^{2+} occupy the octahedral sites in place of half the Fe^{3+} ions, a more stable structure should result. In fact, $NiFe_2O_4$ does assume an inverse spinel structure: the Ni^{2+} ions occupy one-fourth of the octahedral holes, while the Fe^{3+} occupy one-fourth of the octahedral and one-eighth of the tetrahedral holes.

Note that the above calculations and conclusion assume that all other energetic factors remain the same. This would seem to be a case where a large oversimplification has been made, yet the correlation between the predicted and actual structures is remarkably good. Here is one more case where the remarkably (*a*) Ni^{2+} in Weak Field (O^{2-}) Tetrahedral and Octahedral Holes

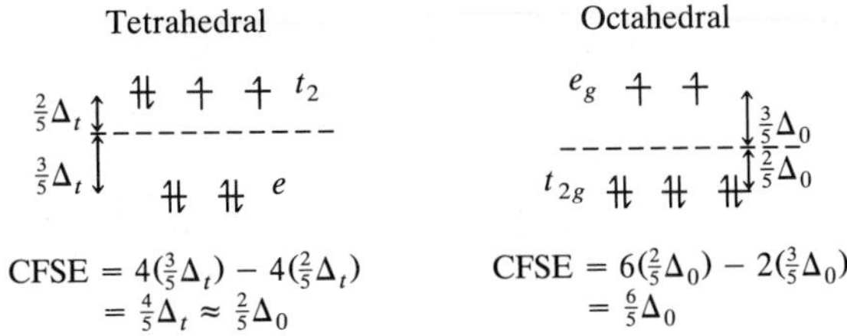

Ni^{2+} prefers the octahedral hole

(*b*) Fe^{3+} in Weak Field (O^{2-}) Tetrahedral and Octahedral Holes

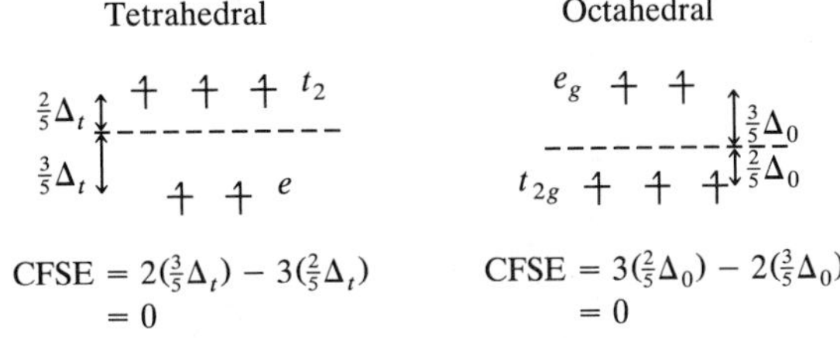

Fe^{3+} shows no preference for either hole

FIGURE 7.27
The calculation of the crystal field stabilization energy (CFSE) of (*a*) Ni^{2+} (d^8) and (*b*) Fe^{3+} (d^5) in weak-field tetrahedral and octahedral holes.

straightforward crystal field theory does an excellent job at predicting the properties of coordination compounds.

SUMMARY

Ionic, metallic, covalent, and atomic-molecular crystals are categorized by the type of forces holding them together. Ionic crystals are characterized by nondirectional electrostatic forces among ions, while metallic crystals are described as a sea of electrons surrounding a lattice of cations. Covalent crystals exhibit interlocking, directional covalent bonds, while atomic-molecular crystals are held together by intermolecular forces among the atoms or discrete molecules.

Solid state structures are adequately described by assuming that the atoms, ions, or molecules act like hard spheres packed together in various ways. The geometries of the structures are summarized in terms of the unit cells of the overall space lattices. In an A-type lattice all the atoms, ions, or molecules are identical. Simple cubic, body-centered cubic (bcc), cubic close-packed (ccp), and hexagonal close-packed (hcp) lattices are among the most common for A-type lattices. Cubic close-packed corresponds to an ABCABC packing scheme and a face-centered cubic unit cell. Hexagonal close-packed corresponds to an ABABAB packing scheme. The geometry of a lattice is characterized by the coordination number of the spheres, the number of spheres per unit cell, the fraction of space occupied, and a density expression.

In AB_n-type structures there are two types of atoms, ions, or molecules. The larger spheres are usually visualized to form an A-type lattice, while the smaller ones occupy some fraction of the holes (cubic, octahedral, or tetrahedral) in that lattice. Which holes are occupied is predicted by the radius ratio of the two spheres. For ionic crystals, the radius ratio is usually calculated as the radius of the cation over that of the anion. Ionic radii are derived from high-resolution x-ray studies.

Common AB lattices include the sodium chloride, or rock salt, structure, the CsCl structure, and the zinc blende and wurtzite forms of zinc sulfide. AB_2 lattices include the fluorite, cadmium iodide, and rutile structures. Each of these is characterized by a radius ratio and the structure and coordination number of both the anions and the cations. Many structures involving polyatomic molecules and ions can also be described by the above lattice types.

Various imperfections in solid state structures include the Schottky and Frenkel defects. The latter can lead to nonstoichiometric compounds.

The structures of spinels, $A^{II}B_2^{III}O_4$, are determined not only by the radius of the ions involved but by the crystal field stabilization energies of the cations which occupy octahedral or tetrahedral holes in the cubic close-packed lattice of oxide ions. These structures offer an opportunity to combine a knowledge of crystal field theory obtained in earlier chapters with the knowledge of solid state structures covered in this chapter.

PROBLEMS

7.1. Carefully define and give an example of metallic, covalent, and atomic-molecular crystals.

7.2. What types of crystals are formed by the following solid elements and compounds: C, Na, CO_2, Na_2O, NH_3, NH_4Cl, Kr, KrF_6, Br_2, BrF_3, LiBr? In each case, state the type of forces between the particles that make up the crystal lattice.

7.3. Briefly define and show a diagram which illustrates (*a*) London dispersion forces, (*b*) dipole-dipole forces, and (*c*) hydrogen bonds.

7.4. When ice melts, some individual water molecules become non-hydrogen-bonded. Using this information and Figure 7.4*b*, speculate on why the density of liquid water is greater than that of ice.

7.5. Given the following two-dimensional array of points and possible unit cells, discuss the advantages and disadvantages of each possible unit cell. Which do you think is the unit cell of choice?

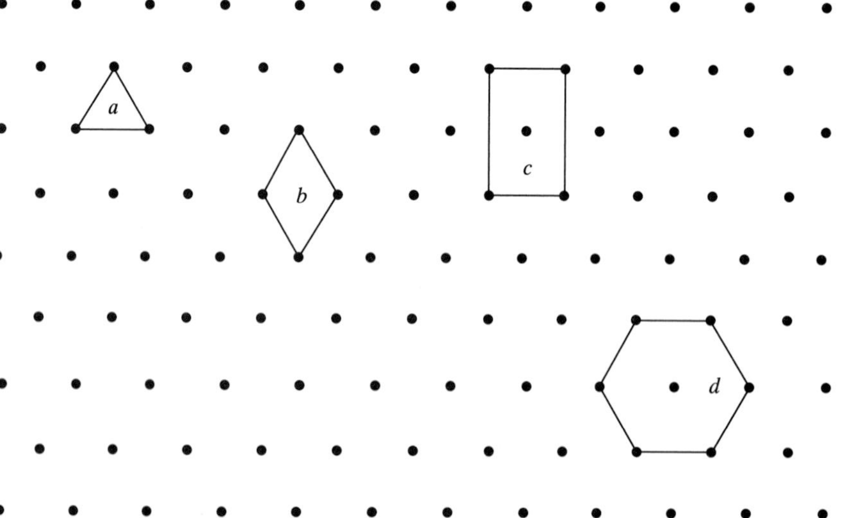

7.6. In describing the body-centered cubic unit cell, it was noted that the atom in the center has a coordination number of 8. Using Figure 7.8*a* as a guide, draw an extended diagram of the body-centered space lattice and demonstrate that a corner atom also has a coordination number of 8.

7.7. Given a face-centered cubic arrangement of atoms in which the corner atoms are type A and those at the face centers are type B, what is the empirical formula of the compound in terms of A and B? Is face-centered cubic the most accurate description of this arrangement? Why or why not?

7.8. Given the base-centered orthorhombic unit cell found in Figure 7.16, how many particles are there per unit cell?

7.9. Briefly recount how to calculate the number of atoms there are in a unit cell of a hexagonally close-packed array of atoms.

7.10. Suppose you have a mixed hexagonal close-packed arrangement where the unit cell has all corner and face atoms of type A and all interior atoms of type B. What is the empirical formula?

***7.11.** Suppose you have hard spheres of radius r in contact with each other in hexagonal close-packing.
(*a*) Calculate the volume of the unit cell of such an array.
(*b*) Calculate the fraction of that volume which is actually occupied by the hard spheres. (*Hint:* First determine the height of a unit cell and then its cross-sectional area.)

7.12. Gold crystallizes in a face-centered cubic arrangement. The observed unit cell length is 4.070 Å.
(*a*) Calculate the radius of a gold atom.
(*b*) Calculate the density of gold in grams per cubic centimeter.

7.13. Aluminum crystallizes in the face-centered cubic arrangement. If the observed density of aluminum metal is 2.70 g/cm^3, what would you predict for the unit cell dimension (in Ångstom units)?

7.14. Europium crystallizes in a body-centered cubic lattice. The density of europium is 5.26 g/cm^3.
(*a*) Calculate the unit cell edge length.
(*b*) Calculate a value for the atomic radius of europium.

7.15. The x-ray diffraction powder pattern for solid krypton shows that this substance exhibits the cubic close-packed structure. The density of solid krypton is 3.5 g/cm^3.
(*a*) How many atoms of krypton are present in a unit cell?
(*b*) What are the dimensions of the unit cell for solid krypton?
(*c*) Estimate the radius of the krypton atom.

7.16. Magnesium metal is very close to being a hexagonal close-packed array of atoms. X-ray studies show that the Mg–Mg distance is 3.203 Å. Calculate a value for the density for metallic magnesium. (The observed density is 1.745 g/cm^3.)

7.17. Show that the (*a*) triangular and (*b*) tetrahedral holes in a close-packed arrangement can accommodate spheres with radius ratios of 0.155 and 0.23, respectively.

7.18. Draw an appropriate diagram and calculate the radius ratio for a cubic hole.

7.19. How many MX formula units are there per unit cell in the (*a*) zinc blende and (*b*) wurtzite structures?

***7.20.** How many MX_2 formula units are there per unit cell of the (*a*) fluorite and (*b*) rutile structures?

7.21. Given the density of fluorite to be 3.18 g/cm^3 and the Ca–F interionic distance to be 2.37 Å, calculate a value for Avogadro's number.

7.22. Using radius ratios, suggest a probable structure for each of the following AB-type ionic crystals:

(*a*) BeO	(*e*) AgCl
(*b*) BeS	(*f*) AgBr
(*c*) MgO	(*g*) AgI
(*d*) MgS	(*h*) TlCl

Using Table 7.8, determine which of your suggestions are correct.

7.23. Using radius ratios, suggest a probable structure for each of the following AB_2-type ionic crystals:

(*a*) $SrCl_2$ (*d*) SnS_2
(*b*) Li_2O (*e*) MgF_2
(*c*) K_2O (*f*) MgI_2

Using Table 7.10, determine which of your suggestions are correct.

7.24. Cesium and gold form an ionic compound, Cs^+Au^-, with a cesium-gold distance of 3.69 Å. What type of lattice will CsAu adopt?

7.25. A part of the NaCl structure is reproduced below Problem 7.26. Which A-type lattice do the Cl^- anions assume? Describe this lattice in terms of a layering scheme ABCD, etc. What type of holes are occupied by the Na^+ cations? Label all the unoccupied tetrahedral and octahedral holes in the figure.

7.26. A part of the ZnS zinc blende is reproduced below. Which A-type lattice do the S^{2-} anions assume? Describe this lattice in terms of a layering scheme ABCD, etc. What type of holes are occupied by the Zn^{2+} cations? Label all the unoccupied tetrahedral and octahedral holes in the figure.

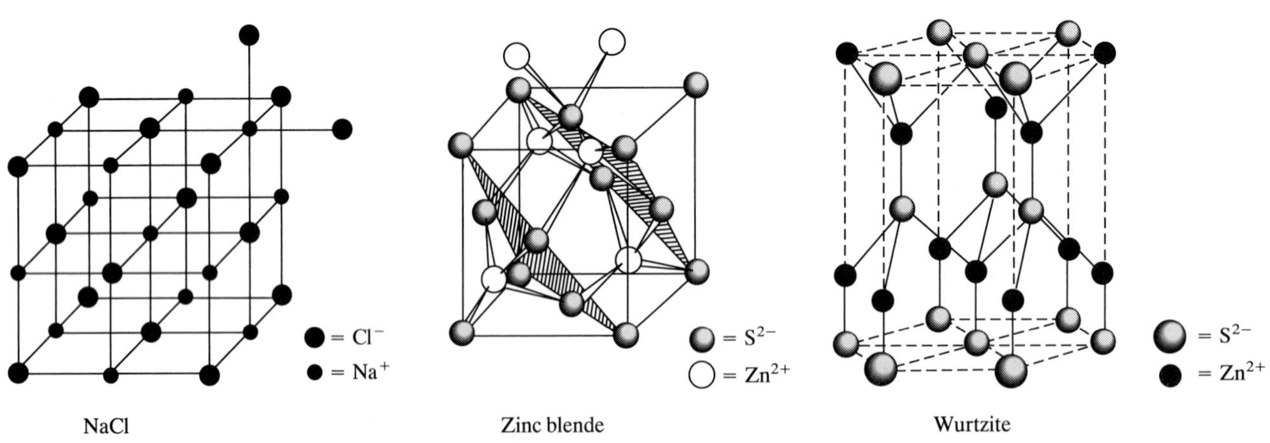

7.27. In the zinc blende structure, the occupied tetrahedral holes form a tetrahedron. Is this also true of the wurtzite structure? If so, outline the tetrahedron in the wurtzite structure found above. If not, do the occupied tetrahedral holes form another common three-dimensional solid? Which one?

7.28. In discussing the fluorite and cadmium iodide structures, we cited a stoichiometric requirement as the reason only half the cubic or octahedral holes were occupied by the cations. Determine the stoichiometry which would result if the cations occupied:

(*a*) All the cubic holes in fluorite
(*b*) All the octahedral holes in cadmium iodide.

7.29. Many MX type compounds with C.N. = 6 have the NaCl structure, whereas few have the NiAs structure in which the As atoms assume a hexagonal close-packed A-type lattice and the Ni atoms occupy the octahedral holes. How many NiAs formula units are there per unit cell in this structure?

7.30. The ionic radii (for C.N. = 6) for Cs^+ and F^- are 1.81 and 1.19 Å, respectively. Using radius ratios as an aid to your thinking, draw a diagram showing a reasonable structure for CsF. Can you think of another structure which is consistent with the radius-ratio concept?

7.31. Potassium fluoride, KF, has a rock salt type of crystal structure. The density of KF is 2.468 g/cm^3.

(*a*) Calculate the dimensions of the KF unit cell.
(*b*) Write an expression whereby the fraction of space occupied in the above KF crystal could be calculated.
*(*c*) Would you be surprised if the result of this calculation showed a fraction greater than 0.75? Why or why not?

7.32. If high pressure is applied to a compound MX, which structure, CsCl or NaCl, would be favored?

7.33. A particular solid has a structure in which W atoms are located at cube corners, O atoms at the centers of the cube edges, and Na atoms at cube centers. The cube edge is 3.86 Å.
(*a*) What is the formula of this material?
(*b*) What is its theoretical density?

7.34. Calculate the volume and the density of KCl. (*Hint:* Work with a single unit cell.)

7.35. Estimate the density of MgO using radii to determine the cell dimensions and the number of formula units per unit cell.

***7.36.** Calculate the density of CaF_2. The experimental density is 3.180 g/cm^3.

***7.37.** Consulting models of the nickel arsenide and sodium chloride structures if possible, speculate on why most MX-type compounds with C.N. = 6 assume the NaCl structure rather than the NiAs structure. Specifically, what features of the NiAs structure limit its occurrence?

***7.38.** Using radius ratios, predict what structure would be assumed by RbBr. Briefly justify your prediction. Using the structure you have predicted, calculate a value for the density of RbBr (*s*). The experimental value is 3.35 g/cm^3. Calculate the percentage difference between your value and the actual. Suppose the radius-ratio rules are violated for RbBr and it assumes a rock salt structure. Calculate a density for this structure of RbBr.

7.39. The ammonium cation is almost spherical and has an estimated radius of 1.37 Å. Suggest probable structures for ammonium iodide, chloride, and fluoride, NH_4X. Using Table 7.8, determine which of your suggestions is correct.

7.40. The hydroxide anion is almost spherical and has an estimated radius of 1.33 Å. Suggest probable structures for magnesium and calcium hydroxide, $M(OH)_2$. Using Table 7.10, determine which of your suggestions is correct.

7.41. $NaSbF_6$ (density = 4.37 g/cm^3) assumes the rock salt structure. Assuming the anion to be spherical, calculate the radius of SbF_6^-. (*Hint:* Use the radius of Na^+, the density, the formula weight, and the number of formula units per unit cell.)

7.42. How might a compound of stoichiometry $Cu_{1.77}S$ come about? The common oxidation states of copper are +1 and +2.

7.43. Speculate on the reason(s) that transition metal oxides are more frequently nonstoichiometric as compared to non-transition metal oxides.

7.44. The equation which determines whether a given compound will form under standard conditions is $\Delta G = \Delta H - T\Delta S$. Knowing that the formation of an ionic compound such as NaCl is almost always exothermic, how would the formation of a Schottky or Frenkel defect affect the value of ΔS, ΔH, and therefore ΔG? If the ΔH term were only a relatively small negative number, would the formation of impure crystals be more or less likely? Why?

7.45. In a spinel structure ($A^{II}B_2^{III}O_4$) the oxides are cubic close-packed, while the A(II) cations occupy one-eighth of the tetrahedral holes and the B(III) cations occupy one-half of the octahedral holes. How many oxide anions, tetrahedral holes, and octahedral holes are there per unit cell? If one-eighth of the tetrahedral and one-half of the octahedral holes are occupied as above, verify the above stoichiometric ratios.

7.46. What type of structure would you predict for the compounds Cr_2CoO_4 and Fe_2CoO_4? Explain any differences.

7.47. Would you expect Mn_3O_4 to form a *normal* or *inverse* spinel? Carefully define these terms and justify your answer.

CHAPTER 8

SOLID STATE ENERGETICS

Ionic and covalent bonds as well as metallic and intermolecular forces are the principal interactions among the particles that make up a solid. In this chapter, we will concentrate on the nature of ionic crystals in which the bonding is primarily characterized by electrostatic forces among oppositely charged ions. As in Chapter 7, we will see that the model in which ions are treated as hard spheres will serve as a good starting point. Nevertheless, it will quickly be complicated by (1) the intricate three-dimensional geometry of space lattices, (2) the necessary realization that ions are not point charges but rather electron clouds which can exert powerful if short-range repulsive forces, and (3) the contribution of covalent interactions among ions.

We start this second and last chapter on the solid state with a description of the theoretical determination of lattice energy. Next will be a consideration of how the lattice energy can be determined experimentally using the principles of thermochemistry you learned in previous chemistry courses. A discussion of such topics as the degree of covalent character in ionic crystals, the source of values for electron affinities, the estimation of heats of formation of unknown compounds, and the establishment of thermochemical radii of polyatomic ions will follow. We will conclude with a special section on the effects of crystal fields on transition metal radii and lattice energies

8.1 LATTICE ENERGY: A THEORETICAL EVALUATION

Lattice energy is the energy change that accompanies the process in which the isolated gaseous ions of a compound come together to form 1 mole of the ionic solid. For a solid composed of monoatomic, singly charged ions, such as sodium chloride, the lattice energy is the energy corresponding to the reaction represented in Equation (8.1):

$$Na^+(g) + Cl^-(g) \longrightarrow NaCl(s) \tag{8.1}$$

We will see shortly that energy is released in such a process; the energy of the products is lower than the energy of the reactants. We will always use the familiar thermochemical convention (that exothermic thermochemical quantities carry a negative sign), from which it follows that lattice energy is always a negative number. (It turns out that some sources of thermodynamic data do *not* follow the usual convention and define the lattice energy as the *magnitude* of the energy *release* and show it as a positive number.)

How can we approach a theoretical evaluation of the lattice energy? The simplest starting point is the electrostatic interaction within an "ion pair" made up of one sodium cation and one chloride anion. Assuming the energy of the isolated gaseous ions is taken to be zero, the potential energy of the ion pair is given by Coulomb's law and is shown in Equation (8.2):

$$E = \frac{(Z^+e)(Z^-e)}{r} \tag{8.2}$$

where Z^+ = integral charge on cation
Z^- = integral charge on anion
e = fundamental charge on an electron
= 1.602×10^{-19} C
r = interionic distance as measured from center of cation to center of anion

If we want the interionic distance to be in Ångstrom units and the energy to be in kilojoules (kJ), the exact relationship for Coulomb's law for one Na^+Cl^- ion pair is as given in Equation (8.3):

$$E = \frac{AZ^+Z^-}{r} \tag{8.3}$$

where $A = 2.308 \times 10^{-21}$†
E = energy, kJ
r = interionic distance, Å

The total coulombic energy associated with a given sodium cation, however, must account for all the charged species surrounding that cation. Figure 8.1*a* shows a portion of the NaCl rock salt lattice with a given sodium cation highlighted. Figure 8.1*b* shows the distances from the cation to various sets of neighbors, both anions and other cations. Note that there are 6 anions at a distance of r, 12 other cations at a distance of $r\sqrt{2}$, 8 anions at a distance of $r\sqrt{3}$, 6 cations at a distance of $2r$, and so forth. The sum of all these coulombic interactions will be the total coulombic energy for one sodium cation, E_{coul}, and is given by Equation (8.4):

$$E_{\text{coul}} = 6\frac{AZ^+Z^-}{r} + 12\frac{AZ^+Z^+}{r\sqrt{2}} + 8\frac{AZ^+Z^-}{r\sqrt{3}} + 6\frac{AZ^+Z^+}{2r} + \cdots \tag{8.4}$$

First noting that in NaCl, $Z^+ = -Z^-$, rearrangement yields Equation (8.5):

$$E_{\text{coul}} = \frac{AZ^+Z^-}{r}\left(6 - \frac{12}{\sqrt{2}} + \frac{8}{\sqrt{3}} - \frac{6}{2} + \cdots\right) \tag{8.5}$$

†Using SI units, Coulomb's law takes the form $E = Z^+Z^-e^2/(4\varepsilon r)$, where ε, the dielectric constant or permittivity, is 8.854×10^{-12} C^2 m^{-1} J^{-1}, r is in meters, and E is in joules. If r is converted to Ångstrom units and E is in kilojoules, Equation (8.3) results.

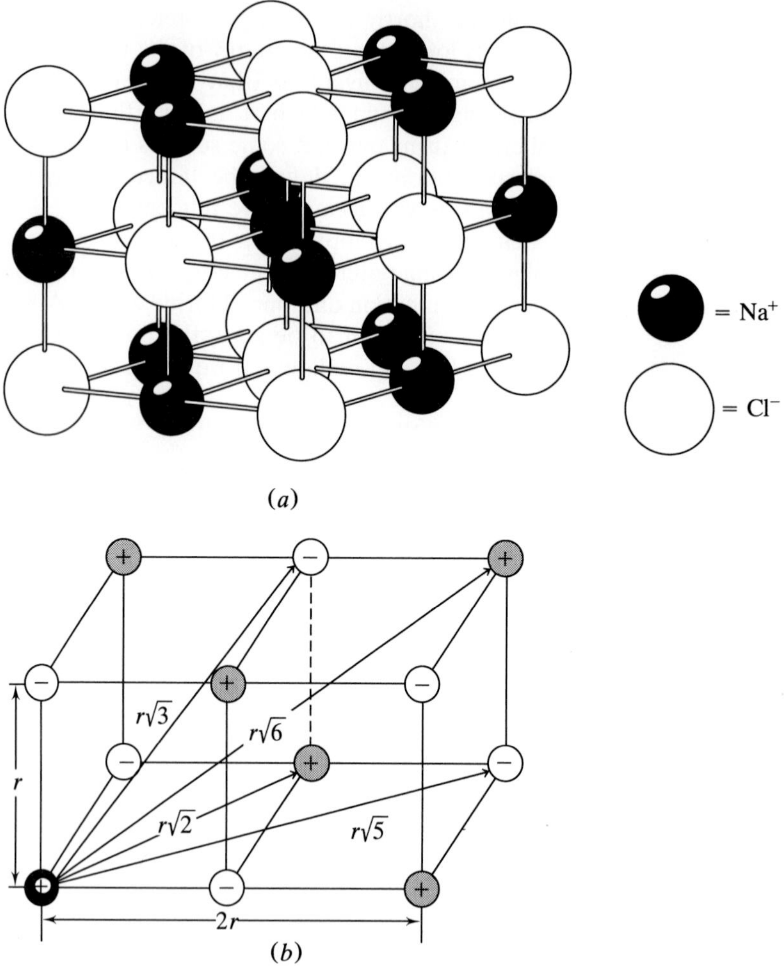

FIGURE 8.1
The ions surrounding a given sodium cation (black) in NaCl. (*a*) A view showing three complete sets of nearest neighbors. [Ref. 19.] (*b*) A schematic showing the distances to the first six sets of nearest neighbors. [Ref. 13.]

The geometric series in the parentheses is a constant which depends on the crystal structure. In other words, if the sodium cations and chloride anions assumed the CsCl, zinc blende, wurtzite, or some other crystal structure, the series in the parentheses would be different from that given above. These unique series for each crystal structure are known as *Madelung constants* (M) and are listed in Table 8.1. Using the symbol M_{NaCl} for the Madelung constant unique to the sodium chloride structure, Equation (8.5) can be simplified to Equation (8.6):

$$E_{\mathrm{coul}} = \frac{AZ^{+}Z^{-}M_{\mathrm{NaCl}}}{r} \tag{8.6}$$

Note that the E_{coul} of Equation (8.6) is the total coulombic energy for one sodium cation, assuming that all the ions are *point charges*, charges that act as if they are at the very center of the hard spheres representing the ions. In Chapter 7 we noted the utility of the hard-sphere model but also that it must be used with caution. Ions are not really hard spheres but electron clouds surrounding a

TABLE 8.1
Madelung constants for some common crystal structures

Crystal structure	Madelung constant
Sodium chloride	1.748
Cesium chloride	1.763
Zinc blende	1.638
Wurtzite	1.641
Fluorite	2.519
Rutile	2.408
Cadmium iodide	2.191

nucleus. Nevertheless, continuing with the analogy, if these hard spheres approach too closely to each other, a strong repulsive force will result. This would correspond to the hard spheres (actually, the filled electron clouds) beginning to interpenetrate each other. The energy of this interaction, E_{rep}, was modeled by Born to be of the form shown in Equation (8.7). The lattice energy of one cation in the crystal will, to a first approximation, be the sum of E_{coul} and E_{rep} and represents a balance between the attractive (negative) E_{coul} term and the repulsive (positive) E_{rep} term.

$$E_{rep} = \frac{B}{r^n} \tag{8.7}$$

where B = a constant
r = interionic distance
n = Born exponent, ranging from 5–12

Compared to a simple inverse (r^{-1}) dependence on r for the E_{coul} term, the relatively high value of n, the Born exponent (ranging from 5 to 12), indicates that E_{rep} will be very small at large interionic distances but significant at small interionic distances. In other words, E_{rep} will be a sensitive function of interionic distance and important only at very short distances—at "short range." n can be estimated from compressibility measurements in which the pressure needed to change the volume of an ionic substance is determined. Plots such as that shown in Figure 8.2 result. Note that at point (b) the crystal becomes very difficult to

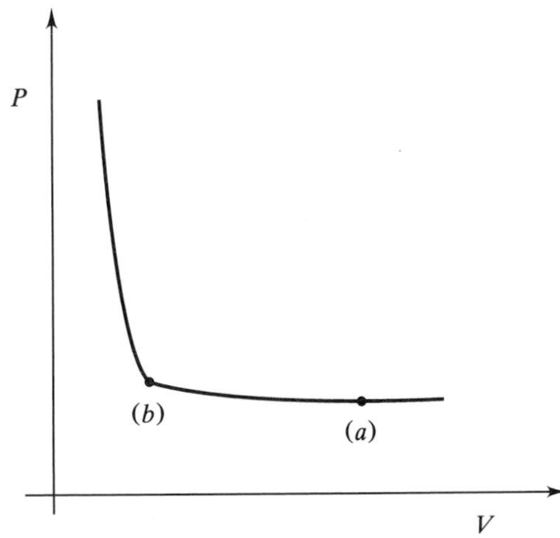

FIGURE 8.2
A representation of a plot of pressure versus volume for a crystal when it is compressed. At point (a) a small increase in the pressure results in a decrease in the volume. At point (b) a large increase in pressure is needed to compress the crystal any further. Data from such measurements can be related to the Born exponent for a given ion.

TABLE 8.2
Values of Born exponents for various electronic configurations

Atom / ion	Principle quantum no. of outermost electron	Electronic configuration	n
He	1	$1s^2$	5
Ne	2	$[He]2s^2 2p^6$	7
Ar	3	$[Ne]3s^2 3p^6$	9
Cu^+	3	$[Ne]3s^2 3p^6 3d^{10}$	9
Kr	4	$[Ar]4s^2 3d^{10} 4p^6$	10
Ag^+	4	$[Kr]4d^{10}$	10
Xe	5	$[Kr]5s^2 4d^{10} 5p^6$	12
Au^+	5	$[Xe]5d^{10}$	12

compress; that is, a sharp increase in the pressure is needed to compress or decrease the volume of the crystal any further. It has been found that the Born exponents derived from such measurements are related to the value of the principal quantum number of the outermost electron of an ion, as shown in Table 8.2.

Having established that the two principal components of the lattice energy U will be E_{coul}, an attractive (negative) term derived from the electrostatic interactions of point charges, and E_{rep}, a short-range repulsive (positive) term, we can write Equation (8.8), which should approximate the total lattice energy for an ionic crystal such as NaCl. Note that in order to put the energy on a per mole basis, both E_{coul} and E_{rep} have been multiplied by Avogadro's number N.

$$U = NE_{\text{coul}} \qquad + NE_{\text{rep}}$$

$$U = \frac{NAZ^+Z^-M_{\text{NaCl}}}{r} + \frac{NB}{r^n} \tag{8.8}$$

A plot of the E_{coul}, E_{rep}, and the resulting U is shown in Figure 8.3. As we anticipated in the beginning of this section, the process of forming the ionic solid from its constituent gaseous ions is exothermic. Note also that U goes through a minimum at r_0, known as the equilibrium interionic distance. Depending on your knowledge of calculus, you are more or less familiar with the idea that the derivative of a function is equal to zero when that function goes through a minimum (or maximum). Taking the derivative of U with respect to r, setting it equal to zero at $r = r_0$, solving for B, and substituting the result back into Equation (8.8) yields Equation (8.9), which is known as the *Born-Landé* equation. Note that we have combined N and A to yield a constant of 1389. The symbol U_0 indicates that this is the lattice energy evaluated at r_0.

$$U_0 = 1389\frac{Z^+Z^-M}{r_0}\left(1 - \frac{1}{n}\right) \tag{8.9}$$

where U_0 = lattice energy, kJ/mole, evaluated at r_0
Z^+, Z^- = integral charges of cation and anion
M = Madelung constant (Table 8.1)
r_0 = equilibrium interionic distance, Å
n = Born exponent (Table 8.2)

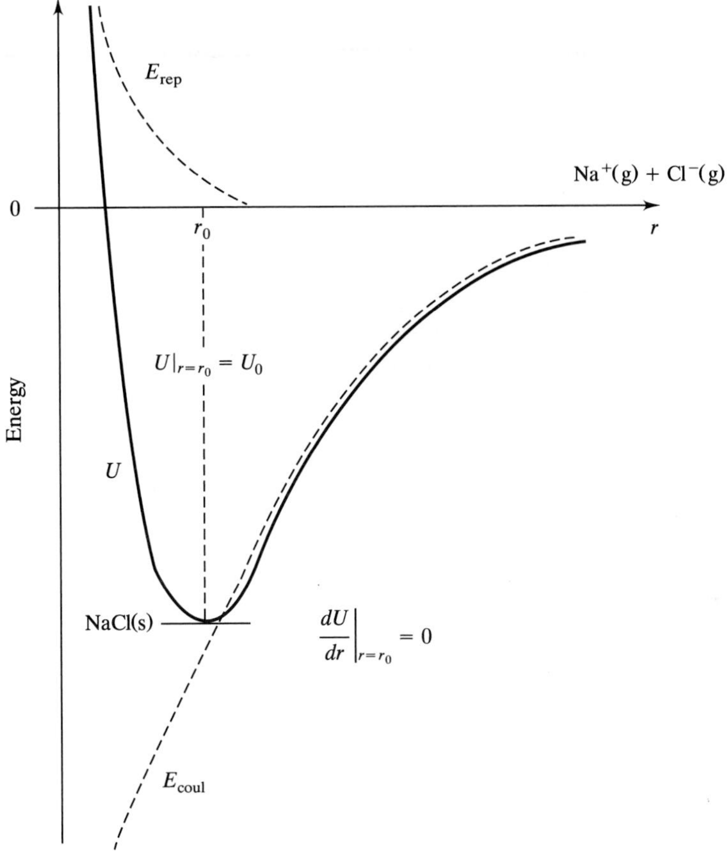

FIGURE 8.3
The lattice energy U as a function of interionic distance (solid line). The contributions of the short-range repulsive and coulombic energies are shown in dashed lines. Note that the derivative of U with respect to r can be set to zero at the minimum at which $r = r_0$, the equilibrium interionic distance. The constant B is evaluated using the result of this calculation.

Note again that because Z^+ and Z^- are of different signs, the lattice energy is a negative number. This means that the ionic solid, with its arrangement of ions in a crystal lattice, is of lower energy than the gaseous ions from which it is formed.

Having come this far, we are now in a position to actually calculate the lattice energy of NaCl using the Born-Landé equation. Z^+ and Z^- are $+1$ and -1, respectively, and $M_{NaCl} = 1.748$. r_0 is taken to be sum of the ionic radii, r_{Na^+} and r_{Cl^-}, found in Tables 7.4 and 7.6. The Born exponent n is determined by averaging the values for each ion. From Table 8.2 we see that the Na^+ ion corresponds to the Ne configuration for which $n = 7$, while the Cl^- ion corresponds to the Ar configuration for which $n = 9$. Accordingly, n for the crystal is 8. The substitution of these values into Equation (8.9) is shown in Equation (8.10):

$$U_0 = 1389\frac{(+1)(-1)(1.748)}{1.16 + 1.67}\left(1 - \frac{1}{8}\right) = -751 \text{ kJ/mol} \tag{8.10}$$

We see that the formation of 1 mole of NaCl from its constituent gaseous ions is a *highly* exothermic process. While there are other refinements to the estimation of lattice energy such as a small contribution from the van der Waals forces among ions and what is known as the *zero-point energy*, these are fairly small factors which we will not take into consideration in this text.

Analysis of the Born-Landé equation shows that there are two factors which affect the magnitude of the lattice energy. One is ionic charge. The more highly charged the ions, the greater the lattice energy. The second is the interionic distance r_0. The smaller this distance, the greater the lattice energy. For a given cation or anion, we can combine these two factors into what is known as the *charge-to-radius* (Z/r) *ratio*, or the *charge density*. As Z^+/r^+ or Z^-/r^- increases, so also does lattice energy. A plot of the magnitude of lattice energy $|U|$ versus equilibrium interionic distance, shown in Figure 8.4, illustrates this dependency for a variety of ionic compounds, all of which assume a rock salt structure (except CsCl added for comparison). Note that the 2 + /2 − compounds have lattice energies roughly four times the 1 + /1 − compounds. Also note that for the 1 + /1 − LiCl, NaCl, KCl, RbCl series (or the shorter MgO, CaO 2 + /2 − series) in which the ionic radius of the cation increases, the lattice energy decreases. The effect of increasing anion radius can be seen by comparing the values for NaF, NaCl, NaBr, NaI in the 1 + /1 − series or the MgO, MgS values in the 2 + /2 − series.

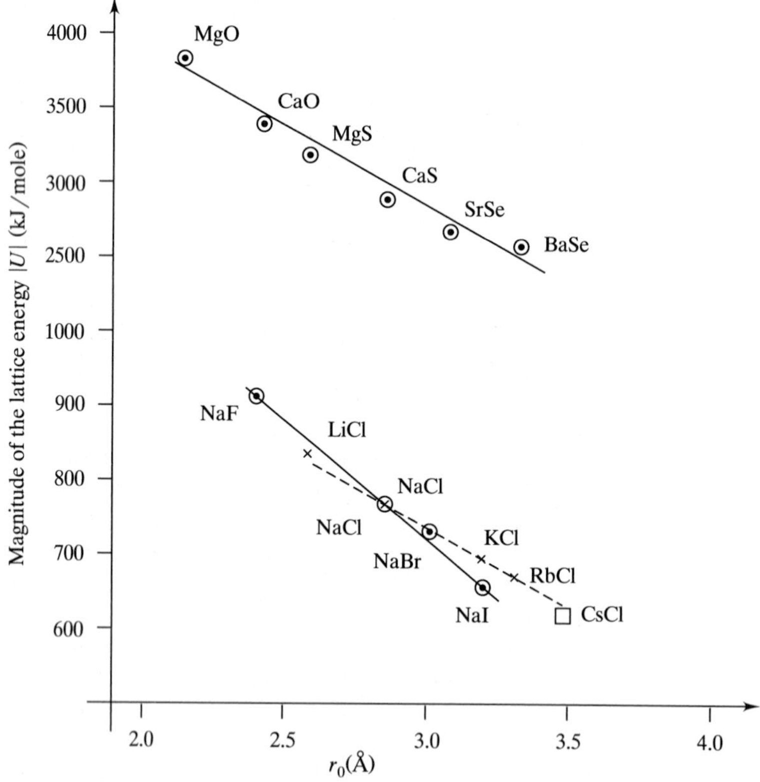

FIGURE 8.4
The magnitude of lattice energy $|U|$ (as calculated using the Born-Landé equation) plotted against equilibrium interionic distance r_0 for compounds with the sodium chloride structure. For unipositive-uninegative salts, the lower solid line shows the effect of increasing anion size, and the dashed line shows the effect of increasing cation size. The lattice energy for CsCl (open square) is shown for comparison. The upper line shows the effect of the variation in charge and in both cation and anion size for salts involving doubly positive and negative charges. [Adapted from Ref. 20, p. 76.]

There is a second theoretical approach to the calculation of lattice energies. A. Kapustinskii suggested that in the absence of specific knowledge about the crystal structure of a compound, the lattice energy could be estimated using Equation (8.11):

$$U = \frac{1202\nu Z^+Z^-}{r_0}\left(1 - \frac{0.345}{r_0}\right) \tag{8.11}$$

where U = lattice energy, kJ/mol
ν = number of ions per formula unit of compound
Z^+, Z^- = integral charges of cation and anion
r_0 = equilibrium interionic distance, Å

In an empirically derived equation such as this, the nature of the dependence of the lattice energy on Z^+, Z^-, ν, and r_0 is known qualitatively and then the constants are chosen so as to provide the best general quantitative agreement with the experimentally available lattice energies. Applying the Kapustinskii equation to NaCl (in which $\nu = 2$, $Z^+ = +1$, $Z^- = -1$, and $r_0 = 1.16 + 1.67$ Å), the lattice energy comes out to be -746 kJ/mol, which is quite similar to the value calculated using the Born-Landé equation.

8.2 LATTICE ENERGY: THERMODYNAMIC CYCLES

We would like to compare the above calculated values of lattice energies with those derived from experiment. Unfortunately, the lattice energy corresponding to the forward reaction shown in Equation (8.12) cannot be directly measured because it is not possible to produce isolated gaseous ions.

$$M^{n+}(g) + X^{n-}(g) \underset{-U}{\overset{U}{\rightleftharpoons}} MX(s) \tag{8.12}$$

The vaporization of an ionic solid results in ion pairs and other more complicated aggregates. If a direct measurement of lattice energy is not possible, how can we confirm or disprove the results of the Born-Landé (or the Kapustinskii) equation? The answer lies in thermodynamic cycles.

Thermodynamic cycles are applications of *Hess' law*, which states that the total enthalpy change for (or *heat of*) a reaction is independent of the pathway followed from reactants to products. Max Born and Fritz Haber are credited for first applying Hess' law to an ionic solid. The Born-Haber cycle for a general alkali metal halide (M^IX) is shown in Figure 8.5. Equation (8.13) at the top of the cycle shows the formation of MX(s) from its constituent elements in their standard states and therefore corresponds to the standard enthalpy of formation. The reactions in the box of the figure constitute a series of steps (a second reaction pathway) which add up to the same overall reaction. According to Hess' law, sometimes known as the *law of heat summation*, the sum of the energies of these steps should be equal to the standard heat of formation ΔH_f° as shown in Equation (8.14) in the figure. With the exception of the lattice energy, all the quantities in Equation (8.14) are known, and therefore a value for lattice energy based on experimental thermochemical values can be calculated. (See Problem 8.18 for some historical perspective on this process.)

Table 8.3 shows the data and resulting experimental lattice energies, labeled $U_{\text{B-H}}$, for the alkali-metal halides. The Born-Landé ($U_{\text{B-L}}$) and the very similar Kapustinskii (U_{Kap}) lattice energies are also tabulated. Note that the $U_{\text{B-H}}$ for

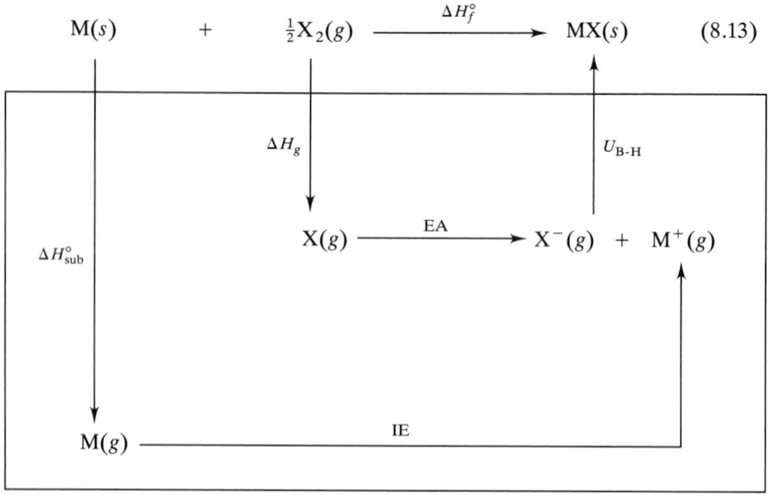

$$\Delta H_f^\circ = \Delta H_{\text{sub}}^\circ + \text{IE} + \Delta H_g + \text{EA} + U \qquad (8.14)$$

where ΔH_f° = standard enthalpy of formation
$\Delta H_{\text{sub}}^\circ$ = heat of sublimation of M(*s*)
IE = ionization energy of M
ΔH_g = enthalpy of formation of gaseous X
EA = electron affinity of X
$U_{\text{B-H}}$ = lattice energy of MX

FIGURE 8.5
The Born-Haber cycle for an alkali-metal halide. Equation (8.13) corresponds to the standard enthalpy of formation of MX(*s*). The equations in the box represent a second pathway for the formation of MX(*s*) from its constituent elements. Equation (8.14) represents Hess' law of summation for these two pathways.

NaCl (−787 kJ/mol) is only 4.6 percent greater than the −751 kJ/mol obtained from the Born-Landé equation, which we know assumes an electrostatic interaction among ions regarded as hard spheres. This agreement, although not proof of the ionicity of these compounds, is certainly an encouraging sign that our assumptions regarding the contributions to lattice energy are consistent with thermochemical observations. In general, the experimental lattice energies ($U_{\text{B-H}}$) in Table 8.3 are 4 to 6 percent greater (less if van der Waals forces and zero-point energy corrections are applied) then the theoretical values, the difference usually being attributed to covalent contributions to the lattice energy.

Under what conditions should we expect significant covalent contributions to the lattice energy? You learned in earlier courses that simple ionic compounds are assumed to result when the two constituent atoms have significantly different electronegativities. As a rule of thumb we take a "significant difference" to be any value greater than about 1.5. For example, the difference in electronegativity, Δ(EN), between sodium and chlorine is 2.1 [= 3.0 − 0.9]. Accordingly, when atoms of these elements are combined (see Figure 7.1), we expect electrons to be transferred and an ionic compound to result. Now you may recall that while there is a range of bond types from pure covalent to pure ionic, no bond is truly 100 percent ionic. There will always be some degree of covalent character, that is, electron cloud overlap and electron sharing. We would anticipate that the degree

TABLE 8.3
Thermochemical data and lattice energies for alkali-metal halides[a]

	ΔH_f° 298	ΔH_{sub}° [b]	IE	ΔH_g [c]	EA[d]	$U_{B\text{-}H}$ [e]	$U_{B\text{-}L}$ [f]	% diff.[g]	U_{kap} [h]
LiF	−616.0	159.4	520.3	79.0	−328.0	−1046.7	−968	7.5	−960
NaF	−573.6	107.3	495.9	79.0	−328.0	−927.8	−886	4.5	−873
KF	−567.3	89.2	418.9	79.0	−328.0	−826.4	−784	5.1	−774
RbF	−557.7	80.9	403.1	79.0	−328.0	−792.7	−752	5.1	−741
CsF	−553.5	76.1	375.8	79.0	−328.0	−756.4	−724	4.3	−709
LiCl	−408.6	159.4	520.3	121.7	−349.0	−861.0	−810	5.9	−810
NaCl	−411.1	107.3	495.9	121.7	−349.0	−787.0	−751	4.6	−746
KCl	−436.8	89.2	418.9	121.7	−349.0	−717.6	−677	5.7	−672
RbCl	−435.3	80.9	403.1	121.7	−349.0	−692.0	−652	5.8	−647
CsCl	−443.0	76.1	375.8	121.7	−349.0	−667.6	−637	4.6	−622
LiBr	−351.2	159.4	520.3	111.9	−324.7	−818.1	−774	5.4	−772
NaBr	−361.1	107.3	495.9	111.9	−324.7	−751.5	−719	4.3	−713
KBr	−393.8	89.2	418.9	111.9	−324.7	−689.1	−650	5.6	−645
RbBr	−394.6	80.9	403.1	111.9	−324.7	−665.8	−628	5.7	−622
CsBr	−395.0	76.1	375.8	111.9	−324.7	−634.1	−613	3.3	−599
LiI	−270.4	159.4	520.3	106.8	−295.2	−761.7	−724	5.0	−718
NaI	−287.6	107.3	495.9	106.8	−295.2	−702.6	−675	3.9	−667
KI	−327.9	89.2	418.9	106.8	−295.2	−647.6	−614	5.2	−607
RbI	−333.8	80.9	403.1	106.8	−295.2	−629.4	−593	5.8	−586
CsI	−346.6	76.1	375.8	106.8	−295.2	−610.1	−580	4.9	−566

[a]All data from *Handbook of Chemistry and Physics*, 67th ed., 1986–1987, Chemical Rubber Company Press, West Palm Beach, Fla. (All values in kJ/mol.)

[b]This quantity is also referred to as the heat of atomization.

[c]For fluorine and chlorine, ΔH_g, the enthalpy of formation of gaseous X corresponds closely to $\frac{1}{2}D$, where D is the X—X bond energy; for bromine and iodine, it does not.

[d]The sign convention used for electron affinity is the normal thermodynamic one in which a negative sign corresponds to an exothermic process.

[e]$U_{B\text{-}H}$ is the lattice energy calculated from the Born-Haber thermodynamic cycle.

[f]$U_{B\text{-}L}$ is the lattice energy calculated from the Born-Landé equation, Equation (8.9). Shannon-Prewitt ionic radii (C.N. = 6) used throughout. No corrections for van der Waals forces or zero-point energy have been applied.

[g]% diff = $[(U_{B\text{-}H} - U_{B\text{-}L})/U_{B\text{-}H}] \times 100$.

[h]U_{Kap} is the lattice energy calculated from the Kapustinskii equation, Equation (8.11). Shannon-Prewitt ionic radii (C.N. = 6) used throughout.

of covalent character should increase as electronegativities become more similar and therefore expect to find more covalent character in NaI [Δ(EN) = 1.6] than in NaF [Δ(EN) = 3.1]. Yet, as shown in Table 8.4, the percentage difference between $U_{B\text{-}H}$ and $U_{B\text{-}L}$ for these compounds is essentially the same. It appears that the correlation between $U_{B\text{-}H}$ and $U_{B\text{-}L}$ is not a particularly sensitive measure of covalent character when Δ(EN) is much above 1.5. It could very well be that the ionic contribution to lattice energy decreases as the covalent contribution increases.

When Δ(EN) falls below 1.5, however, Table 8.4 shows that we begin to see some correlation between the expected high degree of covalent character and the percentage difference between $U_{B\text{-}H}$ and $U_{B\text{-}L}$. For example, note that the electronegativity difference between silver and iodine is only 0.6 and $U_{B\text{-}H}$ is 30.4 percent greater than $U_{B\text{-}L}$. AgCl and AgBr show similar results as do the significantly covalent thallium chloride, bromide, and iodide salts.

TABLE 8.4
A comparison of Born-Haber and Born-Landé lattice energies for sodium, silver, and thallium halides

	$U_{\text{B-H}}$* kJ / mol	$U_{\text{B-L}}$† kJ / mol	Diff. EN‡	% diff.§
NaF	−927.8	−886	3.1	4.5
NaCl	−787.0	−751	2.1	4.6
NaBr	−751.5	−719	1.9	4.3
NaI	−702.6	−675	1.6	3.9
AgCl	−915.5	−734	1.1	19.8
AgBr	−903.3	−703	0.9	22.2
AgI	−889.1	−619	0.6	30.4
TlCl	−748.4	−669	1.2	10.6
TlBr	−732.0	−643	1.0	12.2
TlI	−707.1	−607	0.7	14.2

*$U_{\text{B-H}}$ is the lattice calculated from the Born-Haber thermodynamic cycle.

†$U_{\text{B-L}}$ is the lattice energy calculated from the Born-Landé equation, Equation (8.9). Shannon-Prewitt ionic radii (C.N. = 6) used throughout. No corrections for van der Waals forces or zero-point energy have been applied.

‡Using Pauling electronegativities.

§% diff. = $[(U_{\text{B-H}} - U_{\text{B-L}})/U_{\text{B-H}}] \times 100$.

This is a good place to point out that we also expect a significant degree of covalent character for zinc sulfide, ZnS, for which Δ(EN) is only 0.9. [A value for the second electron affinity for sulfur ($S^- \rightarrow S^{2-}$) independent of a Born-Haber cycle is not available, and therefore we cannot compare $U_{\text{B-H}}$ and $U_{\text{B-L}}$ in this case.] This is consistent with the reasoning (see p. 174) that the covalent character of ZnS is at least partially responsible for it forming the zinc blende and wurtzite structures in which the Zn^{2+} ions occupy tetrahedral holes instead of the octahedral holes predicted by radius ratios. The latter, we have seen, are predicted on an ionic model. The preference for tetrahedral hole occupancy is consistent with the presence of directional covalent bonds (using sp^3 hybrid orbitals) among the zinc and sulfur atoms rather than nondirectional ionic bonds.

Electron Affinities

While Born-Haber cycles in conjunction with estimates of lattice energy from the Born-Landé or Kapustinskii equations were some of the first sources of electron affinities, a variety of methods are now available to produce more reliable values. The newer methods, however, only afford values for the first electron affinity, as represented in Equation (8.15):

$$X(g) + e^-(g) \longrightarrow X^-(g) \tag{8.15}$$

If we want a value for the second electron affinity, we must still use thermodynamic cycles and the Born-Landé and Kapustinskii equations. For example, suppose we wish a value for the second electron affinity of oxygen, depicted in Equation (8.16):

$$O^-(g) + e^-(g) \longrightarrow O^{2-}(g) \tag{8.16}$$

We need to construct a thermodynamic cycle involving a metal oxide for which all quantities except the second electron affinity are either known experimentally or,

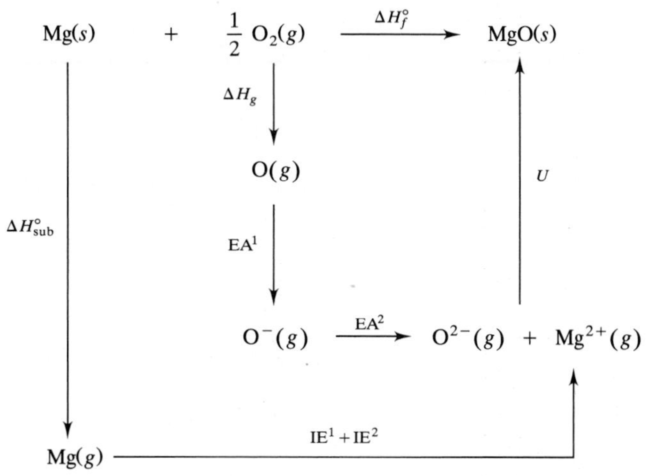

$Mg(s)$ + $\frac{1}{2}O_2(g)$ $\xrightarrow{\Delta H_f^\circ}$ $MgO(s)$

ΔH_g

$O(g)$

U

ΔH_{sub}°

EA^1

$O^-(g)$ $\xrightarrow{EA^2}$ $O^{2-}(g)$ + $Mg^{2+}(g)$

$IE^1 + IE^2$

$Mg(g)$

$$\Delta H_f^\circ = \Delta H_{sub}^\circ + IE^1 + IE^2 + \Delta H_g + EA^1 + EA^2 + U \tag{8.17}$$

$$\begin{aligned} EA^2 &= \Delta H_f^\circ - \Delta H_{sub}^\circ - IE^1 - IE^2 - \Delta H_g - EA^1 - U \\ &= -601.7 - 147.7 - 737.8 - 1450.8 - 249.1 - (-141.0) - (-3930) \\ &= 880\ \text{kJ/mole} \end{aligned} \tag{8.18}$$

FIGURE 8.6
The Born-Haber cycle for magnesium oxide used to calculate the second electron affinity of oxygen.

in the case of lattice energy, available from the Born-Landé or Kapustinskii equation. Such a cycle for magnesium oxide is shown in Figure 8.6. MgO assumes a rock salt structure, and its lattice energy is obtained from the Born-Landé equation as shown in Equation (8.19):

$$U = 1389\frac{(+2)(-2)(1.748)}{(0.86 + 1.26)}\left(1 - \frac{1}{7}\right) = -3930\ \text{kJ/mol} \tag{8.19}$$

Substitution into Equation (8.18) yields a value of 880 kJ/mol for the second electron affinity. (A better value of the second electron affinity of oxygen would be obtained by averaging the results of a number of calculations of the above type for a variety of oxide salts.) Recall that a negative value of electron affinity corresponds to an anion being more stable than the neutral atom. It follows that this positive value for the second electron affinity of oxygen indicates that the $O^{2-}(g)$ ion is unstable relative to $O^-(g)$. However, since MgO readily forms and is a stable compound, it appears that the energy needed to add an electron to $O^-(g)$ is more than compensated for by the highly exothermic lattice energy of a compound like MgO.

Heats of Formation for Unknown Compounds

Have you ever considered why it is always assumed that sodium and chlorine form NaCl rather than $NaCl_2$ or why calcium and chlorine form $CaCl_2$ rather than CaCl or $CaCl_3$? Such results are usually discussed in beginning courses in terms of the special stability of a noble-gas electronic configuration. We are now in a position to calculate the heat of formation for these various possibilities and analyze the

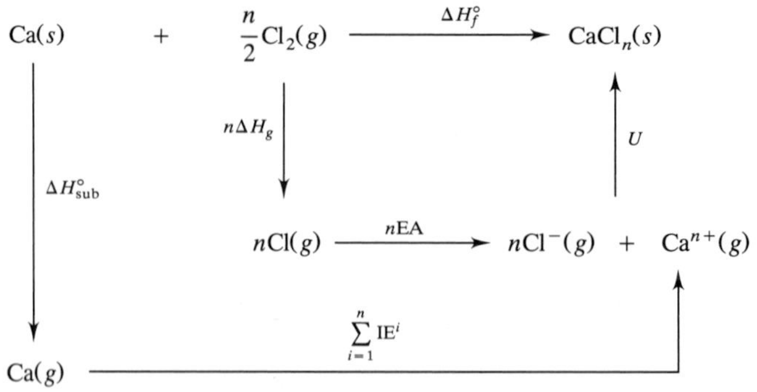

$$
\begin{array}{ccccccc}
Ca(s) & + & \frac{n}{2}Cl_2(g) & \xrightarrow{\Delta H_f^\circ} & & & CaCl_n(s) \\
\downarrow \Delta H^\circ_{sub} & & \downarrow n\Delta H_g & & & & \uparrow U \\
 & & nCl(g) & \xrightarrow{nEA} & nCl^-(g) & + & Ca^{n+}(g) \\
 & & & & & & \uparrow \\
Ca(g) & & \xrightarrow{\sum_{i=1}^{n} IE^i} & & & &
\end{array}
$$

$$\Delta H_f^\circ = \Delta H^\circ_{sub} + \sum_{i=1}^{n} IE^i + n\Delta H_g + nEA + U \tag{8.20}$$

where $\Delta H^\circ_{sub} = 178.2$ kJ/mole
$IE^1 = 589.8$ kJ/mole
$IE^2 = 1145.5$ kJ/mole
$IE^3 = 4912.4$ kJ/mole
$\Delta H_g = 121.7$ kJ/mole
$EA = -349.0$ kJ/mole
U from Kapustinskii equation

FIGURE 8.7
The Born-Haber cycles for $CaCl_n$, $n = 1, 2, 3$, used to calculate the ΔH_f° for these three compounds.

results. Figure 8.7 shows a Born-Haber cycle for the formation of $CaCl_n$, where $n = 1$ to 3. We can use the Kapustinskii equation to estimate lattice energies. The Shannon-Prewitt radii (C.N. = 6) for Ca^{2+} and Cl^- are 1.14 and 1.67 Å, respectively. Ca^+ should be significantly larger than Ca^{2+} due to the addition of a $4s$ electron. For the purpose of making a rough calculation, we will use a rather arbitrarily chosen but reasonable value of 1.5 Å for the radius of Ca^+. As the removal of one of the six $3p$ electrons of Ca^{2+} should not produce a particularly large effect, Ca^{3+} should be only slightly smaller than Ca^{2+}, so we will use a value of 1.1 Å. The calculation of U_{Kap} is shown in Equation (8.21):

$$U_{Kap_n} = \frac{1202(n+1)(+n)(-1)}{r(Ca^{n+}) + 1.67}\left[1 - \frac{0.345}{r(Ca^{n+}) + 1.67}\right] \tag{8.21}$$

The results for both U_{Kap} and the ΔH_f°'s are shown in Table 8.5. Note that CaCl is a thermodynamically feasible compound but that $CaCl_2$ is more stable and

TABLE 8.5
Ca^{n+} radii, U_{Kap_n}, and ΔH_f° for $CaCl_n$

$CaCl_n$	$r(Ca^{n+})$, Å	U_{Kap_n}, kJ/mol	ΔH_f°, kJ/mol
CaCl	1.5	−670*	−130*
$CaCl_2$	1.14	−2250	−792†
$CaCl_3$	1.1	−4500*	1600*

*The number of significant figures in the estimated radii allow only two significant figures in this result.
†Actual value of $\Delta H_f^\circ[CaCl_2] = -795.8$ kJ/mol.

TABLE 8.6
Selected lattice energies for salts containing polyatomic ions*

Name	Formula	Calculated lattice energy, kJ / mol	Thermo-chemical lattice energy, kJ / mol
Sodium tetraborohydride	$NaBH_4$	−703	
Sodium tetraborofluoride	$NaBF_4$	−657	−619
Sodium carbonate	Na_2CO_3	−2301	−2030
Sodium cyanide	NaCN	−738	−739
Sodium bromate	$NaBrO_3$	−803	−814
Sodium chlorate	$NaClO_3$	−770	−770
Sodium hydride	NaH	−782	
Sodium hydroxide	NaOH	−887	−900
Sodium nitrate	$NaNO_3$	−755	−756
Sodium nitrite	$NaNO_2$	−774	−748
Sodium oxide	Na_2O	−2481	
Sodium perchlorate	$NaClO_4$	−643	−648
Sodium sulfide	Na_2S	−2192	−2203
Sodium sulfate	Na_2SO_4	−1827	−1938
Sodium tetrachloroaluminate	$NaAlCl_4$	−556	
Sodium thiocyanate	NaSCN	−682	−682
Potassium hexachloroplatinate(IV)	K_2PtCl_6	−1468	
Ammonium nitrate	NH_4NO_3	−661	−676
Ammonium perchlorate	NH_4ClO_4	−583	−580

*From H. D. B. Jenkins, *Handbook of Chemistry and Physics*, 67th ed., 1986–1987, Chemical Rubber Company Press, West Palm Beach, Fla., p. D-100.

therefore more feasible. We see that the extra ionization energy needed to produce Ca^{2+} is more than repaid in the additional lattice energy released in forming the $CaCl_2$ lattice. $CaCl_3$ is not thermodynamically feasible due to the inordinately high third ionization energy of calcium which is a result of having to extract an electron from a filled Ne shell wherein the effective nuclear charge is very high. Even though the lattice energy for $CaCl_3$ is more than twice that of $CaCl_2$, it is not enough to make up for the third ionization energy. Given the above discussion, it is not surprising that neither CaCl nor $CaCl_3$ exist.

Thermochemical Radii

With the availability of calculated lattice energies (see Table 8.6 for some representative values) of a series of alkali-metal salts involving polyatomic anions and other salts involving polyatomic cations, Kapustinskii's equation can be used to estimate the radii of these ions. Using the listed value for the lattice energy of $NaClO_4$ (−643 kJ/mol) and the Shannon-Prewitt radius of the sodium cation in Kapustinskii's equation, a value of 2.19 Å for the radius of the perchlorate anion can be calculated. Radii derived in this manner are called "thermochemical radii." Some representative values are presented in Table 8.7. Interpretation of these results must be approached with caution, but they do give some indication of the effective size of polyatomic ions.

TABLE 8.7
Selected thermochemical radii*

Ion	Radius, Å	Ion	Radius, Å
BH_4^-	1.93	NO_2^-	1.92
BF_4^-	2.32	O^{2-}	1.49
CO_3^{2-}	1.78	ClO_4^-	2.40
CN^-	1.91	S^{2-}	1.91
BrO_3^-	1.54	SO_4^{2-}	2.58
ClO_3^-	1.71	$AlCl_4^-$	2.95
H^-	1.73	SCN^-	2.13
OH^-	1.33	$PtCl_6^{2-}$	3.13
NO_3^-	1.79	NH_4^+	1.37

*From H. D. B. Jenkins and K. P. Thakur, *J. Chem. Edu.*, **56**(9):577 (1979).

8.3 LATTICE ENERGIES AND IONIC RADII: CONNECTING CRYSTAL FIELD EFFECTS WITH SOLID STATE ENERGETICS

If you have already covered crystal field theory (Chapter 4), we are now in a position to apply it to a discussion of the lattice energies of solid state compounds containing transition metals. We start with the ionic radii of the first-row transition metals.

The dashed lines of Figure 8.8 show the expected radii of the M^{2+} and M^{3+} ions. From general chemistry (also see the discussion of Chapter 9) you know that the radii (atomic or ionic) would be expected to decrease across a period. Briefly, this is because as electrons are added to the $3d$ subshell, these electrons, being on the average the same distance from the nucleus, do not shield each other particularly well from the nucleus. The number of shielding electrons (the [Ar] shell in the first-row transition metals) remains constant, while the nuclear charge increases from left to right. It follows that the effective nuclear charge (the actual nuclear charge minus the number of shielding electrons) increases from left to right and serves to pull the partially filled $3d$ electron cloud closer in to the nucleus. Accordingly, the radii decrease across the period.

Figure 8.8 also shows the actual M^{2+} and M^{3+} radii for both high- and low-spin electronic configurations in octahedral fields. Octahedral crystal fields result when the transition metal ion occupies an octahedral hole in the A-type lattice of anions. To pick two often-cited examples, such a situation is common in the oxides and chlorides. These anions lie fairly low in the spectrochemical series, and therefore a weak-octahedral-field–high-spin case results. We know that the splitting of the d orbitals in an octahedral field is as shown in Figure 8.9 and that the first three $3d$ electrons will enter the t_{2g} set of orbitals which point in between the anionic ligands (Figure 8.9a). The repulsions among the ligands and the metal $3d$ electrons will be less than normal (that is, less than if the metal electrons were in individual or a set of spherically symmetric orbitals; see pp. 61–63 for details), and therefore the ligands will be able to approach the metal ion more closely. Since the radius of the ligand is a constant, the radius of the metal ion will decrease more than expected in these cases. When the fourth and fifth electrons enter the e_g orbitals which point directly at the ligands (Figure 8.9b), the repulsion between these electrons and the ligands will be greater than normal and the metal radius will increase. The resulting $t_{2g}^3e_g^2$ (d^5) electronic configuration is spherically symmetrical, and therefore its radius lies on the line representing the expected

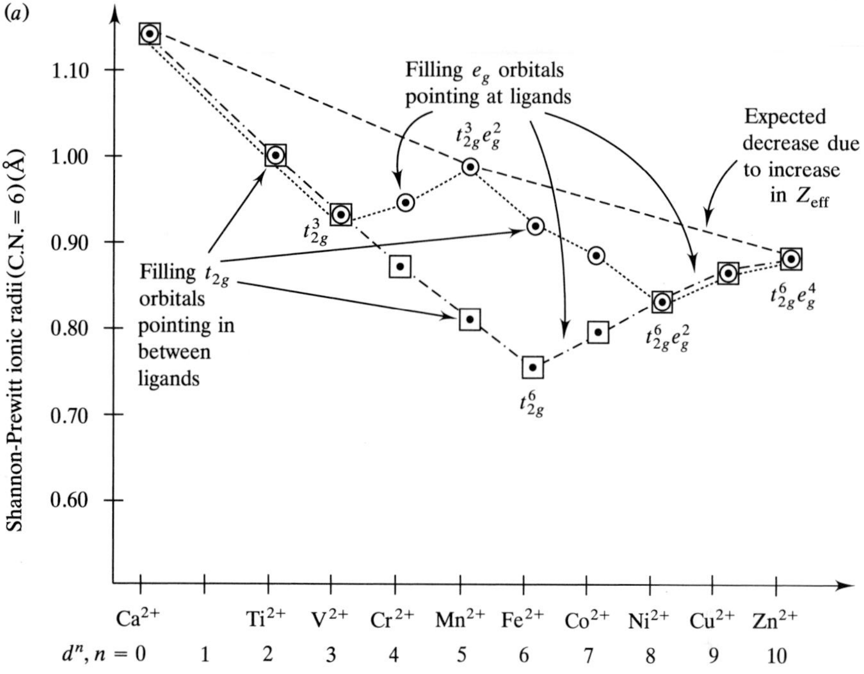

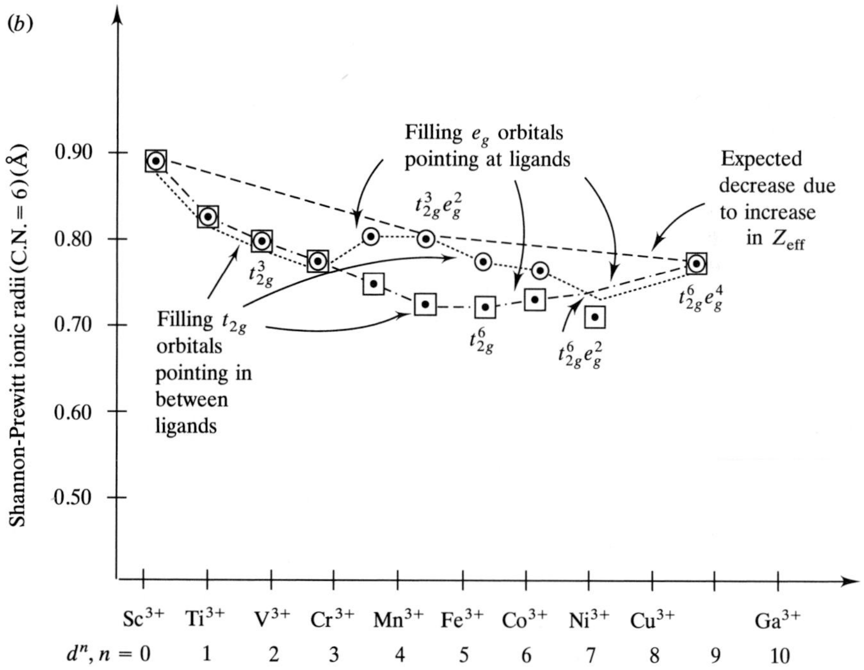

FIGURE 8.8
Shannon-Prewitt ionic radii for (*a*) M^{2+} and (*b*) M^{3+} cations having electron configurations of $3d^n$, $n = 0$ to 10. Open circles = high-spin cases; open squares = low-spin cases; dashed lines = trend for spherically symmetric sets of orbitals; dotted lines = trend for weak-field–high-spin cases; dashed-dotted lines = trend for strong-field–low-spin cases.

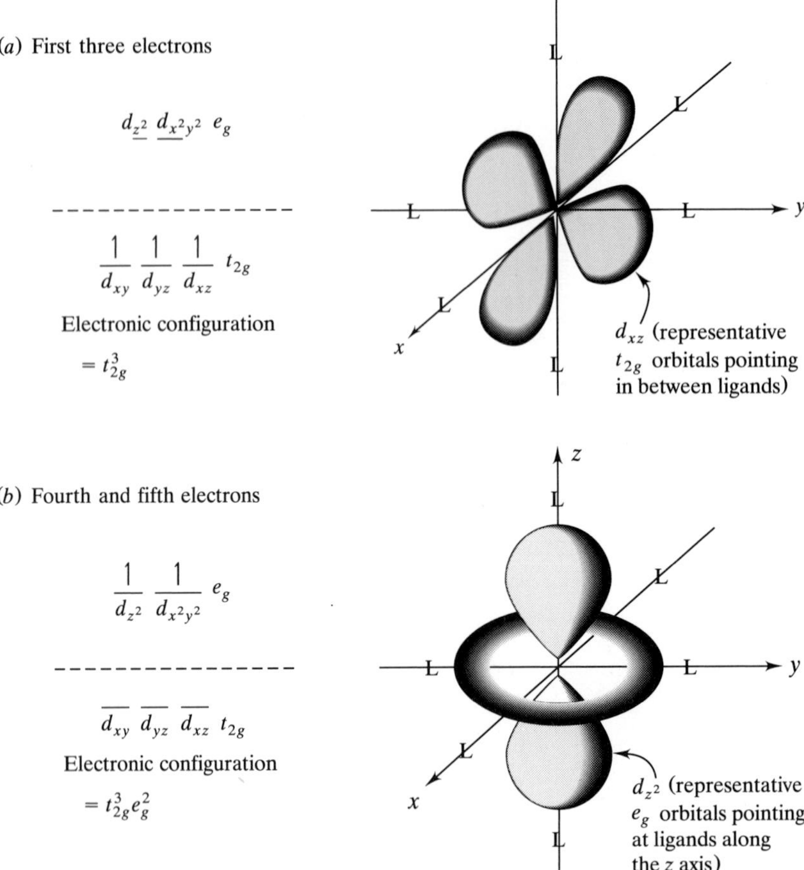

FIGURE 8.9
The placement of the first five d electrons in a weak-octahedral-field–high-spin case. (a) The first three electrons occupy the t_{2g} orbitals pointing in between the ligands. The ionic radii of ions with these configurations are smaller than expected. (b) The fourth and fifth electrons occupy e_g orbitals pointing directly at the ligands. The ionic radii of ions with these configurations increase due to electron-electron repulsions.

trend in the absence of crystal field effects. This decrease (while the t_{2g} orbitals are filled) and increase (while the e_g orbitals are filled) will be repeated as the sixth to tenth $3d$ electrons are added. Again, the $t_{2g}^6e_g^4$ (d^{10}) configuration is spherically symmetric.

In a strong-octahedral-field–low-spin case, the first six electrons occupy the t_{2g} set, and the radii decrease more than expected. The last four electrons occupy the e_g set, and the radii increase until the spherically symmetric $t_{2g}^6e_g^4$ configuration is again attained.

Given these trends in radii for transition metal ions in octahedral fields, we can turn to a discussion of lattice energies in these compounds. Figure 8.10 shows the lattice energies for M^{2+} chlorides. In the absence of crystal field effects (dashed line) we expect an increase in lattice energy as the radius decreases. This is a consequence of the dependence of lattice energy on the charge-to-radius (Z/r) ratio of the cation which we discussed in connection with the Born-Landé equation. Since the charge on the metal cation is always $+2$, the lattice energy

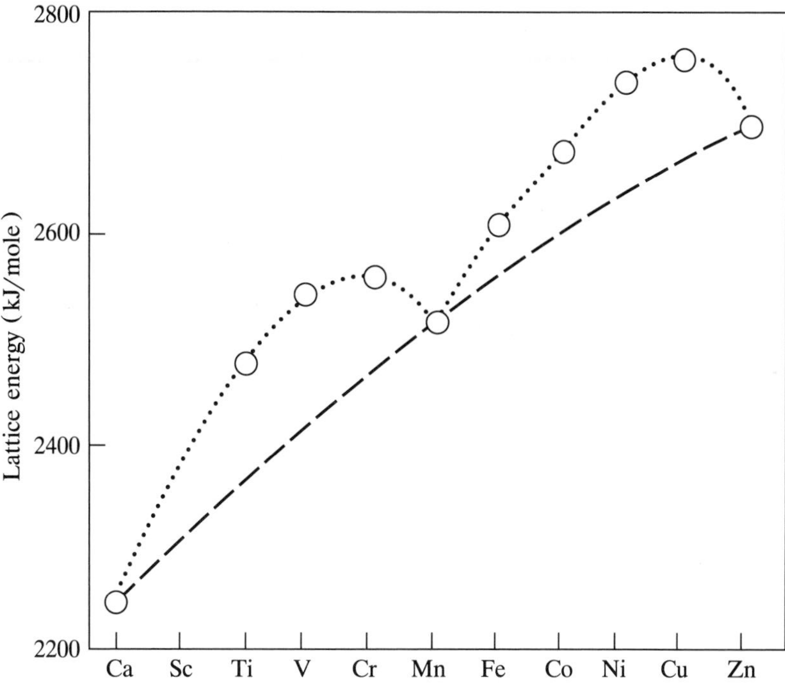

FIGURE 8.10
The lattice energies for the chlorides of the M^{2+} transition-metal cations. The dashed line connects the d^0, d^5, d^{10} cases; the dotted line shows the double-humped curve which reflects the trend for ionic radii of the M^{2+} ions in a weak octahedral field. [Adpated from Ref. 4, p. 683.]

increases as the ionic radius decreases. Given the "double-humped" curve for the metal radii which applies to the weak-field chlorides, it is not surprising that the actual lattice energies reflect the trend of radii, increasing where the radius decreases, and vice versa (dotted line). Quantitative estimates of crystal field stabilization energies can be obtained by comparing the trends in lattice energies with and without crystal field effects (see Problem 8.39).

SUMMARY

The lattice energy of a compound can be estimated theoretically using the Born-Landé and Kapustinskii equations and determined experimentally using thermochemical cycles. The theoretical Born-Landé model considers (1) the series of coulombic or electrostatic interactions among the various ions in a crystal (E_{coul}) and (2) the strong but short-range repulsions (E_{rep}) among interpenetrating filled electron clouds. Summing these contributions produces the Born-Landé equation for the lattice energy of a predominantly ionic compound. Analysis of this equation shows that lattice energies are directly dependent on the charge-to-radius (Z/r) ratio (or the charge density) of the ions involved. For compounds for which we do not know the crystal structure, the empirical equation of Kapustinskii serves to estimate the lattice energy.

Lattice energies cannot be determined directly by experiment. Rather, they must be determined by thermochemical Born-Haber cycles based on Hess' law of heat summation. Given various readily available thermochemical quantities, lattice energies can be determined and compared with the results of the theoretical model. In general, the agreement is within 4 to 6 percent for compounds for which a high degree of ionic character is predicted by the analysis of electronegativity differences. For compounds of significant covalent character, higher percentage differences between the theoretical and thermochemical results are observed.

Using a combination of the theoretical and experimental approaches, (1) more reliable values of electron affinities including those of monoatomic anions like O^- and S^-, (2) heats of formation of unknown compounds like CaCl or $NaCl_2$, and (3) thermochemical (or *effective*) radii of polyatomic ions can be calculated.

A consideration of the lattice energies of transition metal compounds affords an explanation of the trends in the ionic radii of these elements as well as estimates of crystal field stabilization energies. This treatment offers a bridge between the sections (Chapters 2–6) on coordination compounds and those (Chapters 7–8) involving solid state structures and energetics.

PROBLEMS

8.1. (*a*) Using the Shannon-Prewitt ionic radii in Tables 7.4 and 7.6, identify and sketch a plausible unit cell for RbBr(*s*).

(*b*) Using the additional information found in Tables 8.1 and 8.2; calculate the lattice energy of RbBr(*s*) using the Born-Landé equation and the Kapustinskii equation.

8.2. Calculate the lattice energy of lithium iodide using both the Born-Landé and the Kapustinskii equations. Compare these values with those found in Table 8.3.

8.3. The lattice energy of francium chloride, FrCl, has been estimated to be −632 kJ/mol. Calculate a value for the radius of the Fr^+ cation. List any assumptions you make in doing the calculation. Does your result make sense relative to the other alkali metals? Briefly explain.

8.4. Assume that the crystal structure of lanthanum(III) fluoride, LaF_3, is unknown. Using the equation of Kapustinskii, estimate the lattice energy.

8.5. Although berkelium is available only in very small quantities, enough has been prepared to determine some structural parameters.

(*a*) Using a value of 0.97 Å for the ionic radii of Bk^{4+} and using a crystal lattice consistent with the radius ratio rule, estimate the lattice energy of berkelium(IV) dioxide, BkO_2.

(*b*) Assume that the radius-ratio rule is violated and that BkO_2 forms a cadmium iodide lattice. How much difference does this make in your answer?

(*c*) Compare the above values with that obtained from the Kapustinskii equation.

8.6. Explain the following observations: Magnesium oxide (MgO) and sodium fluoride (NaF) have the same crystal structure and approximately the same formula weight, but MgO is almost twice as hard as NaF. The melting points of MgO and NaF are 2852 and 993°C, respectively. The boiling points of MgO and NaF are 3600 and 1695°C, respectively.

8.7. Using the Born-Landé equation as a basis for your answer, to what can you attribute the facts that MgS is much harder and higher melting than LiBr? [$r(Li^+) = 0.90$, $r(Mg^{2+}) = 0.86$, $r(Br^-) = 1.82$, $r(S^{2-}) = 1.70$ Å.] Can you assume that they both exhibit the same crystal structure? Why or why not?

***8.8.** There has been some speculation that an ionic noble gas compound with the stoichiometry Xe^+F^- can be prepared. Using both the Born-Landé and Kapustinskii equations, estimate the lattice energy for this hypothetical compound. Carefully state all assumptions that you make to establish a value for the ionic radius of Xe^+. Atomic radii are given in Table 7.2.

***8.9.** Consider a line of alternating cations and anions as shown below. Estimate a value for the Madelung constant of this "lattice."

$\longrightarrow| \; r_0 \; |\longleftarrow$

***8.10.** Using the relationship found in Equation (8.8) for the total lattice energy, differentiate with respect to r and solve for the constant B. Substitute your expression for B into the equation and show that the Born-Landé equation, Equation (8.9), results.

8.11. Given the following data, calculate the lattice energy of LiF: $IE_{Li} = 520.3$ kJ/mol; $\Delta H^\circ_{sub}(Li) = 159.4$ kJ/mol; $EA_F = -328.0$ kJ/mol; $\frac{1}{2}D_{F\text{-}F} = \Delta H(g) = 79.0$ kJ/mol; $\Delta H^\circ_f(LiF) = -616.0$ kJ/mol. Compare your answer with that given in Table 8.3.

8.12. Given the following data, calculate the lattice energy of barium chloride, $BaCl_2$. $IE^1_{Ba} = 502.7$ kJ/mol; $IE^2_{Ba} = 965.0$ kJ/mol; $\Delta H^\circ_{sub}(Ba) = 175.6$ kJ/mol; $EA_{Cl} = -349.0$ kJ/mol; $\frac{1}{2}D_{Cl\text{-}Cl} = \Delta H(g) = 243.0$ kJ/mol; $\Delta H^\circ_f(BaCl_2) = -858.6$ kJ/mol. Compare your answer with that calculated using the Born-Landé equation (*Note:* You will have to determine the cyrstal structure of this compound.)

8.13. Write a thermodynamic cycle for the heat of formation of KX. For the cases where X = F and Cl, compare the electron affinities, X—X bond energies, and lattice energies (from the Born-Landé equation). Comment on the fact that F_2 is more reactive than Cl_2 despite the fact that the $EA_F < EA_{Cl}$. (*Hint:* Assume both compounds form a rock salt structure.)

8.14. Given the following data plus that found in the tables of the text, calculate a value for the standard heat of formation of copper(I) fluoride, CuF. $IE_{Cu} = 745.3$ kJ/mol; $\Delta H^\circ_{sub}(Cu) = 338.3$ kJ/mol. (The lattice energy of CuF should be determined using the Born-Landé equation. CuF assumes a zinc blende structure.)

8.15. The dissocation energy of the ClF molecule is 246.4 kJ/mol, while the standard heat of formation of ClF(g) is -56.1 kJ/mol. Knowing that the dissociation energy of Cl_2 is 243.4 kJ/mol, use a thermodynamic cycle to calculate the dissociation energy of F_2.

8.16. Calculate the proton affinity of ammonia, $NH_3(g) + H^+(g) \rightarrow NH_4^+(g)$, given the following data: NH_4F crystallizes in a ZnS (wurtzite) structure; the Born exponent of the crystal is 8; the ammonium ion–to–fluoride ion distance is 2.56 Å; the enthalpy of formation of $NH_4F(s)$ is -468.6 kJ/mol; the enthalpy of formation of ammonia gas is -1171.5 kJ/mol; $\frac{1}{2}D_{H\text{-}H}$ [$= \Delta H(g)$] is 218.0 kJ/mol; IE_H is 1305.0 kJ/mol; $\frac{1}{2}D_{F\text{-}F}[= \Delta H(g)] = 79.0$ kJ/mol; $EA_F = -328.0$ kJ/mol.

***8.17.** How might you estimate the standard enthalpy change of the process CsCl (ordinary form) → CsCl (NaCl form)?

8.18. When the Born-Haber cycle was first applied to ionic compounds, all the constituent thermochemical quantities except the lattice energies and electron affinities were known. The lattice energies, as we have seen, could be calculated using the Born-Landé equation, and if they were indeed correct, i.e., if the assumptions of the model were correct, the electron affinity of, for example, chlorine, should be fairly constant if calculated using the data from the series of alkali-metal chlorides LiCl, NaCl, KCl, RbCl, and CsCl. To verify this result, calculate the Born-Landé lattice energies of the above chlorides and then, using Born-Haber cycles, calculate a value for the electron affinity of chlorine in each case. Appropriate thermochemical data are found in Table 8.3. Also calculate an average value for the electron affinity and compare it with the accepted value of -349.0 kJ/mol.

8.19. (*a*) Construct a thermochemical cycle which will allow you to calculate the electron affinity of bromine using data from the formation of rubidium bromide, RbBr.

(*b*) Using the Born-Landé lattice energy you calculated in part (*b*) of Problem 8.1 and selected thermochemical data from Table 8.3, calculate the electron affinity of bromine using your Born-Haber cycle. Compare your result with that given in Table 8.3.

8.20. (*a*) Write a thermodynamic cycle for the synthesis of $Na_2O(s)$ from its constituent elements in their standard states. Identify all the thermochemical data needed to complete the cycle.

(*b*) Estimate the value of the lattice energy of $Na_2O(s)$ using the Kapustinskii equation.

(*c*) Using the following thermochemical data and your value of lattice energy determined in part (*b*) above, calculate a value of the second electron affinity of oxygen. $\Delta H^\circ_{sub}(Na) = 107.3$; $IE_{Na} = 495.9$; $\Delta H^\circ_f[Na_2O(s)] = -418.0$; $\Delta H(g)[O(g)] = 249.1$ kJ/mol; EA^1 of oxygen $= -141.0$ kJ/mol.

8.21. The standard heat of formation of ZnS (wurtzite structure) is -192.6 kJ/mol.

(*a*) Estimate a value for the lattice energy of ZnS using the Born-Landé equation.

(*b*) Using these two values and the following thermodynamic data, determine a value for the second electron affinity of sulfur. $\Delta H_{sub}(Zn) = 130.8$ kJ/mol; $\Delta H_g(S) = 278.8$ kJ/mol; $IE^1 = 906.4$ kJ/mol; $IE^2 = 1733$ kJ/mol; EA^1 of sulfur $= -200.4$ kJ/mol.

8.22. The lattice energy and standard heat of formation of NaH are -782 and -56.3 kJ/mol, respectively. The standard heat of formation ΔH°_f of $H(g)$ is 218.0 kJ/mol. Using these data and others found in Table 8.3, calculate a value for the electron affinity of hydrogen.

8.23. Using the data found in Table 8.5 for $CaCl_n$, calculate the heat of the following disproportionation reaction:

$$2CaCl(s) \longrightarrow CaCl_2(s) + Ca(s).$$

Would you predict such a reaction to occur? Why or why not?

8.24. Using a Born-Haber cycle, calculate a reasonable value for the ΔH°_f of $NaCl_2$. Estimate the lattice energy using the Kapustinskii equation. Rationalize the value you use for the ionic radius of Na^{2+}. Discuss the principal reasons why $NaCl_2$ would or would not be thermodynamically feasible [$IE^2(Na) = 4563$ kJ/mol].

8.25. Construct a Born-Haber cycle which will enable a calculation of a reasonable value for the ΔH°_f of Ne^+Cl^-. Estimate the lattice energy using the Kapustinskii equation. Rationalize the value you use for the ionic radius of Ne^+. Discuss the principal reasons why NeCl would or would not be thermodynamically feasible [IE(Ne) = 2080 kJ/mol].

8.26. A lattice energy of Xe^+F^- was estimated in Problem 8.8. Use this information to estimate a value for the ΔH°_f for this compound [IE(Xe) = 1170 kJ/mol].

***8.27.** We saw that ΔH°_f for $CaCl_3$ was $+1585$ kJ/mol, and therefore it is not likely to be formed. Would you expect ΔS°_f to be positive or negative for the formation of $CaCl_3$? Explain your reasoning and indicate if the entropy factor could somehow be used to force the formation of $CaCl_3$.

8.28. The radius for the hexachloroberkelate(IV) ion, $BkCl_6^{2-}$, has been estimated to be 3.61 Å. Using this value, calculate a value for the lattice energy of potassium hexachloroberkelate(IV), K_2BkCl_6.

8.29. The thermochemical radius of the ammonium cation was given to be 1.37 Å in Problem 7.39. The standard heats of formation of $NH_4^+(g)$ and $NH_4Br(s)$ are 630.2 and -270.3 kJ/mol, respectively.

(*a*) Using these data and others found in the text, calculate a value for the lattice energy of NH_4Br using a Born-Haber cycle.

(*b*) Calculate the thermochemical radius of the ammonium cation using the Kapustinskii equation and compare your result with the above value.

8.30. The thermochemical radius of the hydroxide ion was given to be 1.33 Å in Problem 7.40. The standard heats of formation of $OH^-(g)$ and $NaOH(s)$ are -143.5 and -425.6 kJ/mol, respectively.

(*a*) Using these data and others found in the text, calculate a value for the lattice energy of NaOH using a Born-Haber cycle.

(*b*) Verify the above radius of the hydroxide anion using the Kapustinskii equation.

8.31. Using the data found in Problems 8.29 and 8.30 and the Kapustinskii equation, calculate the standard heat of formation of $NH_4OH(s)$.

8.32. The standard heat of formation and lattice energy for sodium borohydride, $NaBH_4$, are -183.3 and -703 kJ/mol, respectively. Calculate a value for the standard heat of formation of $BH_4^-(g)$.

8.33. The standard heat of formation of $PtCl_6^{2-}(g)$ is -774 kJ/mol. The lattice energy of $K_2PtCl_6(s)$ is -1468 kJ/mol. Calculate a value for the standard heat of formation of potassium hexachloroplatinate, K_2PtCl_6.

8.34. Using the data of Problem 8.32, calculate a value for the thermochemical radius of the borohydride ion, BH_4^-.

8.35. The lattice energy of $CaC_2(s)$ is -2911 kJ/mol. Calculate a value for the thermochemical radius of the carbide ion, C_2^{2-}.

8.36. Given that the lattice energy of pentamminenitrocobalt(III) chloride, $[Co(NH_3)_5(NO_2)]Cl_2$, is -1013 kJ/mol, estimate the effective ionic radius of the complex cation and speculate on the crystal structure of this compound ($r_{Cl^-} =$ 1.67 Å).

8.37. Which would have the larger ionic radius, low-spin iron(II) or high-spin iron(II)? Briefly explain your answer. (Assume an octahedral field.)

8.38. In a weak octahedral field which ion, Cr^{3+} or Fe^{3+}, will have the larger ionic radius? Briefly rationalize your answer.

8.39. Using the observed lattice energies tabulated below, estimate the magnitude of Δ_o in octahedrally coordinated crystals of VO, MnO, and FeO. Carefully explain your method of calculation.

Oxide	Lattice energy, kJ / mol	Oxide	Lattice energy, kJ / mol
CaO	−3465.2	FeO	−3922.9
TiO	−3881.9	CoO	−3991.9
VO	−3916.6	NiO	−4076.0
MnO	−3813.3	ZnO	−4035.0

8.40. The heat of hydration is defined as the energy associated with the following general reaction:

$$M^{n+}(g) + xH_2O(l) \longrightarrow M(H_2O)x^{n+}(aq)$$

Heats of hydration, like lattice energies, depend on the charge-to-radius (Z/r) ratio, or charge density. The heats of hydration ΔH_{hyd} at 25°C and the ligand-field splitting Δ_o for some divalent, octahedrally coordinated transition-metal ions are tabulated below.

(*a*) First plot the negative of the heats of hydration ($-\Delta H_{\text{hyd}}$) as the ordinate against the number of d electrons as abscissa.

M^{2+}	$-\Delta H_{\text{hyd}}$, kJ / mol	Δ_o, cm^{-1}
Ca^{2+}	2470	0
V^{2+}	2780	12,600
Cr^{2+}	2794	13,900
Mn^{2+}	2736	7,800
Fe^{2+}	2845	10,400
Co^{2+}	2916	9,300
Ni^{2+}	3000	8,500
Cu^{2+}	3000	12,600
Zn^{2+}	2930	0

(*b*) How do you account for:
- (i) The general increase in $-\Delta H_{\text{hyd}}$
- (ii) The two humps in the curve

(*c*) Calculate the crystal field stabilization energy, in terms of Δ_o, for the various high-spin configuration d^1 to d^9.

(*d*) Use the given values of Δ_o to find the crystal field stabilization energy in kilojoules per mole for each of these ions. (*Note:* 1 cm^{-1} = 0.0120 kJ/mol.)

(*e*) Apply this crystal field stabilization energy as a correction term to the given heats of hydration and plot the estimated heats of hydration in the absence of ligand-field effects against the number of d electrons. What is the final shape of the curve?

PART C

DESCRIPTIVE CHEMISTRY OF THE REPRESENTATIVE ELEMENTS

In this self-standing section, the descriptive chemistry of the representative elements is introduced over the course of 11 chapters. The titles of these chapters are listed below.

CHAPTER 9

BUILDING A NETWORK OF IDEAS TO MAKE SENSE OF THE PERIODIC TABLE

One of the most attractive aspects of the science of chemistry is the way it all fits together. Typically, students first learn the basics of atomic and molecular structure, add some knowledge of thermodynamics, kinetics, and equilibrium, and then quickly start to apply these ideas to more advanced topics. For example, you may have already studied Chapters 2 to 6 of this book and seen how the ideas that you worked so hard to master in your previous chemistry experiences provide the basis for the study of coordination compounds. Or, perhaps you have read Chapters 7 and 8 on the structures and energetics of solid state chemistry. Alternatively, you may have skipped directly from Chapter 1 to this point to start a study of the chemistry of the periodic table and the representative elements. In any case, no matter in what order you have started to make your way through the discipline we call chemistry, the ultimate goal is the development of an interconnected network of ideas that you can use to rationalize and predict a variety of chemical behavior. Nowhere is such a network more essential than in a study of what has become known as *descriptive chemistry*: the properties, structures, reactions, and applications of the elements and their most important compounds.

In the last half of the nineteenth century, the development of the periodic table closely coincided with the emergence of inorganic chemistry and with good reason. Even a century ago, Mendeleev's great empirical masterpiece was the foundation for the organization of any study of the elements and their compounds. Once the quantum revolution of the early twentieth century firmly established its electronic basis, the periodic table became a mighty framework for the study of descriptive chemistry. Now as we stand at the edge of the twenty-first century, the known chemistry of the 110 elements is incredibly rich and diverse and still growing. Mastering even a small part of that chemistry will be no mean task. We need to choose our starting place carefully and move slowly and logically up through this huge field of study.

In this chapter we make that start. Here we will begin to build a network of specific, organizing ideas on which to base our understanding of the periodic table and the chemistry of the elements. The ideas covered in this chapter will be (1) the familiar periodic law, (2) the uniqueness principle, (3) the diagonal effect, (4) the inert-pair effect, and (5) the division of elements into metals, metalloids, and nonmetals.

The next few chapters will lay the groundwork for a systematic study of the chemistry of the elements. In Chapter 10, we look at hydrogen and the hydrides and then, in Chapter 11, oxygen, oxides, and hydroxides. As we go along in these next few chapters, we will also add two additional ideas to our network. In Chapter 11 the acid-base character of oxides and hydroxides will become the sixth component, and in Chapter 12 we will add a knowledge of standard reduction potentials. With this network of seven ideas in place, we will proceed (in Chapters 12 to 19) to discuss the chemistry of the main-group (or representative) elements (usually defined as those in which the *ns* and *np* orbitals are partially filled).

9.1 THE PERIODIC LAW

The first and foremost of the basic ideas in our network must be the *periodic law*. Originally formulated by Dmitri Mendeleev (and independently by Lothar Meyer) in the 1860s and 1870s, the modern version of this law states that a periodic repetition of physical and chemical properties occurs when the elements are arranged in order of increasing atomic number. Recall that when Mendeleev constructed his periodic tables (an example of which is given in Figure 9.1), he ordered the elements by atomic mass rather than by atomic number and tried to place elements with similar valences in the same group. Finding that some elements did not fit well, he left blanks to represent elements yet to be discovered. By interpolation of the properties of the elements surrounding the blanks, Mendeleev predicted the values of many of the properties of these unknown elements. Three of these undiscovered elements were designated as eka-aluminum (AW = 68), eka-silicon (AW = 72), and eka-boron (AW = 44), where, for example, eka-aluminum indicates the first element below aluminum in his table. (Note that both boron and eka-boron are on the left edge of the "Gruppe III" column, while aluminum and eka-aluminum are on the right edge.) Within the next 15 years, these elements (named gallium, germanium, and scandium, respectively, after the homelands of their discoverers) were isolated and characterized. The agreement between the experimental values of the properties of these elements and those predicted by Mendeleev certainly provided dramatic support for his ideas.

Today we take for granted the arrangement of the elements into periods and groups. As expected, there have been many changes from the days of Mendeleev. The ordering by atomic number rather than atomic mass, for example, and the discovery of the noble gases are two of the more significant ones. Figure 9.2 shows the number of known elements versus time. Note that only about 60 elements were known in Mendeleev's time. The discovery of the other 50 coupled with an

Reihen	Gruppe I. — R^2O	Gruppe II. — RO	Gruppe III. — R^2O^3	Gruppe IV. RH^4 RO^2	Gruppe V. RH^3 R^2O^5	Gruppe VI. RH^2 RO^3	Gruppe VII. RH R^2O^7	Gruppe VIII. — RO^4
1	H = 1							
2	Li = 7	Be = 9.4	B = 11	C = 12	N = 14	O = 16	F = 19	
3	Na = 23	Mg = 24	Al = 27.3	Si = 28	P = 31	S = 32	Cl = 35.5	Fe = 56, Co = 59,
4	K = 39	Ca = 40	— = 44	Ti = 48	V = 51	Cr = 52	Mn = 55	Ni = 59, Cu = 63,
5	(Cu = 63)	Zn = 65	— = 68	— = 72	As = 75	Se = 78	Br = 80	
6	Rb = 85	Sr = 87	?Yt = 88	Zr = 90	Nb = 94	Mo = 96	— = 100	Ru = 104, Rh = 104, Pd = 106, Ag = 108,
7	(Ag = 108)	Cd = 112	In = 113	Sn = 118	Sb = 122	Te = 125	J = 127	
8	Cs = 133	Ba = 137	?Di = 138	?Ce = 140	—	—	—	— — — —
9	(—)	—	—		—	—	—	
10	—	—	?Er = 178	?La = 180	Ta = 182	W = 184	—	Os = 195, Ir = 197, Pt = 198, Au = 199,
11	(Au = 199)	Hg = 200	Ti = 204	Pb = 207	Bi = 208	—	—	
12	—	—	—	Th = 231	—	U = 240	—	— — — —

FIGURE 9.1
A periodic table published in 1872 by Dmitri Mendeleev (insert). Eka-aluminum, eka-silicon, and eka-boron (shaded) were later discovered and named gallium, germanium, and scandium after the homelands of their discoverers. [Ref. 16.] (*Photo The Chemical Heritage Foundation.*)

understanding (gained over the last 100 years!) of the role of atomic structure in organizing the elements has resulted in our modern periodic table shown in Figure 9.3. Note, however, that Mendeleev's original idea of groups of elements with similar properties is very much in evidence more than 100 years after his work. Some of these groups, such as the noble gases, have been given special names. Other commonly accepted (and several rather obscure) group and period names are indicated in Figure 9.4.

Recall from earlier courses that the elements of a given group have similar valence-electron configurations. These configurations have been included in Figure 9.3. For example, the Group 1A elements (the alkali metals) have the general electronic structure [noble gas]ns^1, while the Group 7A elements (the halogens) have the [noble gas]ns^2np^5 or the [noble gas]$ns^2(n-1)d^{10}np^5$ configuration. [When filled $(n-1)d$ and sometimes even $(n-2)f$ subshells are present in

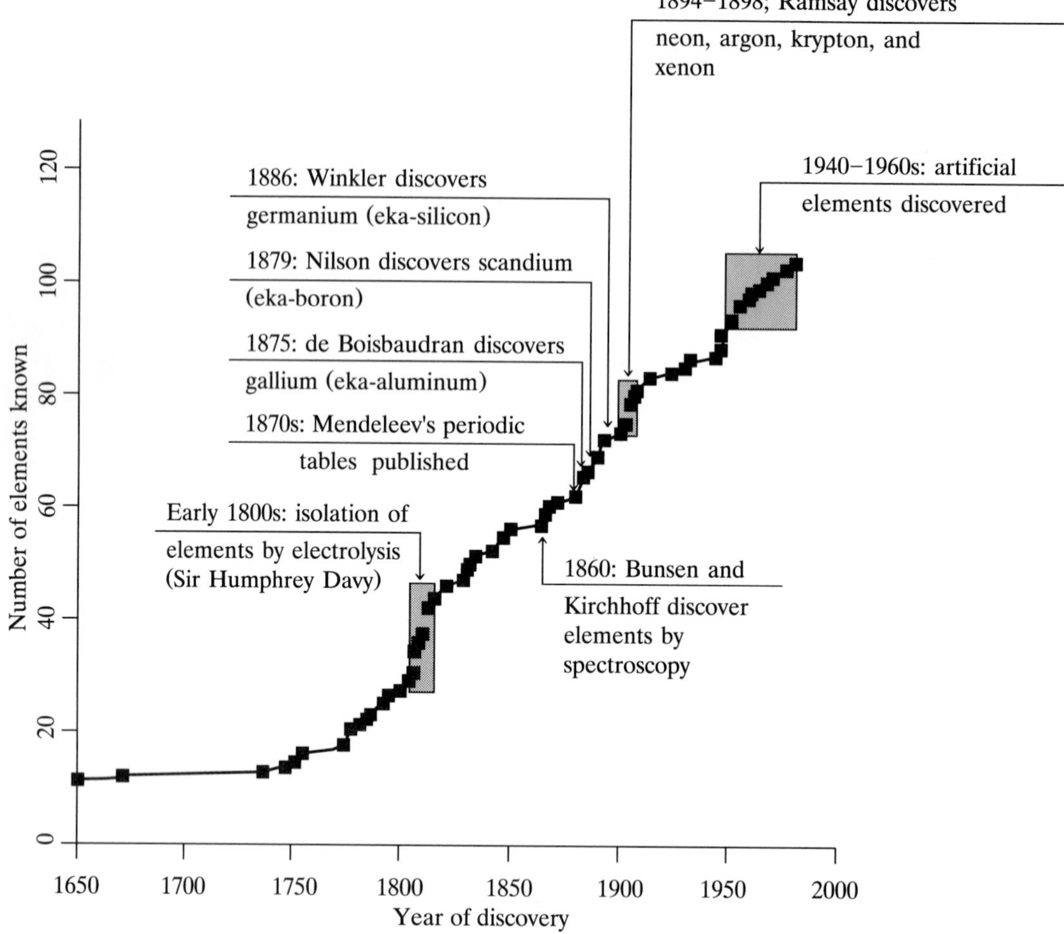

FIGURE 9.2
A plot of the number of known elements versus time.

addition to filled *ns* and *np* subshells, these filled orbitals are sometimes collectively referred to as a *pseudonoble-gas configuration*.] Given these electronic similarities, we expect and in fact do find many corresponding chemical similarities among the members of a given group. Most of the remaining chapters of this book will deal with the descriptive group chemistry of the representative elements.

What characteristics and properties of the elements show periodic trends? You most likely have studied a number of these in previous chemistry courses and may find it helpful to review them at this point. Atomic radii, ionization energies, electron affinity, and electronegativity will be among the most important in our network. Before turning to a discussion of periodicity, a word of warning is in order. *Do not memorize* the trends presented; instead, rely on your developing network to help rationalize and predict them. If you really understand these concepts, try writing down your ideas. When you can produce a well-written paragraph for the trends in each property, you have started to really understand, not just memorize, the periodic table.

	1A	2A	3B	4B	5B	6B	7B	8B	8B	8B	1B	2B	3A	4A	5A	6A	7A	8A
1	1 H $1s^1$																	2 **He** $1s^2$
2	3 **Li** $2s^1$	4 **Be** $2s^2$											5 **B** $2s^22p^1$	6 **C** $2s^22p^2$	7 **N** $2s^22p^3$	8 **O** $2s^22p^4$	9 **F** $2s^22p^5$	10 **Ne** $2s^22p^6$
3	11 **Na** $3s^1$	12 **Mg** $3s^2$											13 **Al** $3s^23p^1$	14 **Si** $3s^23p^2$	15 **P** $3s^23p^3$	16 **S** $3s^23p^4$	17 **Cl** $3s^23p^5$	18 **Ar** $3s^23p^6$
4	19 **K** $4s^1$	20 **Ca** $4s^2$	21 **Sc** $4s^23d^1$	22 **Ti** $4s^23d^2$	23 **V** $4s^23d^3$	24 **Cr** $4s^13d^5$	25 **Mn** $4s^23d^5$	26 **Fe** $4s^23d^6$	27 **Co** $4s^23d^7$	28 **Ni** $4s^23d^8$	29 **Cu** $4s^13d^{10}$	30 **Zn** $4s^23d^{10}$	31 **Ga** $4s^24p^1$	32 **Ge** $4s^24p^2$	33 **As** $4s^24p^3$	34 **Se** $4s^24p^4$	35 **Br** $4s^24p^5$	36 **Kr** $4s^24p^6$
5	37 **Rb** $5s^1$	38 **Sr** $5s^2$	39 **Y** $5s^24d^1$	40 **Zr** $5s^24d^2$	41 **Nb** $5s^14d^4$	42 **Mo** $5s^14d^5$	43 **Tc** $5s^24d^5$	44 **Ru** $5s^14d^7$	45 **Rh** $5s^14d^8$	46 **Pd** $4d^{10}$	47 **Ag** $5s^14d^{10}$	48 **Cd** $5s^24d^{10}$	49 **In** $5s^25p^1$	50 **Sn** $5s^25p^2$	51 **Sb** $5s^25p^3$	52 **Te** $5s^25p^4$	53 **I** $5s^25p^5$	54 **Xe** $5s^25p^6$
6	55 **Cs** $6s^1$	56 **Ba** $6s^2$	57 **La** $6s^25d^1$	72 **Hf** $6s^25d^2$	73 **Ta** $6s^25d^3$	74 **W** $6s^25d^4$	75 **Re** $6s^25d^5$	76 **Os** $6s^25d^6$	77 **Ir** $6s^25d^7$	78 **Pt** $5s^15d^9$	79 **Au** $6s^15d^{10}$	80 **Hg** $6s^25d^{10}$	81 **Tl** $6s^26p^1$	82 **Pb** $6s^26p^2$	83 **Bi** $6s^26p^3$	84 **Po** $6s^26p^4$	85 **At** $6s^26p^5$	86 **Rn** $6s^26p^6$
7	87 **Fr** $7s^1$	88 **Ra** $7s^2$	89 **Ac** $7s^26d^1$	104 **Unq** $7s^26d^2$	105 **Unp** $7s^26d^3$	106 **Unh** $7s^26d^4$	107 **Uns** $7s^26d^5$	108 **Uno** $7s^26d^6$	109 **Une** $7s^26d^7$									

58 **Ce** $6s^24f^15d^1$	59 **Pr** $6s^24f^3$	60 **Nd** $6s^24f^4$	61 **Pm** $6s^24f^5$	62 **Sm** $6s^24f^6$	63 **Eu** $6s^24f^7$	64 **Gd** $6s^24f^75d^1$	65 **Tb** $6s^24f^9$	66 **Dy** $6s^24f^{10}$	67 **Ho** $6s^24f^{11}$	68 **Er** $6s^24f^{12}$	69 **Tm** $6s^24f^{13}$	70 **Yb** $6s^24f^{14}$	71 **Lu** $6s^24f^{14}5d^1$
90 **Th** $7s^26d^2$	91 **Pa** $7s^25f^26d^1$	92 **U** $7s^25f^36d^1$	93 **Np** $7s^25f^46d^1$	94 **Pu** $7s^25f^6$	95 **Am** $7s^25f^7$	96 **Cm** $7s^25f^76d^1$	97 **Bk** $7s^25f^9$	98 **Cf** $7s^25f^{10}$	99 **Es** $7s^25f^{11}$	100 **Fm** $7s^25f^{12}$	101 **Md** $7s^25f^{13}$	102 **No** $7s^2f^{14}$	103 **Lr** $7s^25f^{14}6d^1$

FIGURE 9.3
The modern periodic table with electronic configurations.

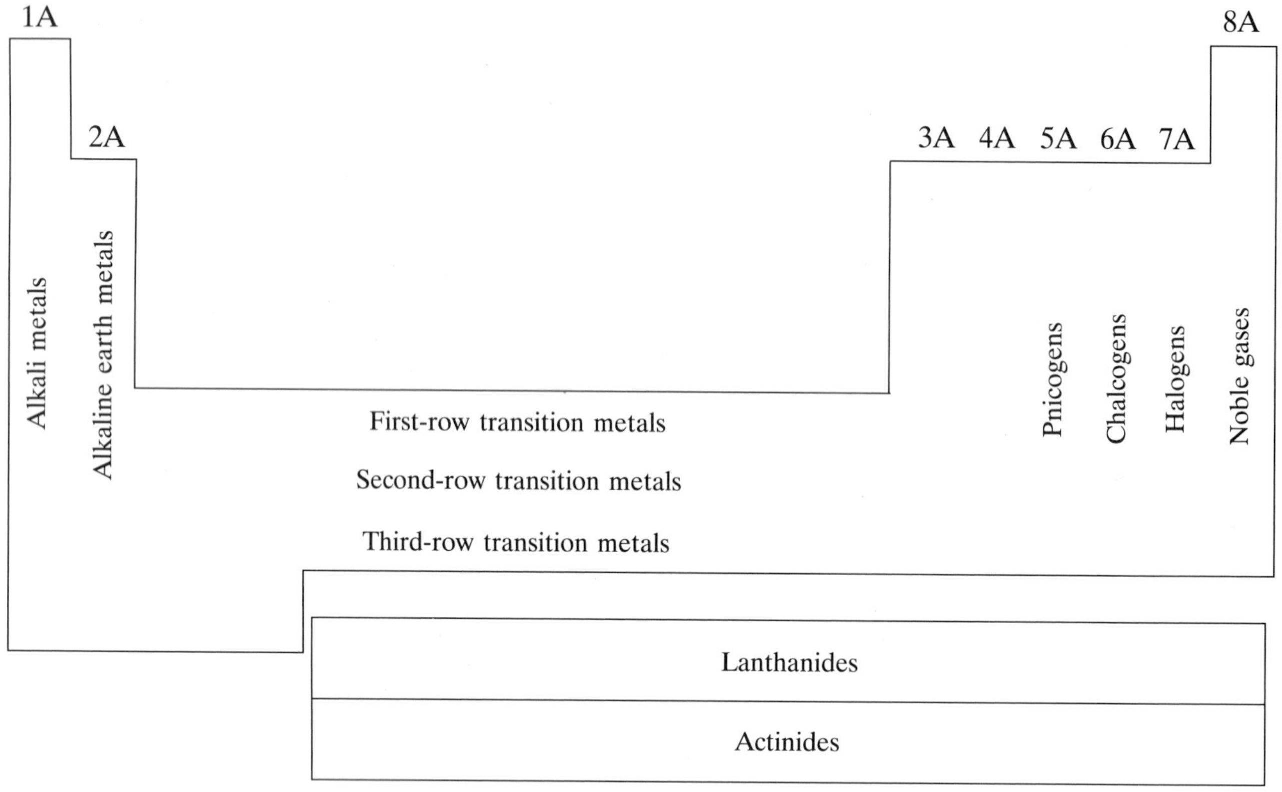

FIGURE 9.4
Common group and period names in the periodic table.

Effective Nuclear Charge

A keystone in the understanding of periodic properties is the concept of effective nuclear charge (Z_{eff}). Although the ideas underlying Z_{eff} are almost always presented in introductory courses, the term Z_{eff} is not always explicitly defined or even discussed. The *effective nuclear charge* acting on a given electron is the actual nuclear charge (atomic number) minus a screening constant that takes into account the effect of *shielding electrons*, electrons closer to the nucleus than the electron under consideration. The relationship among the actual nuclear charge Z, the screening constant σ, and the effective nuclear charge Z_{eff} is shown in Equation (9.1):

$$Z_{\text{eff}} = Z - \sigma \tag{9.1}$$

In many introductory courses, σ is taken to be an integer representing the number of inner-core, or shielding, electrons. For example, consider neon which has an electronic configuration of $1s^2 2s^2 2p^6$. To calculate the effective nuclear charge acting on a valence $2s$ or $2p$ electron, we note that there are two ($1s$) inner-core, or shielding, electrons. Taking σ to be 2 yields an effective nuclear charge of $(10 - 2 =) + 8$. In making this calculation, it is assumed that the two $1s$ electrons in the helium core completely shield the $2s$ and $2p$ valence electrons from the $+10$ nuclear charge, while the $2s$ and $2p$ electrons, all the same distance from the nucleus, do not shield each other at all. Consequently, we estimate that the valence electrons experience a net charge of +8. Note that the assumption that σ equals the number of inner-core electrons yields an effective nuclear charge which is the same as the group number of the element. It follows that the Z_{eff} for any noble gas is +8. Also note that this result requires that the $(n-1)d^{10}$ and $(n-2)f^{14}$ filled subshells (where n is the principal quantum number of the valence electrons) are considered shielding electrons. For example, the Z_{eff} for krypton, [Ar]$4s^2 3d^{10} 4p^6$, also comes out to be (36 − 28 =) + 8, where the $3d$ electrons are considered shielding. This correlation of the effective nuclear charge with group number will be useful throughout our discussion of representative elements. For example, verify for yourself that the Z_{eff} for Ca comes out to be +2.

Slater's Rules—Empirical Rules for Determining σ

In 1930, J. S. Slater formulated an empirical set of rules for determining values of the screening constant σ. These rules were based on calculations in which the energies and sizes of atomic orbitals in many-electron or polyelectronic atoms are estimated by the "self-consistent field (SCF) method." The SCF method considers a given electron to be in the potential field due to the nuclear charge plus the net effect of all the negatively charged clouds of the other electrons. This field is continually refined in the calculations until the results become self-consistent. While the method is beyond the scope of this text, such calculations are usually within a percentage point or two of the energies of atomic orbitals as derived from atomic line spectra.

Slater's rules for determining the screening constant σ are given in Table 9.1. To see how these rules are used, we can calculate the effective nuclear charge acting on a valence $2s$ or $2p$ electron of neon. (Then we can compare our results with those obtained by associating σ with the number of inner-core electrons.) To use Slater's rules, we would first write the electronic configuration in the proper form (rule 1): $(1s^2)\ (2s^2, 2p^6)$. The other seven valence electrons would make a contribution of 7×0.35 (rule 2*b*), while the $1s$ electrons would make a contribution of 2×0.85 (rule 2*c*). The effective nuclear charge would then be calculated as shown in Equation (9.2):

$$Z_{\text{eff}} = 10 - [(7 \times 0.35) + (2 \times 0.85)] = 5.85 \tag{9.2}$$

This result is somewhat less than the value of 8 we calculated assuming that only the

TABLE 9.1
Slater's rules for determining screening constants

1. When writing the electronic configuration of an element, group the orbitals and list them in the following order:

   $(1s)\ (2s, 2p)\ (3s, 3p)\ (3d)\ (4s, 4p)\ (4d)\ (4f)\ (5s, 5p) \cdots$

2. To establish the screening constant for any electron, sum up the following contributions:
   (*a*) Zero contribution from any electrons in groups outside (to the right) of the one being considered.
   (*b*) A contribution of 0.35 for each of the other electrons in the same group (except in the $1s$ group, where a contribution of 0.30 is used).
   (*c*) If the given electron is in an (ns, np) group, a contribution of 0.85 for each of the electrons in the next innermost group (immediately to the left of the group containing the given electron).
   (*d*) If the given electron is in an (nd) or (nf) group, a contribution of 1.00 for each of the electrons in the next innermost group.
   (*e*) A contribution of 1.00 for each electron in the still-lower or farther-in groups.

innermost electrons completely shield the valence electron from the nucleus. A lower value is consistent with the idea that the $2s$ and $2p$ electrons do shield each other from the nuclear charge to a small degree (35 percent). In any case, 5.85 is still a high value and demonstrates why it is so difficult to ionize an electron from an inert gas.

The effective nuclear charge acting on a $1s$ electron of neon would be still higher as shown in Equation (9.3):

$$Z_{\text{eff}} = 10 - [(8 \times 0.0) + (1 \times 0.30)] = 9.7 \tag{9.3}$$

The effective nuclear charge acting on a valence $4s$ or $4p$ electron of a krypton atom is calculated in Equation (9.4) using the following electron configuration:

$$(1s^2)\ (2s^2, 2p^6)\ (3s^2, 3p^6)\ (3d^{10})\ (4s^2, 4p^6)$$

$$Z_{\text{eff}} = 36 - [(7 \times 0.35) + (10 \times 0.85) + (18 \times 1.00)] = 7.05 \tag{9.4}$$

Atomic Radii

Figure 9.5 shows the variation in atomic radii in the representative elements. We should be able to use the concept of effective nuclear charge just presented to rationalize these trends. Note that effective nuclear charge (which we can roughly associate with group number) increases across a period and that this increase in turn predicts a decrease in atomic radii. This decrease makes sense because the electron cloud of the orbitals being filled across a period will contract as the positive effective nuclear charge in the center increases. (See Problem 9.17 for

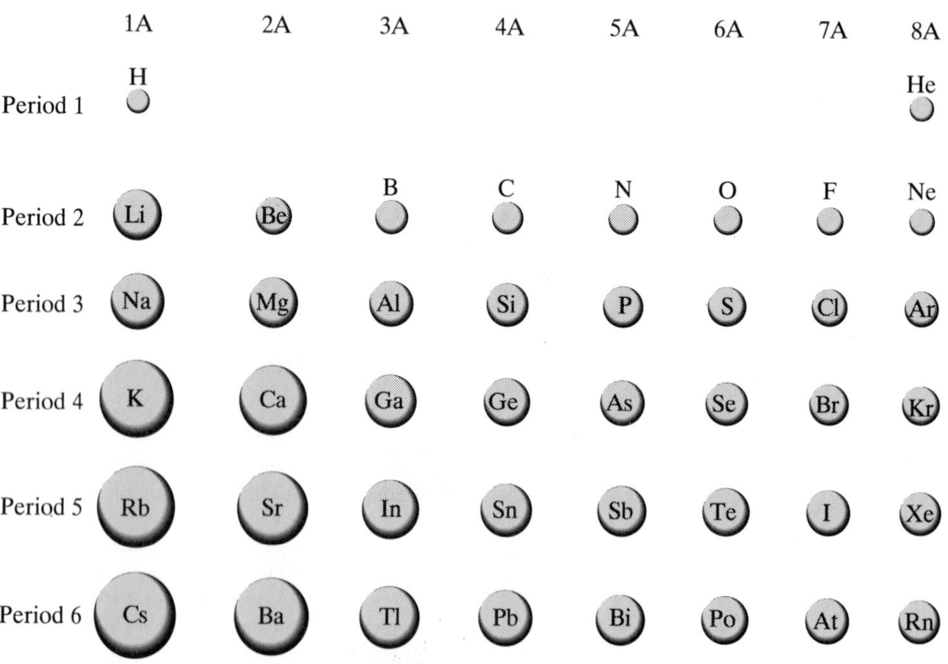

FIGURE 9.5
The variation in atomic radii in the representative elements. [Adpated from Ref. 16.]

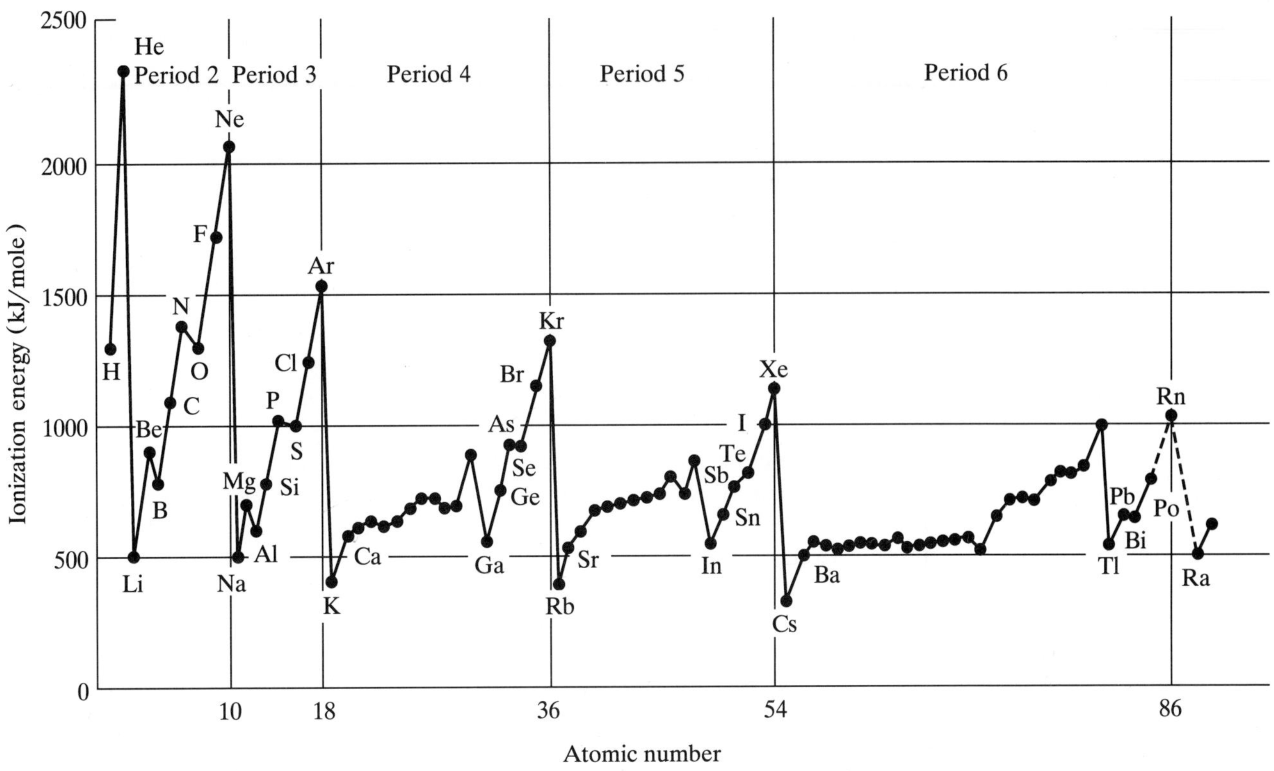

FIGURE 9.6
A plot of ionization energies (in kilojoules per mole) versus atomic number. Ionization energies generally increase across a period and decrease down a group.

results when Slater's rules are used to calculate the effective nuclear charge across a period.) When a new period is started, the atomic radius returns to a still larger value due to the increasing size of atomic orbitals with the value of n, the principal quantum number, and a return to a smaller effective nuclear charge. Note, for example, the atomic radii of the Group 1A elements. The atomic radius of potassium is greater than that of sodium, and so forth, on down the group.

You have probably noticed that many students try to memorize trends such as those presented above. As our network develops, relationships such as that between Z_{eff} and atomic radii will become clearer. As previously suggested, try constructing a paragraph explaining how Z_{eff} varies horizontally and vertically and how trends in atomic radius follow. Having done that successfully, you will not need to bother with consciously memorizing such things. You will know them because you understand them!

Ionization Energy

Similar trends in ionization energy, electron affinity, and electronegativity can be explained by reference to effective nuclear charge and the size of atoms. Concentrate on the general trends shown in these figures. Recall that *ionization energy* is that required to remove an electron from a neutral, gaseous atom. It varies with atomic number, as shown in Figure 9.6. Note that the general increase in Z_{eff} across a given period accounts for the ionization energy increasing from left to right. The exceptions to this general trend, for example, on going from beryllium to boron or from nitrogen to oxygen in the second period, are readily rationalized. You studied these in earlier chemistry courses, and we will not spend much time explaining them here. Suffice it to say that the exception between Groups 2A and 3A occurs because an *np* electron is now being removed rather than an *ns* electron. An electron in an *np* orbital is of higher energy than one in the *ns* orbital and consequently requires less energy for its removal.

The exception between Groups 5A and 6A is readily explained by referring to the orbital diagram for a Group 6A element shown below:

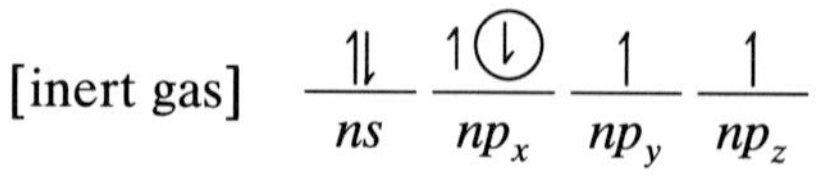

$$[\text{inert gas}] \quad \underset{ns}{\underline{\uparrow\downarrow}} \quad \underset{np_x}{\underline{\uparrow \textcircled{\downarrow}}} \quad \underset{np_y}{\underline{\uparrow}} \quad \underset{np_z}{\underline{\uparrow}}$$

Note that the electron to be removed (circled) occupies the same volume of space (and is spin-paired) with the other electron in that orbital. The circled *np* electron is easier to remove than expected due to electron-electron repulsion between these two electrons.

Vertically, removing the 4*s* electron from a potassium atom should take less energy than removing the 3*s* electron of sodium because the electron being removed is farther from the effective nuclear charge.

Problems 9.27 and 9.28 at the end of this chapter further address the above exceptions in the horizontal trend. As these trends are generally discussed in introductory chemistry courses, you may wish to review them on your own in more complete detail and then try the problems.

Electron Affinity

Electron affinity is the change in energy when an electron is added to a neutral, gaseous atom. An atom of an element on the right side of a period, where the effective nuclear charge is greater, should add an electron more easily than an atom of an element on the left side of the period. (And therefore those on the right will release more energy and have larger negative values for their electron affinities. Recall that the release of energy is a favorable process and results in a more stable final state.) This expected trend is observed in the data as shown in Figure 9.7. [Note, for example, the values for sodium (−53 kJ/mol) and chlorine (−348 kJ/mol) in the third period.] Vertically, adding an electron to a large atom such as iodine should be less favorable (and involve the release of less energy) than the addition of an electron to a smaller atom such as chlorine. This is true because the electron being added to the larger atom is farther away from the effective nuclear charge.

As for ionization energies, there are some exceptions in these general trends for electron affinities, and you more than likely studied these in previous courses. Two horizontal exceptions are encountered (1) on going from Group 1A elements to Group 2A elements (for example, note that the electron affinity of beryllium is +248 kJ/mol instead of being more negative than the value of −58 kJ/mol for lithium) and (2) on going from Group 4A elements to Group 5A (for example, note that the electron affinity of nitrogen is 0 kJ/mol instead of being more negative than that for carbon which is −123 kJ/mol). A vertical exception is found between the second- and third-period elements. [For example, fluorine has a *less* negative electron affinity (−333 kJ/mol) than does chlorine (−348 kJ/mol).]

1A							8A
H −77	2A	3A	4A	5A	6A	7A	He (21)
Li −58	Be (241)	B −23	C −123	N 0	O −142	F −333	Ne (29)
Na −53	Mg (230)	Al −44	Si −120	P −74	S −200	Cl −348	Ar (35)
K −48	Ca (154)	Ga (−35)	Ge −118	As −77	Se −195	Br −324	Kr (39)
Rb −47	Sr (120)	In −34	Sn −121	Sb −101	Te −190	I −295	Xe (40)
Cs −45	Ba (52)	Tl −48	Pb −101	Bi −100	Po ?	At ?	Rn ?

FIGURE 9.7
Variation of the electron affinities (in kilojoules per mole) of the representative elements. [Ref. 21.]

The first exception is explained by referring to the orbital diagrams of 1A and 2A elements shown below. Note that the incoming electron of a 2A element must occupy an *np* orbital which is of higher energy than the *ns* electron to be occupied by the incoming electron of a 1A element. Accordingly, less energy will be released in the case of a 2A element.

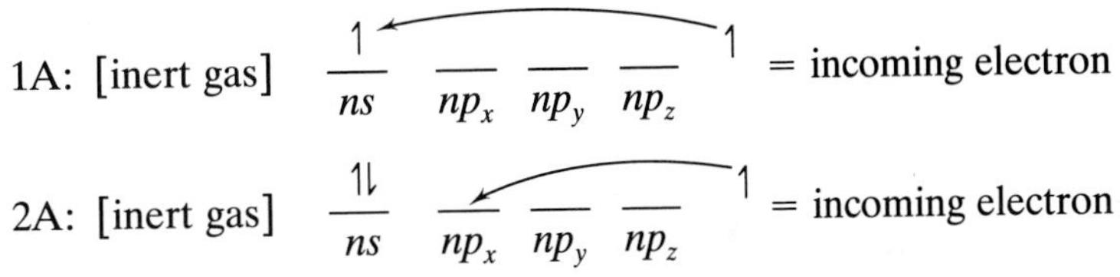

1A: [inert gas] $\underset{ns}{\underline{\uparrow}}$ $\underset{np_x}{\underline{\quad}}$ $\underset{np_y}{\underline{\quad}}$ $\underset{np_z}{\underline{\quad}}$ $\uparrow$ = incoming electron

2A: [inert gas] $\underset{ns}{\underline{\uparrow\downarrow}}$ $\underset{np_x}{\underline{\quad}}$ $\underset{np_y}{\underline{\quad}}$ $\underset{np_z}{\underline{\quad}}$ $\uparrow$ = incoming electron

The second exception is also explained by referring to a set of appropriate orbital diagrams. As shown below, the incoming electron of a 4A element occupies the remaining empty *np* orbital, while for a 5A element the new electron must be put in an *np* orbital already occupied. Accordingly, in the latter, electron-electron repulsions make the incoming electron more difficult to add (and less energy is released) than expected based solely on trends in effective nuclear charge.

4A: [inert gas] $\underset{ns}{\underline{\uparrow\downarrow}}$ $\underset{np_x}{\underline{\uparrow}}$ $\underset{np_y}{\underline{\uparrow}}$ $\underset{np_z}{\underline{\quad}}$ $\uparrow$ = incoming electron

5A: [inert gas] $\underset{ns}{\underline{\uparrow\downarrow}}$ $\underset{np_x}{\underline{\uparrow}}$ $\underset{np_y}{\underline{\uparrow}}$ $\underset{np_z}{\underline{\uparrow}}$ $\uparrow$ = incoming electron

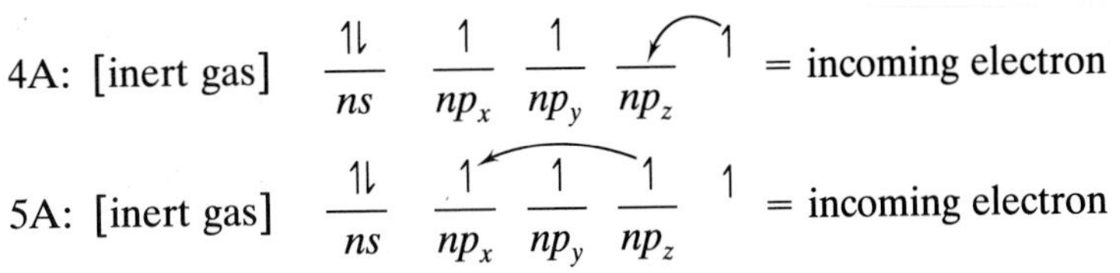

The vertical exception comes from the fact that the second-period elements are much smaller than the third-period elements, and it is much more difficult to add an electron to these smaller elements because of intense electron-electron repulsions. We will discuss this idea in greater detail shortly.

Note that our ability to understand and predict periodic trends is predicated on fairly simple, basic ideas such as electronic configurations, effective nuclear charge, orbital sizes, and coulombic interactions.

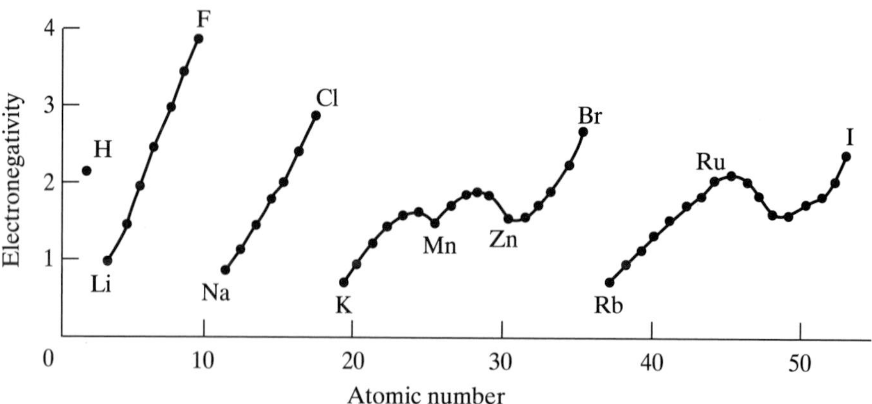

FIGURE 9.8
The electronegativities of the elements. [Ref. 21.]

Electronegativity

The trends for *electronegativity*, the ability of an atom *in a molecule* to draw electrons to itself, are straightforward. Electronegativity should increase across a period due to increasing effective nuclear charge and decrease down a group due to increasing atomic size. These trends are evident in Figure 9.8. Note that the trends in the more qualitative property of electronegativity do not, in general, have the exceptions of the type we encounter for ionization energies and electron

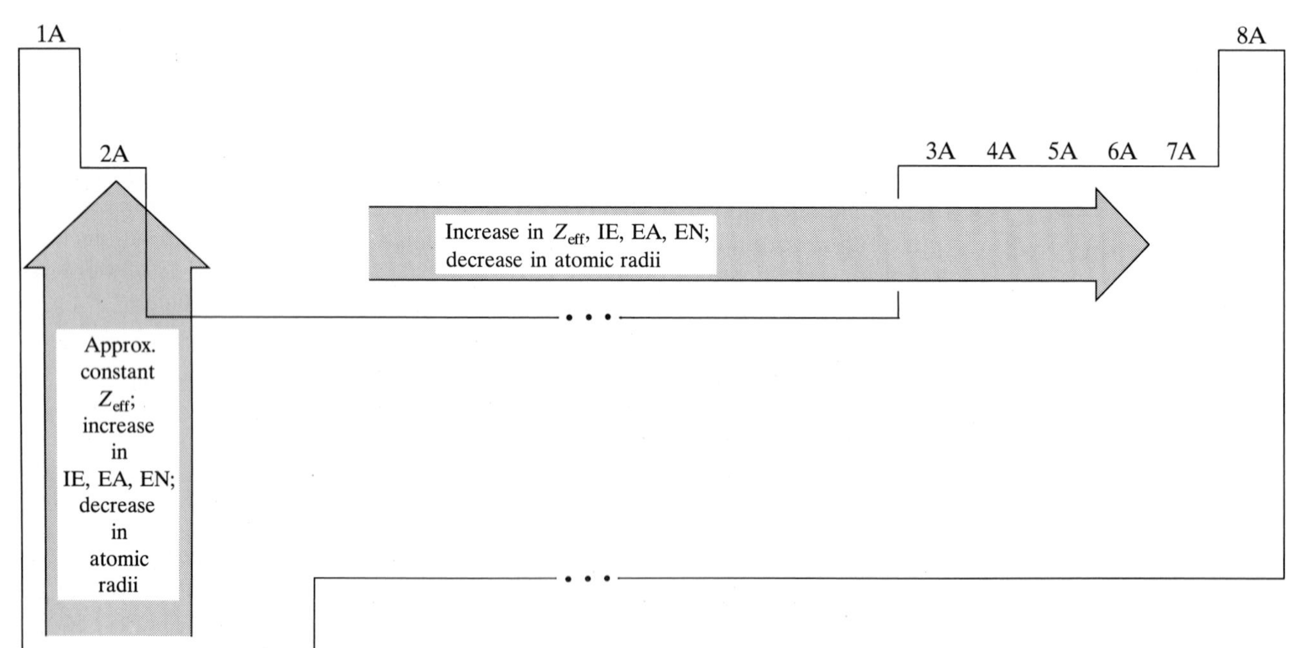

FIGURE 9.9
A summary of the general vertical and horizontal periodic trends in effective nuclear charge, atomic radii, ionization energies, electron affinities, and electronegativities.

affinities. To summarize and emphasize their importance to our network of ideas, the above general trends are summarized in Figure 9.9. This figure should not be memorized.

9.2 THE UNIQUENESS PRINCIPLE

While the familiar periodic law is the fundamental unifying principle for studying descriptive chemistry, it should come as no surprise that there are other organizing ideas which should be a part of our network. One such idea, which will be a recurring theme throughout our consideration of main-group chemistry, is the *uniqueness principle*, which states that the chemistry of the second-period elements (Li, Be, B, C, N, O, F, Ne) is quite often significantly different from that of the other elements in their respective groups. Indeed, these eight elements are so different from their *congeners* (elements of the same group) that some inorganic textbooks have a separate chapter or chapters on them. We will stick by group chemistry as our master organizing principle, but it is certainly important to know that the first element is not the most representative of the group as a whole. Indeed, as we will see, a better case can be made that the second element in each group (Na, Mg, Al, Si, P, S, Cl, Ar) is the more representative. Why is it that the first elements of the groups are so different from their congeners? Essentially, there are three reasons: (1) their exceptionally small size, (2) their enhanced ability to form pi bonds, and (3) the unavailability of *d* orbitals in these elements.

The Small Size of the First Elements

Even a casual reference to Figure 9.5 reveals that the first elements are exceptionally small compared to their congeners. This striking difference in sizes leads to a corresponding difference in electron affinities. In fact, the electron affinities of the first elements are unexpectedly low. (Note, for example, that the electron affinity of fluorine is not greater than that for chlorine as we would expect from our previous discussion. The same is true of oxygen compared to sulfur.) Electrons being added to these small, compact atoms at the top of each group experience more electron-electron repulsions, and thus it is a little more difficult to add an electron to these atoms then it is to their larger congeners.

Another aspect of the uniqueness of the first elements (particularly Li, Be, B, and C) that is related to their unusually small sizes is the relatively high degree of covalent character found in the compounds of these elements. A beginning chemistry student would quite naturally expect, for example, that the compounds of Groups 1A and 2A, elements of low ionization energy, would be primarily ionic. In fact all the congeners of lithium and beryllium do form compounds with a high degree of ionic character, but these two elements tend to form compounds of somewhat greater covalent character than expected. Why should this be? As a representative example, consider lithium chloride, LiCl, which we would assume, based on a low ionization energy for lithium and a highly exothermic electron affinity for chlorine, to be most properly thought of as Li^+Cl^-. Figure 9.10*a* shows the relative sizes of the lithium cation and chloride anion. Note that the very small lithium cation can get very close to the filled electron cloud of the chloride anion. So close, in fact, that the large, rather diffuse chloride electron cloud can be *distorted*, or "polarized," by the compact, positively charged lithium cation as shown in Figure 9.10*b*. This distortion makes orbital overlap between the two ions more likely. Orbital overlap and the resulting sharing of electrons between the two

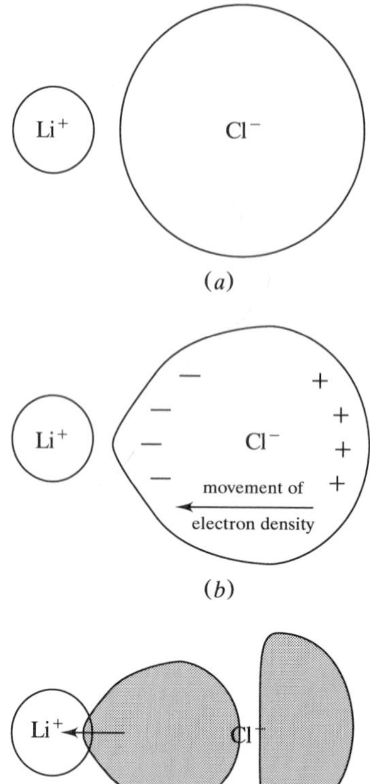

FIGURE 9.10
The polarization of the chloride anion by the lithium cation. (*a*) The small Li^+ is able to get very close to the larger, more diffuse electron cloud of Cl^-; (*b*) the electron cloud of Cl^- is *distorted*, or polarized, by the Li^+; (*c*) the opportunity for overlap between the valence orbitals in Li^+ (empty $2s$) and the Cl^- (filled $3p$) is increased.

species (as shown in Figure 9.10*c*) is, as you know from earlier experience, characteristic of a covalent bond.

The small size of the lithium cation is not the whole story. The larger the positive charge of the cation, the greater its distorting, or polarizing, power and the more covalent the character of the bond. Usually these effects are summarized by saying that the larger the charge-to-radius (Z/r) ratio of the cation, sometimes referred to as the *charge density*, the greater its polarizing power. (See Chapter 8, p. 194, for another context for the term charge density.) We will detail the consequences of this effect in the chapters on the individual groups, but note for now that our network gives us the means to *understand* why LiCl, being significantly covalent in character, is more soluble in less polar solvents (alcohols, for example) than might be expected.

The Increased Likelihood of π Bonding in the First Elements

The small size of the first elements increases the likelihood of π-bond formation among themselves and with other elements. Pi bonds, as you know, involve parallel overlap between, for example, two p orbitals. (Pi bonding can occur using d orbitals and even the antibonding molecular orbitals of some molecules. We discussed this to some degree in Chapter 4, although it is not vital to the argument being presented here.) If the two atoms involved are large, then parallel, π-type overlap will be less effective. This situation is shown in Figure 9.11. On the other hand, sigma bonding, characterized by head-on overlap of orbitals pointing directly

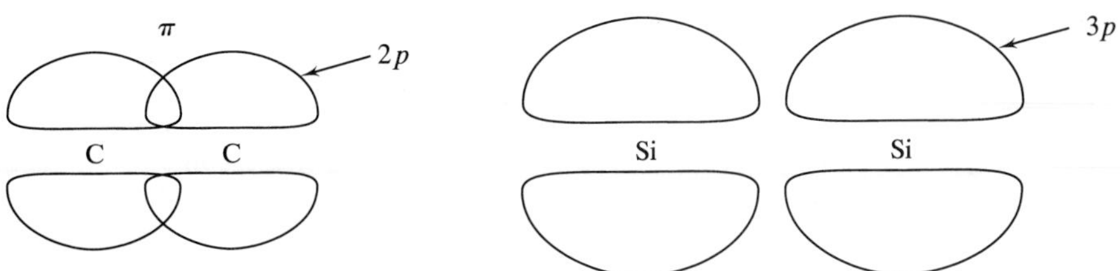

FIGURE 9.11
Parallel orbital overlap, or π bonding, is more effective in (*a*) the smaller first elements, for example, carbon, than it is in (*b*) their larger congeners, for example silicon.

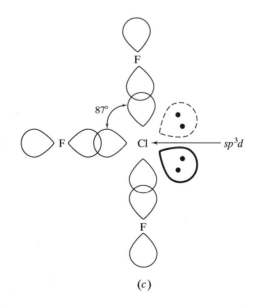

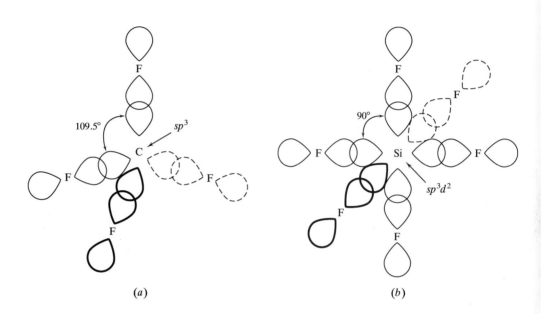

FIGURE 9.12
The availability of *d* orbitals in the heavier congeners of the first elements makes expanded octets possible. (*a*) Carbon is sp^3-hybridized in CF_4. (*b*) Silicon is sp^3d^2-hybridized in SiF_6^{2-}. (*c*) Chlorine is sp^3d-hybridized in ClF_3.

at each other, will be enhanced the greater the size of the participating orbitals. An important consequence of these considerations is a greater incidence of double and triple bonds, utilizing pi as well as sigma bonds, in the chemistry of the first elements (C=C, C≡C, O=O, C=O, C≡O, N≡N, etc.) than in their congeners.

The Lack of Availability of *d* Orbitals in the First Elements

The third reason for the uniqueness of the first elements in each group is the lack of availability of *d* orbitals. Starting with elements such as silicon, phosphorus, sulfur, and chlorine and continuing into the later periods, we find that *d* orbitals

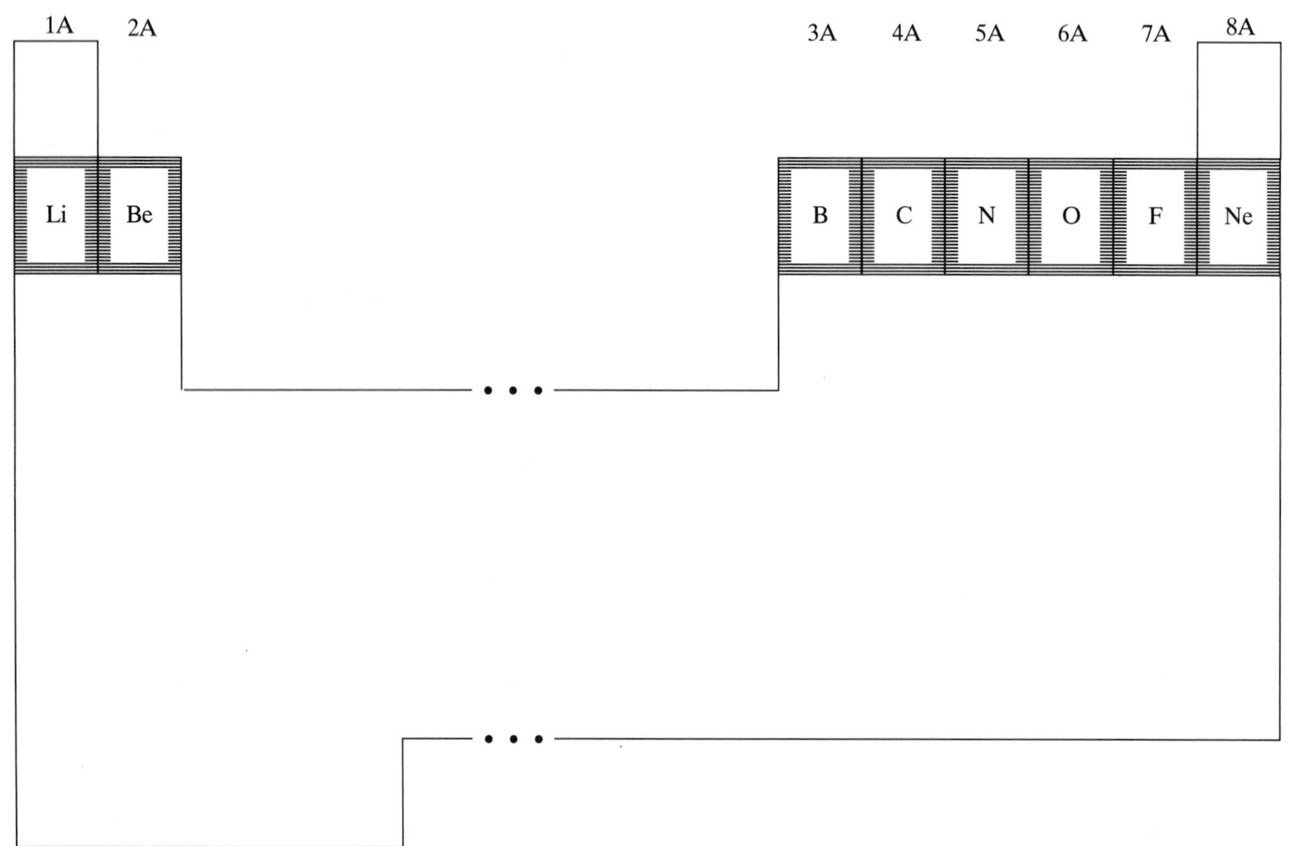

Three aspects of the uniqueness principle:

(*a*) The small size of the elements leading to a high polarizing power and high degree of covalent character in their compounds
(*b*) The greater probability of π bonds ($p\pi$—$p\pi$)
(*c*) The lack of availability of the *d* orbitals

FIGURE 9.13
A summary of three reasons for the uniqueness principle which states that the chemistry of the second-period elements (Li, Be, B, C, N, O, F, and Ne) is significantly different from that of their congeners.

are of low-enough energy that they can be occupied without an undue expenditure of *promotion energy*, the energy needed to "promote" an electron from a lower-energy to a higher-energy orbital. The availability of such *d* orbitals in the heavier congeners of each group makes "expanded octets" possible, whereas in the first element they are not.

There are many consequences of this difference in the availability of *d* orbitals. For example, carbon can form only compounds such as CF_4, but silicon can form the hexafluorosilicate ion, SiF_6^{2-}, as in Na_2SiF_6. In terms of valence-bond theory, the hybridization around the carbon atom in CF_4 would be sp^3 (Figure 9.12*a*), while the silicon atom would be sp^3d^2-hybridized (Figure 9.12*b*). A second example involves halogen chemistry. Fluorine can, in general, only form single bonds with one other element as in F_2 or HF. Chlorine, however, can form a number of compounds in which it is bound to three other atoms, for example, chlorine trifluoride, ClF_3, in which the Cl is sp^3d-hybridized (Figure 9.12*c*). There are many other examples of related differences between the first element in each group and its congeners. Look for these as we start our descriptions of the various groups. (For those who have read Chapters 2 to 6 on coordination chemistry, you should recall that the reason why the phosphines, PR_3, are better ligands than the ammines such as ammonia, NH_3, is because the former can interact with metal atoms and ions through their empty *d* orbitals.)

The three reasons for the uniqueness of the first element in each group are summarized in Figure 9.13. This figure should not be memorized.

9.3 THE DIAGONAL EFFECT

The periodic law and the uniqueness of the first elements in each group are two of the organizing principles in our network. What others are there? A third is the *diagonal effect*, which states that a diagonal relationship exists between the chemistry of the first member of a group and that of the second member of the next group. Actually this effect is important only for the first three groups, where we find that lithium and magnesium are surprisingly similar as are beryllium and aluminum as well as boron and silicon.

There appear to be three principal factors why these pairs, take beryllium and aluminum as a representative example, have so much chemistry in common. One factor is ionic size; the others are charge density (or charge-to-radius ratio Z/r) and electronegativity. Table 9.2 shows the charges, radii, charge densities, and electronegativities of the appropriate eight elements. Note that the ionic radius of Be^{2+} (0.41 Å) is more similar to that of Al^{3+} (0.53 Å) than it is to that of Mg^{2+} (0.71 Å). This means that beryllium and aluminum should be more interchangeable in various crystal lattices than will beryllium and the larger magnesium. (See Section 7.3 for further information on the role of radius ratios in determining the relative stability of various crystal structures.)

Electronegativity differences between the constituent atoms making up a chemical bond are indicative of the relative covalent character of that bond. Since beryllium and aluminum both have an electronegativity of 1.5, it follows that Be—X and Al—X bonds (in which X is typically a nonmetal) should be of similar covalent character on this basis.

We discussed charge density in two earlier contexts: (1) in considering the implications of the Born-Haber equation for lattice energies (Chapter 8, p. 194)

TABLE 9.2
Four relevant properties of elements related to the diagonal effect

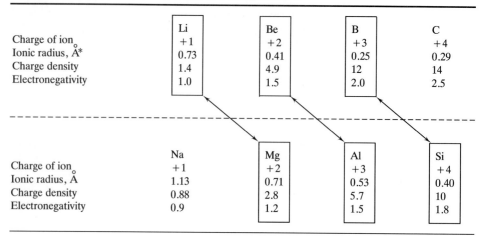

	Li	Be	B	C
Charge of ion	+1	+2	+3	+4
Ionic radius, Å*	0.73	0.41	0.25	0.29
Charge density	1.4	4.9	12	14
Electronegativity	1.0	1.5	2.0	2.5
	Na	**Mg**	**Al**	**Si**
Charge of ion	+1	+2	+3	+4
Ionic radius, Å	1.13	0.71	0.53	0.40
Charge density	0.88	2.8	5.7	10
Electronegativity	0.9	1.2	1.5	1.8

*Shannon-Prewitt radii for C.N. = 4. (See discussion in Section 7.3 for further information.)

and (2) in explaining the first reason why the first elements of each group are unique (p. 226). Note in Table 9.2 that the charge densities of beryllium and aluminum are 4.9 and 5.7, respectively. The two metal ions, then, will similarly polarize the X atom in an M—X bond and give rise to a similar additional covalent character on that basis. We see then that the Be–Al diagonal relationship seems to be due, in large part, to the similarities in ionic size, electronegativity, and charge density of these atoms. In Chapter 13, we will detail the striking similarity in the chemistry of the compounds of these two elements.

Analogous arguments can be made about the other pairs (Li–Mg and B–Si) which show a diagonal relationship. We should note here several warnings. First, keep in mind that group relationships (for example, between beryllium and magnesium) are still the dominant factor. We will certainly see many examples of this in the appropriate group chapters. Second, the ions listed in Table 9.2, particularly the highly charged B^{3+}, C^{4+}, and Si^{4+}, really do not exist as such. Take BCl_3, for example. If it did exist momentarily as the ionic $B^{3+}3Cl^-$, B^{3+} would immediately polarize the chloride ions and primarily covalent bonds would be formed. Nevertheless, even with the above warnings, the diagonal relationship remains a good organizing principle and should now assume its place in our network. The diagonal effect is summarized in Figure 9.14. Do not memorize this figure.

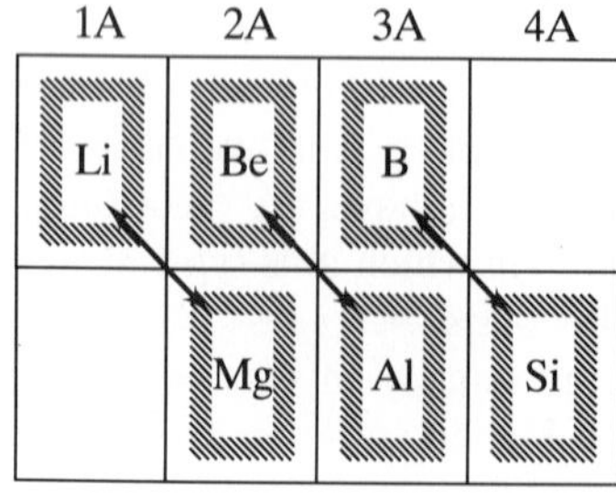

FIGURE 9.14
The elements of the diagonal effect. Lithium and magnesium, beryllium and aluminum, and boron and silicon, each pair diagonally located, have similar properties.

TABLE 9.3
Ionization energies and bond energies of the Group 3A elements

Element	Electron configuration	Sum of IE^2 and IE^3 kJ/mol	$2 \times D_{M-Cl}$*, kJ/mol
B	$[He]2s^2 2p^1$	6087	912
Al	$[Ne]3s^2 3p^1$	4562	842
Ga	$[Ar]3d^{10}4s^2 4p^1$	4942	708
In	$[Kr]4d^{10}5s^2 5p^1$	4526	656
Tl	$[Xe]4f^{14}5d^{10}6s^2 6p^1$	4849	

*M—Cl bond energies are derived from data for the trichloride compounds, MX_3; no value for the Tl—Cl in $TlCl_3$ is available, but presumably it is lower in value than for its lighter congeners.

9.4 THE INERT-PAIR EFFECT

There are various names given to this effect. In addition to just *inert-pair effect*, it has been called the *6s inert-pair effect* and the *inert-s-pair effect*. No matter what we call this idea, it states that the valence ns^2 electrons of metallic elements, particularly those $5s^2$ and $6s^2$ pairs that follow the second- and third-row transition metals, are less reactive than we would expect on the basis of trends in effective nuclear charge, atomic sizes, and ionization energies. This translates into the fact that In, Tl, Sn, Pb, Sb, Bi, and, to some extent, Po do not always show their expected maximum oxidation states but sometimes form compounds where the oxidation state is 2 less than the expected group valence. While this effect is more descriptive and less readily explained than the above ideas in our network, we are learning not to be content with just descriptions. How, then, can we explain or at least partially rationalize this effect?

There seem to be two principal aspects to the explanation. One is the trend in ionization energies going down a given group. A general decrease is expected due to the increase in atomic size. Table 9.3 shows the sum of the second and third ionization energies for the elements of the third group. Note that the expected decrease from boron to aluminum is evident but that gallium and thallium, in particular, have higher values than expected. Why should this be? The best explanation is that the 4*s*, 5*s*, and 6*s* electrons of Ga, In, and Tl, respectively, are not shielded as effectively from the nucleus by the intervening filled *d* and *f* subshells. While we cannot explain this decrease in the efficiency of shielding as well as we might like, we can cite other evidence that it is true. For example, reference to Figure 9.15 shows that there is a general decrease from left to right in the radii of the transition elements as well as in the lanthanides. In fact, the decrease in the size of the lanthanides is sometimes referred to as the *lanthanide contraction*. Also note that the elements La to Hg, following the lanthanide contraction, are very similar in size to their congeners immediately above (Y to Cd). For example, the radius of cadmium (before the lanthanides) is 1.54 Å, while the radius of mercury (after the lanthanides) is 1.57 Å. This is all an indication that the *nd* and *nf* electrons not only do not shield each other from the nucleus (as expected) but that they do not shield succeeding electrons from the nucleus very well either. If they did, the elements after the various contractions would be larger than they are. How is all of this related to the inert-pair effect? It simply means

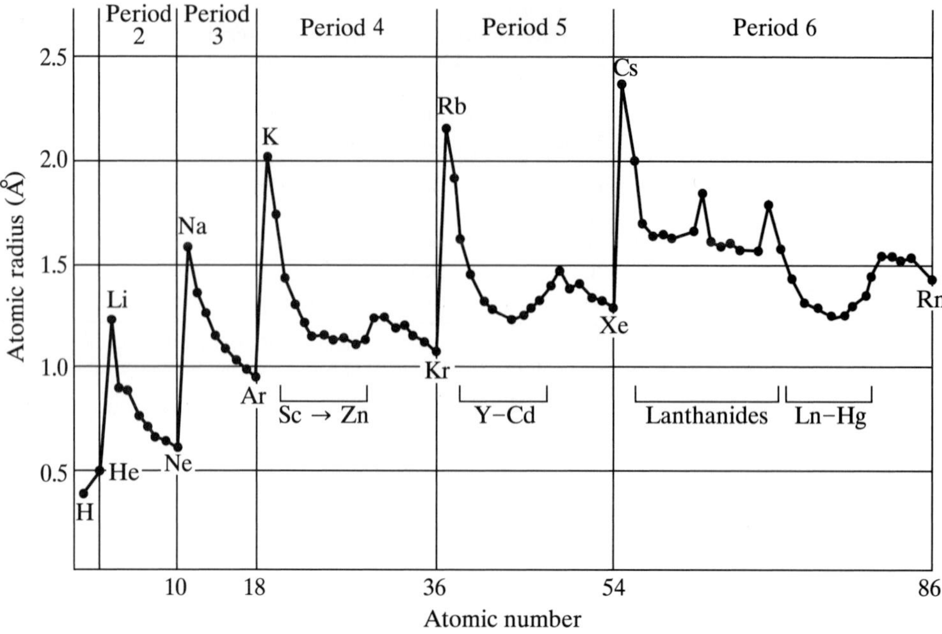

FIGURE 9.15
The atomic radii of the elements versus atomic number. [Adapted from Ref. 16.]

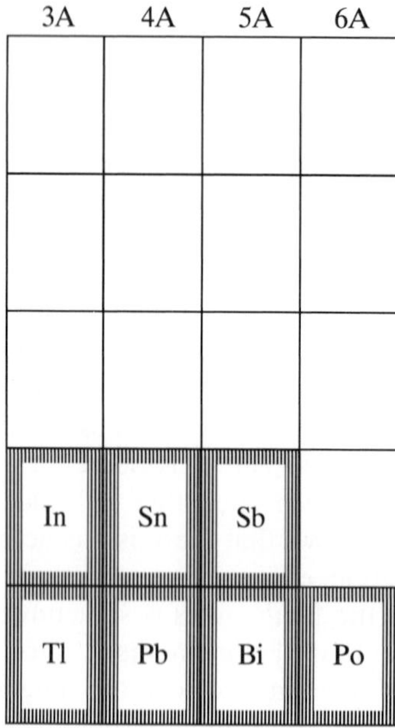

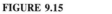

FIGURE 9.16
The elements of the inert-pair effect. The elements shown form compounds where the oxidation state is 2 less than the expected group valence.

that the 4*s*, 5*s*, and 6*s* electrons experience a larger effective nuclear charge than expected and that consequently they are more difficult to ionize.

The second aspect of the explanation of the inert-pair effect is the trends in bond energies going down a group such as Group 3A. The bond energies for the chlorides, which we will take as representative, are also shown in Table 9.3. We expect a decrease in bond energy down the group due to the increase in atomic size and therefore the bond distance. Consequently, the bonding electrons, in the region of overlap of the valence orbitals of these larger atoms, are farther from the nuclei of the atoms and have less ability to hold the two nuclei together.

The combination of these two effects, (1) the higher-than-expected ionization energies for Ga, In, and Tl and (2) the lower bond energies (as expected) for compounds involving these elements, are at least partially responsible for the inert-pair effect. In other words, for thallium, for example, more energy is needed to get the 6*s* electrons to form bonds, but not enough of this energy is recovered upon bond formation. Therefore thallium(I) compounds are more important than one would normally expect from a Group 3A element.

The inert-pair effect is summarized in Figure 9.16. This figure should not be memorized.

9.5 METAL, NONMETAL, AND METALLOID REGIONS

The periodic table is often divided into the metals at the bottom left, the nonmetals at the top right, and the semimetals, or metalloids, in between. Given the status of our network of organizing principles to this point, it is not surprising that the metals (with their low ionization energies and low negative or positive electron affinities) tend to lose electrons to form positive ions. Conversely, the nonmetals (with their high ionization energies and high negative electron affinities) tend to gain electrons to form negative ions. The division between metals and nonmetals, as shown in Figure 9.17, is the familiar stepwise diagonal line found on

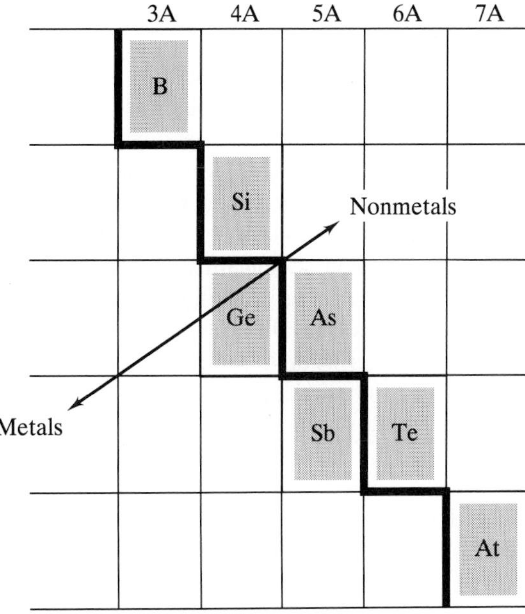

FIGURE 9.17
The metal-nonmetal line in the periodic table. Elements ot the lower left are metals; elements to the upper right are nonmetals. The metalloids, or semimetals, are shaded.

many periodic tables. The elements along this boundary have both metallic and nonmetallic characteristics and are called *metalloids*, or *semimetals*. Since metals are generally conductors of electricity and nonmetals are not, it follows logically that the semimetals are also generally semiconductors. (Such materials will be discussed in detail in Chapter 15.) As we discuss each group in turn, we will find that metals are generally high melting and boiling, lustrous, and exist in close-packed structures of cations surrounded by a sea of electrons (see Chapter 7 for further discussion of crystal structures). Nonmetals, conversely, are low melting and boiling, nonlustrous, and exist in chains, rings, and diatomic molecules. We will also see that metal oxides are basic and nonmetal oxides are acidic.

SUMMARY

To make sense out of the descriptive chemistry of the representative elements, we have started to construct a network of interconnected ideas in this chapter. The first five fundamental, organizing principles of the network are the periodic law, the uniqueness principle, the diagonal effect, the inert-pair effect, and the division of the table into metals, nonmetals, and metalloids. These five components are summarized in Figure 9.18.

The periodic law is the first and most useful network principle. Mendeleev was the first to solidly establish the nature of a table which displayed the periodic repetition of elemental physical and chemical properties. Using his periodic table, Mendeleev was able to organize and verify the similarities and trends of the known elements and also accurately predict the discoveries and properties of elements not yet known. Later, as the structure of the atom was revealed, the electronic configurations of the atoms in a group or period were found to help account for these periodic properties.

The concept of effective nuclear charge Z_{eff} is central to an understanding of the periodic law. Effective nuclear charge, which is the actual nuclear charge Z minus a screening constant σ, can be calculated in two ways. In the first, the screening constant is taken to be just the number of inner-core electrons in an atom. This results in the effective nuclear charge being identified with the group number. In the second method, the screening constant is calculated using Slater's rules. In either case, Z_{eff} is found to increase going from left to right in a period and stay approximately constant within a group. Using these trends in Z_{eff}, the variations in periodic properties such as atomic radii, ionization energy, electron affinity, and electronegativity are readily rationalized. Exceptions in these periodic properties are explained using more detailed analyses of electronic configurations.

The uniqueness principle is the second component of the network. Three reasons are given to rationalize the fact that the chemistry of the second-period elements is quite often significantly different from that of their heavier congeners. (1) The lighter elements are strikingly smaller than their congeners. This size effect leads to smaller electron affinities, larger charge densities, and enhanced degrees of covalent character in their compounds. (2) The first elements in each group, also due to their very small sizes, show an enhanced likelihood of π bonding and are capable of forming strong double and triple bonds. (3) The lighter elements lack the availability of d orbitals and therefore cannot form compounds with expanded octets.

The diagonal effect, the third component of the network, speaks of a strong diagonal relationship between the first member of a group and that of the second member of the next group. This effect relates lithium to magnesium, beryllium to

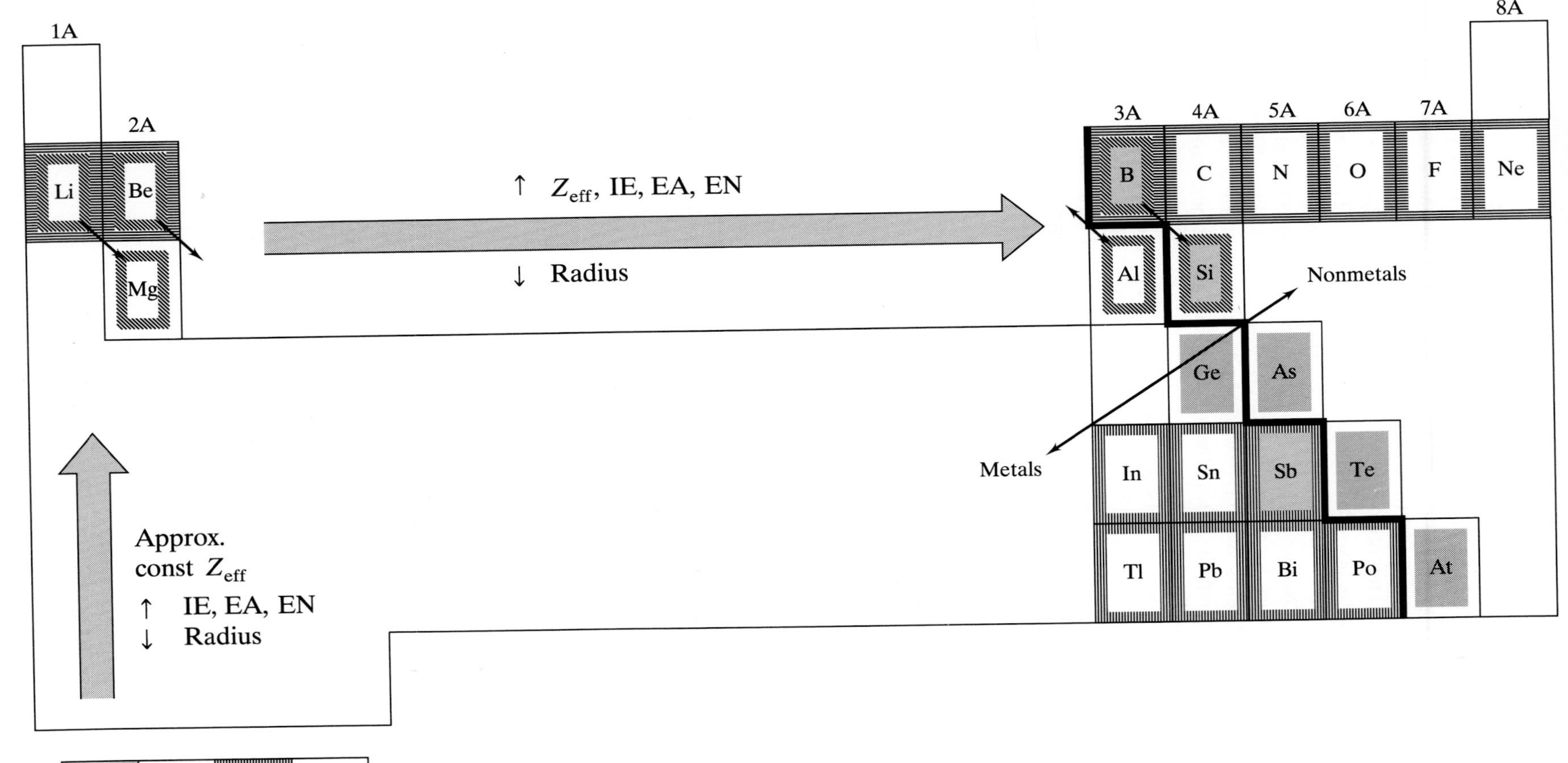

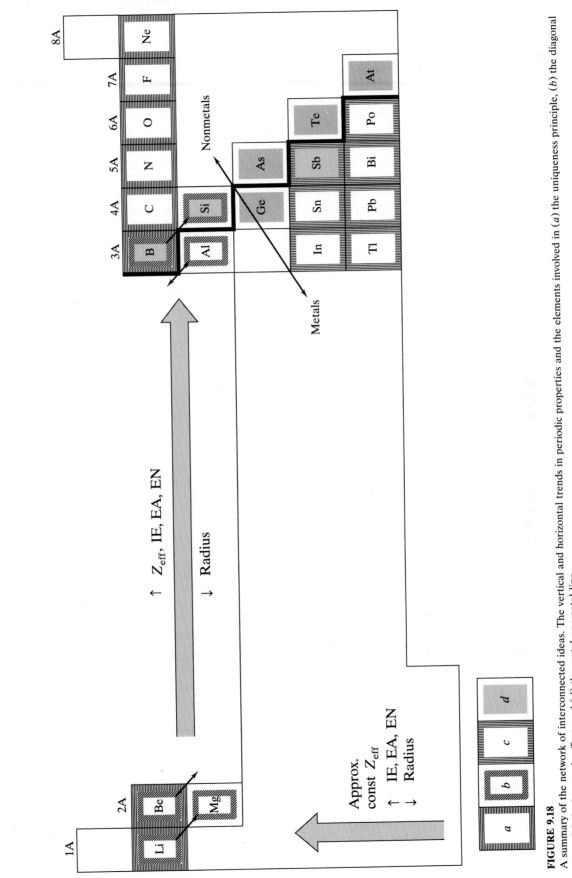

FIGURE 9.18
A summary of the network of interconnected ideas. The vertical and horizontal trends in periodic properties and the elements involved in (*a*) the uniqueness principle, (*b*) the diagonal effect, (*c*) the inert-pair effect, and (*d*) the metal-nonmetal line.

aluminum, and boron to silicon. Ionic size, charge density, and electronegativity are cited as three major factors in the rationalization of the diagonal effect.

The fourth network component, the inert-pair effect, states that the valence ns^2 elements of metallic elements which follow the second- and third-row transition metals are less reactive than expected. These relatively inert ns^2 pairs mean that elements such as In, Tl, Sn, Pb, Sb, Bi, and Po often form compounds where the oxidation state is 2 less than the expected group valence. The two major reasons for this effect are (1) larger-than-normal effective nuclear charges in these elements and (2) lower bond energies in their compounds. The fifth network component is just the division of the periodic table into metal, nonmetal, and metalloid regions.

Figure 9.18 is a representation of the first five components of the network. (Two others will be introduced in the next few chapters.) It is the framework on which we will build our study of descriptive chemistry. Like any superstructure as it is first erected, such as the frame of a house or even a skyscraper, it has many gaping holes. However, this figure represents a mental structure, a mental framework if you like, that we can slowly add to in the succeeding chapters. If we continue to relate additional knowledge to our framework, memorization will be kept at a minimum, and, in the end, we will have an excellent working knowledge of the chemistry of the elements.

PROBLEMS

9.1. When Mendeleev predicted the properties of his "eka" elements, he did so by interpolation; that is, he knew the properties of the elements above and below and before and after the missing element. Suppose that eka-silicon (germanium) was still undiscovered. Using a handbook of chemistry and physics and the information found elsewhere in this textbook, predict values for the following properties of the undiscovered element: atomic weight, density, melting point, boiling point, electronegativity, ionization energy, electron affinity, and atomic radius. For each property, compare your "prediction" with the actual value.

9.2. Gallium, germanium, and scandium were named after the homelands of their discoverers. Can you identify three other elements which were similarly named? When was each discovered?

9.3. Sometimes zinc, cadmium, and mercury are not included as transition metal elements. On what basis might they be excluded from that list?

9.4. Would you classify the noble gases as main-group elements? Give reasons both for and against such a classification.

9.5. The names *pnicogen*, for a Group 5A element, and *pnictide*, for a simple compound of a Group 5A element with an electropositive element, seem to be derived from the Greek word *pnigos*, for "choking" or "strangling." From what you know about nitrogen, phosphorus, and/or arsenic and its compounds, speculate on the appropriateness of these little-used names for the Group 5A elements.

9.6. Look up *chalcogen* in a good dictionary. If you cannot find an entry for it, see if you can find a meaning for both the main root (*chalco*) and the suffix (*-gen*) of the word. Speculate why the Group 6A elements are so designated.

9.7. Look up *halogen* in a good dictionary. Speculate why the Group 7A elements are so designated.

9.8. Restate the periodic law in your own words.

9.9. A plot of atomic volumes versus atomic weight is shown below. Discuss the relationship between this plot and the periodic law. (Figure from Ref. 22, p. 207.)

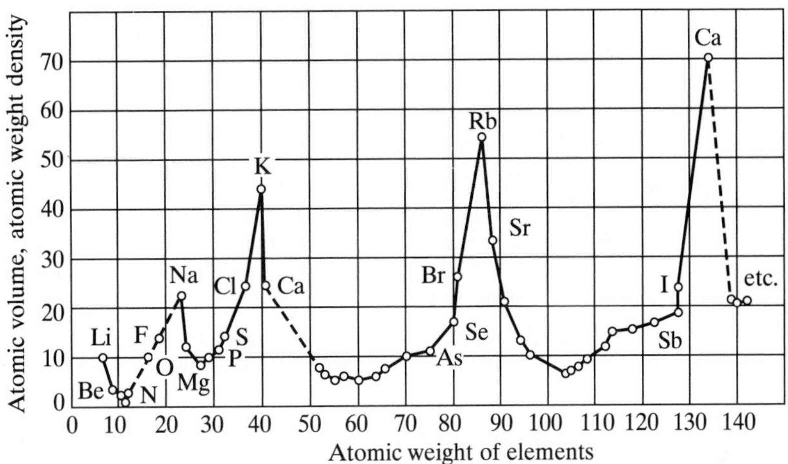

9.10. Suppose that the value of the atomic volume of rubidium were missing in the plot found in Problem 9.9. Could you predict a value for it? On what basis would you make your prediction?

9.11. Write the full electronic configurations of the following elements:
(*a*) Phosphorus
(*b*) Copper
(*c*) Arsenic
(*d*) Thallium

9.12. Write an orbital diagram (noble-gas or pseudonoble-gas core plus valence electrons in boxes) for the elements listed in Problem 9.11. (*Hint:* If you are not sure of the definition of an orbital diagram, look it up in your general chemistry textbook.)

9.13. Discuss the pros and cons of using such a designation as "pseudonoble gas configuration." Is there such a thing as a pseudonoble gas?

9.14. Using the assumption that σ equals the number of inner-shell electrons, calculate the effective nuclear charge felt by a valence electron in the following elements:
(*a*) Calcium
(*b*) Silicon
(*c*) Gallium
(*d*) Argon

9.15. Using the assumption that σ equals the number of inner-shell electrons, calculate a value for the effective nuclear charge felt by (*a*) an electron being added to the $3s$ orbital of a neon atom and (*b*) an electron being ionized from the $2p$ orbital of the neon atom. Comment on your results relative to the stability of the electron configuration of the neon atom.

9.16. Using Slater's rules, calculate a value for the effective nuclear charge felt by (*a*) an electron being added to the $3s$ orbital of a neon atom and (*b*) an electron being ionized from the $2p$ orbital of the neon atom. Comment on your results relative to the stability of the electron configuration of the neon atom.

9.17. Calculate the Z_{eff} for the valence electrons in the atoms Li to Ne using (*a*) the assumption that σ equals the number of inner-shell electrons and (*b*) Slater's rules. Plot both sets of results on the same graph and discuss.

9.18. Write a concise paragraph explaining the trends in effective nuclear charge across the second period (Li → Ne). Your answer should include a concise definition of effective nuclear charge.

9.19. Using Slater's rules, calculate and compare the Z_{eff} for both the $4s$ and $3d$ electrons of a copper atom. Discuss the results relative to the known fact that $4s$ electrons ionize first.

9.20. Using Slater's rules, calculate the Z_{eff} for the $4s$ and then the $3p$ electrons of calcium. Discuss the results relative to the first, second, and third ionization energies.

9.21. Calculate the Z_{eff} for the valence *ns* electron of lithium, sodium, and potassium using the assumption that σ equals the number of inner-shell electrons. Are your results consistent with the trends in ionization energy for these elements? Briefly discuss why or why not.

9.22. Using Slater's rules, calculate the Z_{eff} for Al, Al^+, Al^{2+}, and Al^{3+}. Discuss your results relative to the expected ionization energies for these species.

***9.23.** Recall why the energy of an *ns* orbital is less than that of an *np* orbital. Use this information to discuss the assumption that these orbitals are always considered as a group (*ns*, *np*) in Slater's rules.

***9.24.** A plot of the probability of finding $3s$, $3p$, $3d$, and $4s$ electrons as a function of the radial distance from the nucleus is shown below. Discuss these probabilities relative to rules $2c$ and $2d$ of Slater's rules.

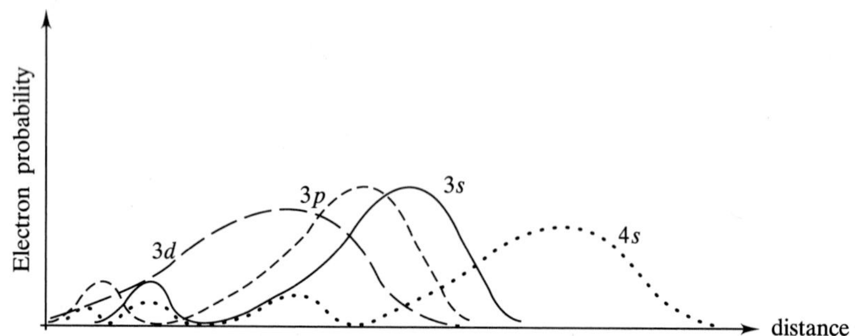

9.25. Write a concise paragraph explaining the general trend in the atomic radii of:
(*a*) The third-period elements
(*b*) The alkali metals

9.26. Write a concise paragraph explaining the general horizontal and vertical trends in ionization energies. Your answer should include a definition of ionization energy.

9.27. Write electron configurations for magnesium and aluminum. Briefly explain why the ionization energy of aluminum is less than that of magnesium when the general trend is the opposite.

9.28. Write electron configurations for phosphorus and sulfur. Briefly explain why the ionization energy of phosphorus is larger than that for sulfur.

9.29. Write a concise paragraph explaining the general trends in electron affinity. Your answer should include a definition of this property.

9.30. Without reference to Figure 9.7, predict which of the following three elements, aluminum, silicon, or phosphorus, would have the most negative electron affinity.

Verify your results by reference to the figure. Briefly explain the relative values of these electron affinities.

9.31. Write electron configurations for sodium and magnesium. Briefly explain why the electron affinities of these two elements are not in accordance with the expected general trends.

9.32. Write electron configurations for carbon and nitrogen. Briefly explain why the electron affinities of these two elements are not in accordance with the general trends expected.

9.33. Sketch a periodic table, indicating the trend of electronegativities from the lowest to the highest. Briefly relate these trends to effective nuclear charge and atomic size. Your answer should include a concise definition of electronegativity.

9.34. Briefly explain why the polarizing power of a cation is directly related to its charge-to-radius ratio, or charge density.

9.35. What are the units of the charge densities given in Table 9.2?

9.36. Which would you expect to be more covalent, $BeCl_2$ or $MgCl_2$? Discuss the reasons for your answer. Would you expect $AlCl_3$ to be more like $BeCl_2$ or $MgCl_2$? Briefly discuss your answer.

9.37. Graphite is a network of hexagonal rings of carbon atoms held together by pi bonds as shown below. The sheets themselves are held together by van der Waals intermolecular forces. Why is their no silicon analog of graphite?

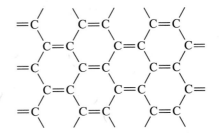

```
      /      \    /      \    /
  =C          C=C          C=C
     \      /    \      /    \
      C=C          C=C          C=
     /    \      /    \      /
  =C          C=C          C=C
     \      /    \      /    \
      C=C          C=C          C=
     /    \      /    \      /
  =C          C=C          C=C
     \      /    \      /    \
```

9.38. Elemental oxygen is characterized by diatomic molecules held together by both sigma and pi oxygen-oxygen bonds, while elemental sulfur is characterized by rings and chains held together by sulfur-sulfur single bonds. Briefly explain the difference in the elemental forms of these elements.

9.39. The most stable state of elemental nitrogen is the N_2 molecule characterized by a strong triple bond. The most stable state of elemental phosphorus, on the other hand, is the P_4 molecule characterized as a tetrahedron of phosphorus atoms (shown below) held together by strong single bonds. Briefly comment on this difference.

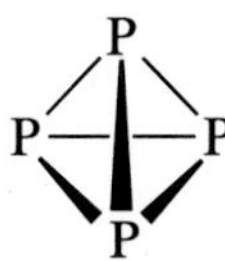

9.40. Nitrogen forms nitrogen trichloride, but phosphorus, arsenic, and antimony form both the trichlorides, XCl_3, and the pentachlorides, XCl_5. Comment on this situation. What is the hybridization of the central atom in the above compounds?

9.41. Unlike phosphorus, arsenic, and antimony, bismuth does not form a pentachloride. Briefly comment.

9.42. Oxygen forms the compound OF_2, but sulfur, selenium, and tellurium form the hexafluorides, XF_6. Comment on this situation. What is the hybridization of the central atom in the above compounds?

9.43. The most stable halides of boron and aluminum are the trihalides, while for thallium the monohalides (TlX) are the most stable. Similarly, the most stable oxides of boron and aluminum are of the formula M_2O_3, while for thallium the most stable oxide is Tl_2O. Comment on the above chemistry.

9.44. Both tin(II) and tin(IV) chloride, $SnCl_2$ and $SnCl_4$, are known as the complex anion $SnCl_6^{2-}$. The single-carbon chlorides, however, are essentially limited to carbon tetrachloride, CCl_4. Briefly discuss these results.

9.45. Using Slater's rules, calculate the effective nuclear charge on the np^1 valence electron of aluminum. Do a similar calculation for gallium, indium, and thallium. Comment on the results relative to the first ionization energies of these elements.

9.46. Write a concise paragraph explaining what is meant by the inert-pair effect and how it can be accounted for.

9.47. Write a concise paragraph explaining what is meant by the diagonal effect and how it can be accounted for.

9.48. Comment on the metallic character of the following elements: mercury, indium, germanium, phosphorus, and oxygen.

9.49. Comment on the metallic character of the following elements:

(*a*) Sodium, magnesium, aluminum, silicon, phosphorus, and sulfur

(*b*) Zinc, gallium, germanium, arsenic, selenium, and bromine

CHAPTER 10

HYDROGEN AND HYDRIDES

10.1 THE ORIGIN OF THE ELEMENTS (AND OF US!)

Where do we come from? Where did the sun and moon, the earth and the other planets, and the stars and galaxies come from? How long has the sun existed? How long will it last? And the universe, has it always existed or did it have a beginning? What about an end? These are all questions that humankind has asked from the moment we could scratch our heads in wonder. In addition, we as chemists (some would say, with a twinkle in their eye, a singular and extraordinary breed of humankind) have more specific questions. Where did the elements we see displayed in the periodic table come from? Were they constructed in some order? Which ones came first? Can we account for the known abundances in the universe and on the earth? Why are there only about 100 of these elementary substances? Some of these questions may have been around for several hundreds of thousands of years (*Homo sapiens* are thought to have evolved about 100,000 years ago), and humankind has constructed some imaginative stories to try to answer them. Only within the last century have we started to gather some empirical evidence to indicate that the stories (we now call them theories) of the last few generations are on the right track.

The presently accepted story or theory about the beginning of the universe goes by an odd, somewhat trivial sounding name, the "big bang theory," a term popularized by physicist George Gamow in the 1940s. According to this theory, all the universe was once contained in the singularity or the *ylem*, a word coined by Aristotle to designate the substance out of which all the universe was created. In the ylem were a variety of particles. Some, like protons, neutrons, and electron-positron pairs, we as chemists recognize. Others, like hadrons and baryons, belong to the realm of subnuclear particle physics. In addition to these and other particles, the ylem contained all the energy necessary to form the universe; all this in a volume no larger than the head of a pin. At a definite point in time, current thinking indicates perhaps 15 to 20 billion years ago, the ylem exploded in an unimaginable flash of light and heat. As the universe expanded from its violent beginnings, energy was converted, or we could say "condensed," to matter in the form of approximately two parts hydrogen to one part helium. These swirling clouds of light atoms were attracted by gravitational forces to start to form the galaxies and their constituent stars. Hydrogen and helium, therefore, are thought to have been the first elements, and hydrogen is the most abundant in the universe.

The interior of these first clouds of hydrogen and helium gradually heated up due to frictional forces. At about 10 million degrees the force of the collisions among these particles was sufficient that these nuclei, instead of bouncing off each other as we usually imagine in the kinetic theory of gases, started to merge together to form bigger nuclei, a process known as *nuclear fusion*. To represent this fusion and other nuclear processes, recall that the standard format for representing the nucleus of an element is ${}^{A}_{Z}X$, where X is the symbol of the element, Z is the *atomic number* (the number of protons), and A is the *mass number* (the total number of protons and neutrons, known collectively as *nucleons*).

Using the above notation, we can represent the simplest fusion reaction, the one requiring the least kinetic energy (and therefore the lowest temperature), as shown in Equation (10.1):

$$ {}^{1}_{1}\text{H} + {}^{1}_{1}\text{H} \longrightarrow {}^{2}_{1}\text{H} + {}^{0}_{+1}e + \text{energy} \qquad (10.1) $$

Note that the first product is also a hydrogen nucleus but one that contains a neutron as well as a proton and therefore has a mass number of 2. Recall that species which have the same atomic number but different mass numbers are *isotopes*. The second product is a positron, sometimes referred to as a *beta plus particle*. A positron has the same mass as an electron but the opposite charge. Note that the symbol for a particle other than an element is given as a lowercase letter. Some other subnuclear particles and their symbols are given in Table 10.1. In such cases the subscript represents the charge on the particle, -1 in the case of an electron, $+1$ for a positron, zero for a neutron, and so forth.

Equation (10.1) is a balanced nuclear equation; that is, the sum of the atomic numbers on either side is the same, as is the sum of the mass numbers. Where then, does the energy come from? A straightforward calculation shows that the sum of the masses of the reactants is just the slightest bit greater than that of the products, the difference being known as the *mass defect*. Converting the mass defect to energy using the well-known Einstein relationship $E = mc^2$ shows that

TABLE 10.1
The symbols and masses for the most common nuclear and subnuclear particles

Particle	Symbol	Mass, u
Nuclear		
Proton (hydrogen nucleus)	${}^{1}_{1}\text{H}$	1.00728
Alpha particle (helium nucleus)	${}^{4}_{2}\text{He}$	4.0015
Subnuclear		
Electron (beta particle)	${}^{0}_{-1}e$	5.488×10^{-4}
Positron (positive electron)	${}^{0}_{+1}e$	5.488×10^{-4}
Neutron	${}^{1}_{0}n$	1.008665

the simple fusion reaction of Equation (10.1) would yield 9.9×10^7 kJ/mol! Nuclear fusion reactions, then, yield four or five orders of magnitude (i.e., powers of 10) more energy than do chemical reactions.

Equations (10.2) and (10.3) show two additional fusion reactions that are thought to occur in an ordinary star such as our sun:

$$^{2}_{1}H + ^{1}_{1}H \longrightarrow ^{3}_{2}He + \text{energy} \qquad (10.2)$$

$$^{3}_{2}He + ^{3}_{2}He \longrightarrow ^{4}_{2}He + 2\,^{1}_{1}H + \text{energy} \qquad (10.3)$$

These three reactions occur sequentially, and the net result [2(10.1) + 2(10.2) + (10.3)] is known as the *proton-proton cycle* shown in Equation (10.4):

$$4\,^{1}_{1}H \longrightarrow ^{4}_{2}He + 2\,^{0}_{+1}e + \text{energy} \qquad (10.4)$$

The result of these nuclear reactions is the "burning" of hydrogen (at an estimated rate of 4 million tons of hydrogen/s) to produce helium. The energy is released in the form of heat and light which we experience as sunshine. Our sun, calculated to have formed almost 5 billion years ago, should continue to burn hydrogen for another 5 billion years and then, after a brief but spectacular red giant stage in which the inner planets would be consumed, die away to nuclear cinders.

But what about the rest of the elements? Even a moderately complete description of the process of *nucleosynthesis*, the process by which all the other elements are synthesized from hydrogen and helium, is outside the scope of this text, and what follows is but the barest of sketches. In a heavy star, one at least 8 to 10 times the mass of the sun, helium can also burn to produce beryllium, carbon, oxygen, neon, and magnesium (all multiples of helium 4), and these and other elements build up in the interior of the star. Other nuclear reactions can produce all the elements up to and including iron. Fusion reactions involving iron are endothermic (not exothermic like all the above) and do not occur spontaneously. With the production of iron, fusion stops, and the star, its structure now resembling that of an onion composed of layers of different elements with iron in the center and helium and hydrogen on the surface, contracts due to the force of gravity. This contraction results in an intolerably high density (the nuclei are postulated to actually touch each other) and such high temperatures and pressures that the star erupts in a massive supernova explosion. Supernovas are rare but certainly spectacular occurrences. The brightest in almost 400 years was observed in the southern latitudes in February 1987. It is in these supernovas that elements heavier than iron are thought to be synthesized. The remains of such exploding stars are swept up with interstellar hydrogen and helium to produce second- and third-generation stars like our sun. The spectroscopic observation of elements such as calcium and iron in the sun are evidence that it is at least a second-generation star.

10.2 THE DISCOVERY, PREPARATION, AND USES OF HYDROGEN

Take a moment to think carefully about the time scale outlined in the previous section. Hydrogen was the first element to form after the big bang 15 to 20 billion years ago. It is the principal component of our sun synthesized almost 5 billion years ago and a primary component of life, the penultimate expression (so far) of which is the species *Homo sapiens* that evolved perhaps 100,000 years ago. Yet these creatures have only been aware of the existence of hydrogen for about 300 years! In 1671, Robert Boyle prepared a gas, later identified as hydrogen, by

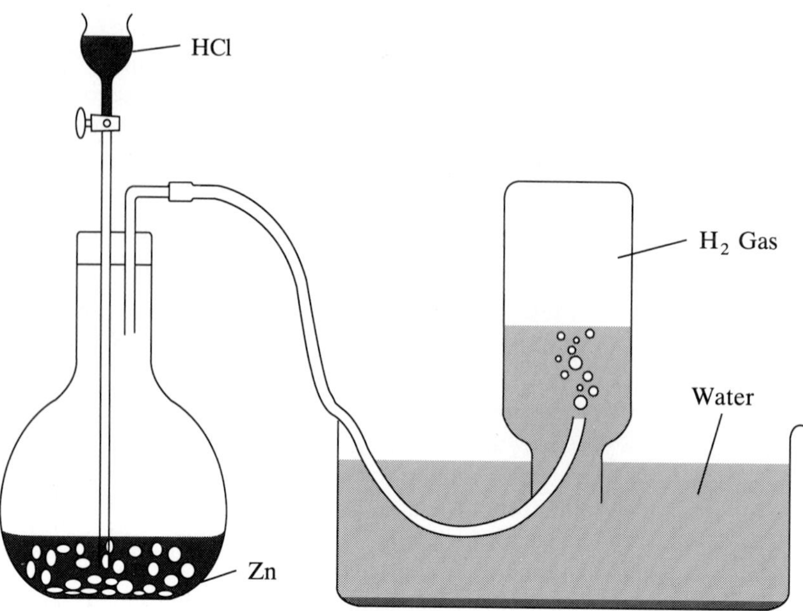

FIGURE 10.1
Apparatus for the laboratory production of hydrogen gas.

dissolving iron in dilute hydrochloric or sulfuric acid. Such a procedure is still used for the laboratory production of the gas. A schematic of the apparatus usually employed is shown in Figure 10.1.

Nicolas Lémery, in the early eighteenth century, also prepared hydrogen as described above and noted that the "vapor will immediately take fire and at the same time produce a violent, shrill fulmination." Although we know better today, it's not surprising to anyone who has exploded a balloon filled with hydrogen that Lémery thought he had discovered the origin of thunder and lightning!

In 1766, Henry Cavendish, in a rare report to the Royal Society of London, described his thorough investigations of the properties of hydrogen gas. Cavendish, painfully shy to the point that he rarely if ever spoke to a group of people, built a separate entrance to his house so that he would not encounter anyone (particularly a woman) as he came in and out. Although he carried out productive research for more than 60 years, Cavendish only published about 20 articles. However, he did report his work on hydrogen and is usually given the credit for its discovery. He showed that hydrogen is lighter than air and that water is produced when hydrogen and oxygen are reacted together. Both hydrogen (water producer) and oxygen (acid producer) were named by Antoine Lavoisier in the late 1770s and early 1780s.

The reaction of a metal such as iron or zinc with a strong acid such as hydrochloric or sulfuric [shown in Equation (10.5)] is still the most common way to produce hydrogen in the laboratory. It is also common to reverse the reaction between hydrogen and oxygen by carrying out the electrolysis of water as represented in Equation (10.6). The sulfuric acid is necessary to carry a current through the otherwise insulating water:

$$Zn(s) + 2HCl(aq) \longrightarrow Zn^{2+}(aq) + 2Cl-(aq) + H_2(g) \quad (10.5)$$

$$2H_2O(l) \xrightarrow[\text{electrolysis}]{H_2SO_4} 2H_2(g) + O_2(g) \quad (10.6)$$

The most common industrial preparation of hydrogen is the *catalytic steam-hydrocarbon reforming process* [Equations (10.7) and (10.8)]:

$$C_3H_8(g) + 3H_2O(g) \xrightarrow[\text{Ni}]{\text{heat}} \underbrace{3CO(g) + 7H_2(g)}_{\text{Synthesis gas}} \tag{10.7}$$

$$CO(g) + H_2O(g) \xrightarrow[\text{Fe}_2\text{O}_3]{\text{heat}} CO_2(g) + H_2(g) \tag{10.8}$$

Aptly named, this process treats a mixture of hydrocarbons (propane, C_3H_8, is shown as a representative example) from natural gas or crude oil with steam (700–1000°C) over a nickel catalyst and reforms it into a mixture of CO and H_2 gases, which is called *synthesis gas*, or just *syngas*. The second reaction is known as the *water-gas shift reaction* perhaps because it shifts oxygen from one reactant to another and thereby adjusts the composition of the syngas. It is carried out at elevated temperatures (325–350°C) over an iron oxide catalyst.

The gasification of coal was seriously considered as an alternate source of energy during the 1970s, but with the apparent (and assuredly temporary) abatement of a crisis atmosphere, it is a technology waiting in the wings. The first step is the reaction of coal, a complex mixture of primarily graphitelike carbon (as well as troublesome impurities such as sulfur and nitrogen compounds, but here represented as just carbon) with hot steam. This *carbon-steam* reaction is represented in Equation (10.9):

$$\text{C(coal)} + H_2O(g) \xrightarrow[\text{Fe or Ru}]{\text{heat}} CO(g) + H_2(g) \tag{10.9}$$

At one time, the resulting syngas (sometimes referred to as *water gas*) was used directly as a fuel before it was realized just how dangerous carbon monoxide is. The water-gas shift reaction (10.8) is employed as above to convert the syngas to carbon dioxide and hydrogen. If hydrogen is the desired product, it can be used for a variety of applications at this point. If hydrocarbons are desired, a *methanation* to a product much like natural gas [Equation (10.10)] might follow:

$$CO(g) + 3H_2(g) \xrightarrow[\text{Ni}]{\text{heat}} \underset{\text{Methane}}{CH_4(g)} + H_2O(g) \tag{10.10}$$

Other catalysts afford different hydrocarbons. Note that all the above reactions require the use of metal catalysts. The research and development of more efficient catalysts and a better understanding of their roles is a topic of considerable interest in the evolving realm of inorganic chemistry. More about the use of these above processes can be found in Section 10.6 concerning alternative energy sources.

The largest single use of hydrogen is for the production of ammonia by the Haber process [Equation (10.11)] that, in turn, provides the starting product for a number of useful nitrogen compounds such as in fertilizers and explosives. More on these compounds can be found in Chapter 6 on Group 5A chemistry.

$$\text{Haber process: } N_2(g) + 3H_2(g) \xrightarrow[\text{iron}]{\text{high } T \text{ and } P} 2NH_3(g) \tag{10.11}$$

Hydrogen gas is used to partially hydrogenate unsaturated oils to produce solid fats used as shortenings and margarines as well as to make soaps. The use of various low-cholesterol oils (corn, olive, peanut, canola, safflower, sesame, soybean) produces low-cholesterol shortenings and margarines useful in minimizing heart disease. Hydrogen is also used to convert various oxides (such as those of silver, copper, lead, bismuth, mercury, molybdenum, tungsten) to the free metals as represented in Equation (10.12) for an M^{II} oxide:

$$MO(s) + H_2(g) \longrightarrow M(s) + H_2O(l) \qquad (10.12)$$

As we have discussed, one of the most spectacular reactions of hydrogen is with oxygen. (It certainly impressed Lémery and continues to impress every new generation of chemists.) Most everyone is familiar with the dramatic burning of the hydrogen-filled dirigible *Hindenburg* in 1937. Since then dirigibles or lighter-than-air aircraft, like the Goodyear blimps, are kept aloft by helium instead of hydrogen. Liquid hydrogen and oxygen are commonly used as rocket fuels, for example, in the Saturn-Apollo launch vehicles of the space shuttle program. The main engines contain several hundred thousand liters of liquid hydrogen and oxygen which are kept separate in a "bipropellant rocket unit" and only mixed at lift-off. The technology supporting the use of these gases is very sophisticated. Only through good engineering and perhaps just a little luck have we avoided more accidents such as the explosion of the *Challenger* space shuttle in early 1986.

The extreme heat generated in nuclear plant accidents such as Three Mile Island (1979) and Chernobyl (1987) is sufficient to split water into hydrogen and oxygen gases. A hydrogen-oxygen explosion ignited by a relatively small chemical blast is thought to have been responsible for the massive devastation of the Chernobyl plant. At Three Mile Island a hydrogen-oxygen explosion was avoided by slowly bleeding off the hydrogen gas bubble which formed at the top of the containment structure and storing it until it could be safely disposed of.

10.3 ISOTOPES OF HYDROGEN

Hydrogen forms three isotopes which are listed in Table 10.2. Deuterium was discovered in 1931 by Harold Urey and his associates. They allowed 4 liters of liquid hydrogen to evaporate down to 1 mL of a liquid that displayed several faint absorption lines (in the visible Balmer series) in addition to those of ordinary hydrogen. Quantum-mechanical calculations confirmed that doubling the mass of the hydrogen accounted for the new spectral lines. Tritium could not be detected in the above manner and was first prepared by Marcus Oliphant in 1934 by nuclear transmutation (see Section 10.4 on nuclear processes). Note that while most elements have a variety of isotopes, only those of hydrogen are given special names and symbols. (The names *protium*, *deuterium*, and *tritium* come from the Greek

TABLE 10.2
The isotopes of hydrogen

Name	No. of neutrons	Symbol	Mass, u	Percent abundance	Half-life, yr
Hydrogen (H) (or protium)	0	^1_1H	1.007825	99.985	
Deuterium (D)	1	^2_1H	2.0140	0.015	
Tritium (T)	2	^3_1H	3.01605	10^{-17}	12.3

prōtus, *deuteros*, and *tritos*, meaning "first," "second," and "third," respectively.) Special names were assigned to hydrogen 2 and 3 because the significant differences in their masses relative to ordinary hydrogen (2:1 and 3:1 ratios) cause them to have correspondingly significant differences in such physical and chemical properties as density, melting point, boiling point, and heats of fusion and vaporization (see Problems 10.20 and 10.22).

The above differences in atomic properties carry over to hydrogen compounds. For example, "heavy water," or deuterium oxide, D_2O, can be separated from ordinary water by electrolysis. One reason for this is that ordinary hydrogen ions (H^+) move to the negative electrode more rapidly than the twice-as-heavy deuterium ions (D^+). It follows that ordinary hydrogen gas, H_2, is the preferred product, and the concentration of deuterium oxide in the water left behind increases as the electrolysis is carried out. The remaining water becomes heavier and heavier (per unit volume), so, not surprisingly, it is often called heavy water. During World War II, when D_2O suddenly became important (as a moderator, see Section 10.4) in primitive nuclear fission reactors, the German control of Norwegian D_2O plants (built to take advantage of cheap and plentiful hydroelectric power) became strategically important. (Anyone who has seen the movie *The Heroes of Telemark*, staring Kirk Douglas and Richard Harris, knows only too well the extent of which the Allies went to destroy those Norwegian heavy water production plants.)

Deuterium oxide is found in greater concentrations in the Dead Sea and other bodies of water that have no outlets other than evaporation. The light water, of lower molecular weight, has a higher mean velocity at a given temperature and escapes from the surface of the sea more rapidly than heavy water. (Recall from the kinetic molecular theory that two substances at the same temperature have the same average kinetic energy $\frac{1}{2}mv^2$, and therefore the molecules of the lighter substance, in this case H_2O, have a greater average or mean velocity than do the molecules of the heavier substance, in this case D_2O.) Put another way, H_2O has a higher vapor pressure than does D_2O, and evaporation increases the concentration of deuterium oxide in the remaining water.

Soon after D_2O was produced, an intriguing and inevitable question was posed. Will heavy water support life in the same manner that ordinary water does? It does not. If large quantities of D_2O are given to mice, they first show signs of extreme thirst and then die. It seems that D_2O, again because it has a lower mean velocity, has a lower rate of diffusion into cells. Another contributing factor may be that the transfer of D^+, catalyzed by various enzymes, is slower than that of the lighter H^+. Independent of the details of the biological mechanism, no matter how much D_2O the mouse drinks, it still dies of dehydration!

Both deuterium and tritium can be incorporated into a variety of hydrogen-containing compounds and used to follow (or *trace*) the course of reactions involving these compounds. For example, one can follow the rate of absorption and excretion of water in the body by using small amounts of D_2O. Some D_2O is almost immediately excreted, but after 9 or 10 days, half of it still remains. It is calculated that the average water molecule stays in the human system for about 14 days. Or one can follow the uptake, storage, and excretion of fats by using a fat labeled with deuterium. Do fat deposits stay immobile in a living system until they are needed? Or is there a turnover, with ingested fats being deposited and stored fats being used as part of a dynamic interchange? It turns out that the dynamic model is correct.

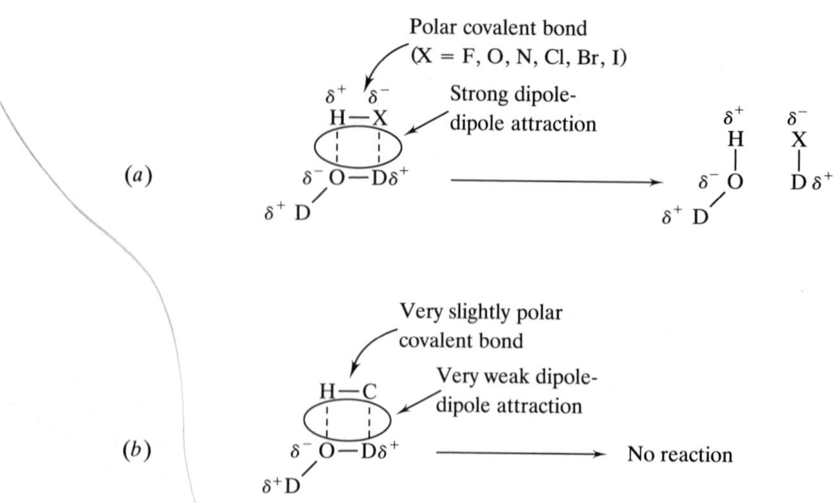

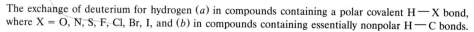

FIGURE 10.2
The exchange of deuterium for hydrogen (*a*) in compounds containing a polar covalent H—X bond, where X = O, N, S, F, Cl, Br, I, and (*b*) in compounds containing essentially nonpolar H—C bonds.

When various hydrogen-containing compounds are dissolved in heavy water, the hydrogens bound to electronegative atoms such as oxygen, nitrogen, sulfur, or one of the halogens are replaced with deuteriums while hydrogens bound to carbon are not. Why should this be? If X represents an electronegative atom, the situation is represented as shown in Equations (10.13) and (10.14):

$$H—X + D_2O \longrightarrow D—X + D—O—H \qquad (10.13)$$

$$H—C + D_2O \longrightarrow \text{no reaction} \qquad (10.14)$$

The replacement of an ordinary hydrogen atom with a deuterium, as shown in Equation (10.13), is known as *deuteration*. The reason why H—X bonds can be deuterated while C—H bonds cannot is outlined in Figure 10.2. Deuterium oxide, like ordinary water, contains polar O—D bonds which can interact via dipole-dipole forces with the polar H—X. C—H bonds are not polar and so do not interact with the polar O—D bonds of the deuterium oxide.

Tritium, as noted above, is not available in nature in sufficient quantities to isolate chemically. Nevertheless, once available from the nuclear processes to be outlined in the next section, it can be used to *tritiate* compounds in a manner analogous to deuteration. In other words, the light hydrogen atoms in compounds containing H—X bonds can be replaced with tritium atoms when dissolved in "superheavy water," T_2O (or ordinary water enriched in T_2O). It follows that tritium-labeled compounds can also be used as tracers to follow the progress of various reactions. For example, tritium oxide can be used to follow water from a source into the water table and then to an outlet. There is one significant difference, however, between deuterium and tritium labeling. Hydrogen atoms bound to carbon can be exchanged with the tritium atoms simply by storing the compound under tritium gas for a few days or weeks. This is not possible with deuterium gas. The beta decay of the radioactive tritium evidently facilitates the exchange. This brings us to a discussion of radioactive processes as they relate to hydrogen chemistry.

10.4 RADIOACTIVE PROCESSES INVOLVING HYDROGEN

Alpha and Beta Decay, Nuclear Fission, and Deuterium

In Section 10.1 we discussed the simplest fusion reaction [Equation (10.1)] in which a beta plus particle, that is, a positron ($_{+1}^{0}e$), is a product. Other common nuclear and subnulcear particles were given in Table 10.1. Having discussed the discovery and some of the chemistry of deuterium and tritium, we are ready to take a closer look at nuclear processes, particularly those related to hydrogen.

Most introductory chemistry courses discuss alpha, beta, and gamma radiation in the context of the history of atomic structure. Recall that Rutherford and his colleagues (including undergraduate students) investigated the effect of shooting alpha particles (later identified as helium nuclei, $_{2}^{4}\text{He}$) at thin pieces of metal foils. From the deflections observed, Rutherford concluded that the atom must have a heavy, dense, positively charged nucleus. Rutherford, Becquerel (who discovered radioactivity in 1896), and the Curies all contributed to the discovery of alpha and beta particles and the high-energy electromagnetic radiation called gamma rays.

An isotope is considered to be *radioactive* if it spontaneously breaks down or decays by emission of a particle and/or radiation. Alpha radiation is denoted by placing an α over the arrow of the equation showing the process. The alpha decay of uranium 238 is shown in Equation (10.15):

$$_{92}^{238}\text{U} \xrightarrow{\alpha} {}_{2}^{4}\text{He} + {}_{90}^{234}\text{Th} \qquad (10.15)$$

Note that the nuclear reaction is balanced as previously described. It turns out that there are two types of beta decay, beta plus decay (which produces a positron) and beta minus decay which produces an ordinary electron. Note, however, that this electron, while indistinguishable from any other electron, is the product of the decay or the falling apart of a nucleus. Beta minus and beta plus decays are denoted by placing a β^- or a β^+ over the arrow. Tritium decays by beta minus decay as shown in Equation (10.16):

$$_{1}^{3}\text{H} \xrightarrow{\beta^-} {}_{-1}^{0}e + {}_{2}^{3}\text{He} \qquad (10.16)$$

An example of an isotope which decays by beta plus emission is boron 8, the longest-lived radioactive isotope of boron. The reaction for its decay is represented in Equation (10.17):

$$_{5}^{8}\text{B} \xrightarrow{\beta^+} {}_{+1}^{0}e + {}_{4}^{8}\text{Be} \qquad (10.17)$$

How long do radioactive isotopes last? This is measured by a quantity familiar to anyone who has studied chemical kinetics. You should recall studying first- and second-order reactions and discussing the concept of half-life in each case. (Incidentally, radioactive decay follows first-order kinetics.) *Half-life* is defined as the time required for the concentration of reactant to decrease to half its initial concentration. Figure 10.3 shows the concentration of a radioactive isotope as a function of time. Note that after one half-life, one-half of the isotope remains, after two half-lives, there is one-quarter of the original remaining, and so forth. The half-lives of uranium 238 with respect to alpha decay, tritium with respect to beta minus decay, and boron 8 with respect to beta plus decay are 4.51×10^9 (4.51 billion) years, 12.3 years, and 0.77 s, respectively.

Two other important radioactive processes are fusion and fission. Fusion, in which lighter nuclei join together to form heavier nuclei, was introduced in Section 10.1. *Fission* is the process whereby a large nucleus splits to form smaller nuclei

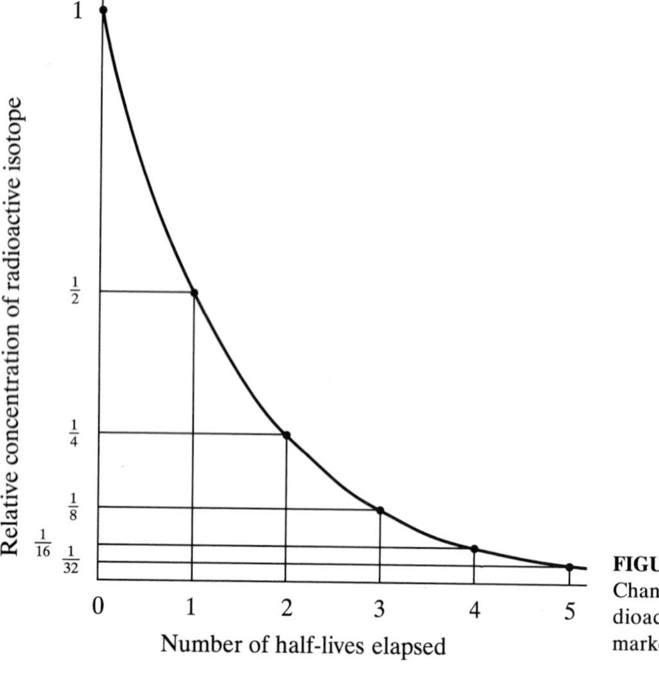

FIGURE 10.3
Change in concentration of a radioactive isotope versus time marked in number of half-lives.

and one or more neutrons. During the late 1930s (the time period, as we will discuss shortly, is especially significant here) uranium 235 was found to undergo fission when bombarded by neutrons (^1_0n). There are a large number of possible reaction schemes for this process, one of which is shown in Equation (10.18) and also in Figure 10.4*a*:

$$^{235}_{92}\text{U} + ^1_0n \longrightarrow \left[^{236}_{92}\text{U}\right] \longrightarrow ^{90}_{38}\text{Sr} + ^{143}_{54}\text{Xe} + 3^1_0n \qquad (10.18)$$

As in fusion, the mass of the reactants is just slightly greater than the mass of the products (although, of course, the equation is balanced) with the difference, the mass defect, being converted to energy. The energy derived from fission, while of great-enough magnitude to be the basis of the atomic bomb and present-day nuclear reactors, is several orders of magnitude less than that derived from fusion.

The neutrons given out as products can go on to strike other fissionable U 235 nuclei and quickly create a *nuclear chain reaction*, a self-sustaining sequence of nuclear fission reactions. Such a process is shown in Figure 10.4*b*. The minimum mass of fissionable material necessary to produce a self-sustaining chain reaction is known as the *critical mass*. Now we come to the special property of deuterium oxide, D_2O, that so interested the Germans and the Allies during World War II. To maximize the efficiency of U 235 fission which was being developed by both sides as the basis of the atomic bomb, neutrons must be slowed down; that is, their velocities must be moderated. A *moderator* is a substance that reduces the kinetic energy or velocity of neutrons. Ordinary light water is a good moderator, but it tends to absorb neutrons to the extent that it reduces the efficiency of the chain reaction. D_2O slows down the neutrons but does not absorb as many. This is significant because using D_2O as a moderator allows the use of natural uranium (a mixture of the fissionable U 235 and predominantly the nonfissionable U 238) in

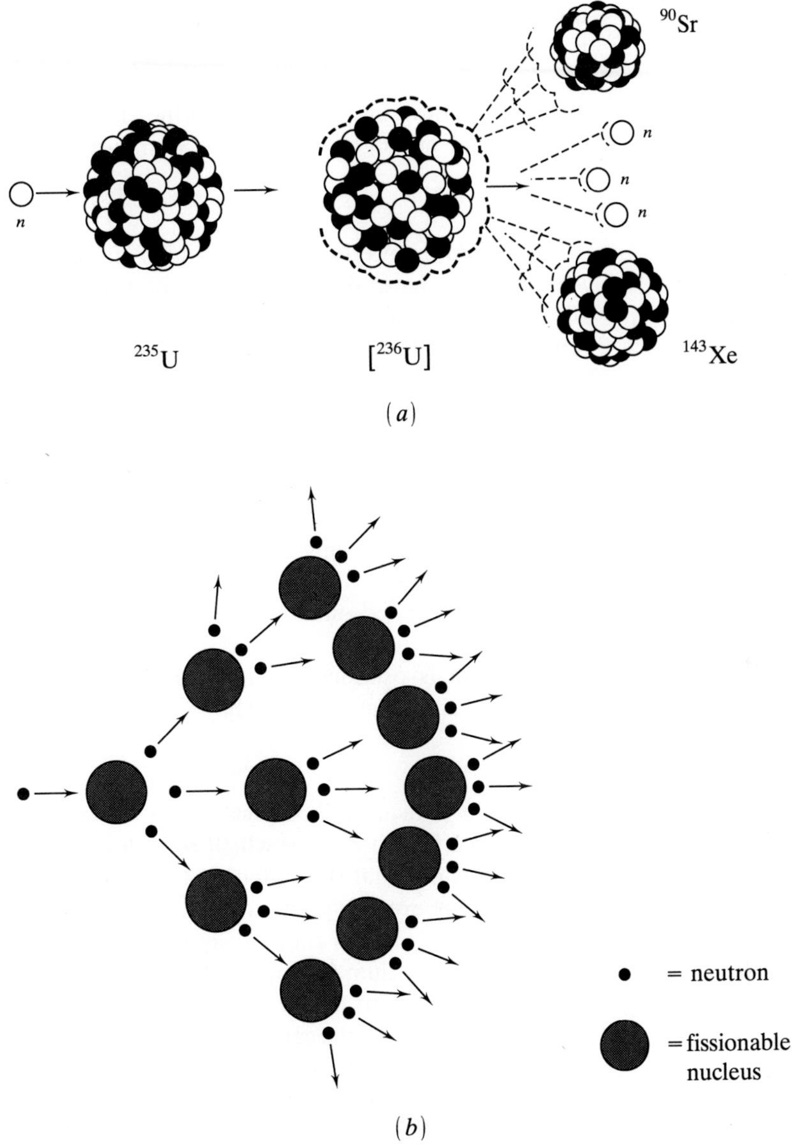

FIGURE 10.4
(*a*) A uranium 235 nucleus absorbs a neutron and than *fissions*, or splits, into two of the many possible fission products and three more neutrons. (*b*) A self-sustaining chain reaction is set up when a critical mass of the fissionable isotope is present. [Ref. 23.]

nuclear reactors rather than uranium laboriously and expensively enriched in U 235.

Tritium

Tritium was first prepared by *nuclear transmutation*, defined as the conversion of one element into another by a nuclear process. Rutherford, in addition to all his other contributions to chemistry and physics, was the first to carry out the alchemists' dream. In 1919, Rutherford was still working with his alpha particles, this time shooting them into various gases. When he used nitrogen gas, the results indicated that hydrogen was produced. Where did hydrogen come from? Further work revealed that the alpha particles were striking the nitrogen atoms, perhaps first fusing into an unstable fluorine 18 nucleus, which then instantaneously decayed into hydrogen and oxygen as shown in Equation (10.19):

$$^{14}_{7}N + ^{4}_{2}He \longrightarrow [^{18}_{9}F] \longrightarrow ^{17}_{8}O + ^{1}_{1}H \tag{10.19}$$

Tritium is produced in a similar fashion both in nature and artificially. Although the natural concentration of tritium is exceedingly small (10 in about 10^{18} hydrogen atoms), it is produced in the upper atmosphere by the reaction represented in Equation (10.20):

$$^{14}_{7}N + ^{1}_{0}n \longrightarrow ^{12}_{6}C + ^{3}_{1}H \tag{10.20}$$

Here neutrons from cosmic rays bombard the nitrogen of the upper atmosphere, producing carbon 12 and tritium. The tritium makes its way down to the earth's surface as precipitation, probably in the form of compounds such as HTO. As the half-life of tritium is only 12.3 years, its concentration in the oceans is much less than it is in rainwater. During the early 1950s when atmospheric testing of nuclear weapons was carried out, the concentration of tritium in rainwater reached about 500 atoms/10^{18} atoms.

Tritium is produced in nuclear reactors by the bombardment of lithium 6 with neutrons [Equation (10.21)]:

$$^{6}_{3}Li + ^{1}_{0}n \longrightarrow ^{3}_{1}H + ^{4}_{2}He \tag{10.21}$$

The lithium is usually incorporated into a magnesium or aluminum alloy that is placed into a fission reactor which provides the neutrons.

It turns out that the radiation from tritium is relatively harmless because the beta emissions are fairly low in energy. Also, the beta particles are not accompanied by gamma rays as they are quite commonly in other nuclear reactions. The radiation from tritium is stopped by about 6 mm of air. For these reasons, using tritium as a tracer to follow various chemical reactions is quite safe, although all the proscribed safety precautions should be followed. In another application of its low-energy beta decay, tritium is used with a mixture of zinc sulfide, a phosphor which lights up when hit by a charged particle, in the production of luminous paints for watch dials and so forth. In this capacity, tritium has replaced the dangerous radium formerly used in such products.

10.5 HYDRIDES AND THE NETWORK

In Chapter 9, we started to develop a network of ideas to make sense of the periodic table. The five components developed at that time were (1) the periodic law (trends in effective nuclear charge, radii, ionization energy, electron affinity, and electronegativity), (2) the uniqueness principle, (3) the diagonal effect, (4) the inert-pair effect, and (5) the metal, nonmetal, and metalloid regions. These components and others to be developed in the next few chapters will come into full play when we start discussing the eight groups of the representative elements, but we can certainly put some of them to use in a discussion of hydrogen compounds, known commonly as *hydrides*. Incidentally, more compounds of hydrogen are known than for any other element, even carbon. Table 10.3 shows some selected properties of hydrogen in comparison to those of lithium and fluorine.

TABLE 10.3
Some selected properties of hydrogen, lithium, and fluorine

	Lithium	Hydrogen	Fluorine
Valence electrons	$1s^2 2s^1$	$1s^1$	$2s^2 2p^5$
Atomic radius, Å	1.55	0.32	0.72
Ionic radius, Å (C.N. = 6)	1.16	1.53 (H^-) 0.000015 (H^+)	1.19
Pauling EN	1.0	2.1	4.0
Z/r (r = ionic)	0.86	0.65 (H^-) 67000 (H^+)	0.84
Oxidation states	+1	+1 (covalent) −1 (ionic)	−1
Ionization energy, kJ/mol	520	1312	1680
Electron affinity, kJ/mol	−58	−77	−333
X—X bond energy, kJ/mol	......	436.4	150.6

Note first that hydrogen has one valence electron like lithium but is also one electron short of completing an inert-gas configuration like fluorine and the other halogens. Not shown is the fact that hydrogen has an electronegativity very close to that of carbon. This implies that hydrogen might have some properties in common with all three of these elements. In fact, hydrogen can be viewed as having a split personality, if you will pardon the use of that phrase in a chemical context, and this idea is reflected in the various positions to which hydrogen is assigned in the periodic table. Most commonly, for example in Figure 9.3, it is placed over the alkali metals (Group 1A), but occasionally it will also be placed over the halogens (Group 7A). More rarely, it will be found above carbon. Some tables even show it placed above all three groups or totally segregated from any one group.

Regardless of where hydrogen is placed, it is not particularly reactive. Remember that while its reaction with oxygen is certainly impressive, a flame, spark, or catalyst is necessary to set it off. Part of the reason for this chemical lethargy is the high H—H bond energy of 436.4 kJ/mol. (This is the highest homonuclear single bond energy known.) Once it does react with another element to form a *binary hydride*, a compound in which hydrogen is bound to one other different element, its resulting oxidation state will usually be either +1 or −1. (The oxidation state of hydrogen in some transition-metal hydrides is not well characterized.) Given that the electronegativity of hydrogen is 2.1, very close to the midpoint of the scale and therefore in between that of a metal and a nonmetal, we are not surprised to find that if the second element is a nonmetal, the hydrogen will assume a +1 oxidation state, while if the second element is a metal, the hydrogen will assume a −1 oxidation state. We start our discussion of hydrides with the +1 oxidation state.

Covalent Hydrides

Note that the ionization energy of hydrogen is high, higher than lithium as expected (because the hydrogen electron is much closer to its effective nuclear charge) but also almost as high as that of fluorine, which is characterized by a large

TABLE 10.4
Classifications of binary hydrides

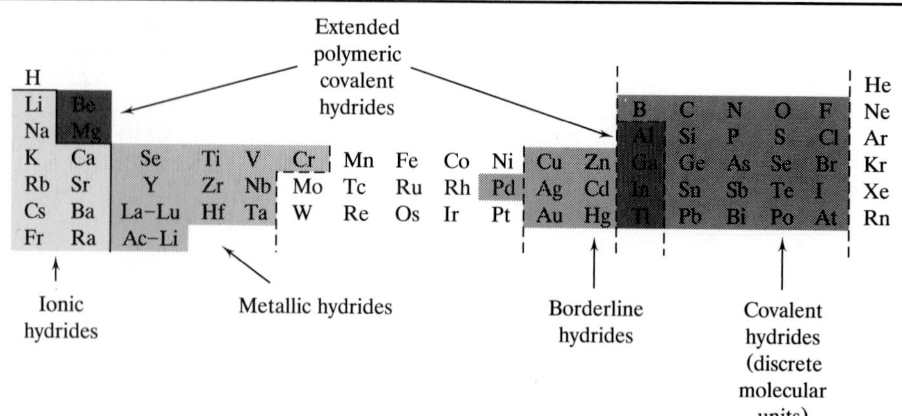

effective nuclear charge. It follows that it is difficult to ionize hydrogen to the full +1 state. Note also that if hydrogen were to lose an electron, it would form the hydrogen cation, which is just a proton. Because of the very small radius of this species, its charge density is extremely large, larger than any other ionic species. We learned in Chapter 9 (p. 226) that a large charge density leads to a high polarizing power and a tendency to form covalent bonds. Given its high ionization energy and extremely high polarizing power, hydrogen in the +1 oxidation state is never found as a bare positively charged proton; rather it is always covalently bound to some other atom or molecule.

Examples of covalent hydrides are given in Table 10.4. They can be subdivided into two types: those which form discrete, self-contained neutral or positively charged molecular units, like HCl, H_2O, H_3O^+, NH_3, and NH_4^+, and those which assume an extended, polymeric structure, such as BeH_2 or AlH_3. The chemistry of these covalent hydrides will be explored in the appropriate group chapters.

Ionic Hydrides

Table 10.4 shows that the ionic hydrides, characterized by hydrogen in the −1 state, occur only with the least electronegative metals, those of Groups 1A and 2A. Note in Table 10.3 that the electron affinity of hydrogen is low compared to the halogens (as represented by fluorine), meaning that relatively little energy is released when H^- is formed. This low electron affinity is consistent with the very small size of the hydrogen atom so that when the second 1*s* electron is added, it experiences significant electron-electron repulsion from the original electron. Electron-electron repulsion also provides the rationale for the atomic radius of hydrogen (0.32 Å) increasing to an ionic radius of about 2.08 Å in the free unassociated H^- ion. In any case, if not much energy is released upon H^- formation, then not much energy can be put into the formation of these compounds; otherwise, the heat of formation would be positive, which tends to make the compounds relatively unstable. [If you have already covered solid state energetics (Chapter 8), then you are prepared to deal with these energy-balance considerations in more detail and may want to attempt Problems 10.39 to 10.46 at the end of the chapter.]

Having noted the rather restricted range of ionic hydrides (sometimes called the *saline* hydrides for their saltlike character), we should note that there is good evidence that these compounds really are significantly ionic. (1) Molten ionic hydrides, like salts, conduct electricity well, implying the existence of charged species. (2) The melt releases hydrogen at the positive anode upon electrolysis, consistent with an H^- species. (3) X-ray analysis shows that in LiH, 80 to 100 percent of the lithium electron density is transferred to the hydrogen. On the other hand, x-ray results also show that the effective radius of the hydride ion in such compounds is only about 1.5 Å instead of the 2.08 Å calculated for the free H^- ion. So either not all the electron density is transferred to the hydrogen and/or perhaps this ion, containing only two electrons in a relatively large volume, is quite compressible.

The ionic hydrides, usually white or gray solids formed by direct combination of the metal and hydrogen at elevated temperatures, are used as drying and reducing agents, as strong bases, and some as safe sources of pure hydrogen. CaH_2 is particularly useful as a drying agent for organic solvents, reacting smoothly with water as represented in Equation (10.22). CaH_2 also acts to reduce metal oxides to the free metal as shown in Equation (10.23):

$$CaH_2(s) + 2H_2O(l) \longrightarrow Ca^{2+}(aq) + 2H_2(g) + 2OH^-(aq) \quad (10.22)$$

$$CaH_2(s) + MO(s) \xrightarrow{\text{heat}} CaO(s) + M(s) + H_2(g) \quad (10.23)$$

NaH reacts violently with water to produce hydroxide in solution and, like the other ionic hydrides, is a strong base. LiH and CaH_2 are convenient portable sources of pure hydrogen. LiH also reacts with aluminum chloride, as discussed in Chapter 14, to form the *complex hydride* lithium aluminum hydride, $LiAlH_4$, that is extremely useful as a reducing agent in organic chemistry.

Metallic Hydrides

Hydrogen reacts with a variety of transition metals, including the lanthanides and actinides (see Table 10.4), to produce a third type of hydride which is rather poorly understood. These brittle solids generally are metallic in appearance, good conductors of electricity, and of variable composition. Their hydrogen-metal ratios are often not ratios of small whole numbers, and so they are referred to as *nonstoichiometric compounds*. Examples include $TiH_{1.7}$, TiH_2, $PdH_{0.65}$, $LaH_{1.86}$, and UH_3.

These metallic hydrides were formerly thought to be *interstitial* compounds with atomic hydrogen fitting into the holes (the interstices, to use the fancier word) left in the crystal structure of the pure metal. However, in many cases, the arrangement of the metal atoms in the hydride has been found to be different from that in the pure metal. While still needing clarification and support, a better model seems to be emerging. We can use the titanium hydride as an example to describe this model. The crystal structure of TiH_2 is well characterized, but the exact nature of the bonding is not. The titanium may very well be in the +4 oxidation state with two of the ionized electrons having been transferred to the hydrogen atoms to produce H^- ions, while the other two electrons are able to flow freely and therefore account for the conducting ability of the compound. (More background on the structure of metallic crystals and their ability to conduct electricity is given in Chapter 7, p. 151.) There may be significant covalent character between the metal and the hydride ions, but more clarification is needed to know for sure.

These metallic hydrides have several important applications. First, note that the hydrides are fairly easily formed from a direct combination of hydrogen gas and the metal. The hydrogen uptake is reversed at higher temperatures, yielding finely powdered metals and hydrogen gas. Accordingly, these compounds are a good way to store and purify hydrogen as well as produce finely divided metals. The palladium-hydrogen system is particularly adaptable. At red heat hydrogen diffuses right through a piece of palladium foil, affording an excellent way to separate and purify hydrogen. The uranium hydride, UH_3, is a good starting point for a number of uranium compounds and also a convenient form in which to store the tritium isotope.

10.6 THE ROLE OF HYDROGEN IN VARIOUS ALTERNATIVE ENERGY SOURCES

The Hydrogen Economy

In these days of high energy demands, dwindling supplies of fossil fuels, political volatility in the Middle East, and concern for the environment, hydrogen has been touted as the energy source of the future. In fact, there are those who advocate that we should work toward what has become known as the *hydrogen economy*, or the production, storage, transportation, and utilization of hydrogen as the primary energy source of the world economy. Such a system is presented schematically in Figure 10.5.

The hydrogen production facility could be based on (1) conversion of crude oil or coal to hydrogen gas, (2) the use of solar energy in a photochemical process to split water into hydrogen and oxygen, or (3) the development of various thermochemical cycles to use heat energy to split water into its component elements. The production of hydrogen from oil or coal has been described in Section 10.2. It involves the steam-hydrocarbon reforming process [Equation (10.7)] and the carbon-steam reaction [Equation (10.9)]. In both cases, the resulting syngas can be converted to hydrogen and carbon dioxide by the water-gas reaction [Equation (10.8)]. Carbon dioxide, however, is a primary cause of the greenhouse effect which may already be starting to warm the earth's atmosphere to the point that, if not reversed, could cause catastrophic flooding of the world's coastal cities by melting the polar ice caps. (See Section 11.6 for further details on this effect and its consequences.) The use of a photochemical process or one or more thermochemical cycles could be used to avoid the production of carbon dioxide.

The development of a new generation of catalysts or semiconductors will be needed if a photochemical process is to be viable, but research and development in this area is growing rapidly. Ruthenium catalysts and various arsenide, selenide, and telluride semiconductors (see Chapter 15) may lead the way to an efficient system. Apart from photochemical processes, any cyclic thermochemical system that starts with water and produces hydrogen and oxygen with the simultaneous regeneration of the reactants will work. Such a scheme is shown schematically in Equation (10.24):

$$H_2O(g) + A \xrightarrow{\text{heat}} H_2(g) + C \qquad (10.24a)$$

$$B + C \xrightarrow{\text{heat}} \tfrac{1}{2}O_2(g) + D \qquad (10.24b)$$

$$D \xrightarrow{\text{heat}} A + B \qquad (10.24c)$$

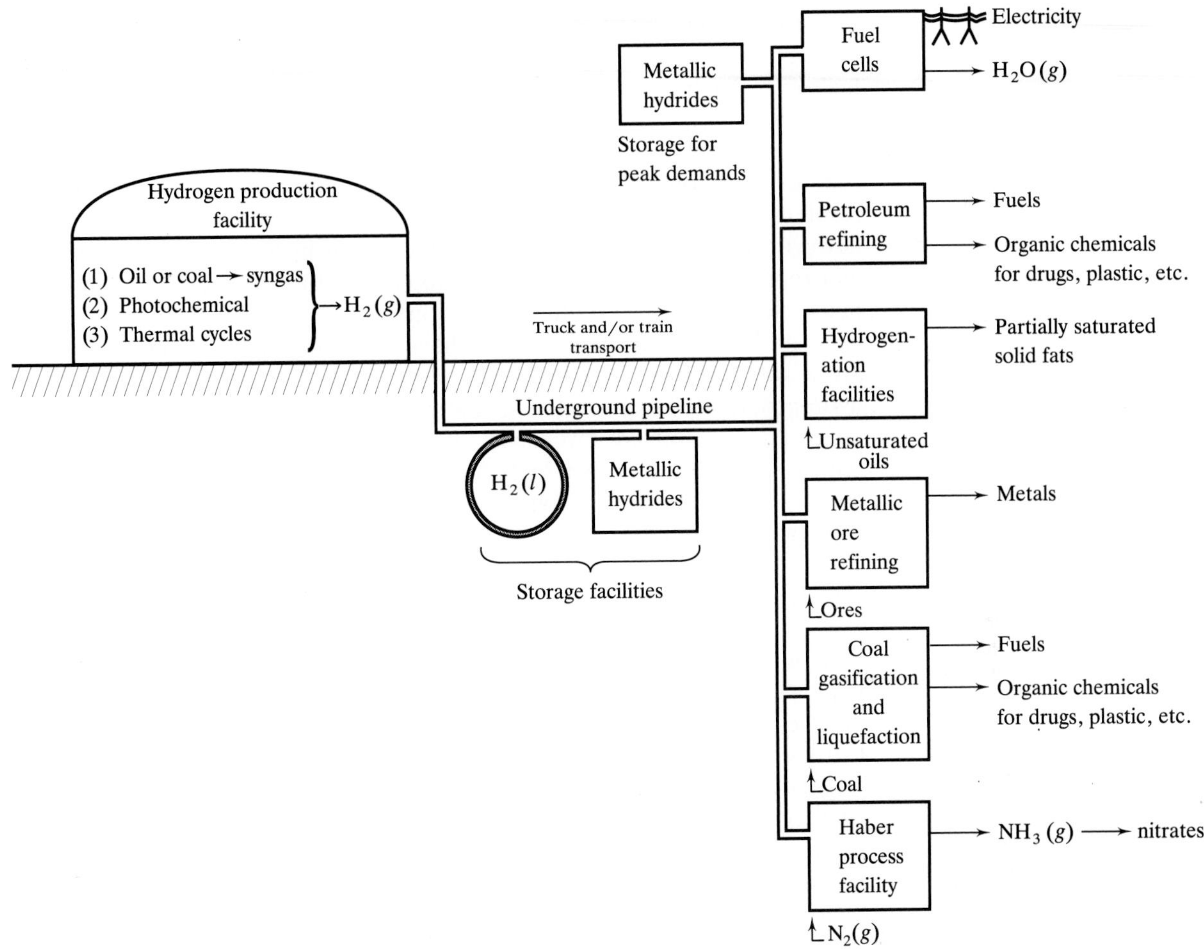

FIGURE 10.5
The hydrogen economy.

Note that A and B are regenerated in the end of the process, while C and D are produced and then consumed. The net result is that water is "split" into hydrogen and oxygen gases.

The storage of liquid hydrogen gas has been done routinely in vacuum-insulated low-temperature facilities as part of the space program. As previously noted, metals have a large capacity to absorb and then regenerate hydrogen by forming hydrides of various compositions. At one time it was thought that automobiles might be redesigned to run on hydrogen generated as needed from a solid metal-hydride fuel. Using hydrogen as a fuel, however, even if it is generated only as needed, is fraught with hazards.

The transportation of liquid hydrogen or solid hydrides could be done by road or rail or it might be pumped by pipeline in the same way that natural gas is now. Care would have to be taken, however, because, as we have seen, many metal hydrides are brittle and pipelines might be susceptible to rupture. Some problems with hydrogen leakage might also be encountered. The transportation of hydrogen or hydrides produced (and perhaps stored) at remote coal mining sites would be more efficient than the transmission of electricity produced by burning that coal.

Electricity cannot be stored, so expensive electrical generation facilities have to be available to meet the peak demands of hot summers and cold winters. With hydrogen, such problems are solved by storing hydrogen or hydrides for such occasions.

The huge advantage of using hydrogen as the basis of an economy is that its combustion produces only water. There are no sulfur or nitrogen oxides which result in acid rain (see Chapter 17) and not even carbon dioxide. [CO_2 might, however, be a by-product of the production of hydrogen gas as discussed earlier and represented in Equations (10.7) to (10.9).] In any case, we could not devise a much cleaner energy-producing reaction.

As Figure 10.5 shows and we have covered in part in earlier sections, there are a variety of uses for hydrogen at the end of a pipeline. Petroleum refining, hydrogenation plants, gasoline or methanol plants, metallic-ore reduction facilities all could be supplied. Even coal gasification and liquefaction plants producing syngas, gasoline, and other fuels could be set up. One added benefit to this use of coal is that sulfur and nitrogen could be removed more easily by hydrogenation than by existing technologies.

Nuclear Fusion

One other use of hydrogen as a future energy source depends on its physics rather than its chemistry. As we discussed in Section 10.1, the sun is powered by the nuclear fusion of hydrogen into helium [Equation (10.4)]. In the 1950s, humans learned how to use the same forces to produce a hydrogen bomb; now in the 1990s we hope to harness them to produce a clean, safe source of energy for the next century. To make two nuclei fuse together, they must strike each other at tremendous velocities to overcome the electrostatic forces between them. This requirement translates into two goals for fusion research. The nuclei must be at extremely high temperatures (100 million degrees), and the resulting *plasma* (nuclei stripped of their electrons) must be at a high density. To achieve such temperatures, great quantities of energy must be applied, but once fusion starts, the reaction will be exothermic and self-sustaining. There are a number of proposed techniques to achieve these two goals simultaneously. Some include using lasers to achieve the high temperatures, containing the plasma in magnetic bottles and mirrors, and injecting the fuel in the form of tiny pellets.

A few of the more promising fusion reaction systems are shown in Equations (10.25) to (10.27):

$$^{2}_{1}H + ^{2}_{1}H \longrightarrow ^{3}_{1}H + ^{1}_{1}H + 3.8 \times 10^{8} \text{ kJ/mol} \tag{10.25}$$

$$^{3}_{1}H + ^{2}_{1}H \longrightarrow ^{4}_{2}He + ^{1}_{0}n + 1.7 \times 10^{9} \text{ kJ/mol} \tag{10.26}$$

$$^{6}_{3}Li + ^{2}_{1}H \longrightarrow 2\,^{4}_{2}He + 2.2 \times 10^{9} \text{ kJ/mol} \tag{10.27}$$

Note that they involve deuterium and tritium. Deuterium can be derived in great quantities from seawater by techniques discussed in Section 10.3, while tritium is prepared in nuclear fission reactors as shown in Equation (10.21) and described in Section 10.4.

SUMMARY

Hydrogen is believed to be the first element synthesized in the big bang and the most abundant element in the universe. Through nuclear fusion it provides sunshine which makes life on earth possible. First isolated and characterized in the eighteenth century by Henry Cavendish, hydrogen is now generated in large quantities for the production of syngas and other fuels, organic chemicals, ammonia and nitrates, metals from their ores, and electricity.

Hydrogen forms three isotopes: protium, deuterium, and tritium. Protium and deuterium can be separated by a variety of physical and chemical processes. Ordinary light hydrogen in H—X bonds can be replaced by deuterium which then provides a means of following the progress of a variety of reactions. Heavy water is used as a moderator in fission reactors. Tritium, produced naturally in the upper atmosphere and artificially in fission reactors, is a mild beta emitter and is used as a tracer and to make luminous paints.

Although diatomic hydrogen gas is only mildly reactive, hydrogen forms more compounds than any other element; the resulting hydrides are usually classified as covalent, ionic, or metallic. Covalent and ionic hydrides are best understood by referring to the interconnected network of ideas developed in Chapter 9. The bonding in metallic hydrides is still poorly understood, although these compounds can be used to purify and store hydrogen.

The production, storage, transportation, and utilization of hydrogen may come to power the entire world economy. Nuclear fusion may be the ultimate power source of the future. A knowledge of the chemistry and physics of hydrogen is fundamental to an understanding of the future as well as the past and present.

PROBLEMS

10.1. Determine the number of protons and neutrons in the following nuclei:

(*a*) $^{11}_{5}B$

(*b*) $^{17}_{8}O$

(*c*) $^{60}_{27}Co$

(*d*) $^{239}_{94}Pu$

10.2. In additional to the reactions which add up to the proton-proton cycle [Equation (10.4)], there are several other possible solar fusion reactions, two of which are shown below. In each case, fill in the missing reactant or product and balance the reaction.

(*a*) $^{2}_{1}H + ^{2}_{1}H \longrightarrow$ ______ $+ ^{1}_{1}H$

(*b*) $^{3}_{2}He + ^{2}_{1}H \longrightarrow ^{4}_{2}He +$ ______

10.3. The following reactions are part of the *carbon cycle*, a different pathway for the conversion of hydrogen into helium. In each reaction, one product or reactant has been omitted. By balancing each reaction, fill in the missing reactant or product and verify that the result of the cycle is the conversion of hydrogen to helium.

(*a*) $^{12}_{6}C + ^{1}_{1}H \longrightarrow$ ______

(*b*) $^{13}_{7}N \longrightarrow ^{13}_{6}C +$ ______

(*c*) $^{13}_{6}C +$ ______ $\longrightarrow ^{14}_{7}N$

(*d*) $^{14}_{7}N +$ ______ $\longrightarrow ^{15}_{8}O$

(*e*) $^{15}_{8}O \longrightarrow ^{15}_{7}N +$ ______

(*f*) $^{15}_{7}N + ^{1}_{1}H \longrightarrow ^{12}_{6}C +$ ______

10.4. The synthesis of the elements with odd atomic numbers has always been a difficult problem for cosmologists to account for. Write a balanced nuclear equation for the following two proposed reactions.

(*a*) Boron 11 (and two protons) is produced from the collision of a proton and carbon 12.

(*b*) Lithium 7 (and two helium 4 nuclei and a proton) is produced from the collision of a helium 4 nucleus and carbon 12.

10.5. A mass defect of 1 atomic mass unit (u) corresponds to the release of 8.984×10^{10} kJ/mol. Given the masses of nuclei in Tables 10.1 and 10.2, calculate the mass defect (in atomic mass units) and the energy released per mole for:
(*a*) ${}^{1}_{1}H + {}^{1}_{1}H \longrightarrow {}^{2}_{1}H + {}^{0}_{1}e$
(*b*) The proton-proton cycle

10.6. Lavoisier named oxygen with the belief that it was a necessary component of acids. Name four acids which are consistent with Lavoisier's belief and at least two exceptions.

10.7. Look at the entry for Robert Boyle in *Asimov's Biographical Encyclopedia of Science and Technology* (2d rev. ed., Doubleday, New York, 1982). Does it mention Boyle's preparation of hydrogen? Now look up Henry Cavendish in the same reference. Why is Cavendish given the credit for discovering this element?

10.8. Use standard heats of formation data (at 298 K) found in an introductory chemistry textbook to calculate the heat of the steam-hydrocarbon reforming process [Equation (10.7) reproduced below]. Briefly explain why this reaction is best carried out at high temperatures. Would the presence of the nickel catalyst alter the heat of reaction?

$$C_3H_8(g) + 3H_2O(g) \xrightarrow{Ni} 3CO(g) + 7H_2(g)$$

10.9. Using standard heats of formation (at 298 K) found in an introductory chemistry textbook, calculate the heat of reaction of the carbon-steam reaction [Equation (10.9) reproduced below with carbon from coal assumed to be pure graphite]. Why might it best be carried out at elevated temperatures? Would it be best to use high pressure or low pressure to increase the yield of this reaction?

$$C(\text{graphite}) + H_2O(g) \xrightarrow{\text{Fe or Ru}} CO(g) + H_2(g)$$

10.10. Syngas (CO and H_2) was formerly used directly as a fuel. Write a reaction for the combustion of these two gases and calculate the heat of the reaction using standards heats of formation found in an introductory chemistry textbook.

10.11. For the reaction of a metal(II) oxide with hydrogen gas [Equation (10.12) reproduced below], what is being oxidized? Reduced? Write a similar equation for the reduction of silver and bismuth oxides.

$$MO(s) + H_2(g) \longrightarrow M(s) + H_2O(l)$$

10.12. The principal oxide of molybdenum is MoO_3. Write the reaction for the reduction of this oxide with hydrogen gas to free molybdenum and water.

10.13. The atomic weight of hydrogen is listed as 1.008 u. Rationalize this value in terms of the information given in Table 10.2.

10.14. Oxygen, like hydrogen, has three major isotopes O 16, O 17, and O 18 with atomic masses 15.995, 16.999, and 17.999 u, respectively. Compare the percentage increase in mass from O 16 to O 17 to O 18 with the increase from hydrogen to deuterium to tritium. How might the differences in rates of diffusion or effusion among oxygen isotopes differ from the rates of the hydrogen isotopes? (*Hint:* Recall Graham's law of effusion.)

* **10.15.** The melting and boiling points of D_2O are 3.8 and 101.4°C, respectively. Briefly rationalize these values as compared to H_2O. Assume that an O—H bond has the same polarity as an O—D bond.

10.16. Using the molecular weights of D_2O and H_2O, calculate the ratio of mean velocities, v_{H_2O}/v_{D_2O}, which is the same as the ratio of rates of effusion or diffusion, of these two molecules. Briefly explain why H_2O evaporates more rapidly than D_2O. (*Hint:* Recall Graham's law of effusion.)

10.17. When water is electrolyzed, its density gradually increases. Write a short paragraph that accounts for this observation.

10.18. When H. C. Urey and his associates first isolated heavy water, they were not quite sure whether the "heaviness" was caused by the hydrogen or the oxygen. They could electrolyze the heavy water to diatomic hydrogen and oxygen but still were not sure of the composition of the gases; for example, the hydrogen could have been H_2 or D_2, the oxygen could be O_2^{16} or O_2^{18}. Suppose they reacted the products of the electrolysis of heavy water with ordinary hydrogen and oxygen, how could they decide the exact cause of the heaviness of heavy water?

10.19. There is only a cup of D_2O in 400 gal of water. How many grams of heavy water are there in the world's oceans? (The total volume of the oceans is estimated at 3.2×10^8 km^3.) (1 cup = 0.236 liters; 1 gal = 3.78 liters; density of D_2O = 1.10 g/mL)

* **10.20.** On the basis of what you can recall from earlier chemistry experiences, rationalize the fact that the boiling points of H_2, D_2, and T_2 are 20.4, 23.7, and 25.0 K, respectively.

10.21. Assuming that H_2 and D_2 are ideal gases, calculate values for the densities of these two gases and compare your results with the known values of 0.0899 and 0.180 g/liter at standard temperature and pressure.

* **10.22.** Which would have the higher vapor pressure, $H_2(l)$ or $D_2(l)$? Explain your results in terms of the kinetic molecular theory of gases.

10.23. The structure of glucose (sugar) is shown below. Which hydrogen atoms would be replaceable with deuterium atoms? Why?

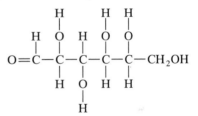

```
           H       H   H
           |       |   |
       H   O   H   O   O
       |   |   |   |   |
   O=C—C—C—C—C—CH₂OH
           |   |   |   |
           H   O   H   H
               |
               H
```

10.24. When methylammonium chloride, $CH_3NH_3^+Cl^-$, is dissolved in deuterium oxide, only half the hydrogen atoms are replaced with deuterium atoms. Explain this result.

10.25. (Extra credit.) After reviewing the movie *The Heroes of Telemark*, write a short paragraph summarizing the plot as seen through the eyes of a chemist.

10.26. Carbon 14 decays via beta minus emission with a half-life of 5730 years. Write a nuclear equation for this process.

10.27. Potassium 40 decays via both beta minus and beta plus emission. Write a nuclear equation for each process.

10.28. Cobalt 60 decays via beta plus emission. Write a nuclear equation for this process.

10.29. When U 238 is bombarded with a neutron, the unstable U 239 is produced. U 239 decays via beta minus emission. Write a nuclear equation for this process.

10.30. Radium 226 decays via alpha emission. Write a nuclear equation for this process.

10.31. Some fission reactions give out three neutrons as products, while others, such as the following, give only two neutrons. Fill in the missing product in this reaction.

$$^{235}_{92}\mathrm{U} + ^{1}_{0}n \longrightarrow ^{96}_{37}\mathrm{Rb} + \underline{\qquad\qquad} + 2\,^{1}_{0}n$$

10.32. Plutonium 239 is also a fissionable isotope. In one of the possible reactions, it produces three neutrons and strontium 90. Write a balanced nuclear equation for this process.

10.33. To control a nuclear fission reaction, cadmium or boron rods are inserted into the reactor. Each cadmium 113 nucleus absorbs one neutron and emits gamma rays, while each boron 10 absorbs one neutron and emits an alpha particle. Write balanced nuclear reactions for these processes.

10.34. Lead 210 decays via alpha emission and mercury 206 decays via beta minus emission. Is it possible to transmute lead into gold using these two decay schemes? Show equations to support your answer.

10.35. In the production of tritium in the upper atmosphere, there is some evidence that high-energy protons from cosmic rays first interact with nitrogen atoms in the upper atmosphere to produce neutrons and oxygen. These neutrons then react with the nitrogen to produce tritium as shown in Equation (10.20). Write a nuclear reaction for the action of the protons with nitrogen to produce neutrons.

10.36. The ionization energy of hydrogen is higher than that of lithium even though the electron in each case experiences an effective nuclear charge of approximately +1. Given this, why would we expect the ionization energy to be large for hydrogen?

10.37. Assume for a moment that a proton could exist independently in water. How might the proton interact with an H_2O molecule? Draw a diagram as part of your answer.

10.38. The value of the ionic radius of H^- in various hydrides varies as shown below. Briefly rationalize these values in terms of the electronegativity of the metal.

Compound	MgH_2	LiH	NaH	KH	RbH	CsH
$r(H^-)$, Å	1.30	1.37	1.46	1.52	1.54	1.52
EN (metal)	1.2	1.0	0.9	0.8	0.8	0.7

* **10.39.** Unlike the halides which occur over a wide range of the periodic table, the ionic or saline hydrides are restricted to the Group 1A and 2A elements. To demonstrate quantitatively why this is the case, use the values of the bond energies and the electron affinities of hydrogen and fluorine found in Table 10.3 to calculate the energy of the following process in which X = H and F:

$$\tfrac{1}{2}X_2(g) + e^-(g) \longrightarrow X^-(g)$$

Discuss how these results imply that ionic hydrides will form only for the most electropositive (or least electronegative) metals.

10.40. Write a Born-Haber cycle for the formation of potassium hydride, KH, and use it to calculate a value for the lattice energy of this compound. (The standard heat of formation of KH is −57.8 kJ/mol; other thermochemical quantities can be found in Tables 8.3 and 10.3.)

10.41. Assuming KH to assume a NaCl structure, calculate a lattice energy using the Born-Landé equation and compare your answer to that obtained in Problem 10.40. (Ionic radii can be found in Tables 7.4 and 10.3.)

10.42. Write a Born-Haber cycle for the formation of CuH and use it to calculate a value for the lattice energy of this compound. (The standard heat of formation of CuH is +21 kJ/mol, the heat of sublimation and first ionization energy of copper are 338.3 and 745 kJ/mol, respectively; other thermochemical quantities can be found in Table 10.3.)

10.43. Without knowledge of the crystal structure of CuH, calculate a value for its lattice energy using the Kapustinskii equation. Radii can be found in Tables 7.4 and 10.3.

10.44. Write a Born-Haber cycle for the formation of CaH_2 and use it to calculate a value for the lattice energy of this compound. (The standard heat of formation of CaH_2 is −186 kJ/mol, the heat of sublimation and the first and second ionization energies of calcium are 178.2, 589.8, and 1145 kJ/mol, respectively; other thermochemical quantities can be found in Table 10.3.)

10.45. Without knowledge of the crystal structure of CaH_2, calculate a value for its lattice energy using the Kapustinskii equation. Radii can be found in Tables 7.4 and 10.3.

*10.46. Using the data given below and a Born-Haber cycle for the formation of both the hydride and the chloride of sodium, calculate the standard heats of formation in each case. Which type of compound is more stable? Briefly discuss the principal reasons for the differences in stabilities as measured by the standard heat of formation. Would you expect differences in entropy of formation to change your conclusions?

	ΔH_{sub}	IE	ΔH_g	EA	$U_{B\text{-}H}$
NaH	107.3	495.9	218.2	−77	−808
NaCl	107.3	495.9	121.7	−349	−787

10.47. Calcium hydride is used to reduce tantalum oxide, Ta_2O_5, to the pure metal. Write an equation for this process.

*10.48. The following thermochemical cycle has been proposed as a method to chemically split water into its constituent elements. Show that the net result of this process is the production of hydrogen and oxygen and that all the reactants needed are regenerated in the necessary stoichiometric quantities. Using standard heat of formation data, calculate the heat of each reaction and the net reaction.

$$CaBr_2(s) + H_2O(g) \xrightarrow{\text{heat}} CaO(s) + 2HBr(g)$$
$$Hg(l) + 2HBr(g) \xrightarrow{\text{heat}} HgBr_2(s) + H_2(g)$$
$$HgBr_2(s) + CaO(s) \longrightarrow HgO(s) + CaBr_2(s)$$
$$HgO(s) \xrightarrow{\text{heat}} Hg(l) + \tfrac{1}{2}O_2(g)$$

*10.49. The following thermochemical cycle has been proposed as a method to chemically split water into its constituent elements. Show that the net result of this process is the production of hydrogen and oxygen and that all the reactants needed are regenerated in the necessary stoichiometric quantities. Using standard heat of formation data, calculate the heat of each reaction and the net reaction.

$$3FeCl_2(s) + 4H_2O(g) \xrightarrow{\text{heat}} Fe_3O_4(s) + 6HCl(g) + H_2(g)$$
$$Fe_3O_4(s) + \tfrac{3}{2}Cl_2(g) + 6HCl(g) \xrightarrow{\text{heat}} 3FeCl_3(s) + 3H_2O(g) + \tfrac{1}{2}O_2(g)$$
$$3FeCl_3(s) \xrightarrow{\text{heat}} 3FeCl_2(s) + \tfrac{3}{2}Cl_2(g)$$

*10.50. The following thermochemical cycle has been proposed as a method to chemically split water into its constituent elements. Show that the net result of this process is the production of hydrogen and oxygen and that all the reactants needed are regenerated in the necessary stoichiometric quantities. Using standard heat of formation data, calculate the heat of each reaction and the net reaction.

$$2HI(g) \xrightarrow{\text{heat}} I_2(g) + H_2(g)$$
$$2H_2O(g) + SO_2(g) + I_2(g) \xrightarrow{\text{heat}} H_2SO_4(l) + 2HI(g)$$
$$H_2SO_4(l) \xrightarrow{\text{heat}} SO_2(g) + H_2O(g) + \tfrac{1}{2}O_2(g)$$

10.51. The requirement that plasma be at very high temperatures is related to the collision theory of chemical kinetics. Briefly describe the similarities and one significant difference between chemical and nuclear reactions such as fusion.

10.52. Hydrogen bombs contain very tightly packed solid lithium deuteride, LiD. The detonation is achieved by using a fission bomb to provide the high temperatures necessary to initiate the fusion reaction between lithium 6 and deuterium which results in helium 4. Write an equation for this fusion reaction.

CHAPTER 11

OXYGEN, AQUEOUS SOLUTIONS, AND THE ACID-BASE CHARACTER OF OXIDES AND HYDROXIDES

In the last chapter we discussed hydrogen, which forms more compounds than any other element. In this chapter, we have three major objectives: (1) to introduce the chemistry of oxygen, which forms compounds with every element except the lighter noble gases, (2) to describe the structure of water and aqueous solutions, and (3) to discuss oxides and hydroxides, with particular emphasis on their aqueous acid-base chemistry. This last subject is so important that it will become the sixth component of our network of interconnected ideas with which to understand the periodic table. It is hoped that the network and the descriptions of oxides and hydroxides in this chapter combined with those of the hydrides covered in the last chapter will serve us well as we make our way through Chapters 12 to 19 describing the group chemistry of the representative elements.

11.1 OXYGEN

Discovery

As we have seen, the credit for the discovery of hydrogen goes to Henry Cavendish. Both Boyle and Lémery prepared the gas but neither described it in any detailed manner nor recognized it as a new element as Cavendish did. The discovery of oxygen, on the other hand, is not as clear-cut and raises the question of what it really means to "discover" an element. Does it occur when an element is first prepared in a reasonably pure form? Or, must it be characterized by various chemical and physical properties? Must the scientist recognize the substance as a new element to be credited with the discovery? What happens if a person is the first to actually prepare and characterize a new element but is not the first to publish his or her results? All these questions play a role in the assignment of the discoverer of oxygen.

The Englishman Joseph Priestley and the Swedish-German Karl Wilhelm Scheele are usually listed as independent codiscoverers of oxygen. Scheele, a pharmacist who concentrated on research, had a hand in the discovery of a number of elements (chlorine, manganese, barium, molybdenum, tungsten, and nitrogen in addition to oxygen) but does not receive the sole credit for the discovery of any of them. He also was the first to prepare a number of acids and, astoundingly, actually documented the taste (resembling rotten almonds) of the deadly poisonous hydrogen cyanide gas! As recorded in his laboratory notebook in 1771, Scheele was able to isolate reasonably pure oxygen (what he called "fire air") from various compounds and went on to characterize it well enough that under normal circumstances, he might very well be listed as the sole discoverer. Tragically, but through no fault of his own, there was a delay in the publication of Scheele's work so that when it did appear in 1777, Priestley had beaten him into print by about 3 years. (Can you image the agony and disappointment Scheele must have felt when he saw Priestley's article?) Due to his careful documentation, however, Scheele is usually listed as the codiscoverer of the element. One moral of this story is to take it to heart when your instructors tell you how important it is to keep an accurate and complete laboratory notebook.

As a Nonconformist or Dissenter in eighteenth century England, Joseph Priestley refused to swear allegiance to the Church of England and was consequently forbidden from attending any of the great English universities like Oxford and Cambridge. In fact, his radical religious and political beliefs (for example, as an Englishman, he supported the American Revolution in the 1770s) tagged him as somewhat of an outcast all his life, at least until he came to America in 1794. Although he was educated as a clergyman and held pastorates in a variety of churches throughout England, Priestley is probably most remembered for his work in the experimental sciences. He was encouraged to pursue his interest in science by Benjamin Franklin, whom he met in London while Franklin was there to discuss the dispute about taxes between the American colonies and the British crown.

When Priestley started to work on gases, or "airs" as they were called then, only common air, carbon dioxide, and hydrogen were known. His work in collecting gases over mercury as well as water enabled him to isolate many water-soluble gases such as ammonia, sulfur dioxide, and hydrogen chloride. Priestley generated oxygen (what he called "dephlogisticated air," discussed in the next paragraph) by focusing sunlight through a powerful lens onto a sample of red mercuric oxide. He noted that the resulting water-insoluble gas supported combustion and that mice lived longer in it. He found it easy to breathe and wrote that "I fancied that my breast felt peculiarly light and easy for some time afterwards. . . . Hitherto only two mice and myself have had the privilege of breathing it."

So both Scheele and Priestley prepared and characterized a gas that Priestley called dephlogisticated air. The phlogiston theory of combustion, first postulated by Georg Stahl about 1700, held that burning objects gave off a substance called phlogiston and that when the air was saturated with it, combustion could no longer occur. Air which supported combustion particularly well was said to lack phlogiston, or be dephlogisticated. Cavendish, Scheele, and Priestley were devoted phlogistonists all their lives. Several months after Priestley's isolation of his new air, he had the opportunity to discuss it at a dinner party with Antoine Lavoisier in Paris. It was Lavoisier who recognized that the phlogiston theory was inadequate to explain combustion and that Priestley's air, which Lavoisier called *oxygen*, was a new element that combined with materials during the combustion process.

Lavoisier maintained that since he was the first to recognize oxygen as a new element, he should receive the credit as its true discoverer, but he is not so recognized today. He named two elements, hydrogen and oxygen (the latter because it was an "acid producer"), formulated the present theory of combustion, and made so many contributions to the new science of chemistry that he is sometimes referred to as the father of the discipline. Continuing in that tradition, the introductory chapter of this book uses Lavoisier's lifetime to mark the beginning of chemistry as an independent science. However, despite all his accomplishments, he never realized his burning ambition to discover an element. In 1794, Lavoisier was beheaded as a result of his activities in prerevolutionary France.

Priestley went on to make a number of other contributions to chemistry. When he took a pastorate in Leeds which happened to be next door to a brewery, he worked with the "fixed air," or carbon dioxide, which is given off as a by-product of fermentation. Taking advantage of the opportunity presented to him, he soon found a way to "impregnate" water with this gas and obtained a tart, bubbly, rather appealing beverage that we know as soda water. With this discovery, he can properly be called the father of the modern soft drink industry. One can only wonder if he would prefer Coke or Pepsi! He also was the first to recognize that breathing air soon made it unsuitable for further use in supporting animal life but that this same air could be restored by growing green plants in it.

In 1794, after some especially unpleasant incidents attributed to his political and religious beliefs, Priestley emigrated to the United States. He left Great Britain only 1 week before Lavoisier was executed in France. He settled in Northumberland, Pennsylvania, where he discovered carbon monoxide, was considered as a presidential candidate, and did much to further the cause of Unitarianism. His house and laboratory are preserved for all to see. In 1874, many chemists gathered in Northumberland to celebrate the one-hundredth anniversary of his discovery of oxygen and lay the groundwork for the foundation of the American Chemical Society.

Occurrence, Preparation, Properties, and Uses

The naturally occurring isotopes of oxygen are listed in Table 11.1. Note that the percent abundances vary somewhat with differing sources. The accuracy of the atomic mass of the element is somewhat limited by this variation, but nevertheless the atomic mass of oxygen served as the standard of comparison for other elements

TABLE 11.1
The stable isotopes of oxygen

Designation	Symbol	Mass, u	Percent abundance	Range of % abundance
Oxygen 16	$^{16}_{8}O$	15.994915	99.763	
Oxygen 17	$^{17}_{8}O$	16.999134	0.037	0.035–0.041
Oxygen 18	$^{18}_{8}O$	17.999160	0.200	0.188–0.209

until 1961 when the International Union of Pure and Applied Chemistry (IUPAC) adopted carbon 12 as the new standard.

Oxygen is the most abundant element in the crust of the earth and in its oceans. The atmosphere is now about 21 percent oxygen, but this has not always been the case. The original atmosphere of the newly formed earth contained very little. All the oxygen in the present atmosphere is the result of biological activity, that is, through the action of photosynthesis [represented in Equation (11.1)] carried on by the members of the plant kingdom:

$$6CO_2(g) + 6H_2O(l) \xrightarrow{h\nu} C_6H_{12}O_6 + 6O_2(g) \tag{11.1}$$

It is significant that Priestley, by recognizing that green plants can restore the life-supporting property of air, was already onto the basic idea of the oxygen cycle and photosynthesis more than 200 years ago. Incidently, oxygen 18 can be used as a tracer to show that the atoms in the diatomic oxygen product of photosynthesis both come from the water and none from the carbon dioxide. The photosynthesis reaction has a positive standard free energy but is driven by the input of solar energy.

Industrially, oxygen is usually prepared by separation from liquid air by fractional distillation. The boiling points of liquid nitrogen and oxygen are -195.8 and $-183.0°$ C, respectively, and therefore the nitrogen boils off first, leaving mostly oxygen (and some argon) behind. Oxygen is usually stored and transported as a liquid. (Have you ever noticed liquid oxygen tank trucks on the highway?)

In the laboratory, oxygen is commonly produced by the decomposition of potassium chlorate, $KClO_3$, as shown in Equation (11.2):

$$2KClO_3(s) \xrightarrow[MnO_2]{heat} 2KCl(s) + 3O_2(g) \tag{11.2}$$

This reaction used to be commonly employed in first-year chemistry laboratory exercises, but reports of isolated explosions have forced its elimination from such situations. Great care should be used in preparing oxygen in this manner. A safer method is to oxidize a mild solution of hydrogen peroxide with potassium permanganate in acid solution as shown in Equation (11.3):

$$5H_2O_2(aq) + 2MnO_4^-(aq) + 6H^+(aq) \longrightarrow 5O_2(g) + 2Mn^{2+}(aq) + 8H_2O(l) \tag{11.3}$$

Oxygen is, not surprisingly, an excellent oxidizing agent. (A quantitative treatment of oxidizing and reducing agents and redox reactions is given in the next chapter and will become a part of our network.) What to the layperson is ordinary combustion is more properly called oxidation by the chemist. One of the most common combustion-oxidation reactions is that of hydrocarbons, for example, those found in natural gas, crude oil, and gasoline. The products of the complete combustion or oxidation of a hydrocarbon are just carbon dioxide and water vapor as shown in Equation (11.4) for hexane, a component of gasoline.

$$C_6H_{14}(l) + \tfrac{19}{2}O_2(g) \xrightarrow{spark} 6CO_2(g) + 7H_2O(l) \tag{11.4}$$

For a long time it was assumed that if complete combustion of fossil fuels could be obtained, air pollution would be a thing of the past. More recently, however, carbon dioxide has been implicated as one of the driving forces of the greenhouse effect first mentioned in the last chapter. Details about this effect and its possible environmental consequences can be found in Section 11.6.

Oxygen is one of the most important industrial chemicals. It is used in the *basic oxygen process* (BOP) to convert pig iron to steel by oxidizing various silicon, manganese, phosphorus, and sulfur impurities. (Iron obtained directly from a blast furnace used to be cast into long oblong masses called *pigs*.) It is also used in making a variety of organic chemicals, as a respiration aid (for example, oxygen tents), as a rocket propellant, and in oxygen-methane, oxyhydrogen, and oxyacetylene torches. The oxygen-methane torch [Equation (11.5)] is usually used to melt the borosilicate glasses used in scientific laboratories, while oxyhydrogen torches are higher in temperature and are used to cut and weld various metals. An oxyacetylene torch [Equation (11.6)] reaches still higher temperatures and is used in construction work.

$$CH_4(g) + 2O_2(g) \longrightarrow CO_2(g) + 2H_2O(g) \qquad (11.5)$$

$$2C_2H_2(g) + 5O_2(g) \longrightarrow 4CO_2(g) + 2H_2O(g) \qquad (11.6)$$

11.2 WATER AND AQUEOUS SOLUTIONS

So far we have included in our network of interconnected ideas the periodic law, the uniqueness principle, the diagonal effect, the inert-pair effect and the metal-nonmetal line (see Figure 9.18). Now we wish to add to these a sixth component, a knowledge of the aqueous acid-base chemistry of oxides and their corresponding hydroxides. To do this we first have to discuss water and aqueous solutions.

Water is without doubt the most important compound of oxygen, that is, the most important oxide. An understanding of its structure and properties is central to an understanding not only of chemistry but of all nature. When we are tempted to think that humankind has a huge and long-standing storehouse of knowledge about nature, it is humbling to note that until Cavendish's experiments in the 1760s proved that water was a compound involving hydrogen, there was still a tendency to regard water (along with the familiar fire, earth, and air) as one of the four fundamental elements composing all (nonliving) substances.

The Structure of the Water Molecule

Although you should be familiar with the structure of an individual water molecule from earlier courses, it is so central to the arguments which follow that we should briefly review it here. The Lewis structure, the geometry as rationalized by the valence-shell electron-pair repulsion (VSEPR) theory, and a valence-bond theory (VBT) representation of water are shown in Figure 11.1. The six electrons of the oxygen atom together with the two electrons from the hydrogen atoms are arranged to give the familiar Lewis structure of Figure 11.1*a* wherein oxygen achieves an octet of electrons.

We all know that water is a bent molecule. (How long do you suppose this has been known? The answer is only about 75 years!) How do we account for its nonlinear structure? We cannot use a Lewis structure because it only accounts for the electrons and rationalizes a particular formula but does not predict a shape. A quick and reliable estimate of molecular geometry is usually established using the VSEPR theory. The four electron pairs (two bonding and two nonbonding or lone pairs) arrange themselves in a tetrahedron in order to minimize electron-electron repulsions as shown in Figure 11.1*b*. The basic tetrahedral angle of 109.5° is compressed slightly by the presence of the two lone pairs so that the H—O—H angle of water is somewhat less than 109.5°. Recall that the lone pairs, being

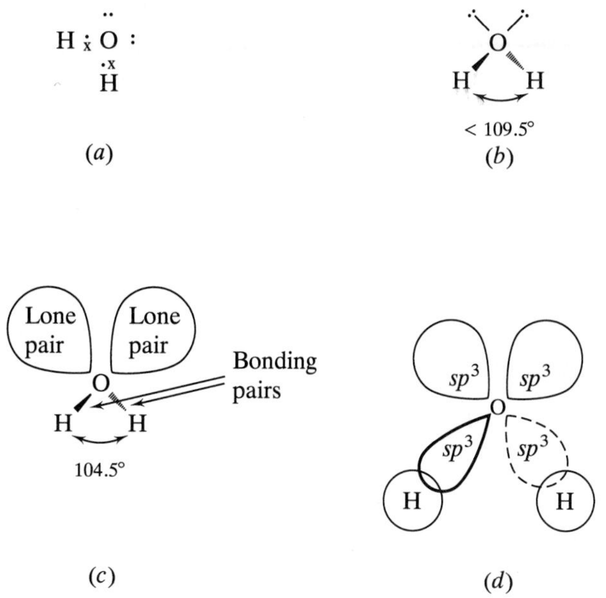

FIGURE 11.1
Various representations of the bonding in a water molecule. (*a*) The Lewis structure showing an octet of electrons around the oxygen atom. (*b*) The VSEPR theory representation with the four pairs of electrons around the oxygen arranged tetrahedrally to minimize electron-electron repulsions. (*c*) The lone pairs spread out and force the bonding pairs together compressing the H—O—H angle from 109.5 to 104.5°. (*d*) The valence-bond theory representation showing four sp^3 orbitals around the oxygen. Two of these contain nonbonding electron pairs, while the other two contain bonding pairs.

confined by only one nucleus, tend to spread out and take up more volume than the bonding pairs which are confined by both the hydrogen and oxygen nuclei. The resulting arrangement with the experimental H—O—H bond angle of 104.5° is shown in Figure 11.1*c*.

In the valence-bond theory (VBT) approach, the overlap of valence atomic orbitals of the constituent atoms result in the structure shown in Figure 11.1*d*. Here the $2s$ and the three $2p$ orbitals of oxygen are hybridized to form four sp^3 hybrid orbitals. Two of these hybrids contain the lone-pair electrons and two contain the bonding pairs which form the bonds with the hydrogen atoms. Again, the lone pairs take up more space than the bonding pairs, resulting in an H—O—H bond angle of 104.5°.

Water is a polar compound. Since the electronegativity of oxygen is greater than that of hydrogen, each O—H bond is polar-covalent; that is, the pair of electrons is shared between the two atoms but not equally so. The pair is closer to the oxygen, resulting in a partial negative charge (δ^-) on the oxygen atom and a corresponding partial positive charge (δ^+) on the hydrogen atom as shown in Figure 11.2*a*. Recall that the resulting polarity of a bond is represented by an arrow pointing toward the more electronegative atom. Such a bond is said to possess a dipole moment in which the centers of positive and negative charge are

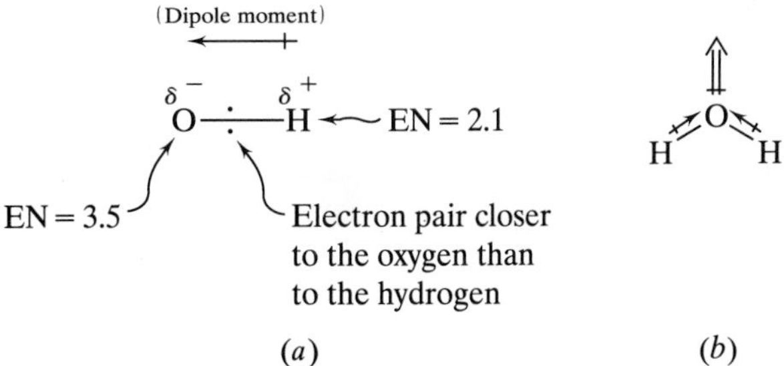

FIGURE 11.2
(*a*) An O—H bond showing the relative electronegativities, the position of the shared electron pair, the resulting partial charges, and the vector representation of the dipole moment. (*b*) The two individual O—H dipole moments add vectorially to produce a nonzero net dipole moment and therefore a polar molecule.

at two different poles, or points in space. Accordingly, the O—H bond is said to possess a dipole moment pointing toward the oxygen atom as shown in the figure. Water is confirmed to be a polar compound when the two O–H dipole moments are added vectorially to produce the net dipole moment represented by the double arrow shown in Figure 11.2*b*. If the molecule had been linear, the two individual dipole moments would have canceled each other out, the net dipole moment would have been zero, and the molecule would be nonpolar.

Ice and Liquid Water

What happens when a large collection (as in Avogadro's number) of water molecules is present in the solid, liquid, or gas phase? To answer this question, we start with a consideration of the structure of ice, which is shown in Figure 11.3. The individual water molecules of ice are held together by hydrogen bonds (shown as dashed lines). Again we call upon material which is covered in most introductory chemistry courses. Recall that intermolecular forces among polar molecules are called dipole-dipole forces. If the polar molecules involve a hydrogen atom covalently bound to an electronegative atom such as fluorine, oxygen, nitrogen, or even chlorine, the dipole-dipole forces among such molecules are strong enough to be dignified by the special name *hydrogen bonds*. Given a collection of molecules containing H—X bonds, the strength of the intermolecular hydrogen bonding among them is greatest when X is fluorine (F) and decreases in the order F, O, N, to Cl. To introduce a slightly whimsical but effective mnemonic (memory device), we might refer to this order as the "FONCl (pronounced fon′-cul) rule" and define it as follows. Hydrogen bonds occur among molecules containing H—X bonds (where X = F, O, N, and Cl) and decrease in strength in the order F to O to N to Cl.

Note that the combination of the bent structure of individual water molecules and the linear or almost linear nature of hydrogen bonds, H---X—H (where X = O in water), leads to an ice structure characterized by rather large hexagonally shaped holes. The shape of snowflakes, one example of which is shown in Figure 11.4, reflects this overall hexagonal symmetry on the molecular level. When ice melts, some of the water molecules break off from the ice structure and fill the hexagonal holes, leading to a greater density (at the melting point) for liquid water

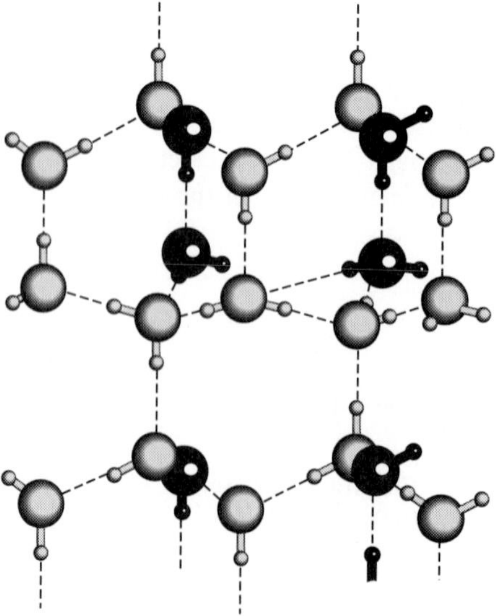

FIGURE 11.3
The structure of ice. Hydrogen bonds shown as dashed lines. [Ref. 14.]

FIGURE 11.4
The familiar hexagonal shape of a snowflake reflects the structure of ice on the molecular level. (*Richard B. Hait/Photo Researchers.*)

than for ice. This increase in density upon liquefaction is highly unusual. (You may also recall that this difference in density leads to an unusual negative slope for the solid-liquid equilibrium line in the phase diagram of water and, not incidentally, to an enhanced popularity of such sports as figure skating, ice hockey, and ice fishing.)

Liquid water is pictured as a mixture of molecules which are (1) hydrogen-bonded together in an icelike cluster and (2) free or nonhydrogen-bonded. At a given moment, an individual water molecule may be in the middle of some cluster, while at the next moment, the same molecule may be found at the edge of a different cluster. Still later it may be "free" of any cluster. This model of clusters which flick on and off throughout the liquid is sometimes known as the *flickering cluster model of liquid water* and is represented in Figure 11.5. The number of clusters as well as the number of molecules per cluster decrease as temperature increases. Some clusters persist even in the vapor phase of water.

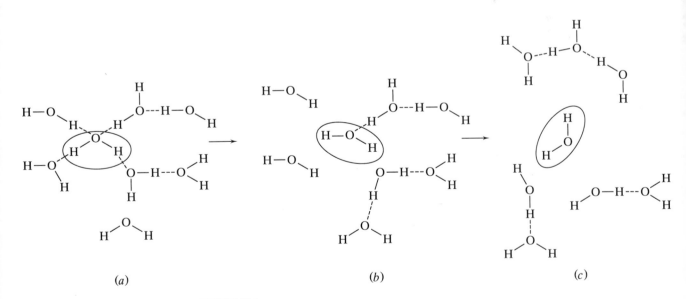

(*a*) (*b*) (*c*)

FIGURE 11.5
The flickering cluster model of liquid water. A given water molecule (circled) is (*a*) in the middle of a cluster at time t_1, (*b*) at the edge of a different cluster at time t_2, and (*c*) free at time t_3. The various water molecules are shown relatively stationary in space from one moment to another. In reality, each would move (translate and rotate) more than is shown here.

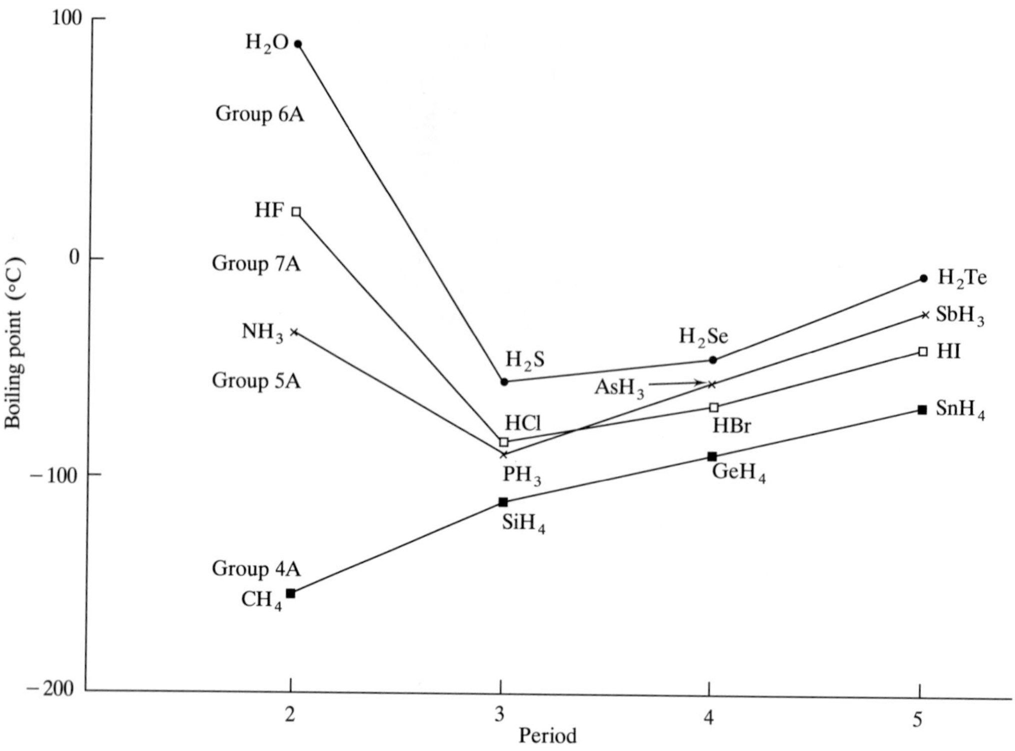

FIGURE 11.6
The boiling points of the hydrides of the Group 4A, 5A, 6A, and 7A elements. Note that the boiling points of HF, H_2O, NH_3, and HCl are elevated over what is expected on the basis of molecular weight alone.

Hydrogen bonds are the strongest of the various intermolecular forces. It takes more energy to break these bonds than to overcome other types of intermolecular forces, and therefore water has an unusually high heat of fusion, heat of vaporization, and heat capacity. In addition, you may already be familiar with the dramatic presentation of the effect of hydrogen bonds on the boiling points of various hydrides shown in Figure 11.6. The unusually high boiling points of water, ammonia, and hydrogen fluoride are all due to the presence of hydrogen bonds which need to be broken to convert the liquid to the gas. Note that the boiling point of HCl is also slightly higher than predicted, a fact that supports the inclusion of Cl in the FONCl rule.

Solubility of Substances in Water

Why was Priestley successful in preparing so many gases that had eluded those before him? Remember from the first section of this chapter that only common air, carbon dioxide, and hydrogen were known before Priestley started his work. On the other hand, he was able to generate, collect, and characterize a number of gases (NH_3, HCl, H_2S, NO_2, and SO_2 among others) that, as it turned out, were soluble in water but insoluble in the liquid mercury that Priestley quite often used. This leads us to a consideration of the question of why water is such a good solvent for so many substances and how we can rationalize the solubility characteristics of various solutes in water.

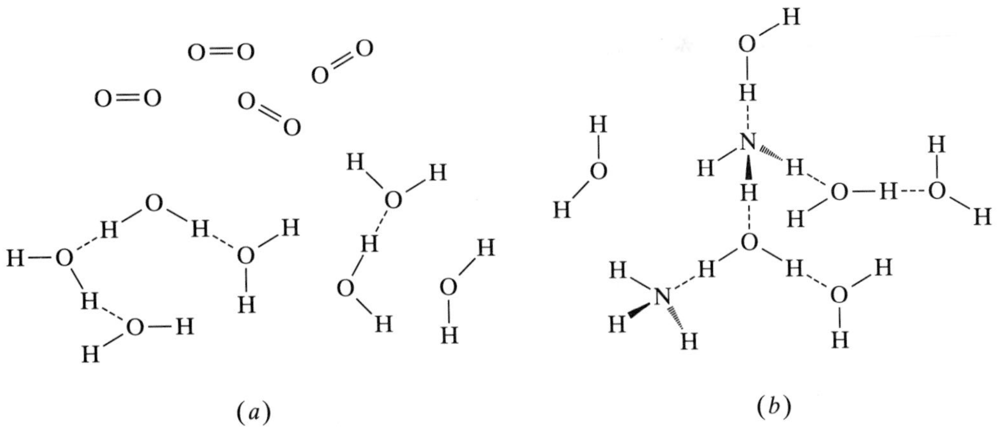

(*a*) (*b*)

FIGURE 11.7
A diagram showing (*a*) the relative insolubility of O_2 molecules and (*b*) the solubility of NH_3 molecules in liquid water. The O_2 does not hydrogen-bond with water molecules, whereas the NH_3 does.

The general rule for predicting the solubility of various covalent solutes is the familiar "like dissolves like." Polar solutes dissolve in polar solvents (like water) and nonpolar solutes dissolve in nonpolar solvents, but polar does not dissolve in nonpolar, and vice versa. Why is this? (We had a hint of what is to come in the last chapter when we discussed the deuteration of various compounds.) Hydrogen bonds are strong intermolecular forces, and individual water molecules will associate together to the exclusion of the solute molecules unless the solute can form hydrogen bonds or strong dipole-dipole forces with the water molecules. The situation is exemplified using oxygen and ammonia gases in Figure 11.7. In Figure 11.7*a* the nonpolar oxygen molecules cannot hydrogen-bond with the polar water molecules which therefore keep to themselves, associating by hydrogen bonds. Oxygen molecules are excluded, and the gas is therefore insoluble. In Figure 11.7*b*, the polar ammonia molecules (you should be able to verify the structure of NH_3) are polar and are capable of participating in hydrogen bonds with the water. Accordingly, ammonia is taken into the water structure and is therefore soluble.

The solubility of solids is somewhat more involved. Nonpolar solids will not dissolve well in polar water as expected. We might expect that polar solids, including ionic solids, would be universally water-soluble. Such is not the case. The solubility of an ionic solute is determined by a rather difficult-to-predict balance of the energy needed to break the lattice of ions in the solid (the lattice energy) and the energy released when the ions associate or are hydrated by the water molecules (the hydration energy). While both these energies depend on the charge density (see Chapter 8, p. 194, and Chapter 9, p. 226), the balance between them is very sensitive, and its details are more the province of physical chemistry than inorganic chemistry. It is important, however, to have some feel for what solids are soluble in water and which are not. Table 11.2 provides some guidelines to keep in mind.

What happens when an ionic solid does dissolve in water? Consider the dissolving (or dissolution, to use the fancier word) of sodium chloride as shown in Figure 11.8 as a representative example. The solid NaCl(*s*) is in the form of a regular array or crystal lattice of alternating sodium and chloride ions. (See Chapter 7 for details on crystal lattices.) Water molecules align their partially positive hydrogen atoms with a chloride anion at a corner of the lattice and

TABLE 11.2

Some general solubility rules for common ionic solutes in water

1. All Group 1A (alkali-metal) hydroxides (LiOH, NaOH, KOH, RbOH, and CsOH) are soluble. Of the Group 2A (alkaline-earth metal) hydroxides, only barium hydroxide [$Ba(OH)_2$] is soluble. Calcium hydroxide [$Ca(OH)_2$] is slightly soluble. All other hydroxides are insoluble.
2. Most compounds containing chlorides (Cl^-), bromides (Br^-), or iodides (I^-) are soluble. The exceptions are those containing Ag^+, Hg_2^{2+}, and Pb^{2+} ions.
3. Most sulfates (SO_4^{2-}) are soluble. Calcium sulfate ($CaSO_4$) and silver sulfate (Ag_2SO_4) are slightly soluble. Barium sulfate ($BaSO_4$), mercury(II) sulfate ($HgSO_4$), and lead(II) sulfate ($PbSO_4$) are insoluble.
4. All compounds containing nitrate (NO_3^-), chlorate (ClO_3^-), and perchlorate (ClO_4^-) are soluble.
5. All carbonates (CO_3^{2-}), phosphates (PO_4^{3-}), sulfides (S^{2-}), and sulfites (SO_3^{2-}) are insoluble, except those of ammonium (NH_4^+) and alkali metals.
6. All ammonium (NH_4^+) compounds are soluble.
7. All alkali-metal compounds are soluble.

Source: R. Chang, *General Chemistry*, 3d ed., McGraw-Hill, New York, 1988, p. 456.

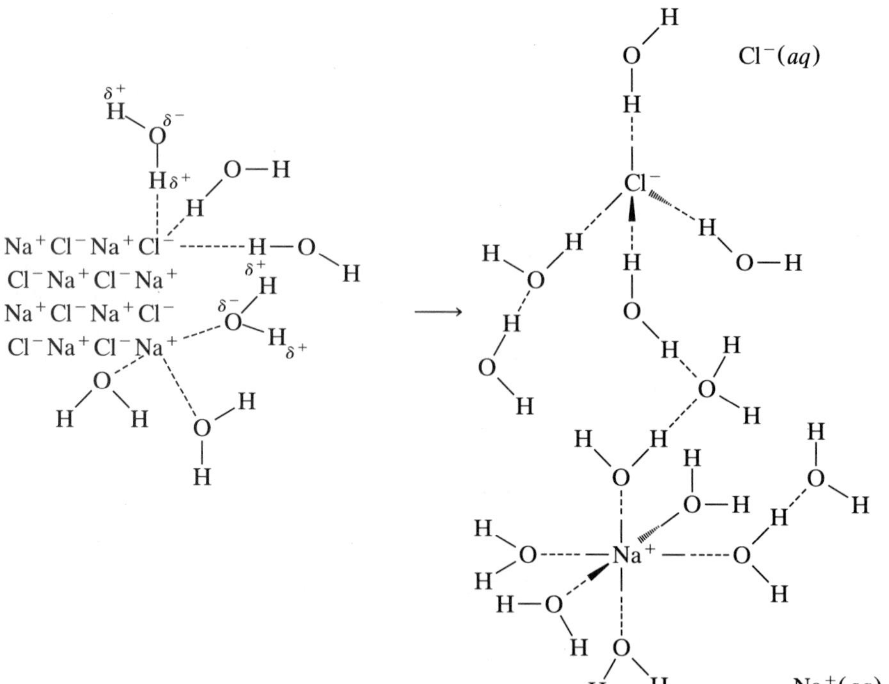

$Cl^-(aq)$

$Na^+(aq)$

$$NaCl(s) \xrightarrow{H_2O} Na^+(aq) + Cl^-(aq)$$

FIGURE 11.8
The interaction of water with a section of the crystal lattice of sodium chloride. The cations and anions are torn away from the lattice and incorporated into the water structure. The hydrated sodium and chloride ions are represented as $Na^+(aq)$ and $Cl^-(aq)$, respectively.

overcome the ionic interactions between the chloride and its neighboring sodium cations. The chloride anion is *hydrated*, that is, surrounded by water molecules, and assumes its place in the water structure. At another corner of the lattice, the sodium cations are similarly attacked by water molecules which now orient their partially negative oxygen ends toward the cation. Hydrated sodium ions are also incorporated into the water structure. For NaCl, it must be that the hydration energy released when sodium and chloride ions are surrounded by water molecules is greater than the energy needed to overcome the lattice energy which holds the sodium chloride crystal together.

Self-Ionization of Water

Before leaving a discussion of the properties of water and aqueous solutions, we should discuss its self-ionization. Recall that the Brønsted-Lowry definition of acids and bases is in terms of proton donors and acceptors, respectively. Water is an amphoteric substance; that is, it can act as either an acid or a base as shown in Equation (11.7). Note that water can act either as a proton donor (acid) or as a proton acceptor (base). As shown in Figure 11.9, the resulting hydronium ion (H_3O^+) and hydroxide ion (OH^-) are hydrated in a fashion similar to that of sodium and chloride ions:

$$\underset{\substack{\text{Proton}\\\text{acceptor}\\\text{(base)}}}{H_2O} + \underset{\substack{\text{Proton}\\\text{donor}\\\text{(acid)}}}{H_2O} \xrightarrow{H^+} H_3O^+(aq) + OH^-(aq) \tag{11.7}$$

Equation (11.7) has an equilibrium constant (often represented as K_w) of approximately 1.0×10^{-14} (at 298 K), and a simple acid-base equilibrium calculation yields the result that in pure water, $[H_3O^+] = [OH^-] = 1.0 \times 10^{-7}$. It follows that the pH ($-\log [H_3O^+] = -\log [H^+]$) of water is 7.00 at room temperature. Equation (11.7), then, is the basis of the pH scale as commonly presented in general chemistry. Any substance which raises the concentration of H_3O^+ ions

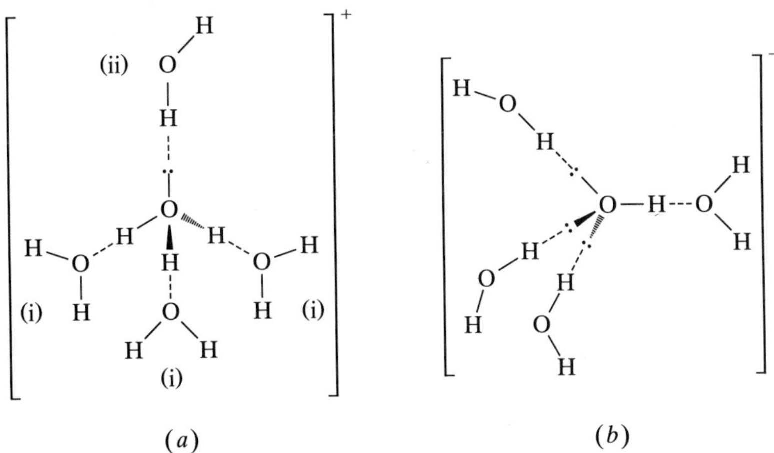

FIGURE 11.9
Hydrated (*a*) hydronium (H_3O^+) and (*b*) hydroxide (OH^-) ions. The water molecules labeled (*i*) can sometimes be isolated with the hydronium ion to form cations of formula $H_9O_4^+$.

produces a pH less than 7 and is an acid. Any substance which lowers the concentration of the hydronium ion or raises the concentration of the hydroxide ion produces a pH greater than 7 and is a base. Note that the small value of K_w indicates that the self-ionization process occurs only to a very small extent. Another way to appreciate just how few hydronium ions there are in pure water is to realize that the concentration of water molecules in pure water is 55.6 *M*, and therefore for every hydronium ion there are ($55.6/1.0 \times 10^{-7} =$) 556×10^6, or 556 million, water molecules.

You may have seen Equation (11.7) in a slightly different form. Sometimes the self-ionization of water is represented as shown in Equation (11.8):

$$H_2O \longrightarrow H^+(aq) + OH^-(aq) \tag{11.8}$$

This is a shorthand notation for what actually happens in solution. Note that no hydrogen ion or bare proton ever exists alone in aqueous solution. We discussed the reasons for this when considering covalent hydrides in Chapter 10. The charge density of a bare proton (and therefore its polarizing power) is so large that it immediately forms a covalent bond with whatever else is present, in this case water molecules. The result is the H_3O^+ ion or something still more complex. From solutions of strong acids such as HBr and $HClO_4$, the cations $H_5O_2^+$, $H_7O_3^+$, and $H_9O_4^+$ have been isolated. These correspond to 1 to 3 water molecules of the type labeled (i) in Figure 11.9*a* being isolated with the hydronium ion. We will quite often use H_3O^+ to represent the acid species in aqueous solution, but it is important to realize that even that is more than likely a drastic oversimplification.

11.3 THE ACID-BASE CHARACTER OF OXIDES AND HYDROXIDES IN AQUEOUS SOLUTION

Oxides: General Expectations from the Network

Oxygen forms compounds with all the elements except helium, neon, argon, and krypton. It *reacts directly* with all the elements except the halogens, a few noble metals such as silver and gold, and the noble gases. Table 11.3 shows the major binary oxides of the representative elements. As expected from our earlier discussions of the network of ideas (Chapter 9) and the hydrides (Section 10.5), we are not surprised to find a tremendous variability in the types of bonding encountered in oxides. As Table 11.3 indicates, metal oxides are ionic solids, while nonmetal oxides are discrete, molecular covalent gases and liquids. The oxides of the heavier nonmetals and the semimetals tend to be covalent polymeric solids. The transition metal oxides are occasionally found to be nonstoichiometric. (See Chapter 7, p. 180, for further discussion of the latter types of compounds.)

In what other ways do we expect our network to be applicable to the oxides? Think first about the uniqueness principle. Even our limited experience with it to this point leads us to anticipate that oxygen compounds will demonstrate a high degree of covalent character and an enhanced tendency toward π bonding (via oxygen *p* but not *d* orbitals). On the basis of the inert-pair effect, we expect that the heavier elements of Groups 3A, 4A, 5A, and 6A will tend to form oxides with an oxidation state 2 less than the group number. Finally, we will not be surprised to find that the variable properties of oxides, as shown in Table 11.3 and discussed below in the section on the acid-base properties of the oxides in aqueous solution, will be strongly correlated with the position of the metal-nonmetal line. For the most part, the details about the oxides are left to the group-by-group discussions of the representative elements found in Chapters 12 to 19.

TABLE 11.3
Classifications of binary oxides

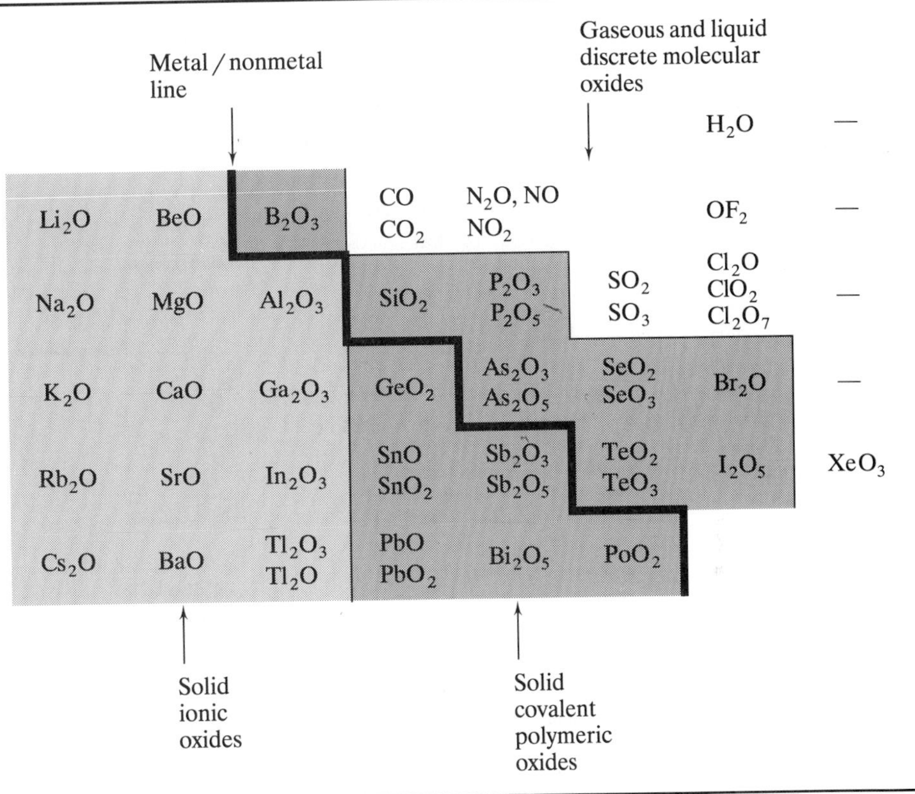

Oxides in Aqueous Solution (Acidic and Basic Anhydrides)

One of the most important aspects of the chemistry of oxides is their acid-base properties. Many oxides are *basic or acidic anhydrides* compounds that are formed by the removal of water from a corresponding base or acid. A comparison of Tables 11.3 and 11.4 shows that the ionic oxides are usually basic anhydrides, while the covalent oxides are usually acidic anhydrides. Some of the oxides of the semimetals are amphoteric anhydrides, capable of acting as either an acid or a base depending on the circumstances. Accordingly, if we list the oxides of a given period, say the third, we find an orderly progression of their acid-base characteristics as shown below:

$$\underbrace{Na_2O \quad MgO}_{\text{Basic}} \longrightarrow \underbrace{Al_2O_3}_{\text{Amphoteric}} \longrightarrow \underbrace{SiO_2 \quad P_2O_5 \quad SO_3 \quad Cl_2O_7}_{\text{Acidic}}$$

The ionic oxides are characterized by the presence of the oxide ion, O^{2-}, that, like the H^+ ion, cannot exist alone in aqueous solution. The reaction between the oxide ion and a water molecule is shown in Equation (11.9):

$$:\ddot{\underset{..}{O}}:^{2-} + \overset{\delta+}{H} \wr \overset{\delta-}{O} \diagdown^{\delta+} H \longrightarrow 2\ :\ddot{\underset{..}{O}} - H^- \tag{11.9}$$

Note that the -2 charged oxide attacks and forms a bond with a partially positive

TABLE 11.4
Oxides as basic, amphoteric, and acidic anhydrides

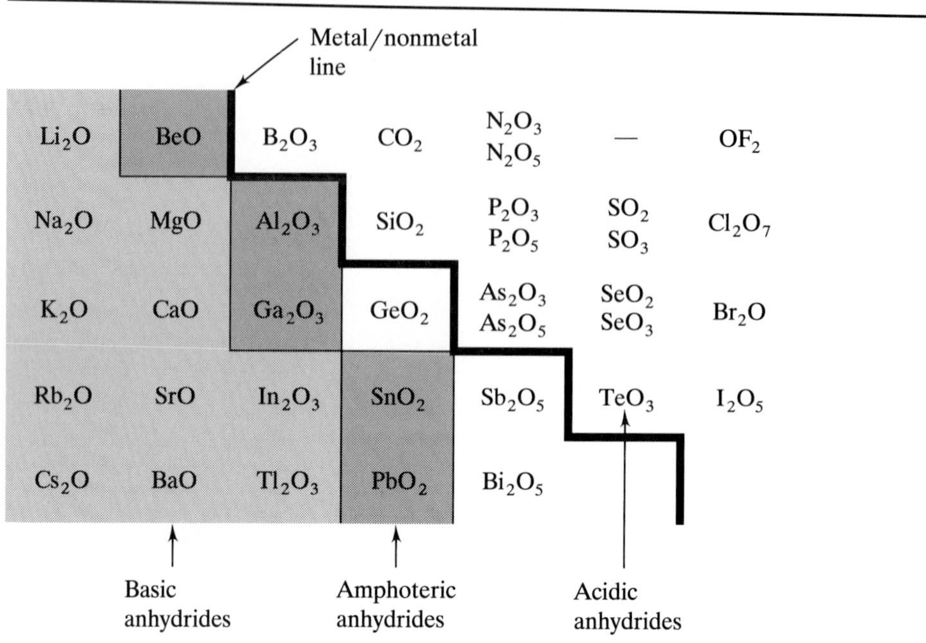

hydrogen atom of the water molecule. The subsequent breaking (represented by a ⌇ in the bond) of the O—H bond produces two hydroxide ions. The equilibrium constant for Equation (11.9) is greater than 10^{22}, so this reaction lies far to the right. Taking sodium oxide as an example, the full reaction of an ionic oxide with water is represented in Equation (11.10):

$$Na_2O(s) + H_2O(l) \longrightarrow 2NaOH(aq) \longrightarrow 2Na^+(aq) + 2OH^-(aq) \quad (11.10)$$

One way to think of the process is in terms of a sequence from the metal oxide to the metal hydroxide which then dissociates into the aqueous hydroxide and metal ions. It follows that sodium oxide is a basic anhydride; it produces the base sodium hydroxide in aqueous solution. Table 11.5 shows some of the more common basic anhydrides and their corresponding bases. Note that the greater the degree of ionic character in the oxide, the more basic it is. In the next section we will discuss the nature of metal hydroxides and explore why they split to produce aqueous metal cations and hydroxide anions.

TABLE 11.5
Some common basic anhydrides and their corresponding bases

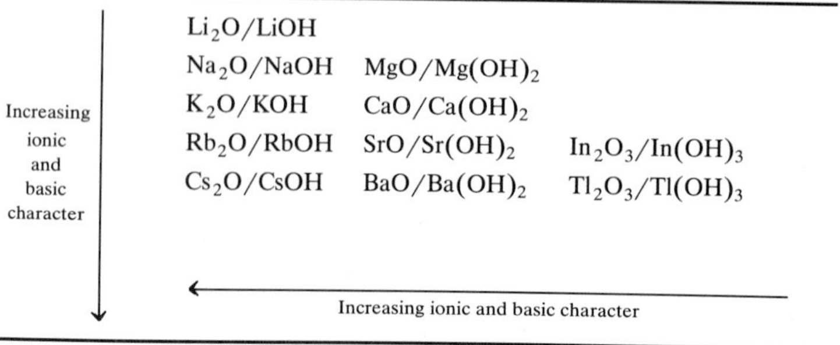

Increasing ionic and basic character ↓			
	$Li_2O/LiOH$		
	$Na_2O/NaOH$	$MgO/Mg(OH)_2$	
	K_2O/KOH	$CaO/Ca(OH)_2$	
	$Rb_2O/RbOH$	$SrO/Sr(OH)_2$	$In_2O_3/In(OH)_3$
	$Cs_2O/CsOH$	$BaO/Ba(OH)_2$	$Tl_2O_3/Tl(OH)_3$

← Increasing ionic and basic character

TABLE 11.6
Some common nonmetal oxides and their corresponding oxoacids

$\mathbf{B_2O_3}$ $B(OH)_3$ or H_3BO_3 Boric acid	$\mathbf{CO_2}$ $CO(OH)_2$ or H_2CO_3 Carbonic acid	$\mathbf{N_2O_3}$ $NO(OH)$ or HNO_2 Nitrous acid	
		$\mathbf{N_2O_5}$ $NO_2(OH)$ or HNO_3 Nitric acid	
	$\mathbf{SiO_2}$ $Si(OH)_4$ or H_4SiO_4 Silicic acid	$\mathbf{P_2O_3}$ $HPO(OH)_2$ or H_3PO_3 Phosphorus acid	$\mathbf{SO_2}$ $SO(OH)_2$ or H_2SO_3 Sulfurous acid
		$\mathbf{P_2O_5}$ $PO(OH)_3$ or H_3PO_4 Phosphoric acid	$\mathbf{SO_3}$ $SO_2(OH)_2$ or H_2SO_4 Sulfuric acid

Nonmetal oxides react with water to produce what are known as the oxyacids or *oxoacids*, acids containing an NM–O–H unit, where NM = nonmetal. A nonmetal oxide is usually characterized by polar covalent bonds rather than the ionic bonds of a metal oxide. As represented generally in Equation (11.11), the partially negative oxygen atom of a water molecule will attack the partially positive nonmetal atom at the same time that the oxygen of the oxide is attracted to one of the hydrogen atoms of the water. The breaking of the O—H bond of the water will produce an oxoacid that, as we will explain shortly, splits to produce the corresponding aqueous anion and hydronium ions:

$$\delta^{-}\!:\!\ddot{O}\!:\ \text{—}\ \delta^{+}\mathrm{NM} \;+\; \mathrm{H}\delta^{+}\text{—}\ddot{O}\delta^{-}\text{—H} \longrightarrow \underset{\text{Oxoacid}}{\mathrm{H{-}\ddot{O}{-}NM{-}\ddot{O}{-}H}} \xrightarrow{2H_2O} NMO_2^{2-} + 2H_3O^+ \qquad (11.11)$$

Equations (11.12) show a specific example starting with sulfur trioxide as the nonmetal oxide which produces sulfuric acid as the oxoacid which, in turn, dissociates into sulfate and hydronium ions. In this case sulfur trioxide is the acid anhydride of sulfuric acid. Table 11.6 shows some of the more common acid anhydrides and their corresponding oxoacids.

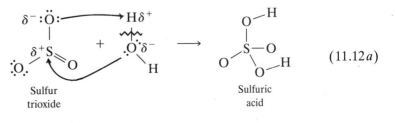

$$\underset{\substack{\text{Sulfur}\\ \text{trioxide}}}{\delta^{-}\!:\!\ddot{O}\!:\text{—}\delta^{+}S(=O)(\ddot{O})} \;+\; \mathrm{H}\delta^{+}\text{—}\ddot{O}\delta^{-}\text{—H} \longrightarrow \underset{\substack{\text{Sulfuric}\\ \text{acid}}}{\mathrm{(H{-}O)(H{-}O)S(=O)_2}} \qquad (11.12a)$$

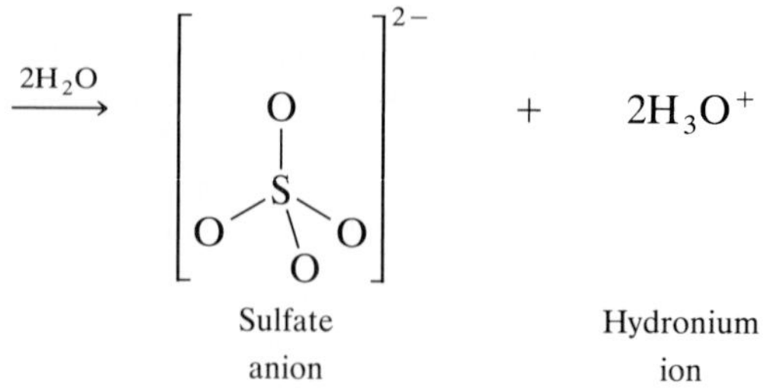

$$\xrightarrow{2H_2O} \left[SO_4 \right]^{2-} + 2H_3O^+ \qquad (11.12b)$$

Sulfate anion — Hydronium ion

Amphoteric anhydrides are often oxides of the semimetals. While the oxides themselves are often insoluble in water, the corresponding hydroxides can act as either an acid or a base. Aluminum hydroxide is shown in Equation (11.13) as an example. Note that it can either act as a base (when it reacts with hydrogen ions) or as an acid (when it reacts with hydroxide ions). More on the exact nature of these amphoteric substances will be presented in the appropriate group chapters.

$$Al(OH)_3(aq) \xrightarrow{+3H^+} Al^{3+}(aq) - + 3H_2O$$
$$Al(OH)_3(aq) \xrightarrow{+OH^-} Al(OH)_4^-(aq) \qquad (11.13)$$

The acid-base character of the oxides in aqueous solution is an important property. In subsequent chapters, these properties will be listed for the elements of a given group.

The E–O–H Unit in Aqueous Solution

As explained above, both metal and nonmetal oxides react with water to produce compounds with an E–O–H unit. If E is a metal (M), the unit acts as a base, releasing hydroxide ions in solution, whereas if E is a nonmetal (NM), hydronium

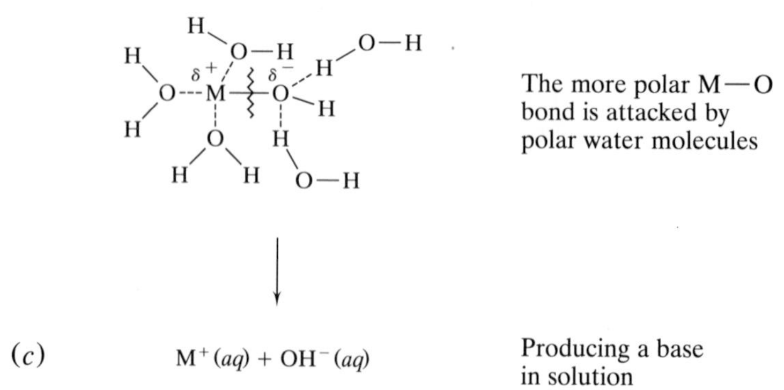

(*a*) $\overset{\delta+}{M}$—$\overset{\delta-}{O}$—$\overset{\delta+}{H}$ ENs 0.7–1.5 3.5 2.1 — The M—O bond is the more polar due to a greater difference in electronegativities

$\downarrow H_2O$

(*b*) The more polar M—O bond is attacked by polar water molecules

$\downarrow$

(*c*) $M^+(aq) + OH^-(aq)$ — Producing a base in solution

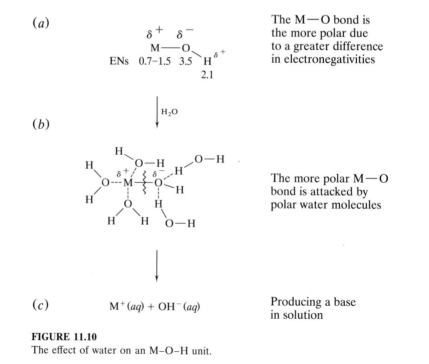

FIGURE 11.10
The effect of water on an M–O–H unit.

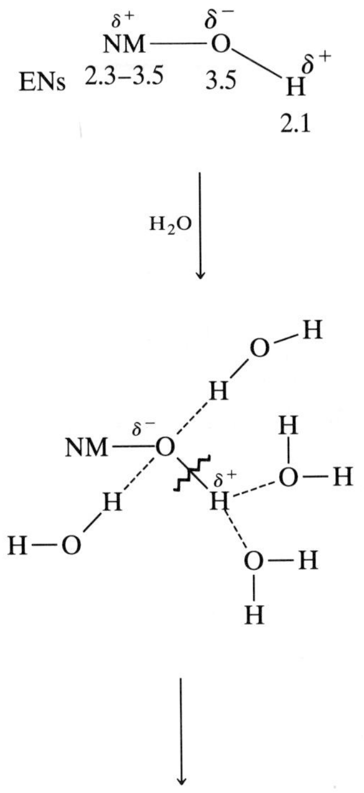

(a) $\overset{\delta^+}{\text{NM}}$—$\overset{\delta^-}{\text{O}}$—$\overset{\delta^+}{\text{H}}$
ENs 2.3–3.5 3.5 2.1

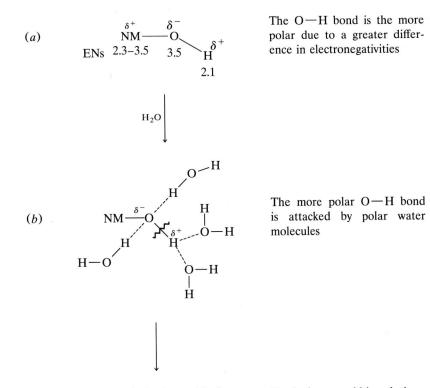

The O—H bond is the more polar due to a greater difference in electronegativities

↓ H_2O

(b) NM—O (attacked by water molecules)

The more polar O—H bond is attacked by polar water molecules

↓

(c) $NMO^-(aq) + H_3O^+(aq)$

Producing an acid in solution

FIGURE 11.11
The effect of water on an NM–O–H unit.

ions are released. Why the difference? How does the nature of the E atom determine whether the unit will be an acid or a base or be amphoteric? To see this, we need to look closely at the relative electronegativities of the unit.

If E is a metal, the relative electronegativities are as shown in Figure 11.10. Part (*a*) of the figure shows that the greater difference in electronegativity is between the metal and the oxygen, making the M—O bond the more polar (to the point that it may properly be classified as predominantly ionic). This highly polar (even ionic) nature is indicated by making the partial charge symbols bigger across the M—O bond. As shown in part (*b*), the more polar M—O bond is more susceptible to attack by polar water molecules, and this results in the breaking of the M—O bond to produce the aqueous metal cation and the hydroxide anion in solution as shown in part (*c*).

If E is a nonmetal, Figure 11.11 shows that now the O—H bond is the more polar and is preferentially attacked by water molecules, resulting in the oxoanion and hydronium ions in solutions.

If E is a semimetal, the two bonds of the E–O–H unit are approximately of the same polarity, and either can be broken depending on the circumstances. In this case, the unit is amphoteric.

An Addition to the Network

Figure 9.18 presented a summary of the five components of the interconnected network of ideas developed in that chapter to begin to make sense of the periodic

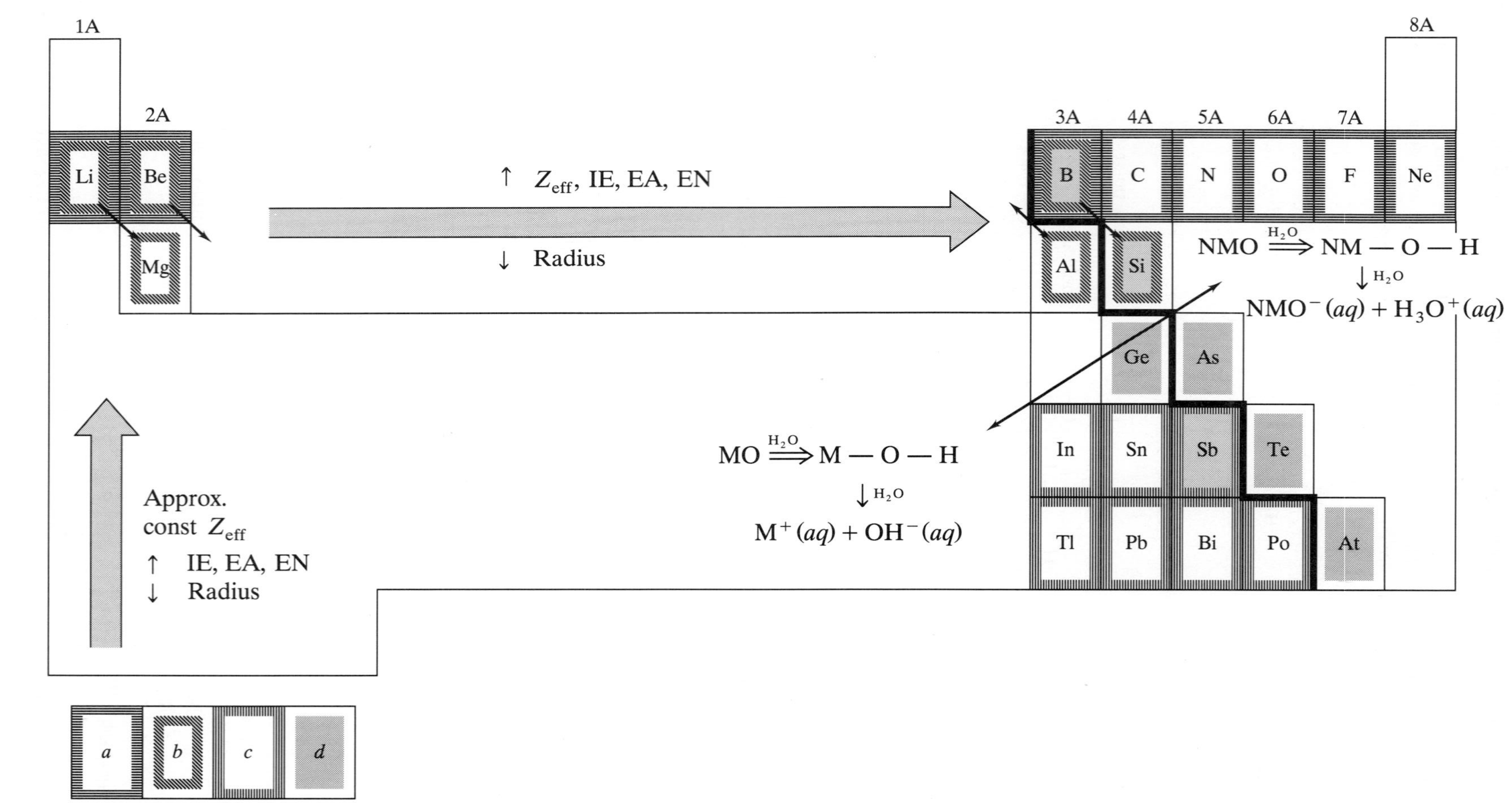

FIGURE 11.12
The six components of the interconnected network of ideas. The original five components are represented by the trends in periodic properties (effective nuclear charge, radius, ionization energy, electron affinity, and electronegativity) and the elements involved in (*a*) the uniqueness principle, (*b*) the diagonal effect, (*c*) the inert-pair effect, and (*d*) the metal-nonmetal line. To this has been added the sixth component: a summary of the acid-base character of metal (M) and nonmetal (NM) oxides in aqueous solution.

table. The acid-base character of metal and nonmetal oxides and their corresponding hydroxides is of central-enough importance that it now becomes the sixth component of the network. We will call this component the acid-base character of oxides. Figure 11.12 shows the latest summary of the network. Note that an indication of the basic character of metal oxides and the acidic character of nonmetal oxides has been added to either side of the metal-nonmetal line and that the splitting of the E–O–H unit as a function of the nature of E has also been indicated in each case.

11.4 THE RELATIVE STRENGTHS OF OXOACIDS AND HYDROACIDS IN AQUEOUS SOLUTION

Oxoacids

Table 11.7 shows the structures of some of the more common oxoacids. In the above section we discussed how these compounds all contain an NM–O–H unit that is split between the oxygen and hydrogen atoms by the attack of polar water. Now we can go on to investigate two additional factors which determine their relative acid strengths. These factors are (1) the electronegativity of the central atom and (2) the number of nonhydroxyl (OH) oxygens bound to the central atom.

As the electronegativity of the central atom increases, so too does its ability to withdraw electron density from its neighboring atoms. The surrounding oxygen atoms have a high electronegativity, so the central atom cannot withdraw electron density from them. Where, then, does the electron density come from? The answer is from the O—H bonds. The hydrogen atom (or perhaps atoms) has a relatively low electronegativity and cannot compete with the other more electronegative atoms in the molecule. As the electron density is withdrawn from the O—H bond, the hydrogen becomes more partially positive, and therefore the O—H bond is more polar and more susceptible to attack by water molecules. The result of all this is that the acid strength of an oxoacid increases as the electronegativity of the central atom increases. For example, sulfuric acid is a stronger acid than selenic acid, phosphoric acid is a stronger acid than arsenic acid, and perchloric acid is stronger than perbromic acid (although the exact acid strength of the latter is not well known).

As the number of nonhydroxyl oxygens (that is, oxygen atoms not in O—H groups) increases, they withdraw more electron density from the central atom, making it more partially positive. In turn, the central atom now withdraws more electron density from the only source available to it, the O—H bond. Again the hydrogen atom becomes more partially positive, and the O—H becomes more polar and therefore more susceptible to attack by water. The result is that as the number of nonhydroxyl oxygens increases, so too does the acid strength of the oxoacid. For example, acid strength increases from hypochlorous to chlorous to chloric to perchloric acid and from nitrous to nitric acid and from sulfurous to sulfuric acid.

Nomenclature of Oxoacids and Corresponding Salts (Optional)

Although you may have been introduced to the nomenclature of the oxoacids and their corresponding salts in earlier courses, there is a large and often mind-boggling array of names to be mastered. It certainly helps to have a systematic approach to this nomenclature. The system to be presented here starts with what

TABLE 11.7
The structures of the more common oxoacids

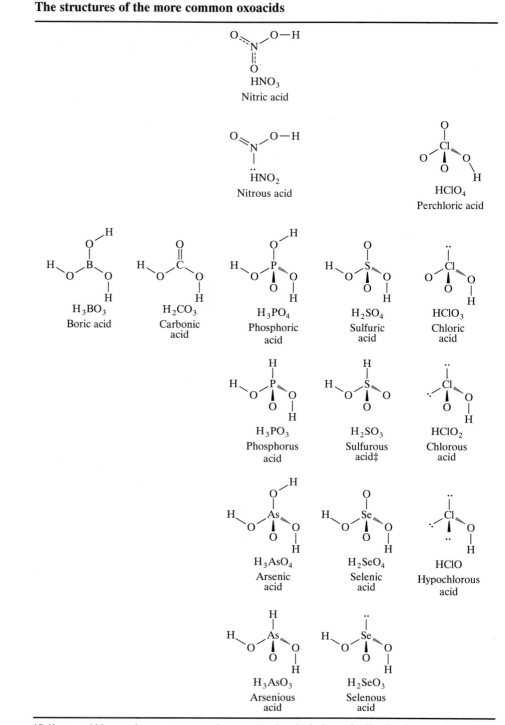

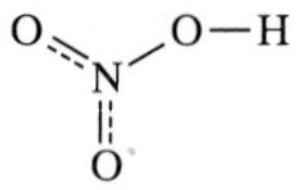

HNO_3
Nitric acid

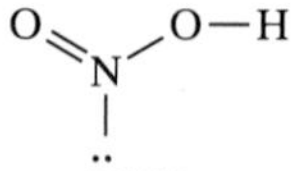

HNO_2
Nitrous acid

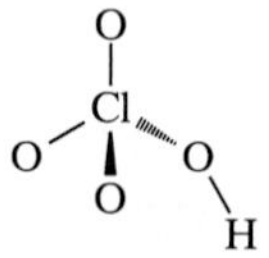

$HClO_4$
Perchloric acid

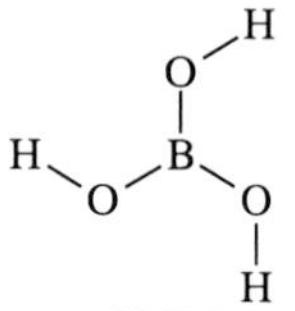

H_3BO_3
Boric acid

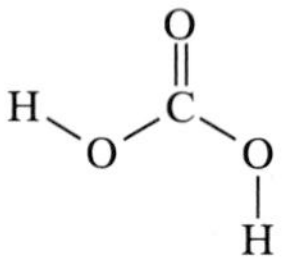

H_2CO_3
Carbonic acid

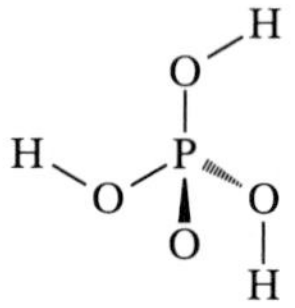

H_3PO_4
Phosphoric acid

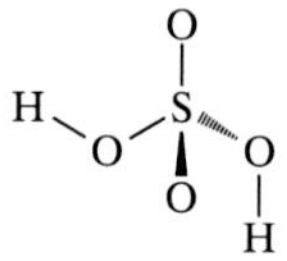

H_2SO_4
Sulfuric acid

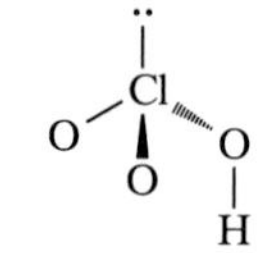

$HClO_3$
Chloric acid

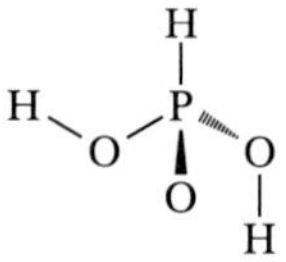

H_3PO_3
Phosphorus acid

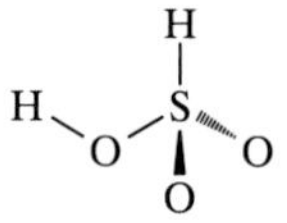

H_2SO_3
Sulfurous acid‡

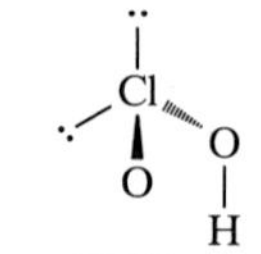

$HClO_2$
Chlorous acid

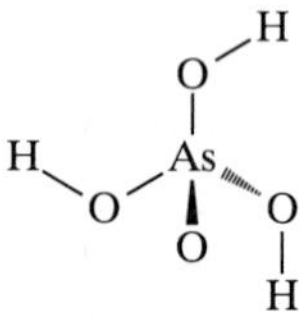

H_3AsO_4
Arsenic acid

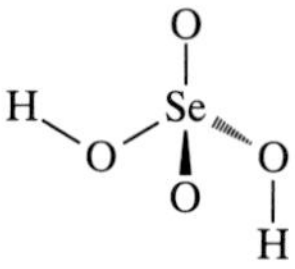

H_2SeO_4
Selenic acid

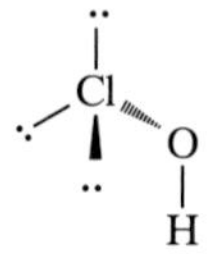

HClO
Hypochlorous acid

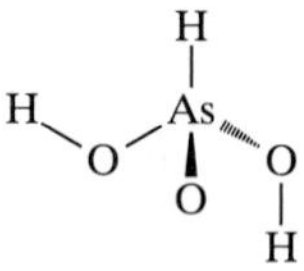

H_3AsO_3
Arsenious acid

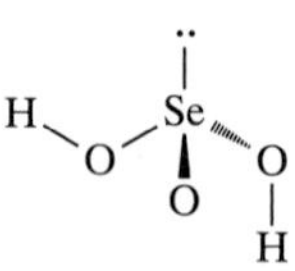

H_2SeO_3
Selenous acid

‡Sulfurous acid is sometimes more accurately represented as the hydrate of sulfur dioxide, $SO_2 \cdot H_2O$.

TABLE 11.8
A system for naming the oxoacids and their corresponding salts

Representative *-ic* acids in each group of the periodic table				
3A	**4A**	**5A***	**6A**	**7A**
H_3BO_3 Boric acid	H_2CO_3 Carbonic acid	H_3PO_4 Phosphoric acid	H_2SO_4 Sulfuric acid	$HClO_3$ Chloric acid

Rules for naming other oxyacids

1. Addition of 1 oxygen to the acid: per . . . -ic acid
2. Subtraction of 1 oxygen from the acid: . . . -ous acid
3. Subtraction of 2 oxygens from the acid: hypo . . . -ous acid
4. All other analogous acids in the same group of the periodic table are named similarly.

Rules for naming anions of oxyacids

1. Acids ending with *-ic* are associated with anions ending with *-ate.*
2. Acids ending with *-ous* are associated with anions ending with *-ite.*
3. The anion is obtained by removing all the hydrogen ions.
4. Anions in which one or more but not all the hydrogen ions have been removed are named with the number of remaining hydrogens indicated before the name of the anion as determined using rules 1 to 3. If one of two hydrogen ions is removed, the anion is commonly referred to using the *bi-* prefix.

*HNO_3 is an exception to the other acids in Group 5A and is called nitric acid.

are called the "representative *-ic* acids" listed in Table 11.8. Note that these five (six, counting nitric acid) acids are the only ones that you have to know outright. All other common oxoacids are named by using the rules given in the table. Another way to visualize these rules is given in the nomenclature "roadmap" shown in Figure 11.13. For example, starting with chloric acid, we can name the other chlorine-containing oxoacids by following the roadmap. Adding an oxygen to chloric acid yields perchloric acid, $HClO_4$. Subtracting oxygens yields chlorous acid, $HClO_2$, and then hypochlorous acid, HClO.

The procedure is the same for any acid with a central atom in any of the groups shown in the table. For example, by analogy with sulfuric acid, H_2SeO_4 is selenic acid, and if one oxygen is subtracted, we get H_2SeO_3, which is selenous

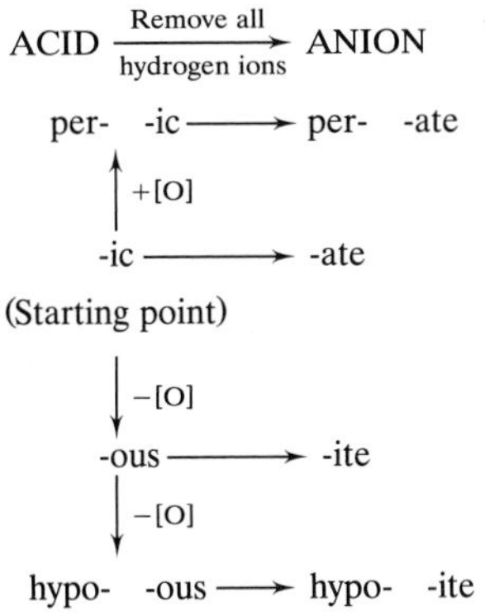

ACID —Remove all hydrogen ions→ ANION

per- -ic ⟶ per- -ate

↑ +[O]

-ic ⟶ -ate

(Starting point)

↓ −[O]

-ous ⟶ -ite

↓ −[O]

hypo- -ous ⟶ hypo- -ite

FIGURE 11.13
A nomenclature "roadmap" for the oxoacids and their salts.

acid. Or by analogy with chloric acid, $HBrO_3$ is bromic acid. Adding one oxygen produces $HBrO_4$, perbromic acid, and subtracting oxygens produces $HBrO_2$, bromous acid, and HBrO, hypobromous acid.

The system also organizes the names of the corresponding anions (and therefore the salts) of the oxoacids. The anion is obtained by removing all the hydrogen ions from the acid. For example, when both the hydrogens are removed from carbonic acid, H_2CO_3, the result is the species CO_3^{2-}. As stated in the table, *-ic* acids become *-ate* anions and *-ous* acids become *-ite* anions. It follows that CO_3^{2-} must be the carbonate anion. Similarly, if we want a name for the NO_2^- anion, we start with the representative nitrogen-containing acid HNO_3, which is nitric acid. Removing one oxygen yields HNO_2, which is nitrous acid. Removing the hydrogen ion produces the NO_2^- ion, which is nitrite. We can even name compounds we have never seen before. For example, what would you call the IO_2^- ion? Start with the appropriate representative acid of the group which in this case is chloric acid. Then proceed to move toward the "mystery species" as shown below:

$$\underset{\text{Chloric acid}}{HClO_3} \longrightarrow \underset{\text{Iodic acid}}{HIO_3} \longrightarrow \underset{\text{Iodous acid}}{HIO_2} \longrightarrow \underset{\text{Iodite}}{IO_2^-}$$

There is one other point to make regarding nomenclature. What happens if not all the hydrogen ions are removed? For example, how about if we remove only one hydrogen from H_2CO_3 to produce HCO_3^-? The rule for naming anions indicates that we would call this either the hydrogen carbonate or the bicarbonate ion. $H_2PO_4^-$ is phosphoric acid with one hydrogen ion removed and two remaining, so it would be dihydrogen phosphate. HPO_4^{2-} would be hydrogen phosphate.

Hydroacids

To this point we have been considering acids that contain oxygen atoms. Lavoisier would have been pleased with the above section since he renamed Priestley's dephlogisticated air as oxygen, meaning "acid producer." But we know that there are other acids which do not contain oxygen. They are called the hydroacids. Table 11.9 shows some representative hydroacids along with the radii and electronegativity of the central atom. Bear in mind that the electronegativity of hydrogen is 2.1.

What should happen to the acid strength as we move horizontally from left to right? Our network (specifically, periodic properties based on effective nuclear charge) and reference to the table shows that the electronegativity of the central atom increases. This should cause the H—X bond to be more polar and more susceptible to attack by water. The size of the central atom does not change appreciably and is not a factor. Thus we would expect that hydrofluoric acid, HF, is a better acid than water, which it is.

What should happen to the acid strength as we move down in a given group? Here we have two effects to investigate. The electronegativity of the central atom decreases going down a group which would make the H—X bond less polar and less susceptible to attack by water. This would indicate a decreasing acid strength. However, the size of the central atom increases significantly going down a group. What effect would this have on acid strength? You may recall that the longer the bond length, the weaker the bond strength. Does this make sense? Yes, because as shown in Figure 11.14, as the bond length increases, the bonding pair of electrons is farther away from the effective nuclear charges of the two atoms and cannot hold them together as well. As the bond strength decreases down a group, the

TABLE 11.9
Acid strengths of common hydroacids

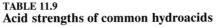

H—N(—H)—H $r_N = 0.75$ A $EN_N = 3.0$	O(—H)—H $r_O = 0.73$ A $EN_O = 3.5$	F—H $r_F = 0.72$ A $EN_F = 4.0$
H—P(—H)—H $r_P = 1.06$ A $EN_P = 2.1$	S(—H)—H $r_S = 1.02$ A $EN_S = 2.5$	Cl—H $r_{Cl} = 0.99$ A $EN_{Cl} = 3.0$
H—As(—H)—H $r_{As} = 1.20$ A $EN_{As} = 2.0$	Se(—H)—H $r_{Se} = 1.16$ A $EN_{Se} = 2.4$	Br—H $r_{Br} = 1.14$ A $EN_{Br} = 2.8$
		I—H $r_I = 1.33$ A $EN_I = 2.5$

Increasing acid strength (→, across rows); Increasing acid strength (↓, down columns)

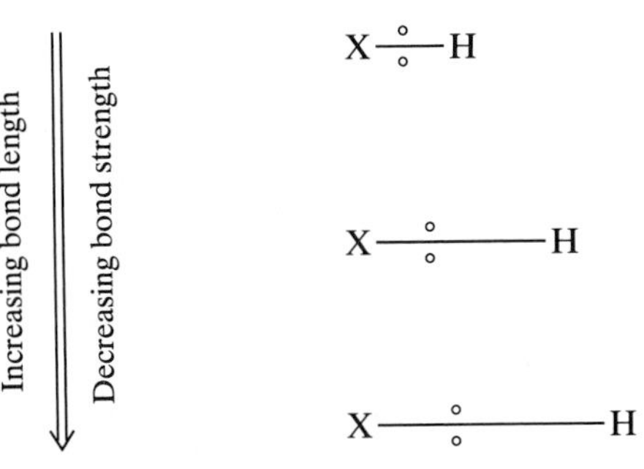

FIGURE 11.14
The increasing length of an H—X bond going down a group. As the bond length increases, bond strength decreases because the bonding pair is farther away from the effective nuclear charges of the two atoms.

bond would be easier to break, thus making for stronger acids. Which effect, decreasing electronegativity or increasing radius, is the dominant one? The answer turns out to be the latter. The decreasing bond strength which causes stronger acids going down the group is the dominant effect. Hydrochloric acid is a stronger acid than hydrofluoric, hydrosulfuric (H_2S) is a stronger acid than water, and so forth.

11.5 OZONE

An *allotrope* is a different molecular form of a given element. Diatomic oxygen, O_2, is the most familiar allotrope of the element oxygen, but ozone, O_3, is a second one of great importance. Ozone was discovered by Christian Schönbein in 1839 when he started to investigate the peculiar odor produced by electrical equipment in his laboratory. In fact, the word *ozone* comes from the Greek *ozein*,

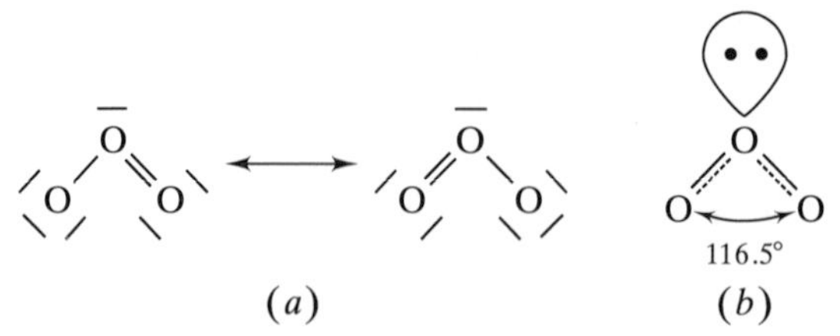

FIGURE 11.15
(*a*) The Lewis structure of the ozone molecule includes two resonance structures. (*b*) The geometry of the resonance hybrid as determined by the VSEPR theory. The O—O—O angle is less than 120° due to the lone pair on the central oxygen atom.

meaning "to smell." Ozone is a bluish, somewhat toxic gas with the pungent odor we can readily detect (in quantities as low as 0.01 part per million) near a subway train, after an electrical storm, or in the vicinity of electrical equipment. It is also a major component of photochemical smog.

The Lewis structure of ozone is shown in Figure 11.15*a*. Note that there are two resonance forms of the molecule. The VSEPR picture showing the geometry of the resonance hybrid, Figure 11.15*b*, has an O—O—O angle less than 120° due to the presence of the one lone pair.

Ozone is prepared by the action of an electrical discharge or ultraviolet (uv) light on ordinary diatomic oxygen as represented in Equation (11.14):

$$3O_2(g) \xrightarrow[\text{discharge}]{\text{electrical}} 2O_3(g) \qquad (11.14)$$

Since it is too reactive to be transported safely, it is usually generated as needed. Ozone is a powerful oxidizing agent, more powerful than ordinary diatomic oxygen. For example, mercury does not react with O_2 at an appreciable rate at room temperature but does with ozone as shown in Equation (11.15). In fact, this reaction is the basis for a common test for ozone because the presence of the mercury oxide causes the liquid metal to become dull in appearance and stick to the sides of glass tubing:

$$3Hg(l) + O_3(g) \longrightarrow 3HgO(s) \qquad (11.15)$$

In the United States, ozone is currently being considered as a replacement for chlorine gas, $Cl_2(g)$, for water purification and treatment. (It is already being used for such purposes in many European countries.) Ozone has two advantages over chlorine: (1) it does not react with various hydrocarbons to produce chlorinated products (such as chloroform, $CHCl_3$) that have been implicated as carcinogens, and (2) it avoids the unpleasant taste and smell of chlorinated water. Its principal disadvantage is cost. Ozone is also used as a bleach, deodorizer, and preservative in a variety of settings.

In 1913, ozone was discovered in the upper atmosphere. Today we hear many discussions about the "ozone layer," but this expression is somewhat misleading. The maximum concentration in the stratosphere (15 to 50 km above the earth) is only 12 parts per million at about 30 km. This hardly constitutes a "layer"! Although the concentration is low, it is of critical importance because both the formation and destruction of ozone in the stratosphere result in the absorption of large amounts of dangerous uv solar radiation. Ozone is formed when short-wavelength (240–300 nm) uv radiation strikes diatomic oxygen to produce single oxygen atoms as shown in Equation (11.16*a*). These reactive atoms then can combine with other molecules of diatomic oxygen to produce ozone as shown in Equation (11.16*b*). The destruction of ozone is also accomplished via the absorption of uv radiation, but this time of longer wavelength (200–360 nm) as shown in Equation (11.17).

$$\left.\begin{array}{rl} O_2(g) & \xrightarrow[\text{(240–300 nm)}]{\text{uv light}} 2O(g) \quad (11.16a) \\ O(g) + O_2(g) & \longrightarrow O_3(g) \quad (11.16b) \end{array}\right\} \text{Ozone formation}$$

$$O_3(g) \xrightarrow[\text{(200–360 nm)}]{\text{uv light}} O(g) + O_2(g) \quad \Big\} \text{ Ozone destruction} \qquad (11.17)$$

The oxygen atom generated here can produce still more ozone by repeating Equation (11.16*b*), or it can strike a second oxygen atom (or an ozone molecule) to reform the diatomic molecule as shown in Equation (11.18). Note that the overall effect of these various processes is (1) the absorption of uv radiation that otherwise would strike the surface of the earth and (2) no net change in stratospheric ozone concentrations.

$$O(g) + O(g) \longrightarrow O_2(g) \qquad (11.18a)$$

$$O(g) + O_3(g) \longrightarrow 2O_2(g) \qquad (11.18b)$$

A variety of compounds can destroy ozone. The identification and assessment of their net effect constitute an ongoing debate. The principal threats to stratospheric ozone involve (1) chlorofluorocarbons (CFCs) used as foaming agents, refrigerants, solvents, and propellants for aerosols and (2) compounds of bromine used in fumigants and fire extinquishers. (These compounds and their potential effects on stratospheric ozone concentrations are discussed in detail in Chapter 18 on the Group 7A elements.)

While we wish to preserve the ozone in the stratosphere, we want to minimize its production in the troposphere, that part of the atmosphere where we live. Photochemical smog produced as a result of the action of solar radiation on the effluents from automobiles is the primary source of ozone in the troposphere. The trigger for photochemical smog is based on nitrogen oxide chemistry, so we will defer our discussion of this problem until Chapter 16 on Group 5A chemistry.

11.6 THE GREENHOUSE EFFECT

Lavoisier established that combustion is the reaction of a substance with oxygen. Complete combustion of hydrocarbons as fuels produces carbon dioxide and water. Now we have long known that continued pollution of the air with various nitrogen and sulfur oxides, not to mention particulates like smoke and soot, could cause severe problems for future generations. Only recently, however, have we started to worry about carbon dioxide. In fact, the aim of many antipollution devices has been to convert hydrocarbon fuels completely to CO_2 (and water vapor). Now we have come to recognize that this gas may be one of the most dangerous air pollutants of all.

Why is carbon dioxide not as harmless as it once seemed? It and several other gases like ozone (O_3), water, methane (CH_4, from animal wastes, rice paddies, termite mounds, and other sources), nitrous oxide (N_2O, from chemical fertilizers and car emissions), and chlorofluorocarbons (CFCs) are responsible for the greenhouse effect. Of these gases, carbon dioxide is the biggest culprit, thought to be responsible for about 50 percent of the effect. The largest sources of CO_2 are coal- and oil-burning electric utilities, transportation, and industry in order of decreasing output.

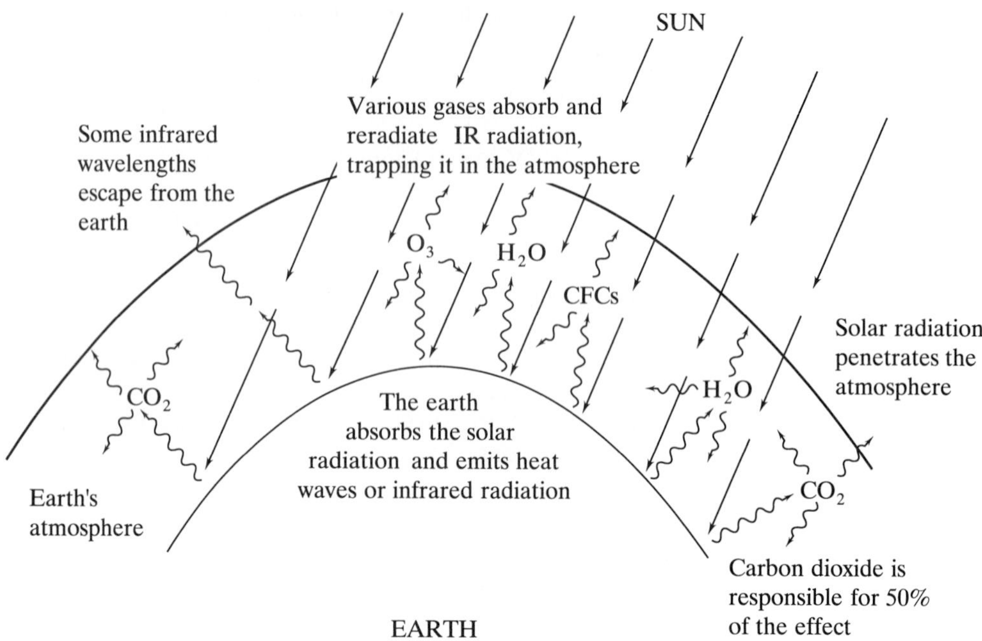

FIGURE 11.16
The greenhouse effect. Solar radiation (straight arrows) penetrates through the earth's atmosphere. The earth absorbs this radiation and reradiates infrared wavelengths (wavy arrows). Various gases absorb and reradiate the infrared light, effectively trapping most of it in the atmosphere. (CFCs indicate the chlorofluorocarbons.)

What is this greenhouse effect? Figure 11.16 demonstrates the idea. Solar radiation penetrates through the atmosphere and warms the earth which reradiates infrared radiation. We feel comfortably warm. Most of this infrared radiation, however, cannot get back out through the atmosphere because it is absorbed by the above gases. The heat (infrared radiation) is trapped in the earth's atmosphere in a process analogous to that which occurs in a greenhouse.

Of course the greenhouse effect works mostly to our benefit. Without it the earth would be 50 to 60° F cooler! However, if we continue to pour carbon dioxide into the atmosphere at the present rate, the mean temperature of the earth could increase by as much as 3 to 8° F over the next 50 years. The consequences could be dramatic. Droughts could be more common. There could be large shifts in weather patterns and climate zones. The polar ice caps could partially melt, causing an increase in sea level (perhaps 1 to 4 ft) which could result in the flooding of major coastal cities and lowland countries.

How can we minimize the amount of carbon dioxide released to the atmosphere? The most obvious action is to decrease the amount of fossil fuels, particularly coal and oil, we consume. The problem is that a drastic cut in their use would cause critical problems and perhaps throw the world economy into an enormous tailspin. One alternative is to reverse the trend away from nuclear power. The United States has backed off from pursuing the nuclear alternative in recent years. (All the plants ordered after 1974 were canceled or abandoned during construction.) Perhaps we will need a new generation of safer nuclear facilities to fill this gap. A second alternative is to accelerate the research and development of the hydrogen economy and nuclear fusion (see Section 10.6) as well as other energy sources such as solar and geothermal power.

Another way to help alleviate the problem is to reverse the massive deforestation which has occurred over the centuries and has accelerated in the last decade. Trees and other plants, as Priestley knew, absorb carbon dioxide through the process we know today as photosynthesis, Equation (11.1). But we have been cutting down trees at an alarming rate. Tropical rain forests are disappearing at a rate of about 25 million acres/year, and it is estimated that the total inventory of trees has been halved since the onset of agriculture.

Has the greenhouse effect taken hold yet? Some experts say that we have already released to the atmosphere enough excess CO_2 (over and above the natural levels already present) to cause a significant temperature increase. Of course, it may turn out that they are wrong. Perhaps there are other countermechanisms. For example, perhaps much larger amounts of carbon dioxide can be absorbed by the oceans as they gently heat up. Or perhaps the larger amounts of water vapor in the atmosphere will produce a heavier cloud cover which will reflect some solar radiation from the earth. Yet we would be foolish to leave such matters to chance. Deserts are already being produced at a rate of some 15 million acres/year. The amount of carbon dioxide in the atmosphere has risen dramatically. Some reliable evidence indicates that the earth is nearly 1° F warmer today on the average than it was 100 years ago (although this may be due to a natural variation).

We need information. We need chemists, physicists, biologists, geologists, and environmental scientists to keep a constant vigil on this potential problem. At this writing, no one can claim for sure that the greenhouse effect has started, but we must continue to monitor the environment and develop potential alternative energy sources. In the meantime, it may come down to a matter of population. Many would argue that we are putting more carbon dioxide into the atmosphere, using up our forests at alarming rates, and so forth, because there are so many people in the world. Humankind may ultimately have to realize that one way to get a handle on the greenhouse effect and its potential (if still unproven) role in global warming is to stem the tide of our increasing numbers. Perhaps it will come down to recognizing the well-known wisdom of Pogo, who said, "We have met the enemy and he is us."

SUMMARY

Oxygen was discovered by Karl Wilhelm Scheele and Joseph Priestley, but Antoine Lavoisier was the first to correctly describe its role in combustion. Industrially, oxygen is usually separated from air, while in the laboratory it is generated from potassium chlorate or hydrogen peroxide. Oxygen has a number of applications, most of them based on its role in combustion.

Water is the most important oxide. The bent structure of the individual molecule, the resulting large dipole moment, and the strong intermolecular hydrogen bonds in the solid and liquid phases give it unique properties as a solvent and as a medium for acids and bases.

Our network of ideas can be applied to oxides, which divide into metal ionic and nonmetal covalent types. Ionic oxides are basic anhydrides which produce metal hydroxides and hydroxide ions in aqueous solution. Nonmetal oxides are acidic anhydrides which produce oxoacids and hydronium ions in solution. These above correlations have become the sixth component of our network of ideas. The relative strengths of oxoacids and hydroacids can be rationalized by using other parts of the network. A systematic approach to the nomenclature of the oxoacids is based on the five representative *-ic* acids.

Ozone is an important allotrope of oxygen and absorbs uv radiation in the upper atmosphere. Various threats to the ozone "layer" will be detailed in later chapters. Carbon dioxide is the primary culprit in the greenhouse effect which threatens to warm the earth and produce a variety of dramatic effects.

PROBLEMS

11.1. Write an equation to represent the method by which Priestley generated oxygen gas. Note that he did not know it was a diatomic molecule. This was not finally established until the first International Chemical Congress in 1860.

11.2. Priestley noted that breathing air soon made it unsuitable for further use in supporting animal life but that this same air could be restored by growing green plants in it. Briefly explain what he observed; use an equation or two in the course of your explanation.

11.3. Priestley tested the "goodness" of air, or what we would call its oxygen content, using nitric oxide, NO, a colorless gas that combines with oxygen to form the reddish-brown gas nitrogen dioxide, NO_2. Why was it that "dephlogisticated air" consumed four or five times as much nitrogen dioxide as "common air"? Write an equation as part of your answer.

11.4. Even in Stahl's time, it was known that when metals rust they gain weight. Given what you know about the phlogiston theory, what conclusions could you draw about the mass of the element phlogiston?

11.5. The longest-lived radioactive isotope of oxygen is O 15 which decays via beta plus emission with a half-life of 124 s. Write a nuclear equation for this decay process.

11.6. Oxygen 15 is prepared by bombarding oxygen 16 with helium 3 particles accompanied by the emission of alpha particles. Write a nuclear equation for this process.

11.7. Using the percentage abundances given in Table 11.1, calculate an average atomic weight for oxygen. Compare your result with that listed on the periodic table.

11.8. Suppose that a sample of oxygen is 0.041% oxygen 17 and 0.196% oxygen 18 with the remainder being oxygen 16. Calculate an average atomic weight for oxygen for this sample.

11.9. Igneous rocks crystallized during the cooling of the primitive earth usually have elements in the lower possible oxidation states, while sedimentary rocks, which were laid down later and subject to weathering, have elements in higher oxidation states. Propose a rationale for these differences in oxidation states.

11.10. Briefly rationalize why the boiling point of diatomic nitrogen ($-195.8°$ C) is less than that of diatomic oxygen ($-183.0°$ C).

11.11. Write an equation for the complete combustion of pentane, C_5H_{12}.

11.12. Briefly, what is the difference between iron and steel? What function would oxygen serve in converting iron to steel? (*Hint:* You might want to consult a dictionary or encyclopedia to answer this question.)

11.13. Draw (i) Lewis structures, (ii) diagrams showing the molecular geometries (including approximate bond angles), and (iii) valence-bond theory diagrams showing the overlap of atomic and/or hybrid orbitals for the following molecules and ions:
(*a*) NH_3
(*b*) SO_3
(*c*) CO_2

(*d*) OF_2
(*e*) NO_2^-
(*f*) CS_2
In each case, state whether the species is nonpolar or polar and for those that are polar, show the net dipole moment on the diagram of the molecular geometry.

11.14. List the type of hybrid orbitals assumed by the central atom in each of the molecules and ions listed in Problem 11.13.

11.15. Explain in your own words why a lone pair of electrons on a central atom takes up more space than a bonding pair.

11.16. When describing how we arrive at a rationalization of water as a polar compound, we started with a statement that the electronegativity of oxygen is greater than that of hydrogen. Without resorting to a memorized trend of electronegativities, account for the difference in electronegativity between the two atoms.

* **11.17.** If you have already covered Chapter 7 on solid state structures, you should be able to comment on the relationship between the ice structure shown in Figure 11.3 and the wurtzite structure shown in Figure 7.22*d*.

11.18. What is the difference between *intra*molecular and *inter*molecular forces? Give an example of each. (*Hint:* An ordinary dictionary may be useful here.)

11.19. Which, if any, of the following molecules would be held together by hydrogen bonds:
(*a*) CH_4
(*b*) CHF_3
(*c*) CH_3OH (like water but a methyl group, $-CH_3$, replaces one of the hydrogen atoms)
(*d*) Glucose (see figure shown with Problem 11.29)
(*e*) CH_3COOH
Show a diagram in each case. (*Hint:* Be careful. Some of the hydrogen atoms in some of these compounds may participate in hydrogen bonding, while others in the same molecule may not.)

11.20. Explain, in your own words, the FONCl rule.

11.21. Using references to molecular structures, explain to an English major why ice floats. (Extra points may be given for documentation that you did this orally as well as in writing.)

11.22. With reference to Figure 11.5 showing the involvement of a given water molecule in the liquid, write a short paragraph describing what is meant by the phrase "flickering cluster model" as applied to the structure of liquid water.

11.23. With reference to Figure 11.6, why would the boiling points of, for example, the hydrides of Group 7A be expected to increase with molecular weight? Ignore the effects of hydrogen bonding in answering this question.

11.24. Explain why water has an unusually high heat of vaporization.

11.25. The solubilities of $CO_2(g)$, $SO_2(g)$, and $NH_3(g)$ are 88, 3937, and 70,000 liters of gas per 100 liters of water, respectively. Rationalize these relative values in terms of the structure and nature of the solutes.

11.26. Cavendish was able to collect $H_2(g)$ over water. Is this surprising or not? Be specific.

11.27. As noted in Problem 11.3, Priestley measured what he called the "goodness" of a newly prepared gas by reacting it with nitric oxide, NO, to form NO_2, a brown gas, which is readily soluble in water. Today we would represent the process as follows:

$$NO(g) + \tfrac{1}{2}O_2(g) \longrightarrow NO_2(g)$$

NO_2, nitrogen dioxide, is much more soluble in water than either $O_2(g)$ or $NO(g)$. Rationalize these solubilities in terms of molecular structure and the nature of water. Draw a molecular-level diagram (of the type shown in Figure 11.7) showing

NO_2 dissolved in water. Your diagram should include a small section of water structure.

11.28. Consider the entropy change which accompanies the dissolution of sodium chloride in water. Is this factor favorable or unfavorable to the dissolution process? Briefly discuss your answer.

11.29. Glucose, $C_6H_{12}O_6$ (MW = 180 u), has the structure shown below. The linear hydrocarbon $CH_3(CH_2)_{11}CH_3$ has a similar molecular weight (184). Which would be more soluble in water? Briefly explain your answer.

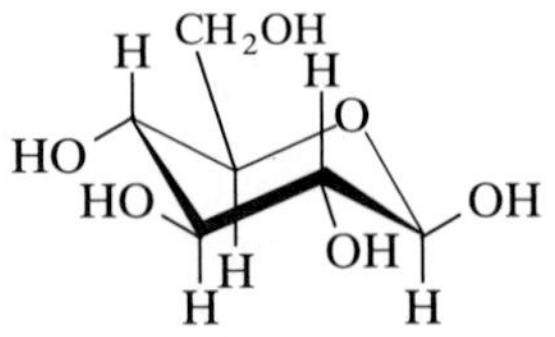

11.30. Draw diagrams of the following hydrated ions in water:
(*a*) $Co^{3+}(aq)$
(*b*) $SO_4^{2-}(aq)$
(*c*) $NH_4^+(aq)$
(*d*) NO_3^-

11.31. If K_w for the self-ionization of water is 1.0×10^{-14}, calculate the concentrations of the hydronium and hydroxide ions. Verify that the pH of water is 7.00.

11.32. The value of K_w for the self-ionization of water is 5.47×10^{-14} at a temperature of 50° C. Calculate the pH of water at this temperature.

11.33. Why is it that raising the concentration of hydronium ions in aqueous solution is equivalent to lowering the concentration of hydroxide ions? Assume a constant temperature.

11.34. The self-ionization of heavy water, D_2O, has an equilibrium constant of 1.4×10^{-15} at 25° C. Write an equation for the self-ionization. Supply a definition of pD such that it is equivalent to the pH defined for ordinary water. Calculate a value of pD for pure heavy water.

11.35. Account for the species $H_{11}O_5^+$ being isolated from solutions of strong acids.

11.36. Starting with the density of water as 1.00 g/mL, calculate the molarity of water in the pure substance.

11.37. Write the formulas for the oxides of potassium, gallium, arsenic, and selenium. Classify each of these as acidic, basic, or amphoteric.

11.38. Write the formula for the most stable oxides of aluminum and thallium.

11.39. Would you expect to find a compound S_3, equivalent to the ozone allotrope found in oxygen chemistry? Why or why not?

11.40. For each of the following oxides, write a two-part equation showing its reaction when placed in water. The equation should show (1) the corresponding hydroxide or oxoacid in molecular form and (2) the ionized products in aqueous solution:
(*a*) K_2O
(*b*) In_2O_3
(*c*) B_2O_3
(*d*) N_2O_5
(*e*) SO_3

11.41. Show a *reaction mechanism*, that is, the molecular-level details of how water attacks nitrogen trioxide, N_2O_3, whose structure is shown below, to produce nitrous acid in solution. Also show the aqueous ionization products of nitrous acid.

11.42. Show the mechanism (sequence of molecular events) of the reaction which occurs when carbon dioxide is placed in water. Show the final reaction products as aqueous bicarbonate and hydronium ions.

11.43. Show the mechanism (sequence of molecular events) of the reaction which occurs when sulfur dioxide is placed in water. Show the final reaction products as aqueous bisulfite and hydronium ions.

11.44. Show the mechanism (sequence of molecular events) of the reaction which occurs when strontium oxide is placed in water.

11.45. In Chapter 10 we discussed the reaction of ionic hydrides in water. A representative reaction is that of sodium hydride as shown below. Apply what you have learned about reactions that occur in water to explain how this reaction might proceed.

$$Na^+H^- + H_2O \longrightarrow Na^+ + H_2 + OH^-$$

11.46. In Chapter 10 we discussed why compounds in which hydrogen was bound to a relatively highly electronegative atom could be deuterated but hydrocarbons could not be. In terms of what you learned in this chapter, explain the above phenomenon.

11.47. Explain in your own words why the E–O–H unit produces hydronium ions if E is a nonmetal but hydroxide ions if E is a metal.

11.48. Explain why sulfuric acid, normally written H_2SO_4, is also found written as $SO_2(OH)_2$.

11.49. Identify the following as acidic, basic, or amphoteric:
(*a*) NO(OH)
(*b*) $Be(OH)_2$
(*c*) $Ti(OH)_4$
(*d*) $Si(OH)_4$

11.50. Given the following pairs of acids in aqueous solution, identify the stronger in each case. For each pair, briefly rationalize your answer:
(*a*) $HClO_3$, HIO_3
(*b*) H_3AsO_4, H_3SbO_4
(*c*) H_3PO_3, H_3AsO_3
(*d*) HSO_4^-, $HSeO_4^-$

11.51. Given the following pairs of acids in aqueous solution, identify the stronger in each case. For each pair, briefly rationalize your answer:
(*a*) HIO_3, HIO_4
(*b*) H_3PO_3, H_3PO_4
(*c*) $HSeO_4^-$, $HSeO_3^-$
(*d*) HClO, $HClO_2$

11.52. Given the following sets of acids in aqueous solution, identify the strongest in each case. For each set, briefly rationalize your answer:
(*a*) $HBrO_2$, HBrO, $HClO_2$
(*b*) H_2SO_3, H_2SeO_3, $HClO_4$
(*c*) H_3PO_4, H_3AsO_4, H_3AsO_3

11.53. Name the following oxoacids:
(*a*) HBrO
(*b*) HIO_4
(*c*) H_3AsO_3
(*d*) HNO_2

11.54. Give the formula for the following oxoacids:
(*a*) Iodous acid
(*b*) Perbromic acid
(*c*) Persulfuric acid

11.55. Name the following anions derived from oxoacids:
(*a*) HSO_3^-
(*b*) $H_2AsO_4^-$
(*c*) HPO_3^{2-}
(*d*) $HSeO_4^-$
(*e*) BrO^-

11.56. Give a formula for the following anions:
(*a*) Biselenite
(*b*) Bromite
(*c*) Dihydrogen arsenite
(*d*) Periodate

11.57. Which is the stronger acid, H_2O or H_2S? Briefly explain why.

11.58. Which is the stronger acid, H_2S or HCl? Briefly explain why.

11.59. Arrange the following in order of increasing acid strength: PH_3, H_2Se, HBr.

11.60. Criticize the following statement: Hydrofluoric acid is the strongest acid of all because fluorine is the most electronegative element.

11.61. Cite two articles from the media which show the current status of the ozone debate.

11.62. Cite two articles from the media which show the current status of the greenhouse effect.

CHAPTER 12

GROUP 1A: THE ALKALI METALS

Having established the first five components of our network in Chapter 9, discussed basic nuclear processes and the chemistry of hydrides in Chapter 10, and the acid-base character of oxides (and their corresponding hydroxides and oxoacids) in Chapter 11, we are now ready to start our tour of the eight groups of representative elements. Our aim is to discuss the history of the discovery of the elements, how the network can be applied to predict and rationalize the chemistry of the group, what special characteristics and practical applications these elements have, and to explore at least one special topic in depth for each group. In this chapter we discuss the alkali metals (lithium, sodium, potassium, rubidium, cesium, and francium), but we also are going to add one final component to the network, a knowledge of reduction potentials. The special topic in depth is liquid ammonia solutions of the Group 1A and 2A metals.

12.1 DISCOVERY AND ISOLATION OF THE ELEMENTS

In Figure 9.2 we saw a plot of the number of known elements versus time. In Figure 12.1, information about the discoveries of the alkali metals is superimposed on this same plot. Note that it took a little more than 130 years to complete Group 1A as we know it today.

Sodium and potassium were discovered and isolated by Humphry Davy, whom we might characterize today as a "wild and crazy guy." Not particularly good at schoolwork, Davy became a teenage apprentice in an apothecary where he was soon fired for carrying out work that led to chemical explosions! He also had the rather foolhardy habit of breathing the gases he produced in the laboratory. Once he reportedly took in 4 quarts (!) of hydrogen. At another time he experimented with nitrous oxide (laughing gas). You can imagine the fun (albeit totally inadvisable) he had with that! It is said that he became a chemist after reading

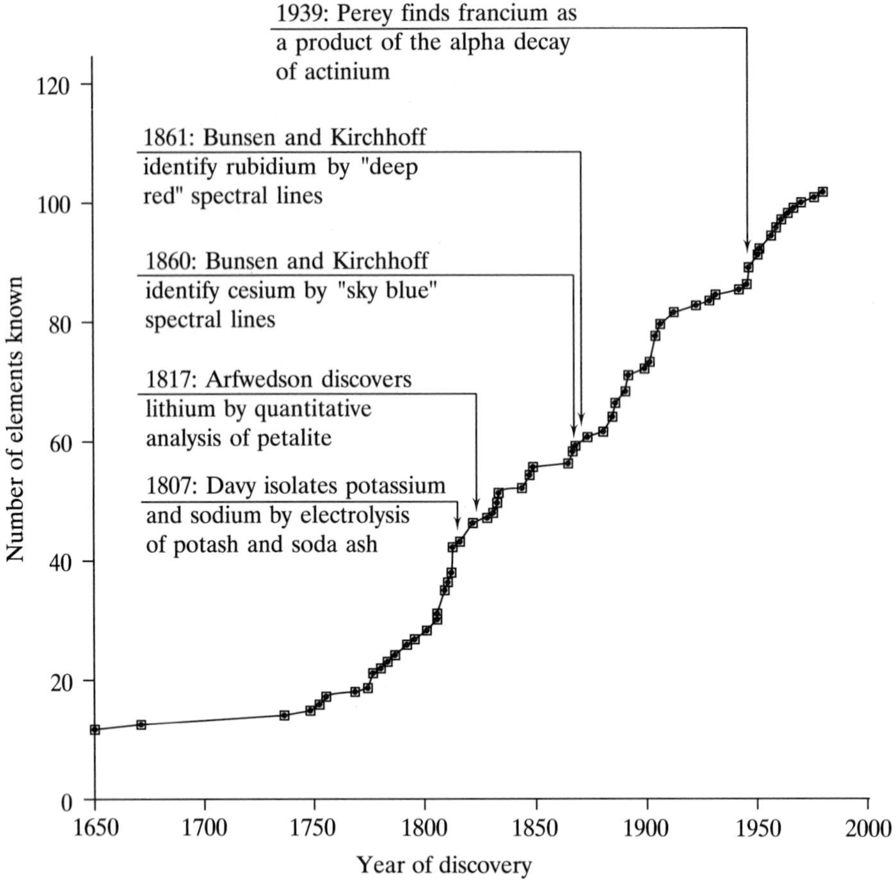

FIGURE 12.1
The discovery of the alkali metals superimposed on the plot of the number of known elements versus time.

Lavoisier's *Elements of Chemistry* in 1797. The same flair and daredevil attitude about chemistry made him an excellent lecturer. Add to this that he was an extremely handsome man with a gift for working audiences and we can appreciate why he attracted so many people to his lectures, including many of the young ladies of that day. In the end, however, Davy paid the price for unwise practices. He became an invalid at the age of 33 and died soon after his fiftieth birthday.

Recall from our earlier discussions about hydrogen, oxygen, and water that many scientists were investigating the effect of electricity on matter. Davy experimented with passing electrical currents through molten samples of potash and soda ash, what we know today to be potassium and sodium carbonates. At that time, however, the exact chemical nature of these "alkalis," literally substances derived from wood and other plant ashes, was unknown. When Davy applied his voltaic pile (battery) to molten potash in 1807, he produced at the negative electrode a soft, bright metal that immediately burst into flame. When this metal was placed in water, it danced excitedly around, producing a hissing sound and a beautiful lavender flame. It is said that upon seeing this Davy *himself* danced excitedly around the room! The analogous experiment with soda ash produced the corresponding sodium metal with albeit not quite as spectacular results.

Sodium and potassium are still produced today in similar ways. As shown in Equation (12.1), sodium is obtained from the electrolysis of molten sodium chloride to which some calcium chloride is added to lower the melting point. Although potassium can be prepared from the electrolysis of potassium chloride, it is more convenient to react sodium metal with potassium chloride at elevated temperatures and continuously remove the potassium as a gas, as shown in Equation (12.2). Both sodium and potassium are soft, highly reactive metals that are easily cut with a knife. They react with water as shown in Equation (12.3) to produce hydrogen gas which can ignite in a manner that Henry Cavendish would find familiar. (See the discussion of reduction potentials in Section 12.3 for further details on this reaction.) These metals must be stored under inert substances such as mineral oil or even kerosene to keep them away from the water of the atmosphere. As you might expect, they are never found as the free metals in nature.

$$2NaCl(l) \xrightarrow[600°C]{\text{with } CaCl_2(l)} 2Na(l) + Cl_2(g) \tag{12.1}$$

$$KCl(l) + Na(g) \xrightarrow{800°C} NaCl(s) + K(g) \tag{12.2}$$

$$2M^{I} + 2H_2O(l) \longrightarrow 2M^{+}(aq) + H_2(g) + 2OH^{-}(aq) \tag{12.3}$$

Young chemists take heart! Lithium was discovered by Johan Arfwedson, a Swede, while performing only his second mineralogical analysis. In 1817, at the age of 25, Arfwedson was an assistant working in the laboratory of Jöns Jakob Berzelius, one of the most influential chemists of the first half of the nineteenth century and the inventor of the modern symbols of the elements, when he started to quantitatively analyze the mineral petalite, which had been discovered by a Brazilian metallurgist a few years earlier. Although Arfwedson could not isolate the metal itself, which he named for the Greek word *lithos*, meaning "stone," he did make a half dozen of its salts. He noted a close similarity between these salts and those of magnesium. Davy was able to isolate a small amount of lithium from lithium carbonate by electrolysis, but today lithium chloride is used. Lithium is much less reactive with water than either sodium or potassium.

Rubidium and cesium, as shown in Figure 12.1, were discovered by the German chemists Robert Bunsen and Gustav Kirchhoff in the early 1860s. Prior to this time, flame tests were routinely used to identify elements. For example, sodium salts characteristically gave yellow flames, while potassium salts produced lavender. Bunsen had earlier invented his famous burner which was useful for these experiments because it did not put out any color of its own and therefore did not interfere with these flame tests. Kirchhoff, some 13 years younger than Bunsen but the source of several key insights in the collaboration, suggested that the light from these tests be passed through a prism. This was the first Bunsen-Kirchhoff spectroscope that subsequently played a role in the discovery of a number of different elements. (Tl, In, Ga, He, Yb, Ho, Tm, Sm, Nd, Pr, and Lu are some others.)

Cesium was identified in 1860 in a salt isolated from a mineral water. It was characterized by several sky blue spectral lines which at that time had not been identified in any other substance. Its name is derived from the Latin *caesius*, meaning "bluish gray," and the word used, Bunsen and Kirchhoff said, to "designate the blue of the upper part of the firmament." The metal itself was not isolated until some 20 years later. It is one of few room-temperature liquid metals along with mercury and gallium. Rubidium was identified a year later than cesium in the mineral lepidolite. It gave two dark-red lines and was named for the Latin *rubidus*, meaning "deepest red." In this case, Bunsen and Kirchhoff were able to isolate the element by electrolysis. Kirchhoff, incidentally, went on to identify six elements in the sun with the aid of the newly developed spectroscope.

There were many chemical attempts to isolate the heaviest alkali metal, which was referred to by the mendeleevian term eka-cesium for many years. It was found using radiochemical techniques by Marguerite Perey of France in 1939. Working at the Curie Institute in Paris, she found that about 1 percent of actinium 227 decays via alpha emission, as shown in Equation (12.4) to produce what she named francium, after her homeland:

$$^{227}_{89}\text{Ac} \xrightarrow{\alpha} {}^{223}_{87}\text{Fr} + {}^{4}_{2}\text{He} \qquad (12.4)$$

Francium 223, with a half-life of 22 min, is the longest-lived isotope of the element. Since then it has been found to occur naturally, but there is probably less than an ounce of the element in the earth's crust at any given time. Francium, by extrapolation from the melting points of the other alkali metals, would most likely be a liquid if enough of it could be isolated. So far no weighable quantities of the element have been produced.

12.2 FUNDAMENTAL PROPERTIES AND THE NETWORK

Figure 12.2 shows the alkali metals superimposed on the network of interconnected ideas as we left it at the end of the last chapter. Table 12.1 shows a number of the properties of the Group 1A elements. A quick reference to the table shows that these elements illustrate many classic and expected variations in periodic properties.

Note first of all that the radii, ionization energies, electron affinities, and electronegativities of the group vary as we expect based on effective nuclear charge and the size of the valence orbitals occupied. There is a slight irregularity in the above properties at rubidium, the first element after the first row of transition metals. For example, note that the ionization energy decreases less on going from potassium to rubidium as compared to any other pair of elements in the group. We discussed this briefly in Chapter 9 (p. 231) and concluded that the *nd* and *nf* electrons do not shield each other or succeeding electrons from the nucleus quite as well as expected.

Hydrides, Oxides, Hydroxides, and Halides

As forecast in the earlier chapters on hydrogen and oxygen, we are now in a position to look at the chemistry of the hydrides, oxides, and hydroxides of this and other groups using the network. For example, from our discussions about relative effective nuclear charges, ionization energies, electron affinities, and electronegativities, we expect both the hydrides and the oxides of the alkali metals to be ionic in nature. The oxides, not all of which are produced by direct reaction with diatomic oxygen (see Table 12.1 and the discussions of peroxides and superoxides in Section 12.4), are basic anhydrides as expected. The hydroxides, characterized by M–O–H units, are bases because the M—O bond is more polar than the O—H bond and therefore more susceptible to attack by polar water molecules. As the metals become less electronegative down the group, the M—O bonds become even more polar, and the hydroxides more basic going down the group.

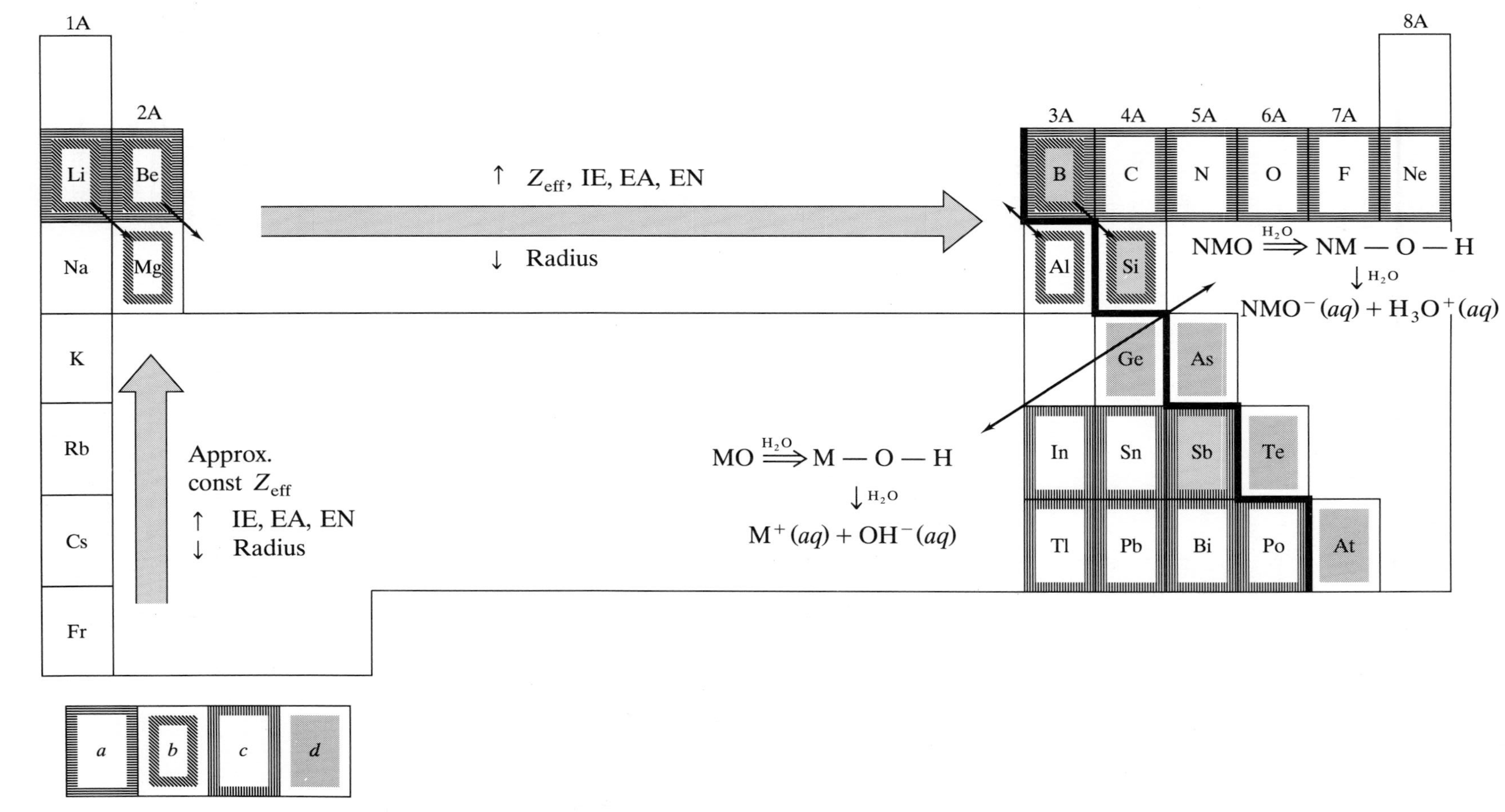

FIGURE 12.2
The alkali metals superimposed on the interconnected network of ideas including the trends in periodic properties, the acid-base character of metal and nonmetal oxides, (*a*) the uniqueness principle, (*b*) the diagonal effect, (*c*) the inert-pair effect, and (*d*) the metal-nonmetal line.

TABLE 12.1
The fundamental properties of the Group 1A elements. [Ref. 24.]

	Lithium	Sodium	Potassium	Rubidium	Cesium	Francium
Symbol	Li	Na	K	Rb	Cs	Fr
Atomic number	3	11	19	37	55	87
Natural isotopes, A/% abundance	6/7.42 7/92.58	23/100	39/93.1 40/0.0118 41/6.88	85/72.15 87/27.85	133/100	223/100
Total no. of isotopes	5	7	9	17	21	20
Atomic weight	6.941	22.99	39.10	85.47	132.9	(223)
Valence electrons	$2s^1$	$3s^1$	$4s^1$	$5s^1$	$6s^1$	$7s^1$
mp/bp, °C	186/1326	97.5/889	63.65/774	38.89/688	28.5/690	27/677*
Density, g/cm^3	0.534	0.971	0.862	1.53	1.87	
Atomic radius, (metallic), Å	1.57	1.91	2.35	2.50	2.72	
Ionic radius, Shannon-Prewitt, Å (C.N.)	0.73(4)	1.13(4)	1.51(4)	1.66(6)	1.81(6)	
Pauling EN	1.0	0.9	0.8	0.8	0.7	0.7
Charge density (charge/ionic radius), unit charge/Å	1.4	0.88	0.66	0.60	0.55	
$E°$,† V	−3.05	−2.71	−2.92	−2.93	−2.92	
Oxidation states	+1	+1	+1	+1	+1	
Ionization energy, kJ/mol	520	496	419	403	376	
Electron affinity, kJ/mol	−58	−53	−48	−47	−45	
Discovered by/date	Arfwedson 1817	Davy 1807	Davy 1807	Bunsen-Kirchhoff 1861	Bunsen-Kirchhoff 1860	Perey 1939
rpw‡ O_2	Li_2O	Na_2O Na_2O_2	K_2O_2 KO_2	RbO_2	CsO_2	
Acid-base character of oxide	Base	Base	Base	Base	Base	
rpw N_2	Li_3N	None	None	None	None	
rpw halogens	LiX	NaX	KX	RbX	CsX	
rpw hydrogen	LiH	NaH	KH	RbH	CsH	
Crystal structure	bcc	bcc	bcc	bcc	bcc	

*Given by *CRC Handbook of Chemistry and Physics* even though no weighable quantities of the element have been isolated.
†$E°$ is the standard reduction potential, $M^+(aq) \rightarrow M(s)$.
‡rpw = reaction product with.

Reactions between the hydroxides (MOH) and the appropriate hydrohalic acids (HX) readily yield the alkali-metal halides, MX. As you may recall, halides are characterized by a -1 oxidation state corresponding to the addition of one electron to produce a filled valence shell. Accordingly, the formulas of the various halides give a strong indication of the predominant valences encountered in a given group. The alkali-metal halides all show a 1 : 1 metal-to-halide ratio consistent with the almost universal +1 oxidation state of these metals. This is consistent with the relatively low values of first ionization energies of the group corresponding to the removal of the ns^1 electrons. The second and subsequent ionization energies (not shown in Table 12.1) are much larger because the electrons of the remaining filled shells all experience much higher effective nuclear charges.

So there is much about alkali-metal chemistry that we expect from our earlier discussions of periodic properties, the metal-nonmetal line, and the acid-base character of oxides. What is left to discuss about these elements? What about the uniqueness principle, the diagonal effect, and the inert-pair effect? The latter does not apply here because we do not yet have a pair of electrons in the *ns* subshell, but the uniqueness principle and diagonal effect do apply. We take these up together in the next section. What else? What about the reduction potentials listed in Table 12.1? We need to discuss (and/or review) this concept, and, in fact, a knowledge of reduction potentials will become the seventh and last component of the network. Finally, what about those somewhat strange formulas for the products of the reactions of the alkali metals with oxygen? These are unexpected on the basis of our network, so they will need to be discussed in some detail. We start with the uniqueness principle and diagonal effect.

Application of the Uniqueness Principle and Diagonal Effect

In Chapter 9 in the section concerning the uniqueness principle (pp. 225ff), we discussed why lithium compounds are expected to be somewhat less soluble in water and other polar solvents than we might ordinarily anticipate. The small lithium cation has a strong ability to polarize a wide variety of anions, and the resulting bonds are of greater covalent character than we might otherwise expect. Since "like dissolves like" (see Chapter 11, p. 273), these somewhat nonpolar lithium compounds have a stronger tendency to dissolve in relatively nonpolar solvents like alcohols. We are not surprised, then, when we find that various lithium salts such as the hydroxide, fluoride, carbonate, and phosphate are all less soluble in water and more soluble in nonpolar solvents than their corresponding sodium salts.

On a related matter, you may have already noticed in Table 12.1 that lithium is the only alkali metal to form a nitride. This further highlights the uniqueness of lithium, but it is not at all clear why lithium should be special in this regard. Presumably (be careful, this word usually foreshadows "hand waving") the high polarizing power of the lithium cation results in a greater degree of covalent character in the predominantly ionic nitride than for the other alkali metals. This may also be related to the enhanced lattice energy and therefore a greater overall stability for this compound due to its covalent character.

The diagonal effect (Chapter 9, pp. 229ff) is based upon the similar ionic radii, charge densities, and electronegativities of the Li–Mg, Be–Al, and B–Si pairs of elements. (Remember that group similarities are still the name of the game, that is, the primary way to organize your thoughts about the periodic table; the diagonal effect is a secondary effect.) Specifically we concern ourselves here with the similarities between lithium and magnesium.

As noted in Section 12.1, the similarity between lithium and magnesium salts was cataloged by Arfwedson himself. As the full mineralogy of lithium has been investigated over the years, its association with magnesium has become even more obvious. The lower solubility of lithium salts in water is paralleled in the corresponding magnesium salts. Both metals form the nitride as well as the carbide by direct reaction with the element. Both form normal oxides rather than the more esoteric peroxides and superoxides found with the heavier alkali metals. The stabilities of their salts when heated, that is, their *thermal stabilities*, are similar. For example, both lithium and magnesium nitrates decompose to the oxide and

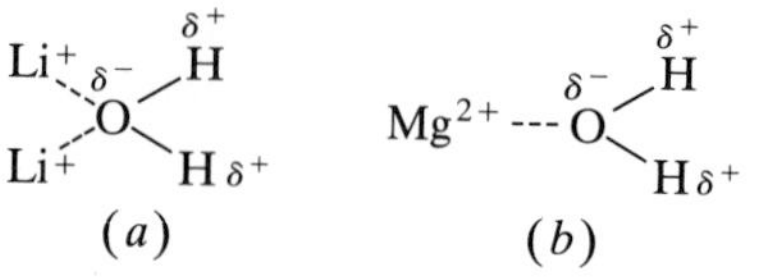

FIGURE 12.3
Lithium and magnesium ions interact with water molecules. (*a*) In $LiClO_4 \cdot 3H_2O$, two small, highly polarizing lithium cations interact with each of the three water molecules. (*b*) In $Mg(ClO_4)_2 \cdot 6H_2O$, one similarly small, highly polarizing magnesium cation interacts with the six water molecules.

dinitrogen tetroxide as shown in Equation (12.5), while both carbonates decompose to the oxide and carbon dioxide as shown in Equation (12.6). Both lithium and magnesium salts are usually highly hydrated, while sodium salts are not:

$$2LiNO_3 \longrightarrow Li_2O + N_2O_4(g) + \tfrac{1}{2}O_2(g) \qquad (12.5a)$$

$$2Mg(NO_3)_2 \longrightarrow 2MgO + 2N_2O_4(g) + O_2(g) \qquad (12.5b)$$

$$Li_2CO_3 \longrightarrow Li_2O + CO_2(g) \qquad (12.6a)$$

$$MgCO_3 \longrightarrow MgO + CO_2(g) \qquad (12.6b)$$

What is the reason for these similarities? Primarily it is a size effect. The lithium and magnesium ions (ionic radii = 0.73 and 0.86 Å, respectively) are of appropriate size to fit into a lattice formed by oxide ions (ionic radius = 1.26 Å) such that the attraction of cations and anions is maximized and the repulsions between the larger anions is minimized. (See Chapter 8 for more details on solid state energetics.) The carbonate and nitrate anions (radii = 1.78 and 1.79 Å, respectively) are just large enough that they get in the way of each other, and their mutual repulsions become more of a dominant factor. Accordingly, the carbonates and nitrates decompose to the more energetically favorable oxides.

The tendency toward enhanced hydration for lithium and magnesium salts is based on the same type of explanation. These small, highly polarizing ions form a stronger interaction with water molecules such that the latter become part of the lattice of the solid. Consequently, lithium perchlorate is found as the trihydrate, $LiClO_4 \cdot 3H_2O$, in which each of the three water molecules is associated with two lithium cations as shown in Figure 12.3*a*. Similarly, magnesium perchlorate is found as the hexahydrate, $Mg(ClO_4)_2 \cdot 6H_2O$, which is structurally similar to the lithium salt but with one Mg^{2+} replacing the two Li^+ cations as shown in Figure 12.3*b*. These strong ion-water interactions make lithium salts among the most *hygroscopic* (having the tendency to absorb water from the air) salts known. For example, both LiCl and LiBr are used in dehumidifiers for just this reason. We will also see in the next section that this exceptionally strong lithium-water interaction is partially responsible for the trends of reduction potentials within the group.

12.3 REDUCTION POTENTIALS AND THE NETWORK

Before we discuss the reduction potentials of the alkali metals, we should recall some definitions related to oxidation-reduction, or redox, reactions. (This will be but a sparse review. You may need to consult your introductory textbook or lecture notes for further clarification.) Recall that when a substance is reduced, its oxidation state (or the oxidation state of some constituent element) is decreased. When a substance is oxidized, the oxidation state increases. Reduction is associated with a gain of electrons, while oxidation corresponds to a loss. (Some students keep these definitions straight by using the mnemonic or memory device "LEO goes GER," which stands for *L*oses *E*lectrons *O*xidized, *G*ains *E*lectrons *R*e-duced.)

Another set of definitions used to describe redox reactions are those for the terms oxidizing and reducing agents. A *reducing agent* causes another substance to be reduced and supplies the electrons to that substance. Since the reducing agent *loses electrons* to the substance it reduces, it must be *oxidized* (LEO). Similarly, an *oxidizing agent* causes another substance to be oxidized and absorbs the electrons from that substance. Since the oxidizing agent *gains electrons* from the substance it oxidizes, it must be *reduced* (GER). (We can add this information to the mnemonic and say LEORA goes GEROA, which translates to *L*oses *E*lectrons *O*xidized, *R*educing *A*gent and *G*ains *E*lectrons *R*educed, *O*xidizing *A*gent.)

As an example of the use of these terms, consider the redox reaction represented in Equation (12.7):

$$Zn(s) + Cu^{2+} \longrightarrow Zn^{2+}(aq) + Cu(s) \qquad (12.7)$$

In this reaction, zinc metal is oxidized to Zn(II), while the Cu(II) ions are reduced to metallic copper. Additionally, we refer to the zinc metal as the reducing agent and copper(II) as the oxidizing agent.

Suppose you wanted to measure the tendency of the above reaction to occur. What would you do? Although it is not crucial to the arguments presented here, you may recall from earlier experiences that an electrochemical cell could be set up with a zinc electrode and a copper electrode connected to each other by a wire and a salt bridge. When the cell is completely assembled, the difference in voltage (E) between the two electrodes could be measured and then related to the tendency of the reaction to occur.

To analyze and tabulate the results of such an experiment, the overall oxidation-reduction equation is often separated into what are known as half-equations or half-reactions, one representing the reduction part and the other the oxidation part. For the $Zn\text{–}Cu^{2+}$ reaction, Equation (12.8*a*) shows the oxidation half-equation and (12.8*b*) the reduction half-equation. Notice that the zinc metal releases or loses two electrons (and so is oxidized) and that these electrons are, in turn, transferred to the Cu^{2+} ion which is reduced. Note also that these two half-equations, as shown, can be added together again to yield Equation (12.7), the overall equation for the reaction.

$$\text{Oxidation:} \qquad Zn(s) \longrightarrow Zn^{2+}(aq) + 2e^- \qquad (12.8a)$$

$$\text{Reduction:} \qquad Cu^{2+}(aq) + 2e^- \longrightarrow Cu(s) \qquad (12.8b)$$

$$\cancel{2e^-} + Zn(s) + Cu^{2+} \longrightarrow Zn^{2+}(aq) + Cu(s) + \cancel{2e^-} \qquad (12.7)$$

(The two electrons on either side of the overall equation cancel each other out because the number of electrons lost by the zinc would be equal to the number gained by the Cu^{2+}. You may recall that getting the number of electrons lost to be equal to the number gained is at the very heart of various methods of balancing redox equations.)

With the above as a brief review of redox reactions, we can now turn to a discussion of the standard reduction potentials of the alkali metals, which are listed in Table 12.1. Specifically, we want to know what information they can provide and how such information can be put to use to better understand the characteristics of not only the alkali metals but other groups of the periodic table as well. Take lithium as an example. The half-equation for the reduction of aqueous lithium ions to lithium metal is shown in Equation (12.9):

$$Li^+(aq) + e^- \longrightarrow Li(s) \tag{12.9}$$

Now we would like to *compare* the tendencies of lithium and the other alkali metals to be reduced. To do this systematically, the reduction potentials must be measured under certain *standard state conditions*. We need not concern ourselves with the details of standard states; it is enough to note that as a first approximation, the standard state for an aqueous solution specifies that all solutes are at a concentration of 1 molar (M) and all gases are at 1 atm of pressure. Under these conditions we can refer to the *standard reduction potential* as the measure of the tendency of a substance to be reduced under standard conditions. The symbol for this is $E°$, where the degree sign specifies the standard conditions.

Now it would be very convenient if we could independently measure the voltage associated with Equation (12.9) or any other individual half-reaction, but it is not possible to measure such absolute voltage differences. Why not? Because Equation (12.9) is only a *half*-reaction and cannot occur on its own. Free electrons cannot be dumped into a beaker and subsequently combine with the lithium ions. The electrons must come from a substance which has lost its electrons, that is, has been oxidized.

To assign a value to the $E°$ for a given half-equation, experiments are set up and/or calculations carried out so that the particular half-reaction under study is paired with a so-called standard electrode, usually taken to be the *standard hydrogen electrode* which has the half-reaction shown in Equation (12.10):

$$2H^+(aq) + 2e^- \longrightarrow H_2(g) \tag{12.10}$$

The tendency for this reaction to occur is *arbitrarily* assumed to be zero and becomes the standard by which the voltages of all other half-reactions are measured. In other words, we take the standard reduction potential ($E°$) for the standard hydrogen electrode to be exactly 0.000 V, and the $E°$ for all other half-reactions are therefore known relative to this standard but arbitrary value.

Now the standard reduction potential given for the reduction of aqueous lithium ions to lithium metal is given in Table 12.1 to be -3.05 V. What does this mean? Does it mean that lithium has a greater or lesser tendency to be reduced relative to hydrogen ions of the standard hydrogen electrode? The answer is a lesser tendency. In general, if $E°$ is less than zero, the substance has less of a tendency to be reduced than hydrogen, while if $E°$ is greater than zero, it has more of a tendency to be reduced. Such a table is called the *electromotive series*, and a reasonably complete version of it is given in Table 12.2. ($E°$ is sometimes referred to as the *electromotive force*, a term relating to the force or the tendency to move electrons.) Note that lithium, specifically the lithium aqueous ion, has the least tendency of all the substances listed to be reduced; that is, it has the least tendency to gain electrons. Conversely, it follows that of all the substances listed, lithium metal has the greatest tendency to lose electrons and is the best reducing agent in the table.

Although you have likely seen these concepts before and keeping in mind that we want to stick pretty close to inorganic chemistry here and not get overly involved with the theory behind electromotive forces, we should briefly relate the reduction potential to a more familiar measure of the tendency of a reaction to

TABLE 12.2
Standard reduction potentials at 25° C

Half-reaction	$E°$, V
$Li^+(aq) + e^- \rightarrow Li(s)$	−3.05
$K^+(aq) + e^- \rightarrow K(s)$	−2.93
$Ba^{2+}(aq) + 2e^- \rightarrow Ba(s)$	−2.90
$Sr^{2+}(aq) + 2e^- \rightarrow Sr(s)$	−2.89
$Ca^{2+}(aq) + 2e^- \rightarrow Ca(s)$	−2.87
$Na^+(aq) + e^- \rightarrow Na(s)$	−2.71
$Mg^{2+}(aq) + 2e^- \rightarrow Mg(s)$	−2.37
$Be^{2+}(aq) + 2e^- \rightarrow Be(s)$	−1.85
$Al^{3+}(aq) + 3e^- \rightarrow Al(s)$	−1.66
$Mn^{2+}(aq) + 2e^- \rightarrow Mn(s)$	−1.18
$N_2(g) + 4H_2O(l) + 4e^- \rightarrow N_2H_4(g) + 4OH^-(aq)$	−1.16
$2H_2O(l) + 2e^- \rightarrow H_2(g) + 2OH^-(aq)$	−0.83
$Fe^{2+}(aq) + 2e^- \rightarrow Fe(s)$	−0.44
$Tl^+(aq) + e^- \rightarrow Tl(s)$	−0.33
$PbSO_4(s) + 2e^- \rightarrow Pb(s) + SO_4^{2-}(aq)$	−0.31
$Sn^{2+}(aq) + 2e^- \rightarrow Sn(s)$	−0.14
$Pb^{2+}(aq) + 2e^- \rightarrow Pb(s)$	−0.13
$2H^+(aq) + e^- \rightarrow H_2(g)$	0.00
$S_4O_6^{2-}(aq) + 2e^- \rightarrow 2S_2O_3^{2-}(aq)$	+0.08
$Sn^{4+}(aq) + 2e^- \rightarrow Sn^{2+}(aq)$	+0.13
$S(s) + 2H^+(aq) + 2e^- \rightarrow H_2S(g)$	+0.14
$SO_4^{2-}(aq) + 4H^+(aq) + 2e^- \rightarrow SO_2(g) + 2H_2O(l)$	+0.20
$Cu^{2+}(aq) + 2e^- \rightarrow Cu(s)$	+0.34
$SO_4^{2-}(aq) + 8H^+(aq) + 6e^- \rightarrow S(s) + 4H_2O(l)$	+0.37
$O_2(g) + 2H_2O(l) + 4e^- \rightarrow 4OH^-(aq)$	+0.40
$CO(g) + 2H^+(aq) + 2e^- \rightarrow C(s) + H_2O(l)$	+0.52
$I_2(s) + 2e^- \rightarrow 2I^-(aq)$	+0.54
$MnO_4^-(aq) + 2H_2O(l) + 3e^- \rightarrow MnO_2(s) + 4OH^-(aq)$	+0.59
$O_2(g) + 2H^+(aq) + 2e^- \rightarrow H_2O_2(aq)$	+0.68
$Fe^{3+}(aq) + e^- \rightarrow Fe^{2+}(aq)$	+0.77
$Ag^+(aq) + e^- \rightarrow Ag(s)$	+0.80
$NO_3^-(aq) + 2H^+(aq) + e^- \rightarrow N_2O_4(g) + H_2O(l)$	+0.80
$OCl^-(aq) + H_2O(l) + 2e^- \rightarrow Cl^-(aq) + 2OH^-(aq)$	+0.89
$NO_3^-(aq) + 4H^+(aq) + 3e^- \rightarrow NO(g) + 2H_2O(l)$	+0.96
$Br_2(l) + 2e^- \rightarrow 2Br^-(aq)$	+1.07
$O_2(g) + 4H^+(aq) + 4e^- \rightarrow 2H_2O(l)$	+1.23
$MnO_2(s) + 4H^+(aq) + 2e^- \rightarrow Mn^{2+}(aq) + 2H_2O(l)$	+1.23
$Cr_2O_7^{2-}(aq) + 14H^+(aq) + 6e^- \rightarrow 2Cr^{3+}(aq) + 7H_2O(l)$	+1.33
$Cl_2(g) + 2e^- \rightarrow 2Cl^-(aq)$	+1.36
$PbO_2(s) + 4H^+(aq) + 2e^- \rightarrow Pb^{2+}(aq) + 2H_2O(l)$	+1.46
$MnO_4^-(aq) + 8H^+(aq) + 5e^- \rightarrow Mn^{2+}(aq) + 4H_2O(l)$	+1.51
$Ce^{4+}(aq) + e^- \rightarrow Ce^{3+}(aq)$	+1.61
$2HOCl(aq) + 2H^+(aq) + 2e^- \rightarrow Cl_2(g) + 2H_2O(l)$	+1.63
$PbO_2(s) + 4H^+(aq) + SO_4^{2-}(aq) + 2e^- \rightarrow PbSO_4(s) + 2H_2O(l)$	+1.70
$BrO_4^-(aq) + 2H^+(aq) + 2e^- \rightarrow BrO_3^-(aq) + H_2O(l)$	+1.74
$H_2O_2(aq) + 2H^+(aq) + 2e^- \rightarrow 2H_2O(l)$	+1.77
$N_2O(g) + 2H^+(aq) + 2e^- \rightarrow N_2(g) + H_2O(l)$	+1.77
$Co^{3+}(aq) + e^- \rightarrow Co^{2+}(aq)$	+1.82
$S_2O_8^{2-}(aq) + 2e^- \rightarrow 2SO_4^{2-}(aq)$	+2.01
$O_3(g) + 2H^+(aq) + 2e^- \rightarrow O_2(g) + H_2O(l)$	+2.07
$XeO_3(aq) + 6H^+(aq) + 6e^- \rightarrow Xe(g) + 3H_2O(l)$	+2.10
$XeF_2(aq) + 2H^+(aq) + 2e^- \rightarrow Xe(g) + 2HF(aq)$	+2.64
$F_2(g) + 2e^- \rightarrow 2F^-(aq)$	+2.87

Increasing strength as oxidizing agent ↓

Increasing strength as reducing agent ↑

Free-energy change	Spontaneity of reaction	Reduction potential
$\Delta G^\circ > 0$	Nonspontaneous reaction as written; spontaneous in reverse direction	$E^\circ < 0$
$\Delta G^\circ = 0$	Reaction at equilibrium, not spontaneous in either direction	$E^\circ = 0$
$\Delta G^\circ < 0$	Spontaneous reaction as written	$E^\circ > 0$

$\Delta G^\circ = -nFE^\circ$

FIGURE 12.4
The connections among reduction potentials, free-energy change, and the spontaneity of a reaction.

occur. This, you should recall, is the change in the free energy of the system ΔG° (where, again, the degree sign specifies a value under standard state conditions). If ΔG° is less than zero, a reaction is spontaneous as written. If ΔG° is greater than zero, the reaction is not spontaneous as written but is spontaneous in the reverse direction. If ΔG° is zero, the reaction is at equilibrium and no net reaction will occur spontaneously. The relationship between the standard reduction potential E° and ΔG° is given in Equation (12.11). The connections among reduction potentials, free-energy change, and spontaneity are summarized in Figure 12.4.

$$\Delta G^\circ = -nFE^\circ \tag{12.11}$$

where ΔG° = change in free energy of reaction, kJ, calculated on a per mole basis
n = number of electrons transferred
F = Faraday constant, 96.5 kJ/V
E° = standard reduction potential, V

Taking all this a step further, we can use half-reactions and the reduction potentials assigned to them to figure out what will happen when various oxidizing and reducing agents are put together. For example, what happens if freshly cut sodium metal is added to water? We start by analyzing the possibilities. The reduction potential of sodium is given in Table 12.1 to be −2.71 V. Writing this out in equation form we see that the reduction of aqueous sodium ions to sodium metal, as shown in Equation (12.12), has a negative E° and a positive ΔG° and therefore does not occur spontaneously:

$$Na^+(aq) + e^- \longrightarrow Na(s) \qquad E^\circ = -2.71\ \text{V} \tag{12.12}$$

$$\Delta G^\circ = -1(96.5\ \text{kJ/V})(-2.71\ \text{V})$$

$$= +262\ \text{kJ/mol}$$

If we reverse the reaction we know that the sign of ΔG° must be changed and that corresponds to a change in the sign of the potential as shown in Equation (12.13). This process has a negative ΔG° (and a positive E°) and therefore tends to be spontaneous:

$$\mathrm{Na}(s) \longrightarrow \mathrm{Na}^+(aq) + e^- \qquad \Delta G^\circ = -262\ \mathrm{kJ/mol} \tag{12.13}$$

$$E^\circ = \frac{\Delta G^\circ}{-nF} = \frac{-262\ \mathrm{kJ}}{-1 \times 96.5\ \mathrm{kJ/V}} = +2.71\ \mathrm{V}$$

Now in aqueous solution what could be reduced by the sodium or, stated another way, what substance could gain the electrons released by the sodium in Equation (12.13)? We know from the earlier discussion in Section 12.1 that the answer is water itself. A quick consultation of Table 12.2 shows that the reduction of water to aqueous hydroxide ions and hydrogen gas has a standard reduction potential of -0.83 V. When we combine these two half-reactions, will the theory bear out what we already know in practice, that sodium spontaneously reacts with water to produce hydrogen gas? That is, will the resulting E° be positive corresponding to a negative ΔG°? The calculation is shown below.

Reduction half-reaction: When water is reduced, it gains two electrons:

$$2\mathrm{H_2O} + 2e^- \longrightarrow \mathrm{H_2}(g) + 2\mathrm{OH}^-(aq) \qquad E^\circ = -0.83\ \mathrm{V} \tag{12.14}$$

$$\Delta G^\circ = -2(96.5)(-0.83) = +160\ \mathrm{kJ/mol}$$

Oxidation half-reduction: To supply two electrons for the reduction of water, we must double the half-reaction for sodium:

$$2 \times [\mathrm{Na}(s) \longrightarrow \mathrm{Na}^+(aq) + e^-] \qquad E^\circ = +2.71\ \mathrm{V} \tag{12.15}$$

$$\Delta G^\circ = -2(96.5)(2.71) = -523\ \mathrm{kJ/mol}$$

(Notice that doubling the half-equation does not affect the value of E°. The tendency to gain or lose electrons does not depend on how many moles of substance are involved. The amount of free energy released, however, *does* depend on the moles of the substance involved.)

Adding the two half-reactions together produces the following net reaction with an accompanying negative ΔG°:

$$2\mathrm{Na}(s) + 2\mathrm{H_2O} \longrightarrow 2\mathrm{Na}^+(aq) + \mathrm{H_2}(g) + 2\mathrm{OH}^-(aq) \tag{12.16}$$

$$\Delta G^\circ = -363\ \mathrm{kJ}$$

$$E^\circ = \frac{\Delta G^\circ}{-nF} = \frac{-363\ \mathrm{kJ}}{-(2)(96.5\ \mathrm{kJ/V})} = 1.88\ \mathrm{V}$$

So we see that the calculation (which produces a negative value of ΔG° and a positive E°) is consistent with the experimental result. Sodium metal will reduce water to produce hydrogen gas and aqueous hydroxide ions (all under standard state conditions). Similar results are obtained for all the alkali metals. They are all good reducing agents.

Now we can move on to consider the *trends* in the reduction potentials of Group 1A. Before we look at the actual E° values in detail, ask yourself what you

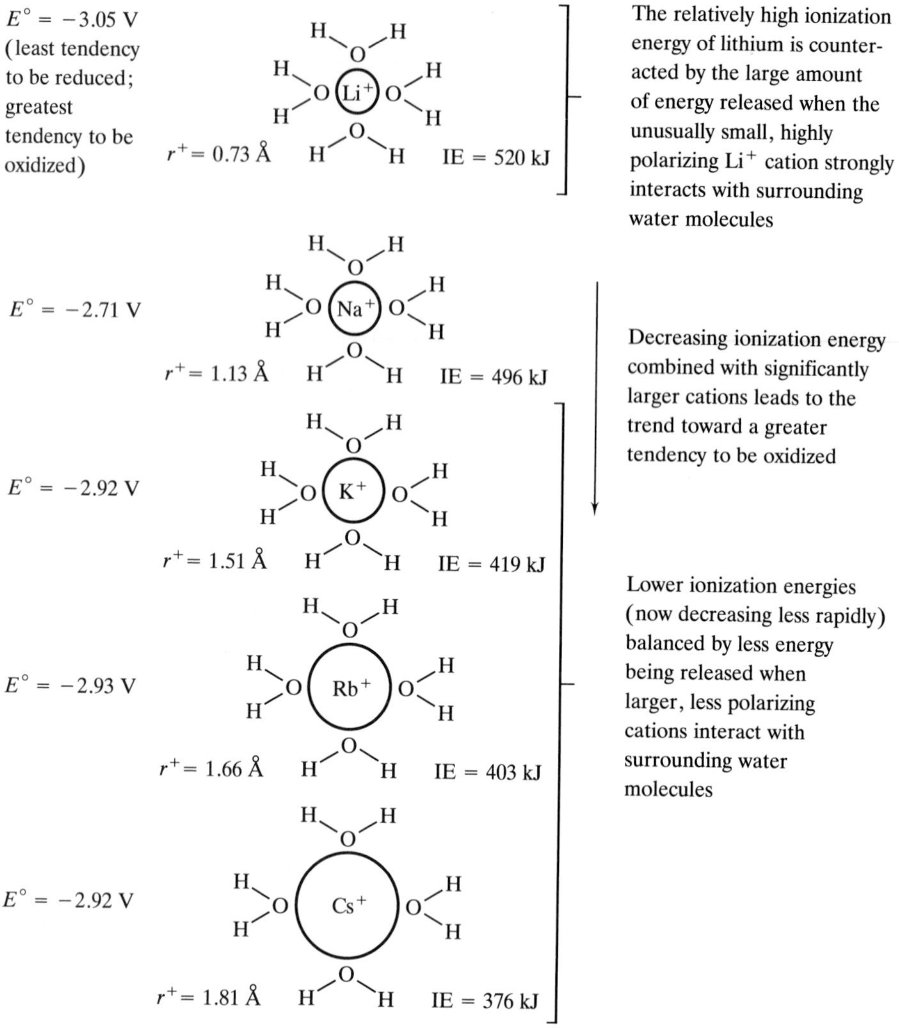

$E° = -3.05$ V (least tendency to be reduced; greatest tendency to be oxidized) — Li^+, $r^+ = 0.73$ Å, IE = 520 kJ

The relatively high ionization energy of lithium is counteracted by the large amount of energy released when the unusually small, highly polarizing Li^+ cation strongly interacts with surrounding water molecules

$E° = -2.71$ V — Na^+, $r^+ = 1.13$ Å, IE = 496 kJ

Decreasing ionization energy combined with significantly larger cations leads to the trend toward a greater tendency to be oxidized

$E° = -2.92$ V — K^+, $r^+ = 1.51$ Å, IE = 419 kJ

$E° = -2.93$ V — Rb^+, $r^+ = 1.66$ Å, IE = 403 kJ

Lower ionization energies (now decreasing less rapidly) balanced by less energy being released when larger, less polarizing cations interact with surrounding water molecules

$E° = -2.92$ V — Cs^+, $r^+ = 1.81$ Å, IE = 376 kJ

FIGURE 12.5
The trends in standard reduction potential as related to ionization energies, ionic radius, polarizing power, and energy released upon interaction with water molecules. (Only four water molecules shown around each ion for clarity.)

expect the trends to be. For example, of the six elements in the group, are you expecting lithium to be the easiest or the hardest to oxidize? On what factors are you basing your expectations? Now look at the $E°$ values on the left side of Figure 12.5. As shown, lithium has the most negative reduction potential. As we know, this means it has the least tendency to be reduced or the greatest tendency to be oxidized. Are you surprised by this result? You very well might be, particularly if you based your expectations primarily on the relative ionization energies of the elements. Lithium has the greatest ionization energy and therefore, it is tempting to conclude, should be the most difficult, not the easiest, to oxidize.

To see why lithium is the most easily oxidized of the alkali metals and to understand the overall trends in the group's standard reduction potentials, we need to look at some factors such as ionic size, charge density, and polarizing power that, in addition to ionization energy, contribute to the variation of reduction potentials. These are summarized in Figure 12.5. Note that as we go down the group, ionization energies do decrease as expected, and it follows that the ns^1 electrons do get easier to dislodge from their gaseous atoms. As indicated above, this one factor may have led you to predict that lithium was going to be particularly difficult to oxidize. It would also lead you to predict (wrongly, as it turns out) that the alkali metals should be easier to oxidize going down the group.

Ionic size, however, is another particularly important factor that has a large influence on the trends in reduction potentials. We are not surprised when reminded that the alkali-metal cations increase in size going down the group. It follows that the charge densities [charge-to-radius (Z/r) ratios] will decrease as will the polarizing power of these cations. This leads to a decrease in the strength of the interactions of these cations with their surrounding water molecules. Consideration of this string of factors would lead us to believe (correctly this time) that lithium should be the easiest to oxidize since its hydrated cation interacts so strongly with the waters of hydration that surround it in aqueous solution. Extrapolated to the entire group, these trends would predict that the alkali metals should become more difficult to oxidize going down the group. So, to summarize to this point, we have a situation where the trends in ionization energies indicate a decreasing tendency to be oxidized, while size considerations lead us to the opposite conclusion.

Figure 12.5 shows how these conflicting predictions are resolved. Three regions are indicated down the right side of the figure. At the top of the group, the unusually small size of the lithium cation dominates. Despite its large ionization energy, the high stability of lithium's hydrated cation makes this element relatively easy to oxidize. In the middle of the group, rapidly decreasing ionization energy is the most important factor, and the elements sodium, potassium, and (to a very small extent) rubidium are successively easier to oxidize. At the bottom of the group, the decrease in ionization energy becomes less pronounced and seems to be just balanced out by the smaller amounts of energy released when the larger, less polarizing cations interact with water.

So we see that a knowledge of standard reduction potentials in aqueous solution gives us another tool with which to analyze the chemistry of the alkali metals. As we go on in the next seven chapters to discuss the group chemistry of the other representative elements, we will see that this knowledge will continue to be of great value. In fact, standard reduction potentials are so valuable that we now make them the seventh and last component of our network. Figure 12.6 shows the network with this final addition.

There is one final note to be made concerning the reactivity of the alkali metals in water. All of the above discussion is certainly valid, but it is based only on *thermodynamics*. For a complete and realistic view of reactivity, we must also consider the kinetics of the situation. It turns out that while lithium reacts with water more spontaneously in a thermodynamic sense, the rate of its reaction with water is less than the rates of the heavier congeners. (The rate is dependent on the degree of contact between the metal and the water. Lithium has a higher melting point than its congeners and therefore does not melt readily and spread out through the water. Therefore, its rate of reaction is smaller.) In fact these rates increase dramatically down the group such that sodium metal reacts more vigorously but not dangerously with water, potassium metal bursts into flame when placed in water, and the heavier congeners react explosively with water.

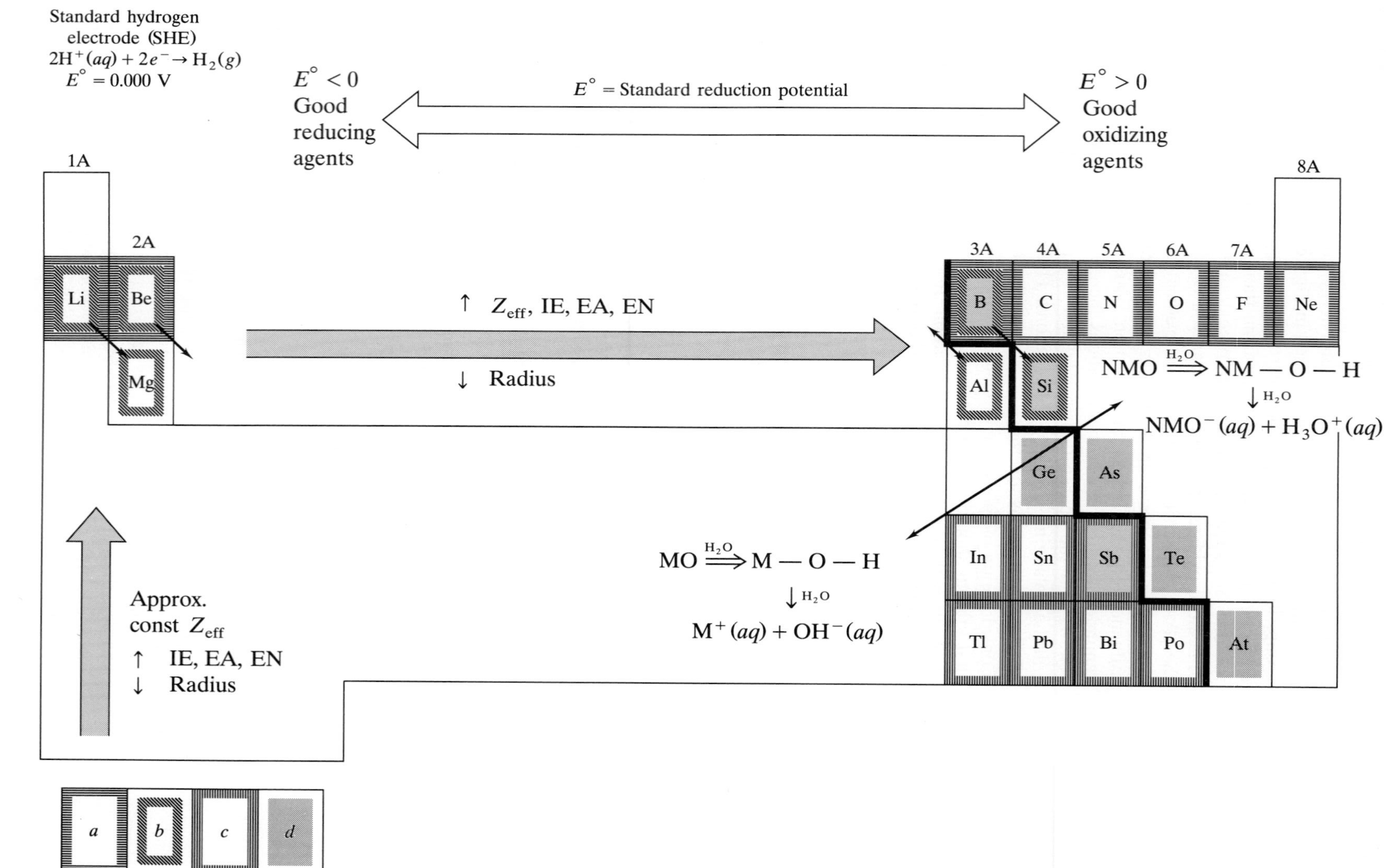

FIGURE 12.6
The seven components of the interconnected network of ideas. Trends in standard reduction potentials join trends in periodic properties, the acid-base character of metal and nonmetal oxides, (*a*) the uniqueness principle, (*b*) the diagonal effect, (*c*) the inert-pair effect, and (*d*) the metal-nonmetal line.

12.4 PEROXIDES AND SUPEROXIDES

Peroxides

Table 12.1 shows some rather unexpected products when the alkali metals react with molecular oxygen. Lithium forms the normal oxide, but when sodium reacts with oxygen, particularly at elevated temperatures, the product is the yellow, hygroscopic sodium peroxide, Na_2O_2. As shown in the following Lewis structure, the peroxide ion, O_2^{2-}, is expected to have a bond order of one:

$$\left[:\ddot{\underset{\cdot\cdot}{O}}:\ddot{\underset{\cdot\cdot}{O}}:\right]^{2-}$$

The most common form of sodium peroxide is the octahydrate, $Na_2O_2 \cdot 8H_2O$. Such hydrates, in which the water molecules are hydrogen-bonded to the peroxide ions, are also common in the alkaline-earth peroxides.

When sodium peroxide is treated with water, hydrogen peroxide is formed as shown in Equation (12.17):

$$Na_2O_2(s) + 2H_2O \longrightarrow H_2O_2(aq) + 2NaOH(aq) \qquad (12.17)$$

Pure hydrogen peroxide is a colorless, syrupy liquid which explodes violently when heated. The individual H_2O_2 molecules have the structure shown in Figure 12.7. Normally we would expect that there would be free rotation about a single bond like the O—O bond in hydrogen peroxide. In this case, however, the O—O bond is short enough that the two lone pairs on each oxygen repel each other if they get too close. Consequently, the most stable configuration of this molecule has a dihedral angle (angle of one OH with respect to the other) ranging from 111.5° in the gaseous phase to 90.0° in the crystalline solid. Hydrogen peroxide is the smallest molecule to show *hindered rotation*, defined here as the interference with the rotation about a single bond caused by lone pairs or other atoms or groups of atoms. The syrupy, viscous nature of this compound is due to the strong hydrogen bonds among the molecules.

As represented in Equation (12.18), hydrogen peroxide decomposes violently to water and oxygen gas when heated:

$$2H_2O_2(l) \xrightarrow[\text{catalysts}]{\text{heat or}} H_2O(l) + O_2(g) \qquad (12.18)$$

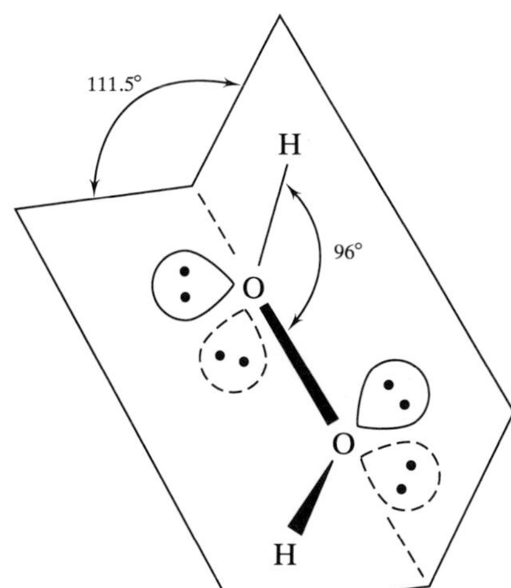

FIGURE 12.7
The structure of hydrogen peroxide. The angle of one O—H group with respect to the other (the dihedral angle) is 111.5°. The O-O-H angle is about 96°. Lone pairs in the plane are shown with solid lines, while those below or behind the plane are shown with dotted lines. H_2O_2 is the smallest molecule to exhibit hindered rotation about a single bond. [Ref. 19.]

The reaction is also catalyzed by a variety of metal (e.g., Fe, Mn, Cu) salts as well as just dust particles or traces of organic matter. This decomposition is an example of a *disproportionation reaction*, one in which an element in a compound is both oxidized and reduced. In this case, the oxygen in the peroxide (-1 oxidation state) is reduced to water (-2 oxidation state) and oxidized to oxygen (0 oxidation state). Due to the unstable nature of this substance, highly concentrated solutions are particularly dangerous to store, and appropriate safety measures should be taken. When they must be stored, solutions of 30% H_2O_2 should be kept refrigerated in polyethylene bottles. The decomposition of H_2O_2 is catalyzed by the trace metals in glass, and the resulting oxygen gas can cause a severe pressure buildup. Even polyethylene bottles used to store these solutions should be equipped with a vented cap to prevent distortion of the bottle. Unvented polyethylene bottles have been known to develop a convex bottom and rock gently back and forth as someone walks by the storage area!

The reduction potential for hydrogen peroxide to water in acid solution, shown in Equation (12.19), is given in Table 12.2 and is consistent with a moderately strong oxidizing agent:

$$H_2O_2(aq) + 2H^+(aq) + 2e^- \longrightarrow 2H_2O \qquad E^\circ = +1.77\text{ V} \qquad (12.19)$$

(Recall from the previous discussion that positive E°'s are consistent with a strong tendency to be reduced and therefore be good oxidizing agents.) For example, hydrogen peroxide can oxidize Fe^{2+} to Fe^{3+} as shown below:

$$\begin{array}{rcll} H_2O_2(aq) + 2H^+(aq) + 2e^- & \longrightarrow & 2H_2O & E^\circ = +1.77\text{ V} \\ 2[Fe^{2+}(aq) & \longrightarrow & Fe^{3+}(aq) + e^-] & E^\circ = -0.77\text{ V} \\ \hline H_2O_2(aq) + 2H^+(aq) + 2Fe^{2+}(aq) & \longrightarrow & 2Fe^{3+}(aq) & E^\circ = +1.00\text{ V} \end{array}$$

Note that when we have a balanced redox reaction, the standard reduction potentials are additive. (If you are not convinced of this, you might want to try calculating the ΔG°'s of each half-equation, adding them together to get a total ΔG° and computing the resulting E°.)

Also from Table 12.2, we see that H_2O_2 can be oxidized to O_2 in acid solution [Equation (12.20)] and can selectively serve as a reducing agent:

$$O_2(g) + 2H^+(aq) + 2e^- \longrightarrow H_2O_2(aq) \qquad E^\circ = +0.68\text{ V} \qquad (12.20)$$

For example, hydrogen peroxide can be oxidized by (or serve as a reducing agent for) permanganate in acid solution as shown below:

$$\begin{array}{ll} 5[H_2O_2(aq) \longrightarrow O_2(g) + 2H^+(aq) + 2e^-] & E^\circ = -0.68\text{ V} \\ 2[MnO_4^-(aq) + 8H^+(aq) + 5e^- \longrightarrow Mn^{2+}(aq) + 4H_2O] & E^\circ = +1.51\text{ V} \\ \hline 5H_2O_2(aq) + 2MnO_4^-(aq) + 6H^+(aq) \longrightarrow & \\ \qquad 5O_2(g) + 2Mn^{2+}(aq) + 8H_2O & E^\circ = +0.83\text{ V} \end{array}$$

In fact, the latter reaction is used to quantitatively determine the concentration of an unknown peroxide solution.

Applications of peroxides include mild (3%) hydrogen peroxide solutions as an antiseptic and as bleach for hair, and 30% solutions as a commercial bleaching agent for flour, fibers, and fats and in methods to artificially age wines and liquors. Hydrogen peroxide and hydrazine (N_2H_2) have even been used as a rocket fuel. (See Chapter 16 for further details on hydrazine.)

Superoxides

When potassium, rubidium, and cesium react with molecular oxygen, the superoxides, MO_2, are produced. There seems to be a rather clear relationship between the size of the alkali-metal cation and the stability of the superoxides. In the salts of the smaller cations, such as those of lithium and sodium, too much contact between the large superoxide ions evidently destabilizes the lattices. The larger cations provide a better match of ionic sizes. Not surprisingly, therefore, the most stable superoxide occurs with cesium.

Superoxides are useful for the removal of carbon dioxide and the production of oxygen in closed areas such as submarines, mines, and space vehicles and in self-contained breathing equipment. The reaction is represented in Equation (12.21). Note that the potassium superoxide reacts as needed with the moisture from the breath to produce the hydroxide and oxygen gas. Subsequently, the KOH absorbs carbon dioxide and generates the carbonate and more water to react with the superoxide once again. The net reaction involves a controlled conversion of carbon dioxide to oxygen. The more often one breathes (producing carbon dioxide and water), the more oxygen is produced.

$$4KO_2(s) + 2H_2O(l) \longrightarrow 4KOH(s) + 3O_2(g) \qquad (12.21a)$$

$$2[2KOH(s) + CO_2(g) \longrightarrow K_2CO_3(s) + H_2O(l)] \qquad (12.21b)$$

$$\text{Sum:} \quad 4KO_2(s) + 2CO_2(g) \longrightarrow 2K_2CO_3(s) + 3O_2(g) \qquad (12.21c)$$

12.5 REACTIONS AND COMPOUNDS OF PRACTICAL IMPORTANCE

Both sodium and potassium are rather abundant (2.6 and 2.4 percent) in the earth's surface, but potassium, due to its greater solubility and subsequent uptake by plant life, is much less prevalent in the seas. Indeed, potassium is so vital to plants that its major use, usually as the chloride or sulfate, is in fertilizers. This is certainly not a newly recognized technology. Even centuries ago, farmers knew that spreading wood ashes on their lands made crops grow better. We now recognize that it was the potassium in these ashes which was primarily responsible for the effect. Both sodium and potassium ions are present in plants and animals and are essential for normal biochemical functions, particularly for the maintenance of the concentrations of ions across various cellular membranes, enzyme functions, and the firing of nerve impulses.

The mode of action of lithium for treatment of *bipolar disorders* (a term which replaces the older *manic depressives*) seems to be related to the function of sodium and potassium in the body. Such patients suffer from alternate fits of joy and rage followed by severe depression. These highs and lows seem to be related to the over- or underfiring of nerve relays which release neurotransmitters. Lithium, administered as the carbonate, is by far the most effective treatment for this condition. While much research remains to be done, it appears that lithium controls and levels out the firing of nerve impulses and thereby severely cuts down the large swings in emotion.

As might be suspected from the earlier discussions, the alkali metals have been highly regarded as reducing agents in various applications. Sodium has been used in the production of chromium, manganese, and aluminum from their oxides and of titanium and zirconium from their chlorides. These routes to the free metals, however, have been largely replaced by other methods. Until fairly recently, a sodium-lead alloy had been used in large quantities to make tetraethyl lead (TEL), an antiknock gasoline additive. Leaded gasolines, however, have been phased out for environmental reasons. Lithium aluminum hydride, $LiAlH_4$, remains a popular reducing agent, particularly in organic synthesis.

Other electrochemical uses of Group 1A metals include lithium in batteries and cesium in photoelectric cells. Lithium batteries, in which the metal serves as the anode and is easily oxidized to lithium ions by various cathode systems in nonaqueous solvents, have the advantage of light weight, small size, good shelf life, and very high currents. Such batteries can be made in such small sizes that they can be mounted on a computer memory chip or used in cardiac pacemakers. Current research focuses on producing a rechargeable lithium battery. Cesium, due to its exceedingly low ionization energy, can be ionized by the use of visible light. This makes the metal useful in photoelectric cells in which a light beam produces an electric current that, in turn, can be used to perform various tasks such as activating a burgler alarm or reading coded or printed information on data cards or packages.

Other applications of these elements take advantage of their nuclear properties. For example, one prevalent method of establishing the age of early humanoids is the potassium-argon dating procedure developed in the 1950s. Potassium 40 has a half-life of 1.3 billion years and decays by either β^- or β^+ emission as shown in Equations (12.22) and (12.23):

$$^{40}_{19}\text{K} \xrightarrow{\beta^-} {}^{0}_{-1}e + {}^{40}_{20}\text{Ca} \qquad (89\%) \tag{12.22}$$

$$^{40}_{19}\text{K} \xrightarrow{\beta^+} {}^{0}_{+1}e + {}^{40}_{18}\text{Ar} \qquad (11\%) \tag{12.23}$$

(Since the human body contains about 1 lb of potassium and a few hundredths of a gram of potassium 40, it turns out that about 500 potassium 40 atoms decay in the body per second! The resulting argon gas leaks through the skin into the atmosphere.) The beta plus decay is responsible for the relatively high amount of argon gas in the atmosphere. When volcanic magma is laid down and cools, argon gas starts to accumulate. Some rocks have a leakproof-enough structure that this argon can be collected and, knowing the above half-life, used to obtain a date for the volcanic event. This in turn can be used to establish a date for any remains found in this stratum of the earth. For example, this was the technique used to date "Lucy," the *Australopithecus afarensis* hominid found in Ethiopia by Donald Johanson in 1974.

Other uses of the alkali metals related to their nuclear properties include the use of lithium 6 and lithium 7 in nuclear reactors to produce tritium as shown in Equations (12.24) and (12.25):

$$^{6}_{3}\text{Li} + {}^{1}_{0}n \longrightarrow {}^{4}_{2}\text{He} + {}^{3}_{1}\text{H} \tag{12.24}$$

$$^{7}_{3}\text{Li} + {}^{1}_{0}n \longrightarrow {}^{4}_{2}\text{He} + {}^{3}_{1}\text{H} + {}^{1}_{0}n \tag{12.25}$$

Tritium would then be readily available for us in fusion reactors. (See Chapter 10, p. 258, for more details on fusion as an alternative energy source of the future.)

12.6 SELECTED TOPIC IN DEPTH: METAL-AMMONIA SOLUTIONS

Although solutions of the alkali and alkaline-earth metals (except beryllium) in liquid ammonia were first studied in the 1860s, only recently have the details of their structure come to light. These solutions have some rather bizarre properties.

For example, the solutions start out as a deep-blue color, but a bronze phase starts to float on top as more metal is added. Eventually, the bronze phase, which has the properties of a liquid metal, predominates. If the ammonia is evaporated off from a solution made with an alkali metal, the original metal is recovered. Evaporation of solutions made with Group 2A metals produce "ammoniates" of formula $M(NH_3)_x(s)$.

The blue color is independent of the metal used. There is a marked increase in the volume and a corresponding decrease in the density of the solution. The conductivity of the solution is much like that of various electrolytes when put in ammonia. The solutions start out highly paramagnetic, consistent with a very large number of unpaired electrons, but become diamagnetic and then mildly paramagnetic again as metal is added.

What is the explanation of these properties? Equation (12.26) represents the formation of the solutions. Note that the (*am*) is used in analogous manner to (*aq*) when discussing these solutions:

$$M(s) \xrightarrow{NH_3(l)} \underset{M(am)^+}{M(NH_3)_x^+} + \underset{e(am)^-}{\left[e(NH_3)_y\right]^-} \qquad (12.26)$$

The formation of the ammoniated cation is not particularly surprising, but the second product, called the ammoniated (or, more generally, the solvated) electron is rather unexpected. The blue color seems to be due to this species. (Does this mean that solvated electrons are blue?) The electron apparently exists in a cavity in the liquid ammonia structure as shown in Figure 12.8. Notice that the cavities are formed by hydrogen bonds among the ammonia molecules and are responsible for the increase in the volume and the decrease in the density of the solution over that of the pure solvent. The individual electron spins start off parallel but eventually form pairs which are represented as $e_2^{2-}(am)$ and lead to a diamagnetic intermediate.

These solutions are not particularly stable and become unusually good one-electron reducing agents. When any impurity, a rusty nail or dust or anything with a high surface area, is added to the above metal-ammonia solutions, hydrogen gas and a solvated amide are produced as shown in Equation (12.27). (Note that in this case hydrogen has been reduced to the elemental state.)

$$NH_3(l) + e(am)^- \longrightarrow NH_2^-(am) + \tfrac{1}{2}H_2(g) \qquad (12.27)$$

If oxygen is added, it may be reduced to a superoxide (oxidation state $= -\tfrac{1}{2}$) and

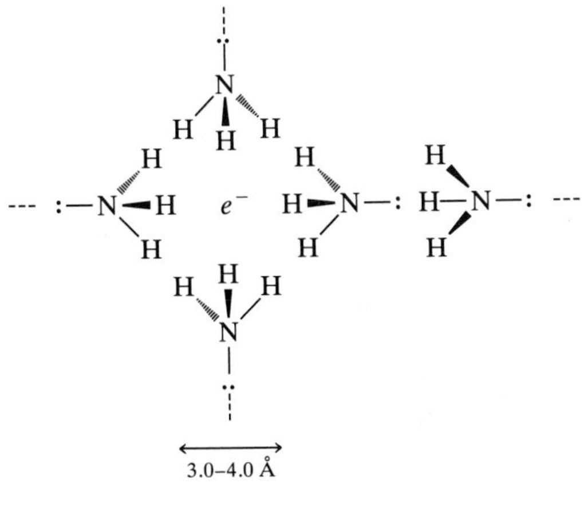

3.0–4.0 Å

FIGURE 12.8
The solvated, or ammoniated, electron is depicted in a cavity in the structure of hydrogen-bonded liquid ammonia.

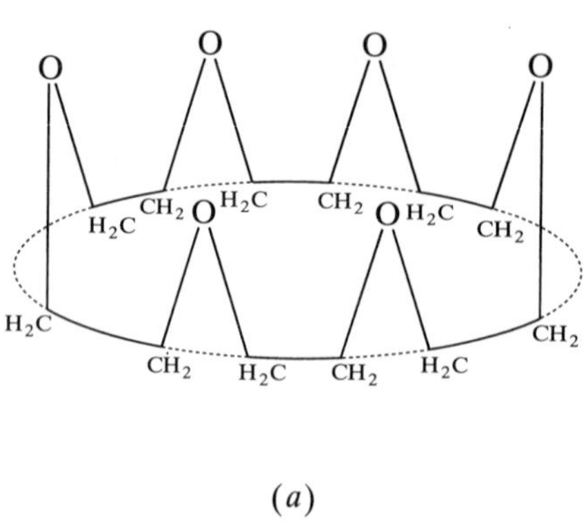

(a)

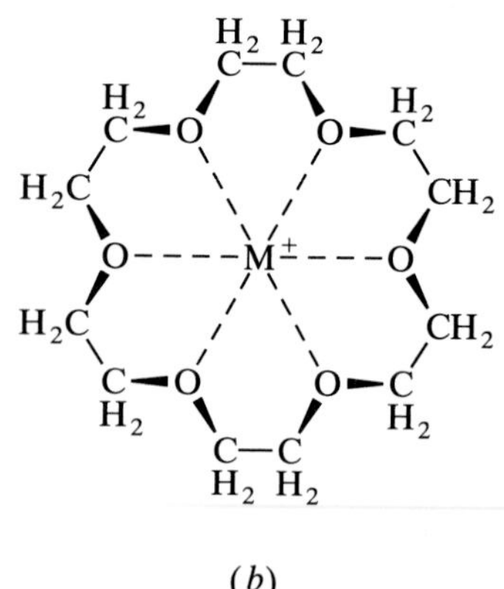

(b)

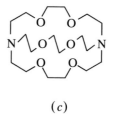

(c)

FIGURE 12.9
Crown ethers and cryptates. (*a*) A representation of 18-crown-6 showing its similarity to a cartoon crown. (*b*) An alkali-metal cation in the cavity of 18-crown-6. (*c*) A cryptate.

then to the peroxide as shown in Equation (12.28):

$$e(am)^- + O_2 \longrightarrow \underset{\text{Superoxide}}{O_2^-(am)} \xrightarrow{e(am)^-} \underset{\text{Peroxide}}{O_2^{2-}(am)} \tag{12.28}$$

If other metals are added, they can be reduced to very unusual negative oxidation states as shown in Equation (12.29) for gold:

$$\text{Au}(s) + e(am)^- \longrightarrow \text{Au}(am)^- \tag{12.29}$$

As if these above results were not strange enough, some truly unique ions have been isolated when some special stabilizing compounds called *crown ethers* and *cryptates* are employed. Crown ethers and the related cryptates are shown in Figure 12.9. Crown ethers are large ring structures with several carbon atoms separating oxygen atoms. The names of the compounds reflect the total number of atoms in the ring and the number of oxygen atoms. Accordingly, 18-crown-6 has 6 oxygen atoms in an 18-member ring. (The name comes from the fact that the free compound looks somewhat like a cartoon crown with the oxygen atoms at the points.) The cavity atop the "crown" varies in size depending on the number of oxygen atoms. Therefore, the cavity size can be matched to accommodate various-size metal cations. Using 18-crown-6 and cesium, the dark-blue liquid ammonia solution yields a dark-blue, paramagnetic solid of formula $[\text{Cs(18-crown-6)}]^+e^-$. This is an example of an *electride*, in which the counteranion is the electron!

Ion-dipole interactions between the metal ion and the partially negative oxygen atoms hold the complex cation together.

Cryptates, containing nitrogen as well as oxygen atoms separated by carbon chains, are only slightly more cryptic (!), but they yield some even stranger and extremely rare compounds. 2,2,2-crypt is just the right size to bind to the sodium cation. Solutions of ethylamine (merely an ammonia molecule with one hydrogen atom replaced with an ethyl group, CH_3CH_2—), sodium, and 2,2,2-crypt yield a compound of formula $[Na(2,2,2\text{-crypt})]^+Na^-$, which has an anionic form of sodium called, logically enough, sodide. The potassium cation is a bit too large for the cryptate so that the analogous $[K(2,2,2\text{-crypt})]^+K^-$, containing a potasside anion, is less stable. In general, anions of alkali metals are called *alkalides*.

SUMMARY

The alkali metals were discovered by several special individuals using some new techniques. The daredevil showman Davy isolated potassium and sodium by electrolysis of molten salts, the young Swede Arfwedson identified lithium by quantitative analysis, rubidium and cesium were identified spectroscopically by Bunsen and Kirchhoff, while Perey prepared minute quantities of francium by radiochemical methods.

Our network of interconnected ideas helps us to account for many expected properties of the alkali metals. The hydrides, oxides, hydroxides, and halides of these elements are ionic. The oxides and hydroxides are basic in character. Lithium, while still an alkali metal with much in common with its congeners, is certainly a good example of the uniqueness principle. It has much in common with magnesium as forecast by the diagonal effect.

After a brief review of oxidation and reduction (LEORA goes GEROA!), standard reduction potentials are seen to be useful in tabulating substances in order of the tendency to be reduced or oxidized. The alkali metals have little tendency to be reduced and are themselves excellent reducing agents. Reduction potentials are related to the change in the free energy of a reaction and therefore can predict thermodynamic spontaneity. Analysis of the standard reduction potentials points out still another criterion by which we regard lithium as unique in its group.

Some of the alkali metals (and alkaline earths) form peroxides and superoxides when combined with molecular oxygen. Sodium and potassium form peroxides, which often are hydrated solids. Sodium peroxide is a ready source of hydrogen peroxide, the smallest molecule to show hindered rotation. Hydrogen peroxide disproportionates violently when heated, is a strong oxidizing agent, but can also be a mild reducing agent. Peroxides have a number of uses ranging from mild antiseptics and bleaches to rocket fuel. The larger alkali metals (potassium, rubidium, and cesium) form superoxides. Potassium superoxide is used in self-contained breathing equipment.

Other practical applications of the alkali metals and their compounds include sodium and potassium in living systems, lithium as treatment for bipolar disorders (manic depression), elemental sodium and lithium compounds as reducing agents, lithium in a new generation of batteries, cesium in photoelectric devices, potassium-argon dating, and lithium as a source of tritium to fuel the hydrogen economy.

Alkali- (and alkaline-earth) metal solutions of liquid ammonia display some bizarre properties which can be accounted for in terms of solvated electrons. The solvated electron is an excellent one-electron reducing agent and can be used to produce peroxides, superoxides, and metals with negative oxidation states. Compounds involving crown ethers and cryptates of the alkali metals in ammonia and related solvents yield the very rare electrides and alkalides.

PROBLEMS

12.1. Given the information in Chapters 10 to 12, arrange the major contributions of Arfvedson, Bunsen, Davy, Cavendish, Kirchhoff, Lavoisier, Mendeleev, and Priestley in chronological order.

12.2. How many alkali metals have been discovered by the time Mendeleev published his first periodic table in 1869?

12.3. Why is a battery referred to as a *voltaic pile* in describing Davy's experiments to produce sodium and potassium? (*Hint*: Start with a good dictionary.)

12.4. What would happen if phenolphthalein indicator were added to the water to which sodium metal was introduced? Specifically, what color, if any, would the phenolphthalein turn? Briefly interpret the result. (*Hint*: The colors of indicators can be found in any good introductory or analytical textbook.)

12.5. Why does adding calcium chloride to sodium chloride lower the melting point of the latter?

12.6. Although francium is the product of the alpha decay of actinium, the majority of actinium nuclei decay by beta minus emission. Write an equation for this process.

12.7. Francium 223 decays via both alpha and beta emission. Write equations for these two processes.

12.8. Plot the values of (*a*) atomic radius, (*b*) ionization energy, (*c*) electron affinity, and (*d*) electronegativity for the Group 1A elements against the period number. Discuss the slight anomalies in these properties which occur at rubidium.

12.9. We have noted that the alkali-metal hydroxides become stronger bases going down the group. However, even sodium hydroxide is considered to be completely ionized in aqueous solution. Could we measure the difference in basicities of the alkali-metal hydroxides in aqueous solution? What other type of solvent might give us additional information about relative basicities of these compounds? Briefly discuss your answer.

12.10. Write an equation representing the reaction between potassium oxide and water.

12.11. Describe, in some detail, giving some experimental conditions, how you might prepare the iodides of lithium, sodium, and potassium. Write a general equation for these syntheses as part of your answer.

12.12. Write an equation representing the reaction between sodium hydride and water.

12.13. Lithium nitride, Li_3N, is often referred to as a *saltlike* nitride. On the basis of previous discussions of hydrides and oxides, briefly discuss this designation.

12.14. If sodium nitride did exist, would you expect it to be more or less saltlike than lithium nitride? Briefly support your answer.

12.15. Based on your own experience, is sodium chloride hygroscopic? Would you expect it to be as hygroscopic as lithium chloride? Briefly discuss your answer.

12.16. Explain in your own words the mnemonic LEORA goes GEROA.

12.17. Why is the standard reduction potential of $2H^+(aq) + 2e^- \rightarrow H_2(g)$ zero volts?

12.18. Consider the reaction between lithium and hydrogen ions in solution:

$$Li(s) + H^+(aq) \longrightarrow H_2(g) + Li^+(aq)$$

(*a*) Break the overall reaction into half-reactions.
(*b*) Label the reducing and oxidizing agents.
(*c*) Balance each half-reaction and then balance the overall reaction.
(*d*) Calculate $E°$ and $\Delta G°$ for the overall reaction.

12.19. Based on standard reduction potentials, would chlorine gas be a reducing agent or an oxidizing agent? Briefly discuss your answer.

12.20. Based on an analysis of standard reduction potentials, should it be more or less difficult to electrolyze lithium chloride to lithium metal or sodium chloride to sodium metal?

12.21. Explain in your own words why (*a*) the standard reduction potentials of potassium, rubidium, and cesium are so similar and (*b*) why the standard reduction potential of lithium is so different from the above three.

12.22. Thermodynamically, lithium is less stable in water than sodium, but kinetically, lithium is more stable. Briefly explain the difference.

12.23. Would you expect hydrogen peroxide to be soluble in water? Why or why not? Briefly discuss your answer.

12.24. Relate the hindered rotation found in hydrogen peroxide to the uniqueness principle.

12.25. The compound H_2S_2, although not particularly stable, does not show hindered rotation. Briefly discuss why this is.

12.26. Hydrogen peroxide can be prepared by reaction of barium peroxide with aqueous sulfuric acid. Write an equation representing this reaction.

12.27. Write the two half-reactions for the disproportionation of hydrogen peroxide to water and diatomic oxygen. Balance the equation. Using Table 12.2, calculate $E°$ for the overall reaction.

12.28. Disproportionation reactions are sometimes called self-oxidation reactions. Briefly describe why.

12.29. Using standard reduction potentials, decide whether hydrogen peroxide could be used to oxidize Co^{2+} to Co^{3+} in an acidic aqueous solution.

12.30. Using standard reduction potentials, decide whether hydrogen peroxide could be used to oxidize cerium(III) to cerium(IV).

12.31. Using standard reduction potentials, decide whether sodium metal could be used to reduce Al^{3+} to aluminum metal.

12.32. Could hydrogen peroxide be used to oxidize sodium metal? Use standard reduction potentials to back up your answer.

12.33. Hydrogen peroxide can be an oxidizing agent or a reducing agent, depending on what reagent it is paired with. Which would it be with (*a*) permanganate, MnO_4^-, and (*b*) iodide, $I^-(aq)$? Write equations to represent these reactions.

12.34. Determine the oxidation states of the oxygen atoms in the superoxide ion.

12.35. Break down the following unbalanced equation for the reaction between potassium superoxide and water into half-reactions and balance the reaction. (*Hint*: Determine the role of the potassium ions in this reaction.)

$$KO_2(s) + H_2O(l) \longrightarrow KOH(s) + O_2(g)$$

12.36. Potassium superoxide is particularly useful in self-contained breathing equipment because it responds to the metabolism of the user. Explain how this response comes about.

12.37. There is some evidence that lithium may be effective in its treatment of bipolar disorders because it regulates the magnesium-calcium balance in the body. Would this appear to make sense from an inorganic chemist's point of view? Why or why not?

12.38. Sodium metal is prepared by the electrolysis of sodium chloride to the metal and chlorine gas. Using the reduction potentials of Table 12.2, calculate the $E°$ and $\Delta G°$ for this reaction.

12.39. Sodium can be used to reduce (*a*) chromium(III) and (*b*) aluminum(III) oxide to the free metal. Write balanced equations for these processes.

12.40. Sodium can be used to reduce (*a*) titanium(IV) and (*b*) zirconium(IV) chlorides to the free metal. Write balanced equations for these processes.

12.41. One reaction system for a promising lithium battery is given by the following overall reaction. Write the half-equations for this reaction and identify the reducing agent and the oxidizing agent. (*Hint*: The product contains the dithionite ion, $S_2O_4^{2-}$.)

$$2Li(s) + 2SO_2(g) \xrightarrow[\text{solvent}]{\text{nonaqueous}} Li_2S_2O_4(s)$$

12.42. Explain in your own words how potassium-argon dating works.

12.43. (*a*) Cesium 137 is a product of nuclear fission. It decays via beta minus emission with a half-life of about 30 years. Write an equation for this process.
(*b*) It is estimated that fission products should be stored for about 20 half-lives before their radioactivity is no longer dangerous. How long would cesium 137 have to be stored?

12.44. List four significant properties of metal-ammonia solutions of the alkali metals and, in a phrase or one or two sentences at the most, sketch out how each property is accounted for by the physical model or theory of these solutions.

12.45. Sketch the structure of the 12-crown-4 crown ether. Speculate on its usefulness in forming the cesium electride in the same manner as 18-crown-6 does.

12.46. Name the following compound. (*Hint*: Be careful, this is a trick question!)

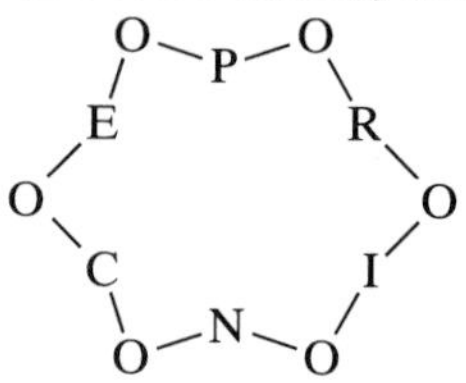

12.47. Neither $[Cs(2,2,2\text{-crypt})]^+$ nor $[Li(2,2,2\text{-crypt})]^+$ exists. Speculate as to why.

12.48. Provide a name for the anions of each of the alkali metals.

CHAPTER 13

GROUP 2A: THE ALKALINE-EARTH METALS

This chapter continues our tour of the representative elements. The alkaline-earth metals (beryllium, magnesium, calcium, strontium, barium, and radium) are similar to the alkali metals in that their compounds were known to the ancients but most of the metals themselves were not isolated until the nineteenth century. In both groups the heaviest congener is rare and radioactive. Again, we start with a history of their discoveries and then move on to see how the network can be applied to predict and rationalize group properties. The periodic law, the uniqueness principle, the diagonal effect, the acid-base character of oxides, and, to a lesser extent, standard reduction potentials are the most important parts of the network needed to organize the chemistry of this group. A section on the practical applications of these elements and their compounds is followed by the selected topic in depth, the commercial use of calcium compounds.

13.1 DISCOVERY AND ISOLATION OF THE ELEMENTS

Figure 13.1 shows information about the discoveries of the alkaline-earth metals superimposed upon the plot of the number of known elements versus time. Note that although their compounds were known to the ancients, most of these elements were actually isolated in the nineteenth century.

The Egyptians used gypsum, which we know today as a hydrate of calcium sulfate, for the mortars and plasters in the pyramids at Giza, the temples of Karnak, and Tutankhamen's tomb. The Romans used limestone and marble,

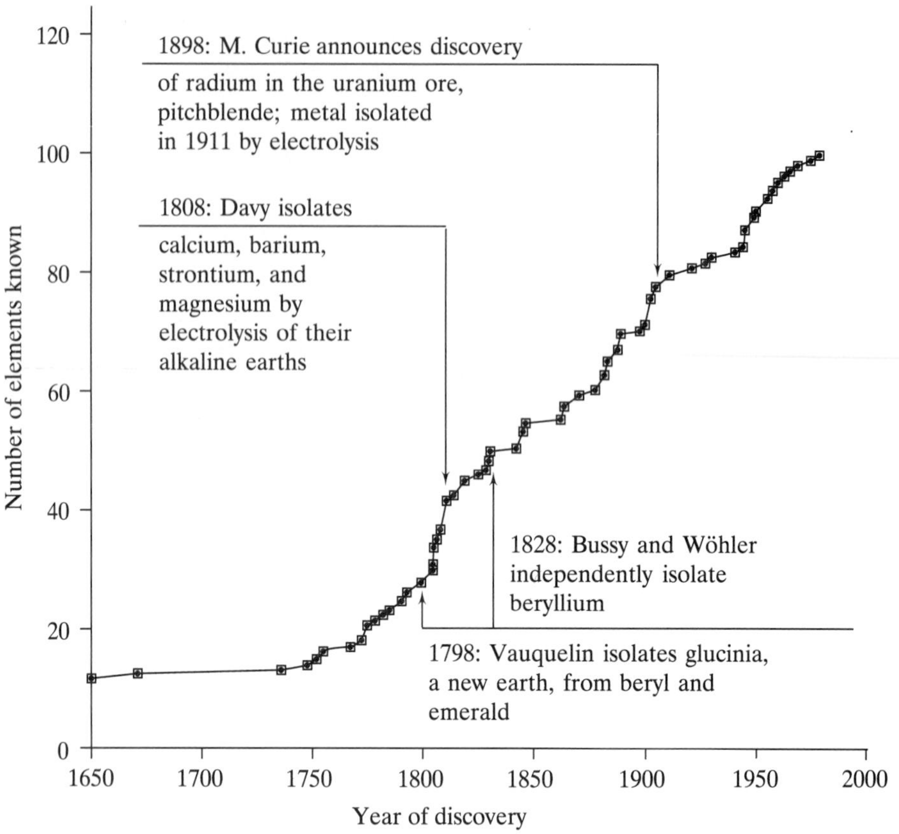

FIGURE 13.1
The discovery of the alkaline-earth metals superimposed on the plot of the number of known elements versus time.

calcium carbonate, for their magnificent buildings. Their mortar was prepared from sand and lime (calcium oxide derived from burning or roasting limestone). When no marble was available, as was the case in northern Europe, they used sandstone blocks constructed with a lime mortar obtained from chalk (also calcium carbonate) that, after mining, left behind huge caves. Today, some of these caves in the Champagne province of France are used to store millions of bottles of the sparkling, bubbling, white wine named after the province.

Lime is an *earth*, which to the mystical alchemists of the Middle Ages (not to be confused with authoritative middle-aged chemists) was a solid substance that did not melt and was not converted into another substance by fire. Alkalis (from the Arabic *al-qili*, meaning "the ashes of saltwort," a plant grown in salt marshes) were known to have a bitter taste and an ability to neutralize acids. Alkalis that did not melt were then, logically enough, called *alkaline earths* and included baryta (BaO), beryllia (BeO), lime (CaO), magnesia (MgO), and strontia (SrO). Most eighteenth century chemists thought these earths to be elements, but Lavoisier surmised (correctly) that they were oxides. Joseph Black, a Scottish chemist, found in the 1750s that limestone, when heated to produce lime, also liberated "fixed air" (air that could be fixed or immobilized into a solid form) which we know today as carbon dioxide. Recall that Priestley used this air to make soda water. The relationships among limestone, lime, and carbon dioxide are shown in Equation (13.1):

$$\underset{\text{Limestone}}{CaCO_3(s)} \xrightarrow{\text{heat}} \underset{\text{Lime}}{CaO(s)} + \underset{\text{“Fixed air”}}{CO_2(g)} \qquad (13.1)$$

$$\xleftarrow{\quad H_2O \quad}$$

This reaction is reversible and shows how moistened lime, when exposed to the carbon dioxide in the air, can be converted to the rocklike limestone, thus serving as a mortar.

Calcium, Barium, and Strontium

Who do you suppose had a large hand in actually isolating the metals from these alkaline earths? Who else but Sir Humphry Davy, fresh from his successful isolation of sodium and potassium. It turned out to be somewhat more difficult to isolate the 2A metals but, with the aid of work done by Berzelius and Pontin, he was able, in 1808, to electrolyze moist lime in the presence of mercuric oxide to make an amalgam, that is, an alloy of mercury, which grudgingly yielded the silvery-white calcium metal. Today, calcium is prepared by the electrolysis of molten $CaCl_2$ in the presence of CaF_2 added to lower the melting point as shown in Equation (13.2). The name *calcium*, coined by Davy, is from the Latin *calx*, meaning “lime.”

$$CaCl_2(l) \xrightarrow[CaF_2(l)]{\text{electrolysis}} Ca(l) + Cl_2(g) \qquad (13.2)$$

The earth or oxide of barium is of high density and hence was called *baryta* (from the Greek *barýs*, meaning “heavy”). It was first distinguished from lime by Karl Scheele in 1774 while he was waiting for the belated account of his discovery of oxygen to be published (see Chapter 11, p. 265). Davy also isolated barium metal by electrolysis of baryta in 1808. The metal itself is quite lightweight, so the name barium is somewhat misleading but is a well-established accident of history. Restoring our faith in the historical record is strontium and its oxide, strontia, that take their names from Strontian, Scotland, where a rare mineral was found in a lead mine. Davy produced the metal (in 1808, his banner year) by the electrolysis of the chloride (obtained by dissolving the Strontian mineral with hydrochloric acid) mixed with potassium chloride.

Magnesium

In the early 1600s a bitter water that promoted the healing of skin sores was found in Epsom, England, which soon became the site of a popular spa. *Epsom salt*, derived from this water and other springs, is primarily $MgSO_4 \cdot 7H_2O$. Throughout history it has often been used as a *purgative*, that is, a laxative, but one very harsh by today's standards. Added to a hot bath, it still provides a popular soaking solution for relief of minor body aches and pains. Magnesium and its oxide, magnesia, take their name from Magnesia, a district of Thessaly, Greece, home of the legendary Achilles and Jason of the Argonauts who searched for the golden fleece. Magnesia was an early source of minerals containing the element. Davy (can you guess in what year?) isolated a small quantity of the grayish-white metal from the electrolysis of its earth. He originally called the metal "magnium" instead of magnesium because he thought the latter would be confused with manganese. He certainly was correct, but it is not apparent that his proposed name would have been much better.

In 1831, Antoine-Alexandre-Brutus Bussy, a French chemist and pharmacist, developed a better method of preparing magnesium by heating the chloride with potassium under isolated conditions as shown in Equation (13.3):

$$MgCl_2(s) + 2K(s) \xrightarrow{\text{heat}} 2KCl(s) + Mg(s) \qquad (13.3)$$

Magnesium chloride is readily available in vast quantities from seawater, and this method is still used today. A second modern method is the ferrosilicon process, in which fused carbonate mixtures are reacted with an alloy of iron and silicon as shown in Equation (13.4):

$$\underset{\text{Dolomite}}{CaCO_3 \cdot MgCO_3} + FeSi \longrightarrow Fe + Ca_2SiO_4 + Mg(l) \qquad (13.4)$$

Beryllium

The mineral beryl was known in ancient times. It comes in a variety of colors, depending on the impurities present. The light-blue or blue-green aquamarine is beryl with trace amounts of iron, while the deep green of emerald results when small amounts of chromium are present. The similarity between emeralds and other forms of beryl was the impetus for the French chemist Louis Nicolas Vauquelin in 1798 to perform a careful quantitative analysis which demonstrated the presence of a new earth. Because of the sweet taste of its salts, Vauquelin called this earth "glucinia" from the Greek *glykys*, "sweet." (Hold it right there! Do not follow in the footsteps of the foolhardy Humphry Davy! Do *not* rush out to taste some beryllium salts. Not only are they sweet, but they are extremely toxic, as we will discuss in Section 13.3.) The metal was first isolated independently by Friedrich Wöhler and Bussy in 1828 by reducing the chloride with potassium as shown in Equation (13.5). Wöhler preferred the name beryllium to glucinium because salts of other metals such as yttrium also have a sweet taste.

$$BeCl_2(s) + 2K(s) \longrightarrow Be(s) + 2KCl(s) \qquad (13.5)$$

Radium

The story of Marie Sklodowska Curie's work to isolate radium from the uranium ore pitchblende is one of the most remarkable in the history of chemistry. Emigrating from Poland to study at the Sorbonne in 1891, she soon met Professor Pierre Curie and married him in 1895. It is significant that Pierre quickly recognized her extraordinary talent and gave up his work on the temperature dependence of magnetism to become her assistant. The existence of rays coming from various uranium minerals had just been established by Becquerel in 1896, but it was Marie Curie who coined the word *radioactivity* from the Latin *radius*, meaning "ray." After the birth of her first child in 1897, Marie started to investigate the radioactivity of various uranium minerals. She found that some of them were much more active than they should have been based solely on their uranium content. She theorized that these minerals must contain very small amounts of an unknown element or elements of extraordinarily intense radioactivity. By 1898, the Curies were able to announce the discovery of two elements, which they called polonium, after Marie's homeland, and radium (also after *radius*).

Seven tons of the uranium ore called pitchblende is needed to produce a single gram of radium. Pitchblende itself was too expensive for the Curies to purchase in sufficient quantities, so they had to work with the tailings left over after the ore had been processed at the mine site. After paying out of their own pockets to have this brown residue shipped from the mines of St. Joachimsthal (in what until recently was Czechoslovakia), they processed it in a poorly equipped shed located in a back alley of Paris. Working with quantities approaching 100 lb, they laboriously reduced the material to insoluble barium and radium carbonates and sulfates until finally, in 1902, they had isolated 0.120 g of nearly pure radium chloride.

In June of 1903, Marie Curie was awarded her doctorate. In December she, Pierre, and Becquerel shared the Nobel Price in physics for their study of radioactivity. In 1911, Marie received the Nobel Prize in chemistry for her isolation of radium and polonium.

In those days no one knew just how dangerous radioactivity was. Radium is an extremely powerful alpha emitter as represented in Equation (13.6):

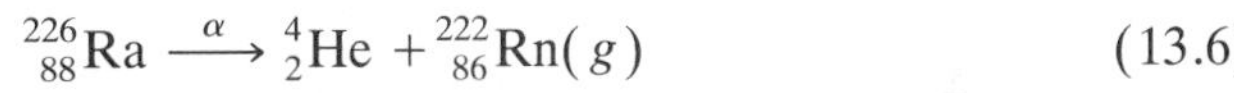

$$^{226}_{88}Ra \xrightarrow{\alpha} {}^{4}_{2}He + {}^{222}_{86}Rn(g) \qquad (13.6)$$

Its emissions are so intense that it constantly ionizes the air around it, producing a glow. (Figure 13.2 shows a photograph of several grams of radium bromide taken

FIGURE 13.2
A photograph of 2.7 g of radium bromide, taken by its own light on October 15, 1922. [*From R. Wolke, "Marie Curie's Doctoral Thesis: Prelude to a Nobel Prize," J. Chem. Educ., 65(7): 561 (1988). Published with permission of the Archives Pierre et Marie Curie, Paris.*]

by its own light!) The Curies have written of returning to their laboratory-shed at night and seeing their various fractions glowing in the dark! Radium remains perpetually ionized, shoots powerful rays through matter, and burns human flesh without feeling hot to the touch! The Curies were constantly subjected to this ionizing radiation, and Marie's laboratory notebook is to this day unavailable for direct consultation because it remains contaminated by the materials she and her husband worked with on an everyday basis. Even reprints of her thesis, published in Britain in 1903, should be checked for radioactivity since they may have been used in contaminated laboratories all over the world. Not surprisingly, Marie Curie died of radiation-induced leukemia at the age of 60.

Radium became the fad of the early 1900s. A number of applications were found for it, ranging from a treatment for cancer to a glow-in-the-dark paint for watch dials. Unfortunately, the workers employed to paint the hands and numerals of the watches with fine brushes (which they inserted in their mouths to obtain a fine point) paid dearly for the ignorance of the effects of radioactivity. As mentioned in Chapter 10, tritium has replaced radium in these paints. Because of the applications found for it and its extreme scarcity, radium soon became extraordinarily expensive. In the early 1920s Madame Curie herself could not afford to purchase it. She toured America twice, and the women she lectured to gave tens of thousands of dollars so that the "radium lady" could purchase 2 g of radium so she could continue to do research on the element she herself had discovered!

13.2 FUNDAMENTAL PROPERTIES AND THE NETWORK

Figure 13.3 shows the alkaline-earth metals superimposed on the completed network of interconnected ideas. Table 13.1 shows a number of the properties of the Group 2A elements.

Briefly, it might be instructive to compare the alkaline-earth metals with the alkali metals. Note that both groups show many of the variations in periodic properties we have come to expect. [However, also note that a few properties of the alkaline earths (melting and boiling point, density) show some irregularities not readily rationalized even by our vast storehouse of interconnected knowledge!] As expected, beryllium, like lithium in its group, is the unique element. Magnesium, much like sodium, is intermediate in character, and the lower congeners (of both groups) form a closely allied series.

The standard reduction potentials, particularly those of the heavier congeners, are similar to those of the heavier alkali metals. These are all good reducing agents. The near constancy of the $E°$ values of calcium, strontium, barium, and radium reflects a balance of the heats of atomization, ionization, and hydration energies. (See Problem 13.22.) Of course, two electrons must be ionized from the alkaline earths, but the resulting +2 cations release more energy when they interact with water. In any case, the approximate balance is maintained. One difference between the two groups is that the standard reduction potential of beryllium is not the most negative in its group as is lithium within the alkali metals. Presumably (here we go again) this is because the energy required to ionize the beryllium to the +2 state is not fully compensated for by the energy released when the Be^{2+} ion is hydrated. (See Problem 13.23 for an opportunity to conduct a more thorough analysis.)

As with the alkali metals, the standard reduction potentials of the alkaline-earth metals indicate that these elements will readily react with water. Given that

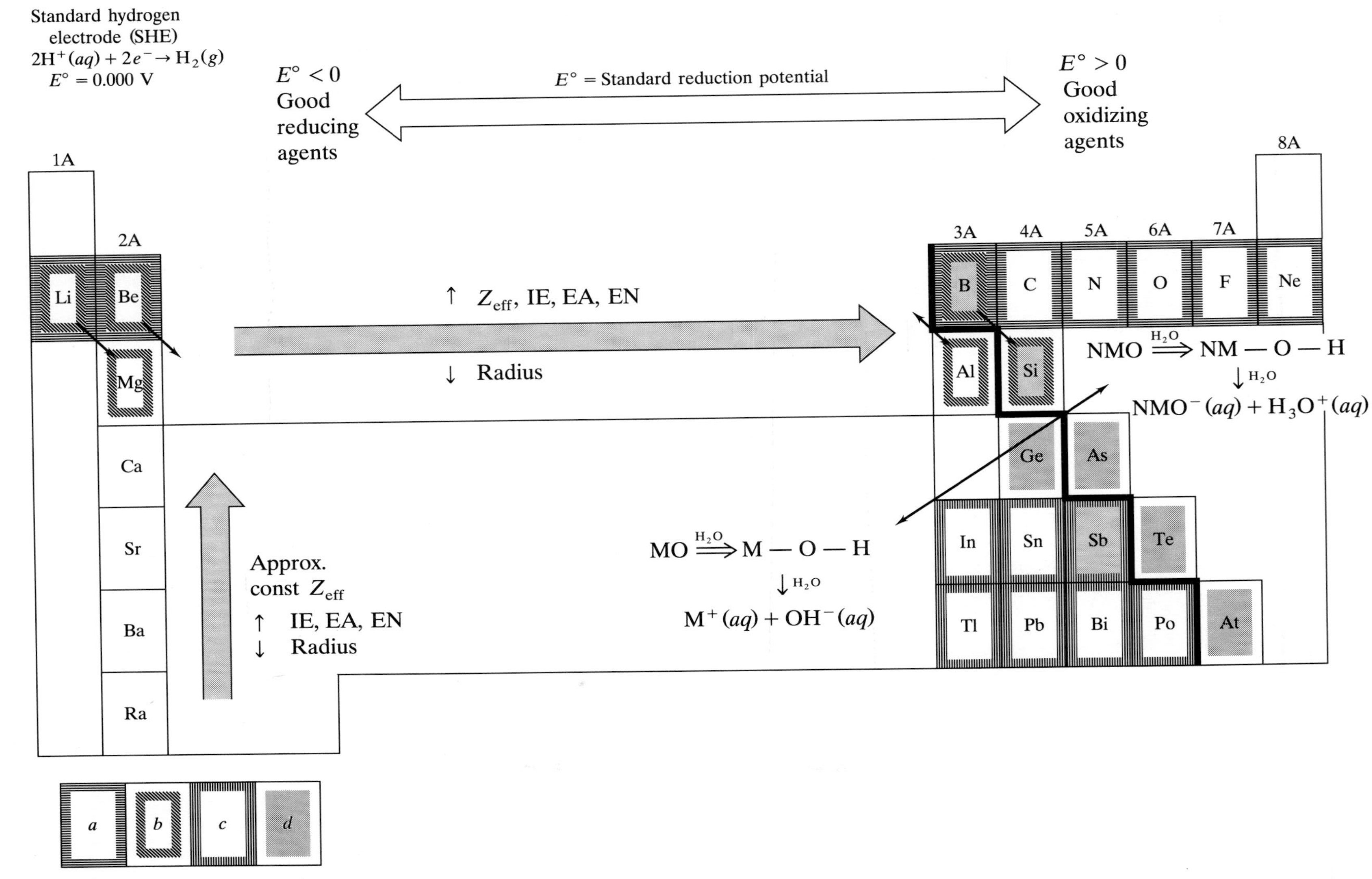

FIGURE 13.3
The alkaline-earth metals superimposed on the interconnected network of ideas including the trends in periodic properties, the acid-base character of metal and nonmetal oxides, trends in standard reduction potentials, (*a*) the uniqueness principle, (*b*) the diagonal effect, (*c*) the inert-pair effect, and (*d*) the metal-nonmetal line.

TABLE 13.1
The fundamental properties of the Group 2A elements. [Ref. 24.]

	Beryllium	Magnesium	Calcium	Strontium	Barium	Radium
Symbol	Be	Mg	Ca	Sr	Ba	Ra
Atomic number	4	12	20	38	56	88
Natural isotopes, A/% abundances	9/100	24/23.98 25/24.99 26/25.98	40/96.95 42/0.646 43/0.135 44/2.08 46/0.186 48/0.18	84/0.56 86/9.86 87/7.02 88/82.56	130/0.101 132/0.097 134/2.42 135/6.59 136/7.81 137/11.32 138/71.66	226/100
Total no. of isotopes	6	8	14	16	20	16
Atomic weight	9.012	24.31	40.08	87.62	137.3	(226)
Valence electrons	$2s^2$	$3s^2$	$4s^2$	$5s^2$	$6s^2$	$7s^2$
mp/bp, °C	1283/2970	650/1120	845/1420	770/1380	725/1640	700/1140
Density, g/cm^3	1.85	1.74	1.55	2.60	3.51	5
Atomic radius, Å	1.12	1.60	1.97	2.15	2.24	
Ionic radius, Shannon-Prewitt, Å (C.N.)	0.41(4)	0.71(4)	1.14(6)	1.32(6)	1.49(6)	
Pauling EN	1.5	1.2	1.0	1.0	0.9	0.9
Charge density (charge/ionic radius), unit charge/Å	4.9	2.8	1.7	1.5	1.3	
E°,† V	−1.85	−2.37	−2.87	−2.89	−2.90	−2.92
Oxidation states	+2	+2	+2	+2	+2	+2
Ionization energy, kJ/mol	899	738	590	549	503	
Estimated electron affinity, kJ/mol	241	230	154	120	52	
Discovered by/date	Wöhler/Bussy 1828	Davy 1808	Davy 1808	Davy 1808	Davy 1808	Curie 1911
rpw‡ O_2	BeO	MgO	CaO	SrO, SrO_2	BaO_2	RaO
Acid-base character of oxide	Ampho.	Weak base	Base	Base	Base	Base
rpw N_2	None	Mg_3N_2	Ca_3N_2	Sr_3N_2	Ba_3N_2	Ra_3N_2
rpw halogens	BeX_2	MgX_2	CaX_2	SrX_2	BaX_2	RaX_2
rpw hydrogen	None	MgH_2	CaH_2	SrH_2	BaH_2	
Crystal structure	hcp	hcp	ccp	ccp	bcc	

†E° is the standard reduction potential, $M^{2+}(aq) \rightarrow M(s)$.
‡rpw = reaction product with.

the E° values become more negative going down the group, we predict that their reactivity should increase. This thermodynamic prediction is borne out in the actual chemistry. The alkaline-earth metals are more electronegative, smaller, and less reactive than the alkali metals, but going down the group they do in fact become progressively more reactive with water. Beryllium is essentially unaffected even by steam, while magnesium reacts slowly with steam (but not with cold water) to produce the oxide and hydrogen gas. Calcium reacts readily and barium violently even with cold water.

Hydrides, Oxides, Hydroxides, and Halides

Following the precedent set in discussing the Group 1A elements, we take a close look now at the hydrides, oxides, hydroxides, and halides of this group.

We expect the hydride of beryllium (based on the uniqueness principle) and to some extent that of magnesium (based on its diagonal similarity to lithium) to be covalent, while the heavier congeners should be ionic. (The classifications of hydrides were presented in Table 10.4.) Indeed this turns out to be the case. Beryllium hydride is not made by the direct combination of the element and hydrogen. It is a polymeric, covalent hydride characterized by bonds of a type that we have not yet encountered. These are called three-center–two-electron electron-deficient bonds. They are more common in Group 3A chemistry, something also consistent with the diagonal effect, and are covered in detail in Section 14.5. Magnesium hydride is intermediate in character with a significant degree of covalent character. Calcium, strontium, and barium hydrides are ionic as discussed in Chapter 10. They are readily formed by reacting the metal with hydrogen gas and react with water to liberate the hydrogen again as shown for calcium in Equations (13.7) and (13.8):

$$Ca(s) + H_2(g) \longrightarrow CaH_2(s) \tag{13.7}$$

$$CaH_2(s) + 2H_2O(l) \longrightarrow Ca(OH)_2(s) + H_2(g) \tag{13.8}$$

These reactions make the ionic hydrides portable sources of hydrogen useful for cooking, in meterological balloons, and perhaps eventually in a hydrogen economy (see Section 10.6). Also based on Equation (13.8), calcium hydride, as mentioned in Chapter 10, serves as a dehydrating reagent for organic solvents.

The oxides, or the "earths," are, as we have seen, difficult to melt. Beryllium oxide is a solid of significant covalent character much like aluminum oxide and rather unlike those of its congeners. Both are amphoteric rather than basic anhydrides (see Chapter 11, p. 277, and Table 11.4). Magnesium reacts vigorously with oxygen to form the oxide as shown in Equation (13.9):

$$2Mg(s) + O_2(g) \longrightarrow 2MgO(s) \tag{13.9}$$

This reaction of magnesium with oxygen produces a brilliant, extremely hot flame and is the basis of the use of magnesium as an incendiary in shells, bombs, etc. Both MgO and BeO find uses as a *refractory*, a material that is hard to melt, reduce, or work. Magnesium oxide is used to cover the heating elements of electric ranges because it is high melting and conducts heat well. Calcium, strontium, and barium oxides are ionic, basic anhydrides as expected. Calcium oxide, lime, is used to lower the acidity of soils and in lakes affected by acid rain. This procedure of adding lime to lower the acidity is sometimes called *liming*. At elevated temperatures, barium reacts with the oxygen of the air to produce the peroxide rather than the oxide. The larger Ba^{2+} forms a more stable lattice with the larger peroxide ion than it does with the smaller oxide. Barium peroxide reacts with water to generate hydrogen peroxide (see Chapter 11) and therefore, like H_2O_2 itself, serves as a powerful oxidizing agent and bleach.

Except for beryllium, all the other hydroxides are basic anhydrides. As represented in Equation (13.10), magnesium oxide reacts slowly with water to make the familiar mild antacid called *milk of magnesia*, $Mg(OH)_2$:

$$MgO(s) + H_2O(l) \longrightarrow Mg(OH)_2(s) \tag{13.10}$$

Lime, CaO, reacts with water to make "slaked lime," $Ca(OH)_2$. These hydroxides are good bases but not particularly soluble in water. $Ba(OH)_2$ is almost as strong a base as the alkali-metal hydroxides and is a useful, somewhat water-soluble base.

The halides of the alkaline-earth metals are readily synthesized by the direct (but often violent) reaction of the elements or, more simply, by treating the oxides (or hydroxides) with the appropriate hydrohalic acid, HX, as shown in Equation (13.11) for magnesium bromide:

$$MgO(aq) + 2HBr(aq) \longrightarrow MgBr_2(aq) + H_2O \qquad (13.11)$$

These halides are all of the general formula MX_2, a fact that demonstrates once again the degree to which the +2 oxidation state dominates the chemistry of these elements. (See Chapter 8 for an extended thermochemical discussion of why these elements do not form either MX or MX_3 salts.)

Uniqueness of Beryllium and Diagonal Relationship to Aluminum

Due to the exceptionally small size and high polarizing power of beryllium, we expect its compounds to be more covalent than those of its congeners. Indeed, as we have just seen, beryllium hydride, oxide, and hydroxide are predominantly covalent in character. In fact, the idea of a separate Be^{2+} ion is really a formality, there being little evidence that such a species ever really exists.

The structure of the beryllium halides, BeX_2, is shown in Figure 13.4 using (*a*) a Lewis structure, (*b*) a diagram of the gas-phase linear geometry as determined by VSEPRT, and (*c*) a valence-bond description. These compounds are characteristically electron-deficient, coming far short of an octet in the Lewis structure. The beryllium halides are often used in introductory courses to first illustrate *sp* hybrids (that account for the linear geometry and the equivalency of the two Be—X bonds in the gaseous forms of these compounds).

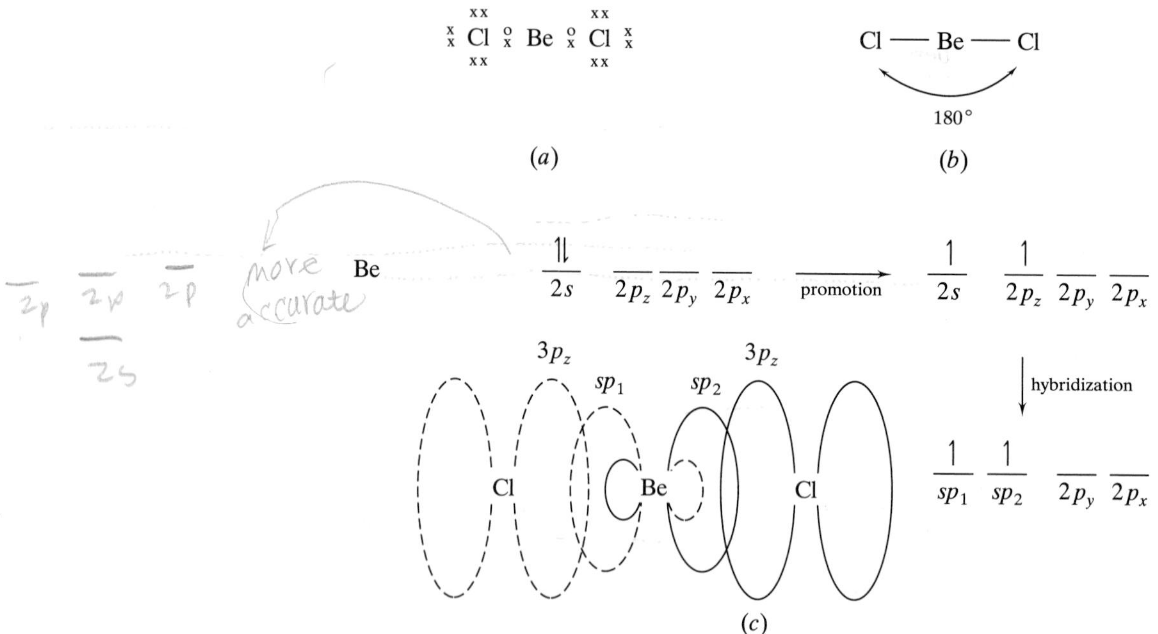

FIGURE 13.4
The bonding in gas-phase beryllium chloride shown as (*a*) a Lewis structure, (*b*) a VSEPRT diagram of the molecular geometry, and (*c*) a valence-bond theory (VBT) description. In the VBT the ground state is shown first, and then one electron is promoted to any one of the three $2p$ orbitals. The $2s$ and $2p$ orbitals are combined to form two equivalent *sp*-hybrid orbitals. The overlap between these *sp* orbitals and the $3p_z$ orbitals of the chlorine atoms gives two equivalent Be—Cl bonds.

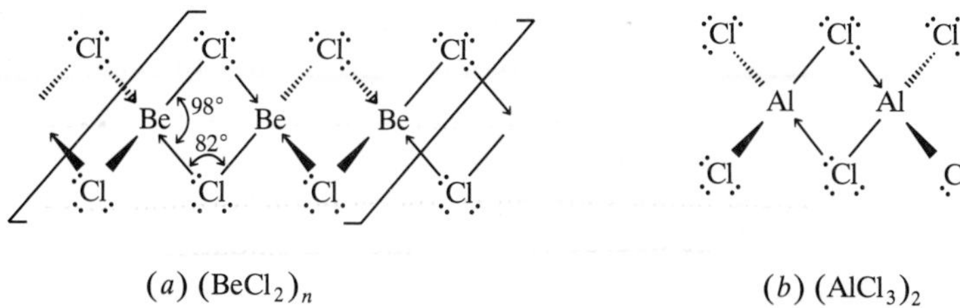

FIGURE 13.5
(a) The chloride bridging between beryllium chloride molecules allows an octet of electrons around each metal and produces a linear infinite-chain polymer, $(BeCl_2)_n$. (b) Similar chloride bridging in aluminum chloride produces a dimer, $(AlCl_3)_2$ or Al_2Cl_6.

Not surprisingly, BeX_2 compounds are excellent Lewis acids or electron-pair acceptors. For example, BeF_2 will react with two additional fluoride ions to produce BeF_4^{2-} as shown in Equation (13.12):

$$BeX_2 + 2X^- \longrightarrow BeX_4^{2-} \qquad X = F, Cl, Br \tag{13.12}$$

In this way, an octet around the beryllium atom is achieved. In such anions, the beryllium atom is sp^3-hybridized. A second way to achieve an octet about beryllium atoms is by forming infinite-chain polymers of the type shown in Figure 13.5a. Note that each beryllium atom is surrounded by four chlorides approximating a tetrahedron.

Another compound of beryllium that has no analog in the heavier congeners is basic beryllium acetate shown in Figure 13.6. Here the oxygen atom in the center is covalently bonded to four beryllium atoms arranged tetrahedrally. Each pair of beryllium atoms is bridged by an acetate. A similar nitrate compound also exists. These predominantly covalent compounds are soluble in nonpolar solvents such as the alkanes and insoluble in polar solvents such as water or even methanol or ethanol.

Recall that in Section 9.3 (p. 229ff) we used beryllium and aluminum to introduce the diagonal effect. We need not repeat that reasoning here. Suffice it to say that these two elements have much in common including (1) beryllium is more easily separated from its congeners than it is from aluminum, (2) both oxides are amphoteric rather than basic, (3) aluminum chloride forms a dimer structurally

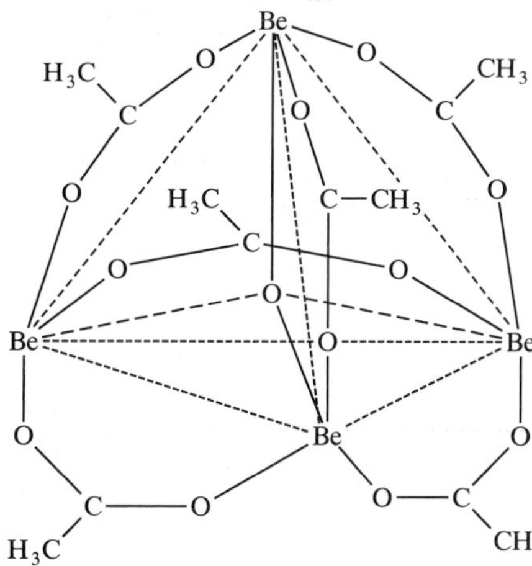

FIGURE 13.6
Basic beryllium acetate, $OBe_4(CH_3COO)_6$, has a central oxygen covalently bonded to four beryllium atoms in the shape of a tetrahedron. The six acetate ions span the sides of the tetrahedron. [Adapted from Ref. 20.]

related to the beryllium halide polymers as shown in Figure 13.5*b*, and (4) the resistance of both of these metals to attack by acids is due to the presence of a strong oxide film. Lest complacency set in about the ability of our network to explain all these things, it should be noted that beryllium shows almost as many similarities to zinc as it does to aluminum.

13.3 REACTIONS AND COMPOUNDS OF PRACTICAL IMPORTANCE

Like sodium and potassium, magnesium and calcium compounds are abundant in nature and essential for life as we know it. The hydrated ions, $[M(H_2O)_6]^{n+}$, where $n = 1$ or 2, can pass over various barriers in the body and thereby produce an electrical current of the type necessary for neurotransmission. Magnesium is the central part of the chlorophyll molecule responsible for photosynthesis. Calcium is the cationic component of bones and teeth and is important in cardiac activity, muscle contraction, as well as nerve impulse transmission.

Beryllium Disease

Beryllium, on the other hand, is toxic. Beryllium disease, originally called berylliosis, was first noted in the 1930s in German and Russian plants set up to extract the metal from beryllium ores such as beryl. As the use of this metal has increased, so also has the incidence of this disease. In occurs in both acute and chronic forms.

The chronic form has been encountered in extracting the metal, in alloy production, in the fluorescent lamp and neon sign industries (before they replaced the beryllium phosphor then in use with a variety of other compounds), and more recently in the nuclear weapons and power industries and in the development of new space- and aircraft. This form develops after a period of months or even many years after exposure. Symptoms include lack of appetite and loss of weight as well as a mild respiratory distress much like the early signs of the miner's black lung. In its most advanced stages, it causes disability and death. Not surprisingly, patients also show a much disturbed calcium balance resulting in the formation of calcium stones in the kidney and various other calcified bodies in the lungs. Perversely, this form also has been known to affect nurses, janitors, and other workers in beryllium plants and even their family members not in direct contact with the metal.

The acute form, which develops immediately after exposure, often shows up as a severe inflammation of the upper respiratory tract. Patients have severe difficulty in breathing, develop a disabling cough, and have a buildup of fluids in the lungs. If the beryllium is lodged in a crack in the skin, an ulcer develops which heals only after the metal is removed. Most patients afflicted with the acute form recover, but some develop the chronic form.

Both forms of beryllium disease are most likely related to the ability of the beryllium to covalently bond with the nitrogen atoms in proteins and therefore interfere with normal body functions of enzymes.

Radiochemical Uses

Radium is not the only element to have a radiochemistry related to practical implications. Beryllium is an excellent moderator and therefore is useful in the nuclear power industry. Mixed with an alpha particle source, commonly radium, beryllium also provides a good laboratory source of neutrons as shown in Equation (13.13). It was the successful interpretation of this reaction that led James

Chadwick to discover the neutron in 1932. Neutrons produced in this manner were also used in Enrico Fermi's early experiments in uranium fission.

$$^{9}_{4}Be + ^{4}_{2}He \longrightarrow ^{1}_{0}n + ^{12}_{6}C \tag{13.13}$$

Strontium 90 is a particularly troublesome fission product due to its (1) long half-life (28.1 years), (2) emission of beta particles [see Equation (13.14)], which penetrate tissue better than alpha particles, and (3) chemical similarity to calcium.

$$^{90}_{38}Sr \xrightarrow{\beta^-} {}_{0-1}e + ^{90}_{39}Y \tag{13.14}$$

During the atmospheric testing of nuclear weapons in the 1950s, there was great concern that this isotope would contaminate the milk supply and thereby endanger the health of children whose bone structure is rapidly developing. (Strontium can replace calcium in bone tissue.) More recently, this same concern has been renewed because the cloud of debris from the Chernobyl disaster contained strontium 90. The more serious problem in the long term is the safe disposal of the long-lived fission products (strontium 90 and cesium 137) derived from spent uranium and plutonium fuel elements. These must be stored for about *six centuries* (20 half-lives) before they can no longer be considered a threat to the environment!

Metallurgical Uses

The Group 2A metals, magnesium in particular and also calcium, are used as reducing agents to isolate metals from their halides and oxides. Two representative reactions are given in Equations (13.15) and (13.16):

$$2Mg(l) + TiCl_4(g) \longrightarrow Ti(s) + 2MgCl_2(l) \tag{13.15}$$

$$5Ca(l) + V_2O_5(s) \longrightarrow 2V(l) + 5CaO(s) \tag{13.16}$$

Magnesium is also used to isolate aluminum, uranium, zirconium, beryllium, and hafnium, among others.

In addition to being strong and of low density, beryllium alloys do not absorb neutrons and therefore are useful in nuclear reactors. Magnesium alloys, also lightweight and strong, are increasingly used for aircraft, naval vessels, luggage, photo and optical instruments, lawnmowers, portable tools, and so forth. For Volkswagen "bug" fans, 20 kg of a magnesium alloy was used in each of its engines. For ICBM fans (!), 1 ton of alloy is used per Titan missile.

Adding 2 percent beryllium to copper increases its strength six times over. These alloys are used for electrical switches, machinery, nonsparking and lightweight tools for the oil industry, and in critical moving parts in aircraft engines, camera shutters, and so forth. It is the rapid expansion of the use of beryllium in alloys which makes an awareness of beryllium disease of increasing importance.

Fireworks and X-Rays

The Group 2A metals play an important role in fireworks and related applications. Fireworks consist of an oxidizer, a fuel, colorizers, and various types of binding agents. Potassium, barium, and strontium salts such as the nitrates, chlorates, perchlorates, and peroxides are common oxidizing agents. (We have already discussed the peroxides; in future chapters we will investigate the oxidizing

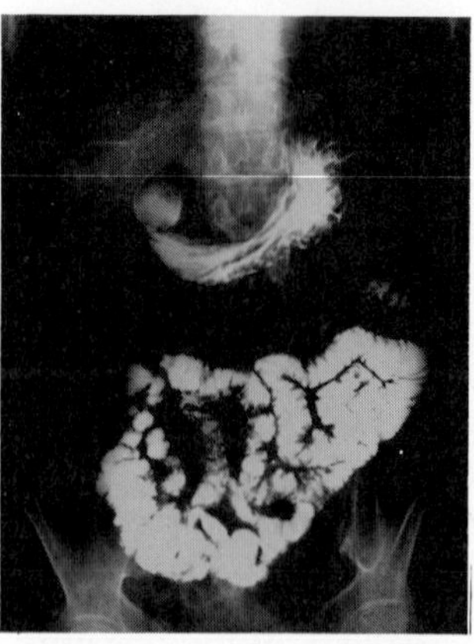

FIGURE 13.7
When a patient drinks a barium meal, an aqueous suspension of $BaSO_4$, his or her digestive tract becomes opaque to x-rays and affords an excellent diagnostic tool. (*Lester V. Bergman & Associates.*)

properties of nitrates, chlorates, and perchlorates.) Fuels include sulfur, charcoal, boron, magnesium, and aluminum. The colorizers are usually salts of strontium (scarlet), calcium (brick-red), barium (green), copper (blue-green), sodium (yellow) as well as elemental magnesium (white). Iron and aluminum granules result in gold and white sparks, while organic dyes are responsible for colored smokes. Powdered titanium results in a booming sound. Related applications include red flares and tracer bullet formulations using strontium and, in years past, flash photography using magnesium. Davy would continue to glory in the elements he helped to isolate!

X-ray technology is aided by the chemistry of the Group 2A metals. Beryllium absorbs x-rays only to a very small extent and so serves as a window in x-ray tubes. Barium sulfate, on the other hand, is opaque to x-rays. In the form of a finely divided suspension of the sulfate, a "barium meal" is given to a patient in order to get a good quality photograph of the digestive tract (see Figure 13.7). Not all barium salts are as nontoxic as the insoluble sulfate. The soluble chloride causes ventricular fibrillation, a dangerous misfiring of the heart muscle.

Hard Water

Hard water contains calcium and/or magnesium ions. Soft water does not. In particular, the calcium ions get into the water supply by the dissolution of calcium carbonate in the presence of carbon dioxide. Carbon dioxide is present in topsoil due to bacteriological activity. As rainwater percolates through the topsoil, it dissolves and carries the CO_2 along with it until it encounters a deposit of $CaCO_3$. $CaCO_3$ is dissolved in such an environment principally by the formation of bicarbonate as shown in Equation (13.17):

$$CaCO_3(s) + CO_2(aq) + H_2O(l) \longrightarrow Ca^{2+}(aq) + 2HCO_3^-(aq) \quad (13.17)$$

If there is a concentrated deposit of the carbonate, it will dissolve away to form a limestone cave or, if that collapses, a sinkhole such as those that have created so much trouble in Florida in recent years.

Hard water causes major difficulties in our society for two reasons. The first is that when it is heated, calcium carbonate can be reprecipitated in the teakettle, water heater, hot water pipe, or boiler. The reaction is represented in Equation (13.18):

$$\underset{\text{Acid}}{HCO_3^-(aq)} + \underset{\text{Base}}{HCO_3^-(aq)} \xrightarrow{\text{heat}} H_2CO_3(aq) + CO_3^{2-}(aq) \qquad (13.18)$$

(The H^+ is transferred from the base-labelled HCO_3^- to the acid-labelled HCO_3^-, and $H_2CO_3(aq) \longrightarrow H_2O(l) + CO_2(g)$.)

Note that the aqueous bicarbonate ion, when heated, reacts as both an acid and a base to produce both the carbonate ion and *carbonic acid*, which is more properly represented as carbon dioxide dissolved in water. (Have you ever noticed the small bubbles of carbon dioxide that form when water is heated but before it boils?) The resulting carbonate ions combine with the calcium ions to precipitate calcium carbonate in the form of a scale or deposit of the type shown in Figure 13.8. Magnesium carbonate is formed by a similar process.

Note that the calcium and magnesium ions that can be precipitated out by the above heating process contribute to what is known as *temporary hardness*, while those that are left over after heating contribute to *permanent hardness*.

The second difficulty posed by the calcium and magnesium ions of hard water is that they react with soap to produce a gummy, white precipitate that adheres to skin, hair, bathtubs, and clothes as well as clogs the small drainage holes in the tubs of automatic washers. The formation of this precipitate can be eliminated in two ways. One is to install an ion exchanger (Hey, Culligan Man!) to eliminate these cations before they get to people or washers. The second is to use a detergent that contains a soaplike *surfactant* molecule as well as a *builder* to

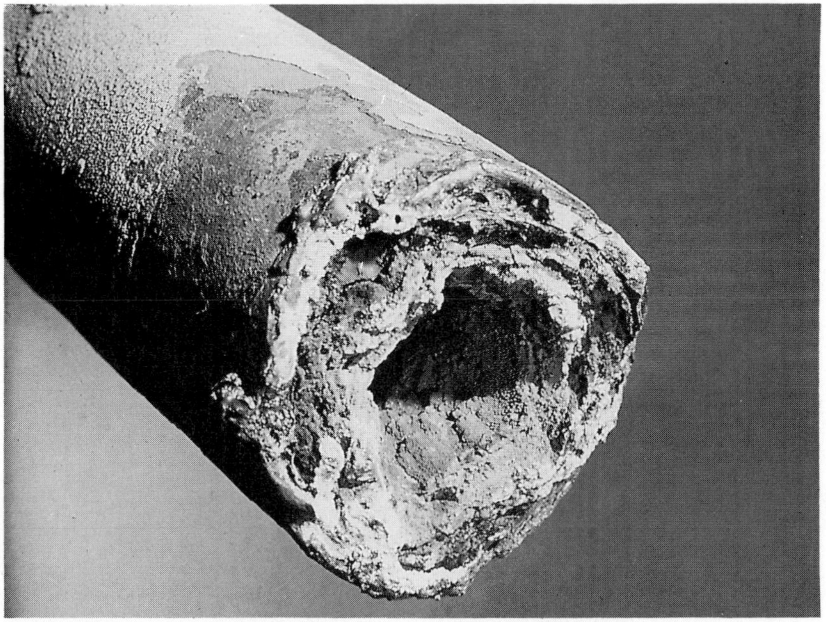

FIGURE 13.8
A hot water pipe filled with a predominantly $CaCO_3$ deposit. (*Betz Industrial.*)

prevent the hard water ions from interfering. In years past, the most common builders were chelating agents like tripolyphosphates which were discussed in Chapter 6. These phosphates, however, have been accused of causing algae growth and, in general, the advanced aging of bodies of water. In the 1970s phosphates in detergents were replaced by *suspended sodium carbonate* which simply precipitates out the calcium and magnesium ions as granular, crystalline $CaCO_3$ and $MgCO_3$ which do not adhere to people, clothing, or washing machines.

13.4 SELECTED TOPIC IN DEPTH: THE COMMERCIAL USES OF CALCIUM COMPOUNDS

$CaCO_3$ (Limestone)

Just as calcium carbonate was one of the cornerstones (pun intended) of the Roman economy, so it remains today. Limestone and marble remain popular for building materials, although they are degraded to the more soluble gypsum by acid rain as represented in Equation (13.19):

$$\underset{\text{Limestone and marble}}{CaCO_3(s)} + \underset{\text{Acid rain}}{H_2SO_4(aq)} + H_2O \longrightarrow \underset{\text{Gypsum}}{CaSO_4 \cdot 2H_2O(s)} + H_2O + CO_2(g) \tag{13.19}$$

The resulting damage to buildings and statues shown in Figure 13.9 is sometimes

FIGURE 13.9
The effect of *stone leprosy* on a limestone statue is apparent in photographs taken 61 years apart in 1908 (left) and 1969 (right). (*Westfälisches Amt für Denkmalpflege, Munich, Germany*.)

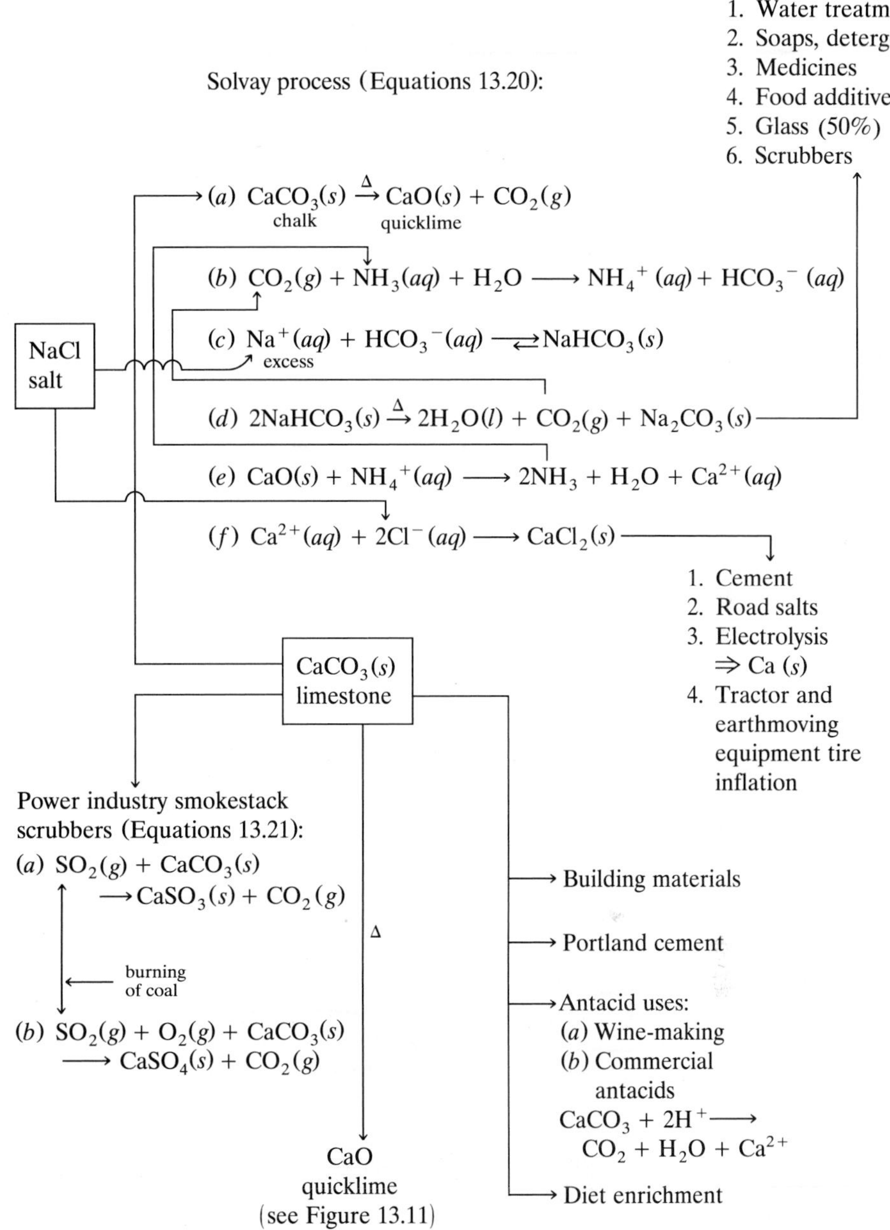

FIGURE 13.10
Some commercial uses of limestone, $CaCO_3$.

referred to as "stone leprosy." We will also see that calcium compounds still figure in mortar formulations.

Figures 13.10 and 13.11 summarize many of the commercial aspects of calcium chemistry. Figure 13.10 shows the uses of limestone, $CaCO_3$. Some of these have been mentioned elsewhere, but the Solvay process, smokestack scrubbers, and Portland cement need further explanation.

The Solvay process [Equations (13.20)] for the production of sodium carbonate starts (*a*) with limestone or chalk ($CaCO_3$) which is heated to produce quicklime (CaO) and carbon dioxide. (*b*) The carbon dioxide is combined with aqueous ammonia to produce the ammonium and bicarbonate ions. Although sodium bicarbonate is fairly soluble, keeping an excess of sodium ions (from NaCl) keeps the equilibrium of (*c*) shifted to the right to precipitate the salt. In (*d*) heating the sodium bicarbonate produces the desired primary product, sodium carbonate, and carbon dioxide gas again which is recycled. The quicklime product of (*a*) is combined with ammonium ion from (*b*) to produce more ammonia which is recycled to (*b*). In (*f*) the chloride ions from the salt and the calcium ions produce the by-product, calcium chloride. The net result is that NaCl and $CaCO_3$ are combined to produce $CaCl_2$ and Na_2CO_3. About half the sodium carbonate is used for the manufacture of glass, while the rest goes to the uses shown in the figure. While the Solvay process remains the primary process for the production of sodium carbonate, this important industrial chemical is increasingly derived from minerals such as natron (Na_2CO_3) and trona ($Na_2CO_3 \cdot NaHCO_3 \cdot 2H_2O$). One of the reasons for trying to reduce the reliance on the Solvay process is the difficulty in finding enough uses for the large amounts of calcium chloride it produces.

The calcium carbonate is also used to "scrub" the emissions from smokestacks of power plants and other industries that burn coal. Coal contains sulfur which in the combustion process is converted to sulfur dioxide and trioxide which go on to become the primary causes of acid rain (see Section 17.6). One solution to this problem is to remove the sulfur oxides before they are emitted from the smokestack. To do this the stack gases are passed through a slurry of calcium carbonate where, as shown in Equations (13.21), the sulfur di- and trioxides are converted to sulfites and sulfates, respectively. Controversy surrounds this process. The power industry claims it is too expensive and inefficient, while environmentalists insist that this is among the best available technologies for preventing acid rain. The more attentive reader may have already noted that this method adds to the amount of carbon dioxide released to the atmosphere which would contribute to the greenhouse effect.

No one chemical equation can come close to representing the complexities of the process by which Portland cement forms concrete. The name Portland cement is derived from the concrete product which resembles the natural limestone found in the Isle of Portland, England. Cement is a mixture of calcium oxide derived from the heating of limestone as well as silicon, aluminum, magnesium, and iron oxides which when mixed with water and sand harden into concrete. The setting process involves hydration of various salts, and, consequently, freshly poured concrete must be kept moist for a few days. The hardening of cement actually continues for years and produces an increasingly strong product over time.

CaO (Quicklime) and $Ca(OH)_2$ (Slaked Lime)

Figure 13.11 shows the commercial uses of quicklime and slaked lime. Note that it is not always clear whether limestone or the product of its burning, quicklime, is the actual raw material. Sometimes the limestone is added directly to a furnace or kiln and immediately converted to the quicklime which is shown as the reactant. In any case, three major uses of quicklime are listed. The calcium oxide combines with various oxides to rid iron ore of its silicon, aluminum, and phosphorus impurities as shown in Equations (13.22). It is a major component (with sodium carbonate from the Solvay process) for the glass industry. (More on this in Chapter 15 when we discuss silicates.) Finally, it is used to make calcium carbide that in turn is used to produce acetylene [Equation (13.23)].

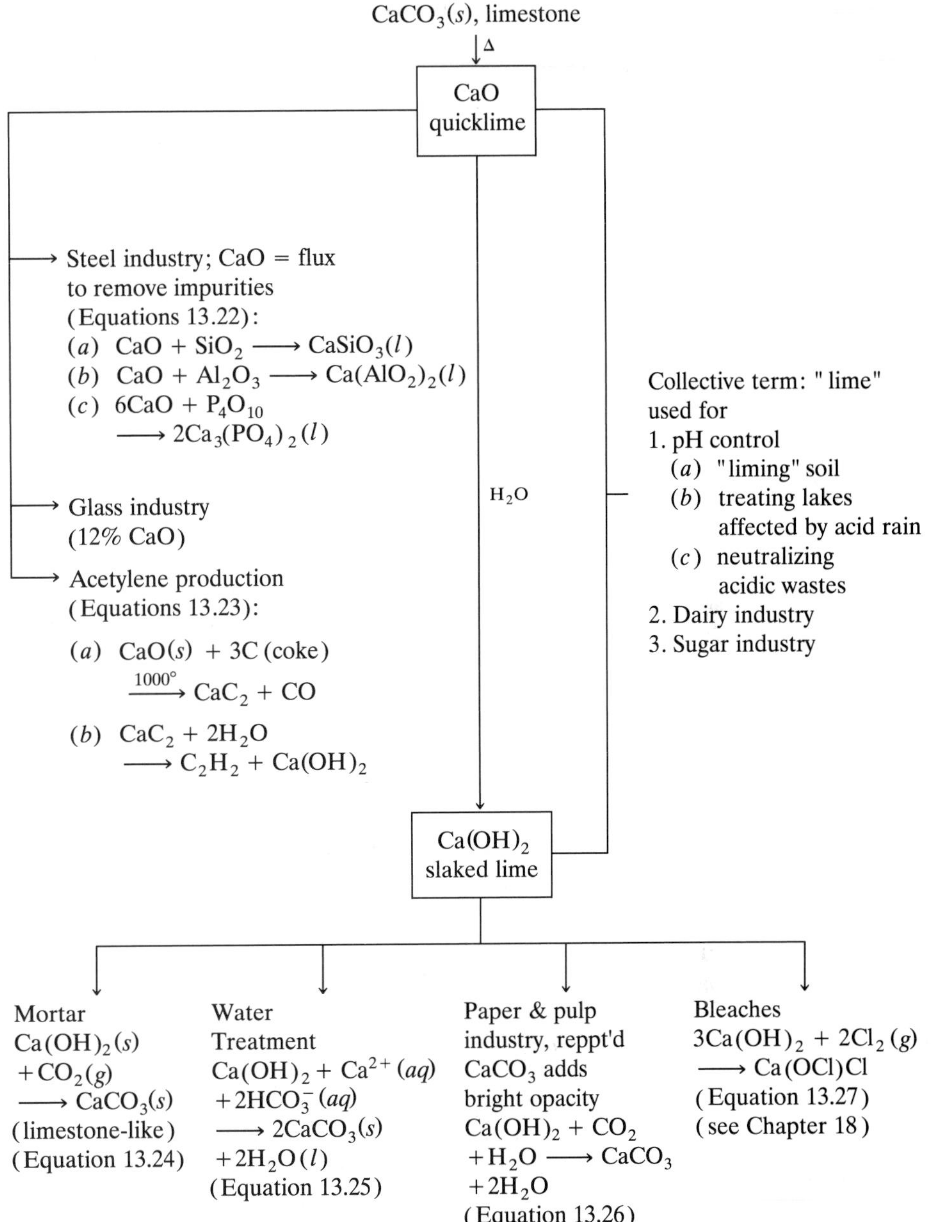

FIGURE 13.11
Some commercial uses of quicklime, CaO, and slaked lime, $Ca(OH)_2$.

When water is added to quicklime (CaO), slaked lime, or $Ca(OH)_2$, is produced. The figure lists four major uses of slaked lime including as mortar, for water treatment (softening temporarily hard water), in the paper and pulp industry, and for bleaches. The major equations pertaining to these uses, Equations (13.24) to (13.27), are listed in the figure.

Finally the figure shows some uses for *lime*, a collective term for quicklime and slaked lime together. It is used for pH control as well as in the dairy and sugar industries.

Calcium sulfate is also used in a variety of commercial applications. The dihydrate, mined as gypsum, is used in Portland cement and for gypsum wallboard. When the dihydrate is heated, it yields the fine hemihydrate powder, $CaSO_4 \cdot \frac{1}{2}H_2O$, known as plaster of paris. When plaster of paris is mixed with water, it makes a paste or slurry which hardens to form accurate molds. These reactions are summarized in Equation (13.28):

$$\underset{\text{Gypsum}}{CaSO_4 \cdot 2H_2O(s)} \xrightarrow[\text{"calcining"}]{250^\circ C} \underset{\text{Plaster of paris}}{CaSO_4 \cdot \tfrac{1}{2}O(s)} \qquad (13.28)$$

(Gypsum ← H_2O → Plaster of paris)

SUMMARY

The Egyptians and Romans used various forms of limestone for building materials and mortars. Earths that were alkalis have been known since the Middle Ages, but it was not until 1808 that Humphrey Davy isolated calcium, magnesium, barium, and strontium from their oxides or chlorides. Beryllium was found in beryl and emerald by Vauquelin in 1798 and isolated electrolytically by Bussy and Wöhler in 1828. Radium chloride was obtained from uranium ores by Marie Curie in 1902. Radium metal was prepared electrolytically in 1911.

The alkaline-earth metals have many similarities to the alkali metals. In both groups, the lightest element is unique, the second element is intermediate in character, the third, fourth, and fifth elements form a closely allied series, and the sixth element is rare and radioactive. The network helps us account for and predict the properties of both groups. The network components of particular importance are the periodic law, the uniqueness principle, the diagonal effect, and the acid-base character of oxides. The alkaline-earth metals have similar reducing properties, are higher boiling and melting, and are less electropositive and reactive than the alkali metals.

Except for beryllium and to some extent magnesium, the hydrides and oxides of the Group 2A elements are ionic and follow the general descriptions set forth in Chapters 10 and 11. The hydroxides are strong bases, and the halides demonstrate the dominance of the $+2$ oxidation state.

Beryllium compounds are covalent and similar to those of aluminum. BeO is amphoteric, while $BeCl_2$ is linear (in the gas phase) and best described in terms of *sp*-hybrid orbitals about the metal. Beryllium compounds are Lewis acids and form complex anions such as BeF_4^{2-} and halide-bridged polymers.

Topics under the practical importance of the alkaline-earth metals include their significance in living systems, the incidence of beryllium disease, radiochemical applications as components and residues of nuclear power, metallurgical uses in preparing pure metals and alloys, use in fireworks and x-ray technology, and the definition and properties of hard water.

Calcium compounds such as limestone, $CaCO_3$; quicklime, CaO; slaked lime, $Ca(OH)_2$; and gypsum, $CaSO_4 \cdot 2H_2O$, have a large number of commercial applications in the steel, glass, power, dairy, sugar, and paper and pulp industries. They are used in pH control and closely tied to the production of such diverse materials as soaps, detergents, food additives, antacids, cement, building materials, acetylene, and bleaches.

PROBLEMS

13.1. Use a good dictionary to look up the word *limelight*. How is it related to the alkaline earths?

13.2. Why is the phrase "mercury amalgam" redundant? What other amalgam is commonly employed as a consumer product?

13.3. Write an equation for the method by which Davy isolated strontium.

13.4. Strontium can also be prepared by the reduction of strontium oxide (strontia) with aluminum. Write an equation for this process.

13.5. Both barium and strontium can be prepared from their oxides and aluminum. Write equations for these two methods.

13.6. In Chapter 10, two types of subnuclear particles, the hadrons and the baryons, were noted to be present in the ylem. On the basis of the material presented in this chapter, which of these two particle types do you think would be the heavier? Briefly rationalize your answer.

13.7. How might you prepare Epsom salt from magnesia?

13.8. As mentioned in Section 13.1, the element magnesium takes its name from Magnesia, a district of Thessaly, Greece. Is this the same place that the Apostle Paul wrote to in the middle of the first century? Support your answer.

13.9. Seawater is 0.13% (by weight) in magnesium. If we were to continue to extract magnesium at the present rate (about 100 million tons/year) how long would it take for the percentage to be reduced to 0.12%? The volume of the earth's seas is approximately 1.5×10^9 km^3. Assume the density of seawater to be 1.025 g/cm^3.

13.10. Lebeau, in 1898, was the first to prepare beryllium electrolytically from beryllium fluoride mixed with sodium or potassium fluoride. Write an equation for the reaction. Speculate on the function of the alkali-metal salts.

13.11. Beryllium can be prepared by the reaction of its fluoride with magnesium. Write an equation for the process. Determine the $\Delta G°$ of the reaction using the following thermochemical data: $\Delta G_f^\circ[BeF_2(s)] = -979.4$ kJ/mol; $\Delta G_f^\circ[MgF_2(s)] = -1070.2$ kJ/mol.

13.12. Radium metal was first isolated by the electrolysis of the chloride in 1911. Write an equation for this reaction.

13.13. Radium is both an alpha and a beta minus particle emitter. Write an equation for its beta decay.

13.14. Radium is a decay product of uranium 238 which undergoes successive alpha, beta, beta, alpha, and alpha decays. Write the five equations representing this sequential decay. All the beta emissions involve beta minus particles.

13.15. Radium is an alpha emitter with a half-life of 3.82 days. Write an equation for its decay.

13.16. Marie Curie determined the atomic weight of radium by quantitatively converting radium chloride to silver chloride. She started with 0.10925 g of the radium chloride and finished with 0.10647 g of silver chloride. Determine the atomic weight of radium using this data.

13.17. Rationalize the trends in radii, ionization energies, electron affinities, and electronegativities of the Group 2A elements based on the network of ideas.

13.18. Would you be surprised to learn that barium emits electrons so easily when heated that its alloys are used in spark plug electrodes? Briefly explain.

* **13.19.** The estimated electron affinities of the alkaline-earth metals are positive, whereas those of the alkali metals are negative. Briefly rationalize this difference between these two groups of elements.

13.20. Although the trends of melting and boiling points within the Group 2A elements are irregular and difficult to explain, they are certainly considerably higher than the corresponding values for the alkali metals. Briefly rationalize this point. (*Hint*: Recall the concept of metallic bonding covered in Chapter 7.)

13.21. Analyze what happens when calcium is added to water using the standard reduction potentials given in Table 12.2. Calculate both the overall $E°$ and the $\Delta G°$ for the reaction.

***13.22.** Set up thermochemical cycles for the reduction of calcium, strontium, and barium ions to the free metals. Determine the enthalpy of these reductions, ΔH_{red}(Ca), ΔH_{red}(Sr), and ΔH_{red}(Ba), using the data provided below. Briefly relate your results to the standard reduction potentials of these three elements.

	IE[1], kJ / mol	IE[2], kJ / mol	ΔH_{atom}, kJ / mol	ΔH_{hyd}, kJ / mol	$E°$, V
Calcium	590	1158	178	−2469	−2.87
Strontium	549	1077	164	−2336	−2.89
Barium	503	977	180	−2198	−2.90

***13.23.** Set up thermochemical cycles for the reduction of $Li^+(aq)$ to $Li(s)$ and $Be^{2+}(aq)$ to $Be(s)$. Determine the enthalpy of these reductions, ΔH_{red}(Li) and ΔH_{red}(Be), by using the thermochemical data given below. Briefly relate your results to the standard reduction potentials of these elements.

	IE[1], kJ / mol	IE[2], kJ / mol	ΔH_{atom}, kJ / mol	ΔH_{hyd}, kJ / mol	$E°$, V
Lithium	520	7298	159	−964	−3.05
Beryllium	899	1757	324	−2494	−1.85

13.24. Calcium hydride is used in the Hydrimet process to reduce various oxides to the corresponding metals. Write an equation for the reactions of CaH_2 with rutile (TiO_2) and baddeleyite (ZrO_2).

13.25. Show a mechanism (that is, a sequence of molecular-level events) for the reaction of calcium hydride with water to produce aqueous calcium cations and hydroxide anions as well as hydrogen gas.

13.26. Show a mechanism (that is, a sequence of molecular-level events) of the reaction of BaO with water to produce aqueous barium cations and hydroxide anions.

13.27. Explain why calcium hydroxide is written as $Ca(OH)_2$ rather than H_2CaO_2 and serves as a base rather than as an acid.

13.28. How might the calcium halides be readily synthesized starting from the oxides? Give a general equation for these reactions as part of your answer.

13.29. Look up the word *formality*. Relate it to the existence of the Be^{2+} ion.

13.30. The beryllium cation in aqueous solution is surrounded by four water molecules, while the corresponding heavier cations are surrounded by six. That is, $Be^{2+}(aq)$ is more likely $Be(H_2O)_4^{2+}$, while its congeners are $M(H_2O)_6^{2+}$. Speculate on a reason for this difference.

13.31. Write an equation involving the reaction of an aqueous beryllium ion with a strong base to form $Be(OH)_4^{2-}$. Draw structures which represent the Lewis, VSEPRT, and VBT representations of the tetrahydroxoberyllate ion. Estimate the bond angles in the anion.

13.32. Give a rationalization for the fact that BeI_4^{2-} is not formed whereas similar compounds with the other halogens are as shown in Equation (13.11).

13.33. Even though we have not covered Group 3A yet, speculate on the nature of aluminum chloride. Will it be covalent or ionic? Will it be soluble in water? In nonpolar organic solvents? Will it be a Lewis acid or Lewis base?

13.34. Speculate on why the ingestion of beryllium sometimes results in the formation of calcium-containing stones in the kidney.

* **13.35.** In the immediate aftermath of the Chernobyl disaster there was concern about the amount of strontium 90, cesium 137, as well as iodine 131 in the resulting radioactive plume. Iodine 131 is a beta minus emitter with a half-life of 8.0 days. Write an equation for its decay and discuss the dangers it might pose as compared to strontium 90.

13.36. Aluminum used to be made by reducing aluminum chloride with magnesium metal. Write an equation for this process.

13.37. Write equations to represent the use of magnesium to "win" zirconium and hafnium metals from their chlorides.

13.38. When Madame Curie was separating radium salts from pitchblende, they were often precipitated with barium salts. She kept recrystallizing until her radium chloride was spectroscopically pure. Speculate on what that might mean in this context.

13.39. Plumbers can rid home plumbing of calcium carbonate deposits by the judicious use of hydrochloric acid. Write an equation for this remedy.

13.40. In Chapter 12 we defined *disproportionation* as a reaction in which an element is both oxidized and reduced. A reaction such as that represented in Equation (13.18) and repeated below is also sometimes referred to as disproportionation.

$$2HCO_3^-(aq) \longrightarrow H_2O(l) + CO_2(g) + CO_3^{2-}(aq)$$

Is bicarbonate oxidized or reduced in this reaction? If not, in what sense do you imagine it disproportionates? Equation (11.7), the self-ionization of water, is also sometimes referred to as a disproportionation. In which sense is this designation made?

13.41. Propose a mechanism (a sequence of molecular-level events) for the self-ionization or disproportionation of bicarbonate when heated.

13.42. Another way to eliminate the formation of the gummy, white precipitate which forms between hard water ions and soap is to add washing soda, $Na_2CO_3 \cdot 10H_2O$. Explain why this procedure works.

13.43. Calcium carbonate is the active ingredient in a number of popular antacids including Tums and Rolaids. Write an equation for the reaction of $CaCO_3$ with stomach acid (HCl).

13.44. Why do many people burp when they use Tums or Rolaids?

13.45. Sodium bicarbonate is also an antacid. Write an equation for its reaction with stomach acid (HCl).

13.46. Why is sodium carbonate less soluble in a solution saturated with sodium ions than it is in pure water? Give a name to this phenomenon that you probably studied in earlier chemistry experiences.

13.47. Sodium carbonate is listed in the *CRC Handbook of Chemistry and Physics* to be soluble to the extent of 7.1 g/100 cm^3 of cold water. Calculate a value of K_{sp} for this salt.

13.48. Using the K_{sp} calculated in Problem 13.47, set up an equation that, when solved, would yield a value for the solubility of sodium carbonate in a solution which is 5.0 *M* in sodium chloride. Is this solubility closer to 5.0, 0.50, 0.050, or 0.0050 *M*? Briefly rationalize your answer.

13.49. Are the reactions represented in Equations (13.22) (in Figure 13.11) redox equations? If so, which reactants are oxidized and which are reduced?

CHAPTER 14

THE GROUP 3A ELEMENTS

Group 3A does not have a name like the alkali metals, the alkaline-earth metals, or the halogens, in part because boron, aluminum, gallium, indium, and thallium are not as chemically similar to each other as are the elements of the above three groups. After the usual section on the discovery and isolation of the elements, we will see that *all* the components of the network significantly contribute to an understanding of this group. These contributions are particularly evident in a study of the properties of the oxides, hydroxides, and halides. A discussion of the special structural aspects of boron chemistry is followed by descriptions of aluminum metal, its alloys, alums, and alumina. The selected topic in depth is electron-deficient compounds which are prevalent in but certainly not restricted to Group 3A chemistry.

14.1 DISCOVERY AND ISOLATION OF THE ELEMENTS

Figure 14.1 shows the discovery dates for these elements superimposed on the plot of the total number of known elements from 1650 to the present.

Boron

Boron was known to the ancients in the form of borax that was used for glazes and various types of glass. Boron is almost always found directly bound to oxygen and is difficult to prepare in pure form. In 1808, the ebullient and foolhardy English chemist Sir Humphry Davy, whom we encountered as the discoverer of potassium and sodium (Chapter 12, pp. 297–298) as well as magnesium, calcium, strontium, and barium (Chapter 13, pp. 325–326), was just barely beaten (by 9 days) to the discovery of boron by the French chemists Joseph Louis Gay-Lussac and Louis Jacques Thénard. Yes, this is the same Gay-Lussac who proved (in 1802) that the volume of a gas is directly proportional to the temperature. (Jacques Charles, a

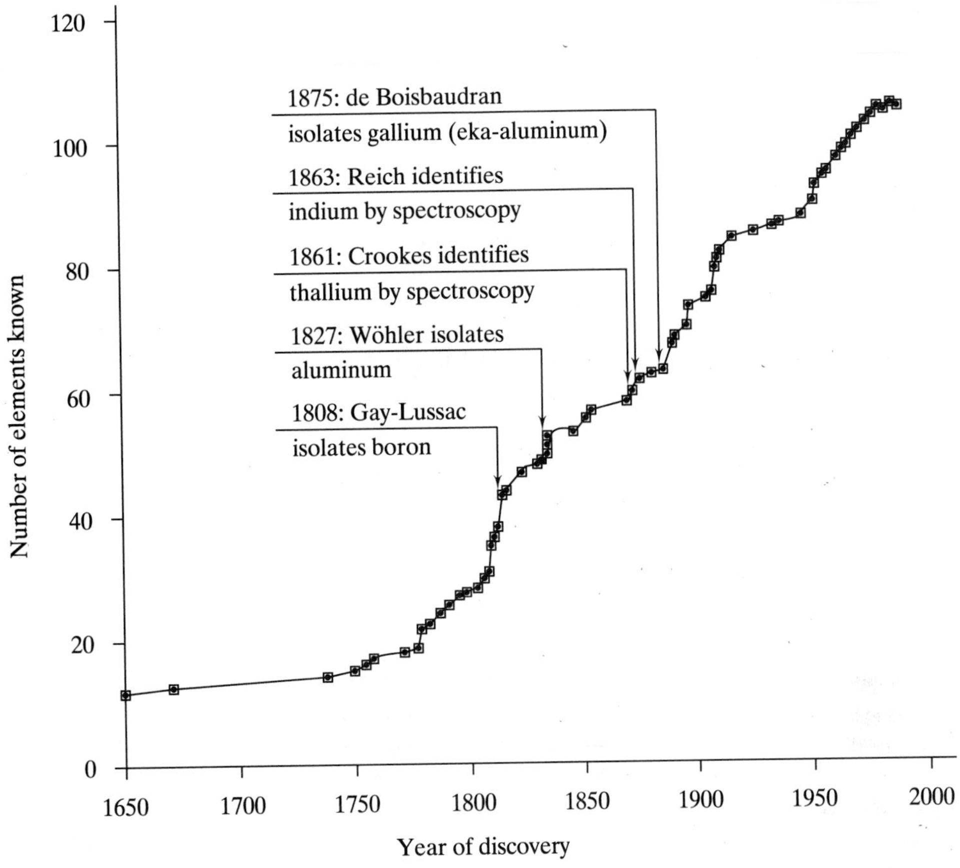

FIGURE 14.1
The discovery dates for the Group 3A elements superimposed on the plot of the number of known elements versus time.

French physicist, actually formulated this relationship some 15 years earlier, but he did not publish his results.) Somewhat ironically, Gay-Lussac and Thénard reacted Davy's potassium with boron oxide to isolate the elemental boron.

The name *boron* (*bor*ax + carb*on*) was proposed by Davy, as the element comes from the mineral borax and is similar in appearance to carbon. Borax, found in deserts and formerly volcanic areas, remains the principal source of boron compounds. One of the largest open-pit borate mines is located in a California mining town appropriately called Boron. Boron is one of the few elements whose percent isotopic abundance varies significantly from one region to another. For example, the borates from California are somewhat low in ^{10}B, while those from Turkey are high in this isotope. In fact, there is more relative uncertainty in the atomic weight of boron than in that for any other element.

Aluminum

Aluminum takes its name from alum, $KAl(SO_4)_2 \cdot 12H_2O$, which the ancients used as an astringent, a substance that shrinks tissues and checks the flow of blood by contracting blood vessels. This time Davy was unable to isolate the element but in

1807 proposed the name *alumium* and later *aluminum* for the yet-undiscovered element. Friedrich Wöhler, who isolated beryllium, is also generally credited with the isolation of what was finally called *aluminium* (pronounced al-u-min′-i-um) in 1828. Not surprisingly from what we have seen before, he reduced $AlCl_3$ with metallic potassium to produce the free metal. The rest of the world continues to call the element aluminium, but in 1925, the American Chemical Society decided that the United States would use Davy's aluminum (pronounced a-lu′-mi-num) for the spelling and pronunciation of the element. Even today, this disparity leads to some initial confusion as well as some good-natured ribbing among the English-speaking peoples of the world.

Efficient methods of preparing this metal, the most abundant in the earth's crust, were slow to develop. It was still considered a precious metal in the 1850s and was displayed as such at the Paris Exposition of 1855. It is reported that Emperor Louis Napoleon III and his special guests used aluminum cutlery and dinner service on state occasions (while lesser guests had to content themselves with the silver and gold versions) and that his son played with an aluminum rattle. A 100-oz slab of this precious metal was also placed atop the newly completed Washington Monument in 1885. How did aluminum become the common, everyday metal we know today? In 1886, Charles Hall, a chemistry major recently graduated from Oberlin College, found a way to prepare aluminum electrolytically (using carbon electrodes) from a mixture of bauxite, an impure ore of the hydrated oxide, dissolved in melted cryolite, Na_3AlF_6. The net reaction for this process is shown in Equation (14.1):

$$3C(s) + 4Al^{3+} + 6O^{2-} \xrightarrow{\text{electrolysis}} 4Al(s) + 3CO_2(g) \qquad (14.1)$$

In France, Paul Louis Héroult discovered the same energy-intensive process at about the same time, and the method is consequently called the Hall-Héroult process. In one of those amazing coincidences, Héroult was born 8 months before Hall and also died 8 months before him. Both men were 23 years old when they discovered a process which changed human history not to mention their own lives.

We cannot leave a discussion of the Hall-Héroult process without emphasizing a growing concern. Almost 5 percent of the electricity produced in the United States is used to make aluminum. As we have noted elsewhere, the burning of fossil fuels to make electricity has a large environmental impact. The resulting carbon dioxide feeds the greenhouse effect (see Chapter 11, pp. 289–291), while the sulfur oxides are responsible, in large part, for acid rain (see Chapter 17, pp. 464ff). The need to maximize the recycling of aluminum products and therefore minimize the energy demands of their production is obvious. In a twist of fate, the resulting acidity in our lakes and ponds caused by acid rain is accompanied by the "mobilization" of various metals such as mercury, cadmium, and, yes, aluminum. Aluminum, at the pH levels now being recorded in some Canadian and American bodies of water, has been found to be toxic to fish and other aquatic organisms. Perhaps the next generation of chemists, with the same investigative and enterprising spirit demonstrated by Charles Hall and Paul Héroult, will find a answer for some of these problems.

Gallium

Gallium, Mendeleev's eka-aluminum, was discovered by Paul Émile Lecoq de Boisbaudran in 1875, five years after the Russian chemist left a blank in his table for it and predicted a number of its properties. De Boisbaudran identified the

TABLE 14.1
A comparison of the properties of eka-aluminum predicted by Mendeleev with the properties of gallium found by Lecoq de Boisbaudran

	Mendeleev's predictions for eka-aluminum (1871)	Lecoq de Boisbaudran's observed properties of gallium (1875)
Atomic weight	≈ 68	69.9
Valence	3	3
Density, g/cm^3	5.9	5.93
Melting point	Low	30.1°C
Volatility	Low	Low
Formula of oxide	M_2O_3	Ga_2O_3
Formula of chloride	MCl_3	$GaCl_3$
Mode of discovery	Spectroscopy	Spectroscopy

element in a sample of zinc ore by using the still new field of spectroscopy that Bunsen and Kirchhoff had employed to detect the "fingerprints" of elements such as cesium and rubidium (see Chapter 12, p. 299). Within a month he had completed a series of conversions ending in the electrolysis of an aqueous solution of the hydroxide and potash (potassium carbonate). He was able to isolate enough of the easily liquefied metal (mp = 29.78°C) to measure its properties and even present some to the French Academy of Sciences. A comparison of the properties of gallium and those predicted for eka-aluminum is given in Table 14.1.

De Boisbaudran originally determined the density of the metal to be 4.7 g/cm^3, but after Mendeleev wrote to him suggesting that he measure the value again (!), the Frenchman found it to be approximately as given in the table. De Boisbaudran named the new element after *Gallus*, the Latin name for the territory that included what we know as modern France. Of course, the name Lecoq means "rooster" and the Latin equivalent of rooster is *gallus*, so perhaps there was a dual reason that de Boisbaudran chose this name.

Gallium is now obtained as a by-product of the aluminum industry. As a solid at room temperature, it is a soft, bluish-white metal which easily melts in the hand. It remains a liquid up to 2403°C, giving it the largest liquid range known.

Indium and Thallium

Indium and thallium were also discovered using spectroscopy. Indium was first identified in 1863 by Ferdinand Reich and his student Hieronymus Theodor Richter who identified an indigo-colored emission line in a zinc ore. (It is interesting to note that although Richter identified his new element by color, he was, in fact, color-blind.) Richter later tried to take sole credit for the discovery. The color of the emission line prompted the name of the element. Thallium was discovered in 1861 by Sir William Crookes, whom you may associate with Crookes' tubes and the initial work that led J. J. Thomson to discover the electron. Crookes, while "scooped" on the electron, analyzed selenium ores to find a beautiful green line previously unidentified. The element takes its name from *thallos*, the Greek word for "green twig."

We now know that thallium is extremely toxic and should be handled with great care. In the first 50 years after its discovery, however, thallium was found to be an effective treatment for syphilis, gonorrhea, gout, dysentery, and tuberculosis. Someone should have taken a clue from the location of this element between lead and mercury in the periodic table. Its use as a medicine had many toxic not to mention unpleasant side effects such as alopecia, the loss of hair. Later thallium compounds such as the colorless and odorless Tl_2SO_4 (note this compound contains thallium in the $+1$ oxidation state) was used as a rat poison and ant killer, but these and similar uses have been prohibited in many countries due to the high toxicity of thallium.

14.2 FUNDAMENTAL PROPERTIES AND THE NETWORK

Figure 14.2 shows the Group 3A elements superimposed on the network, and Table 14.2 shows the usual tabulation of the properties of the group.

Look carefully at Figure 14.2. Compare it with Figure 12.6 presented when the network had just been completed. Note that these two figures are extraordinarily similar. (Only the symbol for gallium has been added to Figure 14.2.) This similarity is significant because the Group 3A elements are intimately tied to our network. Boron is involved in the uniqueness principle, the diagonal effect, and is a metalloid located on the metal-nonmetal line. Aluminum is diagonally related to beryllium (both of which are metals on the metal-nonmetal line), while indium and thallium are involved in the inert-pair effect. With the aid of the network, we know much about Group 3A even before we begin to look at the specifics.

A quick inspection of Table 14.2 shows many of the expected periodic variations. As always, there are a few anomalies. Note that gallium has an atomic radius only 0.10 Å greater than that of aluminum and also a higher-than-expected electronegativity (as do indium and thallium). In addition, the ionization energy of gallium is virtually the same as that for aluminum rather than lower as expected. All of these properties can be traced back, albeit somewhat simplistically, to the supposition that the nd^{10} and nf^{14} subshells are not particularly good at shielding succeeding electrons from the nuclear charge. (See discussion, Chapter 9, pp. 231–233.)

We are certainly not surprised to learn that boron is unique in the group. The theoretical B^{3+} ion is a small, highly charged formal species with a high polarizing power. Boron compounds are all predominantly covalent. Aluminum is intermediate in character but still, like beryllium, a bona fide metal. The remainder of the elements are clearly metals, and we expect their oxides and hydroxides to react accordingly. The $+3$ oxidation state predominates in aluminum, but the $+1$ state becomes more important going down the group until it is actually the most common oxidation state for thallium. Note that TlX and Tl_2O are the common reaction products of thallium with the halogens and oxygen.

Hydrides, Oxides, Hydroxides, and Halides

To further solidify our understanding of this group based on the network, we would normally turn to an inspection of the hydrides, oxides, hydroxides, and halides of these elements. The hydrides, however, are rather special compounds, and a consideration of them is deferred to the selected topic in depth at the end of the chapter. As expected, they do start off predominantly covalent and become more ionic going down the group.

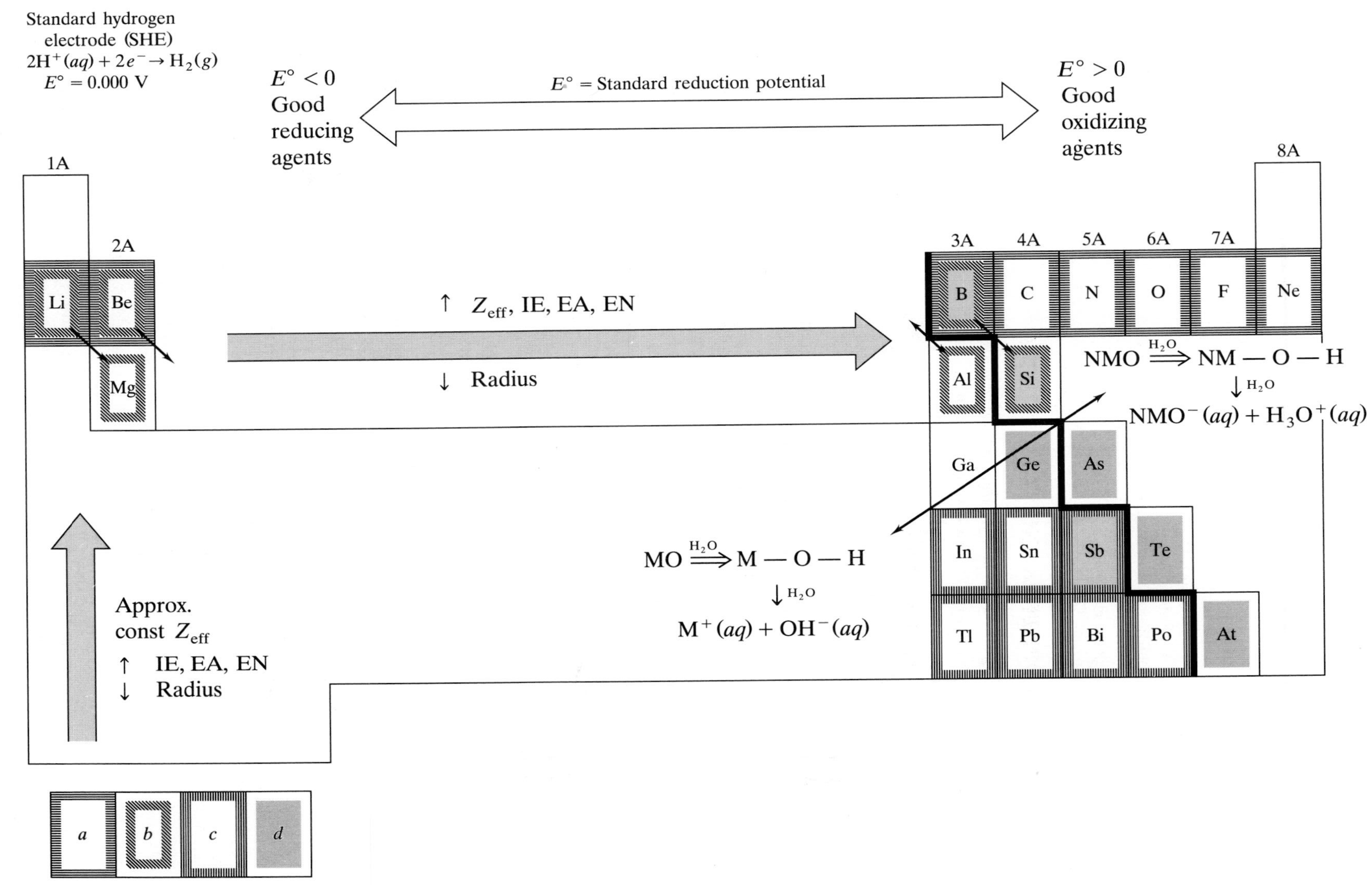

FIGURE 14.2
The Group 3A elements superimposed on the network of interconnected ideas including the trends in periodic properties, the acid-base character of metal and nonmetal oxides, trends in standard reduction potentials, (*a*) the uniqueness principle, (*b*) the diagonal effect, (*c*) the inert-pair effect, and (*d*) the metal-nonmetal line.

TABLE 14.2
The fundamental group properties of the Group 3A elements. [Ref. 24.]

	Boron	Aluminum	Gallium	Indium	Thallium
Symbol	B	Al	Ga	In	Tl
Atomic number	5	13	31	49	81
Natural isotopes, A/% abundances	$^{10}B/19.78$ $^{11}B/80.22$	$^{27}Al/100$	$^{69}Ga/60.4$ $^{71}Ga/39.6$	$^{113}In/4.28$ $^{115}In/95.72$	$^{203}Tl/29.50$ $^{205}Tl/70.50$
Total no. of isotopes	6	7	14	19	21
Atomic weight	10.81	26.98	69.72	114.82	204.37
Valence electrons	$2s^22p^1$	$3s^23p^1$	$4s^24p^1$	$5s^25p^1$	$6s^26p^1$
mp/bp, °C	2300/2550	660/2467	29.78/2403	156.6/2080	303.5/1457
Density, g/cm^3	2.34	2.70	5.90	7.30	11.85
Atomic radius, metallic, Å	—	1.43	1.53	1.67	1.71
Ionic radius, Shannon-Prewitt, Å (C.N.)	(0.25)(4)	0.53(4)	0.76(6)	0.94(6)	(+3)1.02 (6) (+1)1.64 (6)
Pauling EN	2.0	1.5	1.6	1.7	1.8
Charge density (charge/ionic radius), unit charge/Å	12.0	5.7	4.0	3.2	(+3)2.9 (+1)0.61
$E^{\circ\dagger}$, V	−0.90	−1.66	−0.56	−0.34	−0.33
Oxidation states	(+3) covalent	+3	+1, +3	+1, +2, +3	+1, +3
Ionization energy, kJ/mol	801	578	579	558	589
Electron Affinity, kJ/mol	−23	−44	−35	−34	−48
Discovered by/date	Gay-Lussac 1808	Wöhler 1827	de Boisbaudran 1875	Reich 1863	Crookes 1861
rpw‡ O_2	B_2O_3	Al_2O_3	Ga_2O_3	In_2O_3	Tl_2O
Acid-base character of oxide	Acid	Ampho.	Ampho.	Ampho.	Base
rpw N_2	BN	AlN	GaN	InN	None
rpw halogens	BX_3	Al_2X_6	Ga_2X_6	In_2X_6	TlX
rpw hydrogen	—	—	—	—	
Crystal structure	Hexagonal	fcc	Cubic	Tetragonal	Hexagonal

$^\dagger E^\circ$ is the standard reduction potential in acid solution, $B(OH)_3 \longrightarrow B$; $M^{3+}(aq) \longrightarrow M(s)$ for M = Al, Ga, In, Tl.
‡rpw = reaction product with.

Consistent with the position of the metal-nonmetal line, boron oxide is an acid anhydride, while the oxides of the heavier elements progress from amphoteric to basic in behavior. Boron oxide, then, reacts with water as shown in Equation (14.2) to produce boric acid, $B(OH)_3$ or H_3BO_3:

$$B_2O_3 + 3H_2O \longrightarrow 2B(OH)_3 \tag{14.2}$$

Boric acid, which (as an aqueous solution) you may recognize as a mild antiseptic and an eye- and mouthwash, has the planar structure shown in Figure 14.3. The boron atom assumes sp^2 hybridization, while the oxygens are sp^3. The valence-shell electron-pair repulsion theory predicts that the O—B—O bond angle is 120°, while the B—O—H angle is somewhat less than 109.5° due to the presence of the two lone pairs on each oxygen. There is free rotation about all the sigma bonds of the molecule. In the solid, the single $B(OH)_3$ units are hydrogen-bonded into a planar, sheetlike structure (shown in Figure 14.4) which accounts for the flaky texture of this compound.

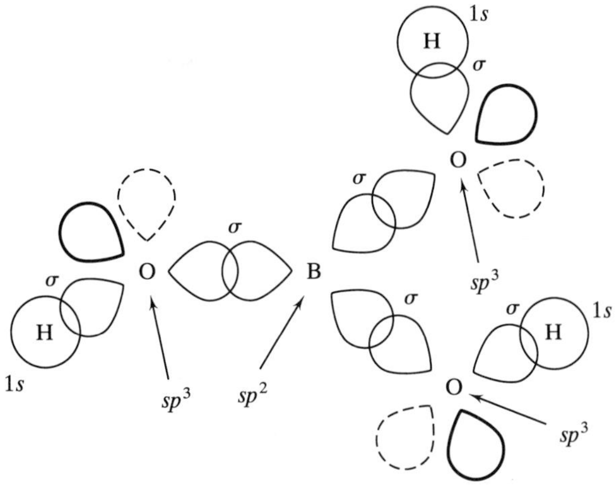

FIGURE 14.3
The planar molecular structure of boric acid, $B(OH)_3$.

Even though $B(OH)_3$ is an acid ($K_a = 5.9 \times 10^{-10}$), it is best to so regard it using the Lewis definition rather than those of Brønsted-Lowry or Arrhenius. As shown in Equation (14.3), $B(OH)_3$ acts as an acid by accepting a pair of electrons from an incoming OH^- ion rather than transferring a proton to the hydroxide:

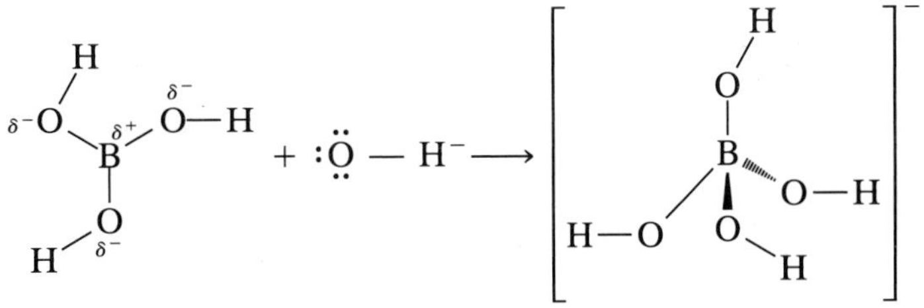

$$B(OH)_3 + :\ddot{O}-H^- \longrightarrow [B(OH)_4]^- \tag{14.3}$$

Note that the negative end of the hydroxide, the oxygen atom, attacks the partially positive boron atom of the acid to produce a fourth B—OH bond. In the process, the boron atom assumes sp^3 hybridization in the product. Only one OH^- group is accepted so that this compound is *monobasic* and not tribasic as one might suspect from the formula, H_3BO_3. Boric acid, by the way, is similar to silicic acid, $Si(OH)_4$, while $Al(OH)_3$, which you might recall from introductory chemistry is aluminum hydroxide, not "aluminic acid," is essentially a base with some amphoteric charac-

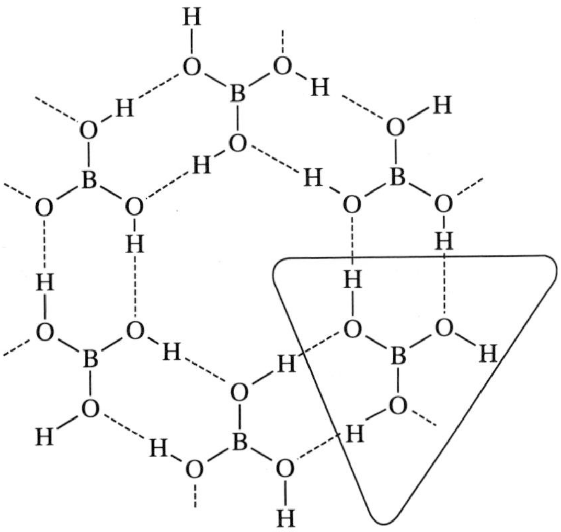

FIGURE 14.4
The planar, sheetlike structure of boric acid showing the individual molecules (one highlighted) held together by hydrogen bonds (dashed lines). [Adapted from Ref. 25.]

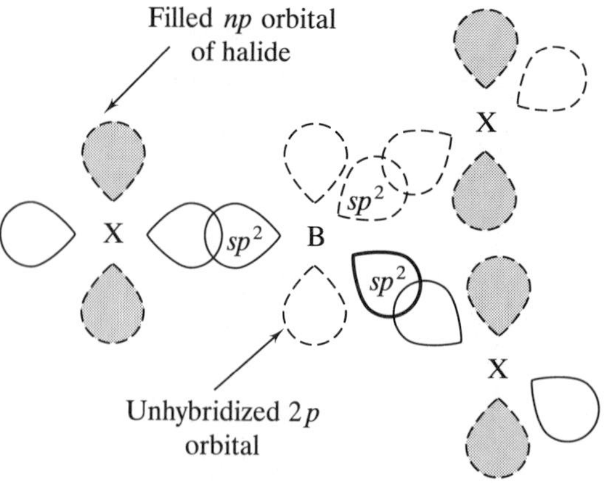

FIGURE 14.5
The structure of BX_3 molecules. The boron atom is sp^2-hybridized. Some evidence exists that some pi bonding may occur between the unhybridized $2p$ orbital of the boron and the filled (shaded) np orbitals of the halides.

ter. The remaining oxides and hydroxides, as expected from our earlier discussion in Chapter 11 (pp. 276ff), become progressively more basic down the group. Consistent with the inert-pair effect, the most stable oxide and hydroxide of thallium are Tl_2O and TlOH, not Tl_2O_3 and $Tl(OH)_3$. Thallium(I) oxide is a mild basic anhydride, while thallium(I) hydroxide is a moderately strong base.

The halides of boron, BX_3, are volatile, highly reactive, covalently bonded, molecular compounds. BF_3 is a gas, BCl_3 and BBr_3 are liquids, while BI_3 is a white solid. These phases reflect melting and boiling points which increase with increasing molecular weight and strength of van der Waals intermolecular forces. Not unexpectedly, they are, as shown in Figure 14.5, nonpolar, trigonal planar molecules in which the boron atom is sp^2-hybridized. The B—X bond distances are shorter than might be expected, while the B—X bond energies are correspondingly higher. Indeed, the B—F bond energy (646 kJ/mol) is the highest known for a single bond. All this has led to the proposal that some pi bonding may occur between the unhybridized $2p$ orbital of the boron and the filled np orbitals of the halides. Such a proposal is consistent with one of the reasons for the uniqueness of the first elements in each group (see Chapter 9, p. 226).

The highly corrosive gaseous boron trifluoride is prepared by the action of concentrated sulfuric acid on a mixture of the oxide and calcium fluoride [Equations (14.4)]:

$$CaF_2(s) + H_2SO_4(l) \longrightarrow CaSO_4(s) + 2HF(g) \qquad (14.4a)$$

$$B_2O_3(s) + 6HF(g) \longrightarrow 2BF_3(g) + 3H_2O(l) \qquad (14.4b)$$

Sulfuric acid is a dehydrating agent and removes the excess water produced in Equation (14.4*b*). The chloride and bromide are prepared by the interaction of the elements at elevated temperatures, while the iodide can be prepared either by direct reaction of the elements or by treating BCl_3 with HI also at high temperatures.

The boron halides, like boric acid, are Lewis acids; that is, they can readily accept a pair of electrons into their vacant, unhybridized p orbitals. Some representative reactions of BF_3 are shown in Equations (14.5) to (14.7):

$$BF_3 + F^- \longrightarrow BF_4^- \qquad (14.5)$$

$$BF_3 + CH_3Cl \longrightarrow [CH_3^+][BF_3Cl]^- \qquad (14.6)$$

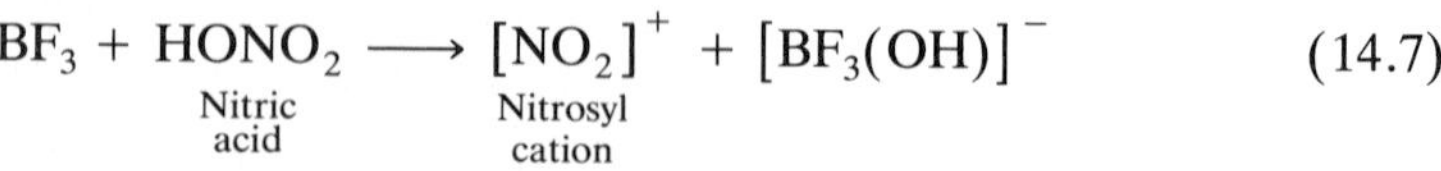

$$\underset{\text{Nitric acid}}{BF_3 + HONO_2} \longrightarrow \underset{\text{Nitrosyl cation}}{[NO_2]^+} + [BF_3(OH)]^- \qquad (14.7)$$

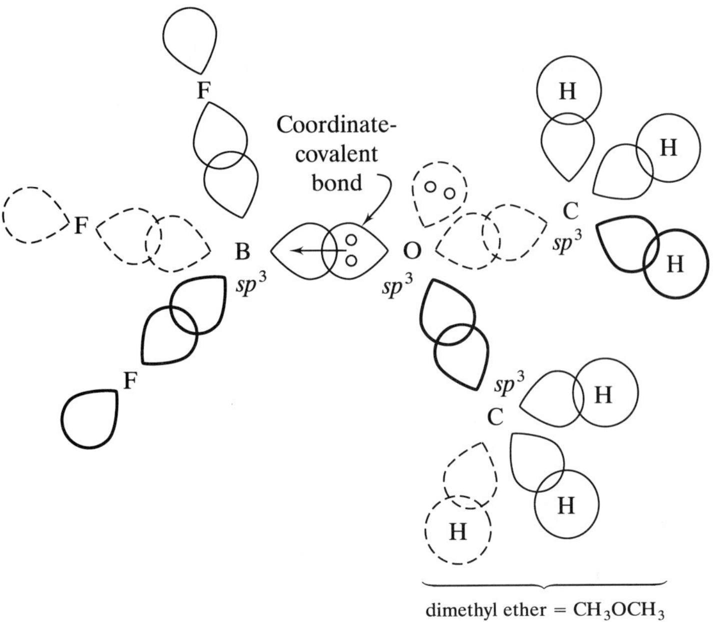

FIGURE 14.6
The "etherate" of BF_3 is a Lewis salt. Dimethyl ether (the Lewis base) contributes a lone pair of electrons to the unfilled, unhybridized $2p$ orbital of the BF_3 (the Lewis acid). The boron atom is rehybridized to sp^3. A molecule characterized by a bond in which both the bonding electrons come from one atom is known as an *adduct* or sometimes an *addition compound*. The bond itself is called a *dative bond*, or sometimes a *coordinate-covalent bond*.

Note that in each case, the negative reactant or the most partially negative part of the reactant (the F^-, the chlorine of CH_3Cl, or the OH oxygen of the nitric acid) attacks the central boron of the trifluoride. The resulting CH_3^+ and NO_2^+ are extremely useful in organic synthesis because they provide economical pathways to a huge variety of organic chemicals, including a large number of polymers.

Since BF_3 is such a reactive, corrosive gas, it is difficult to handle directly. Often it is used in the form of the dimethyl ether adduct (shown in Figure 14.6) which is a Lewis salt. The "etherate" is a liquid which boils at 126°C. All the boron halides, but particularly the chloride and the bromide, react with water to produce boric acid and the corresponding hydrohalic acid as shown in Equation (14.8):

$$BX_3 + 3H_2O \longrightarrow B(OH)_3 + 3HX \tag{14.8}$$

The anhydrous aluminum, gallium, and indium halides can be prepared by the direct reaction of the elements. (Treatment of the hydroxides or the oxides with aqueous hydrohalic acid, HX, often yield the trihydrates, $MX_3 \cdot 3H_2O$.) Anhydrous AlF_3 is produced by adding HF(g) to Al_2O_3 at 700°C. You may have noticed in looking over Table 14.2 that the anhydrous halides are of the general formula M_2X_6. Compounds of empirical formula AlF_3 and $AlCl_3$ do exist in the solid state, but no AlX_3 units are present, only rather complicated layer structures involving distorted AlX_6 octahedra. When solid $AlCl_3$ melts, the structure changes to the Al_2Cl_6 dimer which persists through vaporization until finally, at high temperatures, the $AlCl_3$ trigonal planar unit (like the boron halides) is formed. The aluminum bromides and iodides and gallium and indium chlorides, bromides, and iodides all assume the dimeric structure, particularly in the low-temperature vapor phase.

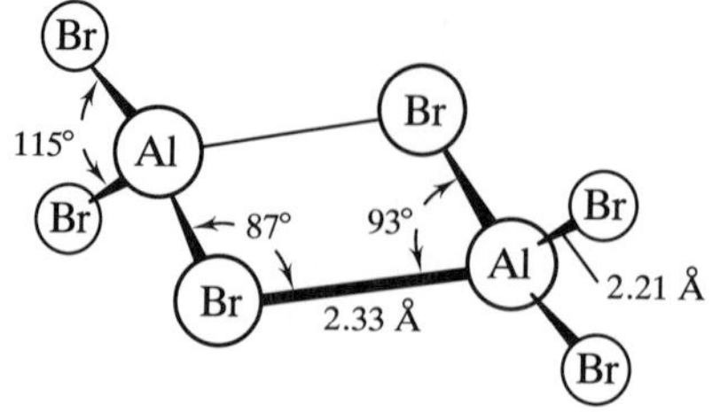

FIGURE 14.7
The structure of Al_2Br_6 showing bromine bridges between the two aluminum atoms.

The structure of the dimeric form is shown in Figure 14.7. It is characterized by halogen bridges between the metal atoms which assume an approximate tetrahedral configuration. The structure seems to result from a tendency of the metal to achieve an octet of electrons. As we saw in Section 13.2, beryllium chloride forms polymers in an analogous manner (p. 333) which is not unexpected given the diagonal effect. Since these compounds are Lewis acids in the same manner as the boron halides, it is not surprising that these dimers are split by reaction with a donor molecule as represented in Equation (14.9):

$$Al_2Cl_6 + 2Cl^- \longrightarrow 2AlCl_4^- \tag{14.9}$$

$AlCl_3$, like BF_3, is used extensively in synthetic organic chemistry. As shown in Equation (14.10), Al_2Cl_6 also reacts with lithium hydride to produce the "complex hydride," $LiAlH_4$:

$$Al_2Cl_6 + 8LiH \xrightarrow{\text{ether}} 2LiAlH_4 + 6LiCl(s) \tag{14.10}$$

$LiAlH_4$, or "lithyl" as organic chemists have been known to call it, as well as the similar $NaBH_4$, are versatile reducing agents. When $LiAlH_4$ reacts with the aluminum chloride dimer, the aluminum hydrides, or alanes, result. These are among the many electron-deficient compounds to be discussed in Section 14.5.

The chemistry of the thallium halides is dominated by the +1 oxidation state as previously discussed above and in Section 9.4 under the inert-pair effect. Note in Table 14.2 that the direct reaction of thallium with halogens produces TlX, not TlX_3 as expected. [Thallium(I) halides, TlX, can also be prepared by adding the appropriate hydrohalic acid to various other thallium(I) salts such as the nitrate or the sulfate.] Analysis of standard reduction potentials gives additional insight into the importance of the +1 oxidation state. The $E°$ of Tl^{3+} relative to Tl^+ is 1.25 V, indicating that Tl^{3+} readily gains electrons (is reduced to form Tl^+) and is therefore a relatively strong oxidizing agent in aqueous solution (GEROA, see Chapter 12, p. 305). The standard reduction potential of Tl^+ relative to $Tl(s)$ is −0.33 V, indicating that $Tl(s)$ is a mild reducing agent in aqueous solution as shown in Equations (14.11). (See Problems 14.23 to 14.25 for additional insight into the utility of the standard reduction potentials in thallium chemistry.)

$$Tl^+(aq) + e^- \longrightarrow Tl(s) \qquad E° = -0.33 \text{ V} \tag{14.11a}$$

$$Tl(s) \longrightarrow Tl^+(aq) + e^- \qquad E° = +0.33 \text{ V} \tag{14.11b}$$

$$\Delta G° = -(1)(96.5)(0.33)$$

$$= -32 \text{ kJ/mol}$$

In summary, the chemistry of the Group 3A oxides, hydroxides, and halides traces a progression from the unique metalloid boron, whose compounds are typically covalent, down to the rather typical metals, aluminum to thallium. At the top of the group, the +3 oxidation state is predominant, but toward the bottom, the +1 state becomes more stable until it predominates in thallium chemistry.

14.3 STRUCTURAL ASPECTS OF BORON CHEMISTRY

Allotropes

The allotropic forms of boron are structurally complex and make it truly unique among all elements. One of the principal recurring structural units in elemental boron is the B_{12} icosahedron shown in Figure 14.8*a*. These and related more complex units such as B_{84}, shown in Figure 14.8*b*, do not pack at all well and result in many open spaces leading to a low percentage of space occupied and a comparatively low density for the element.

Borides

At first glance, there appears to be a mind-boggling array of binary-metal borides, ranging from MB, MB_2, and M_2B up to progressively stranger combinations such as MB_6, MB_{12}, $B_{13}C_2$, and even MB_{66}. These compounds are prepared by a wide variety of methods including most commonly (1) the direct combination of the elements, (2) the reduction of the corresponding metal oxide or chloride (sometimes mixed with boron oxide or the chloride) with boron, carbon, boron carbide, or hydrogen, and (3) the electrolysis of fused salts. Briefly, what are the structures of these compounds and of what use are they?

Borides come in isolated atoms, isolated pairs of atoms (B_2), single-strand, branched, and double chains of atoms, two-dimensional ("chicken wire") networks, and three-dimensional networks. All except the last of these are represented schematically in Figure 14.9. Figures 14.11 and 14.12 show three-dimensional examples. There is no other element which shows this variety of structures in its binary compounds. However, as unique as these compounds are, they are relatively straightforward, particularly so if you have already read Chapter 7 on solid state structures. For example, the diborides, MB_2, as shown in Figure 14.10, have a structure consisting of alternating, parallel, hexagonal layers of metal atoms and boron atoms. Each boron atom is in contact with 6 metal atoms, that is, has a

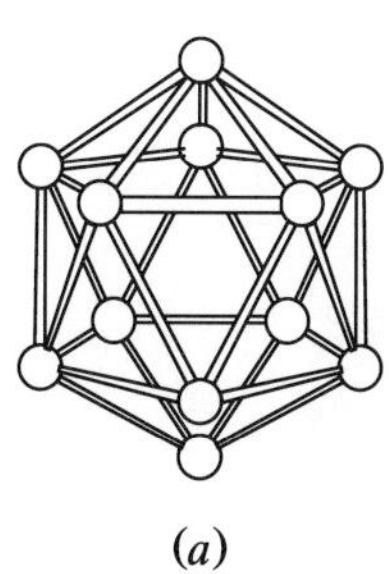

(*a*)

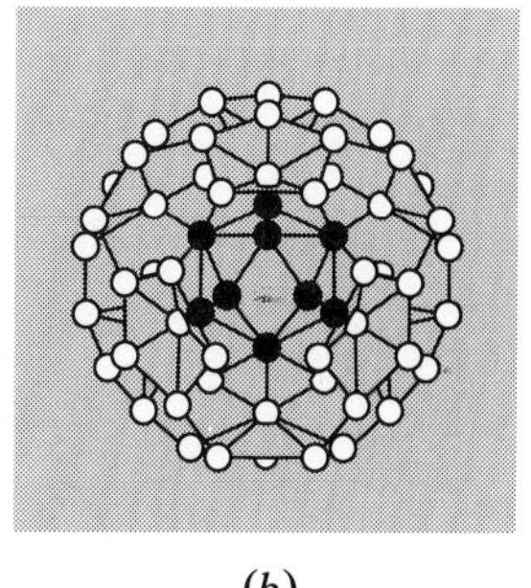

(*b*)

FIGURE 14.8
(*a*) The B_{12} icosahedron unit common in elemental boron (the icosahedron has 12 vertices and 20 triangular faces). (*b*) The B_{84} unit found in some forms of boron is composed of a central B_{12} unit with an outwardly directed pentagonal pyramid associated with each of the 12 atoms of the central unit.

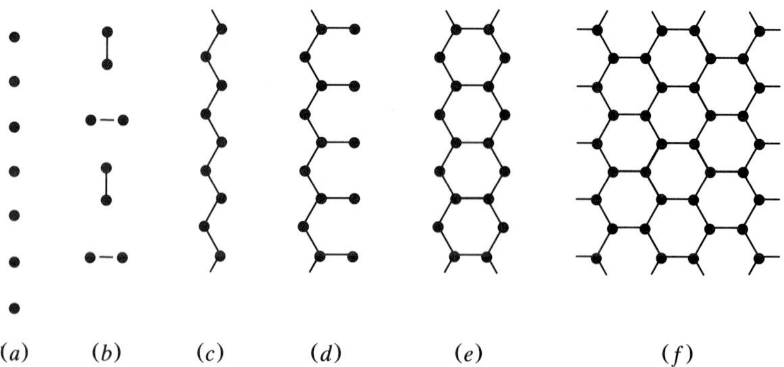

FIGURE 14.9
A schematic representation of the variety of structures found in metal borides: (*a*) single atoms; (*b*) isolated pairs of atoms; (*c*) single-strand chains; (*d*) branched single-strand chains; (*e*) double chains of atoms; (*f*) two-dimensional ("chicken wire") infinite networks. [Adapted from Ref. 26.]

coordination number (C.N.) of 6. Each metal is in contact with 12 boron atoms (C.N. = 12).

The three-dimensional networks of the borides are closely related to the structures of the simple ionic salts covered in Chapter 7 and perhaps in one of your previous chemistry experiences. The structure of MB_6, shown in Figure 14.11, is a body-centered CsCl structure (see Figure 7.22*e*) where the B_6 units take the place of the Cl^- ions. Note, however, that the B_6 units themselves are connected so that the boron atoms form an infinite, three-dimensional, and extremely stable array.

In MB_{12}, shown in Figure 14.12, the B_{12} units are cubooctahedra which replace the Cl^- in the NaCl, or rock salt, structure (see Figure 7.22*a*). In other words, the B_{12} units form a face-centered array with the M atoms in the octahedral holes. Once again the B_{12} units, like the B_6 units above, are linked together to form a huge, beautiful, and orderly array.

Given their extended interwoven nature, it's not surprising that the borides are very hard, high-melting, nonvolatile, and chemically inert materials. They can

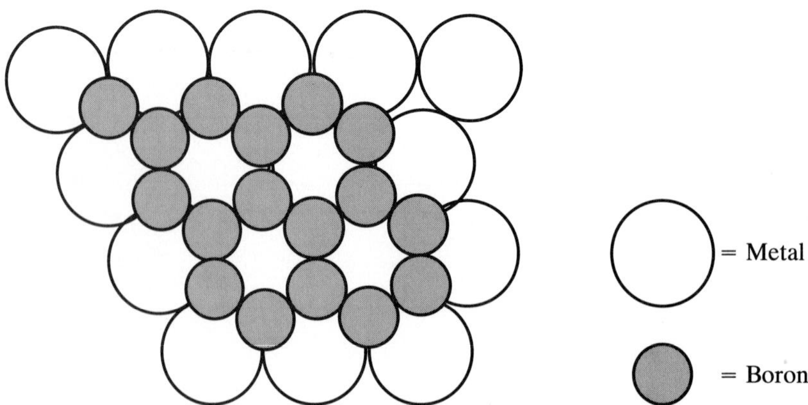

FIGURE 14.10
The structure of metal diborides, MB_2, consists of alternating parallel hexagonal layers of metal atoms and boron atoms. [Ref. 27.]

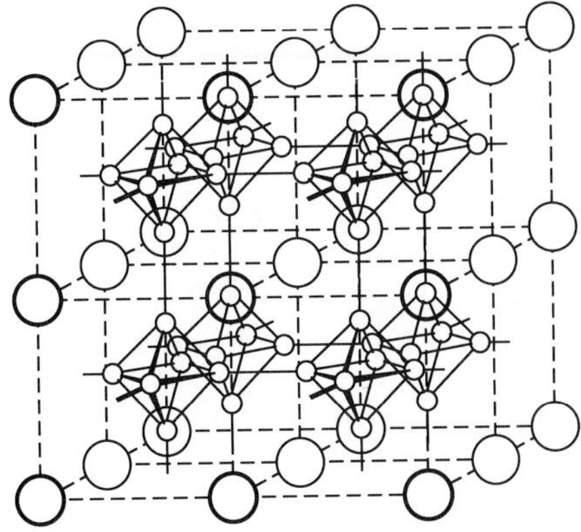

FIGURE 14.11
The three-dimensional network structure of MB_6. Note its resemblance to the CsCl structure shown in Figure 7.22*e*. [Ref. 27.]

withstand a huge variety of forces including very high temperatures and degrees of friction as well as the action of strong acids and molten metals and salts. Name a situation where this type of physical and chemical resistance is desired and borides are probably in use. These applications include turbine blades, rocket nozzles, high-temperature reaction vessels ("boats," crucibles, etc.), polishing and grinding grits, brake and clutch linings, even a new generation of lightweight protective armor. The borides are also good conductors of electricity and find use as hard and tough industrial electrodes.

In addition to these physical and chemical properties, boron has nuclear properties which add to its usefulness. The boron 10 isotope absorbs neutrons efficiently. Note that the products of this absorption, shown in Equation (14.12), are nonradioactive isotopes of helium and lithium:

$$ {}^{1}_{0}n + {}^{10}_{5}\text{B} \longrightarrow {}^{7}_{3}\text{Li} + {}^{4}_{2}\text{He} \qquad (14.12) $$

This ability to absorb neutrons, together with their inertness and structural stability, makes the borides suited to a large number of nuclear applications, such as control rods and neutron shields. Equation (14.12) is also the basis of a neutron

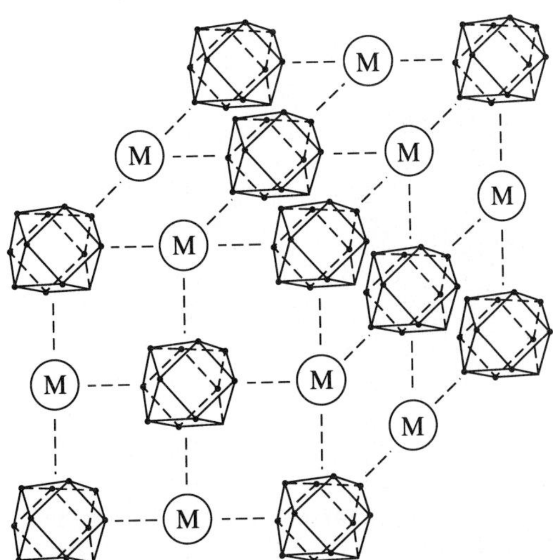

FIGURE 14.12
The three-dimensional network structure of MB_{12}. The cubooctahedron B_{12} clusters replace the chloride anions, while metal atoms replace the sodium cations in the NaCl structure shown in Figure 7.22*a*.

counter which either has boron on its inner walls or is filled with a gas such as boron trifluoride.

Borates

Boric acid, H_3BO_3, as we discussed earlier (p. 353), accepts one OH^- ion in the Lewis sense and is therefore monobasic rather than tribasic. The borate anion, BO_3^{3-}, on the other hand, is named as if boric acid could, in fact, lose all three of its protons. (Sometimes BO_3^{3-} is referred to as orthoborate, with the prefix *ortho*- derived from the Greek for "straight" or "correct.") While BO_3^{3-} is the most straightforward example of a borate, it turns out, not surprisingly given the above discussion of borides, that there are a large number of borates with a variety of structures, only a few of which are shown in Figure 14.13.

Metaborates occur when the BO_3^{3-} units link together to form chains or rings. (*Meta*- is derived from the Greek *meta* for "with" or "after" and means "between" or "among.") Three-dimensional borates, of which $B(OH)_4^-$ (tetrahydroxoborate) and $[B_2(O_2)_2(OH)_4]^{2-}$ (perborate) are fairly simple examples, culminate in borax, $Na_2[B_4O_5(OH)_4]$-$8H_2O$, which contains the $[B_4O_5(OH)_4]^{2-}$ anion

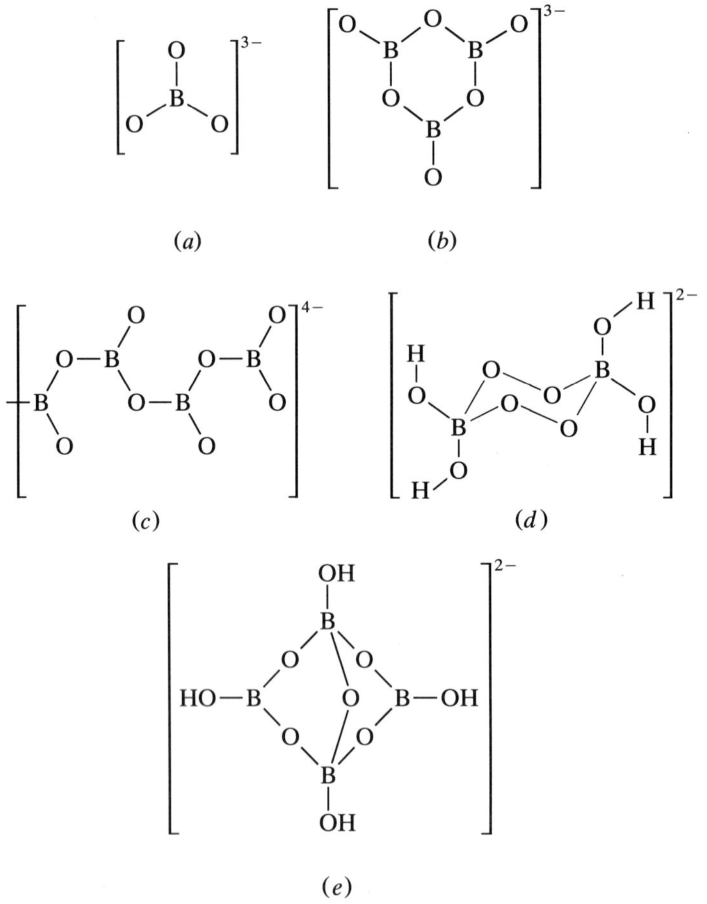

(*a*) (*b*)

(*c*) (*d*)

(*e*)

FIGURE 14.13
Variety of borate structures: (*a*) orthoborate, BO_3^{3-}; (*b*) cyclic metaborate, $B_3O_6^{3-}$; (*c*) chain metaborate, BO_2^-; (*d*) perborate, $[B_2O_2(OH)_4]^{2-}$; (*e*) borax, $[B_4O_5(OH)_4]^{2-}$.

shown in Figure 14.13*e*. Some detergents and "safe bleaches" (for colored clothing) contain sodium perborate which hydrolyzes to form hydrogen peroxide in very hot water as shown in Equation (14.13):

$$\underset{\text{Perborate}}{[B_2(O_2)_2(OH)_4]^{2-}} + 4H_2O \xrightarrow{\text{heat}} 2H_2O_2 + 2H_3BO_3 + 2OH^- \quad (14.13)$$

Sodium perborate is also used as a disinfectant and as a topical antiseptic. The various borates are quite similar to the silicates, as we will discuss in the next chapter.

14.4 ALUMINUM, GALLIUM, INDIUM, AND THALLIUM: REACTIONS AND COMPOUNDS OF PRACTICAL IMPORTANCE

Aluminum Metal and Alloys

As most everyone knows, aluminum is a lightweight, durable, corrosion-resistant metal. You might be surprised to learn, however, that the pure metal is structurally weak (particularly at temperatures above 300°C), reacts readily with water, and is more easily oxidized than iron! (Note in Table 14.2 that the standard reduction potential of aluminum is -1.66 V. The $E°$ for iron is -0.44 V. All of which means it is easier to oxidize aluminum to Al^{3+} than it is to oxidize iron to Fe^{2+}.) Aluminum does have a high thermal as well as electrical conductivity and, since it is much lighter and less expensive than copper, may well become the metal of choice in long-distance power lines.

To improve its structural properties, aluminum is mixed with various metals such as copper, silicon, magnesium, manganese, and zinc to form alloys of varying properties and uses, some of which are shown in Table 14.3. Aluminum evidently forms zones of structural stability with these other metals which harden the alloy in a process somewhat like the formation of steel from iron. The corrosion resistance of aluminum is due to the formation of a hard, tough film of Al_2O_3. A thicker layer of the oxide can be placed on the aluminum or aluminum alloy by an electrolytical process known as *anodization*. These layers of the oxide can be made to absorb various colored dyes which result in a pleasant-appearing product used in siding, furniture, awnings, and a variety of other applications.

Aluminum foil, by the way, was developed in the 1940s by Richard S. Reynolds, the nephew of tobacco king R. R. Reynolds, to protect cigarettes and hard candies from moisture. Soon it was found everywhere. Food was sold in attractive aluminum packaging, cooked in aluminum boats and dishes or on barbecue grills or oven broilers wrapped in foil. When the meal was done, the leftovers were again wrapped and preserved in yet another layer of aluminum. Few products have been so quickly and universally absorbed into the American home as aluminum foil.

Finally, a rather unexpected use of aluminum is in the drain cleaner Drāno, which contains small pieces of the metal mixed with sodium hydroxide. The aluminum reacts with the hydroxide ion as shown in Equation (14.14) to provide hydrogen gas and a great deal of heat which can melt grease deposits and agitate the contents of the clogged drain:

$$2Al(s) + 6H_2O(l) + 2OH^-(aq) \longrightarrow \underset{\text{Aluminate}}{2Al(OH)_4^-(aq)} + 3H_2(g) \quad (14.14)$$

TABLE 14.3
Compositions, properties, and uses of some aluminum alloys

Principal alloying material	Properties	Uses
None or trace amounts of Cu	Good electrical and thermal conductivity; excellent workability	Sheet, foil, wire rod, and tubing; electrical wires
Copper	High strength; high resistance to fracture	Aircraft; fuel and oxidizer tanks; primary structure of Saturn space vehicle boosters
Manganese	Moderate strength; good workability; good weldability	Cooking utensils; tubes; packaging; storage tanks; beer and beverage cans
Silicon	Low melting; low coefficient of expansion	Welding wire; castings; can be anodized to attractive gray color; architectural applications
Magnesium	High corrosion resistance; good weldability and workability	Marine applications; armor plate for military vehicles; beverage can ends; automobile trim
Magnesium and silicon	Heat-treatable; good formability; good corrosion resistance	Trailers, trucks, and other transportation applications; furniture and architectural applications
Zinc	Heat-treatable; high strength	Airframe and other highly stressed parts
Lithium	Low density; heat-treatable; moderate strength	Aerospace applications
Silicon and copper	Castable; high tensile strength; moderate strength	Architectural applications; automotive die-cast engine cylinder blocks and pistons

Aluminum (and gallium) also reacts with acids as shown in Equation (14.15):

$$2Al(s) + 6H^+(aq) \longrightarrow 2Al^{3+}(aq) + 3H_2(g) \qquad (14.15)$$

Although the aluminate ion is shown as $Al(OH)_4^-(aq)$ in Equation (14.14), the coordination number of the aluminum cation is characteristically 6 in this aqueous anion as well as in the $Al^{3+}(aq)$ of Equation (14.15). Their structures are shown in Figure 14.14. Also shown in the figure is a more accurate representation of the amphoteric nature of aluminum hydroxide, which is actually $Al(OH)_3(H_2O)_3$. Note that the $H^+(aq)$ protonates a hydroxide of the hydrated hydroxide, while the $OH^-(aq)$ removes a proton from a coordinated water molecule in the complex. No Al—O bonds are broken during the process. Other amphoteric hydroxides, including those of beryllium (note the diagonal relationship here), gallium, zinc, tin(II), and lead(II), react in the same manner. (If you have covered Chapter 5, you will probably recognize these reactions as those of a coordinated ligand.)

Alums

Hydrated salts such as $AlX_3 \cdot 6H_2O$ (X = Cl, Br, I) and the double salts or alums, $MAl(SO_4)_2 \cdot 12H_2O$ (M = a variety of unipositive cations including NH_4^+), contain the hexaaquoaluminum ion and other hydrated M^+ cations. The astringent properties of alums used since ancient times has already been mentioned. In a way, some things never change; "aluminum chlorhydrate," one of the listed active ingredients of a variety of antiperspirant deodorants, utilizes the same property of aluminum compounds. Evidently these salts are astringents by virtue of the fact that they close the openings of sweat glands by altering the hydrogen bonding among protein molecules.

Saturated solutions of alums kept at a constant temperature yield large, beautiful crystals of various colors which are produced by partially substituting various transition-metal cations for the aluminum in the alum. The addition of the ammonium alum (where $M^+ = NH_4^+$) to a water supply produces a light, loose precipitate, or *floc*, of aluminum hydroxide which purifies drinking water. The fluffy, loosely associated nature of the $Al(OH)_3$ is often attributed to hydrogen bonding between the precipitate and water molecules.

Alumina

The pure form of Al_2O_3, called alumina, is a white solid which exists in a variety of polymorphs. *Polymorph* refers to a particular crystalline form of a compound, whereas allotrope, first defined back in Chapter 11, refers to a different molecular form of an element. (*Morph* comes from the Greek *morphē*, meaning "form.") Al_2O_3, then, crystallizes in several different forms which are usually designated by Greek letter prefixes.

In α-Al_2O_3 the oxide ions form a hexagonal close-packed array and the aluminum ions occupy octahedral holes. γ-Al_2O_3 is a defect spinel structure in which there are not enough cations to occupy the usual fraction of tetrahedral and octahedral holes. The Al_2O_3 formed on the surface of aluminum metal is a third

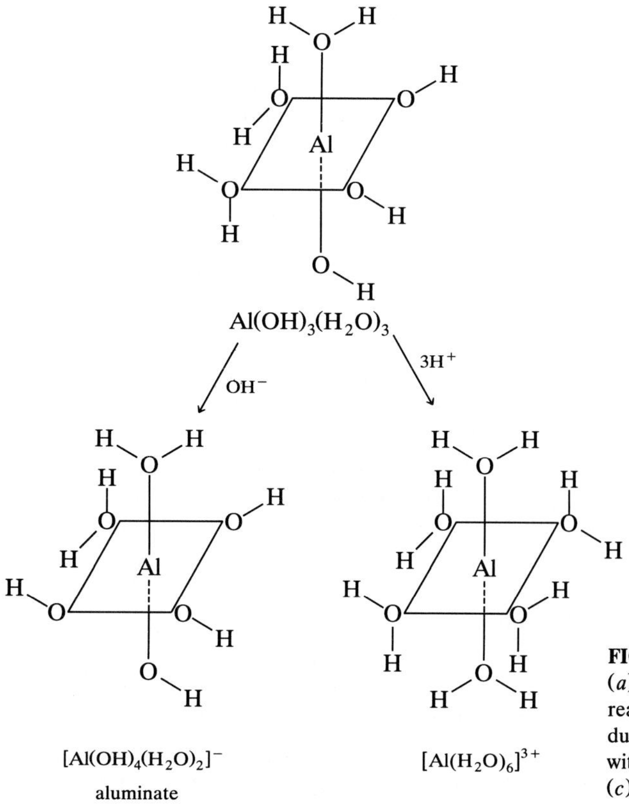

$Al(OH)_3(H_2O)_3$

OH^-

$3H^+$

$[Al(OH)_4(H_2O)_2]^-$

aluminate

$[Al(H_2O)_6]^{3+}$

FIGURE 14.14
(*a*) Amphoteric $Al(OH)_3(H_2O)_3$ reacts with a base (OH^-) to produce (*b*) $[Al(OH)_4(H_2O)_2]^-$ or with an acid (H^+) to produce (*c*) $[Al(H_2O)_6]^{3+}$.

polymorph based on the rock salt (NaCl) structure with every third aluminum ion missing in the cubic close-packed array of oxides. (See Chapter 7 for more details on these crystal structures.)

α-Al_2O_3 occurs naturally as corundum which is used as an abrasive in products such as grinding wheels, sandpaper, and toothpaste. Impure crystalline forms of Al_2O_3 (impurities given in parentheses) are valuable and beautiful gems including red ruby (Cr^{3+}), blue sapphire (Fe^{2+}, Fe^{3+}, and Ti^{4+}), green oriental emerald (Cr^{3+} and Ti^{3+}), violet oriental amethyst (Cr^{3+} and Ti^{4+}), and yellow oriental topaz (Fe^{3+}). Ruby and sapphire are also produced industrially on a large scale as is the gem-quality corundum sometimes called *white sapphire*.

The extreme stability of Al_2O_3 makes aluminum useful for reducing a variety of metal oxides to their corresponding metals using the *thermite* reaction. Several examples of the reaction are given in Equations (14.16) and (14.17):

$$2Al(s) + Fe_2O_3(s) \longrightarrow Al_2O_3(s) + 2Fe(s) \tag{14.16}$$

$$2Al(s) + Cr_2O_3(s) \longrightarrow Al_2O_3(s) + 2Cr(s) \tag{14.17}$$

A thermite reaction is a violent and highly impressive exothermic reaction producing enough heat to weld together the huge steel reinforcing rods used in constructing giant concrete buildings. It also is the basis of incendiary devices of various kinds. The combustion of aluminum when mixed with ammonium perchlorate is forceful enough to be used as solid propellant for the rockets used in the U.S. space shuttle program.

Gallium, Indium, and Thallium Compounds

There are comparatively few uses for gallium, indium, and thallium compounds. Gallium arsenide is an important semiconductor (see Section 15.5) and is used in the light-emitting diodes formerly found in the readout of calculators. Although now replaced by liquid crystal displays, LEDs are still used in applications where an actual light source is desirable. $MgGa_2O_4$ is the brilliant-green phosphor used in the Xerox process, $YSr_2Cu_2GaO_7$ is a fairly recent example of a high-temperature superconductor containing gallium, In_2O_3 and Tl_2O are used in glass manufacture, and crystals of TlBr and TlI have the special ability to transmit infrared light. Nevertheless, these three elements remain rare, widely scattered, and underutilized.

14.5 SELECTED TOPIC IN DEPTH: ELECTRON-DEFICIENT COMPOUNDS

The principal characteristic that distinguishes the Group 3A elements from the rest of the representative elements is the existence of so-called electron-deficient compounds. You may recall earlier references, in this book and elsewhere, to compounds of this type. It is not unusual for boron, aluminum, gallium, and occasionally beryllium and lithium to form compounds in which the metal is surrounded by less than an octet of electrons. Of course one should be wary of such phrases as "electron-deficient." It seems to imply that there is something wrong with such compounds. In fact, it is more likely that there is something wrong with the theories we have traditionally used to picture chemical bonding.

So far in this book, we have explained a great deal of chemistry in terms of what we might more specifically call *two-center–two-electron* (2c-2e) bonds. In these "normal" single (2c-2e) bonds, two atoms (centers) are held together in a single bond by sharing two electrons between them. The two electrons are "localized" between two centers. Now you may recall resonance structures and delocalized pi bonding in such species as nitrate, sulfur trioxide, and the like, but with the discovery of electron-deficient compounds such as those encountered in the borohydrides, or *boranes*, chemists had to radically change their views of chemical bonding. First we will discuss the discovery of these compounds and then proceed to describe the bonding in them.

The first examples of these compounds were synthesized by Alfred Stock, a German chemist active in the first part of this century. He prepared a series of borohydrides by reacting magnesium boride, MgB_2, with various mineral acids as shown in Equation (14.18):

$$MgB_2 + HCl \longrightarrow \text{mixture of } B_4H_{10}, B_5H_9, B_5H_{11}, B_6H_{10}, \text{ and } B_{10}H_{14} \quad (14.18)$$

Note that no BH_3 (which would be called *borane* consistent with compounds such as methane, CH_4) was formed as might be expected. Upon heating B_4H_{10} to 100°C, it decomposes to a compound of that empirical formula, but its molecular formula is B_2H_6 and consequently it is called *diborane*.

Stock found these boranes unpleasant materials to work with. As listed in Table 14.4, they are often volatile, highly reactive, moisture-sensitive, air-sensitive, and/or spontaneously flammable. Consequently, he devised special vacuum techniques which used a variety of mercury-containing switches, valves, "float bottles," etc., so that he could handle these compounds in an inert atmosphere and therefore prevent their decomposition. As it turned out, he and his coworkers developed mercury poisoning using these techniques and experienced symptoms such as headaches, numbness, tremors, anxiety, indecision, depression, loss of memory and the ability to concentrate, as well as more serious mental deterioration. Indeed, at one point in his life, Stock thought that he was losing his mind. Not until 1924, 15 years after he started his work on these compounds, was this mysterious set of symptoms correctly diagnosed as being due to mercury.

Stock devoted much of his later life to investigating the causes and prevention of mercury poisoning, which he called *Quecksilbervergiftung*. Alfred Stock, who never received the Nobel Prize, took a great interest in nomenclature and is honored by having the *Stock nomenclature* of modern inorganic chemistry named after him. This is the system you learned in introductory chemistry in which

TABLE 14.4
Properties of some boranes

Formula	Name†	Physical state at 25°C	Reaction with air	Thermal stability at 25°C
B_2H_6	Diborane(6)	Gas	Spont. flammable	Fairly stable
B_4H_{10}	Tetraborane(10)	Gas	Stable if pure	Decomposes rapidly
B_5H_9	Pentaborane(9)	Liquid	Spont. flammable	Stable
B_5H_{11}	Pentaborane(11)	Liquid	Spont. flammable	Decomposes rapidly
B_6H_{10}	Hexaborane(10)	Liquid	Stable	Slowly decomposes
$B_{10}H_{14}$	Decaborane(14)	Solid	Very stable	Stable

†The number of boron atoms is indicated by a prefix; the number of hydrogen atoms is given by an Arabic numeral in parentheses.

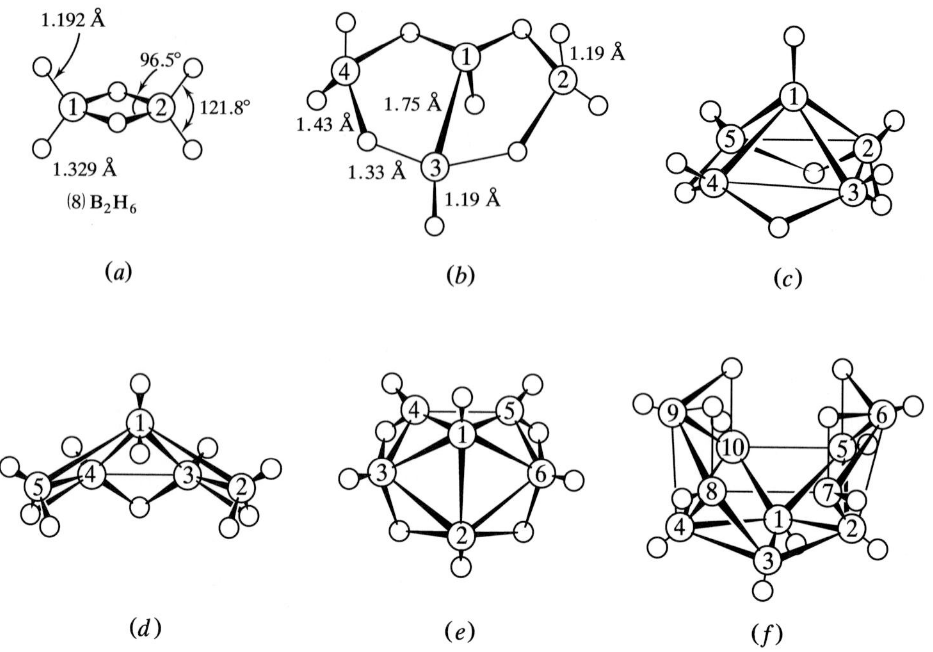

FIGURE 14.15
The structures of the simpler boranes: (*a*) diborane, (*b*) B_4H_{10}, (*c*) B_5H_9, (*d*) B_5H_{11}, (*e*) B_6H_{10}, (*f*) $B_{10}H_{14}$.

compounds such as TlI and $SnCl_4$ are called thallium(I) iodide and tin(IV) chloride, respectively, rather than the old names, thallous iodide and stannic chloride.

The structures of some of the simpler boranes are shown in Figure 14.15. Once these structures were well established, chemists quickly realized that the bonding in them could not be explained by the normal types of schemes. A little electron counting reveals the problem. Take diborane as the prototype. Each of the two boron atoms provides three valence electrons, and the six hydrogens provide one each. This yields a total of 12 electrons, two short of the number needed for conventional two-center–two-electron (2c-2e) bonds. In other words, we need 14 electrons for a structure such as that found in Figure 14.16*a*, and we do not have enough. The compound is electron-deficient. How is this problem to be solved?

The solution, for which William Lipscomb received the Nobel Prize in 1976, turns out to be the existence of multicenter bonds. While most (single) bonds are of the 2c-2e variety, some may have more than two centers; that is, they are *multicenter*. Assuming each boron in diborane to form sp^3 hybrids, picture two "terminal" hydrogen atoms bonded to each boron using regular 2c-2e bonds formed between an sp^3-hybrid orbital of boron and a 1*s* atomic orbital of hydrogen as shown in Figure 14.16*b*. Note that only four electrons remain. These are used to form two *three-center*–two-electron (3c-2e) bonds as shown in Figure 14.16*c*. The three-center bond results from the overlap of one sp^3-hybrid orbital on each boron and the 1*s* atomic orbital of the hydrogen. Thus the two electrons in each of these three-center bonds are spread out over the three atoms and hold them together.

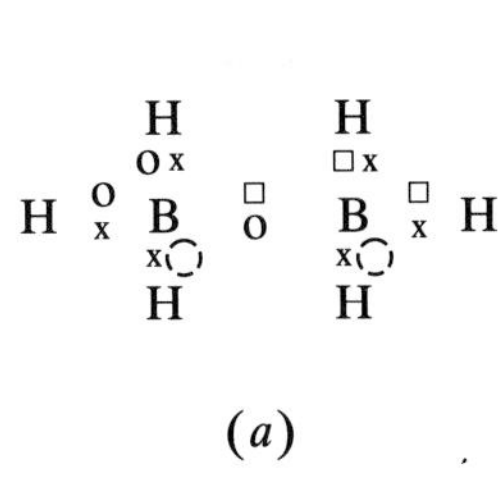

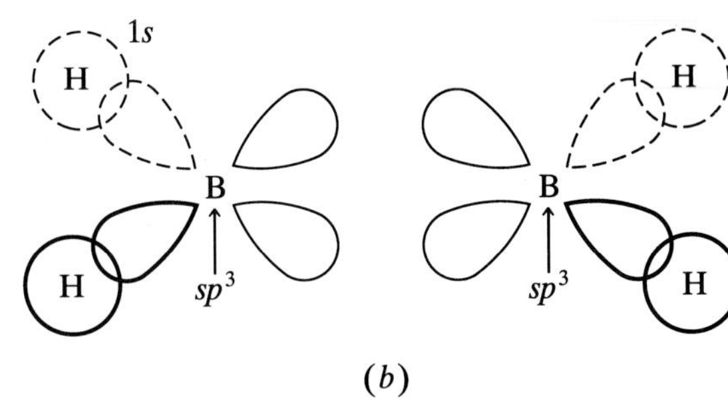

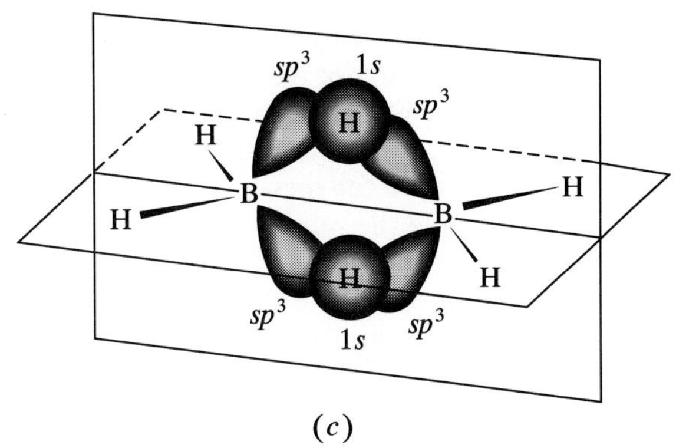

FIGURE 14.16
Three views of the bonding in diborane. (*a*) The 12 electrons available (3 from each boron, squares and circles; 6 from 6 hydrogens, x's) is two short (dashed circles) of that needed to form two-center–two-electron bonds; (*b*) sp^3 boron atoms form regular two-center–two-electron bonds with the "terminal" hydrogen atoms; (*c*) the pair of three-center–two-electron bonds each involving two boron atoms and one hydrogen atom. [(*c*) adapted from Ref. 4, p. 179.]

The overall distribution of electron density in the molecule is shown in Figure 14.17*a*. Understandably, these bonds are sometimes referred to as *banana bonds*.

Since the electron density is spread out over three nuclei rather than two, the three-center bonds should be weaker than conventional 2c-2e bonds. The longer B—H bond lengths in the three-center bonds (see Figure 14.15*a*) would seem to confirm this expectation.

There is another way to represent the electron distribution in diborane and its heavier relatives. These are Lipscomb's *semitopological* diagrams which involve the four structure-bonding elements depicted in Table 14.5. We have already seen two of these elements, the 2c-2e B—H bond and the 3c-2e B—H—B bond, in diborane. Using the symbols for these elements, the semitopological diagram for

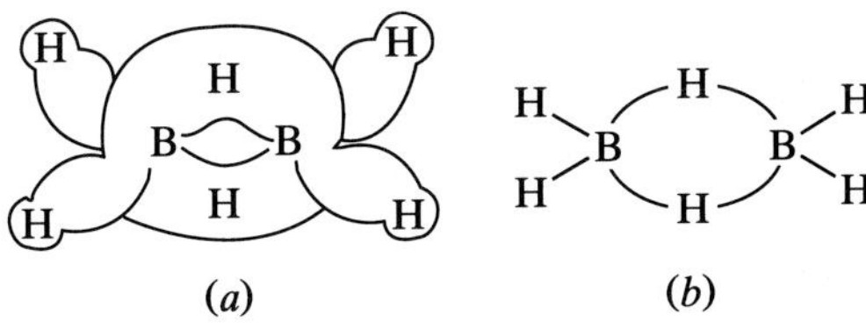

FIGURE 14.17
(*a*) The overall distribution of electron density in diborane; (*b*) the semitopological diagram for diborane.

TABLE 14.5
Structure-bonding elements in boranes

Structure-bonding element	Symbol
Terminal 2c-2e B—H bond	B—H
3c-2e B—H—B bond	B ⁄H∖ B
2c-2e B—B bond	B—B
3c-2e B—B—B bond	B (above), B B (below), joined at a central point

diborane is as shown in Figure 14.17*b*. Note that these diagrams are designed to show the distribution of electrons and not to depict the geometry of the molecule. Using the above ideas, we can go on to picture the bonding in the heavier boranes. Before we do, be sure you have taken note, in Table 14.4, of the special nomenclature used for these compounds. The number of boron atoms is indicated by a prefix, while the number of hydrogen atoms is given by an Arabic numeral in parentheses.

The semitopological diagram for tetraborane(10), B_4H_{10}, is given in Figure 14.18 along with indications of the number of electrons available and how they are distributed among the various structure-bonding elements. It would be instructive to compare this diagram with the actual structure given in Figure 14.15*b*. Note that the bent nature of this molecule is not portrayed in the semitopological diagram. This molecule contains a normal 2c-2e B—B bond.

In pentaborane(9), B_5H_9, shown in Figure 14.19, the fourth structure-bonding element, the 3c-2e B—B—B bond, is present. This involves an overlap of an sp^3-hybrid orbital from each of the three boron atoms. Note that the structure shown is only one of four possible resonance structures. These are required to account for the fact that all the B–B distances involving the one boron at the peak of this square pyramidal molecule (see Figure 14.15*c* for actual structure) are equal. There are molecular orbital pictures of the bonding in these molecules which are quite useful, but they are beyond the scope of this book.

There are a very large and growing number of more complex boranes. In fact, it is fair to say that borane and related carborane chemistry (in which one or more carbon atoms replace the same number of boron atoms) have constituted one of the major areas of growth in inorganic chemistry over the last three decades. Most of the numerous boranes, both neutral and anionic, can best be organized into four

Distribution of electrons			
6	2c-2e	B—H:	$12e^-$
4	3c-2e	B⌒H⌒B:	$8e^-$
1	2c-2e	B—B:	$2e^-$
			$22e^-$

Number of electrons available	
4B:	$12e^-$
10H:	$10e^-$
	$22e^-$

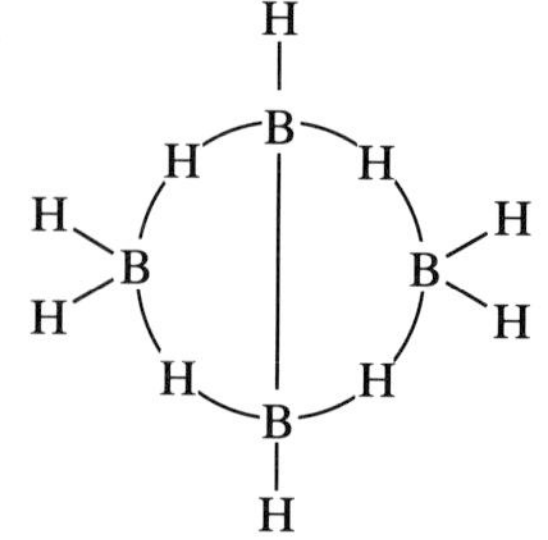

FIGURE 14.18
The semitopological diagram for B_4H_{10}.

	Distribution of electrons:		
5	2c-2e	B—H:	$10e^-$
4	3c-2e	B(H)B:	$8e^-$
2	2c-2e	B—B:	$4e^-$
1	3c-2e	B(B)B:	$2e^-$
			$24e^-$

Number of electrons available	
5B:	$15e^-$
9H:	$9e^-$
	$24e^-$

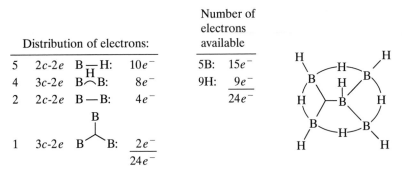

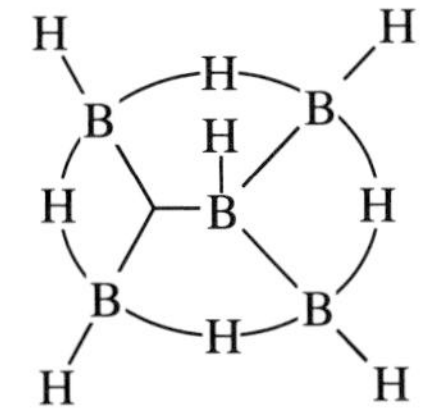

FIGURE 14.19
The semitopological diagram for B_5H_9. (This is one of four possible resonance structures.)

classes (closo, nido, arachno, and conjuncto) of compounds as shown in Table 14.6. Carboranes can be organized in the same manner. The clusters of atoms in both boranes and carboranes can often be thought of as derived from one of the nine polyhedral structures found in Figure 14.20.

As Stock noted, the boranes are dangerously reactive, and proper precautions must be taken when synthesizing them. Diborane, which results when one might expect BH_3, can be prepared from various boron(III) compounds as shown in Equations (14.19). Note that in Equation (14.19*b*) the starting material is the "etherate" of BF_3 which, as mentioned earlier and shown in Figure 14.6, is a more convenient material to work with than the gaseous BF_3.

$$8BF_3 + 6NaH \longrightarrow 6NaBF_4 + B_2H_6 \qquad (14.19a)$$

$$4BF_3 \cdot O(CH_2CH_3)_2 + 3LiAlH_4 \longrightarrow 3LiAlF_4 + 2B_2H_6 + (CH_3CH_2)_2O \qquad (14.19b)$$

$$2KBH_4(s) + 2H_3PO_4(l) \longrightarrow B_2H_6(g) + 2NaH_2PO_4(s) + 2H_2(g) \qquad (14.19c)$$

BH_3, like BF_3 as we saw earlier, can be prepared as an adduct as shown in Equations (14.20):

$$B_2H_6 + 2CO \longrightarrow 2H_3B \leftarrow :CO \qquad (14.20a)$$

$$B_2H_6 + 2R_3N \longrightarrow 2H_3BNR_3 \qquad (R = CH_3, CH_2CH_3, \text{etc.}) \qquad (14.20b)$$

Diborane is the starting material for the preparation of the higher boranes [see Equations (14.21) and (14.22)] which Stock prepared from the borides. Diborane is also used extensively in synthetic organic chemistry, where the hydroboration of olefins (hydrocarbons containing one or more double bonds) is the starting point for a whole new series of compounds. Much of this pioneering work was done by H. C. Brown who subsequently received the 1979 Nobel Prize in chemistry.

$$B_2H_6 \xrightarrow[\text{10 days}]{\text{pressure}} B_4H_{10} + H_2 \qquad (14.21)$$

$$2B_4H_{10} + B_2H_6 \longrightarrow 2B_5H_{11} + 2H_2 \qquad (14.22)$$

Given that B—H bonds (particularly the 3c-2e multicenter bonds) are fairly weak (in other words, it does not take too much energy to break them) and B—O bonds are very strong (in other words, much energy is released when they are

TABLE 14.6
Four structural classes of neutral and anionic boranes

Class (type)	Description	Examples (1)	(2)	(3)	(4)
Closo (cage)	Complete, closed cluster of n boron atoms	$B_6H_6^{2-}$	$B_7H_7^{2-}$	$B_{11}H_{11}^{2-}$	$B_{12}H_{12}^{2-}$
Nido (nest)	Nonclosed B_{n-1} clusters formed by removing 1 B from a B_n polyhedron	B_5H_9	B_6H_{10}	$B_{10}H_{14}$	$B_{11}H_{14}^-$
Arachno (web)	B_{n-2} clusters formed by removing 2 B's from a B_n polyhedron	B_4H_{10}	B_5H_{11}	$B_9H_{14}^-$	$B_{10}H_{14}^{2-}$
Conjuncto (joined)	Formed by linking two or more of the above types of clusters together	$B_{10}H_{16}$ (2 B_5H_9 units)		B_8H_{18} (2 B_4H_9 units)	

formed), it is not surprising that the reactions of diborane with oxygen and water [Equations (14.23) and (14.24)] are among the most exothermic combustion reactions known.

$$B_2H_6 + 3O_2 \longrightarrow B_2O_3 + 3H_2O \quad (14.23)$$

$$B_2H_6 + 6H_2O \longrightarrow 2B(OH)_3 + 6H_2 \quad (14.24)$$

In fact, in the 1950s the United States initiated programs called Project ZIP and Project HERMES (Hermes was the Greek messenger of the gods) which investigated the possibility of using the boranes as *superfuels* for rockets and jet planes. Decaborane(14), the most stable of the lighter boranes, was once produced in ton quantities for these programs. By the early 1960s, however, the boranes were dropped as potential fuels because of high cost, storage difficulties, and toxicity.

The boranes are not the only compounds to exhibit multicenter bonds. Gaseous aluminum hydride, or alane, AlH_3, comes as both the monomer and the dimer, Al_2H_6. In the solid phase it is more properly formulated as the polymeric $(AlH_3)_n$ held together by three-center bridging Al—H—Al bonds. Gallane,

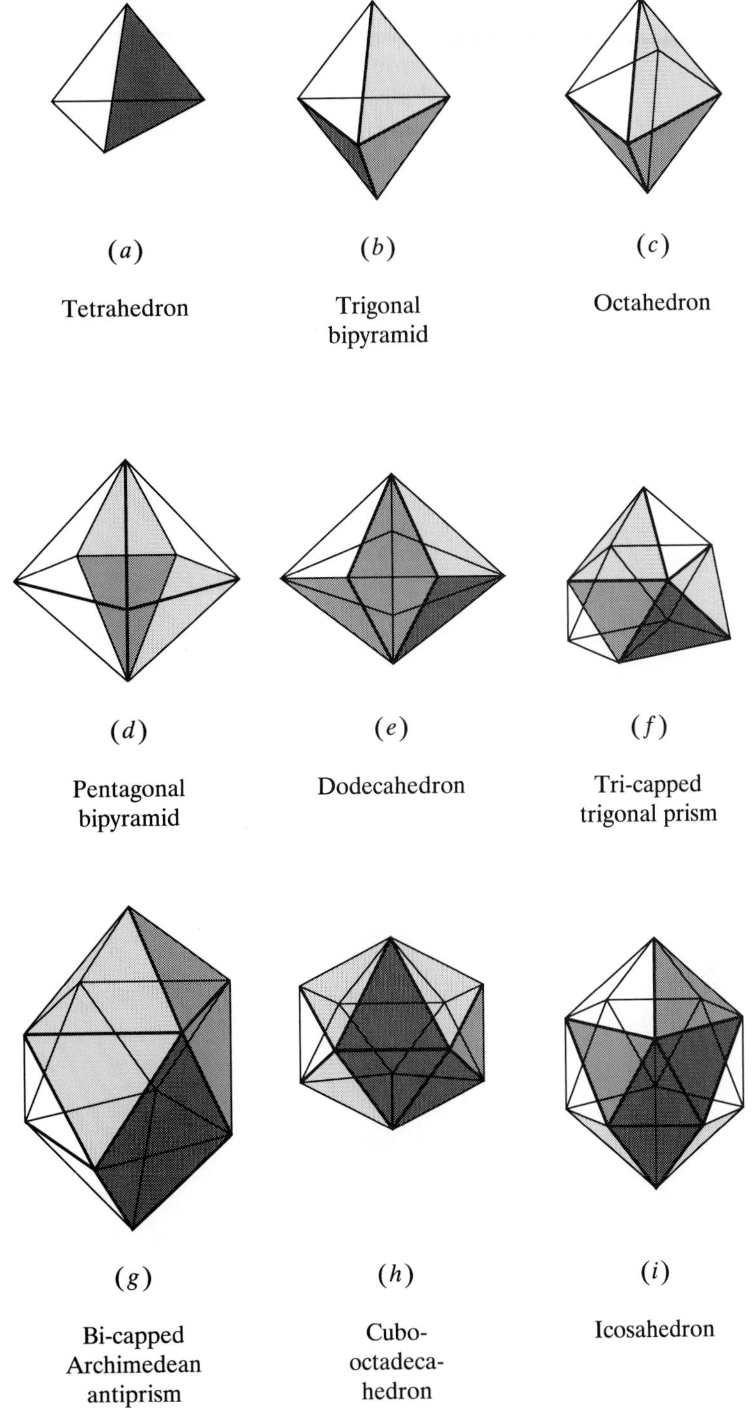

FIGURE 14.20
The nine polyhedra from which the structures of the boranes and carboranes are derived. [Ref. 28.]

GaH_3, is a viscous liquid presumably of similar composition. There are a variety of aluminum, magnesium, beryllium, and lithium compounds, many with CH_3 (methyl) groups, which form multicenter bonds. Some of these are shown in Figure 14.21. $Al_2(CH_3)_6$ is similar to Al_2H_6 with the sp^3-hybridized methyl groups forming both terminal and bridging bonds with aluminum atoms. In $[Be(CH_3)_2]_n$ and $[Mg_2(CH_3)_2]_n$, the bridging three-center bonds extend indefinitely to form infinite chains. The methyllithium tetramer, $[Li(CH_3)]_4$, is perhaps the strangest of them

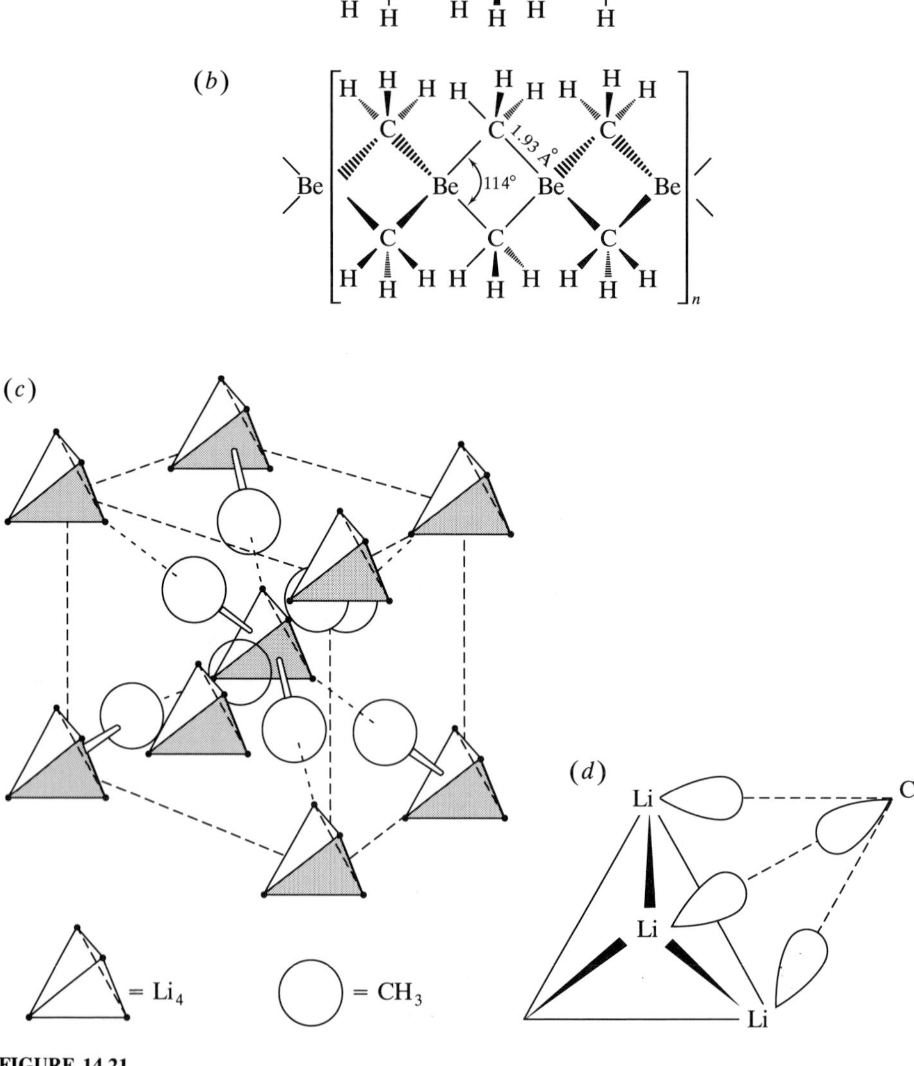

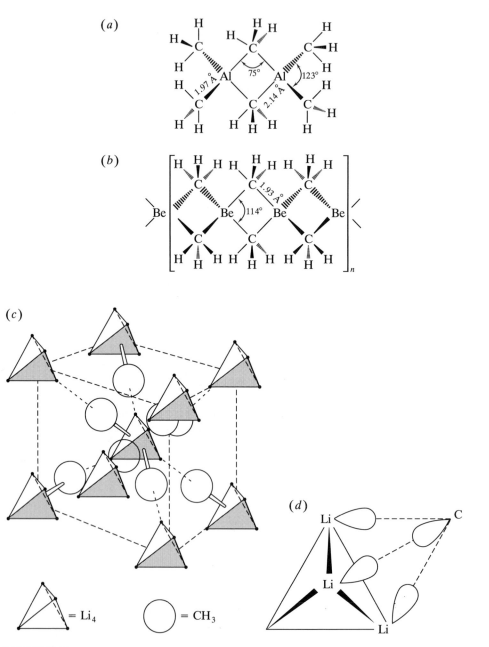

FIGURE 14.21
Other electron-deficient compounds which contain multicenter bonds: (*a*) $Al_2(CH_3)_6$, (*b*) $[Be(CH_3)_2]_n$, (*c*) $[Li(CH_3)]_4(s)$, and (*d*) the four-center–two-electron bonds in $[Li(CH_3)]_4$. [(*c*), (*d*) from Ref. 29.]

all. It is composed of a tetrahedron of lithium atoms held together by four methyl groups which sit in the middle of each triangular face to form four four-center–two-electron (4c-2e) bonds (Figure 14.21*d*).

SUMMARY

Like so many of the alkali and alkaline-earth metals, the Group 3A elements were isolated by reduction (boron) or electrolysis (aluminum and gallium). Indium and thallium were first identified by spectroscopy. Unlike the alkali and alkaline-earth metals, there is no sixth element with striking radioactive properties.

The Group 3A elements have connections to all the major components of the network. Boron, a metalloid located above the metal-nonmetal line, is unique and diagonally related to silicon. Aluminum is intermediate in character with similarities to beryllium. Indium and thallium show oxidation states 2 less than the group number as expected by the inert-pair effect. The predominance of the +1 state in the heavier congeners is underscored by a consideration of standard reduction potentials. The oxides of these elements vary from the acid anhydride B_2O_3 to the amphoteric Al_2O_3 to the basic Tl_2O. The hydroxides range from the mild boric acid to the amphoteric aluminum(III) hydroxide to the strongly basic thallium(I) hydroxide. The halides vary from the covalent trivalent boron variety to the aluminum, gallium, and indium dimers to the ionic univalent thallium halides, TlX.

Boron displays some strikingly different structures ranging from the B_{12} allotrope of the metal through the variety of binary borides to the various ortho-, meta-, and three-dimensional borates. The nuclear properties of boron 10 add to the general usefulness of these structures.

By itself not particularly strong or durable, aluminum in combination with other metals becomes one of the most versatile materials available. It does take vast amounts of electrical energy to manufacture, however, and presently this involves the production of both carbon dioxide and the sulfur oxides. Alums are astringents and the sources of beautiful crystals. Alumina, the oxide, comes in a variety of polymorphs used as various abrasives and in beautiful gem-quality minerals. It is also the product of the extremely exothermic thermite reaction. Gallium, indium, and thallium compounds do not have many applications.

Electron-deficient compounds are formed by boron, aluminum, gallium, and indium as well as lithium, beryllium, and magnesium. Such compounds are characterized by multicenter bonds in which two electrons hold three and sometimes four centers (atoms) together. The boranes, or borohydrides, are the most representative of these compounds. First prepared and characterized by Alfred Stock, these compounds are described by Lipscomb's four *structure-bonding elements* arranged into semitopological diagrams. The structures of these compounds and the related carboranes are derived from nine polyhedral structures ranging from the tetrahedron to the icosahedron. The boranes are most reactive and were once considered as rocket superfuels.

PROBLEMS

* **14.1.** (Extra credit) Calling on your vast storehouse of knowledge of Group 3A chemistry as well as any knowledge you have of Latin and Greek (or of any other language), propose a name for this group of the periodic table. Briefly justify your choice.

14.2. Write equations to represent the reactions by which (*a*) Gay-Lussac and Thénard prepared elemental boron and (*b*) Wöhler prepared aluminum.

14.3. The Group 3A elements do not have a highly radioactive sixth element like the alkali and alkaline-earth metals, but new superheavy transactinide elements are still being synthesized. What atomic number would the sixth element of the Group 3A elements have? Briefly describe some of its chemical properties.

14.4. The Turkish samples of boron typically contain 20.30% boron 10, while Californian samples typically contain 19.10%. Calculate an atomic weight for boron for each sample. Masses: boron 10 = 10.01294 u; boron 11 = 11.00931 u.

* **14.5.** The structure of cryolite, Na_3AlF_6, is shown below (from Ref. 4, p. 335). Describe the positions of the the AlF_6^{3-} anions and the sodium cations in the lattice. Is this structure consistent with the given stoichiometry of the compound? Draw a diagram showing the geometry and bonding in the AlF_6^{3-} octahedral anion.

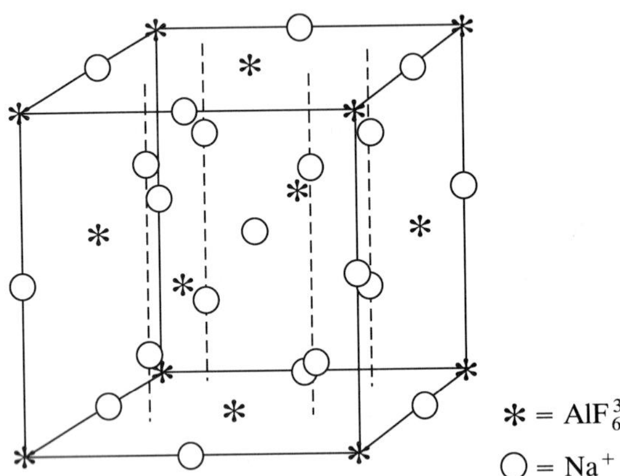

14.6. Thallium 204 is a beta minus emitter with a half-life of 3.81 years. Write a nuclear chemical equation for this process.

14.7. Like boron 10, lithium 6 also captures neutrons and emits an alpha particle. Write a nuclear equation for this process.

14.8. List two examples from this chapter which illustrate each of the five principal components (periodic law, uniqueness principle, diagonal effect, inert-pair effect, and metal-nonmetal line) of the network of interconnected ideas developed in Chapter 9.

14.9. Briefly state and explain the general trend expected for the ionization energies of the Group 3A elements. Do the actual values support that general trend? What, if any, are the exceptions to the general trend? Briefly present a rationale for these exceptions.

14.10. Briefly state and explain the general trend expected for the atomic radii of the Group 3A elements. Do the actual values support that general trend? What, if any, are the exceptions to the general trend? Briefly present a rationale for these exceptions.

14.11. Briefly state and explain the general trend expected for the electronegativities of the Group 3A elements. Do the actual values support that general trend? What, if any, are the exceptions to the general trend? Briefly present a rationale for these exceptions.

14.12. Carefully explain why boron does not form ionic salts such as $B^{3+}Cl_3^-$.

14.13. Explain why H_3BO_3 is, despite its formula, only a monoprotic acid.

14.14. Using the acid-base character of metal and nonmetal oxides component of the network, explain why boron oxide is acidic while indium oxide is basic.

14.15. Briefly describe how $TlCl_3$ and TlCl might be prepared. Write equations for these preparations as part of your answer.

14.16. Write equations representing the preparations of $AlCl_3$ and $AlCl_3 \cdot 3H_2O$.

14.17. In your own words explain why aluminum hydroxide is $Al(OH)_3$ while thallium hydroxide is TlOH.

14.18. Why is the B—F bond length so much longer in BF_4^- than it is in BF_3? As part of your answer, draw a well-labeled diagram showing the geometry and hybridization used in each species.

14.19. Write a mechanism (that is, a sequence of molecular-level events) for the reaction of boron trifluoride and methyl chloride which is represented in Equation (14.6).

14.20. Draw a diagram showing the hybridization of all atoms in the dimethyl ether adduct of boron trifluoride.

14.21. Explain the mechanism (sequence of molecular-level events) of the hydrolysis of a boron trihalide shown in Equation (14.8), repeated below.

$$BX_3 + 3H_2O \longrightarrow B(OH)_3 + 3HX$$

14.22. Briefly explain why aluminum, gallium, and indium form halides of the type M_2X_6 but thallium does not.

14.23. Consider the borohydride ion BH_4^-. Describe its relationship to methane, CH_4. Draw a diagram showing the molecular geometry and hybridization of the boron atom in this anion.

14.24. Use standard reduction potentials to determine the ΔG° of the reaction between $Tl^{3+}(aq)$ and $Tl(s)$ to produce $Tl^+(aq)$. Would this reaction be thermodynamically spontaneous under standard state conditions? Why or why not?

14.25. Use standard reduction potentials to determine the E° and ΔG° of the half-reaction in which $Tl^{3+}(aq)$ would be reduced to $Tl(s)$. [The standard reduction potential for the reaction $Tl^{3+}(aq) + 2e^- \longrightarrow Tl^+(aq)$ is 1.25 V.]

14.26. Use standard reduction potentials to determine the ΔG° of the reaction of $Tl(s)$ and $H^+(aq)$ to produce (*a*) $Tl^+(aq)$ and $H_2(g)$ and (*b*) $Tl^{3+}(aq)$ and $H_2(g)$. Which reaction is more thermodynamically spontaneous under standard state conditions? Use your answer for Problem 14.25 as one piece of data for part (*b*).

14.27. CaB_6 has the crystal structure shown in Figure 14.11. How many Ca^{2+} and B_6^{2-} ions are there per unit cell? Explain your reasoning.

14.28. ScB_{12} has the structure shown in Figure 14.12. How many scandium atoms and B_{12} clusters are there per unit cell? Explain your reasoning.

14.29. Carefully explain in your own words why the metal borides are such inert, hard, and high-melting structures.

14.30. Draw structures for the cyclic anion in the salt $K_3B_3O_6$ and the chain anion in CaB_2O_4. Estimate the value of all bond angles and indicate the hybridization of the boron atoms in each structure.

14.31. Draw a well-labeled diagram showing the geometry and orbitals used in cyclic metaboric acid, $H_3B_3O_6$.

14.32. Draw a well-labeled diagram showing the geometry and orbitals used in (*a*) the perborate anion, $[B_2O_2(OH)_4]^{2-}$ and (*b*) the anion found in borax, $[B_4O_5(OH)_4]^{2-}$.

14.33. Draw a well-labeled diagram showing the geometry and the hybridization of all boron and oxygen atoms in the tetrahydroxoborate ion, $B(OH)_4^-$.

14.34. Why is pure aluminum metal not used for aircraft frames or automobile engines? In what form and composition is aluminum usually used in these applications?

14.35. Why is aluminum resistant to air and water even though it is rather easily oxidized?

14.36. Analyze the standard reduction potentials of aluminum and iron to demonstrate that aluminum is easier to oxidize than is iron. Assume that iron is oxidized to the +2 state.

14.37. Using standard reduction potentials determine the $\Delta G°$ of Equation (14.15), repeated below.

$$2Al(s) + 6H^+(aq) \longrightarrow 2Al^{3+}(aq) + 3H_2(g)$$

14.38. Describe what happens when anhydrous $AlCl_3$ is dissolved in water and the solution made progressively more alkaline to pH 11.

14.39. Speculate on why $Al(OH)_3$ must be freshly precipitated to show the amphoteric behavior described on p. 362.

14.40. "Aluminum chlorhydrate" is actually aluminum hydroxychloride, $Al(OH)_2Cl$, or perhaps $Al_2(OH)_5Cl \cdot 2H_2O$. Speculate on the structure corresponding to this latter molecular formula.

14.41. Conduct a survey of antiperspirant deodorants used by your friends. List four different products and their active ingredients.

14.42. The standard heats of formation of $Fe_2O_3(s)$, $Cr_2O_3(s)$, and $Al_2O_3(s)$ are −822.2, −1128.4, and −1675.7 kJ/mol, respectively. Calculate the heats of reaction of the two thermite reactions represented in Equations (14.16) and (14.17), repeated below.

$$2Al(s) + Fe_2O_3(s) \longrightarrow Al_2O_3(s) + 2Fe(s)$$

$$2Al(s) + Cr_2O_3(s) \longrightarrow Al_2O_3(s) + 2Cr(s)$$

14.43. Without reference to the text, write a one-paragraph description, accompanied by well-labeled diagrams, of what is meant by a *multicenter* bond as found in diborane. Give several other examples.

14.44. Draw a semitopological diagram for the hexaborane(10), B_6H_{10}.

14.45. Draw a semitopological diagram for the pentaborane(11), B_5H_{11}.

14.46. The $B_3H_6^+$ cation is not known, but it is possible to speculate on its structure. Draw a semitopological diagram representing a "reasonable" structure for this cation. Indicate the hybridization of each boron in your structure.

14.47. The closo borane anion $B_{12}H_{12}^{2-}$ has the icosahedral structure shown in Table 14.6. By counting available valence electrons, does it seem reasonable that the carborane $B_{10}C_2H_{12}$ should exist? Speculate on the possible geometric isomers of this molecule.

14.48. Describe the structural modifications which occur in the sequence $B_6H_6^{2-}$, B_5H_9, to B_4H_{10} as found in Table 14.6.

14.49. There are at least two other possible structural isomers of the conjuncto $B_{10}H_{16}$ which are not shown in Table 14.6. Draw diagrams and describe these two isomers.

14.50. What is the best way to make diborane in the laboratory? Write an equation for the reaction. Contrast this method with that used by Stock to obtain diborane.

* **14.51.** Draw the structures of $Be(BH_4)_2$ and $Al(BH_4)_3$. Each involves three-center bonds of the type B—H—M, where M = beryllium or aluminum.

14.52. BeB_2H_8 is a known compound, but its structure has proven difficult to determine. By counting electrons, write two possible semitopological diagrams for this compound. One should be a linear species in which the boron and two beryllium atoms are in a straight line and the second a triangular species held together by 3c-2e bonds.

14.53. Speculate on the structure and isomers of $(CH_3)_2B_2H_4$. Draw a well-labeled diagram showing estimated bond angles and hybrids used by all boron and carbon atoms. This molecule contains 3c-2e B—H—B bonds.

14.54. In Chapter 9 it was asserted that the first elements in each group are not the most representative of the group as a whole and that, in fact, a better case can be made that the second element is more representative. Do you agree that this is the case for Group 3A? Be specific.

CHAPTER 15

THE GROUP 4A ELEMENTS

The Group 4A elements (carbon, silicon, germanium, tin, and lead) continue the trend of an increasingly wide variety of properties within a single group. Like Group 3A, these elements are so diverse that the group has no descriptive name like the halogens or alkali metals. Following our common practice, we start with a discussion of the discovery and isolation of the elements. Next is the application of our network. As was the case for the Group 3A elements, all seven components significantly contribute to our understanding of the group. In particular, the description of the hydrides, oxides, hydroxides, and halides reveals a wide range of properties. The many and diverse practical applications of these elements include the lubricating and cutting abilities of the graphite and diamond allotropes, respectively, radiochemical dating methods, and the lead storage battery. A discussion of the structure and properties of silica, silicates, and aluminosilicates merits its own separate section. The selected topics in depth are semiconductors and glass.

15.1 DISCOVERY AND ISOLATION OF THE ELEMENTS

Figure 15.1 shows the usual information about the discoveries of these elements superimposed on our chronological plot of the number of known elements. Note that for the first time we encounter elements (carbon, tin, and lead) known to the ancients.

Carbon, Tin, and Lead

Carbon has been known in the forms of coal, oil, petroleum, natural gas, and charcoal for thousands of years. For example, lampblack (a fine carbon soot) was used as an ink pigment six centuries before the time of Christ. Given that free carbon was known to the ancients, no discoverer is listed for it. The recognition

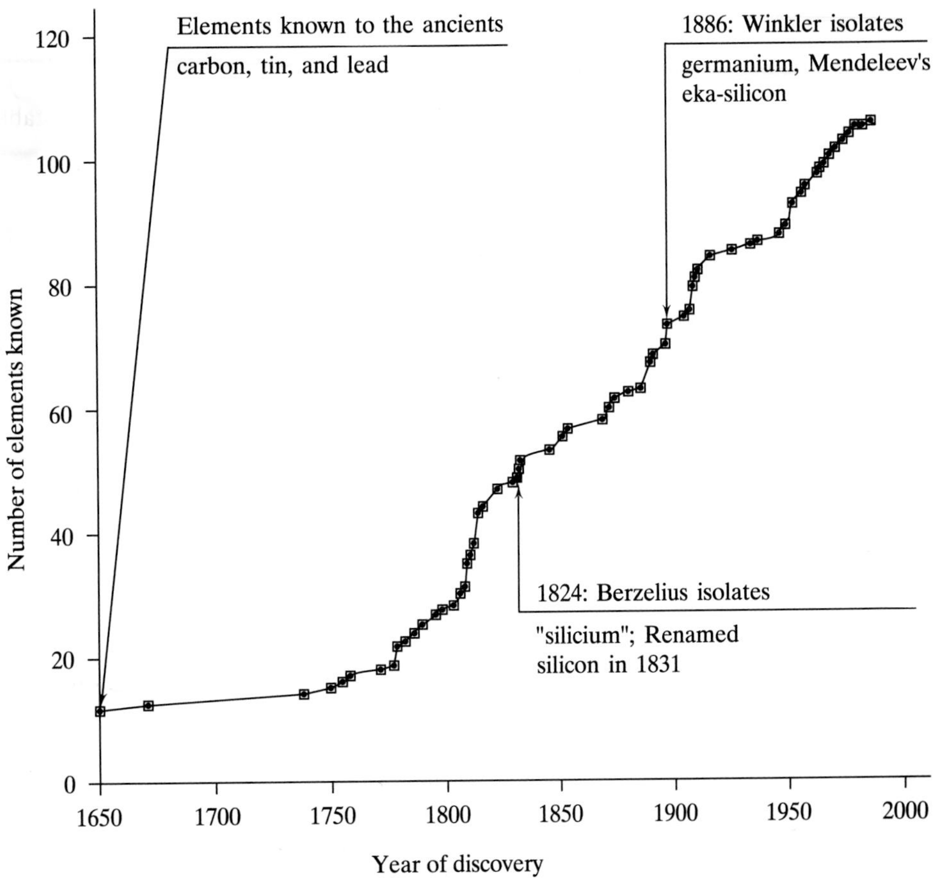

FIGURE 15.1
The discovery of the Group 4A elements superimposed on the plot of the number of known elements versus time.

that it was an element in the modern sense, however, dates back only to the late eighteenth century. The independently wealthy Lavoisier carried out many experiments on the combustion of diamonds in the 1770s, and the books of that time start to refer to "carbone" as an element. By the end of that century it had been shown that diamonds and graphite were but two forms of the very same element.

As mentioned in Chapter 1, it was in the first part of the nineteenth century that Berzelius divided all chemical compounds into either organic (derived from living tissue) or inorganic. Initially he thought that there might even be different laws governing these two types of compounds, but when this was disproven by Wöhler in the 1820s, the distinguishing element (literally in this case) seemed to be whether the compounds contained carbon or not. Today we know that the boundary between organic and inorganic is a very blurry one indeed but nevertheless a demarcation that, it appears, will persist for some time to come.

The production of tin can be dated back to 3000 B.C., most likely because its oxide could be readily reduced to the metal by the glowing coals of a wood fire. The production and use of bronze (an alloy of copper and tin) is still older. Tin dishes were common in the 1700s as were tin-plated materials. In fact, the plating of iron with a thin layer of tin was a booming industry in New York and New

England in colonial times. Today the United States continues to be a major user of tin but must import almost all it consumes. Much of it is used in various alloys including solder (with lead), pewter (with antimony and copper), and bronze (with copper). Tin has the distinction of having more stable isotopes (10 in fact) than any other element. (See Table 15.1 for a listing.)

Lead is perhaps the oldest metal known. The Book of Job, probably written about 400 B.C., has its author wishing to have his devotion to God be recorded forever "with an iron pen and lead" (Job 19:24). Lead was easy to hammer into sheets for writing and flooring material, into vessels for cooking and storing foods, and into pipes for plumbing. Not only were lead pipes the original plumbing materials (the insignia of Roman emperors can be found in lead pipes still in service), but both the words *plumbing* and *plumber* are derived from the same Latin word, *plumbum*, for "lead." This is also the origin of the symbol Pb for the element.

Silicon

Glass, of which silicon in the form of silica is a prime component, has been known since about 1500 B.C. As discussed in Chapter 12, the "earths" including silica were regarded by many to be elements, but Davy did not agree even though he could not isolate the corresponding metalloid from silica using his voltaic pile or by reacting it with potassium. In 1811, Gay-Lussac and Thénard, who together had isolated boron 3 years earlier, tried reacting the gaseous silicon tetraflouride (isolated by Scheele) with potassium but obtained only an impure form of silicon—not pure enough to be credited with the discovery. In 1824, Berzelius, by patiently purifying the products, succeeded in isolating amorphous silicon where others before him had failed. He named his new element "silicium" from the Latin *silex*, for "flint," a major source of silica. The name *silicon* was proposed in 1831, with the suffix *-on* replacing *-ium* to establish the parallel with boron and carbon. Shiny, blue-gray crystalline silicon was not prepared until nearly 25 years later. Today, reasonably pure silicon is prepared by reacting silica with carbon in an electric furnace as represented in Equation (15.1):

$$SiO_2(s) + 2C(s) \xrightarrow[\text{furnace}]{\text{electric}} Si(s) + 2CO(g) \tag{15.1}$$

Germanium

Silicon, having taken its place in the list of elements in the 1820s, appeared in Mendeleev's periodic table of the 1870s. Directly under silicon the Russian chemist left one of his famous blanks for undiscovered elements. (See Chapter 9, p. 214, for a recounting of this process and Chapter 14, p. 348, for an account of de Boisbaudran's discovery of eka-aluminum, or gallium.) Eka-silicon was discovered in 1886 by Clemens Winkler who, in a line of reasoning similar to that of young Arfwedson in isolating lithium in 1817, found that he could not account for 7 percent of a new silver ore. The "missing" percent turned out to be a new element, which he named germanium in honor of his fatherland. Winkler, incidentally, did not identify his element with eka-silicon but rather with eka-stibium, an element which Mendeleev had predicted to lie between antimony and bismuth. Mendeleev himself thought Winkler's new element to be eka-cadmium which he had placed between cadmium and mercury. It was Meyer (who, you might recall from Chapter 9, independently formulated the periodic law in the same time period as Mendeleev) who correctly identified it as eka-silicon. Eka-stibium and eka-cadmium are two of Mendeleev's *not-so-famous blanks* which we seldom hear about any more.

Not much use was found for germanium (except perhaps to make bad, nongermane puns) until 1942, when the transistor was invented at the Bell labs. And now germanium has returned to relative obscurity having lost out in the semiconductor-transistor market to silicon. It is separated from other elements by the fractional distillation of its chloride that in turn is hydrolyzed and reduced to the gray-white metalloid element as shown in Equations (15.2):

$$GeCl_4(s) + 2H_2O(l) \longrightarrow GeO_2(s) + 4HCl(aq) \qquad (15.2a)$$

$$GeO_2(s) + 2H_2(g) \longrightarrow Ge(s) + 2H_2O(l) \qquad (15.2b)$$

15.2 FUNDAMENTAL PROPERTIES AND THE NETWORK

Figure 15.2 shows the Group 4A elements superimposed on the network and is, in fact, exactly identical to Figure 12.6 presented after the network had just been completed. Group 4A, then, like 3A, is intimately tied to our organizing scheme of ideas. A careful inspection of Table 15.1, the usual tabulation of group properties, reveals the expected periodic trends with only a few irregularities.

What about the other components of the network? Is carbon unique in its group? It most certainly is. No other element is so special that it has an entire branch of chemistry built around it! Carbon is arguably the most striking example of an element being unlike its heavier congeners as one can find in the periodic table. This will be particularly evident when we explore the ability of carbon to catenate (form self-links with itself) and form π bonds.

The metal-nonmetal line passes through the heart of the group with carbon being a bona fide nonmetal and lead a bona fide metal. In between are two metalloids (silicon and germanium) and a borderline metal (tin). The progression in the acid-base character of the oxides of the elements further emphasizes the trend from nonmetal to metal. The formulas of the oxides and halides show the increasing importance of the +2 oxidation state down the group, and this is reinforced by a consideration of standard reduction potentials.

So our network brings nearly its full weight to bear in organizing these elements. Note, however, that the diagonal effect is not as important now as it once was in the earlier groups. We have seen that silicon is much like boron, but carbon is not similar to phosphorus. Further evidence of the full organizing power of the network is shown in the following considerations of the hydrides, oxides, hydroxides, and halides of the Group 4A elements.

Hydrides

We have seen in the earlier chapters (for example, see Table 10.4 and the accompanying discussion) that the binary hydrides of lithium, beryllium, and boron progress from one, LiH, displaying significant ionic character to the polymeric BeH_2 and then to the prototypical borohydride, B_2H_6, the latter two being characterized by multicenter covalent bonds. The lower hydrides of the 1A and 2A elements are predominantly ionic and become more so going down the groups,

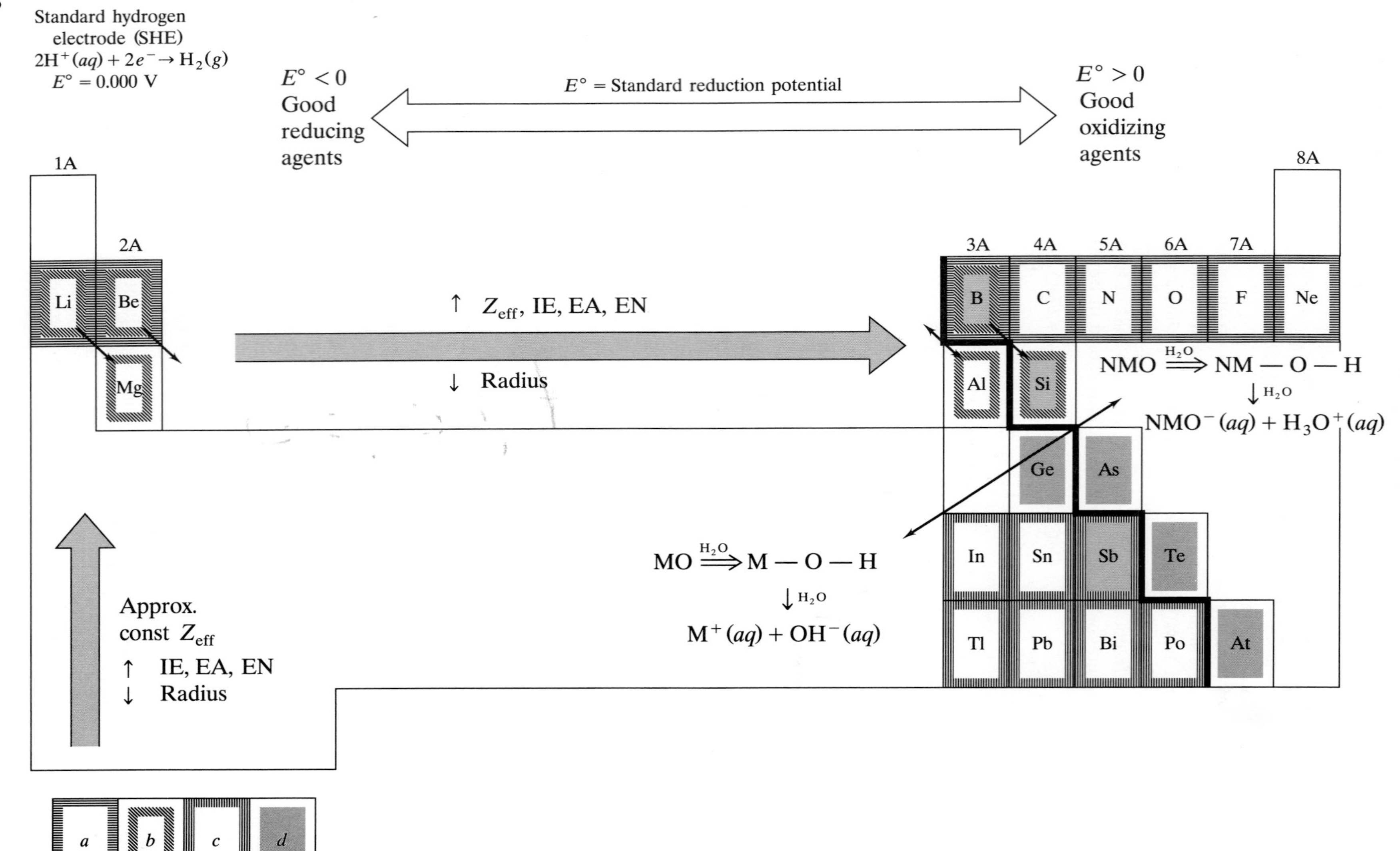

FIGURE 15.2
The Group 4A elements superimposed on the network of interconnected ideas including the trends in periodic properties, the acid-base character of metal and nonmetal oxides, trends in standard reduction potentials, (*a*) the uniqueness principle, (*b*) the diagonal effect, (*c*) the inert-pair effect, and (*d*) the metal-nonmetal line.

TABLE 15.1
The fundamental properties of the Group 4A elements. [Ref. 24.]

	Carbon	Silicon	Germanium	Tin	Lead
Symbol	C	Si	Ge	Sn	Pb
Atomic number	6	14	32	50	82
Natural isotopes, A/ % abundances	$^{12}C/98.89$ $^{13}C/1.11$	$^{28}Si/92.21$ $^{29}Si/4.70$ $^{30}Si/3.09$	$^{70}Ge/20.52$ $^{72}Ge/27.43$ $^{73}Ge/7.76$ $^{74}Ge/36.54$ $^{76}Ge/7.76$	$^{116}Sn/14.30$[a] $^{117}Sn/7.61$ $^{118}Sn/24.03$ $^{119}Sn/8.58$ $^{120}Sn/32.85$ $^{122}Sn/4.72$ $^{124}Sn/5.94$	$^{204}Pb/1.48$ $^{206}Pb/23.6$ $^{207}Pb/22.6$ $^{208}Pb/52.3$
Total no. of isotopes	7	8	14	21	21
Atomic weight	12.01	28.09	72.59	118.7	207.2
Valence electrons	$2s^2 2p^2$	$3s^2 3p^2$	$4s^2 4p^2$	$5s^2 5p^2$	$6s^2 6p^2$
mp/bp, °C	3570/sublimes	1414/2355	937/2830	232[b]/2270	328/1750
Density, g/cm^3	2.25[c]	2.33	5.32	7.30[b]	11.35
Atomic radius (metallic), Å	...	...	1.39	1.58	1.75
Covalent radius, Shannon-Prewitt, Å	0.77	1.17	1.22	1.41	1.47
Ionic radius, Shannon-Prewitt, Å (C.N.)	0.29(4)	0.40(4)	0.67(6)	(+4) 0.83(6) (+2) 1.22(8)	(+4) 0.79(6) (+2) 1.33(6)[d]
Pauling EN	2.5	1.8	1.8	1.8	1.9
Charge density (charge/ionic radius), unit charge/Å	14	10	6.0	(+4) 4.8 (+2) 1.6	(+4) 5.1 (+2) 1.5
$E°$, V ($MO_2 \rightarrow M$, acid soln.)	+0.21	−0.91	−0.07	−0.10	0.74[e]
Oxidation states	−4 to +4	−4, +2, +4	−4, +2, +4	−4, +2, +4	+2, +4
Ionization energy, kJ/mol	1086	786	760	709	716
Electron affinity, kJ/mol	−123	−120	−118	−121	−101
Discovered by/date	Antiquity	Berzelius 1824	Winkler 1886	Antiquity	Antiquity
rpw[f] O_2	CO, CO_2	SiO_2	GeO_2	SnO_2	PbO
Acid-base character of oxide	Acid	Acid	Ampho.	Ampho.	Ampho.
rpw N_2	None	Si_3N_4	None	Sn_3N_4	None
rpw halogens	CX_4	SiX_4	GeX_4	SnX_4	PbX_2
rpw hydrogen	CH_4	None	None	None	None
Crystal structure	Diamond graphite	Diamond	Diamond	Tetragonal	fcc

[a] Tin has three other stable isotopes of less than 1% abundance.

[b] Value for white tin.

[c] Value for graphite.

[d] Value questionable.

[e] $PbO_2 \rightarrow Pb^{2+}$, $E° = 1.46$ V (acid solution).

[f] rpw = reaction product with.

while those of 3A retain an electron-deficient flavor but also become progressively more ionic.

The parent carbon hydride, on the other hand, is the fully covalent methane, CH_4, the simplest of thousands of alkanes of general formula C_nH_{2n+2}. While others had prepared the corresponding silanes before, Stock, in 1916, was the first to carry out a systematic investigation of them with the aid of his vacuum line and its mercury-activated stopcocks and valves. In the absence of air and water, he reacted the metal silicides (as he had the borides) with various acids as represented in Equation (15.3):

$$Mg_2Si(s) + HX \longrightarrow Si_nH_{2n+2}(g) + MgX_2(s) \qquad (15.3)$$

It was also Stock who suggested that we call silicon hydrides the silanes by analogy with the alkanes. However, while we took his suggestion and still name these two classes of compounds analogously, there are some striking chemical differences between them. One difference is that the alkanes can form much longer and more stable chains. Specifically, n for the alkanes can reach values of 100 or more, while for the silanes n is restricted to the single digits. A second striking difference has to do with the relative reactivities of the silanes and their corresponding alkanes. For example, the alkane in which $n = 4$ is butane, C_4H_{10}, a very stable gas, while tetrasilane, Si_4H_{10}, is a violently reactive liquid.

Catenation is defined as the self-linking of an element to form chains and rings. Carbon, then, given the above discussion, is the all-time champion catenator, much better than silicon (or sulfur, boron, phosphorus, germanium, and tin, the other elements that show this ability). Why should this be so? A comparison of the relevant carbon and silicon bond energies as shown below is helpful:

C—C	356 kJ/mol	Si—Si	226 kJ/mol
C—H	413 kJ/mol	Si—H	298 kJ/mol
C—O	336 kJ/mol	Si—O	368 kJ/mol

Note first that the C—C bond is (more than 50 percent) stronger than the Si—Si bond. This is due to the increased bond length and therefore less effective overlap of p (or sp^3) orbitals to form the σ bonds in the silanes. (The C—C internuclear distance is 1.54 Å, while the Si—Si distance is 2.34 Å, some 50 percent longer.) Secondly, C—H bonds are (about 40 percent) stronger than the corresponding Si—H bonds, again due to less effective orbital overlap in the latter. These shorter and therefore stronger C—C and C—H bonds are certainly major factors in the enhanced stability of the chains found in the alkanes as compared to the silanes.

Now let us consider the related issue of why the silanes are more reactive than the alkanes. Note that the C—C bond energy is *greater* than that for C—O, while the Si—Si bond energy is considerably *less* than that for Si—O. This means that C—C bonds tend to be stable relative to C—O (single) bonds, while Si—Si bonds are less stable than and tend to give way to the stronger Si—O bonds. It follows then that since oxygen is almost always present, except under rigidly controlled conditions, the silanes will spontaneously react (in the thermodynamic sense) to form silicon-oxygen compounds.

But, you may very well be wondering, *why* are Si—O bonds so strong relative to Si—Si bonds? The answer to this question is part of a trend that will become more and more important now that our tour of the representative elements has moved almost halfway across the periodic table. From our considerations of the uniqueness principle, we know that the very small second-period elements (particularly C to O) readily form π bonds among themselves. (This type of pi bond between p orbitals is often called a $p\pi$-$p\pi$ bond.) We also know that the heavier elements in each group are unlikely to π-bond among themselves due to the relatively ineffective overlap between the larger p orbitals of these atoms. But now, as we move further to the right in the table, we come to the point where we must consider the possibility of π bonding between a second-period element (using its often *filled p* orbitals) and a heavier element (using its *empty d* orbitals). Such a pi bond is often designated a $p\pi$-$d\pi$ bond with both electrons coming from the smaller element. The contribution of such a π interaction in a Si—O bond (in addition to the normal sigma Si—O bonds still of primary importance) is the reason that this bond is stronger than that between two silicon atoms. The $p\pi$-$d\pi$ overlap possible only in the Si—O bond (and not in the Si—Si) enhances its strength and makes it more stable.

One more point should be made here that, although somewhat incidental to the argument at hand, is certainly important in the long run. It should make sense that as the larger elements get smaller in a given period (for example, Si to P to S in the third period), the $p\pi$-$d\pi$ overlap across a progressively shorter and shorter bond distance will become more and more effective. In the next few paragraphs we will see the effect of $p\pi$-$d\pi$ overlap in Si—N bonds that is quite similar to that just discussed for Si—O bonds. Moreover, in subsequent chapters, we will see how $p\pi$-$d\pi$ interactions play even greater roles in increasing the strengths of P—O, P—N, S—O, S—N, and other bonds. Look for the growing strength and importance of $p\pi$-$d\pi$ bonding as we continue our survey of the representative elements.

The greater reactivity of the silanes is certainly due, in large measure, to the above thermochemical considerations. However, there are related kinetic factors to be considered as well. Take, for example, methane versus silane (or monosilane as it is sometimes called). Methane (the primary component of natural gas) does not react readily with water, and it takes a flame or a catalyst to set off its reaction with oxygen. Silane, on the other hand, quickly ignites in air without any catalyst and violently reacts with water if even small traces of base are present. These reactions are represented in Equations (15.4) and (15.5):

$$SiH_4(g) + O_2(g) \xrightarrow[\text{catalyst}]{\text{no}} SiO_2(s) + H_2O(g) \tag{15.4}$$

$$SiH_4(g) + nH_2O(l) \xrightarrow{OH^-} SiO_2 \cdot nH_2O + 2H_2(g) \tag{15.5}$$

Let's consider why silane reacts so much more rapidly than its distant cousin, methane. Again, the answer has to do with the participation of the $3d$ orbitals of silicon. These orbitals are low enough in energy that they can be used to bind to an incoming molecule or ion. For example, the hydroxide ion (which catalyzes the hydrolysis of silane) or even a water molecule itself might form a fifth bond to the silicon and so facilitate the reaction through the resulting five-coordinate intermediate. In the oxidation of silane with molecular oxygen, the O_2 could form a bond with a $3d$ orbital of silicon in the same manner. In either case such interactions would produce lower-energy transition states, smaller energies of activation, and consequently more kinetically favorable (faster) reactions. On the other hand, such five-coordinate intermediates are not available in carbon because it has no low-lying d orbitals.

We go on now to consider the differences between other carbon and silicon hydrides. For example, carbon forms compounds such as ethylene, $H_2C{=}CH_2$, and acetylene, $HC{\equiv}CH$. These involve bonding between p orbitals of the carbon

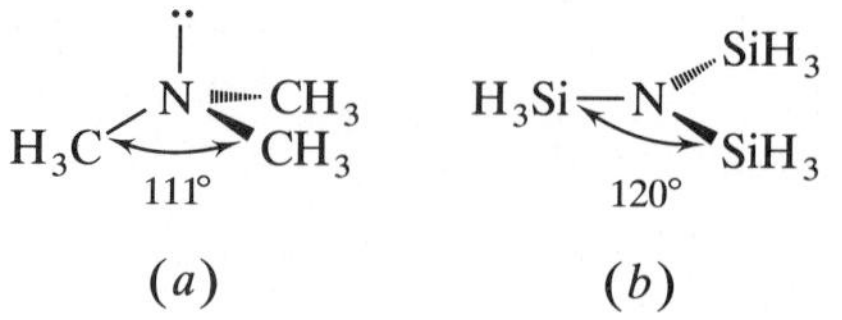

FIGURE 15.3
The structures of (*a*) the pyramidal trimethylamine, $N(CH_3)_3$, and (*b*) the planar trisilylamine, $N(SiH_3)_3$.

atoms. As we know from the discussion in Chapter 9 and in the preceding paragraphs, such $p\pi$-$p\pi$ overlap is extraordinarily difficult to achieve with silicon and its heavier congeners because the internuclear distances are too large to be effectively spanned by the parallel overlap of $3p$ orbitals (see Figure 9.11). Accordingly, although a few examples of compounds containing Si=Si double bonds have been synthesized, no simple analogs of ethylene or acetylene have been found.

Another example of the striking differences between carbon and silicon chemistry occurs in nitrogen-containing compounds. As we have just discussed, silicon does not use its p orbitals to π-bond, but it can use its empty d orbitals to participate in $p\pi$-$d\pi$ bonding. One of the best pieces of evidence for this is the structure and reactivity of trimethylamine, $N(CH_3)_3$, and the analogous silicon compound, trisilylamine, $N(SiH_3)_3$. Based on their similarity to ammonia, we expect these compounds to be pyramidal (around the nitrogen) as shown in Figure 15.3*a*. In the trimethyl compound the nitrogen is evidently sp^3-hybridized with one of the hybrid orbitals containing a nonbonding pair of electrons. The corresponding silyl compound, however, is, surprisingly, not pyramidal but rather planar as shown in Figure 15.3*b*. It appears that in the silyl compound, the nitrogen is sp^2-hybridized (leading to the planar configuration), and the lone pair of electrons in the unhybridized $2p$ orbital of the nitrogen is donated into empty d orbitals of the three silicon atoms to form delocalized $p\pi$-$d\pi$ interactions as shown in Figure 15.4. The acid-base character of these two compounds helps to confirm the above interpretation. We find that the trimethyl amine is a Lewis base as expected (the lone pair of electrons can be donated to an electron-pair acceptor), while the trisilyl is not basic, presumably because of the delocalization of the electron pair into the silicon atoms.

Germanes, Ge_nH_{2n+2} (n = single digits), form in a manner similar to silanes and are a little less reactive. Distannane, Sn_2H_6, has been prepared, but the higher analogs have not. Plumbane, PbH_4, is extremely unstable.

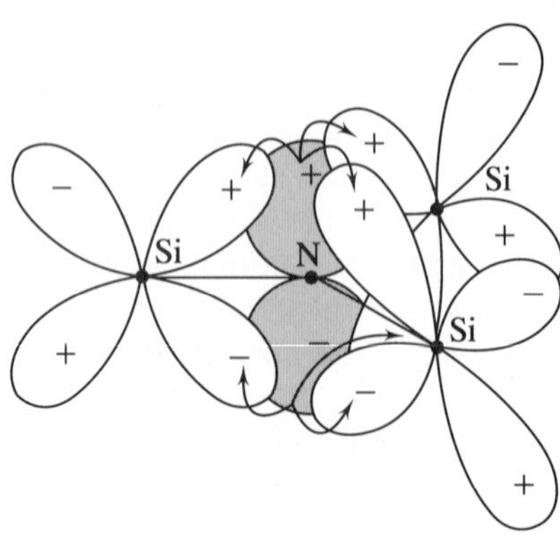

FIGURE 15.4
The $p\pi$-$d\pi$ bonding in trisilylamine, $N(SiH_3)_3$. The nitrogen is sp^2-hybridized and forms sigma bonds (not shown) with each of the three silicon atoms. The filled, unhybridized $2p$ orbital (shaded) of the nitrogen overlaps the unfilled silicon $3d$ orbitals (unshaded), spreading or delocalizing the electron density evenly to all three silicon atoms. [Adapted from Ref. 20, p. 164.]

Oxides and Hydroxides

In Chapter 11 (pp. 276ff) we discussed the general trends in oxides including the network component concerning their acid-base properties. In Groups 1A and 2A, all the oxides are ionic and basic anhydrides except for the partially covalent and amphoteric beryllium oxide. In Group 3A, we encountered our first acidic oxide in B_2O_3 and noted that its heavier congeners move from amphoteric (Al_2O_3, Ga_2O_3, and In_2O_3) to the basic Tl_2O. Now in Group 4A we encounter our first gaseous acidic oxide in carbon dioxide, the acid anhydride of carbonic acid. So we are reminded once again that, moving horizontally from left to right in a period, the oxides progress from ionic basic anhydrides to partially covalent amphoteric and finally to covalent acidic anhydrides.

While the empirical formulas of carbon and silicon dioxides are similar, the bonding and structures of these compounds have little in common. The linear carbon dioxide, O=C=O, with its small central atom capable of forming strong $p\pi$ bonds, is a discrete molecular gas. Silicon dioxide, with its larger central atom incapable of forming $p\pi$-$p\pi$ bonds, is a three-dimensional polymeric solid (see the α-quartz structure shown in Figure 15.9) described more fully in the section on silicates and silica. There is no stable analog of the common, gaseous carbon monoxide (C≡O) in silicon chemistry.

One property that carbon and silicon dioxides *do* have in common is that both are acidic anhydrides. Silicon dioxide, a major component of glass, is acidic enough that strong bases will react with glass bottles. Such bases should be stored in polyethylene and not glass containers. Going down the group the central atom becomes less electronegative, giving the oxides greater ionic and consequently greater basic character.

You may have noted in Table 15.1 that the oxides of the heavier Group 4A elements are characterized by an increasing stability of the +2 oxidation state. The most stable oxide of tin is $Sn^{IV}O_2$, but for lead it is $Pb^{II}O$. Given our earlier discussions of the inert-pair effect, this trend is to be expected. On the other hand, you might be surprised to find that no true neutral tetrahydroxides exist for these elements. The interaction between a Group 4A element in the +4 oxidation state and the oxygen of a hydroxide ion is strong enough that the hydrated oxides, $MO_2 \cdot nH_2O$, are more stable in each case.

Halides

Ionic halides were the rule for the Group 1A and 2A elements, with the covalent, polymeric, electron-deficient $BeCl_2$ being the only exception. In Group 3A, however, halide chemistry became more involved. The boron halides were covalent, electron-deficient compounds that were strong Lewis acids. Aluminum, gallium, and indium halides, M_2X_6, were characterized by bridging halogen atoms. At the bottom of the group, the +1 oxidation state dominated due to the inert-pair effect.

The various carbon halides, more the province of organic chemistry than inorganic, are a logical continuation of the horizontal left-to-right trend toward greater covalent character. Not only are the discrete CX_4 compounds known for all the halides, but catenation is also prevalent, one of the best examples being the polymer Teflon, characterized by long —(CF_2)— chains. The chlorofluorocarbons, CF_mCl_n, or just CFCs, are discussed in Chapter 18 (pp. 493ff).

The silicon halides are prepared by the reaction of elemental silicon or silicon carbide, SiC, with the halogens. The halides exhibit longer Si—Si chains than the hydrides do. Compounds such as $Si_{16}F_{34}$, Si_6Cl_{14}, and Si_4Br_{10} are known. As expected from the discussion of the hydrides, CF_4, with no available carbon *d* orbitals, is relatively inert, while SiF_4 is extremely reactive. For example, silicon tetrafluoride reacts with hydrogen fluoride to produce H_2SiF_6, characterized by an expanded octet around the silicon atom. This can only be possible through the involvement of the *d* orbitals of the silicon atom.

The halides also provide more examples of the increasing stability of the +2 oxidation states down the group. The germanium tetrahalides are more stable than the dihalides, but in lead the dihalides predominate. The germanium and tin tetrahalides are prepared by direct reaction of the elements. Germanium(IV) oxide, GeO_2, treated with the hydrohalic acids, HX, also yields the tetrahalides. However, when GeF_4 or $GeCl_4$ is combined with elemental germanium, the dihalides result. Tin(II) fluoride, which not long ago was a common toothpaste additive (see Chapter 18, p. 492), is prepared by the reaction of SnO with 40% aqueous HF. Elemental tin plus dry hydrogen chloride affords $SnCl_2$, not $SnCl_4$. The element plus iodine in 2 *M* HCl results in SnI_2. Lead tetrahalides are very difficult to prepare, and the tetrafluoride is the only stable one of the four. Lead(II) halides can be readily isolated by adding aqueous hydrohalic acids to a variety of water-soluble lead(II) salts.

The existence of both the +2 and +4 oxidation states in germanium, tin, and lead chemistry leads to the question of how these compounds vary, particularly in ionic versus covalent character. Based on what you know so far, what would you guess? Would the compounds containing the higher oxidation state tend to be more or less covalent than those with the lower oxidation state? Does it help you to mention that the higher oxidation states have a higher charge-to-radius (Z/r) ratio, or charge density? The greater the charge density, the more covalent the compound, right? So it follows that the higher oxidation state compounds are the more covalent. For example, tin(II) fluoride is a rather saltlike ionic compound, while tin(IV) fluoride is covalent. The greater solubility of the ionic tin(II) salt once made it a good fluoride additive in toothpastes, whereas tin(IV) was not soluble enough to be effective.

Before we leave the subject of the relative stability and character of the +2 versus the +4 oxidation states, we should see what can be gained by looking at the standard reduction potentials of these compounds. The standard reduction potential for $Sn^{4+}(aq)$ being reduced to $Sn^{2+}(aq)$ is +0.13 V as shown in Equation (15.6):

$$Sn^{4+}(aq) + 2e^- \rightleftharpoons Sn^{2+}(aq) \qquad E^\circ = +0.13\text{ V} \tag{15.6}$$

This means that Sn^{4+} is a very mild oxidizing agent, or, stated another way, Sn^{2+} is a mild reducing agent. In any case, the Sn^{2+} ion is not particularly stable in aqueous solution being rather easily oxidized to $Sn^{4+}(aq)$. This property is traditionally harnessed to reduce Fe^{3+} to Fe^{2+} in quantitative iron analyses. On the other hand, the standard reduction potential of PbO_2 to the $Pb^{2+}(aq)$ is +1.46 V as shown in Equation (15.7):

$$PbO_2 + 4H^+ + 2e^- \rightleftharpoons Pb^{2+} + 2H_2O \qquad E^\circ = +1.46\text{ V} \tag{15.7}$$

This indicates that lead(IV) oxide is a strong oxidizing agent and that the $Pb^{2+}(aq)$ is a relatively stable species in aqueous solution.

15.3 REACTIONS AND COMPOUNDS OF PRACTICAL IMPORTANCE

Graphite, Fullerenes, and Diamond

At first it was difficult for early chemists to imagine that the graphite and diamond allotropes of carbon were really different forms of the same element. Graphite is metallic in appearance and very soft, while diamond is transparent and one of the hardest substances known. The diamond structure, pictured in Figure 7.3, is a three-dimensional covalent crystal composed of interconnected C—C single bonds. (If you have covered solid state structures, you should be able to see that the diamond unit cell is the same as that for zinc blende except that all the spheres represent carbon instead of alternating between zinc and sulfur. The zinc blende unit cell is shown in Figure 7.22*c*.) Since these interconnected C—C bonds extend throughout the crystal, one could say that diamond is *one* giant molecule. It is this bonding arrangement that makes the hard and high-melting diamond so useful in cutting tools and abrasives.

The graphite structure is shown in Figure 15.5. It is a layered structure characterized by strong delocalized π bonding within each layer and only van der Waals forces between them. The nature of these forces is reflected in the C—C distances shown in the figure. The soft and lubricating properties of graphite are due to these layers being able to easily slide by each other. Pencil "lead" is also graphite (mixed with clay) which is easily confused with the dark-gray lead sulfide. Pressure on the pencil head causes the layers of graphite to rub off onto a piece of paper. Charcoal, soot, and lampblack are very tiny particles of graphite. The large surface areas of these materials make them useful for adsorbing various gases and solutes.

Many highly advertised pieces of sporting equipment such as golf clubs, tennis and squash rackets, and fishing rods are advertised as being made of lightweight graphite fibers. Can this be the same graphite described above? No, in fact, it is not. Rather, this light, fibrous, and high-strength material is produced by pyrolyzing (heating) organic polymeric fibers.

When graphite is vaporized in a laser, a variety of large clusters with even numbers of carbon atoms are formed. One of the most prevalent of these, C_{60}, is a truncated icosahedron characterized by 60 vertices, 32 faces, 12 of which are

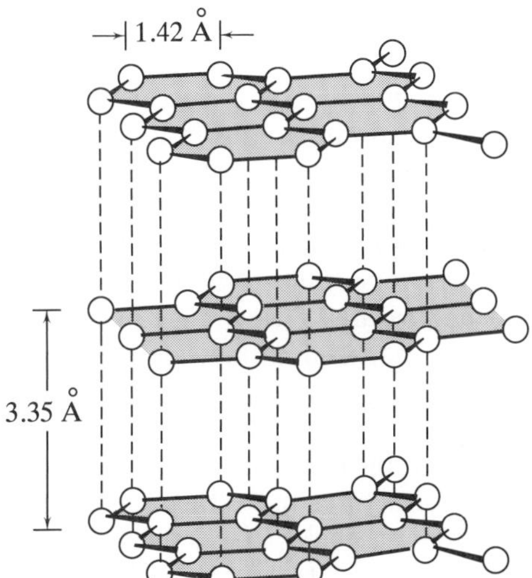

FIGURE 15.5
The structure of graphite. The carbon atoms within each layer are bonded by delocalized $p\pi$ electrons. The layers are held together only by van der Waals forces. [Ref. 27.]

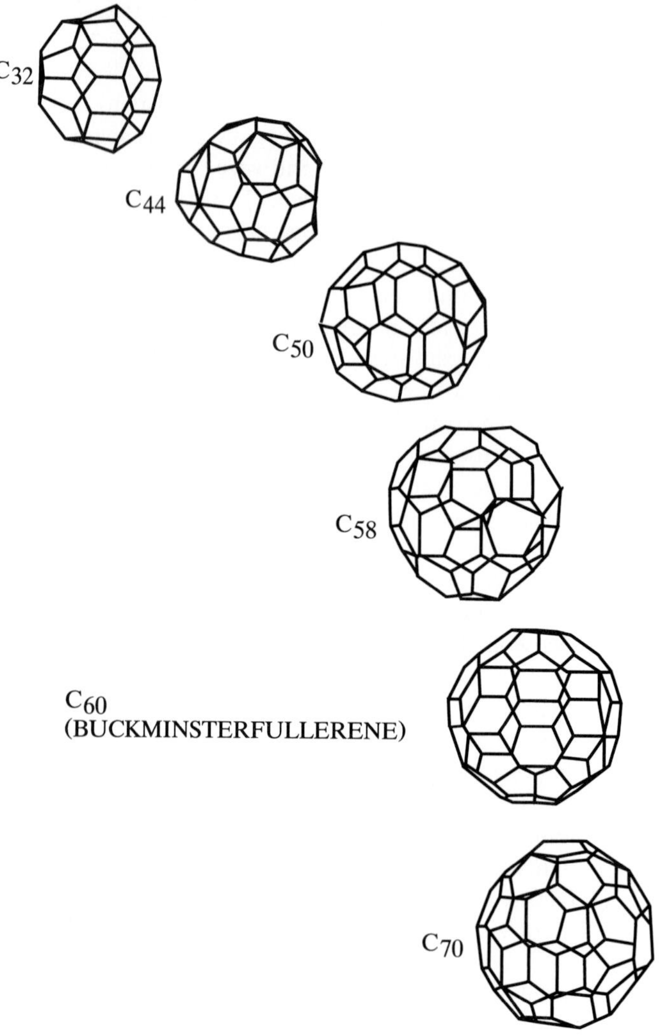

FIGURE 15.6
Some representative fullerenes: C_{32}, C_{44}, C_{50}, C_{58}, C_{60} (buckministerfullerene, or bucky ball), and C_{70}. [Ref. 30.]

pentagons. It looks much like a soccer ball or one of the icosahedra that Buckminster Fuller made so famous in the construction of his geodesic domes. In fact, C_{60} is now called *buckminsterfullerene*, or sometimes *bucky ball* for short. C_{60} is essentially a section of graphite curled up into a ball. It retains the delocalized nature of graphite, and both its interior and exterior are a sea of π electrons. Since the initial discovery of these *fullerenes* in the mid-1980s, an incredible variety of them have been prepared. They range in size from C_{32} to some *giant fullerenes* containing nearly a thousand carbon atoms. Some of these are shown in Figure 15.6.

Since their discovery, the fullerenes have been investigated at a feverish pace. Each is characterized by 12 pentagons and a varying number of hexagons. Their highly symmetrical structures make them extremely stable and resilent. For example, C_{60} can be fired at steel surfaces at velocities approaching that of the space shuttle (17,000 mi/h) and just bounce back unharmed.

The interior diameter of the C_{60} cluster is about 7 Å and can accommodate a variety of types of small ions, for example, helium and potassium, as in $C_{60}K$. Other atoms can be placed in the holes created by the packing of the nearly spherical C_{60}'s, or they can be chemically attached to the exteriors of the spheres or even incorporated into the very fabric of their structures. These possibilities lead to a truly astounding number of possible compounds, many of which may turn out to possess some extremely useful chemical and physical properties. For example, K_3C_{60} has been found to be a three-dimensional superconductor that may be useful in fabricating superconducting wires.

Although the laser vaporization of graphite was the first method of preparing the fullerenes, they have now been made in a variety of ways. Perhaps the simplest method producing the highest yields involves setting up an arc between two graphite electrodes under a stream of helium gas. (Recent evidence shows that when this is done, a few helium atoms are often trapped inside the fullerene. If *n* helium atoms are trapped in a C_{60} fullerene, for example, this situation is represented by writing the "formula" $n\text{He@C}_{60}$.) Even when a candle is burned, fullerenes are present in the carbon soot that results. When these highly symmetrical, nonpolar fullerenes are dissolved in a nonpolar solvent such as benzene, a beautiful magenta solution results.

While the fullerenes are fairly easily formed, diamonds are certainly not. Natural diamonds are formed in rock melts at a depth of 93 mi or more where there is a high pressure and temperatures in excess of 1400° C. Synthetic diamonds, first produced in 1955, are routinely made in small grit sizes used for various grinding applications. Gem-quality diamonds can be made, but the cost is not competitive with the natural variety. Diamonds, by the way, have not "been forever." The custom of giving a diamond engagement ring seems to have been started by the Venetians toward the end of the fifteenth century. Imitation diamonds are yttrium-aluminum garnet, strontium titanate, or cubic zirconium, ZrO_2.

Tin Disease

Tin has two principal allotropes, the white or metallic β-tin and the crumbly gray α-tin. The transition temperature between these forms is 13.2° C, or 55.8° F, as shown in Equation (15.8):

$$\underset{\text{gray}}{\alpha\text{-Sn}} \underset{13.2^\circ\text{C}}{\rightleftharpoons} \underset{\text{white}}{\beta\text{-Sn}} \tag{15.8}$$

The white tin, incidently, displays a unique "tin cry" due, it is said, to the breaking of the crystals when it is bent or deformed. The white tin, from which numerous objects have been fashioned, is the more stable at the higher temperatures. However, if the metallic form is exposed to temperatures below 55.8° F for prolonged periods of time, it will be converted to the crumbly gray tin. This transformation is known as the *tin pest* or *tin disease*, and knowledge of it is certainly of practical importance. It has wreaked havoc with European organ pipes, many of which were fashioned from tin and housed in constantly cold European cathedrals. It is also said to have been partially responsible for the defeat of Napoleon's army because the French soldiers' uniforms came equipped with tin buttons which fell apart during the long Russian winter.

Radiochemical Uses

The discovery that ordinary carbon is composed of three isotopes was made by William Giauque in 1929. Carbon 14 was isolated in 1940, and by 1945 Willard Libby had formulated the carbon 14 dating method, still the most commonly known radiochemical chronometric (time-measuring) method. As represented in Equation (15.9), carbon 14 is formed in the upper atmosphere by the bombardment of nitrogen with neutrons derived from cosmic ray activity:

$$^{1}_{0}n + ^{14}_{7}N \longrightarrow ^{14}_{6}\text{C} + ^{1}_{1}\text{H} \qquad (15.9)$$

It decays by β^- emission as shown in Equation (15.10):

$$^{14}_{6}\text{C} \xrightarrow{\beta^-} {}^{0}_{-1}e + ^{14}_{7}\text{N} \qquad (15.10)$$

Although present in very small amounts, 1.2×10^{-10} percent by weight, carbon 14, as compounds such as $^{14}CO_2$, participates in the normal photosynthesis and carbon-cycle reactions. In this manner, being continually incorporated into all plant and animal life but also steadily decaying, carbon 14 is always present in a small, but detectable, steady-state concentration. Once a living organism dies, however, only the decay continues, and the amount of the isotope steadily decreases with a half-life of 5730 years. Libby realized that analysis of the amount of carbon 14 remaining (or the $^{14}C/^{12}C$ ratio) in any once-living material (parchment, cloth, wood, some paints, and so forth) could be related to its age. Numerous objects such as parchment from various manuscripts and wood from tombs and ships have been dated in this manner. The utility of the method is limited to about eight or nine half-lives, or approximately 50,000 years.

Lead is the most abundant of the heavy metals because it is the end product of three different radioactive series. Starting from uranium 238, uranium 235, and thorium 232, three different lead isotopes are ultimately produced by a series of alpha and beta minus decays. The overall reactions for these series are shown in Equations (15.11) to (15.13):

$$\text{Uranium series:} \quad \text{U 238} \longrightarrow \text{Pb 206} + 8\,^{4}_{2}\text{He} + 6\,{}^{0}_{-1}e \qquad (15.11)$$

$$\text{Actinium series:} \quad \text{U 235} \longrightarrow \text{Pb 207} + 7\,^{4}_{2}\text{He} + 4\,{}^{0}_{-1}e \qquad (15.12)$$

$$\text{Thorium series:} \quad \text{Th 232} \longrightarrow \text{Pb 208} + 6\,^{4}_{2}\text{He} + 4\,{}^{0}_{-1}e \qquad (15.13)$$

The three series have half-lives of 4.5, 0.71, and 13.9 billion years, respectively, and serve as the basis of the *lead isochron method* used to determine the age of meteorites, the moon, the earth, and by implication, the entire solar system. The amounts of lead 206, 207, and 208 in a given sample depend upon the amount of lead originally present plus that generated by one or more of the above series. Since a fourth naturally occurring lead isotope, Pb 204, is not the product of a decay scheme, it affords a measure of the original amount of lead in a sample. In uranium-containing samples, the ratios of $^{206}Pb/^{238}U$ and $^{207}Pb/^{235}U$ both afford an estimate of the age of the specimen. When these ratios agree, the results are said to be concordant and yield the same time (thus the name *isochron* method) period for the origin of the sample. Using such methods, the age of the earth and the moon is estimated to be 4.6 billion years.

Isotopic compositions can also be used as a tracer to determine the sources of lead pollution. For example, lead compounds (mostly bromides and some chlorides) from the tetraethyl lead used as a gasoline antiknock additive have been the primary source of lead pollution. This source of lead can be distinguished from that derived from the combustion of coal (which contains lead impurities and constitutes the distant second leading source of lead pollution) or various lead-smelting operations.

Carbon Compounds

The carbon oxides, particularly the dioxide, have been discussed in earlier chapters, principally Chapter 11. Recall (p. 266) that Joseph Priestley worked with "fixed air" to produce soda water. The carbonation of soft drinks is still the second leading use of CO_2. (The leading use is of the solid as the refrigerant dry ice.) CO_2 is the product of the complete combustion of hydrocarbons (Chapter 11, p. 267) and plays a major role in the greenhouse effect (Chapter 11, pp. 289–291). Earlier we discussed the role of CO and CO_2 in the production of syngas (Chapter 10, p. 245) and the gasification of coal. In Chapter 6 we noted the utility of carbon monoxide (carbonyl) as a ligand in the Mond process for the purification of nickel.

The ionic carbides, containing either the C_2^{2-} or C^{4-} anions, are often prepared by reacting either the elements themselves or a metal oxide with carbon in an electric furnace at elevated temperatures. They can be hydrolyzed to produce various hydrocarbons in industrial processes. For example, calcium carbide, CaC_2, is a major source of acetylene as shown in Equation (15.14) (and briefly discussed in Chapter 13, p. 340), and aluminum carbide, as shown in Equation (15.15), generates methane:

$$CaC_2(s) + H_2O(l) \longrightarrow C_2H_2(g) + CaO(g) \tag{15.14}$$

$$Al_4C_3(s) + 12H_2O(l) \longrightarrow 3CH_4(g) + 4Al(OH)_3(s) \tag{15.15}$$

The covalent silicon carbide, SiC, is known as Carborundum, a name derived from *carbon* + *corundum*. (Corundum, as discussed in Chapter 14, p. 364, is an extremely hard mineral consisting of aluminum oxide, Al_2O_3.) SiC is a giant molecule like diamond, is extremely hard, and finds use as an abrasive, in cutting tools, and as a refractory.

When the lattices of various metals are expanded by the presence of carbon atoms, the interstitial carbides result. These materials retain many metallic properties such as high conductivity and luster and often are harder and higher melting than the pure metals themselves. The addition of carbon to iron makes steel, but too much carbon makes it brittle. Tungsten carbide, WC, is used for cutting and drilling tools.

Other useful carbon compounds include the disulfide, CS_2 (a poisonous, flammable liquid used as a solvent in dry cleaning and other applications); hydrogen cyanide, HCN, and the cyanides, CN^- (the former a poisonous gas smelling like bitter almonds that may have been of great importance in the chemical evolution of life and the latter mentioned as an important ligand in Chapter 6); the cyanamides, CN_2^- (briefly mentioned in Chapter 12 and used as fertilizers, weed killers, pesticides, and in the production of acrylic fibers and plastics); the chlorofluorocarbons, CFCs (as aerosols and refrigerants but implicated as threats to the ozone layer, see Chapter 18); and oxyhalides such as phosgene, $COCl_2$ (a former chemical warfare agent now used in the manufacture of polyurethanes).

Lead Compounds and Toxicology

Lead compounds have been used in paints for years. The familiar red primer paint for iron and steel, most commonly seen on older automobiles, is Pb_3O_4 or $Pb^{II}{}_2Pb^{IV}O_4$, known as *red lead*. Lead monoxide, PbO, can be red (*litharge*), orange, or yellow depending on the method of preparation. Other paint formulations include white lead ($2PbCO_3 \cdot Pb(OH)_2$), blue lead (a combination of basic lead sulfate, $PbSO_4 \cdot PbO$, zinc oxide, and carbon) and the lead chromates (used in the preparation of yellow, orange, red, and green paints).

The lead storage battery was invented in 1859, but when it was used to power automobile self-starters in the early twentieth century, it revolutionized the personal transportation industry. This battery has remained fundamentally the same for nearly a century because it has a long lifetime (3–5 years), can operate in high and low temperatures ($+100$ to $-30°$ F), and holds up well under the beating meted out by road vibrations.

Figure 15.7 shows a schematic of a lead storage battery. Note that both the anode and the cathode are lead grids containing the metal and the dioxide, respectively. The overall reaction, represented in Equation 15.16*c*, that is harnessed to provide the energy to turn the starting motor also produces lead sulfate (which is retained in each grid), but that reaction can be reversed (when the battery is charged by the alternator once the engine is running). It used to be that in the charging process, water was unavoidably electrolyzed and therefore would have to be replaced. Today, the use of calcium-lead alloys in the electrodes so significantly diminishes the electrolysis that water need not be replaced and the batteries are sealed off. When jumping a car battery, however, electrolysis does occur and the resulting hydrogen gas can explode if the proper precautions are not taken.

The myriad uses of lead by the Romans, particularly the ruling class, has been blamed (at least in small part) for the fall of their empire and their way of life. Even today lead is called "the everywhere poison" because it seems to be just that, everywhere. Until recently, almost all gasolines contained tetraethyl lead, $Pb(CH_2CH_3)_4$, or tetramethyl lead, $Pb(CH_3)_4$, as an antiknock additive. Reactions with dibromoethylene and other "lead scavengers" in such fuels produced lead bromide which settled along highways and was incorporated into the roadside trees and grass and whatever else might be growing there. In the United States, residual lead from the 60-some years of leaded gasoline use is a significant environmental and health hazard. In Europe, the use of leaded gasoline continues unabated.

Paint chips from houses painted before the mid-1950s or furniture painted before the mid-1970s are significant sources of lead and particularly affect inner-city children. Furthermore, it is leached from (1) lead pipes into water supplies, (2) lead-soldered cans (by acid foods) into food supplies, and (3) from lead foil wrappers into even the best European wines. Smelting operations, the combustion of coal, and various industrial processes have helped to spread more lead into the environment. It has been found in older, improperly glazed earthenware vessels and even in moonshine whiskey and other alcoholic products distilled from equipment using lead solders. Who knows, perhaps future historians will also have cause to speculate about the role of lead in the decline of *both* the great EuroAmerican civilization as well as that associated with the Roman Empire.

Lead is absorbed either through the lungs or the gastrointestinal tract but, in normal individuals, is excreted almost but not quite as fast as it is absorbed. Accordingly, lead slowly accumulates in bone tissue where it remains relatively

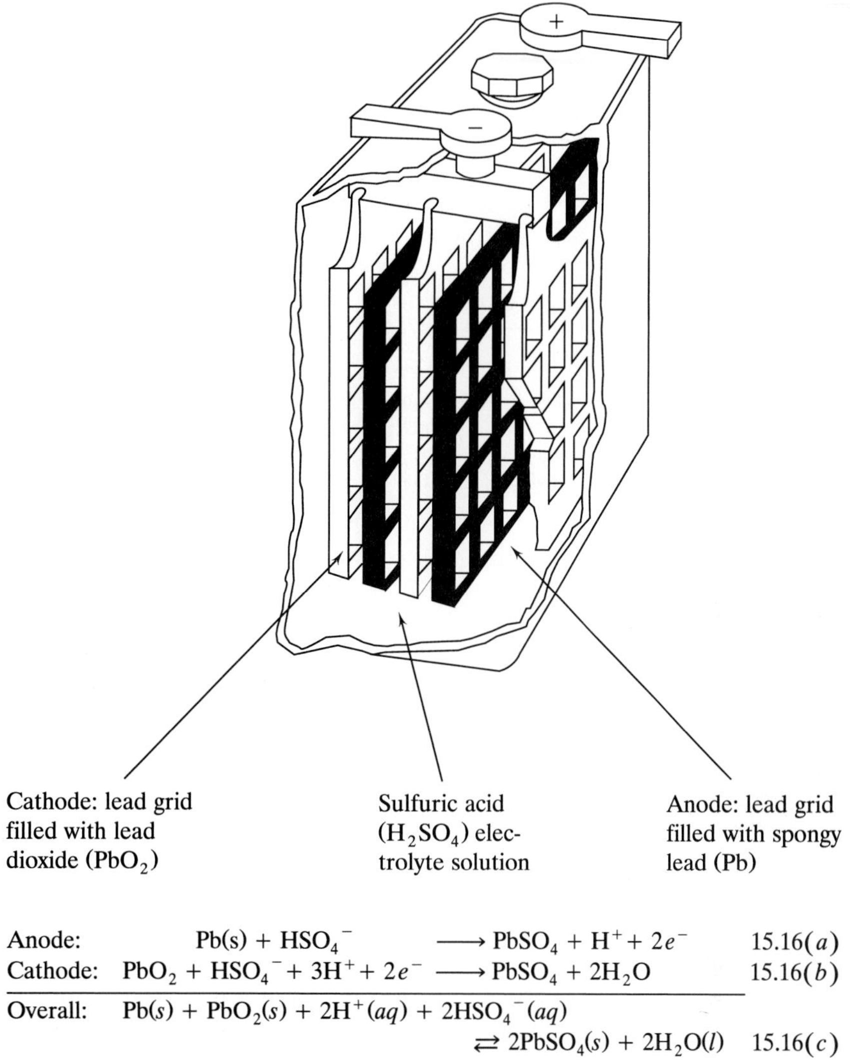

$$\begin{array}{llll} \text{Anode:} & Pb(s) + HSO_4^- & \longrightarrow PbSO_4 + H^+ + 2e^- & 15.16(a) \\ \text{Cathode:} & PbO_2 + HSO_4^- + 3H^+ + 2e^- & \longrightarrow PbSO_4 + 2H_2O & 15.16(b) \\ \hline \text{Overall:} & \multicolumn{2}{l}{Pb(s) + PbO_2(s) + 2H^+(aq) + 2HSO_4^-(aq)} & \\ & & \rightleftarrows 2PbSO_4(s) + 2H_2O(l) & 15.16(c) \end{array}$$

FIGURE 15.7
One of six cells of a 12-V lead storage battery. When the battery discharges providing the voltage to turn the starter motor, it produces lead sulfate which adheres to the surfaces of the anodic and cathodic lead grids. When the battery is charged, the overall reaction is reversed.

inert. People who are exposed to large amounts exhibit higher lead levels in the blood. The early symptoms of lead poisoning in adults include anemia, fatigue, headache, weight loss, and constipation. Low levels of lead have often been misdiagnosed as psychological problems. Symptoms accompanying larger amounts, again in adults, include neurological damage. Children, particularly those between 12 and 36 months old, seem to be more susceptible to lead poisoning, and they show more pronounced neurological damage than adults similarly exposed. The difference between the effects on adults and children is not well understood. (The use of therapeutic chelating agents as partial antidotes for lead and other heavy-metal poisoning is discussed in Section 6.5.)

TABLE 15.2
Types of silica and silicates

Type	Repeating unit	Number of oxygens shared	Examples
Ortho-	SiO_4^{4-}	0	Be_2SiO_4, phenacite Zn_2SiO_4, willemite $ZrSiO_4$, zircon $(M^{2+})_2SiO_4$, olivine $M^{2+} = Mg^{2+}, Fe^{2+}, Mn^{2+}$ $(M^{3+})_2(M^{2+})_3(SiO_4)_3$, garnet $M^{2+} = Ca^{2+}, Fe^{2+}, Mg^{2+}$ $M^{3+} = Al^{3+}, Cr^{3+}, Fe^{3+}$
Pyro-	$Si_2O_7^{6-}$	1	$Sc_2Si_2O_7$, thortveitite
Cyclic-	$Si_3O_9^{6-}$	2	$BaTiSi_3O_9$, benitoite
	$Si_6O_{18}^{12-}$	2	$Al_2Be_3Si_6O_{18}$, beryl
Chain:			
Single	SiO_3^{2-}	2	Pyroxenes, e.g., $MgSiO_3$, enstatite $LiAl(SiO_3)_2$, spodumene $CsAl(SiO_3)_2$, pollucite
Double	$Si_4O_{11}^{6-}$	2 or 3	Amphiboles, e.g., $Ca_2(OH)_5Mg_5[Si_4O_{11}]_2$, tremolite $Na_2(OH)_2Fe_5[Si_4O_{11}]_2$, crocidolite
Sheet	$Si_2O_5^{2-}$	3	$Mg_3(OH)_2[Si_2O_5]$, talc (or soapstone) $LiAl[Si_2O_5]$, petalite
3-d	SiO_2	4	α-Quartz Cristobalite

15.4 SILICATES, SILICA, AND ALUMINOSILICATES

Silicates and Silica

The chemistry of silicon, as we have seen, is dominated by silicon-oxygen bonds. The basic unit of all the silicates is the SiO_4^{4-} tetrahedron. As shown in Table 15.2, the various types of silicates are characterized by the sharing of from zero to all four of the oxygens in this unit. The charge on each repeating unit is determined by recalling that the oxidation state of the silicon is $+4$ while the oxygen is, as expected, -2.

The orthosilicates feature a discrete, self-contained SiO_4^{4-} unit. (See the discussion of borates, Chapter 14, p. 360, for a short discussion of the *ortho-* prefix.) The straightforward tetrahedral structure is shown in Figure 15.8*a*. The rare pyrosilicates (or disilicates) are characterized by the sharing of one oxygen of each SiO_4^{4-} unit to produce the $Si_2O_7^{6-}$ ion, shown in Figure 15.8*b*.

Cyclic silicates find two oxygens of the SiO_4^{4-} being shared to produce six-membered $Si_3O_9^{6-}$ (Figure 15.8*c*) and twelve-membered $Si_6O_{18}^{12-}$ (Figure 15.8*d*) rings. Recalling the earlier (Chapter 13, p. 326) discussion concerning beryl, $Al_2Be_3Si_6O_{18}$, the partial replacement of the aluminum cations (1) with chromium characterizes emeralds and (2) with iron characterizes aquamarine.

Single-chain silicates also find two oxygens of the SiO_4^{4-} unit being shared but not closing to a ring. The repeating unit of these infinite chains is SiO_3^{2-}, outlined in Figure 15.8*e*. Spodumene, $LiAl(SiO_3)_2$, is one of the few important lithium ores.

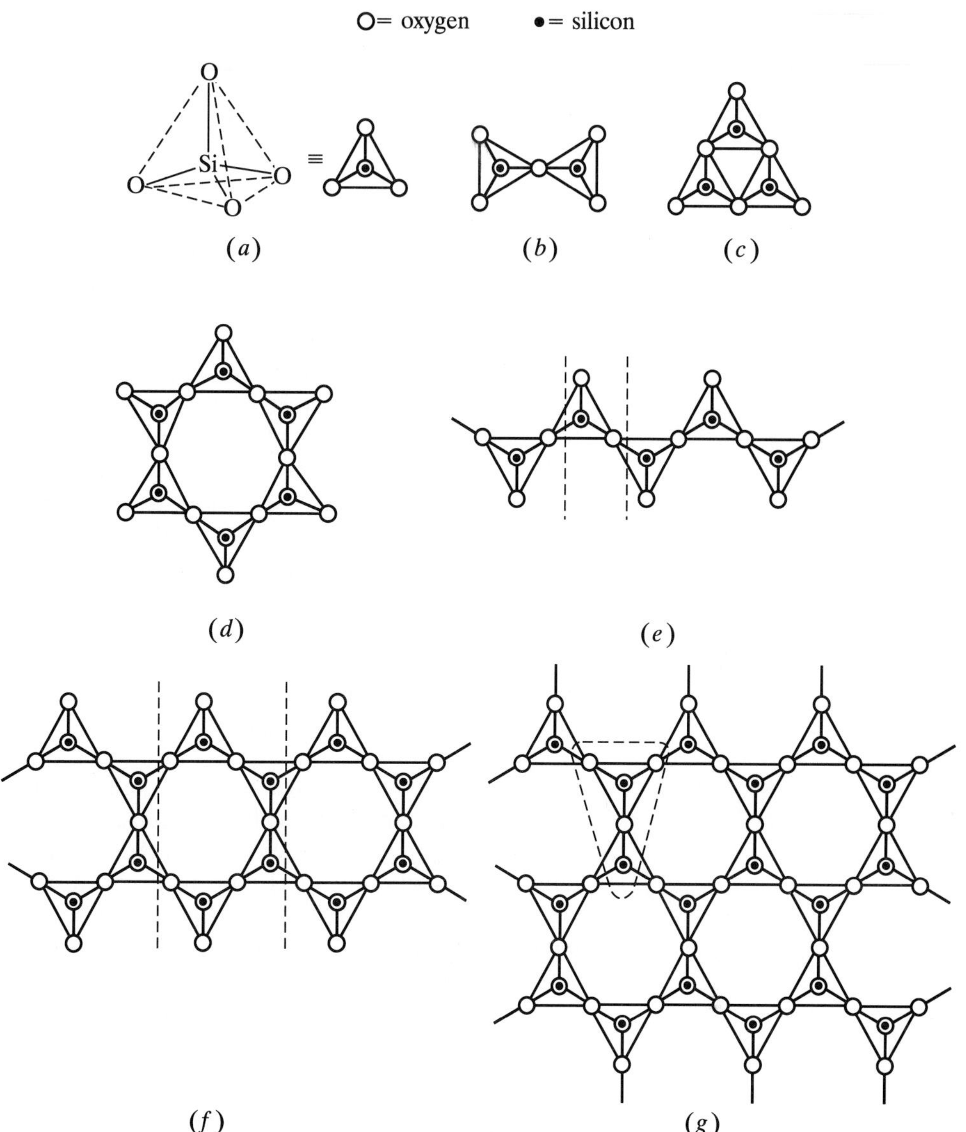

FIGURE 15.8
Silicate structures: (*a*) ortho SiO_4^{4-}, (*b*) pyro $Si_2O_7^{6-}$, (*c*) cyclic $Si_3O_9^{6-}$, (*d*) cyclic $Si_6O_{18}^{12-}$, (*e*) single-chain SiO_3^{2-}, (*f*) double-chain $Si_4O_{11}^{6-}$, and (*g*) sheet $Si_2O_5^{2-}$.

Double-chain silicates, shown in Figure 15.8*f*, find some SiO_4^{4-} units sharing two oxygens while others share three. The repeating unit is $Si_4O_{11}^{6-}$. Five double-chain silicates (tremolite, grunerite, crocidolite, anthophyllite, and actinolite) and the mineral serpentine go under the generic name *asbestos*. The molecular-level structure of these materials is reflected in their fibrous macroscopic structure. Cations located among the negatively charged chains hold the material together, although it can be fairly readily broken apart to form a stringy fibrous material. Asbestos has been used for a variety of heat-insulating applications from building insulation to brake linings. Its fibrous nature also allows it to be woven into a variety of insulating nonflammable garments. It is said that the great Emperor

Charlemagne (742–814), conqueror of the Lombards of northern Italy and the Saxons of northwestern Germany, greatly impressed his barbarian dinner guests by throwing the festive asbestos tablecloths into the fire to clean them. Today, actually wearing or even touching asbestos cloth or occupying a building in which the structural integrity of the asbestos insulation has broken down is considered unsafe due to its carcinogenic nature.

Sheet silicates find the bulk of the SiO_4^{4-} tetrahedra sharing three oxygens to form a $Si_2O_5^{2-}$ repeating unit shown in Figure 15.8*g*. Talc or soapstone is a sheet silicate as is the mineral petalite from which Arfwedson first isolated lithium (see Chapter 12, p. 299).

Silica, SiO_2, in which all four of the oxygens in the SiO_4^{4-} unit are shared, comes in a variety of polymorphs of which the two most stable are α-quartz and β-cristobalite. The structure of β-cristobalite, shown in Figure 15.9*a*, is similar to that of diamond (or zinc blende), with the silicon atoms occupying the positions held by the carbon atoms but with an oxygen bridge (Si—O—Si) between each silicon. α-Quartz, shown in Figure 15.9*b*, comes in both right-handed and left-handed forms. (See Chapter 3 for a discussion of enantiomers, or optical isomers, in inorganic compounds.) Note that the helix formed by —Si—O—Si—O— chains can be an alpha helix as shown or a beta helix. These two forms are analogous with right- and left-handed screws and can be mechanically separated.

Quartz is a major component of granite and sandstone. Colorless forms of quartz are known as rhinestones, while purple forms are amethyst. The pink variety is often called rose quartz.

It was Pierre Curie, in the early 1880s (before he met Marie Sklodowska), who first showed that quartz is able to generate *piezoelectricity*, electricity resulting from the application of mechanical pressure on a crystal. The effect arises when crystals produce electrical charges on their surfaces when compressed in particular directions or vibrated in certain ways. Such crystals can convert vibrations into electrical signals and find applications in such devices as crystal microphones, phonograph pickups, and vibration-detecting devices. The inverse effect, that is, converting electrical signals into specific vibrations, finds use in headphones, loudspeakers, and cutting heads for disk recording. This property is also the basis for quartz watches and clocks. A precisely cut section of quartz crystal is placed in an oscillating electric circuit (generated by a battery or house current) and then vibrates with its own characteristic frequency that can be used to run a timing device. Citizen band (CB) radios contain crystals cut in a way which ensures a given characteristic frequency unique to that station or channel.

Two other materials related to silica should be mentioned here. When silica is heated with sodium carbonate, it forms a melt of variable composition that is water-soluble and often referred to as sodium silicate. Solutions of sodium silicate are known as *water glass* and have been put to a variety of uses over the years. For example, it has been used as a fireproofing agent, for sizing paper, and for preserving eggs. An aqueous solution of water glass just slightly denser than water makes an excellent medium in which to slowly grow large and beautiful crystals in what is sometimes called a *chemical garden*.

Another form of SiO_2 is called *silica gel*. When an aqueous solution of sodium silicate is acidified and then roasted or dried to remove most of the excess water, a white, amorphous (or noncrystalline), high-porosity powder called silica gel is obtained. Because of its anhydrous, highly porous nature, silica gel finds a variety of uses as a desiccant, catalyst, and chromatographic support. For similar

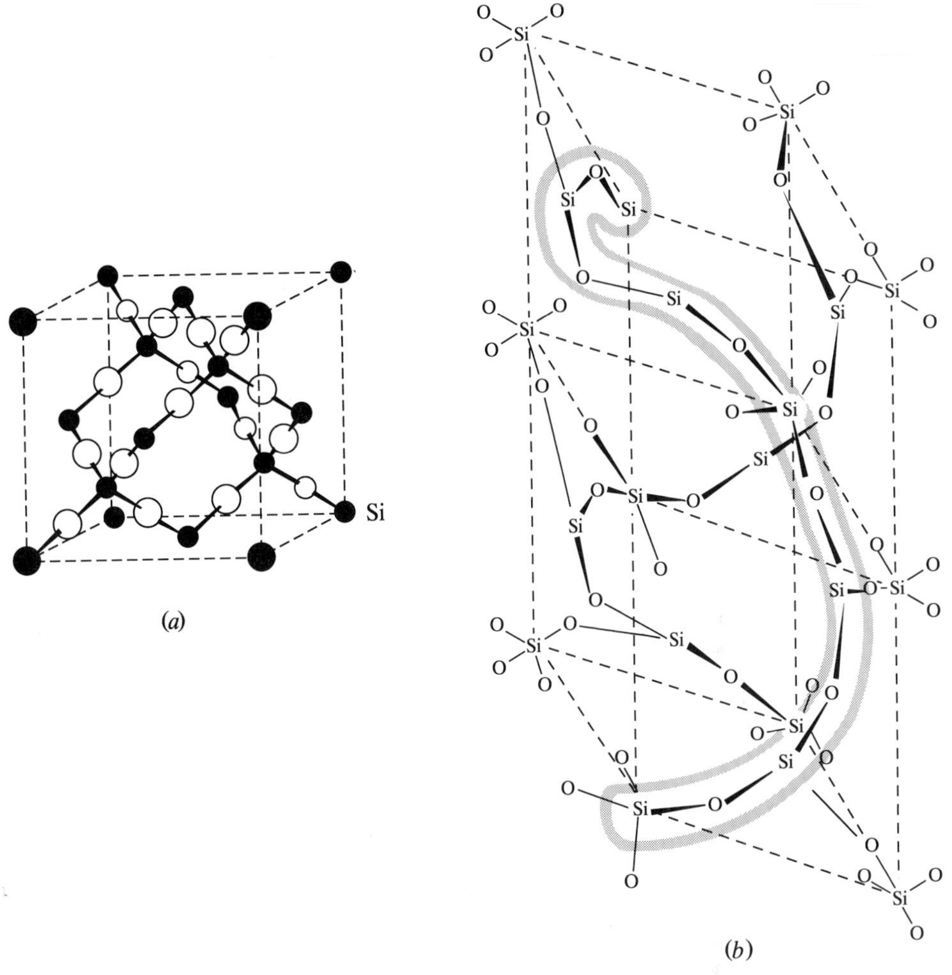

FIGURE 15.9
Two prevalent forms of silica, SiO_2: (*a*) β-cristobalite and (*b*) two unit cells of α-quartz. In the latter, two helices composed of O—Si—O—Si—O— bonds are shown, with one shaded for emphasis. Crystals with the helices rotating in the opposite direction are equally possible, and these two forms can be mechanically separated.

reasons it is used in the marketplace as an anticaking agent in various finely powdered or crystalline food products such as cocoa and powdered fruit juices.

Aluminosilicates

When an aluminum atom replaces a silicon in a silicate, the resulting aluminosilicate carries an additional negative charge. So, for example, suppose there is a section of silica containing 24 formula units of SiO_2. If the Si^{4+} in half of these units are replaced with Al^{3+}, the resulting material has a formula of $[(AlO_2)_{12}(SiO_2)_{12}]^{12-}$. To achieve electrical neutrality, 12 positive charges must be supplied in the form of cations. Often these aluminosilicates have framework structures with cagelike cavities large enough to accommodate various-sized cations and small molecules. Such materials are called *zeolites*. Linde A, shown in Figure

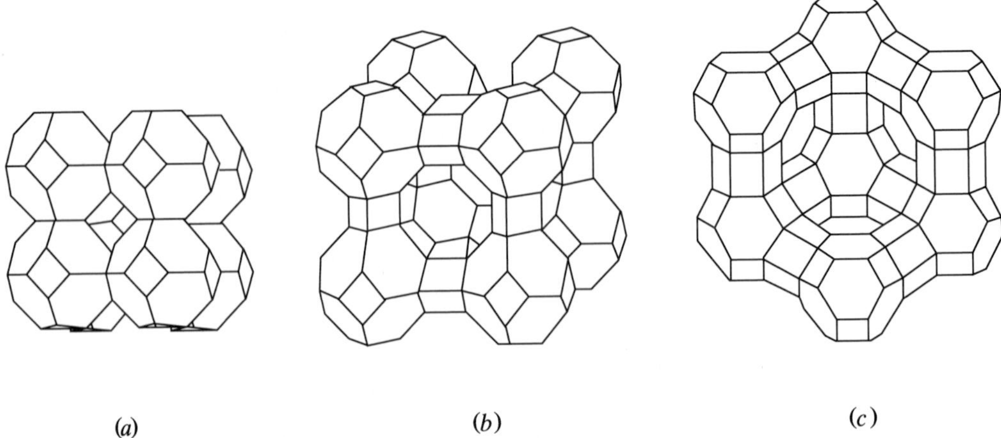

FIGURE 15.10
Several different zeolites structures, including (*a*) sodalite, (*b*) zeolite A, and (*c*) faujasite. [Ref. 18.]

15.10*b*, is a synthetic zeolite with a formula of $Na_{12}[(AlO_2)_{12}(SiO_2)_{12}] \cdot 27H_2O$. When such materials are dehydrated, many of their cavities are evacuated and are available to trap a variety of small molecules such as H_2O, NH_3, and CO_2. Linde A "molecular sieves" can be synthesized with varying cavity sizes. Linde 5A sieves selectively trap water molecules and are used to dehydrate organic solvents. Other zeolites can trap various reactants and serve as heterogeneous catalysts, for example, in the cracking of crude oil to make lighter, more volatile fuels.

Zeolites served as the first ion-exchange water softeners because the sodium ions could be displaced by the higher-charged calcium or magnesium ions of hard water (see Chapter 13, pp. 336–338). A schematic of a water softener is shown in Figure 15.11. Heart disease patients must be wary of using such softeners because

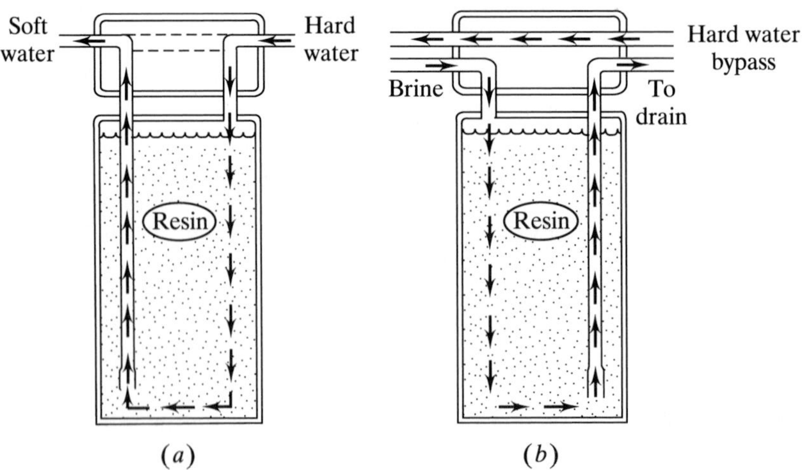

FIGURE 15.11
A commercial water softener. (*a*) The Ca^{2+} and Mg^{2+} ions of hard water are trapped in the ion-exchange resin or a zeolite and replaced by sodium ions. (*b*) The softener is recharged when a brine (concentrated solution of sodium chloride) is passed through the resin to force sodium ions to replace the magnesium and calcium ions which are then discharged. [Ref. 31.]

small amounts of sodium are released into the water supply. After a period of time, the water softener must be recharged by passing a concentrated solution of sodium chloride back through the zeolite to displace the calcium and magnesium ions. For the most part, zeolites have been replaced as water softening agents by various synthetic resins.

Micas are also aluminosilicates. For example, in muscovite, $KAl_2[AlSi_3O_{10}]$-$(OH)_2$, the extra cations are located between the aluminosilicate anionic sheets. This material can be split up into sheets so thin that 1000 are needed to make a pile only 1 in high. Vermiculite is a hydrated mica which splits off into soft flakes, scales, and layers when dehydrated. It makes an excellent packing material or soil conditioner. A variety of materials such as clays, feldspars, and talc are also aluminosilicates as is the absorbent used as cat litter.

15.5 SELECTED TOPICS IN DEPTH: SEMICONDUCTORS AND GLASS

Semiconductors

Consistent with the general inability of silicon and germanium to form $p\pi$-$p\pi$ bonds, these elements are found only in diamond-type structures and not in layer-type structures like graphite. In diamond itself all the electrons are involved in localized covalent bonds and therefore are unavailable for conduction. In germanium or silicon, however, the bonding electrons are held less tightly because they are farther away from the effective nuclear charges. It is not surprising, therefore, that even at room temperature, a small number of these electrons have broken free of the covalent bonds and make these semimetals feeble conductors of an electric current. If these elements are heated or exposed to light (as in photovoltaic solar cells), more electrons are freed and the conductivity increases. Substances whose conductivity is poor at low temperatures but increases with heat, light, or the addition of certain impurities are called *semiconductors*.

Pure germanium and silicon have a natural or built-in ability to be semiconductors and are known as intrinsic semiconductors. To describe these devices we make use of terms from what is known as *band theory*. We start with all the valence electrons involved in covalent bonds as shown in the left side of Figure 15.12*a*. We say these electrons occupy a filled valence band as shown in the right side of part (*a*). In this configuration the material cannot conduct electricity.

When electrons are freed from the covalent bonds as shown in the left side of part (*b*), they are said to occupy a conduction band, because there they are mobile and free to be conducted from one place to another in the lattice. The valence band is shown occupied by a few electrons in the right side of part (*b*). Note also in part (*b*) that when electrons move up to the conduction band, they leave behind some "holes" in the covalent bonds and therefore in the valence band.

Now when a potential difference is applied to the crystal as shown in the left side of part (*c*), electrons can jump from hole to hole and in that sense carry a current. Also note in the left side of part (*c*) that when a valence electron moves to fill a hole, another hole appears where it came from. So as electrons move in one direction, the holes appear to move in the opposite direction. This is also depicted in the right side of part (*c*).

Germanium and silicon have valence electrons which are rather easily excited to the conduction band. As we move toward the upper right of the periodic table, the valence electrons are under the influence of a greater effective nuclear charge and cannot be as easily excited. Accordingly, in the nonmetals including diamond

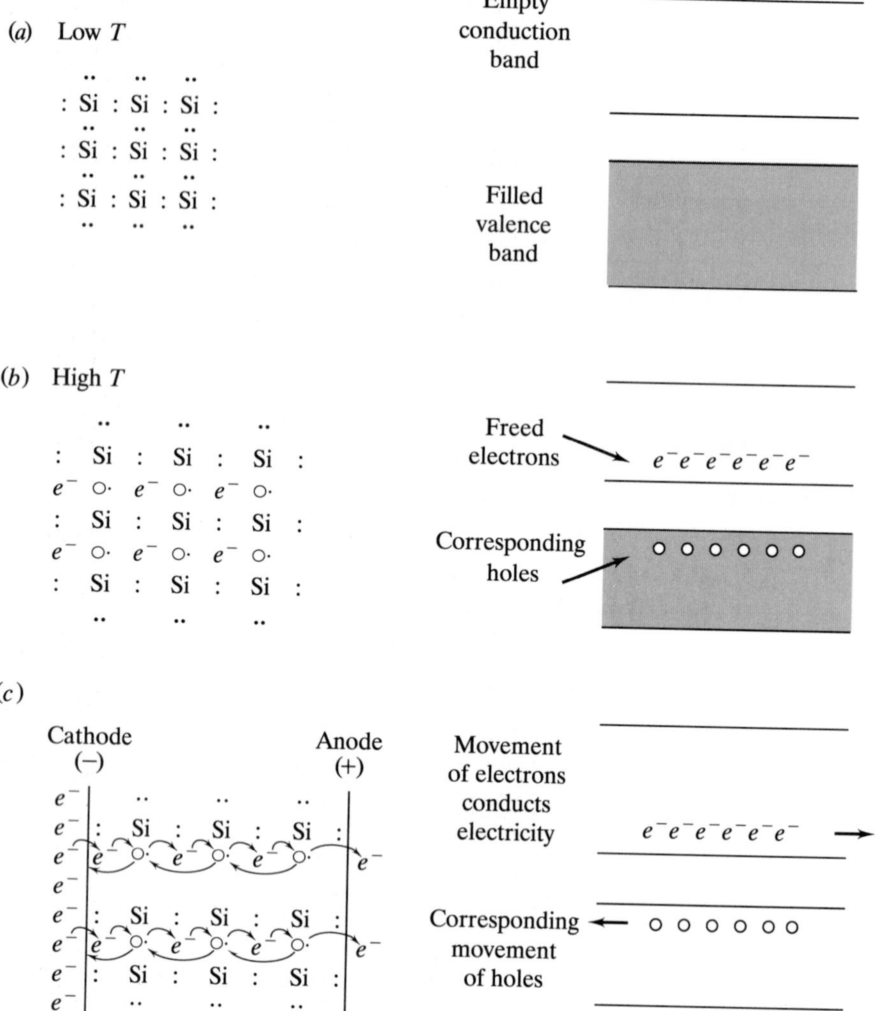

FIGURE 15.12
The position and movement of electrons in an intrinsic semiconductor. (*a*) At a low temperature all the electrons are involved in covalent bonds. The valence band is filled and the conduction band is empty. There are no free electrons and the substance cannot carry a current. (*b*) At a higher temperature some electrons are freed from the covalent bonds and move from the valence band to the conduction band. For each electron liberated from a covalent bond, a hole is left behind. (*c*) When a potential difference is applied, electrons move toward the anode (+) while the holes appear to move toward the cathode (−). The movement of electrons is an electric current. (· = Si electron, e^- = free electron, ○ = hole.)

itself, the gap between the valence and conduction bands is said to be so large that they cannot conduct electricity even at very high temperatures. They are insulators. In metals, on the other hand, the valence band is incompletely filled and becomes one in the same with a conduction band. Metals are conductors even at low temperatures.

Doping, or adding selected impurities to, an intrinsic semiconductor can enhance its conductivity. In an *n-type semiconductor*, a Group 5A element (usually phosphorus or arsenic) replaces a silicon or germanium in the lattice as shown in

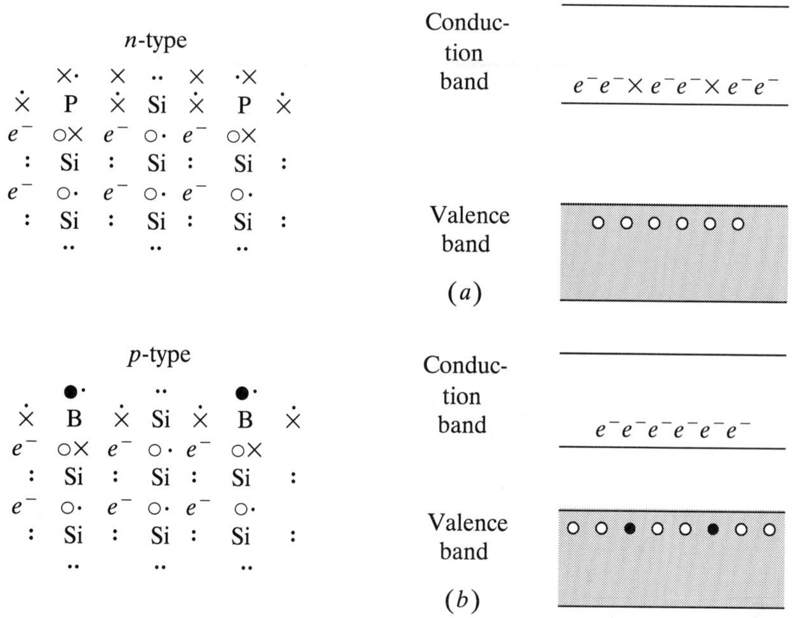

FIGURE 15.13
The position and movement of electrons in doped semiconductors. (*a*) An *n*-type semiconductor results when some silicon atoms are replaced by a Group 5A element. Each phosphorus atom has an additional electron which resides in the conduction band. (*b*) A *p*-type semiconductor results when some silicon atoms are replaced by a Group 3A element such as boron. Each boron atom has one less electron resulting in an additional hole in the valence band. (· = Si electron, × = phosphorus or boron electron, e^-= free electron, ○ = ● = hole.)

Figure 15.13*a*. The P or As atom brings with it one extra negative electron (therefore the term *n*-type) which occupies the conduction band. In a *p-type semiconductor*, a Group 3A element (usually boron or gallium) is inserted in the silicon or germanium lattice as shown in Figure 15.13*b*. This type of impurity is one electron short (and hence is positive, therefore the term *p*-type), and so an extra hole develops in the valence band. In either case, electrons and holes can move under the influence of an applied electric potential, and the material is a semiconductor.

The real utility of these semiconductors is realized when they are combined. *pn* junctions (a joining of an *n*- and *p*-type semiconductor along a common interface) will conduct well in only one direction and therefore converts alternating current (ac) to direct current (dc). A *pnp* or *npn* junction can serve as a selective resistor to transfer and therefore amplify an electronic signal in a given direction. Such a device is called a *transistor* and does in a very small space what formerly was done by very bulky, hot vacuum tubes. Transistors and other such devices have revolutionized and miniaturized the electronics and computer industries. By selectively doping a pure silicon (or germanium) chip, various self-contained or integrated electronic circuits can be produced in a very small area (see Figure 15.14). These chips have become the primary components of microcomputers, pocket calculators, electronic watches, and solar photovoltaic cells not to mention radios, televisions, and other electronic equipment.

The above semiconductor devices require large quantities of silicon which must be pure to 1 part in a billion. To achieve such a purity the crude silicon

FIGURE 15.14
Computer chips on a paper clip. (*C. Falco/Photo Researchers.*)

produced from the reduction of silica with carbon [Equation (15.1)] is converted to the liquid silicon tetrachloride as shown in Equation (15.17):

$$Si(s) + 2Cl_2(g) \longrightarrow SiCl_4(l) \qquad (15.17)$$

The $SiCl_4$ is repeatedly distilled and then reduced with magnesium yielding a substantially purified silicon as shown in Equation (15.18):

$$SiCl_4(l) + 2Mg(s) \longrightarrow 2MgCl_2(s) + Si(s) \qquad (15.18)$$

The final step necessary to produce the required purity is accomplished by a process known as *zone refining* shown in Figure 15.15. Here a silicon rod is passed through a heated electric coil. As the rod becomes molten, the remaining impurities dissolve in the partially melted portion and are carried to the end where they are cut off.

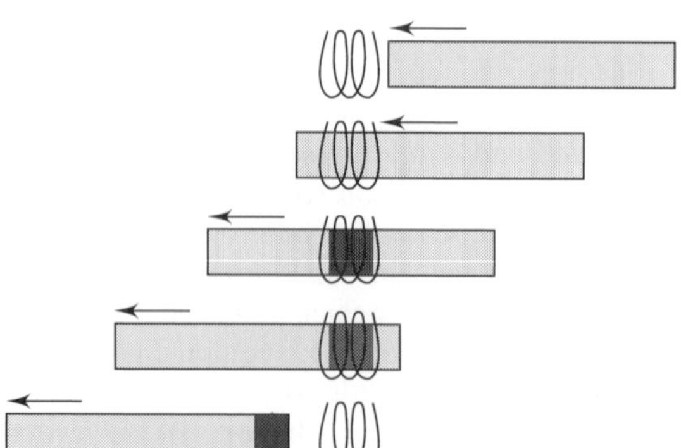

FIGURE 15.15
The zone-refining technique for purifying silicon. A rod of the impure element is passed slowly through a heated coil. Since the impurities are more soluble in the moving molten zone, they build up there until they are disposed of when the end of the rod is removed. [Ref. 21.]

Glass

When solid silica is heated above its melting point (approximately 1700° C), many of its Si—O bonds are broken. Upon cooling, the bonds begin to reform, but even at temperatures significantly below the melting point the highly ordered interconnected crystalline state cannot easily be reestablished. Instead a glass is formed. A *glass* is a homogeneous, noncrystalline state resembling that of a liquid whose rate of flow is so slow that it appears to be rigid over long periods of time. If even small crystalline surfaces should be formed in the glass, light would be reflected and the material would appear to be white (assuming that no visible wavelengths were absorbed). In a glass, however, no such reflecting surfaces are formed, and light is allowed to pass through the material; that is, the material is transparent to visible light.

The most common glasses contain silica, SiO_2, as their principal ingredient. Pure silica itself makes a good glass (quartz glass or fused silica) but it is very high melting. If sodium oxide, Na_2O, is added, the melting point is lowered, but the glass is not durable and is water-soluble. (This material is the water glass referred to earlier.) The further addition of calcium oxide adds durability and makes the glass water-insoluble. These three compounds are the principal constituents of ordinary soda lime-silica glass found in sheet or window glass and in bottles.

Glass with specialized properties is prepared by adding a variety of other ingredients. For example, the addition of potassium oxide produces a harder glass that can be precisely fashioned into lenses for eyeglasses and other optical applications. Cookware is aluminosilicate glass prepared by adding aluminum and magnesium oxides. Pyrex or Kimax laboratory glass contains boron oxide which decreases its coefficient of thermal expansion, yielding a product which can be heated and cooled quickly without breaking. The addition of lead oxide produces a highly refractive flint glass used in fine crystal. (There is some evidence that significant amounts of lead can be leached out of crystal glass when a wine is in contact with it for only a short period of time.) High amounts of lead also produce a glass used to shield radiation equipment. Glass containing strontium oxide absorbs the x-rays emitted by a color TV. Tin(IV) chloride adds a toughness and iridescence to glass bottles and, in the appropriate amounts, reflects infrared light for use in heat-insulating windows. Various metals and metal oxides are added for color [for example, red (small amounts of gold and copper); green (Fe_2O_3 or CuO); yellow (UO_2); orange (Cr_2O_3); dark-blue, visibly nontransparent but uv-transparent "cobalt glass" (CoO)]. Photochromic eyeglasses darken when exposed to sunlight. They contain small amounts of silver chloride that is photochemically and reversibly reduced to metallic silver and elemental chlorine as shown in Equation (15.19):

$$AgCl \underset{\text{dark}}{\overset{\text{sunlight}}{\rightleftharpoons}} Ag + Cl \tag{15.19}$$

Porcelain contains large amounts of aluminum oxide and is heterogeneous, opaque, and more chemically resistant than glass.

Glass fibers include the continuous variety used to weave a cloth and a discontinuous type used as insulation (glass wool) and as a plastic reinforcement. Special optical fibers capable of transmitting light over long distances through a flexible cable have revolutionized the communications industry. These fibers are fashioned from extremely pure ingredients using new fiber-drawing techniques. To keep from losing intensity, the outer core of such fibers is doped with ingredients (such as germanium oxide) which has a higher refractive index. These fibers are used for delivering light to remote places (for example, into the organs of the body for medical diagnostic purposes), transmitting telephone signals, power transmission, and heat sensing and transporting devices.

The above optical properties can be combined with semiconductors on the same silicon chip. Photodetection and transmission in combination with the sophisticated new miniaturized electronics is sure to be applied to an ever-increasing variety of devices.

SUMMARY

For the first time in our survey of the representative elements, we encounter several that were known to the ancients. Carbon, tin, and lead have been known in elemental form for thousands of years. In the 1820s, silicon was isolated by reduction of the fluoride with potassium, while germanium, Mendeleev's eka-silicon, was found in a silver ore some 60 years later.

Group 4A is closely intertwined with all seven components of our network. Carbon may be the most striking example of the uniqueness principle that we will encounter. With the metal-nonmetal line passing through the heart of the group, we expect a large variety of properties including the variation of the acid-base properties of the oxides. The diagonal relationship of silicon with boron represents the last fading influence of this network component. Consistent with the inert-pair effect, the $+2$ oxidation state becomes more stable until it dominates the chemistry of lead. Standard reduction potentials help us quantify the relative stabilities of these oxidation states.

A consideration of the hydrides emphasizes the uniqueness of carbon. It is the champion catenator of all the elements. The silanes are more reactive than the alkanes for both thermochemical and kinetic reasons. The relative ineffectiveness of $p\pi$-$p\pi$ overlap as well as the introduction of $p\pi$-$d\pi$ interactions in the compounds of silicon play a large role in distinguishing them from those of its lighter congener.

The acid-base character of the oxides progresses from the acidic carbon and silicon dioxides to the amphoteric germanium and tin oxides to the almost fully basic lead oxide. Silicon and carbon dioxides are exceedingly different compounds, again due to the inability of silicon to form $p\pi$-$p\pi$ bonds. The oxides also demonstrate the trend toward the stability of the $+2$ oxidation state in the heavier congeners.

Halide chemistry serves to reemphasize both the uniqueness of carbon as well as the dominance of the inert-pair effect in tin and lead. Tin(II) and lead (II) compounds are more ionic than their analogous compounds with $+4$ oxidation states. Standard reduction potentials further document the increased stability of the $+2$ state in tin and lead.

The graphite and diamond allotropes of carbon have a variety of applications. Diamond, characterized by interconnected C—C single bonds, is extremely hard and high melting. Graphite, characterized by delocalized $p\pi$ electrons, is a soft, layered form of the element. When graphite is vaporized in a laser, it forms a variety of clusters called fullerenes, the most important of which is the highly symmetrical and stable C_{60}. Other matters of practical importance concerning these elements include tin disease, carbon 14 dating, and the lead isochron method. Useful compounds include the carbon oxides mentioned in previous chapters, the ionic, covalent, and interstitial carbides, and various lead compounds in paints and the common lead storage battery. Lead is the "everywhere poison" and causes a variety of different symptoms, with children suffering more severe symptoms than adults.

Silicon-oxygen bonds dominate the chemistry of the silicates, silica, and the aluminosilicates. The SiO_4^{4-} unit can be linked together to form rings, chains, sheets, and three-dimensional arrays. Asbestos is a generic name for a class of double-chain silicates that, until lately, have been put to a wide variety of uses. Silica comes in a variety of polymorphs including the optically active α-quartz. Quartz is also piezoelectric. *Water glass* and *silica gel* have a variety of practical applications. When aluminum replaces some silicon atoms in a silicate, the aluminosilicates result. These include the zeolites and the micas.

Silicon and germanium are intrinsic semiconductors. Their ability to carry a current is best described using the *band theory* in which electrons can occupy a valence band or a conduction band. The conduction-band electrons and the corresponding holes in the valence band move in opposite directions when a potential difference is applied to the crystal. *N*-type and *p*-type semiconductors result when silicon or germanium is doped with Group 5A elements or Group 3A elements, respectively. Various *np* junctions are the basis of transistors and have revolutionized the electronics industry. Ultrapure silicon and germanium is produced by a variety of steps ending in *zone refining*.

Glass is commonly composed of silica, sodium oxide, and calcium oxide. Specialized glasses and porcelain are prepared by adding a variety of ingredients. Optical glass fibers have revolutionized the communications industry.

PROBLEMS

15.1. Assuming the coals of a wood fire provide carbon as the reducing agent, write an equation for the reduction of tin(IV) oxide to the metal which was observed by the ancients.

15.2. Consult the *Biographical Encyclopedia of Science and Technology* by Isaac Asimov and briefly recount how Wöhler disproved that there might be two sets of laws, one for nonliving substances and another for living substances.

15.3. Interview at least one organic chemistry professor in your department and discuss the distinctions between inorganic and organic chemistry. Briefly summarize your discussion here.

15.4. Scheele prepared gaseous silicon tetrafluoride from silica and hydrofluoric acid. Write an equation for this process.

15.5. Write an equation for Berzelius' preparation of silicon.

15.6. When we say that Berzelius was able to isolate "amorphous" silicon, what does this mean? Cite a dictionary definition as part of your answer.

15.7. Why, do you suppose, Ge was not called germon or perhaps germanicon instead of germanium? After all, silicium was renamed silicon within 7 years of its discovery.

15.8. Closely examine the trend in the ionic radii of the Group 4A elements. Is this trend as expected? Are there any unexpected values? Briefly discuss your answer.

15.9. Closely examine the trend in the Pauling electronegativities of the Group 4A elements. Is this trend as expected? Are there any unexpected values? Briefly discuss your answer.

15.10. Closely examine the trend in the ionization energies of the Group 4A elements. Is this trend as expected? Are there any unexpected values? Briefly discuss your answer.

15.11. Closely examine the trend in the electron affinities of the Group 4A elements. Is this trend as expected? Are there any unexpected values? Briefly discuss your answer.

15.12. Summarize the trends in the ionic character of the hydrides of lithium, beryllium, boron, and carbon.

15.13. Summarize the trend in the ionic character of the hydrides of the Group 4A elements.

15.14. Based on your reading of boron chemistry in Chapter 14, is it justified to include this element as a catenator? Cite evidence to back up your conclusion.

15.15. Anyone who has read some science fiction or seen the episode of *Star Trek* "The Devil in the Dark" about the Horta monster knows that silicon-based life forms have often been theorized. How realistic is this possibility? Would these "silons" be a threat to humanoids? Would it be possible for them to invade the earth? Briefly comment.

15.16. Draw out as many different structures as possible for:
(*a*) Tetrasilane, Si_4H_{10}
(*b*) Pentasilane, Si_5H_{12}

15.17. Methane is chlorinated with some difficulty, while silane reacts violently with chlorine. Discuss this difference in terms of the network of interconnected ideas.

15.18. Methane does not react readily with alcohols such as methanol ($R = CH_3$) or ethanol ($R = CH_3CH_2$), but silane does as shown below. Comment on this difference.

$$SiH_4 + 4ROH \longrightarrow Si(OR)_4 + 4H_2$$

15.19. When a collision between monosilane and a reacting molecule or ion takes place, will the greater size of the silicon atom tend to make that collision more or less effective as compared with the reaction between methane and the reacting species? Briefly discuss your answer.

15.20. Draw all the possible isomers of hexagermane, Ge_6H_{14}.

15.21. On the basis of the covalent radii listed in Table 15.1, speculate as to why both the silanes and germanes up to $n = 8$ or so have been prepared but the stannanes go up only to $n = 2$.

15.22. Summarize the trends in the ionic character of the oxides of lithium, beryllium, boron, and carbon.

15.23. Summarize the trends in the ionic character of the oxides of the Group 4A elements.

15.24. Carbon forms a suboxide C_3O_2 ($O{=}C{=}C{=}C{=}O$) for which there is no silicon analog. Discuss some reasons for the existence of this unique carbon compound.

15.25. Summarize the trends in the nature of the bond types found in the chlorides of lithium, beryllium, boron, and carbon.

* **15.26.** Considering electronegativity and bond length versus bond strength, give a rationale for greater catenation exhibited by the silicon halides as compared to the silicon hydrides.

15.27. How would you prepare (*a*) $GeCl_2$ and $GeCl_4$, (*b*) $SnCl_2$ and $SnCl_4$? Write equations as part of your answer.

15.28. How would you prepare (*a*) SnF_2 and SnF_4, (*b*) PbF_2 and PbF_4? Write equations as part of your answer.

15.29. Rationalize the fact that the carbon halides do not form bridged, dimeric structures as the Group 3A elements quite often do.

15.30. Which would you expect to be the more covalent, germanium tetrahalides or the germanium dihalides? Briefly explain your answer.

15.31. Which would be more soluble in water, lead(IV) chloride or lead(II) chloride? Briefly rationalize your answer.

15.32. Write a balanced equation to represent the reaction in which tin(II) is used to reduce iron(III) to iron (II). Determine $E°$ and $\Delta G°$ for the reaction.

15.33. Write an equation for the possible oxidation of HCl to Cl_2 by PbO_2. Calculate $E°$ and $\Delta G°$ and decide whether or not this oxidation is possible. (Assume PbO_2 is reduced to Pb^{2+}.)

15.34. Graphite is a good conductor of electricity, whereas charcoal is not. Speculate on the reasons for this difference. Include references to the structures of these compounds in your answer.

15.35. The thermal conductance of diamond is very high. Would you suspect that its electrical conductance would also be high? Briefly explain your answer.

15.36. Carbon 14 cannot be used to date objects much younger than about 1000 years. Speculate as to reasons for this limitation.

15.37. Give at least two reasons why the age of rocks cannot be determined by the carbon 14 method.

15.38. Uranium 238 decays to lead 206 through the following series of α and β^- emissions: α, 2β, 5α, β, α, 3β, and α. Write the sequence of nuclear reactions showing the new isotopes produced at each stage. How many isotopes of lead are produced in the sequence?

15.39. With reference to the sequence of alpha and beta emissions that occur in the transformation of uranium 238 to lead 206 outlined in Problem 15.38, comment on the Curies' discovery of radium in uranium ores and not in barium deposits as one might suspect from their chemical similarities.

15.40. In a sequence of rock samples arranged from the oldest to the youngest, should the $^{206}Pb/^{204}Pb$ ratio increase, decrease, or stay constant? Briefly justify your answer.

15.41. Why should the isotopic trail of lead from the combustion of coal be different from that of a smelter? Why would a lead smelter in Missouri produce a different ratio of lead isotopes than a smelter in California?

15.42. Comment on why the atomic weight of lead depends on the mineral from which it is isolated.

15.43. Write a Lewis structure, VSEPR diagram, and VBT diagram of carbon disulfide.

15.44. Write a Lewis structure, VSEPR diagram, and VBT diagram of phosgene, $COCl_2$.

15.45. Write a Lewis structure, VSEPR diagram, and VBT diagram of hydrogen cyanide, HCN.

15.46. Before lead storage batteries were sealed, the condition of a battery could be determined by measuring the density of the sulfuric acid electrolyte solution with a hygrometer. Would a relatively low density indicate that the battery needed charging or that it was fully charged? Briefly explain your answer.

15.47. Calculate the $E°$ and $\Delta G°$ for the overall reaction of the lead storage battery.

15.48. Two different double silicate chains from the one pictured in Figure 15.8*f* are given below. Determine the repeating unit in each.

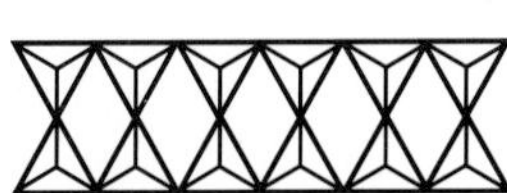

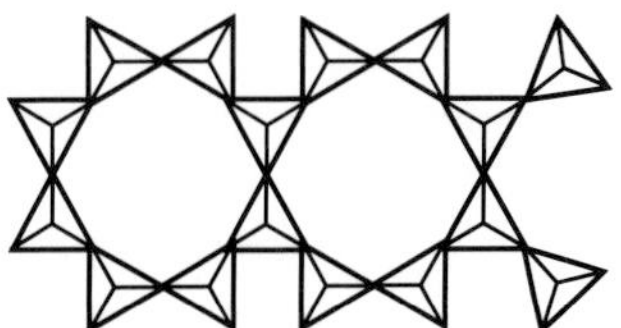

15.49. Briefly explain why aluminum phosphate, $AlPO_4$, forms in quartzlike structures like SiO_2. (*Hint*: How are aluminum and phosphorus related electronically to silicon? How would an AlP pair be related to two silicon atoms?)

*** 15.50.** Briefly explain why BeF_2 is structurally similar to SiO_2.

15.51. On the basis of the structures, briefly explain why some zeolites are called *sieves*. Use a dictionary to first look up the meaning of sieve.

15.52. Briefly explain the difference between an intrinsic and a doped semiconductor.

15.53. Would gallium arsenide be a suitable compound for a semiconductor? How could n- and p-type semiconductors be fashioned from it?

15.54. Why are old window panes thicker on the bottom than they are on the top?

15.55. Instead of sodium oxide and calcium oxide, sodium carbonate and limestone are listed as the common constituents of ordinary glass. Briefly explain why.

CHAPTER 16

GROUP 5A: THE PNICOGENS

Not many chemists refer to the Group 5A elements as the *pnicogens*, a name that means "choking producers" (Greek *pnigein*, "to choke," and *-genēs*, "be born"). Although obscure and not officially accepted, just the existence of a group name indicates that we have started back toward more uniform properties within a group. Still, there is a great deal of diversity among nitrogen, phosphorus, arsenic, antimony, and bismuth.

The first two sections of this chapter address the history of the discoveries of the elements and the application of the network to group properties. The third section highlights the structures, preparations, reactions, and applications of compounds exhibiting the large variety of nitrogen oxidation states from -3 to $+5$. This is followed by a discussion of some other reactions and compounds of practical importance, including nitrogen fixation, nitrites and nitrates, the phosphorus allotropes, and the many phosphates. The chemistry of photochemical smog is the selected topic in depth.

16.1 DISCOVERY AND ISOLATION OF THE ELEMENTS

Figure 16.1 displays the discovery dates of the Group 5A elements. A quick comparison with Figure 15.1 reveals that the 5A elements, while not quite as ancient as carbon, tin, and lead, were all known before the establishment of the United States. Three of them (phosphorus, arsenic, and antimony) have strong connections to alchemy. Alchemists (magicians-chemists of the Middle Ages) were in search of the philosopher's stone which would convert base metals into gold and the elixir of life which would grant immortality. Antimony was a guarded secret of the alchemists, while the discoverers of arsenic and phosphorus conducted alchemical experiments to varying degrees.

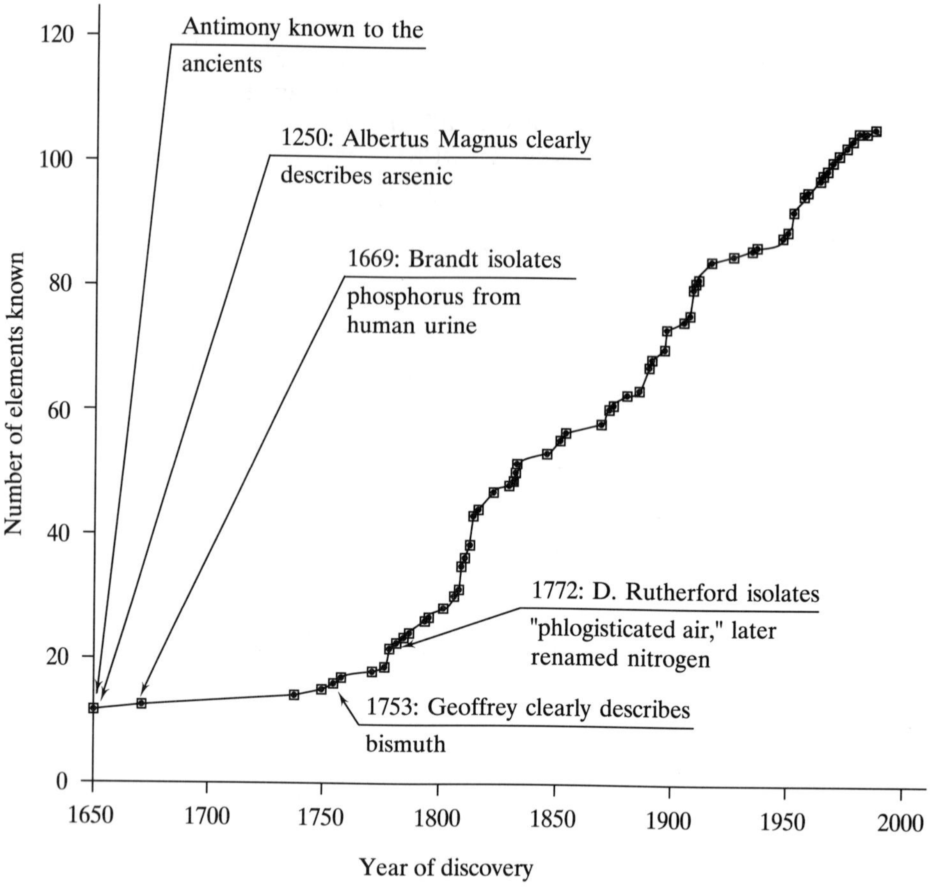

FIGURE 16.1
The discovery of the Group 5A elements superimposed on the plot of the number of known elements versus time.

Antimony and Arsenic

Even the name *antimony* is shrouded in alchemical mystery and, like other such terms, is most likely a corruption of some Arabic word or phrase so as to appear Greek or Latin. The Latin *stibium*, formerly used for the element and still the root for many of its compounds, may very well have the same (but unknown) Arabic root. Stibnite, or the "stibic stone," what we know today as antimony (III) sulfide, Sb_2S_3, was used by the ancients as a cosmetic to darken and beautify eyebrows. Rhazes, the tenth century Persian physician and alchemist, described metallic antimony, but it is not known when this brittle gray metal was first isolated. Early uses included emetics (vomiting inducers), additives to bell metals to impart a silver sound, and type metal alloys.

The name *arsenic* comes from the Greek word *arsenikon* which is a somewhat mystical adaptation of the Persian word for "yellow orpiment," a common sulfide ore of the element. Again, the first isolation of the metal is unrecorded, but Albertus Magnus, or Albert the Great, a thirteenth century German scholar and skeptical alchemist, is commonly credited with its discovery because of the clear descriptions he gives in writings dated to 1250. Early descriptions of the reduction of the mineral orpiment, arsenic(III) sulfide, to the free arsenic involved heating it with eggshells ($CaCO_3$) or lime (CaO) followed by charcoal.

The poisonous nature of arsenic compounds, or *arsenicals*, has long been known. Both accidental and intentional poisonings have been well documented for centuries. For example, some Chinese were poisoned by drinking beverages which had stood in new tin vessels. Arsenic almost always occurs with tin, and proper measures must be taken to separate out the former. Accounts of murder using arsenical poisonings have been the favorite of mystery writers over the years. One of the most famous of these is *Strong Poison* (1930), by Dorothy L. Sayers, in which the murderer, who has gradually built up his immunity to arsenic over time, deflects suspicion by administering the arsenic to his victim and himself in the same doses. With his increased immunity, the murderer is unaffected, but not so his victim!

Phosphorus

The discovery of phosphorus is one of the most bizarre stories we have encountered in accounts of this type. Hennig Brandt, a German alchemist and physician, isolated the element in 1669, making him the very first person to discover an element not known before his time. Brandt was looking for a substance capable of converting silver into gold and chose, of all things, to investigate human urine. One recipe called for letting 50 to 60 pails of urine stand in a tub until it putrefied. It was then boiled down to a paste and the vapors drawn into water, producing a white, waxy substance that glowed in the dark. If the material was removed from water, it burst into flame. Brandt called his product "cold fire." Later it was called *phosphorus* from the Greek word meaning "light-bearing."

The recipe was secretly purchased and otherwise passed among a select number of entrepreneurs until finally in 1737, some 68 years after its initial discovery, the secret was sold to the French Academy of Sciences which made it public. In 1769, Scheele and Gahn found phosphorus in bones, but it is remarkable that for a full century the only way to make this element involved the processing of human urine! By the early 1800s, the manufacture of matches in England made the demand for phosphorus so great that European battlefields were being combed for human remains. Thankfully, it is now manufactured by the relatively mundane procedure of heating phosphate rock with sand and coke as represented in Equation (16.1):

$$2Ca_3(PO_4)_2 + 10C + 6SiO_2 \longrightarrow 6CaSiO_3 + 10CO(g) + P_4(s) \quad (16.1)$$

Bismuth

The exact date of the first isolation of this metal, like several of its congeners, is shrouded in mystery. The German *weisse masse*, later wismuth and the latinized bisemutum, may have been derived from the word for "white metal." Bismuth is indeed a white crystalline metal with a pinkish tinge. It seems to have played a role in the literacy revolution associated with the development of the Gutenberg printing press around 1440. By 1450, a secret method of casting type from bismuth alloys was known.

For centuries, bismuth was confused with tin or lead. Alchemists believed it to be further along in a transformation they foresaw ending in silver. Accordingly, miners, when they chanced to come across bismuth, would lament that "alas, we have come too soon." By 1753, Claude-François Geoffrey made such a thorough and definitive investigation of bismuth that he is often listed as the discoverer.

Early uses of bismuth included its addition to tin to increase its hardness and brilliancy. It was also added to tin and bronze bell metals to produce a deeper, richer sound when struck. Alloys of appreciable bismuth content are low-melting and are used today for safety devices in fire detection and sprinkler systems as well as fuses and relief valves. A teaspoon made of Wood's metal (50% bismuth, 25% lead, 12.5% tin, and 12.5% cadmium) melts in a cup of coffee. (Drinking this brew afterward would be decidedly inadvisable!)

Nitrogen

Like those of so many other elements we have discussed, nitrogen compounds were well known long before the free element was isolated. Ammonium salts such as sal ammoniac (ammonium chloride) have been characterized since the fifth century B.C., while aqua fortis (nitric acid) was described in the thirteenth century and by the end of the sixteenth century was in high demand for the separation of silver and gold. Saltpeter, or niter (potassium nitrate), and Chilean saltpeter (sodium nitrate) have long been prized as fertilizers and for use in gunpowder.

In 1771, Daniel Rutherford, working under the direction of Scottish chemist Joseph Black, announced his isolation of a gas or "air" that was left over after carbonaceous (carbon-containing) substances were burned in it and the carbon dioxide (what Black called "fixed air") removed. Recall from earlier discussions (Chapter 11, pp. 265–266) that in the days prior to Lavoisier's definitive work, all accounts of what we call combustion were couched in terms of phlogiston. Rutherford therefore called his product "phlogisticated air." Three other phlogistonists, Karl Scheele and Joseph Priestley, the codiscoverers of "dephlogisticated air" (oxygen), and Henry Cavendish, the discoverer of hydrogen, also independently produced the same substance. Lavoisier proposed the name "azote," meaning "without life," but a few years later the name *nitrogen* ("nitron producer") was accepted after it was found that this element was a component of nitric acid and nitrates. Nitrogen is still manufactured by liquefying and fractionally distilling common air. It is the most abundant uncombined element known; the atmosphere contains 4 trillion tons of nitrogen!

16.2 FUNDAMENTAL PROPERTIES AND THE NETWORK

Figure 16.2 shows the Group 5A elements superimposed on the network, but this representation is only slightly altered from that given in Figure 12.6, presented when the network had just been completed. The pnicogens are indeed intimately tied to our organizing fabric.

Table 16.1 shows the usual tabulation of group properties that vary in ways we have come to expect. Bismuth 209, incidentally, is the heaviest stable isotope of any element.

Uniqueness Principle

Nitrogen, the first gas encountered in our group tour, is certainly unique as expected but is more closely allied with its congeners than carbon is in its group. For example, nitrogen is not far and away a better catenator than phosphorus like carbon is compared to silicon. Nitrogen-nitrogen single bonds are considerably

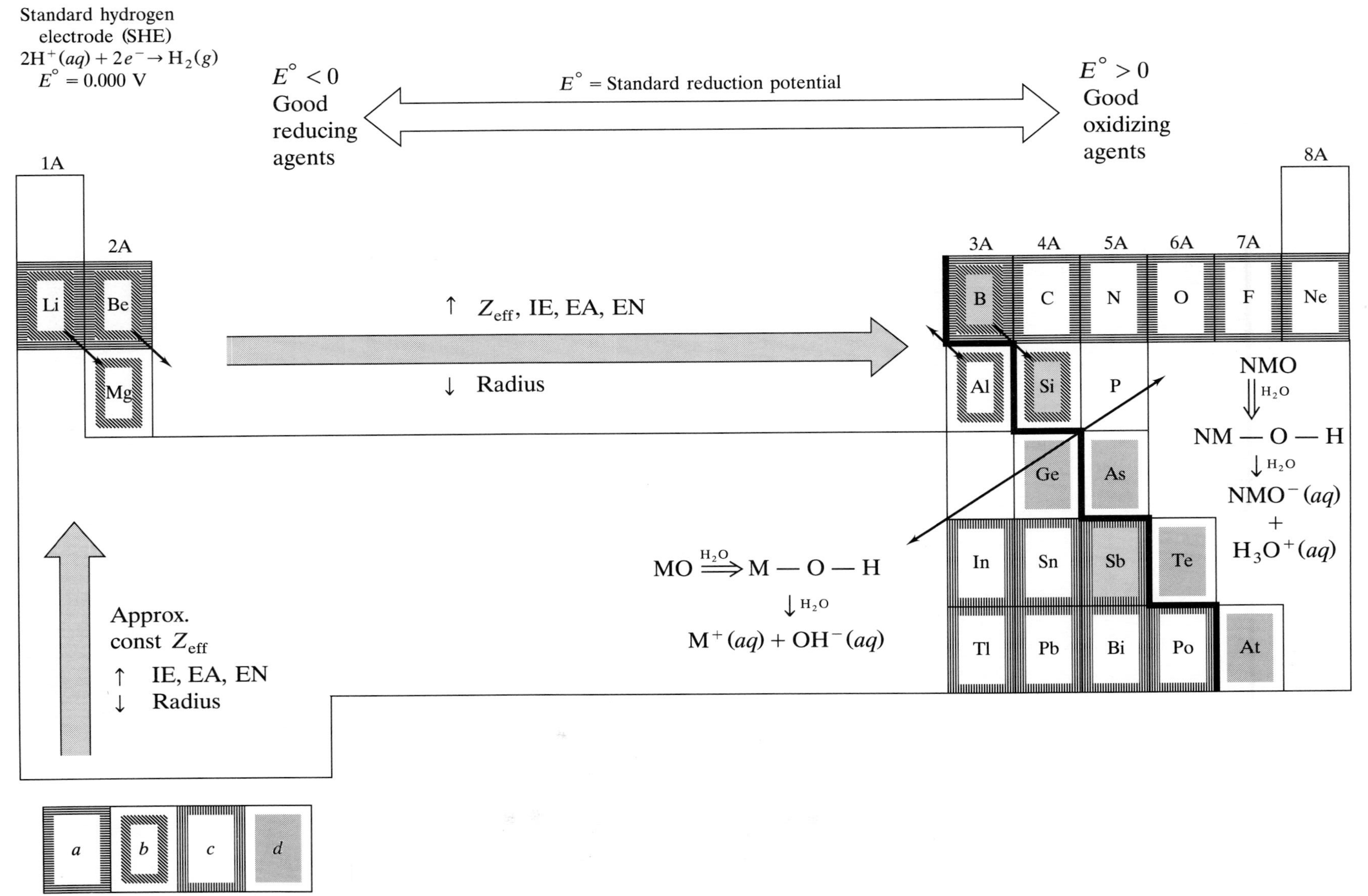

FIGURE 16.2
The Group 5A elements superimposed on the network of interconnected ideas including the trends in periodic properties, the acid-base character of metal and nonmetal oxides, trends in standard reduction potentials, (*a*) the uniqueness principle, (*b*) the diagonal effect, (*c*) the inert-pair effect, and (*d*) the metal-nonmetal line.

TABLE 16.1
The fundamental properties of the Group 5A elements

	Nitrogen	Phosphorus	Arsenic	Antimony	Bismuth
Symbol	N	P	As	Sb	Bi
Atomic number	7	15	33	51	83
Natural isotopes, A/% abundances	$^{14}N/99.63$ $^{15}N/0.37$	$^{31}P/100$	$^{75}As/100$	$^{121}Sb/57.25$ $^{123}Sb/42.75$	$^{209}Bi/100$
Total no. of isotopes	7	7	13	22	18
Atomic weight	14.01	30.97	74.92	121.8	209.0
Valence electrons	$2s^22p^3$	$3s^23p^3$	$4s^24p^3$	$5s^25p^3$	$6s^26p^3$
mp/bp, °C	−210/− 196	44/280†	814/sublimes	631/1380	271/1560
Density, g/cm^3	1.25 (g/liter)	1.83†	5.73‡	6.69	9.75
Atomic radius, (metallic), Å	...	...	...	1.61	1.82
Covalent radius, Å	0.75	1.06	1.20	1.40	1.46
Ionic radius, Shannon-Prewitt, Å (C.N.)	0.27(6)	0.52(6)	(+5) 0.60(6) (+3) 0.72(6)	(+5) 0.74(6) (+3) 0.94(5)	(+5) 0.90(6) (+3) 1.17(6)
Pauling EN	3.0	2.1	2.0	1.9	1.9
Charge density (charge/ionic radius), unit charge/Å	18	9.6	(+5) 8.3 (+3) 4.2	(+5) 6.8 (+3) 3.2	(+5) 5.6 (+3) 2.6
$E°$,§ V	+0.25	−1.49	−0.68	−0.57	−0.45
Oxidation states	−3 to +5	−3, +3, +4, +5	+3, +5	+3, +5	+3, +5
Ionization energy, kJ/mol	1400	1012	947	834	703
Electron affinity, kJ/mol	0	−74	−77	−101	−100
Discovered by/date	Rutherford 1772	Brandt 1669	Albertus 1250	Antiquity	Geoffrey 1753
rpw¶ O_2	NO NO_2	P_4O_6 P_4O_{10}	As_4O_6	Sb_2O_5	Bi_2O_3
Acid-base character of oxide	Acid	Acid	Acid	Ampho.	Base
rpw N_2		None	None	None	None
rpw halogens	NX_3	PX_3, PX_5	AsX_3, AsF_5, $AsCl_5$	SbX_3, SbF_5, $SbCl_5$	BiX_3, BiF_5
rpw hydrogen	NH_3	PH_3	...	...	
Crystal structure	...	Cubic†	Rhomb.	Rhomb.	Rhomb.

†For white phosphorus.
‡For gray arsenic.
§$NO_3^- \rightarrow N_2$; $EO_4^{3-} \rightarrow E$, E = P, As; $Sb(OH)_6^- \rightarrow Sb$; $Bi_2O_3 \rightarrow Bi$ (in basic solution)
¶rpw = reaction product with

weaker than carbon-carbon bonds, a fact attributed to repulsions among lone pairs usually present on small adjacent nitrogen atoms. On the other hand, unlike phosphorus, it does show *many* examples of a variety of oxidation states ranging from −3 to +5 (the subject of Section 16.3), a high ability to form $p\pi$ bonds, and no availability of 3*d* orbitals.

The enhanced ability of nitrogen to form $p\pi$ bonds is most obvious in the special stability of the diatomic molecule, N_2. The high energy associated with the triple bond is responsible for (1) molecular nitrogen constituting nearly 80 percent of the atmosphere, (2) the difficulty in "fixing" nitrogen, that is, converting molecular nitrogen into other nitrogen compounds (see p. 431), and (3) the highly exothermic, often explosive reactions that result when N_2 is a product (see p. 432).

The ability to form $p\pi$ bonds also results in a variety of nitrogen compounds that have no analogs among the heavier Group 5A congeners. These include (1) catenated chains involving alternating single and double bonds such as PhN=N—NPh—N=N—NPh—N=NPh (where Ph = C_6H_5–), (2) a huge variety of oxygen compounds (oxides, oxoacids, and the corresponding oxoanions) involving N=O bonds, (3) hydrocyanic acid, H—C≡N, and the corresponding cyanides, $C{\equiv}N^-$, (4) thiocyanic acid, H—S—C≡N, and the corresponding thiocyanates,

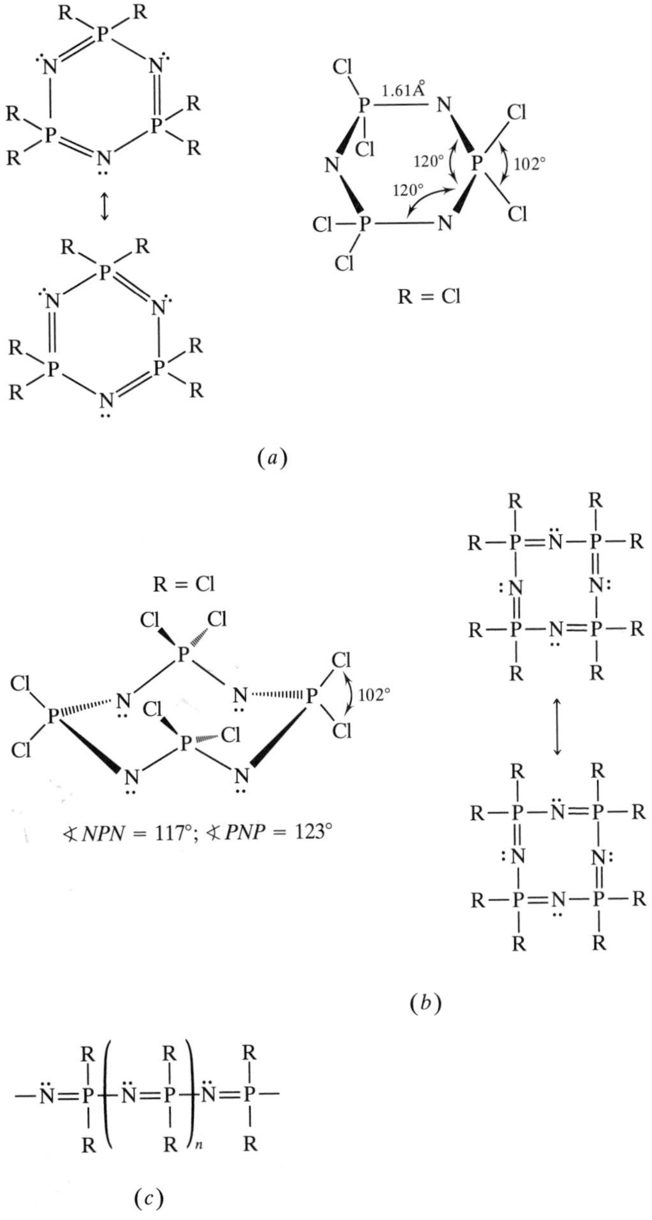

(a)

R = Cl

1.61Å 120° 120° 102°

R = Cl

∢NPN = 117°; ∢PNP = 123°

102°

(b)

(c)

FIGURE 16.3
The structure of (a) cyclotriphosphazenes, (b) cyclotetraphosphazenes, and (c) the linear polymeric phosphazenes.

and (5) a variety of sulfur and phosphorus nitrides involving double bonds to nitrogen.

By contrast, the chemistry of phosphorus and the heavier congeners is dominated by element to element (E—E) single bonds and, particularly in the case of phosphorus, the availability of *d* orbitals to form $p\pi$-$d\pi$ double bonds with a variety of other atoms such as oxygen, nitrogen, and sulfur. *d* Orbital participation results in expanded octets as found in compounds such as PF_5, $SbCl_5$, $X_3P{=}O$ (where X = F, Cl, Br, I), the phosphorus oxoacids and oxoanions, and a class of compounds called the phosphazenes. The latter, formerly called the *phosphonitriles*, contain both N and P atoms in the same molecule.

The *phosphazenes* are cyclic or chain phosphorus-nitrogen compounds of general formula $[NPR_2]_n$, where R = F, Cl, Br, OH, and a variety of organic groups. The best known cyclic phosphazenes are the trimers and tetramers shown in Figure 16.3*a* and *b*. These molecules are pictured with alternating single and double $p\pi$-$d\pi$ P—N bonds, but in fact all the P–N distances are about 1.5 Å, shorter than P—N bonds and longer than P=N bonds. The nitrogen atoms are sp^2-hybridized with the third unhybridized *p* orbital forming the double bonds. The phosphorus atoms, on the other hand, are sp^3-hybridized with a 3*d* orbital forming the double bonds. Linear polymeric chains, as shown in Figure 16.3*c*, are also known. These inorganic polymers, with a variety of substituents that in some cases provide cross-linking reminiscent of rubber, are being increasingly employed in a variety of fibers, rubbers, glasses, and even biodegradable sutures.

Other Network Components

The inert-pair effect continues to be applicable. While phosphorus, arsenic, and antimony show the prevalence of both the +3 and +5 oxidation states to varying degrees, bismuth is dominated by +3 chemistry.

The metal-nonmetal line moves to a lower position in each succeeding group but still is relevant here. Significantly, however, this group is dominated by the nonmetals nitrogen and phosphorus and, in bismuth, contains only one real metal. The acid-base properties of the oxides reflect the placement of the metal-nonmetal line with nitrogen and phosphorus oxides being acidic, arsenic and antimony amphoteric, and, again, only bismuth basic.

The general horizontal trends in standard reduction potentials would indicate that these elements and their compounds are better oxidizing agents than we have previously encountered, but our discussions of individual compounds will show a variety of redox properties.

Hydrides

The difference between nitrogen and phosphorus is aptly represented in their simplest hydrides, ammonia (NH_3) and phosphine (PH_3). Ammonia is the covalent hydride with the familiar pyramidal shape. We have discussed its ability to hydrogen-bond (Chapter 11, p. 273), to complex (Chapter 2), to dissolve in water (Figure 11.7), and so forth, in a variety of sections in this book.

While we all feel at home with ammonia and recognize its distinctive pungent smell, phosphine is a different kettle of fish. In fact, that is exactly what it smells like—rotten fish. It is conveniently prepared by the reaction of water and an ionic phosphide like that of calcium as shown in Equation (16.2). In the open air the deadly poisonous phosphine gas immediately reacts with oxygen (and ignites due to traces of P_2H_4 or elemental phosphorus) as shown in Equation (16.3):

$$Ca_3P_2(s) + 6H_2O(l) \longrightarrow 2PH_3(g) + 3Ca(OH)_2(s) \qquad (16.2)$$

$$4PH_3(g) + 8O_2(g) \longrightarrow P_4O_{10}(s) + 6H_2O(g) \qquad (16.3)$$

Structurally, it is pyramidal (H–P–H angle = 93.7°) like ammonia but does not hydrogen-bond or dissolve in water because the P—H bonds are essentially nonpolar. Phosphine also differs from ammonia in that it is not a good Brønsted-Lowry base (proton acceptor). It can, however, be forced to react with strong acids to form phosphonium (PH_4^+) salts.

Arsine, AsH_3, is much less stable than phosphine. For example, it readily decomposes upon heating to form metallic arsenic which can be deposited as a mirror on hot surfaces. This is the basis of the common criminological Marsh test for the presence of arsenic compounds. In practice, the contents of the victim's stomach are mixed with hydrochloric acid and zinc, producing hydrogen gas. The hydrogen in turn reacts with a variety of arsenic compounds to produce arsine that is thermally decomposed to the arsenic mirror. These processes are represented in Equations (16.4) to (16.6):

$$Zn(s) + 2HCl(aq) \longrightarrow ZnCl_2(aq) + H_2(g) \qquad (16.4)$$

$$4H_2(g) + H_3AsO_4 \longrightarrow AsH_3(g) + 4H_2O(l) \qquad (16.5)$$

$$2AsH_3(g) \xrightarrow{\text{heat}} 2As(s) + 3H_2(g) \qquad (16.6)$$

Stibine, SbH_3, and bismuthine, BiH_3, are still less stable than their lighter analogs.

Oxides and Oxoacids

There are a variety of nitrogen oxides in which the oxidation state of the nitrogen atom ranges from +1 to +5. These oxides and their corresponding oxoacids, including the strong electrolyte nitric acid, will be covered in detail in the next section. They are characterized by a variety of single and double N—O bonds involving both localized and delocalized $p\pi$-$p\pi$ interactions.

The phosphorus oxides and corresponding acids are only mildly acidic. Both the +3 and +5 phosphorus oxides are known. Normally, we would expect these oxides to have formulas of P_2O_3 and P_2O_5, and indeed these compounds are quite often referred to as phosphorus trioxide and phosphorus pentoxide. The molecular formulas, however, are P_4O_6 and P_4O_{10}, respectively. White phosphorus, P_4, and the two oxides are structurally related as shown in Figure 16.4. P_4 is a tetrahedron of phosphorus atoms, and P_4O_6 has six oxygen atoms bridging the sides of the tetrahedron. In P_4O_{10}, four terminal P=O bonds (of the $d\pi$-$p\pi$ type) have been added. These oxides are the acid anhydrides for phosphorous acid, H_3PO_3, and phosphoric acid, H_3PO_4, respectively, as shown in Equations (16.7) and (16.8):

$$P_4O_6(s) + 6H_2O(l) \longrightarrow 4H_3PO_3 \qquad (16.7)$$

$$P_4O_{10}(s) + 6H_2O(l) \longrightarrow 4H_3PO_4 \qquad (16.8)$$

Equation (16.8) is the basis of the excellent desiccant properties of phosphorus pentoxide which is widely employed in glove bags and boxes to ensure that the inert atmospheres in these spaces are free of trace amounts of water.

The structures of phosphoric and phosphorous acid are shown in Table 11.7. Phosphoric acid, H_3PO_4, has three hydroxyl hydrogen atoms and is therefore triprotic. (Recall the discussion in Chapter 11, pp. 283–285.) It is a weaker acid

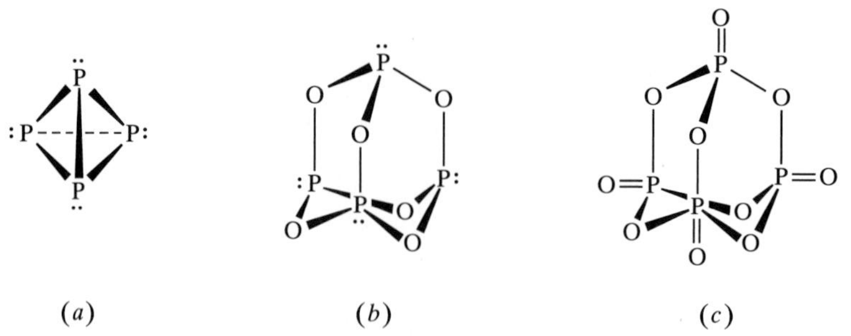

(*a*) (*b*) (*c*)

FIGURE 16.4
The structures of (*a*) P_4 (white phosphorus), (*b*) phosphorus trioxide, P_4O_6, and (*c*) phosphorus pentoxide, P_4O_{10}, are all based on a tetrahedron of phosphorus atoms.

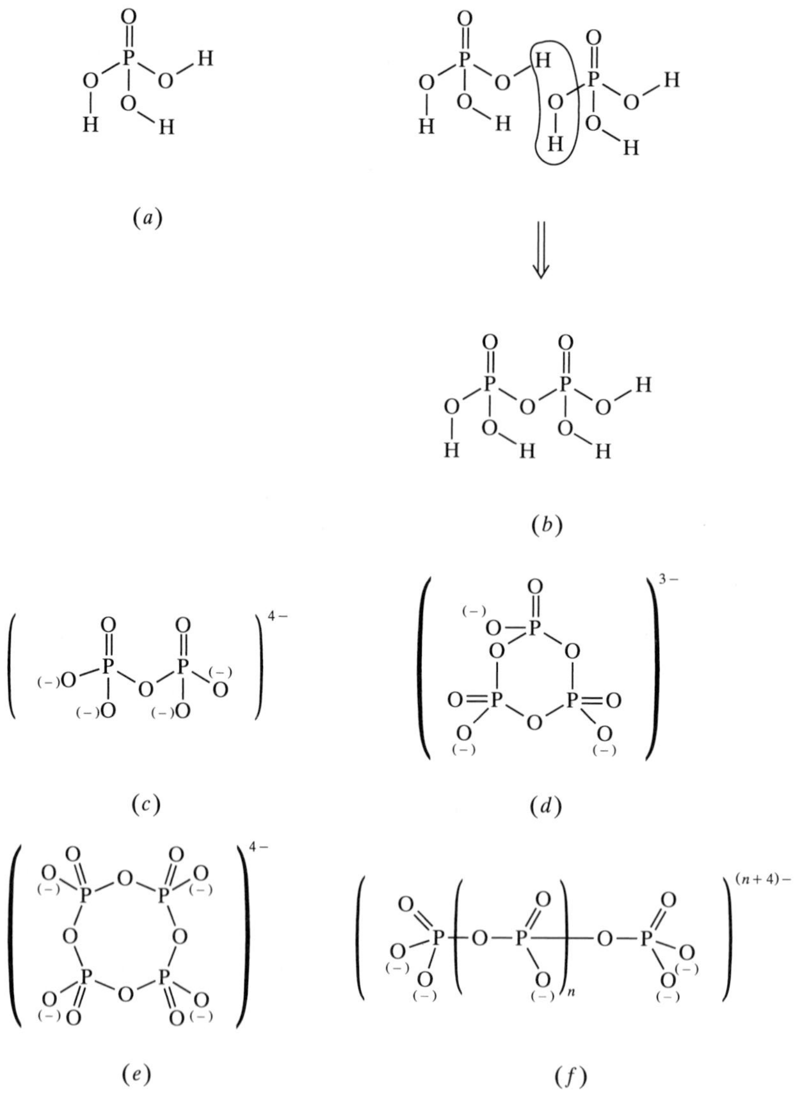

(*a*)

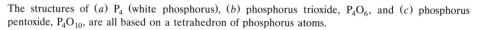

(*b*)

(*c*) (*d*)

(*e*) (*f*)

FIGURE 16.5
Phosphoric acids and phosphates: (*a*) phosphoric acid, H_3PO_4; (*b*) two phosphoric acid molecules split out a molecule of water to form pyrophosphoric acid, $H_4P_2O_7$; (*c*) pyrophosphate; (*d*) tricyclophosphate; (*e*) tetracyclophosphate; (*f*) polyphosphate.

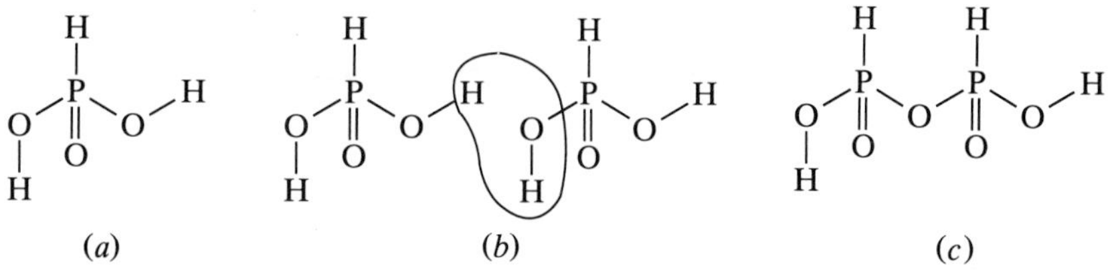

FIGURE 16.6
(*a*) Phosphorous acid, H_3PO_3. (*b*) Two phosphorous acid molecules split out a molecule of water to form (*c*) pyrophosphorous acid, $H_4P_2O_5$.

than nitric due to the lower electronegativity of the central atom. The thick, syrupy nature of the commercially available 85% aqueous solution of this acid is due to the hydrogen bonds among the acid molecules.

The corresponding oxoanion to phosphoric acid is phosphate, PO_4^{3-}, sometimes, similar to what we have discussed in both borate (Chapter 14, pp. 360ff) and silicate (Chapter 15, pp. 396ff) chemistry, called the orthophosphate anion. As might be suspected from the experience gained in earlier chapters, two or more phosphoric acid molecules can come together as shown in Figure 16.5, split out one or more water molecules, and form various chain and cyclic acids and their corresponding phosphate anions. These metaphosphates are sometimes referred to as condensed phosphates. Such acids and anions have been put to a variety of uses, some of which are detailed in Section 16.4.

Phosphorous acid, H_3PO_3, has only two hydroxyl protons and therefore is diprotic. The third hydrogen atom is bound directly to the phosphorus atom and is not an acidic proton. (Since hydrogen and phosphorus have identical electronegativities of 2.1, the P—H bond is nonpolar, not susceptible to attack by the polar water molecule, and therefore not ionized.) Like the phosphates, there are a variety of condensed phosphites as shown in Figure 16.6.

The acidic arsenic oxides and the corresponding oxoacids are fairly similar to the above phosphorus compounds. As_4 (gray arsenic) and As_4O_6 have structures like their phosphorus analogs, but the +5 oxide is polymeric. Arsenious acid, H_3AsO_3, and arsenic acid, H_3AsO_4, as well as the corresponding arsenites, AsO_3^{3-}, and arsenates, AsO_4^{3-}, are known. Some examples of meta- and polyarsenates and arsenites have been isolated, but the rings are smaller and the chains are shorter than their phosphorus cousins.

When antimony and bismuth are burned in air, only the +3 oxides result. The amphoteric Sb_4O_6 is structurally similar to the phosphorus(III) oxide and with some difficulty can be oxidized to a polymeric +5 oxide. The more ionic Bi_2O_3 is very difficult to oxidize and is distinctly basic in character. Bismuth hydroxide, $Bi(OH)_3$, a true base, can be precipitated out of hydroxide solutions of various bismuth(III) salts.

The chemistry of the oxides and their corresponding oxoacids (or in the case of bismuth, the corresponding hydroxide) demonstrates that both (1) the stability of the +5 oxidation state and (2) the acidity of the oxides steadily decrease down the group.

Halides

A huge number of Group 5A halides are known, but only a very few can be detailed here. Nitrogen, as expected, forms only the pyramidal trihalides, there being no *d* orbitals available to expand the octet. Gaseous nitrogen fluoride is the most stable. It is prepared by treating ammonia with fluorine in the presence of a copper catalyst. The chloride, bromide, and iodide are well-known explosives. Nitrogen triiodide, which is prepared as an ammonia adduct of formula $NI_3 \cdot NH_3$, is extremely shock-sensitive and explodes to give purple iodine vapors.

Phosphorus forms both pyramidal trihalides and trigonal bipyramidal pentahalides with all four halogens. Taking the chlorides as representative, the reaction of white phosphorus, P_4, with a limited amount of chlorine gas yields the colorless liquid PCl_3, while if an excess of chlorine is used, the off-white solid PCl_5 is formed. The trihalide is hydrolyzed to phosphorous acid, while the pentahalide forms phosphoric acid, as shown in Equations (16.9) and (16.10). (Note that the oxidation state of phosphorus does not change in either one of these hydrolysis reactions.) In solid form, PCl_5 is actually made up of alternating PCl_4^+ and PCl_6^- ions.

$$P_4(s) + \underset{\text{(Limiting)}}{6Cl_2(g)} \longrightarrow 4PCl_3(l) \xrightarrow{H_2O} H_3PO_3(+\ HCl) \tag{16.9}$$

$$P_4(s) + \underset{\text{(Excess)}}{10Cl_2(g)} \longrightarrow 4PCl_5(s) \xrightarrow{H_2O} H_3PO_4(+\ HCl) \tag{16.10}$$

The decreasing stability of the +5 oxidation state is shown by the fact that all 16 EX_3 compounds (E = P, As, Sb, Bi; X = F, Cl, Br, I) are formed, but only phosphorus forms pentahalides with all four halogens. Arsenic and antimony form only the pentafluoride and pentachloride, while bismuth forms only the pentafluoride. Of the pentafluorides of the lower three pnicogens, only AsF_5 is trigonal bipyramidal. SbF_5 and BiF_5 are polymers of EF_6 octahedra held together by bridging fluorine atoms.

The heavier phosphorus trihalides (PX_3, X = Cl, Br, I) are prepared by direct halogenation. The trifluoride, on the other hand, is best synthesized by treating the trichloride with an ionic fluoride such as CaF_2 or ZnF_2. The arsenic, antimony, and bismuth trifluorides are prepared by adding hydrogen fluoride to the oxides. (If fluorine gas is used, the pentafluorides result.) The other trihalides of arsenic and antimony can be produced by direct halogenation of either the elements or the trioxides, while the bismuth trihalides are best produced by the action of aqueous hydrohalic acids on Bi_2O_3.

16.3 A SURVEY OF NITROGEN OXIDATION STATES

Nitrogen displays each of the nine oxidation states from −3 to +5. Indeed, it is this variety that makes it unique within the group. Phosphorus displays the same range, but only −3, 0, +3, +4, and +5 are of any major importance. Arsenic and antimony are restricted to −3, 0, +3, and +5, while bismuth shows only 0, +3, and +5.

To organize and simplify this survey, the description of many of the major compounds discussed will be accompanied by a *compound profile*. Included in each profile will be (1) any common names, (2) a physical description, (3) the molecular and structural formulas (with bond angles and distances), (4) a note concerning the history and/or application of the compound, (5) equations representing the common preparation(s), and (6) equations representing major reactions. The description of the compounds will amplify these entries as necessary.

Nitrogen (−3) Compounds: Nitrides and Ammonia

The nitrides and ammonia (including the ammonium salts) are the two major types of compounds in which nitrogen has a −3 oxidation state. The nitrides can be divided into ionic, covalent, and interstitial types in the same way as we saw earlier for the hydrides, oxides, and carbides. There are only a few ionic nitrides, the most important being those of lithium, the alkaline-earth metals, and zinc. These are prepared by direct reaction of the elements and readily hydrolyze to ammonia as shown in Equation (16.11). Note that the N^{3-} ion is acting as a Brønsted-Lowry base in this reaction:

$$N^{3-}(aq) + 3H_2O(l) \longrightarrow NH_3(aq) + 3OH^-(aq) \qquad (16.11)$$

Major covalent nitrides include those of boron, sulfur, and phosphorus. Some indication of the nature of the phosphorus nitrides (the phosphazenes) was given in the last section. Boron nitrides have rather similar formulas but involve $p\pi$-$p\pi$ B—N bonds instead of $d\pi$-$p\pi$ P—N bonds. Figure 16.7 shows borazine, sometimes known as *inorganic benzene*. Note that two resonance structures are necessary to describe this molecule using valence-bond concepts. Experimentally, we find that all the B—N bonds have the same bond length, intermediate between that characteristic of single and double bonds. Sulfur nitrides will be discussed in the next chapter.

The transition metals often form interstitial, nonstoichiometric nitrides in which nitrogen atoms occupy holes in the close-packed metal lattices.

Table 16.2 shows a compound profile for ammonia, NH_3. This is the familiar pyramidal molecular hydride. It is capable of forming strong hydrogen bonds with itself and other molecules, most notably water. Much heat is needed to break the hydrogen bonds of liquid ammonia, and this results in a high heat of vaporization. Accordingly, even though ammonia has a boiling point of only −33.3° C, it evaporates rather slowly and can be easily handled in a thermos-type or Dewar flask.

The commercial preparation of ammonia is accomplished in huge quantities by the Haber process [Equation (16.12)], discussed in Section 16.4 under nitrogen fixation. In the laboratory the most common preparation is the treatment of ammonium salts with strong bases as represented in Equation (16.13). The self-ionization of liquid ammonia (analogous to that of liquid water) is shown in Equation (16.14). $NH_3(l)$ is often used as a nonaqueous solvent. As shown in Equation (16.15), NH_3 acts as a base in water and serves as a prototype for a number of nitrogen-containing bases such as methylamine, pyridine, and aniline which you may recall from studying acid-base equilibria in earlier courses.

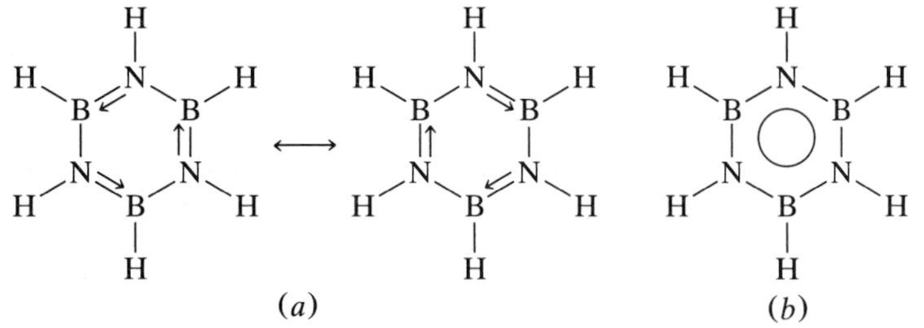

(*a*) (*b*)

FIGURE 16.7
(*a*) The two resonance structures of borazine, $B_3N_3H_6$, sometimes known as inorganic benzene. (*b*) The resonance hybrid of borazine. The circle indicates that all six B—N bonds are equivalent.

TABLE 16.2

Compound profile of NH_3

Nitrogen oxidation state = −3

Name: Ammonia

Physical description: Colorless gas with pungent odor; weak base in aqueous solution

History / application note: First isolated by Priestley in 1774; now prepared commercially by the Haber process; used as fertilizer and to make nitrates

Molecular structure

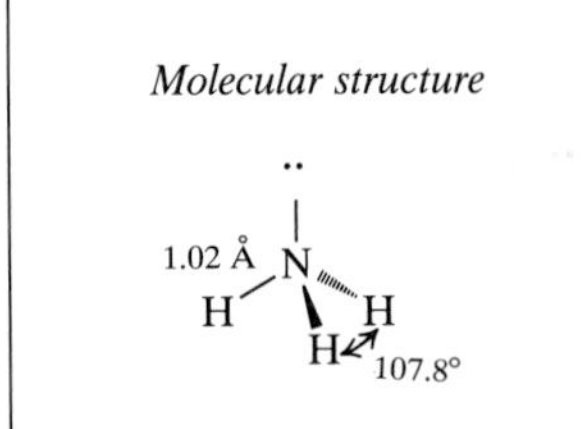

Preparations:

$$N_2(g) + 3H_2(g) \xrightarrow[\text{high } T \text{ and } P]{\text{iron}} 2NH_3(g) \quad \text{(Haber process)} \tag{16.12}$$

$$NH_4Cl(aq) + NaOH(aq) \longrightarrow NaCl(aq) + H_2O(l) + NH_3(g) \tag{16.13}$$

Reactions:

$$2NH_3(l) \rightleftharpoons NH_4^+ + NH_2^- \qquad K = 1 \times 10^{-33} \tag{16.14}$$

$$NH_3 + H_2O \rightleftharpoons NH_4^+(aq) + OH^-(aq) \qquad K = 1.8 \times 10^{-5} \tag{16.15}$$

Household ammonia consists of approximately 2 *M* aqueous NH_3 together with a detergent. It should *never* be mixed with household bleach (containing hypochlorites, OCl^-) because the extremely toxic and explosive chloramines result as represented in Equation (16.16):

$$NH_3(aq) + OCl^-(aq) \longrightarrow OH^-(aq) + H_2NCl(aq) \tag{16.16}$$

There are a large number of ammonium salts containing the tetrahedral NH_4^+ cation. The effective thermochemical radius of the ammonium cation is 1.37 Å (see Table 8.7 and accompanying discussion) which makes it about as large as the potassium (1.52 Å) and rubidium (1.66 Å) ions, and indeed ammonium compounds closely resemble salts of these ions in solution and in structure. One difference, however, is that ammonium salts are acidic in aqueous solution, or, to use the older expression, hydrolyze, as represented in Equation (16.17):

$$NH_4^+(aq) + H_2O(l) \rightleftharpoons NH_3(aq) + H_3O^+(aq) \tag{16.17}$$

$$K = \frac{K_w}{K_b \text{ of } NH_3} = \frac{10^{-14}}{1.8 \times 10^{-5}} = 5.6 \times 10^{-10}$$

Another property that distinguishes ammonium salts is the manner in which they thermally decompose. For example, ammonium chloride decomposes at about 300° C to ammonia and gaseous hydrogen chloride as shown in Equation (16.18):

$$NH_4Cl(s) \xrightarrow{\text{heat}} NH_3(g) + HCl(g) \tag{16.18}$$

If the anion is a good oxidizing agent, the thermal decomposition may be accompanied by the oxidation of ammonia to dinitrogen oxide or elemental nitrogen as shown in Equations (16.19) and (16.20). Equation (16.20) is the basis of the familiar ammonium dichromate volcano demonstration:

$$NH_4NO_3(s) \longrightarrow N_2O(g) + 2H_2O(g) \quad (16.19)$$

$$(NH_4)_2Cr_2O_7(s) \longrightarrow N_2(g) + 4H_2O(g) + Cr_2O_3(s) \quad (16.20)$$

Nitrogen (−2): Hydrazine, N_2H_4

Table 16.3 shows a profile of hydrazine, N_2H_4. It is produced by the Raschig process shown in Equation (16.21). Here ammonia is treated with the mild hypochlorite oxidizing agent in basic solution to first produce chloramine as shown earlier in Equation (16.16). Subsequently the chloramine reacts with excess ammonia to form hydrazine. Hydrazine resembles ammonia in its structure, ability to hydrogen-bond, and its properties as a base [Equation (16.22)]. Unlike ammonia, it is widely used as a reducing agent. In fact, hydrazine and its methyl derivatives are extensively employed as rocket fuels. The reaction represented in Equation (16.23) uses dinitrogen tetroxide as the oxidizing agent and is extremely exothermic, predominantly because of the great amount of energy liberated when the nitrogen-nitrogen triple bond is formed. The methyl derivatives of hydrazine, $(CH_3)NHNH_2$ and $(CH_3)_2NNH_2$, in combination with dinitrogen tetroxide were used in the lunar excursion module (LEM) for both landing and for re-blast-off. These reactions are *hypergolic*, or self-igniting. All the astronauts had to do was open the separate tanks containing the reactants and hang on.

Nitrogen (−1): Hydroxylamine, NH_2OH

The −1 oxidation state is the least stable in nitrogen chemistry. Hydroxylamine is a colorless, thermally unstable, hygroscopic, white solid usually available as aqueous solutions of salts of formula $(NH_3OH)^+X^-$. Its structure in the solid is shown in Figure 16.8.

TABLE 16.3
Compound profile of N_2H_4
Nitrogen oxidation state = −2

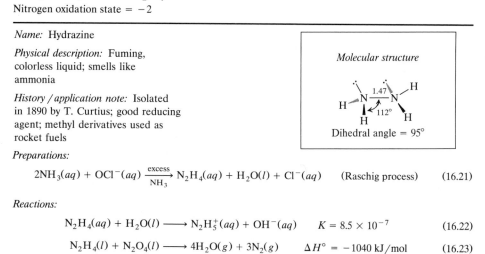

Name: Hydrazine

Physical description: Fuming, colorless liquid; smells like ammonia

History / application note: Isolated in 1890 by T. Curtius; good reducing agent; methyl derivatives used as rocket fuels

Molecular structure

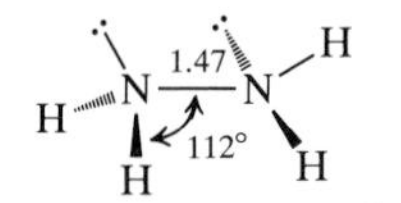

Dihedral angle = 95°

Preparations:

$$2NH_3(aq) + OCl^-(aq) \xrightarrow[NH_3]{excess} N_2H_4(aq) + H_2O(l) + Cl^-(aq) \quad \text{(Raschig process)} \quad (16.21)$$

Reactions:

$$N_2H_4(aq) + H_2O(l) \longrightarrow N_2H_5^+(aq) + OH^-(aq) \quad K = 8.5 \times 10^{-7} \quad (16.22)$$

$$N_2H_4(l) + N_2O_4(l) \longrightarrow 4H_2O(g) + 3N_2(g) \quad \Delta H° = -1040 \text{ kJ/mol} \quad (16.23)$$

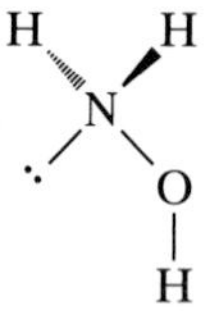

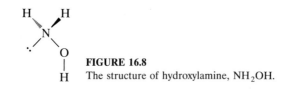

FIGURE 16.8
The structure of hydroxylamine, NH_2OH.

Nitrogen (+1): Nitrous Oxide, N_2O

N_2O goes by a variety of names including the older nitrous oxide and the more modern dinitrogen oxide. Its most famous name, however, is *laughing gas*. It was first discovered by Priestley in 1772, but why are we not surprised to find that it was Humphry Davy who did numerous experiments and demonstrations involving the inhalation of the gas? He reported a euphoric intoxication and a tendency to large swings in emotion, ranging from silliness to rage. A good portrayal of the effects of this gas is given in Figure 16.9, which shows an 1839 painting of some students who had imbibed some nitrous oxide produced in a lecture demonstration. Modern college administrations, encumbered as they are (rightly so, of course) by safety rules and insurance policies, would probably frown on such scenes!

Nitrous oxide was the first modern anesthetic, first being used in surgery in 1837. The gas is somewhat tricky to administer, as different people metabolize it at different rates. N_2O does support combustion; a candle glows brighter in this gas than it does in air itself. However, body temperature is not high enough for the dissociation indicated in Equation (16.25) to take place so that oxygen gas must be administered along with the N_2O when it is used as an anesthetic. There have been anesthesiologists who have been unaware of this caution! Nitrous oxide is no longer the principal anesthetic for surgical procedures, but it is used as an inductant to relax the patient and is still used in dentistry. There is no satisfactory explanation for its unusual effect on humans.

TABLE 16.4
Compound profile of N_2O
Nitrogen oxidation state = +1

Names: Dinitrogen oxide (IUPAC)
Nitrous oxide
Laughing gas

Physical description: Colorless, fairly unreactive gas with pleasing odor and sweet taste; will support combustion once reaction has started

History / application note: Discovered in 1772 by Priestley; surgical anesthetic first used in 1837; now used in dental and other minor surgery as well as a propellant gas

Molecular structure

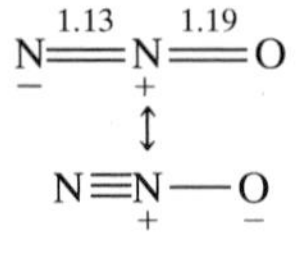

$$\overset{-}{N}\overset{1.13}{=}\overset{+}{N}\overset{1.19}{=}O \leftrightarrow N\equiv\overset{+}{N}-\overset{-}{O}$$

Preparations:

$$NH_4NO_3(s) \xrightarrow{250^\circ} N_2O(g) + 2H_2O(g) \tag{16.24}$$

Reactions:

$$2N_2O(g) \xrightarrow{\text{heat}} 2N_2(g) + O_2(g) \tag{16.25}$$

FIGURE 16.9
The effects of laughing gas being passed around a classroom. A drawing by George Cruikshank from the book *Chemistry No Mystery*, published in London, 1839. [Ref. 32.]

Being fat-soluble, nitrous oxide is also used in large quantities as a propellant gas in cans of whipped cream. This seems an innocent-enough use, but it turns out that N_2O is one of those gases responsible for the greenhouse effect. Perhaps more importantly, N_2O is also produced by soil microorganisms, and there is a growing concern that as Brazilian tropical rain forests are cut down, the rate of production of N_2O by these organisms will increase. Given that it is but slowly oxidized, any increase in the atmospheric concentrations of this gas could have significant consequences. All this is just one more example of the old adage (particularly relevant when discussing environmental concerns), "everything is connected to everything else."

Nitrogen (+2): Nitric Oxide, NO

The compound profile of nitric oxide is given in Table 16.5. Lewis structures leave much to be desired in describing the bonding in NO because it contains an odd number of electrons. Not surprisingly, it readily loses one electron to produce NO^+, the nitrosonium ion, which is isoelectronic with CO and has a bond order of three.

As described in more detail in Section 16.5, the temperatures within an internal combustion engine are high enough to make Equation (16.26) an important source of NO. In the laboratory it can be produced by the action of dilute nitric acid on copper metal as shown in Equation (16.27). Equation (16.28) is the

TABLE 16.5
Compound profile of NO

Nitrogen oxidation state = +2

Names: Nitrogen oxide (IUPAC)
Nitric oxide

Physical description: Colorless, paramagnetic, slightly toxic gas

History / application note: Important product of internal combustion engine leading to photochemical smog

Molecular structure

$$|\dot{N}\overset{1.15}{=\!=}O \leftrightarrow \overset{-}{\underline{N}}=\overset{+}{O}|$$

Preparations:

$$N_2(g) + O_2(g) \longrightarrow 2NO(g) \quad (16.26)$$

$$K_{298\,K} = 4.5 \times 10^{-31} \qquad K_{1800\,K} = 1.2 \times 10^{-4}$$

$$3Cu(s) + \underset{\text{dilute}}{8HNO_3(aq)} \longrightarrow 3Cu(NO_3)_2(aq) + 4H_2O(l) + 2NO(g) \quad (16.27)$$

$$4NH_3(g) + 5O_2(g) \xrightarrow[\text{Pt}]{\text{heat}} 4NO(g) + 6H_2O(l) \quad (16.28)$$

Reactions:

$$2NO(g) + O_2(g) \longrightarrow 2NO_2(g) \quad (16.29)$$

$$2NO(l) \rightleftharpoons N_2O_2(l) \quad (16.30)$$

$$NO \longrightarrow \underset{\text{nitrosonium}}{NO^+} + e^- \quad (16.31)$$

first step in the conversion of ammonia to nitric acid, a reaction that is known commercially as the *Ostwald reaction*. Nitric oxide is easily oxidized in air to the red-brown nitrogen dioxide, NO_2. At temperatures at or less than its boiling point, NO will dimerize to a small extent to form the diamagnetic N_2O_2.

Nitrogen (+3): Dinitrogen Trioxide, N_2O_3, and Nitrous Acid, HNO_2

The compound profiles of dinitrogen trioxide and nitrous acid are given in Table 16.6. Careful oxidation of nitric oxide with molecular oxygen as shown in Equation (16.32) or carefully controlling the equilibrium shown in Equation (16.35) will yield the unstable blue liquid dinitrogen trioxide. N_2O_3, in turn, is the acid anhydride of nitrous acid, HNO_2, as indicated in Equation (16.33). Pure liquid nitrous acid is not known, but it is stable in aqueous solution and the vapor state. One convenient reaction that yields pure aqueous nitrous acid is represented in Equation (16.34). This type of preparation is sometimes called a "milkshake" reaction because of the formation of the fine, milk-white barium sulfate precipitate. Nitrous acid is a weak acid as shown in Equation (16.36). (Also see the discussion of the relative strength of oxoacids as presented in Chapter 11, p. 283.) Nitrites are used as meat preservatives as discussed in the next section.

Nitrogen (+4): Nitrogen Dioxide, NO_2

Nitrogen dioxide, as shown in Table 16.7, is another odd-electron species. Note that the presence of only one electron in the lone-pair position allows the O—N—O bond angle to open up from its normal trigonal planar value of 120°

TABLE 16.6
Compound profile of N_2O_3, HNO_2, NO_2^-
Nitrogen oxidation state = +3

Names: N_2O_3, Dinitrogen trioxide
HNO_2, Nitrous acid
NO_2^-, Nitrite

Physical descriptions: N_2O_3 intensely blue liquid; HNO_2 known in aqueous solution and in vapor state; nitrites white solids

History / application notes: Nitrites used to prevent botulism in hams and other meat products

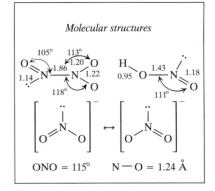

Molecular structures

Preparations:

$$4NO(g) + O_2(g) \longrightarrow 2N_2O_3(l) \tag{16.32}$$

$$N_2O_3(l) + H_2O(l) \longrightarrow HNO_2(aq) \tag{16.33}$$

$$Ba(NO_2)_2 + H_2SO_4(aq) \longrightarrow 2HNO_2(aq) + BaSO_4(s) \tag{16.34}$$

Reactions:

$$N_2O_3 \rightleftharpoons NO + NO_2 \tag{16.35}$$

$$HNO_2(aq) + H_2O(l) \longrightarrow NO_2^-(aq) + H_3O^+(aq) \quad K = 3.2 \times 10^{-4} \tag{16.36}$$

to 134.1°. NO_2 can be prepared from the action of concentrated nitric acid on copper as shown in Equation (16.37) and also from the oxidation of nitric oxide as shown in Equation (16.29) found in Table 16.5.

With one unpaired electron, nitrogen dioxide is ripe for the dimerization represented in Equation (16.39). This equilibrium is highly temperature-dependent. Since the formation of the dimer is an exothermic reaction because a nitrogen-nitrogen bond is formed, increasing the temperature forces this equilibrium to the left. Given that the nitrogen dioxide is reddish-brown while the dinitrogen tetroxide dimer is colorless, the equilibrium can be monitored visually. If a sample in a sealed flask is put into boiling water, it is a deep reddish-brown gas, while in ice it turns to a yellow liquid. At −11.2° C, the colorless N_2O_4 solid freezes out. The equilibrium can also be monitored magnetically since the NO_2 is paramagnetic while N_2O_4 is diamagnetic.

Because nitrogen dioxide dissociates into nitric oxide and atomic oxygen under the action of sunlight as shown in Equation (16.40), it plays a central role in the formation of photochemical smog, as we will discuss in the selected topic in depth. It also serves as a useful nonaqueous solvent.

Nitrogen (+5): Dinitrogen Pentoxide, N_2O_5, and Nitric Acid, HNO_3

The maximum oxidation state displayed by nitrogen is +5. The oxide has an empirical formula of N_2O_5, analogous to that of the corresponding phosphorus(V) oxide, but the two differ considerably. N_2O_5 is the acid anhydride of nitric acid,

TABLE 16.7
Compound profile of NO_2

Nitrogen oxidation state = +4

Name: Nitrogen dioxide

Physical description: Highly toxic, paramagnetic, reddish-brown gas with choking odor

History / application note: Secondary pollutant and trigger for further reactions of photochemical smog

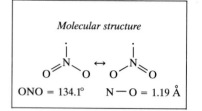

Molecular structure

ONO = 134.1° N—O = 1.19 Å

Preparations:

$$Cu(s) + \underset{\text{conc}}{4HNO_3(aq)} \longrightarrow Cu(NO_3)_2(aq) + 2H_2O(l) + 2NO_2(g) \qquad (16.37)$$

Reactions:

$$2NO_2(g) + H_2O(l) \longrightarrow HNO_2(aq) + HNO_3(aq) \qquad (16.38)$$

$$2NO_2(g) \rightleftharpoons N_2O_4(g) \qquad (16.39)$$

$$NO_2(g) \xrightarrow{h\nu} NO(g) + O(g) \qquad (16.40)$$

TABLE 16.8
Compound profile of HNO_3

Nitrogen oxidation state = +5

Name: Nitric acid

Physical description: Syrupy, colorless pungent liquid usually available as 68%, 15.7 *M* aqueous solution; strong oxidizing agent; often yellow due to small concentrations of NO_2

History / application note: Known since alchemical times; solvent for metals except gold and platinum group; principally used to make NH_4NO_3 for fertilizers and explosives

Molecular structure

102° 0.96 1.41

∢ONO = 114°

∢ON═O = 130° N═O = 1.21 Å

Preparations:

$$NH_3(g) + 2O_2(g) \longrightarrow [NO] \longrightarrow [NO_2] \longrightarrow HNO_3 \quad \text{(Ostwald reaction)} \qquad (16.41)$$

$$2M^{I}NO_3 + H_2SO_4 \longrightarrow 2HNO_3 + M^{I}_2SO_4(aq) \qquad (16.42)$$

Reactions:

$$HNO_3(aq) + H_2O(l) \longrightarrow NO_3^-(aq) + H_3O^+(aq) \qquad (16.43)$$

$$Cr(s) + 6H^+ + 3NO_3^- \longrightarrow Cr^{3+}(aq) + 3NO_2 + 3H_2O \qquad (16.44)$$

$$NO_3^-(aq) + 2H_3O^+(aq) + e^- \xrightarrow{\text{conc.}} NO_2 + 3H_2O \qquad E° = 0.803\ V \qquad (16.45a)$$

$$NO_3^-(aq) + 4H_3O^+(aq) + 3e^- \xrightarrow{\text{dil.}} NO + 6H_2O \qquad E° = 0.96\ V \qquad (16.45b)$$

$$4HNO_3(aq) \xrightarrow{h\nu} 4NO_2(aq) + 2H_2O + O_2(g) \qquad (16.46)$$

although the latter is hardly ever produced in this manner. Instead, this strong mineral acid, one of the most widely used chemicals in the world, is produced commercially by the multistep Ostwald reaction summarized in Equation (16.41) in Table 16.8. Nitric acid does not exist as the pure liquid but rather is most commonly available as a 68% aqueous solution (15.7 *M* HNO_3, 1.42 g/cm^3). In the laboratory, aqueous solutions of HNO_3 can be prepared by the action of sulfuric acid on metal nitrates as shown in Equation (16.42). The reactions which produce insoluble sulfates yield the purest samples of the aqueous acid.

As we discussed in Chapter 11 and as represented in Equation (16.43), nitric acid is a strong acid like sulfuric and hydrochloric. HNO_3, however, has the additional advantage of being an excellent oxidizing agent. Equation (16.44) shows the action of the acid on chromium metal. This is such a strong acid and good oxidizing agent (due to the nitrate ion and not just the hydrogen ion) that it will oxidize and therefore dissolve most metals. In fact, it is used to dissolve the spent fuel rods from nuclear power plants. These rods typically contain some 35 different metals, and nitric acid dissolves each and every one of them. When the concentrated acid is used as an oxidizing agent, the nitrate is reduced to nitrogen dioxide, while using the dilute acid yields nitric oxide as shown in Equation (16.45). When the concentrated aqueous acid is stored in bottles subject to strong sunlight, it decomposes as shown in Equation (16.46) to produce nitrogen dioxide that dissolves to give the solution a yellowish cast and often collects as a reddish-brown gas in the air space above the solution.

16.4 REACTIONS AND COMPOUNDS OF PRACTICAL IMPORTANCE

Nitrogen Fixation

Nitrogen fixation, the conversion of molecular nitrogen into nitrogen compounds of use to agriculture, is particularly difficult because of the strong triple bond in N_2. Essentially, there are three agents that bring about nitrogen fixation: (1) lightning, (2) bacteria, (3) and human beings.

Lightning provides enough electrical energy to bring about the reaction [Equation (16.26)] between nitrogen and oxygen gases in the atmosphere. The resulting nitric oxide is partially oxidized to nitrogen dioxide [Equation (16.29)], and both oxides are washed out of the atmosphere into the soil by precipitation. Lightning, however, is an unreliable and small source of fixed nitrogen.

Certain leguminous (pod-bearing) plants have a symbiotic relationship with nitrogen-fixing bacteria living on their roots. Such crops (alfalfa, clover, soybeans, other beans, peas, and peanuts in decreasing order of fixing ability) provide valuable nitrogen compounds to the soil and are often rotated with other plants and even plowed under for just that purpose. The biochemical mechanism by which these plants accomplish nitrogen fixation has been under investigation for years. It may involve various iron- and molybdenum-sulfur clusters but remains poorly understood.

Lightning and bacteria fall far short of supplying enough fixed nitrogen for modern agricultural needs. This was recognized as early as 1898 by William Crookes, who argued that humankind must come up with artificial means of providing fixed nitrogen. Crookes himself devised a method of blowing air through an electric arc where the action of lightning was duplicated. A second artificial method, summarized in Equations (16.47) and (16.48), involves the absorption of nitrogen by calcium carbide to produce calcium cyanamid that in turn is hydrolyzed to ammonia. Unfortunately, neither one of these methods is particularly feasible on the large scale needed.

$$CaC_2(s) + N_2(g) \xrightarrow{\text{heat}} CaCN_2(s) + C(s) \quad (16.47)$$

$$CaCN_2(s) + 3H_2O(l) \longrightarrow CaCO_3(s) + 2NH_3(g) \quad (16.48)$$

The Haber process is currently the method of choice for the production of ammonia gas that is in turn converted to a variety of nitrogen compounds. Using nitrogen from the atmosphere and hydrogen from syngas (see Chapter 10, p. 245, and Figure 10.5), this process, represented in Equation (16.49), converts nitrogen and hydrogen gases directly to ammonia:

$$N_2(g) + 3H_2(g) \xrightarrow[\text{Fe/FeO}]{\substack{500^\circ\text{C} \\ 500\text{–}1000 \text{ atm}}} 2NH_3(g) \qquad \Delta H^\circ = 92.6 \text{ kJ/mol} \quad (16.49)$$

(See Problem 16.56 for an opportunity to analyze the optimal experimental conditions for the Haber process.) The ammonia, as the liquid or an aqueous solution, can be directly used as a fertilizer or it can be converted to a solid ammonium salt such as $(NH_4)_2SO_4$. Using the Ostwald process, represented in Equation (16.41), ammonia can also be converted to various nitrates.

The timing of the development of the Haber process was of critical importance to the course of World War I. Fritz Haber, a German chemist, had perfected the chemistry of the process by 1908. Karl Bosch improved the technology involved and by 1914 was directing the construction of a huge German plant. During the war, the British navy cut off the German imports of Chilean saltpeter (sodium nitrate) necessary for munitions manufacture. Had it not been for the Haber and Ostwald processes, it is estimated that the German war machine would have run out of ammunition by 1916! As it was, the Kaiser fought on for several more years. Haber also worked on chemical warfare agents (chlorine and mustard gas) used by the Germans during the war. Ironically, being Jewish, he was forced to leave Germany prior to World War II and died on his way to Palestine.

Nitrates and Nitrites

Nitrates are generally soluble salts and are only rarely found in nature. They do occur in arid regions such as deserts and in caves but most are produced by the action of nitric acid on various bases. Nitrates, being strong oxidizing agents, have been prized as explosives starting with gunpowder developed during the Middle Ages. Gunpowder is a combination of potassium nitrate, carbon, and sulfur. Although difficult to detonate, the oxidation of carbon to the monoxide and dioxide and the sulfur to sulfates, combined with the reduction of nitrate to dinitrogen, as shown rather simplistically in Equation (16.50), is exceedingly exothermic because of the formation of the very strong C=O, C≡O, and N≡N bonds:

$$14KNO_3 + 18C + 2S \longrightarrow 5K_2CO_3 + K_2SO_4 + K_2S + 10CO_2(g) + 3CO(g) + 7N_2(g) \quad (16.50)$$

Nitric acid is used to form the nitrated organic compounds TNT (trinitrotoluene) and nitroglycerine (glycerol trinitrate) which also yield very strongly bonded products upon detonation. Somewhat ironically, Alfred Nobel, father of the Nobel Prizes (including, of course, the peace prize), made his fortune by developing dynamite, a relatively inert mixture of nitroglycerine and a packing material made of diatomaceous earth. [Nobel left his entire fortune for the establishment of five prizes in chemistry, physics, physiology (medicine), literature, and peace. The economics prize was added later by the Nobel Foundation.]

Nitroglycerin, dynamite, and TNT have now been replaced by simple mixtures of ammonium nitrate and fuel oil. While ammonium nitrate and most modern explosives are inert unless set off by powerful detonators, there are exceptions. In 1947, the *SS Grandcamp*, a ship being loaded with NH_4NO_3 fertilizer in Texas City, Texas, caught fire and exploded, destroying a Monsanto chemical plant and numerous oil storage tanks and killing nearly 600 people. Detonators are also commonly nitrogen compounds with the most widely used being lead azide, $Pb(N_3)_2$, readily exploded by an electrical current or mechanical shock. The azide ion, N_3^-, is isoelectronic and isostructural with carbon dioxide.

Both nitrates and nitrites are used as meat preservatives. Nitrates retard meat spoilage, produce a characteristic cured-meat flavor, and slow the growth of the microorganisms which cause botulism. Nitrites also slow bacterial growth and, in addition, decompose to nitric oxide that complexes (see Chapter 6) to the blood hemoglobin, imparting an appealing red color to the product. The amount of nitrates and nitrites allowed in the curing process has been limited because there is evidence that in cooking and in the stomach, these ions may react to produce carcinogenic nitrosoamines, $R_2NN{=}O$.

Matches and Phossy Jaw

As shown in Figure 16.4*a*, white phosphorus, P_4, is a nonpolar, tetrahedral molecule with lone pairs sticking out from each phosphorus atom. The severely strained 60° P–P–P angles are thought to be responsible for its high reactivity. Normal body temperatures are high enough to ignite it, and the resulting burns are extremely painful and slow to heal. Due to its nonpolar nature, phosphorus can be safely stored under water without reacting or dissolving. It is soluble in alcohols and carbon disulfide.

When white phosphorus is heated to about 250° C in the absence of air, one or more of the P—P bonds of P_4 are broken, leading to the polymeric, less strained structure of red phosphorus shown in Figure 16.10*a*. Red phosphorus is less reactive, higher-melting, and less soluble in nonpolar solvents. Black phosphorus is rather graphitelike in appearance and is characterized by the puckered sheets shown in Figure 16.10*b*.

The spontaneous combustion of white phosphorus in air makes it an object of great fascination and curiosity. (Can you imagine Humphry Davy's response if he had discovered this element?) Almost immediately it was recognized that phosphorus could be formulated into some type of device to replace the tinderbox still used to start fires in those days. Various matches were devised in which paper or a wood splint was coated with phosphorus and kept in an evacuated glass tube. When the tube was broken, the phosphorus would burst into flame. Other formulations contained ingredients like gum and starch that protected the phosphorus from the air until it was scratched against a rough surface. In time, potassium chlorate was added for its property as an oxidizing agent. Sulfur was added because it could sustain the flame and transfer it to a wooden splint. Nevertheless, all of these early efforts were cumbersome and unreliable as well as dangerous to make and store.

It has always been dangerous to be a matchmaker (in more ways than one!), but in those days it was particularly so. Phosphorus workers were often afflicted with a fatal disease called "phossy jaw." It seems that when phosphorus fumes are inhaled, they can be absorbed through cavities in the teeth and attack and destroy bones, particularly in the jaw. Death was painful and nearly inevitable. Even to this

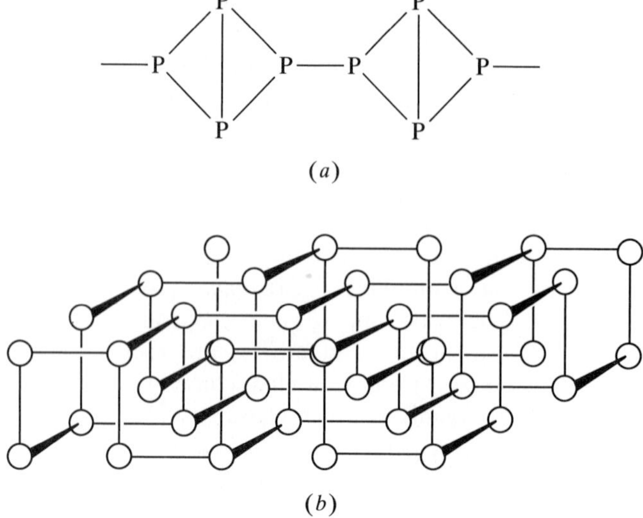

FIGURE 16.10
Representative structures of (*a*) red and (*b*) black phosphorus. Compare these structures with that of white phosphorus, Figure 16.4*a*.

day, phosphorus plant workers are monitored very carefully for the condition of their teeth. Storage also posed problems. Many fires resulted when rats gnawed at matchheads. Babies died when they chewed on match tips. People collected the phosphorus and used it to commit murder and suicide.

One of the answers to these problems was to use the safer and less toxic red phosphorus in place of the white allotrope. The red phosphorus could also be oxidized by potassium chlorate to produce a flame which, being sustained by the burning of a small amount of sulfur, was transferred to the splint. Eventually the phosphorus and sulfur were united in the form of tetraphosphorus trisulfide, P_4S_3,

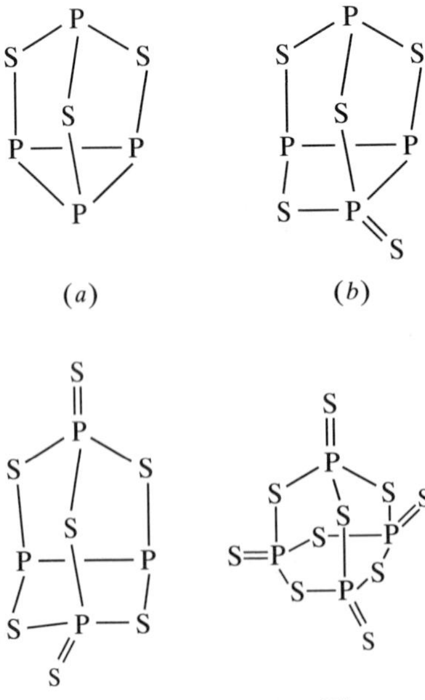

FIGURE 16.11
Some representative phosphorus sulfides: (*a*) P_4S_3, (*b*) P_4S_5, (*c*) P_4S_7, and (*d*) P_4S_{10}. [(*a*), (*b*), (*c*) adapted from Ref. 27; (*d*) Ref. 4.]

which, together with $KClO_3$ would produce a controlled, sustained flame when struck against paper containing powdered glass. (P_4S_3 is only one of many covalent phosphorus sulfides. Figure 16.11 shows the structures of P_4S_3 and several other representative compounds.)

Phosphorus has been used for a variety of incendiaries. One of the most famous is the Molotov cocktail, a combination of phosphorus and gasoline in a bottle. This concoction was first used by the British government to prepare millions of its citizens in the event England was invaded by ground troops in World War II. The cocktails were stored in beer or milk bottles and quite often submerged in a nearby stream. When the bottle broke upon impact, the phosphorus ignited the gasoline to produce an effective and cheap explosive.

Phosphates

Phosphate fertilizers have been known for about 150 years. Phosphate rock contains minerals such as fluorapatite, $Ca_5(PO_4)_3F$, which are generally too insoluble to be of much use for plant uptake. To increase the solubility, phosphate rock can be treated with sulfuric acid as shown in Equation (16.51). The resulting mixture, often called *superphosphate*, was the most common fertilizer of the 1940s. If phosphoric acid is used instead of sulfuric, as shown in Equation (16.52), the inert calcium sulfate of superphosphate is eliminated. This product is called *triple superphosphate*. Both of these products have gradually been replaced by ammonium phosphates or mixtures of potassium and ammonium phosphates, which supply both nitrogen and phosphorus to the soil:

$$2Ca_5(PO_4)_3F + 7H_2SO_4 + 3H_2O \longrightarrow 7CaSO_4 + 3Ca(H_2PO_4)_2 \cdot H_2O + 2HF \quad (16.51)$$

$$Ca_5(PO_4)_3F + 7H_3PO_4 + 5H_2O \longrightarrow 5Ca(H_2PO_4)_2 \cdot H_2O + HF \quad (16.52)$$

Phosphates have been used in a variety of ways for food processing, one of the most common being the leavening of bread products. Baking powder, invented in the middle of the last century by E. N. Horsford, a Harvard chemistry professor, was originally a mixture of calcium dihydrogen phosphate, sodium bicarbonate (baking soda), and starch (to keep the two active ingredients separated until water is added). While the bread, cake, pancake, or biscuit dough is put together and later during baking, the two ingredients react as represented in Equation (16.53) to produce carbon dioxide gas that diffuses and raises the dough into a light and expanded state. Under certain circumstances, the escaping gas can be observed. For instance, the tiny holes that appear when pancakes are first put on the griddle is just carbon dioxide escaping from the batter.

$$Ca(H_2PO_4)_2 + 2NaHCO_3 \longrightarrow CaHPO_4 + Na_2HPO_4 + 2CO_2(g) + 2H_2O(g) \quad (16.53)$$

To minimize the loss of carbon dioxide during the mixing process, "combination" baking powders containing a slow-acting acid such as sodium aluminum sulfate, $NaAl(SO_4)_2$, have partially replaced the dihydrogen phosphate in the original "straight" product. Cornstarch is also added to separate the active ingredients during storage. Similar formulations are used in self-rising flours and even effervescent tablets.

While Priestley was the father of the soft drink industry because he added carbon dioxide to water to produce a carbonated "soda water," these beverages are also refreshing because of their tart taste. This tartness comes from added acids such as citric in orange and grapefruit drinks, tartaric in grape drinks, malic in apple drinks, and phosphoric in colas and root beers. Phosphoric acid content varies between 0.01 and 0.10%; the more acid, the greater the sourness, or tartness, of the product.

In the 1930s, calcium hydrogen phosphate dihydrate, $CaHPO_4 \cdot 2H_2O$, replaced chalk as the principal polishing agent in toothpastes. It is less abrasive than chalk and imparts a much advertised shine to teeth. Unfortunately, it also has a tendency to dehydrate to the gritty, even rock-solid anhydrous $CaHPO_4$, not a good property in a toothpaste. A great deal of research time was expended on methods of preventing this dehydration, resulting in formulations that work but still are not particularly well understood. When fluoride toothpastes (see Section 18.5) were introduced, another problem arose. The fluoride, supplied as the sodium salt, was precipitated out by the polishing agent and therefore was unavailable to the teeth. For a while, other less effective polishing agents were used until, in the late 1960s, sodium monofluorophosphate (MFP), Na_2PO_3F, was found to be an excellent fluorinating agent and compatible with the $CaHPO_4 \cdot 2H_2O$.

In Chapter 6 (pp. 134–135) we detailed the use of tripolyphosphate as a detergent builder. Solutions of phosphoric acid and metal phosphates are also used to clean the surfaces of metals to prevent corrosion. First applied to corset stays more than a century ago, *phosphatization* is now a routine procedure before painting or enameling automobiles and appliances. The finishing touches on such products are also provided by a phosphate product. Dipping aluminum and its alloys in a phosphoric-nitric acid solution gives it a bright chromelike finish. These "bright-dipped," lightweight metals have now replaced chrome in automobile trim and appliance handles.

16.5 SELECTED TOPIC IN DEPTH: PHOTOCHEMICAL SMOG

There are two principal types of smog: London, or classical, and photochemical. The first observed was *London smog*, a poisonous mixture of smoke, fog (*sm*oke + f*og* = smog), air, and other chemicals. It was responsible for an air pollution disaster over London in 1911 that reportedly killed 1150 people. In December 1952, a 3-day smog tragically killed another 4000 people but also triggered the first scientific efforts to understand and control air pollution. London smog has been common in areas where large amounts of coal are burned. Most coal contains appreciable amounts of sulfur and when burned, sulfur dioxide results. The surfaces of the smoke particulates serve to catalyze the oxidation of sulfur dioxide to the trioxide which is then hydrolyzed to sulfuric acid. This sulfuric acid (in aerosol form), combined with various suspended solids like soot, causes a great deal of respiratory stress particularly among the elderly and those afflicted with respiratory disorders. Prolonged exposure to these conditions, for example, when a temperature inversion keeps an air mass localized in one area for days at a time, is what leads to large numbers of fatalities. Since the 1950s, there has been significant progress in controlling London smog.

Photochemical smog, sometimes called Los Angeles smog, is a combination of unburned hydrocarbons, carbon monoxide, and nitrogen oxides from automobile exhaust which reacts under the influence of sunlight to produce a variety of oxidation products including ozone. It was first observed in the early 1940s and characterized as due to a series of photochemical reactions in the 1950s. Our discussion of photochemical smog, which we will simply call smog from here on, will be divided into three sections: primary pollutants, secondary pollutants, and control measures.

Primary pollutants, those produced directly by the polluting source, include "lower" hydrocarbons, carbon monoxide, and nitric oxide. These all are emitted by the internal combustion engine of our beloved automobiles, which to many are not only basic means of transportation but also important symbols of prestige and independence. The internal combustion engine (ICE) derives its power from the combustion of gasoline, a mixture of hydrocarbons containing anywhere from 5 to 10 catenated carbon atoms. A representative but certainly oversimplified equation for the combustion of gasoline is given in Equation (16.54) for the complete oxidation of octane:

$$C_8H_{18}(l) + \tfrac{25}{2}O_2(g) \longrightarrow 8CO_2(g) + 9H_2O(g) \qquad (16.54)$$

Such reactions are exothermic and serve to systematically expand the pistons within the cylinders of the engine, thus converting chemical energy into mechanical. The problem is that the ICE does not have a history of being particularly efficient. Instead of completely oxidizing the components of gasoline, it produces significant amounts of carbon monoxide and, perhaps more importantly, leaves some hydrocarbons only partially burned. These "lower" hydrocarbons of fewer than five carbon atoms, for example, methane (CH_4) and ethane (C_2H_6), are emitted as exhaust products.

The third primary pollutant is nitric oxide which, as discussed in Section 16.3, is produced by the reaction of molecular nitrogen and oxygen. [See Equation (16.26) in Table 16.5.] While at ordinary air temperatures this reaction has a very low equilibrium constant, the value is appreciably higher at the high temperatures of the ICE. In summary, then, the primary pollutants from automobile exhaust are carbon monoxide, lower hydrocarbons, and nitric oxide. These are dangerous, toxic materials in themselves but are just the beginning of the story. The action of sunlight on the primary pollutants is what really makes photochemical smog one of the most difficult problems facing humankind as we move into the twenty-first century.

Nitric oxide is readily oxidized to yellow-brown, choking nitrogen dioxide gas that is in large part responsible for the color of the air over the major cities of the world. Nitrogen dioxide, the first of the secondary pollutants (those produced by subsequent reactions involving the primary pollutants), is the "trigger" for many of the large number of reactions involved in the production of photochemical smog. As shown in Equation (16.40), sunlight serves to dissociate nitrogen dioxide into nitric oxide and the extremely reactive atomic oxygen. These oxygen atoms, with their two unpaired electrons, react with just about everything else present in the atmosphere to produce a mind-boggling number of dangerous products. Much work remains to be done to figure out the most important of the more than 200 chemical reactions that occur under various meteorological and geographic conditions. In the meantime, an insidious cycle is set up because the nitric oxide, NO, can be reoxidized to form more nitrogen dioxide, NO_2, that in turn photochemically dissociates, generating more and more atomic oxygen.

Although the details of the complicated photochemistry of smog formation will take years to completely decipher, it seems clear that one of the most important reactions is that between oxygen atoms, O, and molecular oxygen, O_2 [in the presence of a third atom or molecule—designated M in Equation (16.55)] to produce ozone:

$$O(g) + O_2(g) + M \longrightarrow O_3(g) + M \qquad (16.55)$$

Now this is the same ozone that, in the stratosphere, shields the biosphere from dangerous uv radiation. (See Section 11.5, pp. 287–289, for an extended discussion of ozone and the ozone layer. Chapter 18 discusses threats to the ozone layer from chlorine and bromine compounds.) But this ozone is in the troposphere, the layer of air we breathe and live in. It causes a number of health problems including respiratory irritation, choking, coughing, and fatigue and has also been implicated in damage to forests and crops. It also attacks rubber products and causes cracks in tires.

Ozone makes up about 90 percent of a general class of secondary pollutants called *oxidants*. Atomic oxygen and ozone can react with the lower hydrocarbons to produce a large variety of oxidants. Principal among these are aldehydes ($R{-}\overset{\overset{\displaystyle O}{\|}}{C}{-}H$), ketones ($R{-}\overset{\overset{\displaystyle O}{\|}}{C}{-}R'$), and peroxy acyl nitrates (PANs, $R{-}\underset{\underset{\displaystyle O}{\|}}{C}{-}O{-}O{-}\underset{\underset{\displaystyle O}{|}}{N}{-}O$), where R and R′ stand for a variety of carbon-containing radicals such as methyl (CH_3–), ethyl (C_2H_5–), and so forth. These compounds are mainly eye and lung irritants but may be linked with increased incidents of cancer and heart disease. PANs also contribute to respiratory distress and are also known to cause severe damage to plants and trees. Forests in the Los Angeles area, for example, have been severely damaged.

Figure 16.12 shows the variations in the amounts of the primary and secondary pollutants over the course of a day. As the morning traffic starts, nitric oxide is the first of the primary pollutants to appear. Carbon monoxide, not shown since it does not react photochemically, is also formed in these first hours. The lower hydrocarbons register by midmorning as does the first of the secondary pollutants, nitrogen dioxide. As the day wears on, ozone and other oxidants start to form and peak in the early afternoon. The afterwork traffic often produces another

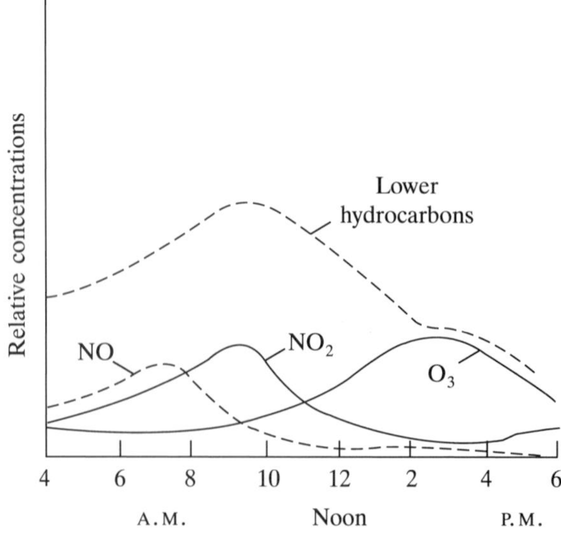

FIGURE 16.12
The daily variation of primary (dashed lines) and secondary (solid lines) pollutants in photochemical smog. [Ref. 21.]

peak of the primary pollutants and perhaps nitrogen dioxide, but with the sun low in the sky, very few additional secondary pollutants are produced.

Measures to control the incidence of smog have been difficult to devise and more difficult to put into practice. The Clean Air Act of 1967 and the subsequent amendments of 1970 and 1977 set limits for exhaust emissions. Modifications to the ICE and its exhaust system have followed. Primary among these is the catalytic converter used to fully oxidize hydrocarbons to carbon dioxide and water. In the past, these have used expensive platinium and palladium compounds, but recent improvements may make this catalytic system obsolete. The palladium-platinum catalytic converter necessitated the removal of lead antiknock compounds from gasolines. (See Chapter 15, p. 392, and pp. 394–395, for a discussion of lead pollution.)

The control of nitrogen oxides in automobile exhaust seems to be a more difficult problem, but various catalytic systems to dissociate NO into molecular nitrogen and oxygen have been devised. Intense basic and applied research on the details of nitrogen chemistry in this setting must continue if the problem of photochemical smog is to be satisfactorily solved.

SUMMARY

While not known as early as carbon, tin, and lead, the Group 5A elements were all discovered before the founding of the United States. Antimony was known to the ancients and was a protected secret of the alchemists. Similarly, arsenic is mentioned in the mystical literature of alchemy, but its discovery is often attributed to Albertus Magnus because of his definitive descriptions of the element. Phosphorus was isolated from human urine for a century before it was discovered in bones and in phosphate rock. Bismuth was probably known well before Geoffrey described it so thoroughly, but he is usually listed as its discoverer. Nitrogen was discovered by the phlogistonist D. Rutherford.

With Group 5A we have started back toward more uniform group properties. Nitrogen, the first gas encountered in the group tour, does not catenate nearly as well as carbon but is unique in the group for its (1) large array of oxidation states, (2) high ability to form $p\pi$ bonds, and (3) inability to use d orbitals to expand its octet. By contrast, phosphorus does expand its octet by using d orbitals. The phosphazenes combine the $p\pi$ ability of nitrogen with the $d\pi$ ability of phosphorus. The inert-pair effect is evident in the increasing stability of the $+3$ oxidation state down the group. The metal-nonmetal line has moved to a still lower position in this group in which bismuth is the only metal.

A survey of the hydrides, oxides, hydroxides, and halides highlights the network components. The hydrides of nitrogen and phosphorus emphasize the uniqueness of the lightest element. Unlike the polar ammonia, the nonpolar phosphine is a poor base and not capable of forming hydrogen bonds. Arsine is less stable than phosphine, and its decomposition is the basis of the criminological Marsh test for the presence of arsenic.

The nitrogen oxides are characterized by $p\pi$-$p\pi$ bonds, but those of phosphorus have strong P—O single bonds and $d\pi$-$p\pi$ P=O double bonds. Diphosphorus pentoxide is the anhydride of phosphoric acid, a thick, syrupy triprotic acid capable of forming the condensation polymers meta- and polyphosphoric acids and their corresponding phosphates. Diphosphorus trioxide is the anhydride of the diprotic phosphorous acid and the corresponding phosphites. Arsenic has similar oxides and corresponding acids, but antimony and bismuth move toward more basic properties. Bismuth hydroxide is the only true base in the group. The halides again display the uniqueness of nitrogen and the increasing stability of the +3 oxidation state down the group.

The survey of the nitrogen oxidation states starts with the nitrides and ammonia (−3). The binary nitrides are much like hydrides and oxides in that they can be ionic, covalent, or interstitial. Ammonia is the familiar pyramidal, hydrogen-bonding base; ammonium salts are much like those of potassium and rubidium except that the ammonium ion is a weak acid. Hydrazine (−2) is closely related to ammonia and is such an excellent reducing agent that it is used in rocket fuels. Hydroxylamine contains nitrogen in the −1 oxidation state. Nitrous oxide (+1), laughing gas, was the first anesthetic. Nitric oxide (+2), is an odd-electron species. Dinitrogen trioxide is the anhydride of nitrous acid (+3), a weak acid that produces the nitrites upon reaction with bases. Nitrogen dioxide (+4), another odd-electron species, is a reddish-brown gas of critical environmental importance. Dinitrogen pentoxide is the anhydride of nitric acid (+5), one of the most important commercial chemicals in the world today.

Reactions and compounds of practical importance include three agents (lightning, bacteria, and humankind) for nitrogen fixation; nitrates as fertilizers, explosives, and food additives; nitrites as food additives; phosphorus allotropes and sulfides in matches; and phosphates in fertilizers, baking products, soft drinks, and toothpastes. The Haber process produces ammonia from nitrogen and hydrogen gases. It altered world history and continues to be of critical importance in food production. Nitrates have been important in explosives such as gunpowder, nitroglycerine, TNT, and in modern devices because of the strongly bonded products they form upon detonation.

The two principal types of smog are London and photochemical. London smog contains smoke, fog, and sulfur oxides but is yielding to current control measures. Photochemical smog starts with the primary pollutants (lower hydrocarbons, carbon monoxide, and nitric oxide) emitted by automobiles. The action of the sun produces the secondary pollutants (nitrogen dioxide, ozone, and other oxidants). The daily rise and fall of the concentrations of these materials can be monitored. Control measures have concentrated on eliminating carbon monoxide and lower hydrocarbons and redissociating nitric oxide to molecular nitrogen and oxygen. Basic and applied research must continue in order to gain further insight into the production and control of photochemical smog.

PROBLEMS

16.1. Based on your reading of this chapter, cite three different examples of elements or compounds which earn the members of Group 5A the name pnicogens.

16.2. Look up the term *iatrochemistry*. What connection would you imagine existed between alchemists and iatrochemists?

16.3. A typical recipe for the production of metallic antimony would be to roast stibnite, Sb_2S_3, to the oxide and then reduce it to the metal. Write equations for this recipe.

16.4. Write equations for the reactions of orpiment, As_2S_3, with eggshells ($CaCO_3$) followed by charcoal.

16.5. Write an equation for the reaction of Brandt's white, waxy "cold fire" with Priestley's dephlogisticated air.

16.6. Daniel Rutherford had some difficulty in completely phlogisticating his new air. He let a mouse live in common air until it died. Then he burned a candle and phosphorus in this air and finally treated the product with a strong alkali. In terms of modern chemistry, describe the reactions he was carrying out.

16.7. Identify and discuss any irregularities in radii, ionization energies, electron affinities, and electronegativities of the pnicogens.

16.8. Draw VSEPRT structures of hydrazine and ethane and discuss the reason that the N—N bond is considerably weaker than the C—C bond.

16.9. Describe the N_2 molecule using the Lewis and valence-bond theories. Draw well-labeled diagrams where appropriate.

16.10. $P(SiH_3)_3$ is pyramidal, while $N(SiH_3)_3$ is planar. Discuss these structures relative to the favorability of $p\pi$-$d\pi$ bonding.

16.11. The phosphine oxides, R_3PO, have the P—O bond distance shorter than that for a single bond. Draw Lewis, VSEPRT, and VBT structures for these compounds. Discuss the unusually short P—O distance in terms of these structures.

16.12. Isocyanates, OCN^-, do not have any phosphorus analogs. Draw Lewis, VSEPRT, and VBT structures for this ion and discuss them in terms of the uniqueness principle.

16.13. Cyclodiphosphazenes, $R_4P_2N_2$, were first synthesized in the mid-1980s. These are planar molecules with nearly equal P–N distances. Draw Lewis, VSEPRT, and VBT diagrams for this class of molecules. What nitrogen and phosphorus orbitals would be used in the bonding?

16.14. Draw Lewis, VSEPRT, and VBT diagrams of the phosphonium ion, PH_4^+.

16.15. Consider ammonia and phosphine acting as acids to produce the corresponding EH_2^- conjugate bases. Which would be the stronger acid, NH_3 or PH_3? Briefly discuss the rationale for your answer.

16.16. As mentioned in Chapter 10, deuterium oxide, D_2O, is prepared by the electrolysis of water. How would you prepare the compound D_3PO_4?

16.17. Which would be the stronger acid, phosphoric or arsenic acid? Include a structure of these acids as part of your answer.

16.18. Which would be the stronger acid, phosphoric or phosphorous acid? Briefly discuss your answer.

16.19. When the concentration of phosphoric acid solutions falls below 50%, hydrogen bonding among the acid molecules is replaced by that among acid and water molecules. Draw a diagram showing both types of interactions.

16.20. Write the formula for:
(*a*) Sodium hydrogen phosphate
(*b*) Ammonium dihydrogen phosphite
(*c*) Potassium dihydrogen pyrophosphate
(*d*) Calcium hydrogen phosphite.

16.21. Name the following acids and salts:
(*a*) KH_2PO_4
(*b*) $Ca(H_2PO_4)_2$
(*c*) $Mg_2P_2O_7$
(*d*) $NaH_3P_2O_5$.

16.22. Determine the formula of the acid that results when one phosphoric and one phosphorous acid condense. How many acidic protons would this molecule have?

16.23. How many acidic protons are there in tripolyphosphoric acid and tripolyphosphorous acids? Draw structures of these acids that verify your answer.

16.24. Hypophosphorus acid, H_3PO_2, is a weak, monoprotic acid. Draw a structure for this molecule and discuss why only one hydrogen is acidic.

16.25. Determine the formula of sodium hypophosphite.

16.26. Draw structures of arsenious and arsenic acids.

16.27. Name the following compounds:
(*a*) NaH_2AsO_4
(*b*) Ag_3AsO_3
(*c*) $(NH_4)_3AsO_4$.

16.28. Write formulas for the following compounds:
(*a*) Antimonious acid
(*b*) Copper(II) hydrogen arsenite
(*c*) Ammonium dihydrogen arsenate

16.29. Solid PCl_5 is actually made up of $[PCl_4^+][PCl_6^-]$ units. Describe the expected shape of both the cation and the anion in the solid.

16.30. Write an equation for the preparation of nitrogen trifluoride from ammonia.

16.31. How would you prepare the trichloride and trifluoride of phosphorus? Write equations as part of your answer.

16.32. How would you prepare the trihalides of arsenic and bismuth? Write equations as part of your answer.

16.33. With reference to borazine, $B_3N_3H_6$, shown in Figure 16.7, identify the hybridization about each boron and nitrogen atom. Would you expect to find a similar bonding scheme in a corresponding boron-phosphorus compound? Why or why not?

16.34. Discuss the similarities and the differences between the trimeric phosphazene, $P_3N_3Cl_6$, and hexachloroborazine, $B_3N_3Cl_6$.

16.35. Draw a diagram showing the structure of liquid ammonia. Be sure that the geometry of the individual ammonia molecules is correct and that hydrogen bonds are labeled clearly.

16.36. Draw a diagram showing the structure of a solution of ammonia in water. Be sure that the geometries of the water, ammonia, and any ions are correct and that any hydrogen bonds are labeled clearly.

16.37. Assign oxidations states to the atoms in Equation (16.20), reproduced below. Which, if any, have been oxidized or reduced?

$$(NH_4)_2Cr_2O_7(s) \longrightarrow N_2(g) + 4H_2O(g) + Cr_2O_3(s)$$

16.38. Hydrazine is useful as a reducing agent. In acid solution, the hydrazinium ion is oxidized to $N_2(g)$ as shown below.

$$N_2H_5^+(aq) \longrightarrow N_2(g) + 5H^+(aq) + 4e^- \qquad E° = +0.23\ V$$

(*a*) Could acidic aqueous solutions of hydrazine be used to reduce Fe^{3+} to Fe^{2+}?
(*b*) MnO_4^- to Mn^{2+}?
Standard reduction potentials are given in Table 12.2.

16.39. The standard reduction potential for hydrazine in basic solution is given below. Could hydrazine be used to reduce MnO_4^- to MnO_2 in basic solution?

$$N_2(g) + 4H_2O + 4e^- \longrightarrow N_2H_4(aq) + 4OH^-(aq) \qquad E° = -1.16\ V$$

16.40. Discuss the likelihood of hydrogen bonding in solid hydroxylamine, NH_2OH. Include a diagram in your discussion.

16.41. Calculate the percentage by weight of oxygen in nitrous oxide. Compare your value with that in air. Do your calculations support the observation that candles glow brighter in nitrous oxide?

16.42. Balance the reaction shown below in which nitrous oxide disproportionates to dinitrogen and nitric oxide in acid solution. Use the standard reduction potentials given to determine if this disproportionation reaction would actually occur under standard state conditions.

$$N_2O \xrightarrow{\text{acid}} N_2 + NO$$

$$2e^- + N_2O + 2H^+ \longrightarrow N_2 + H_2O \qquad E^\circ = +1.77\ \text{V}$$

$$2e^- + 2NO + 2H^+ \longrightarrow N_2O + H_2O \qquad E^\circ = +1.59\ \text{V}$$

16.43. Hyponitrous acid is a symmetrical molecule of formula $N_2(OH)_2$. Draw Lewis, VSEPRT, and VBT diagrams of this molecule. Include estimates of the bond angles in the molecule and the hybridization of the nitrogen and oxygen atoms. Also determine the oxidation state of the nitrogen atoms in this molecule.

16.44. The ΔS° at 298 K for the reaction of nitrogen and oxygen diatomic gases to make nitric oxide is 24.7 J/(mol·K). Does this favor or disfavor the increased production of nitric oxide at high temperatures? Briefly discuss your answer.

16.45. The ΔH° of the dimerization of NO to N_2O_2 is -57.2 kJ/mol. Briefly rationalize the exothermic nature of the reaction by relating it to bond energies.

16.46. Draw Lewis, VSEPRT, and VBT diagrams of N_2O_2. Include estimates of bond angles in the molecule and the hybridization of the nitrogen and oxygen atoms.

16.47. The asymmetrical structure of dinitrogen trioxide given in Table 16.6 can be altered to a symmetrical O—N—O—N—O structure by exposing the ordinary form to light of wavelength 720 nm. Write Lewis, VSEPRT, and VBT structures for the symmetrical structure. Include estimates of bond angles in your VSEPRT diagram.

16.48. Comment on the expected O—N—O bond angles in NO_2^+ (the nitronium ion), NO_2 (nitrogen dioxide), and NO_2^- (nitrite). Justify your answer with VSEPRT diagrams.

16.49. Balance the reaction shown below in which nitrogen dioxide disproportionates to nitrous acid and nitrate ions in acid solution. Use the standard reduction potentials given to determine if this disproportionation reaction would actually occur under standard state conditions.

$$NO_2 + H_2O \longrightarrow HNO_2 + NO_3^- + H^+$$

$$NO_2 + H^+ + e^- \longrightarrow HNO_2 \qquad E^\circ = 1.10\ \text{V}$$

$$NO_3^- + 2H^+ + e^- \longrightarrow NO_2 + H_2O \qquad E^\circ = +0.78\ \text{V}$$

16.50. Write a balanced equation showing the action of water on dinitrogen pentoxide to form nitric acid. Is this a redox equation? Why or why not?

16.51. In addition to oxidizing and therefore dissolving a variety of metals, nitric acid can also oxidize nonmetals such as sulfur (S_8) and phosphorus (P_4) to the corresponding sulfates and phosphates. Write balanced equations representing these processes.

16.52. The network (Figure 16.2) indicates that the good oxidizing agents tend to exist on the right-hand side of the periodic table. Using nitrogen chemistry for examples, discuss why this is an overgeneralization.

16.53. Write balanced reactions for the action of concentrated nitric acid on aluminum metal and of dilute nitric acid on iron.

16.54. Phosphorus pentoxide can be used to dehydrate nitric acid. Write an equation to represent this reaction.

16.55. Draw Lewis, VSEPRT, and VBT diagrams of the symmetrical cyanamide ion, CN_2^{2-}.

16.56. Think carefully about the experimental conditions prescribed for the Haber process in Equation (16.49). Can these high temperatures and pressures be rationalized from a thermodynamic point of view? If not, what other factor must be considered to rationalize such conditions?

16.57. Write a reaction for the oxidation of (*a*) P_4 and (*b*) P_4S_3 by the chlorate ion, ClO_3^-.

16.58. Speculate on the nature of the P═S bonds found in many of the phosphorus sulfides.

16.59. In baking powders and various biscuit mixes, the leavening agents include monocalcium phosphate, $Ca(H_2PO_4)_2$. Rationalize this name. What alternate name could you suggest?

16.60. Before the invention of baking powder, a combination of cream of tartar ($KHC_4H_4O_6$) and baking soda was used to leaven bread. Write an equation for this process.

16.61. Write an equation for the elimination of soluble fluoride when sodium fluoride is mixed with calcium hydrogen phosphate dihydrate.

16.62. Speculate on the structure of the monofluorophosphate ion, PO_3F^{2-}. With what phosphate ion is it isoelectronic?

16.63. Briefly rationalize in your own words the daily variations of concentrations of primary and secondary pollutants in photochemical smog as shown in Figure 16.12.

CHAPTER 17

SULFUR, SELENIUM, TELLURIUM, AND POLONIUM

The Group 6A elements are known collectively as the *chalcogens*, a name meaning "copper-producing." The name certainly is appropriate for sulfur as copper sulfide is a primary copper ore. Selenium and tellurium too are found in conjunction with the coinage metals (copper, silver, and gold) and are consistent with the group name. Polonium and oxygen, on the other hand, are not categorized as well by such a designation. Polonium is another of those highly radioactive, heavy congeners that reminds us of francium and radium in Groups 1A and 2A. Oxygen is one of the best examples of a unique element at the top of a group. In fact, it is so special that we chose to cover it in Chapter 11 as a prelude to the group tour. The chemistry of oxygen will be mentioned here only as a point of departure for the discussion of its heavier congeners.

The first two sections of this chapter are the usual ones on (1) the discovery and isolation of the elements and (2) the application of the network to group chemistry. The third section, necessitated by the great ability of sulfur to catenate, concentrates on the allotropes and compounds that involve element-to-element bonds. Next is a short section on the relatively new and potentially useful sulfur nitrides. The reactions and compounds of practical importance in the fourth section include sodium-sulfur batteries, the photoelectric properties of selenium and tellurium, and the most important commercial chemical in the world, sulfuric acid. The selected topic in depth is the production, effects, and possible control of acid rain.

17.1 DISCOVERY AND ISOLATION OF THE ELEMENTS

Figure 17.1 shows the discovery dates of all the chalcogens including oxygen. Only sulfur was known to the ancients. Oxygen was discovered by Priestley (and

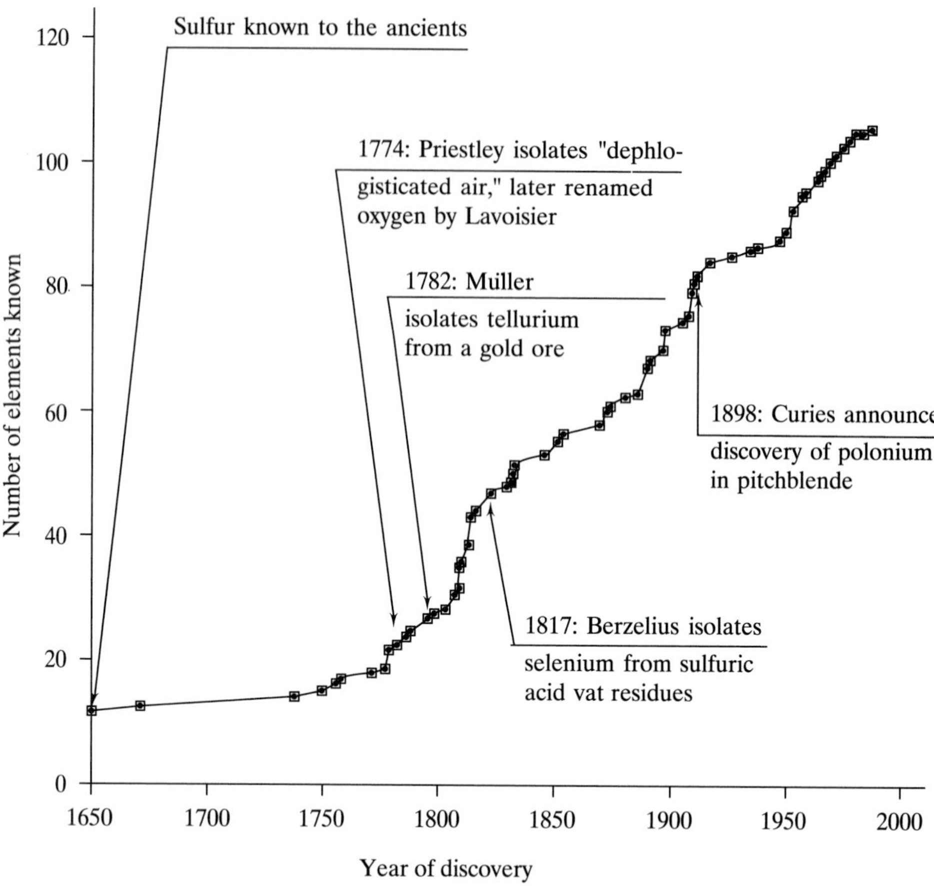

FIGURE 17.1
The discovery dates for the Group 6A elements superimposed on the plot of the number of known elements versus time.

independently by Scheele) in the early 1770s, tellurium in the early 1780s, selenium in 1817, and polonium at the very end of the nineteenth century.

Sulfur

Genesis 19:24 describes the destruction of Sodom and Gomorrah by brimstone, which literally means "burning stone." The term survives in "fire and brimstone" speeches which warn us that unacceptable behavior will result in permanent residence in a very hot and fiery place! Brimstone is the old word for sulfur, and both names, at various times, have been used as general terms for combustible substances. By 1774, Lavoisier had recognized that Priestley's dephlogisticated air was a new element. Similarly, in 1777, the Frenchman claimed the same status for sulfur but was challenged by Humphry Davy who, as late as 1809, maintained that it contained both hydrogen and the new oxygen. The elementary nature of sulfur seems never to have been contested after a study published by Gay-Lussac and Thénard in 1809.

This odorless and tasteless yellow solid can be found in free form in volcanic and hot spring areas. Hydrogen sulfide, on the other hand, as anyone who has been in a high school chemistry laboratory or Yellowstone National Park can testify to, is certainly not odorless. But hydrogen sulfide is not just an inconvenient nuisance. It so significantly "sours" natural gas that this otherwise clean-burning material cannot be used as a fuel. In the late 1880s the young German emigrant Herman Frasch patented a method to recover sulfur from the vast deposits of natural gas found in Louisiana and Texas. In the early 1890s he went on to perfect a method, still called the Frasch process, for recovering sulfur from huge deposits under the swamps and quicksands of Louisiana. He devised a way to pump superheated water down some 500 ft or more to these deposits where it melted the sulfur and forced it to the surface as a bright-yellow liquid. The Frasch method is gradually being replaced by newer methods of recovering sulfur from oil and natural gas.

Sulfur has found a variety of uses over the many centuries it has been known. Homer refers to its use as a fumigant in early Greece. It was a prime component of gunpowder, first introduced in the thirteenth century (see Chapter 16, p. 432), and of matches starting in the 1800s (see Chapter 16, pp. 433–435). In the late 1830s, Charles Goodyear discovered a method of heating a mixture of sulfur and rubber to produce a dry, flexible, immensely versatile product. (Goodyear did not profit from his discovery of what became known as the vulcanization of rubber. He died a pauper. Benjamin Goodrich, in the 1870s, made his fortune producing and marketing vulcanized rubber products.) The most important and widespread use of sulfur is to make sulfuric acid, which was first described about 1300. We will discuss the production and uses of sulfuric acid in Section 17.5.

Tellurium and Selenium (Earth and Moon)

Franz Joseph Müller was a director of various gold mines in Transylvania (western Romania) in the late 1700s. "Aurum problematicum," a bluish-white ore of gold, was, as its name indicates, proving to be particularly difficult to analyze at that time. Müller took on the challenge posed by this ore and, in 1782, isolated from it a metal that, on the basis of specific gravity, he identified as antimony. Strangely, he noted that the metal smelled something like radishes!

In 1798, Martin Klaproth, who had earlier discovered and named uranium, isolated the same silvery-white metal from the same problematical ore. Klaproth, however, recognized that he and Müller had isolated a new element for which he suggested the name *tellurium*, meaning "earth." He properly acknowledge the prior work of Müller, who is consequently listed as its discoverer. Later it was found that tellurium does occasionally exist as the free element but more often than not exists as the gold telluride. Oddly, workers who process this ore and the metal derived from it acquire a garliclike odor to their breath, a condition referred to (rather honestly but certainly not flatteringly) as tellurium breath.

Selenium was discovered in 1817 by Jöns Jakob Berzelius. Recall that we have previously noted some of the other contributions of this most famous Swedish chemist. (He invented the modern symbols for the elements in 1813, was young Arfwedson's mentor when the younger man discovered lithium in 1817, attempted to divide all compounds into either organic or inorganic, and isolated silicon in 1824.) Berzelius discovered a reddish-brown precipitate in the sulfuric acid vats of the plant that he had invested in. This material too had a very strong odor of decaying radishes or cabbage which at first led him to believe that it was tellurium. Later, however, he proved that he had isolated a new element closely related to tellurium which he appropriately called *selenium*, from the Greek *selene* for "moon." It turns out that it is selenium which smells like radishes, and trace quantities of it impart that odor to tellurium.

Berzelius also discovered or codiscovered cerium and thorium and was the first to isolate zirconium. He determined the atomic weights of nearly all the elements known at the time, wrote an extremely well regarded textbook, and generally was the most influential chemist of his time. He was so famous that medallions carrying his portrait were cast from selenium and, while very rare, still exist today.

Polonium

We have already discussed (Chapter 13, pp. 326–328) the circumstances surrounding the discovery of polonium and radium by Pierre and Marie Curie. The polonium work was, in some ways, the more challenging. Polonium was much less abundant in the uranium ores available to the Curies than was radium. In addition, polonium 210 has a half-life of only 138.4 days, while that of radium 226 is 1600 years. Compounding the difficulty still further was the coprecipitation of polonium compounds with those of bismuth. This Po–Bi connection led the Curies to attempt to separate polonium based on Group 5A chemistry. Only later was polonium properly identified with Group 6A. Finally, as if all of the above was not enough, trace quantities of polonium compounds were later shown to adhere to glassware, causing the radioactivity to appear and disappear in what must have been a most frustrating fashion. Nevertheless, in 1898, the Curies felt confident enough of their work that they announced their discovery of polonium, named for Marie's homeland.

Polonium is a strong alpha emitter. A half-gram sample reaches temperatures as high as 500°C due to this radiation. Both the heat and radiation decompose polonium-containing compounds and greatly complicate their characterization. Somewhat incidentally, but still unnerving to contemplate, is the fact that this radiation causes a blue glow in the air around polonium samples. In 1934, a method of preparing polonium 210 by neutron bombardment of bismuth 209 was devised. It is represented in Equation (17.1):

$$^{209}_{83}\text{Bi} + ^{1}_{0}n \longrightarrow ^{210}_{83}\text{Bi} \xrightarrow{\beta^-} ^{210}_{84}\text{Po} + ^{0}_{-1}e \tag{17.1}$$

Twenty-seven polonium isotopes, none of them stable, are now known, but only polonium 210 has been produced in sufficient quantities (milligrams) for chemical investigations. Due to its self-heating ability, polonium is a potential lightweight source of heat for space and lunar stations.

17.2 FUNDAMENTAL PROPERTIES AND THE NETWORK

Figure 17.2 shows the Group 6A elements superimposed on the network. Before considering some representative hydrides, oxides, and halides, we should briefly check on the status of each network component as it relates to the chalcogens. Starting as usual with the periodic law, we are not surprised to see that properties such as ionization energies, electron affinities, electronegativities, and various radii (all shown with other fundamental properties in Table 17.1) vary much as we have come to expect. The metal-nonmetal line has been sinking lower in each group we have considered, so it follows that polonium, although radioactive and difficult to work with, is expected to be the only element which approaches a degree of metallic character. It also follows that the group oxides start out strongly acidic but

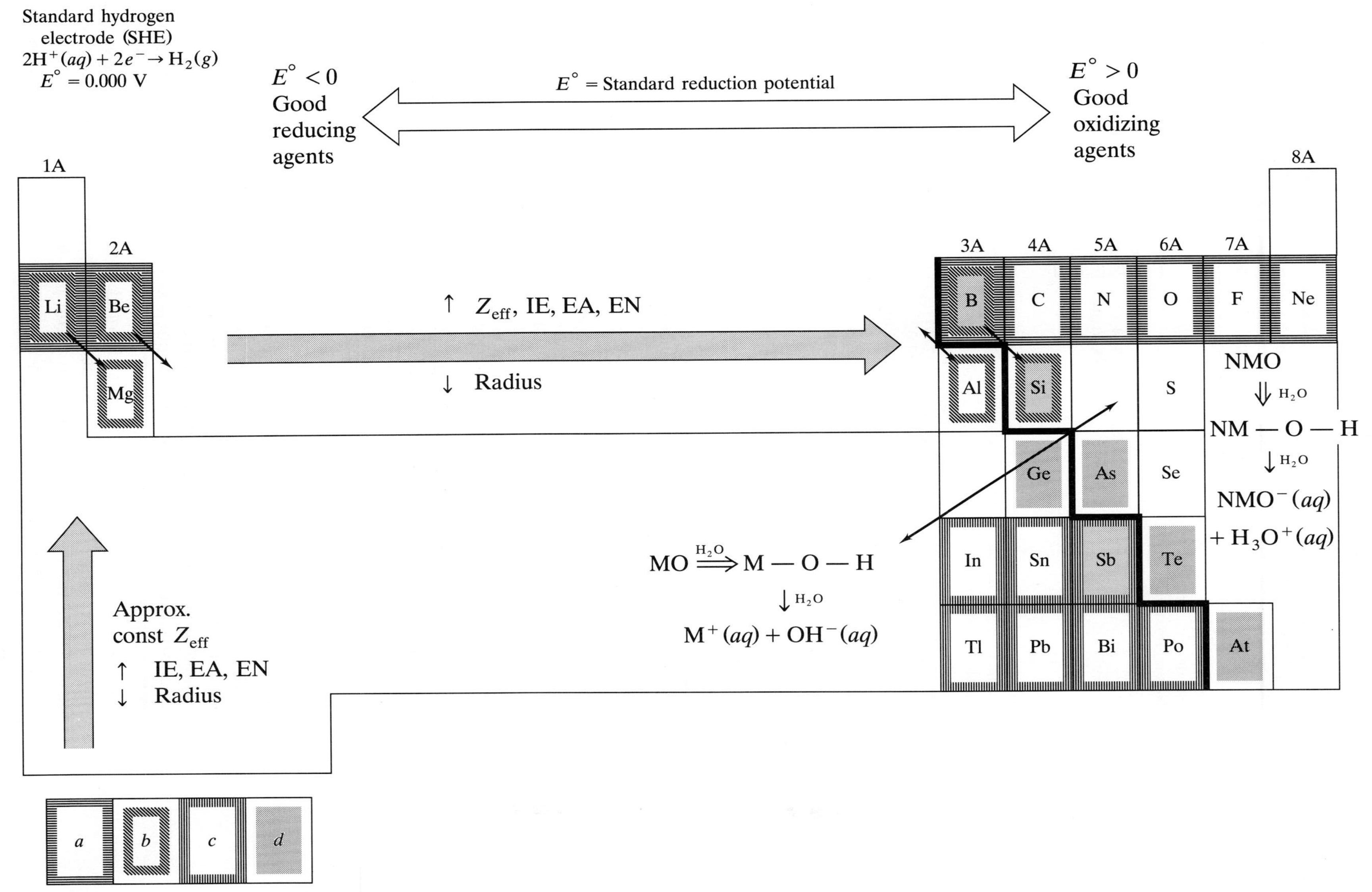

FIGURE 17.2
The chalcogens (Group 6A) elements superimposed on the network of interconnected ideas including the trends in periodic properties, the acid-base character of metal and nonmetal oxides, trends in standard reduction potentials, (*a*) the uniqueness principle, (*b*) the diagonal effect, (*c*) the inert-pair effect, and (*d*) the metal-nonmetal line.

TABLE 17.1
The fundamental properties of the Group 6A elements. [Ref. 24.]

	Oxygen	Sulfur	Selenium	Tellurium	Polonium
Symbol	O	S	Se	Te	Po
Atomic number	8	16	34	52	84
Natural isotopes, AW/% abundances	$^{16}O/99.76$ $^{17}O/0.037$ $^{18}O/0.204$	$^{32}S/95.0$ $^{33}S/0.76$ $^{34}S/4.22$ $^{36}S/0.014$	$^{74}Se/0.87$ $^{76}Se/9.02$ $^{77}Se/7.58$ $^{78}Se/23.52$ $^{80}Se/49.82$ $^{82}Se/9.19$	$^{120}Te/0.089$ $^{122}Te/2.46$ $^{123}Te/0.87$ $^{124}Te/4.61$ $^{125}Te/6.99$ $^{126}Te/18.71$ $^{128}Te/31.79$	$^{210}Po(100)$†
Total no. of Isotopes	8	10	17	21	27
Atomic weight	16.00	32.07	78.96	127.6	(210)
Valence electrons	$2s^22p^4$	$3s^23p^4$	$4s^24p^4$	$5s^25p^4$	$6s^26p^4$
mp/bp, °C	−218/−183	112/444	217/685	450/990	254/962
Density, g/cm^3	1.43 (g/liter)	2.07‡	4.79	6.24	9.32
Covalent radius, Å	0.73	1.02	1.16	1.36	
Ionic radius, Shannon-Prewitt, Å (C.N.)	...	0.43(6)	0.56(6)	(+6)0.70(6) (+4)1.11(6)	(+6)0.81(6) (+4)1.08(6)
Pauling EN	3.5	2.5	2.4	2.1	2.0
Charge density (charge/ionic radius), unit charge/Å	...	14	11	(+6)8.6 (+4)3.60	(+6)7.4 (+4)3.70
$E°$,§ V	1.23	0.14	−0.11	−0.69	−1.0
Oxidation states	−1, −2	−2 to +6	−2 to +6	−2 to +6	−2 to +6
Ionization energy, kJ/mol	1314	1000	941	870	814
Electron affinity, kJ/mol	−142	−200	−195	−190	NA
Discovered by/date	Priestley 1774	Antiquity	Berzelius 1817	Müller 1782	Curie 1898
rpw¶ O_e		SO_2	SeO_2	TeO_2	PoO_2
Acid-Base character of oxide	Acidic	Acidic	Acidic	Ampho.	
rpw N_2	NO, NO_2	None	None	None	None
rpw halogens	O_2F_2	SF_4,SF_6 S_2Cl_2 S_2Br_2	SeX_4 SeF_6 Se_2Cl_2,Se_2Br_2	TeX_4 TeF_6	PoX_4 $PoCl_2$,$PoBr_2$
rpw hydrogen	H_2O	H_2S	H_2Se	None	None
Crystal structure	Cubic	Ortho.	Hexag.	Hexag.	Cubic

†Only radioactive isotope available in milligram quantities.
‡For orthohombic.
§$\frac{1}{2}O_2 \rightarrow H_2O$; $E \rightarrow H_2E$, E = S, Se, Te, Po (in acidic solution).
¶rpw = reaction product with.

become less so going down the group. While Figure 17.2 indicates that only polonium is directly influenced by the inert-pair effect, the +6 oxidation state steadily decreases in stability down the group until, as expected, it is very rare in polonium. Given that the diagonal effect was last heard from in Group 4A chemistry, we are left only with the uniqueness principle and standard reduction potentials to consider.

Since most of Chapter 11 was devoted to the chemistry of oxygen, the uniqueness principle would seem to be very much alive and well. Oxygen's high electronegativity, small size, ability to form $p\pi$ bonds, and inability to use d orbitals all combine to make it distinctly different from its congeners. This leaves sulfur as the more representative chalcogen. (Much like phosphorus was more representative of the Group 5A elements.) As expected, sulfur (again like phosphorus) forms particularly strong $d\pi$ interactions with a variety of other elements, leading to strong double bonds and a large number of compounds with expanded octets. Perhaps a bit surprisingly, sulfur is second only to carbon in the ability to catenate. Indeed, this ability leads to a special section (17.3) in this chapter on compounds with element-to-element bonds.

The standard reduction potentials given in Table 17.1 are for the reduction of the element to the hydride as shown in general in Equation (17.2):

$$E + 2H^+(aq) + 2e^- \longrightarrow H_2E \qquad \text{where E = O, S, Se, Te, and Po} \tag{17.2}$$

The value of +1.23 V for oxygen is consistent with O_2 being an excellent oxidizing agent. (It has a strong tendency to be reduced to water and therefore readily oxidizes other substances.) For sulfur, the standard reduction potential is much smaller (0.141 V) but still a positive number. Sulfur is not regarded as a particularly good oxidizing agent. Proceeding down the group the reduction potentials become increasingly negative, consistent with a decreasing stability of the hydrides relative to the free elements. All of which leads into a more detailed discussion of the noncatenated hydrides to be followed by similar accounts of the oxides and halides. Catenated compounds are discussed in Section 17.4.

Hydrides

Hydrogen sulfide, like water, is a bent, covalent molecule but otherwise has little in common with water, its familiar cousin. It is, however, representative of the other Group 6A hydrides. The trademark smell of H_2S is not idly associated with rotten eggs. When eggs spoil, sulfur-containing proteins decompose to produce small amounts of H_2S gas. Somewhat ironically, getting a whiff of H_2S is a healthy sign. At high concentrations, this odor so overrides the sense of smell that one is no longer aware of it. This is when the concentrations of this gas, every bit as dangerous as hydrogen cyanide, become life-threatening. As offensive as the odor of H_2S is, the other hydrides make it seem rather pleasant by comparison.

Hydrogen sulfide can be prepared directly from the elements, but it is more readily generated by reacting various metal sulfides with strong acids. Reactions such as the one represented in Equation (17.3) for iron sulfide and hydrochloric acid were used for many years to generate H_2S gas for qualitative analysis. Indeed, the old Kipp generators still found in out-of-the-way corners of most chemistry departments were designed expressly for the purpose of safely generating H_2S in this way. (If you have never seen a Kipp generator, ask around your department and see if anyone knows where you could find one.)

$$FeS(s) + 2HCl(aq) \longrightarrow H_2S(g) + FeCl_2(aq) \tag{17.3}$$

H_2S is a weak diprotic acid with a $K_{a1} = 1.0 \times 10^{-7}$ and $K_{a2} = 1.3 \times 10^{-13}$. The analogous H_2Se, also produced by the action of acids on various metal selenides and by direct combination of the elements, is less stable and a stronger acid (K_{a1} = approximately 10^{-4}) than hydrogen sulfide. H_2Te and H_2Po cannot be produced directly because they are so unstable. H_2Te is a still stronger acid (K_{a1} = approximately 10^{-3}) than its lighter analogs. Recall that the increase in the acid strength of the hydroacids going down a group was explained in Chapter 11 (pp. 286–287).

Oxides and Oxoacids

Sulfur dioxide and sulfur trioxide are two of the most important nonmetal oxides shown in Table 11.4 and discussed in Chapter 11 (pp. 279–280). They are both characterized by strong delocalized S—O bonds and are acid anhydrides (see Table 11.6).

Sulfur dioxide is a colorless, toxic gas with a penetrating odor. It is produced when sulfur is burned in moist air and is of pivotal importance in the formation of acid rain, as we will explore later. It is also a common by-product when various metal sulfides are "roasted" (reacted with the oxygen in air) to convert them to the oxides which can in turn be reduced to the free metal as shown in Equations (17.4) and (17.5):

$$2M^{II}S(s) + 3O_2(g) \longrightarrow 2MO(s) + 2SO_2(g) \tag{17.4}$$

$$MO(s) + C(s) \longrightarrow M(s) + CO(g) \tag{17.5}$$

Some sulfides, for example, PbS, HgS, FeS_2, Sb_2S_3, and most importantly Cu_2S, react directly with oxygen to produce the free metal as shown for copper in Equation (17.6). In the laboratory, sulfur dioxide is generated by adding strong acids to sulfites as represented in Equation (17.7). This gas is used as a bleaching agent, a food additive to inhibit browning, and a fungicide and antioxidant in the wine industry.

$$Cu_2S(s) + O_2(g) \longrightarrow 2Cu(s) + SO_2(g) \tag{17.6}$$

$$2HCl(aq) + Na_2SO_3(aq) \longrightarrow SO_2(g) + H_2O(l) + 2NaCl(aq) \tag{17.7}$$

The structure of sulfur dioxide is shown in Figure 17.3*a*. Two resonance structures are necessary to account for the two equal S—O bonds intermediate in energy and length between the corresponding values for single and double S—O bonds which occur in other compounds. The resonance hybrid involves electron density delocalized over the three atoms of the molecule as shown in Figure 17.3*b*.

Selenous dioxide, SeO_2, and tellurous dioxide, TeO_2, are white, polymeric solids produced by burning the free elements in air. SeO_2 is a chain polymer

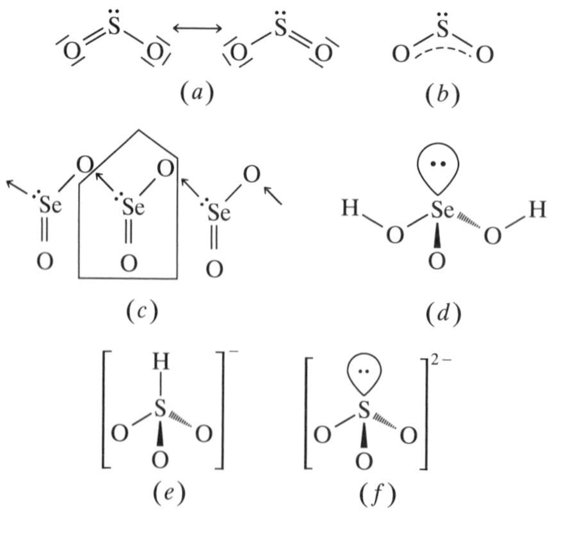

(*a*) (*b*) (*c*) (*d*) (*e*) (*f*)

FIGURE 17.3
Structures of selected sulfur and selenium oxides and oxoacids: (*a*) sulfur dioxide resonance structures, (*b*) sulfur dioxide resonance hybrid, (*c*) selenium dioxide chain polymer, (*d*) selenous acid, (*e*) bisulfite, and (*f*) sulfite.

involving strong dative bonds between the groupings as shown in Figure 17.3*c*. TeO_2 is a three-dimensional network of such bonds in which the tellurium atoms have a coordination number of 4. Polonium dioxide, PoO_2, also produced from the direct combination of the elements at elevated temperatures, assumes a fluorite structure more typical of ionic compounds. (See Figure 7.24 and accompanying discussion for more details on these solid state structures.)

Sulfur dioxide is the anhydride of sulfurous acid, usually represented as H_2SO_3, but more accurately is just the hydrate, $SO_2 \cdot H_2O$. It behaves as a weak diprotic acid ($K_{a1} = 1.3 \times 10^{-2}$, $K_{a2} = 6.3 \times 10^{-8}$). Selenous and tellurous acids are properly written as H_2SeO_3 and H_2TeO_3, respectively. The structure of selenous acid is shown in Figure 17.3*d*, but that for tellurous acid is not known. The acid strength of these oxoacids decreases down the group as expected.

Stepwise neutralization of sulfurous acid produces the bisulfite, HSO_3^-, and sulfite, SO_3^{2-}, anions which have the structures shown in parts (*e*) and (*f*) of Figure 17.3. Sulfite is a mild reducing agent and is used as a fungicide and a fruit and vegetable preservative.

Sulfur trioxide is produced from the catalytic oxidation of the dioxide. Its trigonal planar structure requires three resonance structures to describe. The very short S—O bond distances seem to indicate that there are some $d\pi$-$p\pi$ interactions in addition to the expected delocalized $p\pi$ bonds. It is a powerful oxidizing agent. Although selenium and tellurium are less stable in the +6 oxidation state than is sulfur, both SeO_3 and TeO_3 can be prepared.

Sulfuric acid is obtained by treating sulfur trioxide with water. [The mechanism of this reaction was given in terms of structural formulas in Equation (11.12) and was discussed in detail at that time.] Commonly available as the familiar colorless, corrosive, viscous, concentrated (98%, 18 *M*) aqueous solution, it is a strong acid with a very large K_{a1} and $K_{a2} = 1.3 \times 10^{-2}$. In addition to its well-known acid properties, sulfuric acid is also a powerful dehydrating agent and, in its concentrated form, a good oxidizing agent.

The reaction of concentrated sulfuric acid with water produces large amounts of heat. Care must be taken to always add the denser acid (1.84 g/cm^3) to water and not the other way around so as to avoid dangerous splattering. Splattering results when the less dense water is allowed to sit on top of the heavier acid. Large amounts of heat, released only at the interface of these two liquid layers, vaporizes some water, resulting in a buildup of pressure that is released by blowing the liquid layers apart. On the other hand, if acid is added to water, the denser acid diffuses down through the water, and the heat is diffused throughout the resulting solution.

Gases which do not react with acids can be dehydrated by bubbling them through concentrated sulfuric acid. The reaction with water is so powerful that it can remove hydrogen and oxygen (in a 2:1 mole ratio) from compounds that contain no free water molecules. This is the basis of the familiar lecture demonstration in which adding concentrated sulfuric acid to sugar produces carbon and water vapor as represented in Equation (17.8). Similar reactions cause the familiar blackening of wood, paper, and skin (yipes!) when treated with this powerful acid.

$$C_{12}H_{22}O_{11}(s) \xrightarrow{\text{conc. } H_2SO_4} 12C(s) + 11H_2O \qquad (17.8)$$

Sulfuric acid contains sulfur in its highest possible oxidation state, +6, which, while more stable than the +6 state of selenium and tellurium, makes it a good oxidizing agent, particularly at elevated temperatures and high acid strength. The sulfate may be reduced to sulfur dioxide or to elemental sulfur (or even sulfide, see

Problem 17.23) as shown in Equations (17.9) and (17.10):

$$SO_4^{2-}(aq) + 4H^+(aq) + 2e^- \longrightarrow SO_2(g) + 2H_2O(l) \qquad E^\circ = +0.20\text{ V} \tag{17.9}$$

$$SO_4^{2-}(aq) + 6H^+(aq) + 6e^- \longrightarrow S(s) + 4H_2O(l) \qquad E^\circ = +0.37\text{ V} \tag{17.10}$$

As expected, selenic acid is a weaker acid than sulfuric. However, due to the decreasing stability of the +6 oxidation state going down the group, H_2SeO_4 is a stronger oxidizing agent. The selenate ion is isostructural with sulfate, but telluric acid and the tellurates are quite different. Telluric acid exists as the octahedral $Te(OH)_6$ and is a very weak acid (K_{a1} = approximately 10^{-7}).

Before leaving the noncatenated oxoacids of the chalcogens, mention should be made of peroxodisulfuric acid, $H_2S_2O_8$, and the peroxodisulfate ion, $[O_3S{-}O{-}O{-}SO_3]^{2-}$. The latter is a very strong oxidizing agent as shown by the half-reaction of Equation (17.11). It will oxidize the manganous ion to permanganate and the chromic ion to dichromate (see Problem 17.32).

$$[O_3S - O - O - SO_3]^{2-} + 2e^- \rightleftharpoons 2SO_4^{2-} \qquad E^\circ = 2.01\text{ V} \tag{17.11}$$

Halides

Although there are a large variety of chalcogen halides, many of them are catenated and so are covered in the next section. Sulfur hexafluoride is certainly one of the most important noncatenated compounds. Produced by the direct fluorination of sulfur, this rather inert gas is composed of the familiar octahedral molecules discussed in most introductory courses. Although characterized by strong S—F bonds, the stability of this compound is primarily kinetic rather than thermodynamic. The energy of activation of reactions involving SF_6 is high because of the difficulty in producing a transition state in which a fluorine atom has been removed.

Due to its high molecular weight, SF_6 does not readily effuse or diffuse. (Recall Graham's law.) This combined with its nontoxic, inert, nonconducting nature has made it useful as a gaseous insulator in high-voltage generators. In addition, 50:50 SF_6-air mixtures have also been considered as the pressurizing agent for tennis balls. Oddly, such balls give off a "ping" when put into play. The frequency of the ping is inversely proportional to the molecular weight of the pressurizing gas. Although there are no chloride, bromide, or iodide analogs of SF_6 (six larger halogens would not fit around the sulfur), the corresponding selenium and tellurim fluorides do exist but are rather unstable.

The relatively inert hexafluorides are in sharp contrast to the extremely reactive tetrafluorides. All four tetrafluorides can be prepared by controlled direct fluorination, but their structures vary rather widely. As predicted by VSEPR theory, both SF_4 and SeF_4 exist in the distorted tetrahedral, or seesaw, shape shown in Figure 17.4*a*. The one lone pair of electrons occupies an equatorial position in a trigonal bipyramid. TeF_4 is a one-dimensional polymer made up of the linked square planar TeF_5 groups shown in Figure 17.4*b*, while PoF_4 is not well characterized. Sulfur forms no other tetrahalides, but the larger selenium, tellurium, and polonium do. The "cubane" structure of $TeCl_4$, given in Figure 17.4*c*, is typical of the chlorides and bromides. Iodides form even more extended structures based on octahedral EI_6 units.

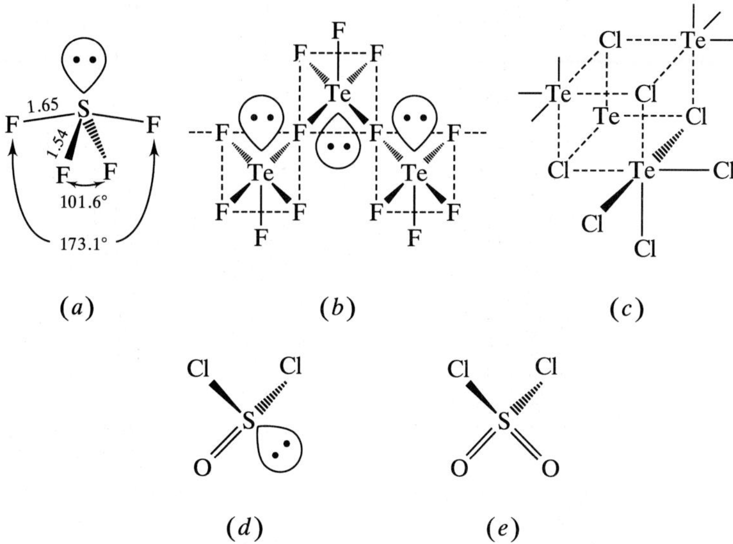

FIGURE 17.4
Structures of selected chalcogen halides: (*a*) sulfur tetrafluoride, (*b*) one-dimensional polymer of tellurium tetrafluoride, (*c*) cubane structure of tellurium chloride, (*d*) thionyl chloride, and (*e*) sulfuryl chloride.

Sulfur tetrafluoride is an important fluorinating agent for both organic and inorganic compounds. For example, it can be used to convert various oxides such as those of thallium(III), carbon (both the monoxide and the dioxide), or selenium (IV) to the corresponding fluorides.

Two important oxyhalides of this group are thionyl chloride, $SOCl_2$, and sulfuryl chloride, SO_2Cl_2. Thionyl chloride, shown in Figure 17.4*d*, is prepared by reacting sulfur dioxide with phosphorus pentachloride as represented in Equation (17.12). It is a fuming, colorless liquid which readily hydrolyzes to sulfur dioxide and hydrogen chloride. This reaction, represented in Equation (17.13), makes $SOCl_2$ useful for preparing anhydrous metal chlorides, as shown for chromium in Equation (17.14).

$$SO_2 + PCl_5 \longrightarrow SOCl_2 + POCl_3 \tag{17.12}$$

$$SOCl_2 + H_2O \longrightarrow SO_2 + 2HCl \tag{17.13}$$

$$[Cr(H_2O)_6]Cl_3 + 6SOCl_2 \longrightarrow CrCl_3(s) + 6SO_2 + 12HCl \tag{17.14}$$

Sulfuryl chloride, SO_2Cl_2, is made by catalytically chlorinating sulfur dioxide. It is in turn a good chlorinating agent. Its structure is given in Figure 17.4*e*. Selenium forms analogous compounds.

17.3 ALLOTROPES AND COMPOUNDS INVOLVING ELEMENT-ELEMENT BONDS

Sulfur is second only to carbon in its ability to catenate, that is, form element-to-element bonds. Sulfur-sulfur single bonds are strong due to good-sized $3p$ orbitals which can effectively overlap. The participation of the $3d$ orbitals may also enhance the strength of such bonds. While thermodynamically stable, S—S bonds are also kinetically labile, that is, they break and reform readily, a property which

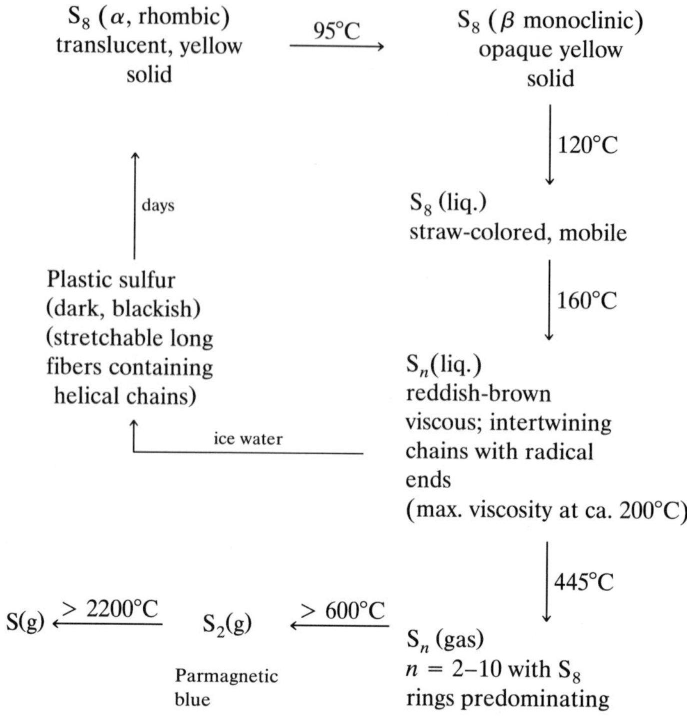

FIGURE 17.5
The major allotropes of sulfur in the solid, liquid, and gaseous phases.

produces an ever-changing equilibrium mixture of chain lengths. Accordingly, sulfur allotropes and compounds are often difficult to characterize with any great certainty. While Se—Se and Te—Te bonds are longer and weaker than S—S bonds, there are still a number of parallels to sulfur structures in the allotropes and compounds of the heavier chalcogens.

Allotropes

Figure 17.5 summarizes the major sulfur allotropes which exist in the solid, liquid, and gaseous phases. The most stable form is α or rhombic sulfur which is composed of the S_8 crown-type molecules shown in Figure 17.6*a*. Heating to 95°C produces the somewhat more disordered β or monoclinic phase, also composed of S_8 rings. Sulfur in rings is sometimes called *cyclosulfur* and S_8, more specifically, *cyclooctasulfur*.

At approximately 120°C, a mobile, straw-colored liquid phase is produced still containing mostly cyclooctasulfur but also a variety of other ring sizes ranging from 6 to over 20. At approximately 160°C, these various rings break apart to form *diradical chains*, meaning that the sulfur atom on each end of a given chain is a radical, that is, it possesses an unpaired electron. As might be expected, the presence of these unpaired electrons makes these chains rather easy to link together to form exceedingly long superchains. The resulting reddish-brown viscous liquid is thought to be composed of intertwining superchains composed of hundreds of thousands of sulfur atoms. As the liquid is heated to its boiling point of 445°C, the average chain length steadily decreases. If the liquid sulfur is

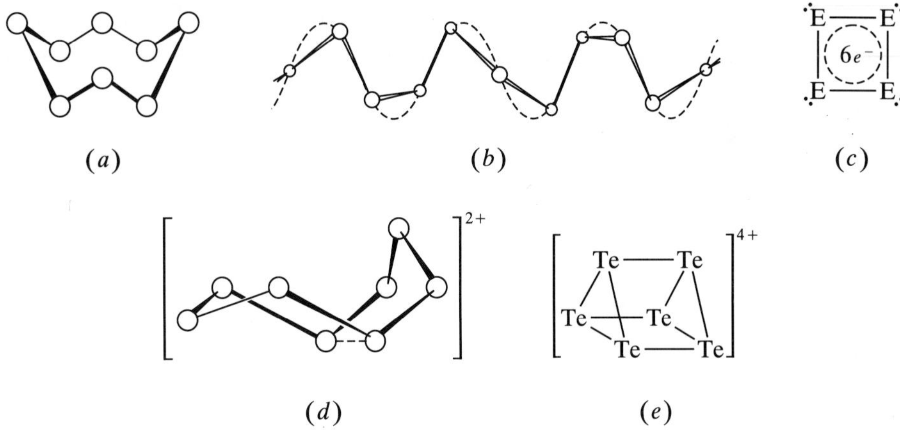

FIGURE 17.6
Catenated chalcogen allotropes and cations: (*a*) cyclooctasulfur, S_8; (*b*) polycatenasulfur, S_n; (*c*) E_4^{2+}, E = S, Se, and Te; (*d*) S_8^{2+}; and (*e*) Te_6^{4+}.

suddenly placed in ice water, long stretchable fibers of "plastic sulfur" are produced. These fibers appear to be made up of right- and left-handed helical chains as represented in Figure 17.6*b* and are one example of a *polycatenasulfur*.

The vapor contains cyclosulfur units again, with ring sizes ranging from 2 to 10 atoms. S_8 rings predominate just over the boiling point, but ring sizes decrease with increasing temperature. At about 600°C, gaseous, paramagnetic, blue S_2 gas is produced, and finally, at low pressures and temperatures over 2200°C, the vapor is composed of free sulfur atoms.

Selenium allotropes also include rings and chains but not nearly in the variety as we have just seen for sulfur. A red Se_8 solid exists and also a black form containing rings with perhaps as many as 1000 selenium atoms. The most stable allotrope, however, is a metallic-gray form composed of polymeric helical chains like that of plastic sulfur. The most stable allotrope of tellurium is a silvery-white metallic form also containing helices. Polonium exists in two metallic lattices, one cubic, the other a simple rhombohedral form. (See Figure 7.17 for figures of these crystal types.)

Polycations and Anions

There are a surprisingly large number of catenated cations and anions in sulfur, selenium, and tellurium chemistry. As early as the end of the eighteenth century when these elements were added to oleum (pure H_2SO_4 with additional SO_3), a variety of brightly colored solutions were produced depending on the chalcogen used, the strength of the oleum, and the time elapsed. We now know that the colors of these solutions correspond to the presence of polycations usually of general formula E_n^{2+}. For example, when E = S, n can be 4 (yellow), 8 (blue), and, rather amazingly, 19 (red). When E is changed to selenium, n can be 4 (yellow), 8 (green), and 10 (red). Solutions containing ions in which E = Te and $n = 4$ are red. Also for tellurium, brown compounds containing the Te_6^{4+} cation are known. Some representative structures of these cations are shown in Figure 17.6*c* to *e*. A variety of methods and counteranions are available for synthesizing these polycations.

Fool's gold, or iron pyrite, FeS_2, is perhaps the most familiar polysulfide, containing discrete anionic S_2^{2-} units in a elongated sodium chloride lattice (see Figure 7.23). Other polysulfide anions of general formula S_n^{2-} ($n = 3$ and 4) can be produced by heating sulfur with a variety of Group 1A and 2A aqueous sulfides. Other more involved methods have yielded species with $n = 5$ and 6. Like the polycations, these compounds are generally highly colored. Polyselenide and polytelluride anions are less common, but the E_3^{2-} anions are known in each case.

Catenated Halides and Hydrides

The catenated sulfur halides have a general formula of S_nX_m. The fluoride, chloride, and bromides for which $m = n = 2$ are well characterized. The chloride and bromide are produced by the direct reaction of the halogen with sulfur. S_2Cl_2, an orange liquid with an indescribably revolting odor, was studied by Scheele in the 1770s. It is now used in huge quantities in the vulcanization of rubber. The reaction of S_2Cl_2 with potassium fluoride can be controlled to produce S_2F_2, which exists as two structural isomers, FSSF and SSF_2. The structures of these two molecules are shown in Figure 17.7*a* and *b*. Both have very short S—S bonds indicating a large degree of double-bond character. Other halides include S_2F_4 and the chlorides with values of n up to 100. S_2F_{10}, a very toxic gas, can be made as a by-product of the fluorination of sulfur. Its structure is shown in Figure 17.7*c*. Catenated halides of selenium and tellurium are much more difficult to prepare; only Se_2Cl_2 and Se_2Br_2 have been well characterized.

The catenated hydrides of sulfur are known as the *polysulfanes*, of general formula H_2S_n, $n = 2$ to 8. Some of these reactive yellow liquids can be prepared by heating sulfides with sulfur and then acidifying as shown in Equation (17.15), but a variety of more sophisticated methods were devised starting in the 1950s and include the treatment of the S_nCl_2 with H_2S as shown in Equation (17.16):

$$Na_2S \cdot 9H_2O + (n-1)S + 2HCl(aq) \longrightarrow 2NaCl(aq) + H_2S_n \qquad (n = 4\text{–}7) \tag{17.15}$$

$$S_nCl_2(l) + 2H_2S(l) \longrightarrow 2HCl(g) + H_2S_{n+2}(l) \tag{17.16}$$

Catenated Oxoacids and Corresponding Salts

Table 17.2 shows the formulas of the catenated sulfur oxoacids and one resonance structure for each of the corresponding anions. Note from the table that not all

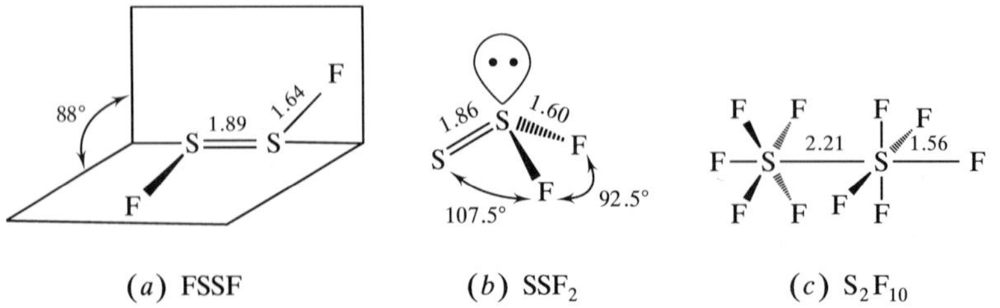

FIGURE 17.7
Catenated sulfur fluorides: (*a*) disulfur difluoride, FSSF; (*b*) thiothionylfluoride, SSF_2; and (*c*) disulfur decafluoride, S_2F_{10}.

TABLE 17.2
Some representative catenated oxoacids and corresponding anions

Acid	Formula	Salt	Structure[†]
Thiosulfuric	$H_2S_2O_3$	Throsulfate, $S_2O_3^{2-}$	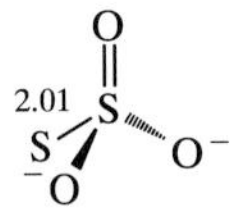 2.01; S — O = 1.47 Å
Dithionous[‡]	$H_2S_2O_4$	Dithionite, $S_2O_4^{2-}$	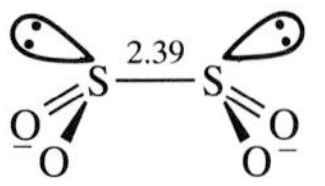 2.39; S — O = 1.51 Å
Dithionic	$H_2S_2O_6$	Dithionate, $S_2O_6^{2-}$	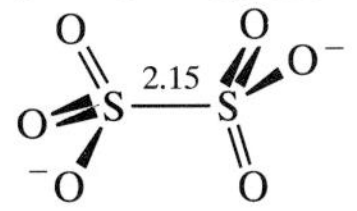 2.15; S — O = 1.43 Å
Polythionic[§]	$H_2S_{n+2}O_6$	Polythionate, $S_{n+2}O_6^{2-}$	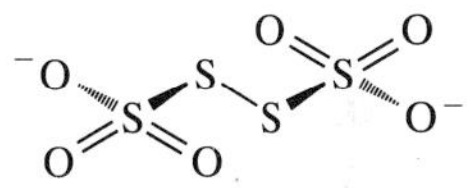 for n = 2, $S_4O_6^{2-}$ (tetrathiomate)

†One resonance structure shown.

‡Free acid unknown.

§Dihedral angle = 15°.

these acids exist in the free state, but all the anions are well known. None of these acids or anions have selenium or tellurium analogs in which *all* the sulfur atoms are replaced by a heavier congener.

Thiosulfuric acid, $H_2S_2O_3$, cannot be prepared as the free acid in aqueous solution, but the thiosulfate ion can be made by boiling elemental sulfur with aqueous sulfite solutions as represented in Equation (17.17):

$$S(s) + SO_3^{2-}(aq) \xrightleftharpoons{\text{boiling}} S_2O_3^{2-}(aq) \qquad (17.17)$$

Note that this reaction is reversible. $S_2O_3^{2-}$ is a tetrahedral anion directly analogous with sulfate but with one oxygen replaced by a sulfur atom. The S—O bond seems to have significant double-bond ($d\pi$-$p\pi$) character as expected, but the S—S is essentially only a single bond. Note that the average oxidation state of the sulfur in thiosulfate is +2, but the central sulfur atom is +5, while the peripheral one is −1.

The use of sodium thiosulfate as the hypo in black-and-white photography is described in Chapter 6 (pp. 127–128). It is also a moderately strong reducing agent and is routinely used to determine the amount of iodine produced in various quantitative analytical procedures. The reaction of thiosulfate with iodine is shown in Equation (17.18):

$$2S_2O_3^{2-}(aq) + I_2(aq) \longrightarrow S_4O_6^{2-}(aq) + 2I^-(aq) \qquad (17.18)$$

Tetrathionate, $S_4O_6^{2-}$, is a polythionate which is described at the end of this section.

The free dithionous acid, $H_2S_2O_4$, is not known, but its salts (sulfur oxidation state = +3) can be prepared by reducing sulfites (S = +4) with zinc dust as shown in Equation (17.19):

$$SO_3^{2-}(aq) + Zn(s) \xrightarrow{SO_2} Zn^{2+}(aq) + S_2O_4^{2-}(aq) \qquad (17.19)$$

The structure of dithionite anion, $S_2O_4^{2-}$, as shown in Table 17.2 is noteworthy for (1) the longest S—S bond (2.39 Å) of any catenated S-S compound and (2) its "playground swing" structure which puts the lone pairs on each sulfur atom in close proximity to each other. Dithionite solutions are rapid and good reducing agents in basic solution. Its standard reduction potential with respect to sulfite in basic solution is −1.12 V.

Dithionate salts may be prepared by oxidizing acidic sulfite solutions with manganese dioxide as shown in Equation (17.20). The dithionate ion, $S_2O_6^{2-}$, is stable in solution and is useful as a moderately large counteranion to precipitate similarly sized cations. The free dithionic acid, $H_2S_2O_6$, can be prepared by treating dithionate solutions with acid as represented in Equation (17.21):

$$MnO_2 + 2SO_3^{2-} + 4H^+ \longrightarrow Mn^{2+} + S_2O_6^{2-} + 2H_2O \qquad (17.20)$$

$$Ba_2S_2O_6(aq) + H_2SO_4(aq) \longrightarrow H_2S_2O_6 + 2BaSO_4(s) \qquad (17.21)$$

There are a variety of polythionates, $[O_3S—S_n—SO_3]^{2-}$, where n can vary from 0 (dithionate) to the 20's. The anions with lower n values are named by reference to the total number of sulfur atoms; for example, $S_3O_6^{2-}$ is trithionate and $S_4O_6^{2-}$ is tetrathionate. The structure of the latter is given in Table 17.2 as a representative polythionate. None of the corresponding acids are stable. While selenium analogs of the polythionates in which all the sulfur atoms are replaced with the heavier congener do not exist, compounds with a general formula $[O_3S—Se_n—SO_3]^{2-}$, $n = 2$ to 6, do.

17.4 THE SULFUR NITRIDES

Sulfur-nitrogen compounds are currently the subject of intense research. Structurally, these compounds call to mind the allotropes of sulfur discussed in the last section, but their preparations and properties are not easily rationalized.

Tetrasulfur tetranitride, S_4N_4, an easily detonated, orange-yellow solid at room temperature, was first prepared in 1835 and remains the starting point for a large number of other S–N compounds. One way S_4N_4 can be prepared is by bubbling ammonia gas through a nonaqueous solution of S_2Cl_2 as shown in Equation (17.22), but the mechanism of this and other synthetic reactions is unclear:

$$6S_2Cl_2 + 16NH_3 \xrightarrow[50°]{CCl_4} S_4N_4 + S_8 + 12NH_4Cl \qquad (17.22)$$

The structure of S_4N_4, shown in Figure 17.8*a*, remained unknown for more than a century after it was first prepared. It turns out to have a structure reminiscent of S_8 and S_8^{2+} but with a square planar arrangement of nitrogens having the same geometric center as a tetrahedron of sulfur atoms. All the S—N bonds are intermediate in length between single (1.74 Å) and double (1.54 Å) $p\pi$-$d\pi$ bonds, implying considerable delocalization around the ring. The S–S distances (shown with dashed lines) are also too short for nonbonded atoms,

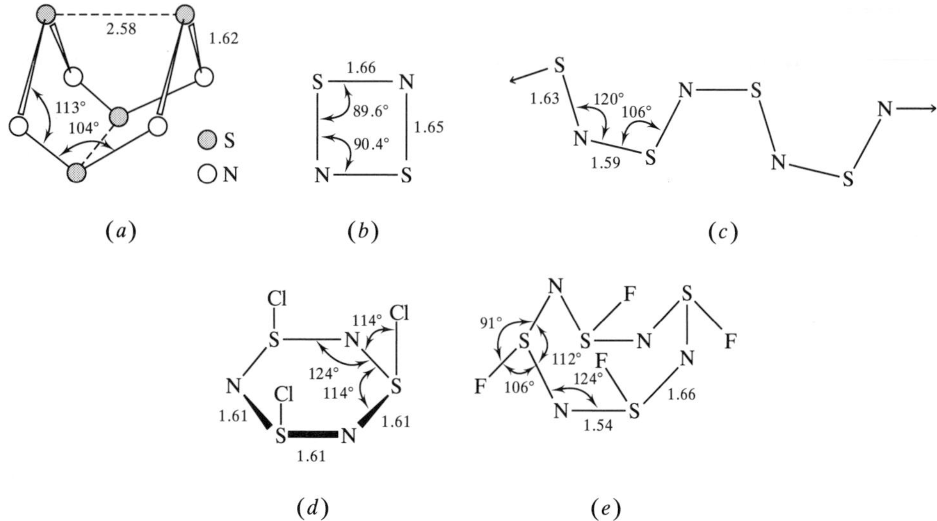

FIGURE 17.8
Sulfur-nitrogen compounds: (*a*) tetrasulfur tetranitride, S_4N_4; (*b*) disulfur dinitride, S_2N_2; (*c*) sulfur nitride polymer, $(SN)_x$; (*d*) trithiazyltrichloride, $(NSCl)_3$; and (*e*) tetrathiazyltetrachloride, $(NSF)_4$.

implying a weak interaction between these *transannular* (across the ring) sulfur atoms.

If S_4N_4 is heated in a vacuum and passed over silver "wool," as represented in Equation (17.23), colorless, explosive crystals of disulfur dinitride, S_2N_2, can be trapped by rapidly cooling the resulting vapor. This compound has the delocalized square planar form, isostructural with S_4^{2+}, shown in Figure 17.8*b*:

$$2S_4N_4 + 8Ag \xrightarrow{250\text{–}300°\,C} 4Ag_2S(s) + 2N_2(g) + 2S_2N_2(s) \qquad (17.23)$$

If S_2N_2 is allowed to warm to about 0°C and left for several days, a metallic, fibrous sulfur nitride polymer, $(SN)_x$, is produced. The structure of this polymer, reminiscent of that of plastic sulfur, is shown in Figure 17.8*c*. Note that it can be viewed as adjacent S_2N_2 units in which the bonds have been rearranged to form a nearly flat chain of alternating sulfur and nitrogen atoms. Again, the S—N bonds are intermediate in length, implying complete delocalization of electron density along the axis of the polymer.

This material acts remarkably like what we might call a one-dimensional metal. For example, it is a golden-bronze metallic color when viewed from the side, but flat-black when viewed end on. In addition, it conducts electricity as well as mercury but only along the polymeric axis. Even more intriguing, when it is cooled to nearly absolute zero, it loses its electrical resistance entirely and becomes a superconductor.

Various derivatives of these compounds have been prepared in great number in the last few years. For example, see the cyclothiazenes (analogous to the cyclophosphazenes) $(NSCl)_3$ and $(NSF)_4$ shown in Figure 17.8*d* and *e*. The chlorine derivative has delocalized S—N bonds, but $(NSF)_4$ has unequal S—N bond lengths, implying more localized S—N and S=N $d\pi$-$p\pi$ bonds. In addition to this growing list of derivatives, there are also a large variety of neutral, cationic, and anionic S_xN_y species. Research continues to unravel the nature of the bonding and the reaction chemistry of these sulfur-nitrogen compounds.

17.5 REACTIONS AND COMPOUNDS OF PRACTICAL IMPORTANCE

Sodium-Sulfide Batteries

Battery technology made the automobile a practical means of transportation. (See Chapter 15, p. 394, for a description of the lead storage battery.) But the generation of automobiles available in the twentieth century is now known to be largely responsible for photochemical smog, and control measures have been difficult and slow to achieve (see Section 16.5). Potential replacements for the gasoline-powered internal combustion engine have always included electric vehicles (EV), but their development has been hampered by the availability of lightweight, efficient, easily rechargeable batteries. The sodium-sulfur battery system has the potential to make EVs competitive.

A sodium-sulfur cell is shown schematically in Figure 17.9. It consists of an inner anode (positive) composed of liquid sodium and an outer cathode (negative) of liquid sulfur (with some added graphite to provide for some electrical conduction). Between the two electrodes is a solid β-Al_2O_3 electrolyte that allows ions to pass through while keeping the two liquid electrodes separated. The operating temperature of the cell is about 350°C.

During the operation of the battery, composed of about 1000 of the above individual cells, sodium atoms are oxidized, that is, they give up their electrons which in turn pass through the external circuit (thereby providing electrical energy) to the sulfur which is reduced to a variety of polysulfides, S_n^{2-}. The sodium ions produced in the liquid sodium pass through the Al_2O_3 electrolyte into the liquid sulfur and serve there as countercations. Therefore, as the battery discharges, the sodium in the cathode is depleted and the concentration of sodium polysulfide in the anode increases. The overall cell reaction is shown in Equation (17.24). The cell can be recharged by reversing the direction of current flow.

$$2Na(l) + \frac{n}{8}S_8(l) \longrightarrow Na_2S_n(l) \qquad E° = 2.08\text{ V} \tag{17.24}$$

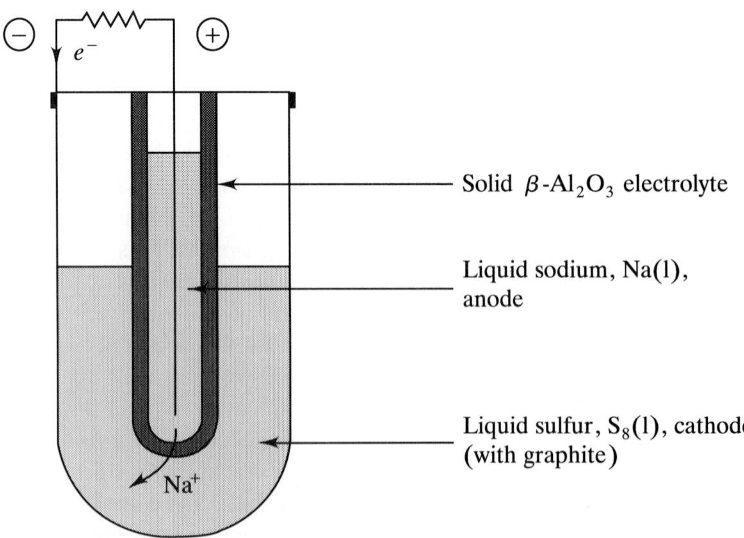

FIGURE 17.9
The sodium-sulfur battery. [Ref. 26.]

Originally developed at the Ford Motor Company in the early 1960s, the sodium-sulfur battery still needs more research and development. Advantages include its light weight, high energy capacity, and long lifetime. A principal disadvantage is the need for frequent and time-consuming (15–20 h) rechargings.

Photoelectric Uses of Selenium and Tellurium

Selenium and tellurium are both photosensitive semiconductors; that is, their electrical conductivity is markedly greater when exposed to light. In the language of band theory, electromagnetic radiation is able to promote electrons from valence bands to conduction bands. (See Section 15.5 for more on band theory and semiconductors.) This property makes both of these elements (and some of their compounds) of interest for such applications as photocells, solar cells, transistors, and rectifiers as we discussed in Chapter 15, and also for the Xerox process, or what is more commonly known as photocopying.

As previously described, germanium and silicon are intrinsic semiconductors. Doping them with Group 5A (or even 6A) elements produces *n*-type semiconductors, while adding Group 3A (or even 2A) elements produces the *p*-type. In addition to these materials, compounds composed of equal moles of 3A and 5A elements, called III-V compounds (due to the older Roman numeral nomenclature for the groups of the periodic table), are good semiconductors. For example, gallium arsenide (GaAs) is a III-V semiconductor of growing importance in supercomputers. II-VI compounds are quite often composed of elements from Group 2B (zinc, cadmium, and mercury) and Group 6A (particularly sulfur, selenium, and tellurium). Zinc sulfide (ZnS), zinc selenide (ZnSe), zinc telluride (ZnTe), cadmium selenide (CdSe), and mercury(II) selenide (HgSe) are all examples of II-VI semiconductors. Cadmium selenide is particularly sensitive to visible light and is used to activate street lights at dusk and in camera exposure meters. Other compounds are more sensitive to infrared light and can be used to monitor heat losses from buildings and even in diagnostic systems for the early detection of breast cancer. (Tumor cells produce more heat than normal cells.)

Xerography (derived from the Greek words for "dry writing"), first developed by C. F. Carlson starting in the 1930s, is dependent upon the light sensitivity of thin films of selenium. In this process, a high-intensity lamp flashes a greatly contrasting light and dark image of the document to be copied onto a thin film of selenium. Areas of the photosensitive selenium exposed to the light are charged to a different extent than those areas not exposed. These gradations of charge are transferred to paper (or perhaps to a piece of transparent overhead projector film), and then a charged toner is added which adheres only to the dark areas. The paper or overhead film is briefly heated to fuse the toner permanently into the fabric of the finished copy. The toner may be black or, in an increasingly common process, of various colors to produce a full-color image.

Sulfuric Acid

We cannot leave a section on the practical importance of compounds of the Group 6A elements without discussing the industrial production and uses of sulfuric acid. Huge amounts (in excess of 10^8 tons) of this multifaceted strong mineral acid are produced worldwide every year. It is so intricately involved in such a large variety of manufacturing processes that the amount produced by a given country is taken as a reliable indicator of economic development.

Sulfuric acid is produced by the contact process. As previously discussed, the burning of sulfur or the roasting of sulfides [Equation (17.4)] are the two most important sources of sulfur dioxide. The oxidation to sulfur trioxide is accomplished by a multistage catalytic system in which the dioxide and oxygen gas (as air) are in contact with a V_2O_5 catalyst at a variety of temperatures. The conversion of the acidic anhydride sulfur trioxide to sulfuric acid is more complicated than might be expected. Just the simple addition of water to the trioxide produces a fine mist or fog of H_2SO_4 which is difficult to condense. So instead the trioxide is passed through sulfuric acid itself to produce oleum, which can be represented as disulfuric or pyrosulfuric acid, $H_2S_2O_7$, as shown in Equation (17.25). The latter can be readily hydrolyzed to a 98% aqueous solution of sulfuric acid as represented in Equation (17.26):

$$SO_3(g) + H_2SO_4(aq) \longrightarrow H_2S_2O_7(aq) \quad (17.25)$$

$$H_2S_2O_7(aq) + H_2O(l) \longrightarrow 2H_2SO_4(aq) \quad (17.26)$$

Sulfuric acid has such a variety of applications it is difficult to gain a good appreciation of its pervasive role in the chemical industry. More than half of it is used to produce superphosphate and other fertilizers (see Chapter 16, p. 435). Other major uses include the manufacture of hydrochloric acid from salt, represented in Equations (17.27) and (17.28), and as the electrolyte in lead storage batteries (see Chapter 15, pp. 394–395):

$$NaCl(s) + H_2SO_4(l) \longrightarrow HCl(g) + NaHSO_4(s) \quad (17.27)$$

$$HCl(g) \xrightarrow{H_2O} HCl(aq) \quad (17.28)$$

Other commercial products produced with the aid of sulfuric acid include detergents, drugs, dyes, explosives, insecticides, metal alloys, paper, petroleum products, pigments, and plastics.

17.6 SELECTED TOPIC IN DEPTH: ACID RAIN

Acid rain is one of the four major atmospheric pollution problems facing modern society. The others are photochemical smog (Section 16.5, pp. 436–439), the depletion of the ozone layer (Chapter 11, pp. 287–289, and Chapter 18, pp. 493–496), and the greenhouse effect (Section 11.6, pp. 289–291). In previous chapters, we have touched on subjects related to acid rain: (1) London smog (Chapter 16, p. 436) which involves sulfuric acid derived from the burning of high-sulfur coal, (2) "stone leprosy" (Chapter 13, pp. 338–339), the effect of acid rain on buildings and statues composed of $CaCO_3$ (marble and limestone), and (3) the use of calcium carbonate in "scrubbers" (Chapter 13, p. 340) designed to control the release of sulfur oxides from the smokestacks of large power plants.

Acid rain can be defined as precipitation (rain, mist, fog, etc.) that has been made acidic by sulfur and nitrogen oxides. The primary source of acid rain starts with the sulfur dioxide (and some sulfur trioxide) released into the atmosphere by various industrial processes. The most important of these is the burning of fossil fuels, mostly coal, to produce electricity. The percentage of sulfur in coal varies from 0.5 to 5.0. In the United States, coal from east of the Mississippi River is significantly higher in sulfur content than coal from the west. Burning sulfur-containing coal accounts for about 60 percent of the humanmade sources of sulfur dioxide. Oil also varies in sulfur content, and refining and burning it accounts for another 25 percent of anthropogenic (human-produced) SO_2. Copper smelting [see

Equations (17.4) and (17.6)] accounts for another approximately 12 percent, while the production of sulfuric acid accounts for only about 2 percent. The net effect of these processes is to release some 50 to 60 million tons of SO_2 into the atmosphere per year. This is almost matched by natural sources such as biodegradation and volcanic activity.

The atmospheric oxidation of sulfur dioxide to the trioxide is still the subject of intense research, but it seems to be accomplished by several complex series of reactions. One of these appears, at first glance, to be the straightforward oxidation of the dioxide to the trioxide, but a large variety of catalytic agents (many of which are themselves air pollutants), operating through still unclear mechanisms, makes this a poorly understood reaction. Another series of reactions, which add up to the overall reaction given in Equation (17.29), may be initiated by the photodissociation of tropospheric ozone to molecular and atomic oxygen:

$$SO_2(g) + O_3(g) \xrightarrow{h\nu} SO_3(g) + O_2(g) \qquad (17.29)$$

These reactions involve the formation of hydroxyl radicals, –OH, produced when atomic oxygen reacts with water molecules. Further complicating the process, other reactions regenerate the hydroxyl radicals and thereby keep the acid-producing cycle going. Once sulfur trioxide is present, it is readily hydrolyzed to sulfuric acid that, in turn, is carried down to the earth's surface mostly by various types of precipitation.

Nitric acid is another important component of acid rain. Although coal does contain some nitrogen that is released during its combustion, the most important source of HNO_3 is derived from the reaction of atmospheric oxygen and nitrogen gases at the high temperatures (approximately 3000° F) required for the efficient combustion of coal. This reaction [see Equation (16.25) and also Equation (16.28) in Table 16.5] produces large amounts of various nitrogen oxides, NO and NO_2, referred to collectively as NO_x. These oxides are then caught up in the same above cycle of reactions involving hydroxyl radicals and ultimately are oxidized to nitric acid.

The above acids build up in the atmosphere, particularly at the base of clouds, and are spread far and wide by the prevailing winds of a given geographical area. In the northeastern United States, the sulfur and nitrogen oxides emitted by coal-burning plants in the midwest are thought to be primarily responsible for the drop in the pH of rainwater from natural values of about 5.5 to 6.0 to values as low as 3.5. (See the contour lines on Figure 17.10.) In northern Europe, similar sources have spread acid rain throughout the region.

In high-altitude areas, such as the Adirondack mountains of New York, forests are bathed directly in high acidity clouds, or "acid fog." The devastating effects on these forests have been well documented. More commonly, when acid rain falls in areas with little natural capacity to neutralize it (see the shading of Figure 17.10), the pH of lakes and ponds has been known to fall to as low as 5.2, and aquatic and plant life have greatly suffered.

How is acid rain to be controlled? Assuming that we will need to continue to burn our vast resources of coal to produce electricity, there are several steps that could be taken to decrease the amounts of sulfur and nitrogen oxides released. These steps include (1) stopping the use altogether of high-sulfur coal to produce electricity, (2) removing sulfur from coal before it is burned, (3) modifying the combustion process to minimize the amounts of nitrogen oxides produced, (4) removing the sulfur and nitrogen oxides from power plant effluents before they are emitted to the atmosphere, or (5) neutralizing the acidity once it has fallen over

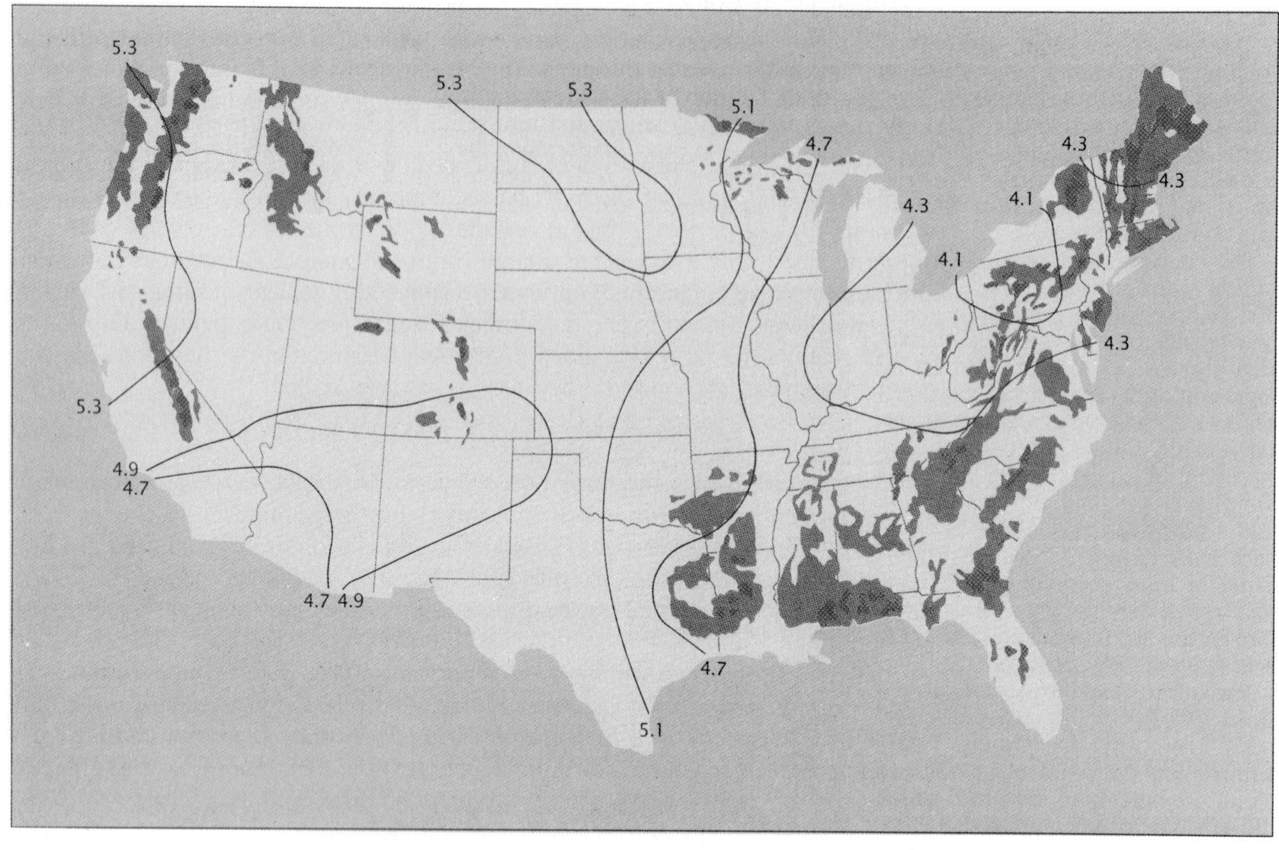

FIGURE 17.10
The effect of acid rain in the continental United States. The average pH of rainfall is shown by contour lines. The alkalinity of surface waters, a reflection of the capacity to neutralize acid rain, is shown by shading, with gray designating areas of moderately low alkalinity and black designating area of severely low alkalinity. The combination of low pH and low neutralizing capacity makes a given area most susceptible to the adverse effects of acid rain. [Ref. 33.]

vast geographical regions. While the first and last options are usually dismissed as unfeasible, a great deal of ongoing research seems to hold out the promise of significant progress in the other areas.

Removing the sulfur from coal before combustion can and is increasingly carried out by physically washing or even centrifuging sulfur-containing iron pyrites, FeS_2, from coal prior to combustion. Modifications to the combustion process that lower boiler temperatures have been proven to cut the amounts of NO_x's produced. One promising way to do this appears to be to burn a small amount (5 to 15 percent) of natural gas with the coal.

Another more complicated technology involves "scrubbing" the sulfur dioxide from the combustion effluents during or after the combustion step. The first generation of *wet scrubbers*, required on power plants built since 1978, spray an aqueous solution of lime (CaO) or limestone ($CaCO_3$) into the smokestack and isolate the sulfur as a difficult-to-manage slurry of calcium sulfite, $CaSO_3$, or sulfate, $CaSO_4$. (See Chapter 13, p. 340, for a more complete discussion of how this technology works.) Unfortunately, these first attempts at scrubbing have been anything but cheap and fraught with an array of technological difficulties.

A second generation of scrubbers now seems to be on the horizon. Several of these are *dry scrubbers* that produce more manageable dry waste products. Others, variations on wet scrubbers, may (1) catalytically reduce NO_x's to molecular nitrogen, (2) tie up nitric oxide, NO, as an easily disposed of iron complex, or (3) change NO_x's to more water-soluble forms that can be more easily handled. One potentially promising technique injects an aqueous solution of yellow phosphorus (P_4) and limestone ($CaCO_3$) into effluents to remove sulfur oxides and change, through a complex series of reactions, the NO_x's into nitrites and nitrates that, along with phosphite and phosphate by-products, could find use as fertilizer materials.

In the long term, newer methods of burning coal, even high-sulfur coal, will need to be devised. Active research in alternative coal-burning technologies continues. The goal is to produce a system which can be retrofitted into existing power plants and incorporated into new ones as they are built. Ultimately, even better technologies or, somewhat ironically, the resurrection of older ones, like coal gasification, will most likely be the answer to this growing global problem.

SUMMARY

Sulfur was known to the ancients but was not recognized as an element until Lavoisier did so in the late 1700s. During the last decade of the nineteenth century, Frasch developed methods of extracting this yellow element from huge underground deposits. Tellurium was isolated from a gold ore by Müller and later by Klaproth. At first Berzelius mistook selenium for tellurium when analyzing sulfuric acid vat residues. The radioactive polonium was discovered in pitchblende by the Curies at the turn of the century.

The properties of the chalcogens are organized by network components including the periodic law, the metal-nonmetal line, and the inert-pair effect. Sulfur is more typical of the chalcogens than is oxygen, which is a good example of the uniqueness principle. Standard reduction potentials reveal that oxygen is the best oxidizing agent of these elements, and this ability decreases going down the group.

The noncatenated hydrides, oxides, and halides further illustrate the rationalization of group properties by the network. The hydrides decrease in thermodynamic stability and increase in acid strength going down the group. Sulfur dioxide, produced from the roasting of metal sulfides or by adding acids to sulfites, is an acid anhydride as are the analogous selenium and tellurium analogs. Sulfurous acid is a weak diprotic acid best thought of as hydrated sulfur dioxide. The corresponding selenous and tellurous acids are weaker acids. Neutralization of sulfurous acid produces bisulfite and sulfite anions.

Sulfur trioxide is a powerful oxidizing agent and the acid anhydride of sulfuric acid, which, in addition to being a strong acid, is a good oxidizing and dehydrating agent. Selenic acid is a weaker diprotic acid than sulfuric, and telluric acid is weaker still. Peroxodisulfuric acid is a very strong oxidizing agent.

The noncatenated halides include the relatively inert sulfur hexafluoride and the highly reactive tetrafluorides of sulfur, selenium, and tellurium. Two important oxyhalides are thionyl and sulfuryl chloride. The former is particularly useful in preparing anhydrous metal halides.

The allotropes of sulfur in the solid, liquid, and gaseous phases include a wide variety of structures dominated by sulfur-sulfur bonds. Cyclooctasulfur, intertwining superchains, helical chains, rings of various sizes, diatomic molecules, and free atoms characterize this element at various temperatures. Although not to the same extent as those of sulfur, the allotropes of selenium and tellurium have similar catenated chains and rings. Sulfur, selenium, and tellurium also exist in a variety of catenated polycations and polyanions.

The simplest catenated sulfur halides have a formula of S_2X_2 (X = F, Cl, and Br). The fluoride is particularly interesting because it is available as two structural isomers. S_2F_2 and S_2F_{10} also contain sulfur-sulfur bonds. Se_2Cl_2 and Se_2Br_2 are the only catenated selenium halides of note. The polysulfanes or sulfur hydrides of formula H_2S_n (n = 2–8) have been prepared. The catenated oxoacids include thiosulfuric, dithionous, dithionic, and the polythionics. The corresponding oxoanions have a variety of properties and structures. Both thiosulfate and dithionite are moderately good reducing agents, while dithionate is stable enough to be a useful counteranion.

Sulfur nitrides include S_4N_4, S_2N_2, and $(SN)_x$. All of these are characterized by delocalized π electrons and structures reminiscent of the sulfur allotropes. The polymeric sulfur nitride is of particular interest because it acts as a one-dimensional metal. The cyclothiazenes are analogous to the cyclophosphazenes.

Reactions and compounds of practical importance include sodium-sulfur batteries, the photoelectric uses of selenium and tellurium, and sulfuric acid. Sodium-sulfur batteries have the potential to make electric vehicles competitive with those powered by the internal combustion engine. Selenium and tellurium are photosensitive semiconductors useful as photocells, solar cells, and transistors. The II-VI semiconductor cadmium selenide is of particular use in detecting visible light. The Xerox process depends on the electrical conductivity of selenium being a sensitive function of light. Sulfuric acid is manufactured by the contact process and plays a pervasive role in the chemical industry.

The production of acid rain starts when atmospheric sulfur dioxide is oxidized to sulfur trioxide by a complex series of reactions. SO_3 is, in turn, hydrolyzed to sulfuric acid. Anthropogenic sources of sulfur dioxide include the burning of coal, the refining and burning of oil, and the smelting of copper ores. The pH of rainwater in the northeastern United States and other areas downwind of such sources has fallen to values between 3 and 4. Particularly at risk are geographical areas which have little natural ability to neutralize this acidic precipitation. In the short term, acid rain control must rely on scrubbing effluent power plant gases to remove the sulfur dioxide. In the long term, new technologies must be developed or old ones, such as coal gasification, revived.

PROBLEMS

17.1. Given what you have read in Chapters 16 and 17, did those who invented gunpowder and matches recognize sulfur as an element? Support your answer.

17.2. Berzelius' scale of atomic weights was all relative to the atomic weight of oxygen to which he arbitrarily assigned a value of 100. Approximately what atomic weight would he have assigned to the elements he helped to discover (silicon, selenium, thorium, cerium, lithium, and zirconium)?

17.3. Compare the densities of antimony and tellurium. Is it understandable that Müller confused them? Would such a comparison distinguish selenium and tellurium?

17.4. Polonium occurs three times in the course of the uranium radioactive decay series. Polonium 210 decays via alpha emission with a half-life of 138.4 days. Write an equation for its decay.

17.5. Discuss the trend in the electron affinities of the chalcogens. Account for this trend using effective nuclear charges, atomic sizes, and so forth.

17.6. Rationalize the trends in atomic radii, ionization energies, and electronegativities of the chalcogens on the basis of effective nuclear charges, atomic sizes, and so forth.

17.7. In one well-written paragraph, make a case that sulfur is the more representative chalcogen rather than oxygen.

17.8. Compare the boiling points of the Group 6A hydrides. Why is the lightest hydride so much higher-boiling than the others? In your answer, refer to the atomic properties (such as electronegativity and effective nuclear charge) of the central chalcogen atoms.

17.9. Is water a better acid than hydrogen sulfide? Cite data which support your answer and rationalize the result.

17.10. Small amounts of hydrogen sulfide are thought to be responsible for the tarnishing of silver, represented in the following equation:

$$4Ag(s) + 2H_2S(g) + O_2(g) \longrightarrow 2Ag_2S(s) + 2H_2O(l)$$

Analyze this process. Identify the oxidizing agent and the reducing agent. Based on what you know about tarnished silver objects, what color do you suspect silver sulfide is?

17.11. Hydrogen selenide, H_2Se, is made by the hydrolysis of aluminum selenide and by the reaction of iron selenide with hydrochloric acid. Write equations corresponding to these reactions.

17.12. On the basis of trends in hydroacids discussed in Chapter 11, would you expect the hydrides of sulfur, selenium, and tellurium to become stronger or weaker in acid strength going down the group? Briefly rationalize your answer.

17.13. Write balanced reactions for the roasting of pyrite (FeS_2) and cinnabar (HgS) to the free metals.

* **17.14.** Sulfur dioxide can serve as either a Lewis acid or a Lewis base. Describe how this might be possible.

17.15. Write an equation which corresponds to the first acid dissociation constant of sulfurous acid as accurately as possible.

17.16. While the most prevalent structure of the bisulfite anion in the solid state is that shown in Figure 17.3*e*, it is also known to exist (particularly in aqueous solution) as a pyramidal structure in which the hydrogen atom is bound to an oxygen atom. Draw a Lewis and VSEPR diagram of this isomer.

17.17. Calculate the base dissociation constant of the sulfite ion given that K_{a1} and K_{a2} of sulfurous acid are 1.3×10^{-2} and 6.3×10^{-8}, respectively.

17.18. On the basis of the discussion of the relative strengths of the oxoacids given in Chapter 11, would you expect sulfurous or selenous acid to be the stronger acid? Briefly rationalize your answer.

17.19. Write a balanced equation for the use of sulfite to reduce chlorine to chloride in acidic aqueous solution. Would you expect this reduction to be more or less likely at very high acid concentrations? Briefly explain.

17.20. Sulfur trioxide can be prepared as the cyclic trimer, $(SO_3)_3$. Draw Lewis and VSEPR diagrams for such a structure.

17.21. Briefly explain why sulfuric acid is a stronger acid than sulfurous.

17.22. Discuss the role of intermolecular forces in determining the viscous nature of concentrated aqueous solutions of sulfuric acid. Draw a diagram as part of your answer.

17.23. Using Tables 12.2 and 17.1, write a balanced half-equation for the reduction of sulfate to hydrogen sulfide. Calculate the reduction potential of this half reaction.

17.24. Cold, dilute sulfuric acid is capable of oxidizing metals above H^+ in the table of standard reduction potentials, but the oxidation state of sulfur does not necessarily change in such reactions. Write an equation representing the oxidation of Mn to the manganous ion by sulfuric acid. Calculate a value for $E°$ for this reaction. (Use Table 12.2.)

17.25. Using a table of standard reduction potentials (for example, Table 12.2), show equations representing (*a*) the oxidation of tin to tin(II) using just the reduction of hydrogen ions from sulfuric acid and also (*b*) the oxidation to tin(IV) using the oxidizing power of the sulfate ion of sulfuric acid.

17.26. By calculating $E°$'s, demonstrate that the reduction of the hydrogen ion (to H_2) of sulfuric acid cannot by itself oxidize copper metal to Cu(II) while the reduction of the sulfate ion of H_2SO_4 to SO_2 can. (Use the standard reduction potentials given in Table 12.2.)

17.27. Fluorosulfuric acid, $FSO_2(OH)$, is a very strong acid. Write a Lewis structure for this compound and discuss the reasons for its high acid strength.

17.28. Sulfamide, $(H_2N)_2SO_2$, has a structure similar to that of sulfuric acid. Draw Lewis and VSEPR diagrams of this molecule.

17.29. Sulfamic acid, $(H_2N)SO_3H$, has a structure similar to that of sulfuric acid.

(*a*) Draw Lewis and VSEPR diagrams of this molecule.

(*b*) Discuss why it is usually found in the ionic form, $H_3^+NSO_3^-$, rather than the above molecular form.

17.30. Speculate on the nature of the intermolecular forces existing in solid telluric acid.

17.31. Draw Lewis, VSEPR, and VBT diagrams of the peroxodisulfate ion.

17.32. Write equations representing the oxidation of the manganous ion to permanganate and the chromic ion to dichromate by the peroxodisulfate ion. Calculate $E°$ and $\Delta G°$ for both these reactions.

17.33. Disulfurous or pyrosulfurous acid, $H_2S_2O_5$, cannot be prepared as the free acid, but the disulfite or pyrosulfite salt is known. Draw a Lewis and VSEPR diagram of this anion. Speculate on how it might be prepared from bisulfites.

17.34. Write a formula for disulfuric or pyrosulfuric acid. Draw corresponding Lewis, VSEPR, and VBT diagrams for this acid. (*Hint*: Recall the nature of pyrophosphoric acid discussed in Chapter 16.)

17.35. In SF_4, there are four bonding pairs (forming the S—F bonds) and one lone pair around the sulfur. Discuss why the lone pair occupies an equatorial position rather than an axial.

17.36. Describe how anhydrous ferric chloride can be prepared from the hexahydrate.

17.37. Polonium does not exist as the free +6 cation and has only a few stable Po(VI) compounds, but there are a variety of compounds containing +4 and +2 cations. Which would you suspect to be the more covalent, $PoCl_2$ or $PoCl_4$? Briefly explain your answer.

17.38. The O—O and O=O bond energies are 146 and 496 kJ/mol, respectively, while those of S—S and S=S are 226 and 423 kJ/mol. Rationalize the following:

(*a*) The single-bond energy of sulfur is greater than that of oxygen.

(*b*) The double-bond energy of oxygen is greater than that of sulfur.

17.39. Using the bond energies found in Problem 17.38, discuss which would be more thermodynamically stable:

(*a*) Cyclooctasulfur, S_8, or four diatomic S_2 molecules?

(*b*) Cyclooctaoxygen, O_8, or four diatomic O_2 molecules?

17.40. Give a formula and speculate on a structure for cyclohexasulfur.

17.41. The structures of S_8 and S_8^{2+} are given in Figure 17.6. Using an analysis of Lewis structures, give a reason that the latter has a transannular S–S interaction but the former does not.

17.42. Analyze the structure of Te_6^{4+} (Figure 17.6*e*) using Lewis electron-dot diagrams. Can the structure be accounted for using all single Te—Te bonds? How might the Lewis structure analysis relate to the fact that the Te–Te distances within a triangular face are shorter (2.67 Å) than the Te–Te distances between the triangular faces (3.13 Å)?

17.43. Speculate on the structure of H_2S_2. Draw a diagram as part of your answer.

17.44. What should happen if aqueous solutions of thiosulfate are acidified? [*Hint*: Refer to Equation (17.17).]

17.45. On the basis of other molecules named in this chapter, justify the name thiothionylfluoride for SSF_2.

17.46. Dithionite can be used in basic solution to reduce lead(II) and silver(I) to their respective metals. Write equations corresponding to these two reactions. Dithionite is oxidized to sulfite in these processes.

17.47. Draw Lewis and VSEPR diagrams of the trithionate anion, $S_3O_6^{2-}$.

17.48. Draw several resonance structures for S_4N_4 which account:
(*a*) For the S—N bond lengths intermediate between those of single and double S—N bonds
(*b*) For the S–S interactions found in the molecule

17.49. How do the number of electrons in S_4N_4 compare with the number in S_8 and S_8^{2+}? How do these numbers compare with the number of transannular S–S interactions? Speculate on the relationship between these results. (*Hint*: An analysis of Lewis structures may be helpful here.)

17.50. Considering the results of the electron counting in Problem 17.49, speculate on the structure of the compound $S_4(NH)_4$.

17.51. Draw resonance structures for disulfur dinitride, S_2N_2, which account for the bond lengths intermediate between those of single and double S—N bonds.

* **17.52.** Oxygen and sulfur are both chalcogens and both form compounds of empirical formula EN (where E = sulfur or oxygen) with nitrogen. Using what you know about the stability, structure, and type of bonding employed in these molecules, comment on the uniqueness principle.

17.53. Analyze the relationship among germanium, gallium arsenide, and zinc selenide. Does it make sense that GaAs and ZnSe should also be good semiconductors?

17.54. Would you suspect that aluminum phosphide might possibly be a semiconductor? Why or why not?

17.55. Discuss the optimum temperature, pressure, and oxygen concentrations (high or low) under which the maximum yield of sulfur trioxide (from the oxidation of sulfur dioxide with molecular oxygen) will be produced. Consider both the thermodynamic and the kinetic points of view.

17.56. Summarize in four major equations the industrial production of sulfuric acid from sulfur.

17.57. As described in Chapter 10, deuterium oxide, D_2O, can be prepared by the electrolysis of water. How would you prepare the compound D_2SO_4?

17.58. In addition to triple phosphate and ammonia itself, another major fertilizer is ammonium sulfate. Speculate on how this might be efficiently produced.

17.59. Why is acid rain more of a problem in the northeastern part of the United States than in any other region?

17.60. Would you expect that in the absence of acid rain, the pH of rainwater would be 7.0? Briefly explain.

17.61. Even if new technologies are developed to rid fossil fuel power plants of sulfur dioxide emissions, one large problem remains. Identify and briefly discuss this problem.

CHAPTER 18

GROUP 7A: THE HALOGENS

The Group 7A elements fluorine, chlorine, bromine, iodine, and astatine are known collectively as the *halogens*, meaning "salt producers." The name was first applied to chlorine because of its ability to combine with metals to form salts. All but the extremely rare and poorly characterized astatine are now known to have this same ability. While there is the usual variation in group properties, the striking similarities among these elements are reminiscent of those of the alkali and alkaline-earth metals.

After the normal sections on the history of the discoveries and the application of the network to the halogens, special sections on (1) the oxoacids and their salts and (2) the interhalogens follow. The usual section on reactions and compounds of practical importance comes next and finally the selected topic in depth, which concerns the threat posed by chlorofluorocarbons to the ozone layer.

18.1 DISCOVERY AND ISOLATION OF THE ELEMENTS

As shown in Figure 18.1, the halogens were discovered in the order chlorine, iodine, bromine, fluorine, and then the radioactive astatine. This section describes the discoveries of these elements in chronological order. Given our experience with other groups, it should not be surprising that chlorine is the most representative halogen. It is appropriate that we should consider it first.

Chlorine

Chlorine was discovered by a chemist we have encountered before, Karl Wilhelm Scheele. Scheele, you will recall, often was in the hunt to discover new elements and is officially listed as the codiscoverer of oxygen with Priestley. This time the Swedish chemist won the day. Sodium chloride, as you might imagine, had been known for centuries. Priestley himself was the first to collect (over mercury) the

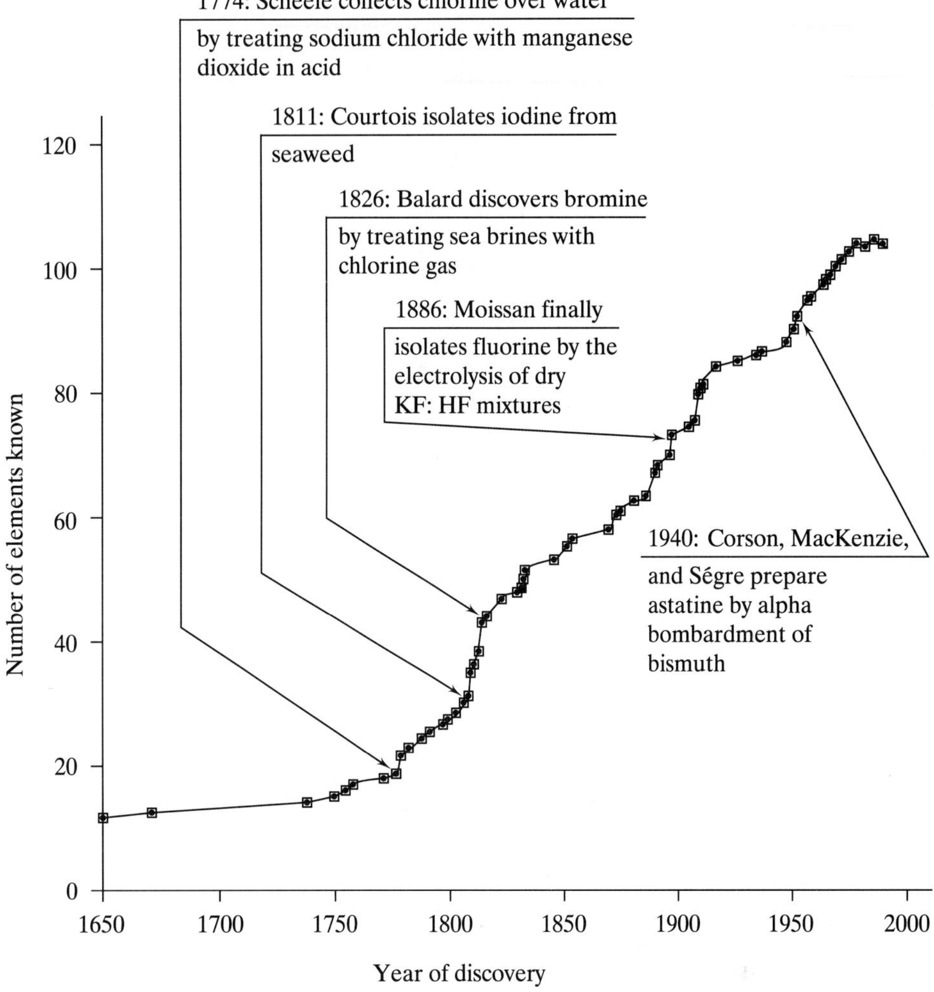

FIGURE 18.1
The discovery of the Group 7A elements superimposed on the plot of the number of known elements versus time.

water-soluble hydrogen chloride gas, obtained by treating sodium chloride with strong acids. Two years later in 1774, Scheele heated sodium chloride with sulfuric acid and manganese dioxide. The acid soon acquired a suffocating odor, and then a pale-green, only slightly water-soluble gas was given off. Scheele's preparation, still a common method of producing small quantities of chlorine, is represented in Equation (18.1):

$$4\text{NaCl}(aq) + 2\text{H}_2\text{SO}_4(aq) + \text{MnO}_2(s) \xrightarrow{\text{heat}} 2\text{Na}_2\text{SO}_4(aq) + \text{MnCl}_2(aq) + 2\text{H}_2\text{O}(l) + \text{Cl}_2(g) \quad (18.1)$$

Scheele found that chlorine gave water an acid taste, and, since Lavoisier had earlier proposed that all acids contain oxygen, the Swede thought he had made a new compound of oxygen. Indeed, a French school of thought (including Gay-Lussac and Thénard) was founded on the idea that oxygen, the "acid producer," was a necessary component of all acids. Humphry Davy, the colorful Englishman we have encountered so many times before, was not of the same opinion regarding acids. In 1810, he announced his contention that the pale-green gas was a new element and named it *chlorine* for the Greek *chlōros*, meaning "pale green" or "greenish yellow." Davy and the English school were, in the end, correct. Davy demonstrated that hydrogen chloride was a nonoxygen containing acid in water, and Gay-Lussac, probably somewhat to his dismay, demonstrated that HCN, then known as prussic acid, also lacked oxygen. Davy suggested that it was hydrogen that was characteristic of acids, a concept that lasted for many years.

Scheele had also noted that chlorine could decolorize flowers and green leaves. This bleaching ability was capitalized upon from the very start. It also was found to be a good disinfectant and, of course, is still used for this purpose. The smell of chlorine gas is not very different from the smell of your local municipal, chlorinated swimming pool. Chlorine has its dark side too. It was the very first war gas, first released by the Germans against the British in 1915. Fortunately, World War I was the only major conflict in which chemical warfare played any significant role. Sodium chloride remains the principal source of chlorine and, ultimately, all chlorine compounds.

Iodine

Bernard Courtois, a French chemist, was involved in manufacturing potassium nitrate (saltpeter) from potassium carbonate (potash), commonly obtained by burning seaweed. Part of the procedure involved eliminating impurities by heating the seaweed with acid. On one occasion in 1811 he added too much sulfuric acid and, much to his surprise, produced a wondrous violet vapor with a penetrating odor similar to that of chlorine. Eventually the beautiful purple gas condensed into dark, lustrous, almost metallic crystals. Suspecting a new element to be in hand, Courtois did some preliminary experiments but, lacking confidence, consulted with others who almost succeeded in stealing the credit for his discovery.

As was the usual custom of the day, Davy was eventually consulted. Again, he made the case that here was another new element which he called *iodine*, from the Greek word *ioeidēs*, meaning "violetlike." The competition between the English and French continued. Gay-Lussac claimed that he, not Davy, had first proposed that iodine was an element. Again, not without irony, the Frenchman prepared hydrogen iodide, another nonoxygen containing acid and showed that when metals were added to it, various iodides and hydrogen gas were produced. The battles between the French and English did not solely concern territory all over the new world but extended to the world of science as well.

Iodine, it turns out, is concentrated in seaweed, which remains one of the primary sources of the element to this day. In the early 1800s, it was also known that "burnt sponge" and kelp were partial remedies for goiter, the name given to an enlargement of the thyroid gland at the base of the throat. People naturally wondered if these two facts might be related. Indeed, we now know that the thyroid gland produces thyroxine, an iodine-containing amino acid responsible for growth regulation. When the thyroid cannot get enough iodine, it grows larger in an attempt to increase the amount of iodine it can biochemically harvest. Today, a small amount of potassium iodide added to table salt, the combination called *iodized salt*, provides the dietary iodine necessary to prevent goiters. A solution of KI and I_2 in alcohol, called *tincture of iodine*, was for many years the common household treatment of choice for minor external wounds.

Bromine

So, by 1820, two halogens were known. Were there others? In 1826, Antoine Jérôme Balard, a young French chemistry assistant interested in the chemistry of the sea, passed a stream of chlorine gas through the mother liquor (the saturated solution left over after the precipitation of a salt) of some sea brines he was working with. From this reaction he isolated a reddish-brown liquid which was easily vaporized to a red vapor with a strong, irritating odor resembling that of chlorine. This liquid had properties seemingly intermediate between those of gaseous chlorine and solid iodine. Indeed, for a while Balard was convinced that he had isolated a compound of iodine and chlorine but quickly thought better of it and announced that he had discovered a new element. (Justus von Liebig, a German chemist we have not previously had occasion to mention, had also isolated this same substance and thought it to be iodine chloride. Liebig merely put it in a bottle labeled ICl and went on with his work. After Balard's announcement he discovered that he too had prepared this new element but had not realized it.) While Balard had called his new element muride, due to its origin in the sea, the third halogen was soon established as *bromine*, from the Greek word *brōmos* for "stench."

After the establishment of bromine, the number of halogens remained at three for 60 years. The properties of bromine were quickly recognized to be intermediate between those of chlorine and iodine and helped to verify the claim that all three were closely allied elements. This triad of elements (along with others such as lithium, sodium, and potassium; calcium, strontium, and barium; and sulfur, selenium, and tellurium) was part of a strong and rapidly building case for the establishment of a periodic table of the elements. Mendeleev first proposed the forerunner of the modern periodic table in the 1870s.

Fluorine

The fourth halogen proved to be more elusive. Fluorspar, or fluorite, (CaF_2) can have a variety of colors depending on the impurities present. The name is derived from the Latin *fluere*, meaning "to flow," as these minerals were used as fluxes, that is, to make a metal ore liquefy and flow at lower temperatures. Later, fluorspars were found to emit a bluish-white light when heated. This property, although the definition has been broadened considerably in today's usage, is still called *fluorescence*. In 1670, a glass cutter named Heinrich Schwanhard treated a fluorspar with strong acid and, much to his surprise we can be sure, found that the lenses of his glasses were no longer transparent. They had become permanently etched. It did not take long before very beautiful patterns could be produced on glass by selective etching.

Scheele studied this process carefully. When he heated various fluorspars with sulfuric acid, the inner surfaces of his glass vessels were corroded, a new acid appeared, and a solid white mass was left over. He reported that the acid was impossible to isolate because it reacted with just about everything. Scheele had prepared what we today call hydrofluoric acid, HF, which in turn had reacted with the glass of his vessels. These reactions are summarized in Equations (18.2) and (18.3):

$$\underset{\text{Fluorspar}}{CaF_2(s)} + H_2SO_4(aq) \xrightarrow{\text{heat}} 2HF(aq) + CaSO_4(aq) \tag{18.2}$$

$$\underset{\text{Glass}}{SiO_2(s)} + HF(aq) \longrightarrow H_2SiF_6(s) + 2H_2O(l) \tag{18.3}$$

There followed a number of tragic attempts to isolate an element from this new acid of fluorspar. Davy showed that the acid did not contain oxygen, but the isolation of its constituent element proved both difficult and dangerous. Davy, Gay-Lussac and Thénard, among many others, suffered greatly from inhaling small amounts of HF during such attempts. A student of Gay-Lussac, Edmond Frémy, attempted to electrolytically decompose fluorite (CaF_2) to produce fluorine, but the elusive fourth halogen reacted too quickly to isolate.

In 1886, Ferdinand Frédéric Henri Moissan, in turn a student of Frémy, succeeded where his mentor had failed. (Moissan also had his work interrupted a number of times while he recovered from HF and F_2 poisoning.) He finally isolated this furiously reactive element by electrolyzing a mixture of anhydrous hydrofluoric acid and potassium fluoride with platinum-iridium electrodes in a platinum vessel. He chilled the apparatus to reduce the activity of the resulting pale-yellow gas. Equation (18.4) summarizes his procedure, which was the only method of preparing fluorine for a century and remains the principal method even today:

$$2\text{HF (in KF melt)} \xrightarrow[-50°\text{C}]{\text{electrolysis}} \text{H}_2(g) + \text{F}_2(g) \tag{18.4}$$

(Great precautions must be taken to keep the product gases separate as they will explosively re-form HF.) Moissan received the Nobel Prize in chemistry in 1906, edging out by one vote another arguably even more worthy chemist, Dmitri Mendeleev.

Astatine

There are no stable isotopes of the heaviest halogen. It was originally prepared in 1940 by D. R. Corson, K. R. MacKenzie, and E. Segrè by alpha bombardment of bismuth as shown in Equation (18.5):

$$^{209}_{83}\text{Bi} + ^{4}_{2}\text{He} \longrightarrow 2\,^{1}_{0}n + ^{211}_{85}\text{At} \tag{18.5}$$

Its name is derived from the Greek *astatos*, meaning "unstable." Although there are 20 known isotopes of astatine, At 211, with a half-life of 7.21 h, remains one of the most stable. Given the instability of this element, it comes as no surprise that there is probably less than an ounce of astatine in the earth's crust, making it the rarest naturally occurring terrestrial element. Little is known about its chemistry. Most of what is known has been derived from work done with extremely dilute (about 10^{-14} M) aqueous solutions.

18.2 FUNDAMENTAL PROPERTIES AND THE NETWORK

Figure 18.2 shows the halogens superimposed on the network. Table 18.1 is the usual listing of group properties that, upon close inspection, are seen to be as regular as those encountered in Groups 1A and 2A.

For the first time since the alkaline-earth metals, we encounter a group which is not divided by the metal-nonmetal line. Consequently, all of the halogens are nonmetals, although iodine and probably astatine do show some signs of metallic character. For example, solid iodine exhibits a metallic luster and, under some conditions, a complexed I^+ cation. At high pressures, iodine is a conductor of electricity. But these are exceptions. The properties of these elements, including iodine, are indeed consistent with their classification as nonmetals.

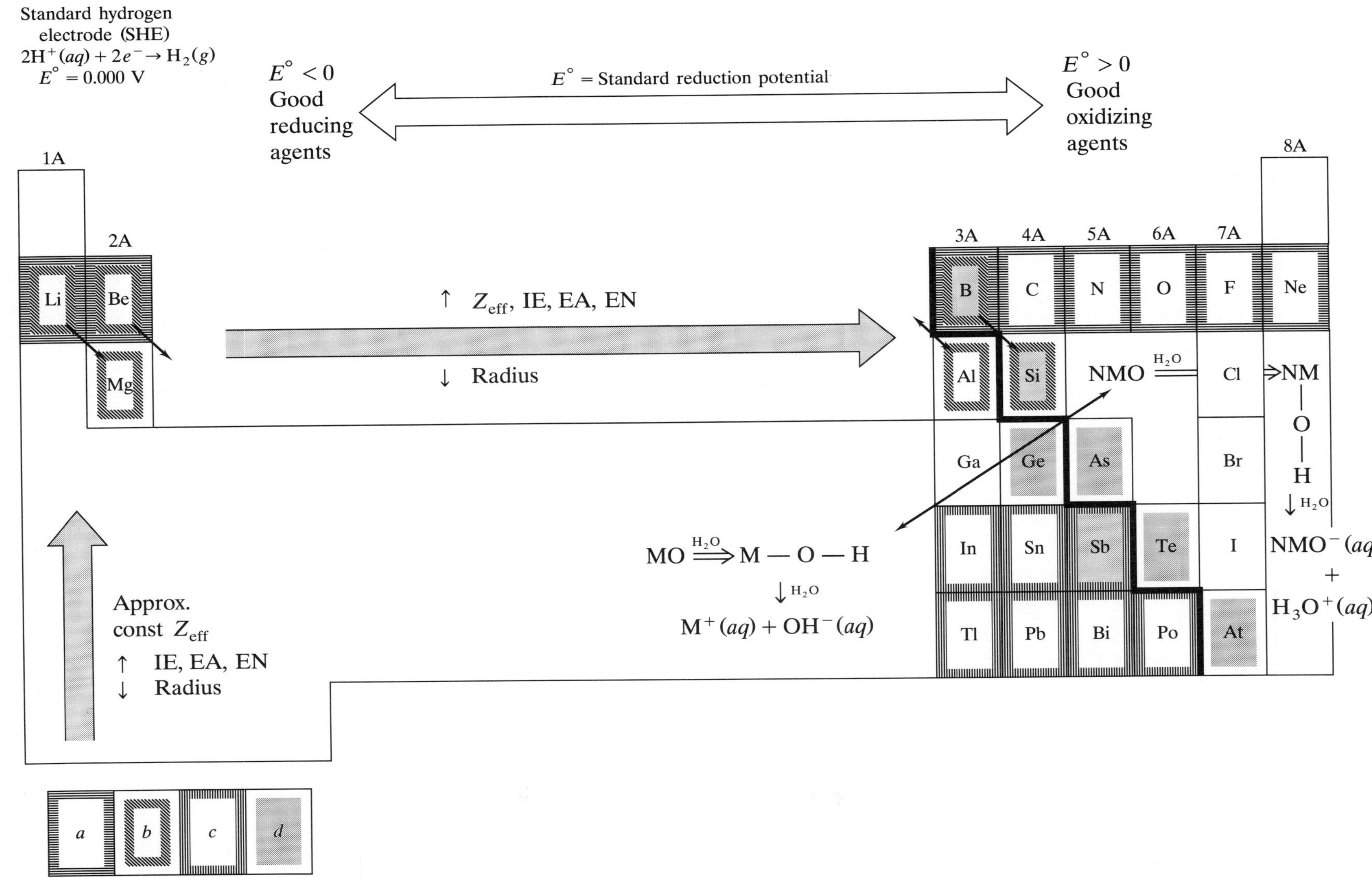

FIGURE 18.2
The halogens (Group 7A) superimposed on the network of interconnected ideas including the trends in periodic properties, the acid-base character of metal and nonmetal oxides, trends in standard reduction potentials, (*a*) the uniqueness principle, (*b*) the diagonal effect, (*c*) the inert-pair effect, and (*d*) the metal-nonmetal line.

TABLE 18.1
The fundamental properties of the Group 7A elements. [Ref. 24.]

	Fluorine	Chlorine	Bromine	Iodine	Astatine
Symbol	F	Cl	Br	I	At
Atomic number	9	17	35	53	85
Natural isotopes, A/% abundances	$^{19}F/100$	$^{35}Cl/75.53$ $^{37}Cl/24.47$	$^{79}Br/50.54$ $^{81}Br/49.46$	$^{127}I/100$	$^{210}At^{\dagger}$
Total no. of isotopes	6	9	17	23	24
Atomic weight	19.00	35.45	79.90	126.9	(210)
Valence electrons	$2s^22p^5$	$3s^23p^5$	$4s^24p^5$	$5s^25p^5$	$6s^26p^5$
mp/bp, °C	-220/-188	-101/-34	-7.3/58.8	114/184	302/337
Density	1.81 g/L	3.21 g/L	3.12 g/cm^3	4.94 g/cm^3	
Covalent radius, Å	0.72	0.99	1.14	1.33	
Ionic radius, Shannon-Prewitt, Å (C.N.)	1.19(6)	1.67(6)	1.82(6)	2.06(6)	
Pauling EN	4.0	3.0	2.8	2.5	2.2
Charge density (charge/ionic radius), unit charge/Å	0.84	0.60	0.55	0.49	
$E°$,‡ V	+2.87	+1.36	+1.07	+0.54	+0.3
Oxidation states	−1	−1 to +7	−1 to +7	−1 to +7	−1, +1, +3(?), +5
Ionization energy, kJ/mol	1680	1251	1143	1009	916
Electron affinity, kJ/mol	−333	−348	−324	−295	
Discovered by/date	Moisson 1886	Scheele 1774	Balard 1826	Courtois 1811	Corson, MacKenzie, Segrè 1940
rpw§ O_2	O_2F_2	None	None	None	None
Acid-base character of oxide	Acid	Acid	Acid	Acid	Acid
rpw N_2	None	None	None	None	None
rpw halogens	(See Section 18.4 on Neutral and Ionic Interhalogens)				
rpw hydrogen	HF	HCl	HBr	HI	HAt
Crystal structure	...	...	Ortho-rhomb.	Ortho-rhomb.	

†Longest half-life (8.3 h).
‡$X_2 \rightarrow X^-$, basic solution.
§rpw = reaction product with.

We have already seen that fluorine was the most difficult of the stable halogens to isolate because of its extremely high reactivity. This is just one of many examples in which fluorine is observed to have properties so special that it cannot be classified as just another halogen, but rather should be called the *superhalogen*. In a group of reactive elements, it is superreactive. It combines with all the elements except helium, neon, and argon. Diatomic fluorine is so reactive that it will remove hydrogen from all its compounds except HF. Wood and paper and even water burst into flames when exposed to a stream of flourine gas. Metals react violently to produce salts, although a few, such as copper, nickel, aluminum, and iron, form a protective fluoride layer reminiscent of aluminum and oxygen.

Why should fluorine be so much more reactive than its congeners? The first factor is the weakness of the F—F bond in the diatomic molecule. The fluorine-fluorine internuclear distance is so small that the lone pairs on each atom are thought to significantly repel each other. A similar line of reasoning, you may recall from earlier discussions, can be applied to the element-element bonds in hydrazine, N_2H_4, and hydrogen peroxide, H_2O_2. (See Chapter 12, p. 313, for more on the structure of hydrogen peroxide.) On the other hand, heteronuclear E—F bonds are exceptionally strong. The high electronegativity of fluorine gives a large polar or ionic component to all E—F bonds in addition to the normal covalent bond strength. The hydrogen-fluorine bond has a particularly high bond energy (568 kJ/mol), making it among the strongest single bonds known. For the above reasons, the reactions of fluorine with nonmetals (producing polar covalent E—F bonds) are most favorable.

When diatomic fluorine reacts with metals, ionic fluorides are formed. These too are exceptionally stable, and the reactions that produce them are correspondingly favorable. The stability of the ionic fluorides can be traced to the exceptionally small size of the fluoride ion, leading to a high charge density compared to other anions. A high charge density, you may recall from our study of the Born-Landé equation (Chapter 8, p. 192), corresponds to a high value of the lattice energy, the energy released when an ionic lattice is formed from its constituent gaseous ions.

The trends in the standard reduction potentials of the halogens are closely related to the uniqueness of fluorine. A quick inspection of Table 18.1 indicates that fluorine has the highest standard reduction potential (2.87 V) of the group and that chlorine is a distant second (1.36 V), with the rest of the values steadily decreasing from 1.07 V to 0.3 V after that. Note that the standard reduction potentials correspond to the half-reaction shown in Equation (18.6):

$$X_2(s, l, g) + 2e^- \longrightarrow 2X^-(aq) \qquad (18.6)$$

The standard state phase (the most stable phase at 25°C and 1 atm) is the gas for fluorine and chlorine, the liquid for bromine, and the solid for iodine. Recall that the more positive the value of the standard reduction potential, the more spontaneous the reduction half-reaction tends to be and the stronger the oxidizing properties of the halogen. Fluorine, then, is by far the strongest oxidizing agent of the halogens, and the oxidizing abilities steadily decrease going down the group.

Why is fluorine such an unusually strong oxidizing agent? The exceptionally low F_2 bond energy makes the F—F bond relatively easy to break and greatly favors the above half-reaction. So also does the high charge density of the small fluoride ion, leading to a large exothermic hydration energy (the energy released when a gaseous ion is surrounded by polar water molecules). Only the somewhat lower electron affinity of fluorine compared to its congeners takes away from the overwhelming spontaneity of the reduction of $F_2(g)$ to $F^-(aq)$. The primary reason for the decrease in the oxidizing power of the heavier halogens is the decreasing hydration energy (as the ions get larger) and the decreasing values of electron affinity.

We can better appreciate now why fluorine, one of the most powerful oxidizing agents known, was so difficult to isolate. When an aqueous solution of HF is electrolyzed, it produces $H_2(g)$ and $F_2(g)$. However, as shown in Equations (18.7), fluorine gas is a better oxidizing agent than oxygen itself and immediately oxidizes water to oxygen gas. Moissan solved this problem by using a dry mixture rather than an aqueous solution of HF and KF.

$$2[F_2(g) + 2e^- \longrightarrow 2F^-(aq)] \qquad E° = +2.87\text{ V} \qquad (18.7a)$$

$$2H_2O(l) \longrightarrow O_2(g) + 4H^+(aq) + 4e^- \qquad E° = -1.23\text{ V} \qquad (18.7b)$$

$$2F_2(g) + 2H_2O(l) \longrightarrow 4F^-(aq) + O_2(g) + 4H^+(aq) \qquad E° = +1.64\text{ V} \qquad (18.7c)$$

Chlorine is produced industrially by the "chlor-alkali process" in which a brine (very concentrated solution of sodium chloride) is continuously electrolyzed using a variety of cells. The overall reaction for this process is given in Equation (18.8):

$$2NaCl(aq) + 2H_2O \xrightarrow{\text{electrolysis}} 2NaOH + Cl_2(g) + H_2 \quad (18.8)$$

Until recently, amalgam cells were used in which molten mercury served as the cathode and a variety of materials as anodes. Unfortunately, small but environmentally significant amounts of mercury are discharged using these cells. For this reason these mercury cells are gradually being replaced by newer ion-selective membrane cells.

The other halogens can be readily produced by oxidizing aqueous halide solutions as shown in Equation (18.9) for bromine:

$$Cl_2(g) + 2e^- \longrightarrow 2Cl^-(aq) \qquad E^\circ = +1.36\text{ V} \quad (18.9a)$$

$$2Br^-(aq) \longrightarrow Br_2(l) + 2e^- \qquad E^\circ = -1.07\text{ V} \quad (18.9b)$$

$$Cl_2(g) + 2Br^-(aq) \longrightarrow 2Cl^-(aq) + Br_2(l) \qquad E^\circ = +0.29\text{ V} \quad (18.9c)$$

This is the type of reaction that Balard first used to isolate liquid bromine and is still the basis of the commercial preparation of the element from seawater. The above equation is certainly an oversimplification of the actual industrial process. Recall, for example, that standard reduction potentials are tabulated for 1 *M* solutions and gas pressures of 1 atm. The above reactions are generally conducted under considerably higher concentrations and pressures.

Hydrides

Our usual practice is to turn now to a consideration of the hydrides, oxides, oxoacids, and halides of the group. However, the chemistry of the halogen oxoacids is so extensive that Section 18.3 is entirely devoted to it.

The hydrides can be produced by direct combination of the halogens with hydrogen. Fluorine and hydrogen combine with explosive force, and chlorine-hydrogen mixtures can become explosive when exposed to light. The products of such direct reactions are called *hydrogen halides* if they are gases, but when put into aqueous solutions they become acids and are called *hydrohalic acids*. Recall from Chapter 11 that the strength of these acids increases down the group. HF is not a strong acid principally because of its very high H—F bond energy.

Hydrofluoric and hydrochloric acids can also be prepared by the action of other strong acids on the halides. For example, Priestley first prepared hydrogen chloride this way, and Scheele's attempts to similarly prepare hydrogen fluoride was thwarted only by the subsequent reaction of HF with his glass vessels. However, when bromides and iodides are treated with sulfuric acid, the elements are produced rather than the hydrides. Indeed, this is how Courtois first produced iodine. Note that the mild oxidizing power of sulfuric acid is sufficient to produce iodine from iodides but not fluorine or chlorine from the corresponding halides. Equation (18.10) illustrates this reaction for bromine:

$$2NaBr(aq) + 2H_2SO_4(aq) \longrightarrow Br_2(l) + SO_2(g) + 4H_2O + Na_2SO_4(aq) \quad (18.10)$$

In addition to the direct reaction of the elements, hydrogen bromide and iodide can also be prepared by the hydrolysis of the phosphorus trihalides.

While HCl was prepared for many years from the reaction of rock salt (NaCl) with other acids, the most important modern source of hydrochloric acid is the chlorination of hydrocarbons using chlorine gas. For example, when benzene is chlorinated as shown in Equation (18.11), chlorobenzene, an important solvent and dye intermediate, is produced with hydrogen chloride as the by-product. Hydrogen chloride is a major industrial chemical and is used to produce ammonium chloride from ammonia, for the synthesis of chlorine dioxide, an important industrial bleach and, most importantly, to "pickle" (remove the oxides from) the surface of steel and other metals:

$$C_6H_6(l) + Cl_2(g) \longrightarrow C_6H_5Cl(l) + HCl(g) \tag{18.11}$$

Hydrogen fluoride, we have seen, is extremely reactive. It etches glass and therefore must be stored in plastic, Teflon, or inert-metal containers. Liquid HF, in many ways, is the universal solvent dissolving (or actually reacting with) a number of oxides, such as those of uranium, silicon, and boron, that water will not. HF is characterized by the strongest possible hydrogen bonds. Indeed, recall that fluorine is the "F" of the FONCl rule discussed in Chapter 11 (p. 270). These strong hydrogen bonds make possible the aqueous bifluoride ion, $F{-}H{-}F^-$, that is present in concentrated hydrofluoric acid or when a fluoride salt is added to the acid. The dry mixture of potassium fluoride and hydrogen fluoride, used by Moissan to prepare fluorine gas, is sometimes referred to as *potassium bifluoride*, KHF_2, as it contains the bifluoride anion. This linear ion has two equal H—F bond lengths and is perhaps best thought of as containing a three-center–four-electron bond. (See Chapter 14, p. 366, for further discussion of multicenter bonds.) In the vapor phase, HF is characterized by polymeric units such as the hexamer $(HF)_6$, held together by F—H --- F hydrogen bonds.

Halides

Throughout our tour of the representative elements, we have considered the halides of each group. Their properties have varied from the ionic, nonvolatile compounds of the early groups to the molecular, volatile compounds of the pnicogens and chalcogens. In each of the sections on the halides we have briefly discussed some of the methods of synthesis. At this point, it is appropriate to summarize some of the general trends we have previously encountered and perhaps extend the discussion slightly.

Although it is difficult to systematize all the synthetic methods we have encountered, they can be broken down into roughly three types of reactions:

1. Direct reaction of the elements
2. Reactions of oxides or hydroxides with the hydrogen halides
3. Reactions of covalent oxides or lower halides with covalent fluorides

1. The direct reaction of various elements with the halogens is the most prevalent method of preparing halides. There is no doubt that this method will work for Groups 1A and 2A, but considerable caution must be exercised. After all, these metals are excellent reducing agents and the halogens, as we have just discussed, excellent oxidizing agents. Reactions between such elements, not sur-

prisingly, are violent, even explosive, and are not particularly safe or practical in many situations. For Groups 3A and 4A, the reactions with chlorine, bromine, and iodine work well, but the fluorides are usually better produced by the alternative procedures described in the next paragraph. For Groups 5A and 6A, reactions with fluorine are more likely to produce the higher oxidation states, while those with the heavier halogens produce the lower.

2. The basic properties of metal hydroxides and oxides (basic anhydrides) make their reactions with the appropriate hydrohalic acid an excellent way to prepare metal halides. These reactions, shown in general in Equations (18.12) to (18.14), also extend to a variety of transition metal halides as well:

$$MOH(aq) + HX(aq) \longrightarrow MX(s) + H_2O \tag{18.12}$$

$$MO(s) + 2HX(g) \longrightarrow MX_2(s) + H_2O \tag{18.13}$$

$$M_2O_3(s) + 6HX(g) \longrightarrow 2MX_3(s) + 3H_2O \tag{18.14}$$

3. The reactions of various oxides with covalent halides present a less precise category of synthetic procedures. Nevertheless, such halogenating agents as chlorine trifluoride (ClF_3), bromine trifluoride (BrF_3), sulfur tetrafluoride (SF_4), and carbon tetrachloride (CCl_4) often are effective in preparing chlorides and particularly fluorides of a variety of elements. Again, this method can be extended to the transition metal halides as well. A few typical reactions are represented in Equations (18.15) to (18.18):

$$6NiO(s) + 4ClF_3(l) \longrightarrow 6NiF_2(s) + 2Cl_2(g) + 3O_2(g) \tag{18.15}$$

$$2SeO_3(s) + 4BrF_3(l) \longrightarrow 2SeF_6(s) + 2Br_2(l) + 3O_2(g) \tag{18.16}$$

$$CO_2(g) + SF_4(g) \longrightarrow CF_4(g) + SO_2(g) \tag{18.17}$$

$$Cr_2O_3(s) + 3CCl_4(g) \longrightarrow 2CrCl_3(s) + 3COCl_2(g) \tag{18.18}$$

4. Halogen exchange reactions have not been covered in the individual sections on the halides. These reactions are often particularly useful in preparing fluorides from the corresponding chlorides. As a general rule they proceed in such a manner that results in the formation of a bond between the least and the most electronegative elements in the reaction. The reason this works is that a polar covalent bond has an ionic component to supplement its covalent character and therefore is stronger. So, in Equations (18.19) and (18.20), the formation of aluminium fluoride and tin(IV) fluoride are favored as the stronger M—F bonds are formed in each case:

$$BF_3(g) + AlCl_3(s) \longrightarrow AlF_3(s) + BCl_3(g) \tag{18.19}$$

$$SnCl_4(s) + 4HF(g) \longrightarrow SnF_4(s) + 4HCl(g) \tag{18.20}$$

One of the more typical reactions of the halides is hydrolysis to the hydroxides, oxides, or the oxoacids. Some representative reactions are given in Equations (18.21) to (18.25):

$$BX_3(s) + 3H_2O(l) \longrightarrow B(OH)_3(aq) + 3HX(aq) \tag{18.21}$$

$$SnX_4(s) + 2H_2O(l) \longrightarrow SnO_2(aq) + 4HX(aq) \tag{18.22}$$

$$PX_3(s) + 3H_2O(l) \longrightarrow HPO(OH)_2(aq) + 3HX(aq) \tag{18.23}$$

$$PX_5(s) + 5H_2O(l) \longrightarrow PO(OH)_3(aq) + 5HX(aq) \tag{18.24}$$

$$SF_4(g) + 2H_2O(l) \longrightarrow SO_2(g) + 4HF(g) \tag{18.25}$$

Before leaving the topic of the halides, a brief mention of the so-called pseudohalides is in order. These are anions which resemble halides in their chemical behavior and generally include azide, N_3^-; cyanide, CN^-; cyanate, OCN^-; thiocyanate, SCN^-; selenocyanate, $SeCN^-$; and tellurocyanate, $TeCN^-$. The similarities between these ions and ordinary halides are remarkable. These include (1) the existence of -1 ions with electronegativities (when averaged over all the atoms in the anion) similar to that of the halides, (2) the formation of volatile dimeric molecules which react with metals to make salts, (3) the ability of these dimeric molecules to serve as oxidizing agents, (4) the ability of HX compounds to serve as acids in aqueous solution, (5) the insolubility of salts of such cations as silver(I), lead(II), and mercury(I), and (6) the tendency to form tetrahedral complexes with a variety of metal ions.

Oxides

The halogen oxides tend to be obnoxious-smelling, highly unstable oxidizing agents with the disconcerting tendency to spontaneously explode. Only dioxygen difluoride, O_2F_2, an unstable orange-yellow solid, can be prepared directly from the elements (by passing an electrical discharge through a mixture of fluorine and oxygen gases). Oxygen difluoride, OF_2, is a pale-yellow gas prepared by the reaction of fluorine gas with aqueous sodium hydroxide. It hydrolyzes to form oxygen gas and hydrogen fluoride. Since fluorine is the only element more electronegative than oxygen, these compounds are among the very few in which oxygen is found in a formal positive oxidation state.

The major chlorine oxides, ClO, Cl_2O, ClO_2, and Cl_2O_7, are rather unpleasant, explosive substances. Chlorine oxide is of central importance in accounting for the depletion of the ozone layer by chlorofluorocarbons and is discussed in Section 18.6. Dichlorine oxide is a powerful chlorinating agent and an important commercial bleach. It is used industrially to produce household bleaching powder, $Ca(OCl)_2$, as shown in Equation (18.26):

$$Cl_2O(g) + CaO(aq) \xrightarrow[\text{lime water}]{} Ca(OCl)_2(s) \qquad (18.26)$$

Chlorine dioxide is used to bleach flour and wood pulps and for water and waste treatment but is always kept well diluted to decrease the chances of explosion. Dichlorine heptoxide, Cl_2O_7 or $O_3ClOClO_3$, is a shock-sensitive, oily liquid. It is formally the acid anhydride of perchloric acid. There are no stable bromine oxides at room temperatures. Of the iodine oxides, diiodine pentoxide, I_2O_5, is the most important. It is used to quantitatively determine carbon monoxide concentrations by the reaction represented in Equation (18.27):

$$5CO(g) + I_2O_5(s) \longrightarrow I_2(s) + 5CO_2(g) \qquad (18.27)$$

18.3 OXOACIDS AND THEIR SALTS

Given that the halogens are nonmetals with an extended list of possible oxidation states, we expect them to exhibit a variety of oxoacids. (You may want to refer to Table 11.2 for an overview of representative oxoacids.) In fact, the number of such acids per group has been growing steadily throughout our tour as more oxidation states have become available. In the metallic main groups (1A and 2A), compounds that contained an E—O—H group were hydroxides. In Groups 3A and

TABLE 18.2
The known chlorine, bromine, and iodine oxoacids

	Oxidation state			
	+1 Hypohalous acid	+3 Halous acid	+5 Halic acid	+7 Perhalic acid
	Chlorine			
	HOCl	HOClO	$HOClO_2$	$HOClO_3$
K_a	3.4×10^{-8}	1.1×10^{-2}	5.5×10^{2}	1×10^{8}
$E°$ (acid)	1.63 V†	1.64 V†	1.47 V‡	1.42 V‡
$E°$ (anion)	0.89 V§	0.78 V§	0.63 V§	0.56 V§
	Bromine			
	HOBr	HOBrO(?)	$HOBrO_2$	$HOBrO_3$
K_a	2×10^{-9}	....	1.0	Large
$E°$ (acid)	1.59 V†	....	1.52 V‡	1.59 V‡
$E°$ (anion)	0.76 V§	....	0.61 V§	0.69 V§
	Iodine			
	HOI	....	$HOIO_2$	$HOIO_3$¶
K_a	2×10^{-11}	....	1.6×10^{-1}	Large
$E°$ (acid)	1.45 V†	....	1.20 V‡	1.34 V‡
$E°$ (anion)	0.49 V§	....	0.26 V§	0.39 V§

†For acid to halogen in acid solution.
‡For oxoanion to halogen in acid solution.
§For oxoanion to halide in basic solution.
¶Also occurs as H_5IO_6; $K_{a1} = 5 \times 10^{-4}$, $K_{a2} = 5 \times 10^{-9}$, $K_{a3} = 2 \times 10^{-12}$.

4A, where only the lighter elements were metalloids or nonmetals, there was one major oxoacid per element, for example, boric or carbonic acid, corresponding to the maximum oxidation state of those groups. For most of the pnicogens and chalcogens, with a greater number of possible oxidation states, we found two major oxoacids per element. Now for the halogens, with the exception of fluorine, there are four possible oxoacids per element corresponding to the +1, +3, +5, and +7 oxidation states. These acids and some of their properties are listed in Table 18.2.

One of the first things you should note from this table is that not all the possibilities are realized. Prominent among the missing oxoacids are those of fluorine which, being the most electronegative element, does not exist in positive oxidation states. (There is a compound of formula HOF. First synthesized in 1968, it is not acidic and should be called hydroxyl fluoride rather than hypofluorous acid.)

Another point to notice in Table 18.2 is that any given halogen oxoacid is a better oxidizing agent, that is, has a higher standard reduction potential, than its corresponding oxoanion. Take the $HOCl–OCl^-$ case as typical. Equation (18.28) shows the half-equation for hypochlorous acid in acid solution:

$$2e^- + H^+ + HOCl \rightleftharpoons Cl^- + H_2O \qquad E° = 1.63\text{ V} \tag{18.28}$$

Suppose base (in the form of hydroxide, OH^-) were to be added to the above system. In which direction would the equilibrium shift? The base would combine not only with the hypochlorous acid to produce hypochlorite but also with the hydrogen ion, thereby decreasing its concentration. Therefore, by Le Châtelier's principle, the reaction should be less spontaneous to the right and the $E°$ less positive. The standard reduction potential for hypochlorite to chloride in basic solution [shown in Equation (18.29)] is 0.89 V, the lower value being consistent with the above reasoning:

$$2e^- + H_2O + OCl^- \rightleftharpoons Cl^- + 2OH^- \qquad E° = 0.89\ \text{V} \qquad (18.29)$$

Hypohalous Acids, HOX, and Hypohalites, OX^-

We are now in a position to examine the oxoacids one oxidation state at a time, starting with the hypohalous acids in which the halogen has a +1 oxidation state. With only one oxygen atom per halogen, these are all weak acids (that become weaker going down the group, see Chapter 11, p. 283) but are good oxidizing agents. The acids are generally prepared by the hydrolysis of the halogen, represented in Equation (18.30). This equilibrium can be shifted significantly to the right by the removal of the halide ion as a mercury- (or silver-) mixed oxide-halide as shown in Equation (18.31). If it is the hypohalite ions that are wanted, a base is added to neutralize the hydrogen ions with the resulting reaction being represented as in Equation (18.32). These preparations are complicated by the disproportionation of the hypohalite, XO^-, to the halide and halate, XO_3^-, as represented in Equation (18.33). The rate of this reaction increases in the order $ClO^- < BrO^- < IO^-$:

$$X_2(g, l, s) + H_2O \rightleftharpoons HOX(aq) + H^+(aq) + X^-(aq) \qquad (18.30)$$

$$2X_2(s, l, g) + 2HgO(s) + H_2O \longrightarrow HgO\cdot HgX_2(s) + 2HOX(aq) \qquad (18.31)$$

$$X_2(s, l, g) + 2OH^-(aq) \longrightarrow XO^-(aq) + X^-(aq) + H_2O \qquad (18.32)$$

$$3XO^-(aq) \rightleftharpoons 2X^-(aq) + XO_3^-(aq) \qquad (18.33)$$

Hypochlorous acid is readily made by the general reaction of Equation (18.31), particularly if conducted at low temperatures to limit the rate of the disproportionation reaction (18.33). Industrially, HOCl is prepared by the gaseous hydrolysis of dichlorine dioxide, Cl_2O, shown in Equation (18.34). Hypochlorites made this way are used in huge quantities as bleaches. (See p. 493 for further details on the mechanism of bleaching.)

$$Cl_2O(g) + H_2O(g) \rightleftharpoons 2HOCl(g) \qquad (18.34)$$

Hypochlorite salts are among the best and most widely used oxidizing agents. Their reactions with a variety of atoms and ions usually result in the net transfer of one or more oxygen atoms to the reactant. Some common examples are given in Equations (18.35) to (18.37). (Although the following equations show hypochlorite being cleanly reduced solely to chloride, these redox systems are complicated by a variety of competing reactions.)

$$NO_2^- + ClO^- \longrightarrow NO_3^- + Cl^- \qquad (18.35)$$

$$S + 3ClO^- + H_2O \longrightarrow SO_4^{2-} + 3Cl^- + 2H^+ \qquad (18.36)$$

$$Br^- + 3ClO^- \longrightarrow BrO_3^- + 3Cl^- \qquad (18.37)$$

The moderately high rate the disproportionation of hypobromite makes it difficult to store and therefore of little use as an oxidizing agent. The rate is so high for hypoiodite that it is unknown in aqueous solution.

Halous Acids, HOXO, and Halites, XO_2^-

Chlorous acid is the only one of the possible three halous acids that exists beyond a shadow of doubt. It is a much stronger acid than hypochlorous (by six orders of magnitude) but is still classified as a weak acid. It cannot be isolated in the free state, but, as shown in Equations (18.38) and (18.39), dilute aqueous solutions are prepared by (1) reducing chlorine dioxide to chlorite with hydrogen peroxide in the presence of barium hydroxide followed by (2) the addition of sulfuric acid which precipitates out barium sulfate, leaving chlorous acid in solution:

$$Ba(OH)_2 + H_2O_2 + ClO_2 \longrightarrow Ba(ClO_2)_2 + 2H_2O + O_2 \qquad (18.38)$$

$$Ba(ClO_2)_2 + H_2SO_4 \longrightarrow BaSO_4(s) + 2HClO_2 \qquad (18.39)$$

Chlorite is also an excellent oxidizing agent and, like hypochlorites, is used in huge quantities as an industrial bleaching agent. It is also used to oxidize and eliminate various smelly, toxic, or environmentally dangerous gases from the effluents of a variety of industrial processes.

Halic Acids, $HOXO_2$, and Halates, XO_3^-

In the hypohalous-hypohalite section above, we saw that the heavier three halogens can be hydrolyzed to hypohalites [Equation (18.32)] and that these in turn can disproportionate with varying rates ($IO^- > BrO^- > ClO^-$) to the halates and halides [Equation (18.33)]. Given these reactions, it is not surprising that hot alkaline solutions of chlorine and bromine go directly to chlorate and bromate, respectively. The overall reaction is represented in Equation (18.40):

$$3X_2(g, l) + 6OH^-(aq) \xrightarrow{\text{heat}} XO_3^- + 5X^- + 3H_2O \qquad (18.40)$$

The same reaction is the basis of a brine electrolysis scheme for the mass production of sodium chlorate in which chlorine and hydroxide are mixed together as they are produced. Bromates and iodates can also be made on a small scale by oxidizing the halide to the halate with hypochlorite. [See, for example, Equation (18.37).] Iodic acid is easily prepared from the oxidation of iodine with hot nitric acid as shown in Equation (18.41). (See Chapter 16, p. 431, for more details on nitric acid as an oxidizing agent.)

$$I_2(aq) + 10HNO_3(aq) \xrightarrow{\text{heat}} 2HIO_3 + 10NO_2 + 4H_2O \qquad (18.41)$$

Iodic acid, a white solid, is the only one of the three halic acids that can be isolated outside of aqueous solution. It exists as pyramidal $IO_2(OH)$ molecules held together by hydrogen bonds. All the halates, XO_3^-, are pyramidal ions with O–X–O angles somewhat less than the tetrahedral 109.5°.

Chlorates have a number of uses based on their oxidizing abilities. $NaClO_3$ is converted to chlorine dioxide used for bleaching paper pulp and as a herbicide and defoliant, while $KClO_3$ is the primary oxidant in fireworks (see Chapter 13, pp. 335–336) and matches (see Chapter 16, p. 433). The catalyzed thermal decomposition of $KClO_3$ was, for many years, used as a convenient laboratory source of oxygen gas. However, see the precautionary note concerning this reaction in Chapter 11, p. 267.

Iodate (and periodate) are the only halogen oxoanions to occur naturally. They are found in great quantities, for example, in Chilean mineral deposits. Iodine is produced from these iodates by reduction with sodium bisulfite as shown in Equation (18.42):

$$2IO_3^-(aq) + 5HSO_3^-(aq) \longrightarrow I_2(s) + 5SO_4^{2-}(aq) + 3H^+(aq) + H_2O \quad (18.42)$$

Potassium iodate is used as a primary standard for thiosulfate solutions in quantitative analysis. Iodate reacts quantitatively with iodide to produce iodine as shown in Equation (18.43):

$$IO_3^- + 5I^-(aq) + 6H^+(aq) \longrightarrow 3I_2(aq) + 3H_2O \quad (18.43)$$

The iodine is then titrated with the thiosulfate solution as detailed in Chapter 17, p. 459. Finally, potassium iodate is used as an oxidizing agent in the intriguing *iodine clock* reaction commonly used as a demonstration of limiting reagents and as a system to exemplify the determination of a rate law (see Problem 18.47).

Perhalic Acids, $HOXO_3$, and Perhalates, XO_4^-

Both perchloric and periodic (per-ī-ōd′-ic) acids can be prepared from the electrolytic oxidation of the corresponding halate followed by the addition of a strong acid. The preparation of perbromic acid proved more difficult and was not accomplished until 1968 (see Problem 18.49). The best route (but still difficult and low yielding) involves the oxidation of bromate to perbromate with fluorine gas as shown in Equation (18.44):

$$BrO_3^- + F_2 + 2OH^- \longrightarrow BrO_4^- + 2F^- + H_2O \quad (18.44)$$

Other possible oxidizing agents, for example, peroxydisulfate or ozone, although thermodynamically feasible, are evidently too slow. The reason for the difficulty in preparing perbromic acid is not satisfactorily understood. This is another example of the anomalous properties of elements that follow the completion of the $3d$ subshell which we have encountered previously (see Chapter 9, pp. 231–233, and Chapter 14, p. 350).

The structures of the perhalic acids and their anions are tetrahedral as expected. There is greater double-bond character to the X—O bonds of perchlorate than there is in the analogous phosphate and sulfate ions. The $d\pi$-$p\pi$ double-bond character increases with the formal charge on the central atom.

In addition to HIO_4, an oxoacid of iodine +7 with an expanded octet is possible. H_5IO_6, usually called orthoperiodic acid, has an octahedral array of oxygen atoms linked together by hydrogen bonds. Orthoperiodic acid is a weak acid as shown by the dissociation constants in Equations (18.45) and (18.46):

$$H_5IO_6 \rightleftharpoons H^+ + H_4IO_6^- \qquad K = 5 \times 10^{-4} \quad (18.45)$$

$$H_4IO_6^- \rightleftharpoons H^+ + H_3IO_6^{2-} \qquad K = 5 \times 10^{-9} \quad (18.46)$$

Its most common sodium salt is sodium dihydrogenorthoperiodate, $Na_3H_2IO_6$, but the fully reacted sodium orthoperiodate salt, Na_5IO_6, can also be prepared. Orthoperiodic acid is the starting point for a small number of polyperiodates, but the tendency to form condensation polymers is much decreased from that found in the phosphoric and, to a lesser extent, sulfuric acid.

These are all strong acids and powerful oxidizing agents. Perchloric acid oxidizes organic materials explosively when heated. Great care should be exercised when employing this acid, particularly in concentrated form. Perchlorates are used as explosives, solid rocket fuels, and in fireworks and flash powders. More than half of the $NaClO_4$ produced is used to make ammonium perchlorate, NH_4ClO_4, the solid rocket fuel used in the space shuttle booster rockets. Each shuttle launch requires about 700 tons of ammonium perchlorate. The white flash and boom that make most of us wince at a display of fireworks is made possible by including separate pockets of potassium perchlorate and sulfur.

Perbromates are surprisingly stable once synthesized. With proper procedures, solutions of perbromic acid, $HBrO_4$, can be stored for extended periods of time. It is a stronger oxidizing agent than perchloric acid (see Table 18.2) but kinetically is slower than its lighter analog.

18.4 NEUTRAL AND IONIC INTERHALOGENS

The interhalogens are generally considered to be compounds composed of two or more different halogen atoms. The neutral binary interhalogens, of formula XX'_n, where X = the less electronegative and X' = the more electronegative atom, are given in Table 18.3. As a general rule the larger, less electronegative atom is in the center surrounded by one, three, five, or seven smaller, more electronegative atoms.

The majority of binary interhalogens are best prepared by direct interaction of the elements at various temperatures. Notable exceptions include bromine pentafluoride, made by fluorination of the trifluoride, and iodine heptafluoride, similarly prepared from the corresponding pentafluoride.

TABLE 18.3
The neutral binary interhalogens

	Gas-phase structure	Color and Phase at STP	Preparation
ClF	Linear	Colorless gas	Direct and $Cl_2 + ClF_3$
ClF_3	Distorted T-shaped	Colorless gas	Direct
ClF_5	Square pyramidal	Colorless gas	$F_2 + ClF_3$
BrCl	Linear	Red-brown gas	Direct
BrF	Linear	Pale-brown gas	Direct and $Br_2 + BrF_3$
BrF_3	Distorted T-shaped	Yellow liquid	Direct
BrF_5	Square pyramidal	Colorless liquid	$BrF_3 + F_2$
IBr	Linear	Black solid	Direct
ICl	Linear	Red solid	Direct
IF	Linear	Brown solid	Direct and $I_2 + IF_3$
ICl_3	Planar dimer	Orange solid	Direct
IF_3	Trigonal planar	Yellow solid	Direct
IF_5	Square pyramidal	Colorless liquid	Direct
IF_7	Pentagonal bipyramidal	Colorless gas	$F_2 + IF_5$

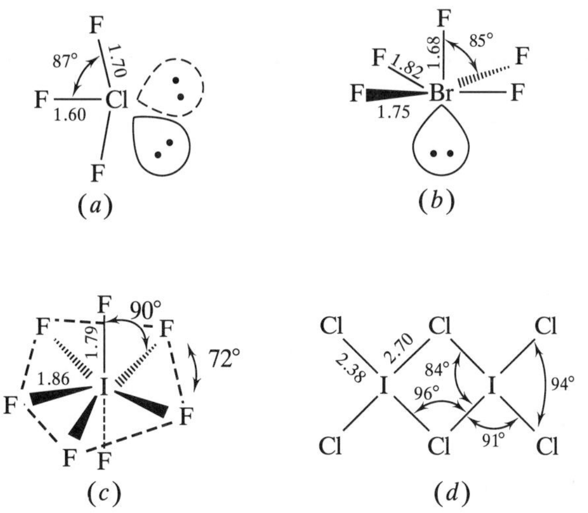

FIGURE 18.3
The molecular structures of some representative neutral interhalogens: (*a*) chlorine trifluoride, ClF_3 (distorted T-shaped); (*b*) bromine pentafluoride, BrF_5 (square pyramidal); (*c*) iodine heptafluoride, IF_7 (pentagonal bipyramidal); and (*d*) diiodine hexachloride, I_2Cl_6 (planar dimer).

The structures of the interhalogens given in Table 18.3 apply to the gas phase and generally follow the rules of the valence-shell electron-pair repulsion theory. (Sometimes AB_7 structures are not covered in a first presentation of this theory, but a rationale for the pentagonal bipyramidal configuration is consistent with VSEPR assumptions.) In various condensed phases, halogen-bridged structures are common. The structures of a few representative interhalogens are given in Figure 18.3.

Some common reactions of the interhalogens (XX'_n) include (1) oxidations resulting in X′ halogenation, (2) hydrolysis to the HX′ and the oxoacid of X, and (3) halide donor-acceptor reactions. Halogen fluorides are usually strong fluorinating agents. The order of decreasing activity is $ClF_3 > BrF_5 > IF_7 > ClF > BrF_3 > IF_5 > BrF > IF_3 > IF$. Both chlorine fluoride and chlorine trifluoride are common, effective fluorinating agents as shown in Equations (18.47) and (18.48). (Recall also the preparations of halides discussed in Section 18.2, pp. 481–483.)

$$W(s) + 6ClF(g) \longrightarrow WF_6(g) + 3Cl_2(g) \qquad (18.47)$$

$$U(s) + 3ClF_3(g) \longrightarrow UF_6(l) + 3ClF(g) \qquad (18.48)$$

Fluorination with bromine trifluoride is exemplified in Equation (18.49). Similarly, iodine chloride is an effective chlorinating agent as shown in Equation (18.50):

$$2B_2O_3(s) + 4BrF_3(l) \longrightarrow 4BF_3(g) + 2Br_2 + 3O_2 \qquad (18.49)$$

$$P_4 + 20ICl \longrightarrow 4PCl_5 + 10I_2 \qquad (18.50)$$

Chlorinetrifluoride is commercially available but is extremely reactive. For example, it reacts explosively with cotton and paper and with hydrazine constitutes a hypergolic (self-igniting) fuel used for short-range missiles.

The hydrolysis of an interhalogen XX'_n commonly produces HX′ and the oxoacid of X as exemplified in Equations (18.51) and (18.52). Note that the oxidation states of the halogens do not change in these reactions:

$$ClF_3 + 2H_2O \longrightarrow 3HF + HClO_2 \qquad (18.51)$$

$$BrF_5 + 3H_2O \longrightarrow 5HF + HBrO_3 \qquad (18.52)$$

Halide donor-acceptor reactions (of XX'_n) are generally those in which X'^- is donated to or accepted from an interhalogen. They include self-ionization reactions such as that of BrF_3 shown in Equation (18.53). This property makes bromine trifluoride a common *aprotic* (without protons) self-ionizing solvent. In addition to its self-ionization, BrF_3 readily accepts fluoride ions from other sources, such as alkali-metal fluorides, to produce salts containing the bromine tetrafluoride ion as shown in Equation (18.54). Conversely, it can donate fluoride ions to produce salts containing the bromine difluoride cation as shown in Equation (18.55):

$$2BrF_3(l) \rightleftharpoons BrF_2^+ + BrF_4^- \tag{18.53}$$

$$BrF_3 + KF \xrightarrow{BrF_3} KBrF_4(s) \tag{18.54}$$

$$BrF_3 + SbF_5 \xrightarrow{BrF_3} [BrF_2^+][SbF_6^-] \tag{18.55}$$

Other self-ionizing interhalogens include iodine trichloride, chlorine trifluoride, and iodine pentafluoride.

In general, most interhalogens can accept halide ions, most commonly from alkali-metal halides, to produce a large variety of interhalogen anions. The preparations of interhalogen cations are generally more difficult as they require specialized solvents and oxidizing agents. A representative list of the simpler interhalogen ions is shown in Table 18.4. Note that some polyatomic monohalogen ions, $X_n^{m\pm}$, are included in this table as well as the binary ions. There is a growing number of ternary halogen ions, but these are not included.

The most common of the polyatomic monohalogen anions is the triiodide ion, I_3^-. The solubility of solid iodine, I_2, in water is greatly enhanced by the addition of an alkali-metal iodide, most often KI. The increased solubility is due to the formation of the aqueous I_3^- ion as represented in Equation (18.56):

$$I_2(aq) + I^-(aq) \longrightarrow I_3^-(aq) \tag{18.56}$$

The linear structure of the triiodide anion is in accord with VSEPR theory ideas. Other I_n^{m-} ions can be thought of as combinations of I^-, I_2, and I_3^- groups held together by weak intermolecular forces. The structures of some of these anions are shown in Figure 18.4.

TABLE 18.4
Binary interhalogen and polyatomic monohalogen ions

Terminal halogen	Central halogen					
	Chlorine		**Bromine**		**Iodine**	
Fluorine	ClF_2^+ Cl_2F^+ ClF_4^+ ClF_6^+	ClF_2^- ClF_4^-	BrF_2^+ BrF_4^+ BrF_6^+	BrF_2^- BrF_4^- BrF_6^-	IF_2^+ IF_4^+ IF_6^+	IF_2^- IF_4^- IF_6^- IF_8^-
Chlorine	Cl_3^+	Cl_3^-	$BrCl_2^+$	$BrCl_2^-$ Br_2Cl^-	ICl_2^+ I_2Cl^+ I_2Cl^+	ICl_2^- Cl^- ICl_4^-
Bromine			Br_2^+ Br_3^+ Br_5^+	Br_3^-	IBr_2^+ I_2Br^+	IBr_2^- I_2Br^-
Iodine					I_2^+ I_3^+ I_5^+ I_7^+ I_4^{2+}	I_3^- I_5^- I_7^- I_9^- I_8^{2-}

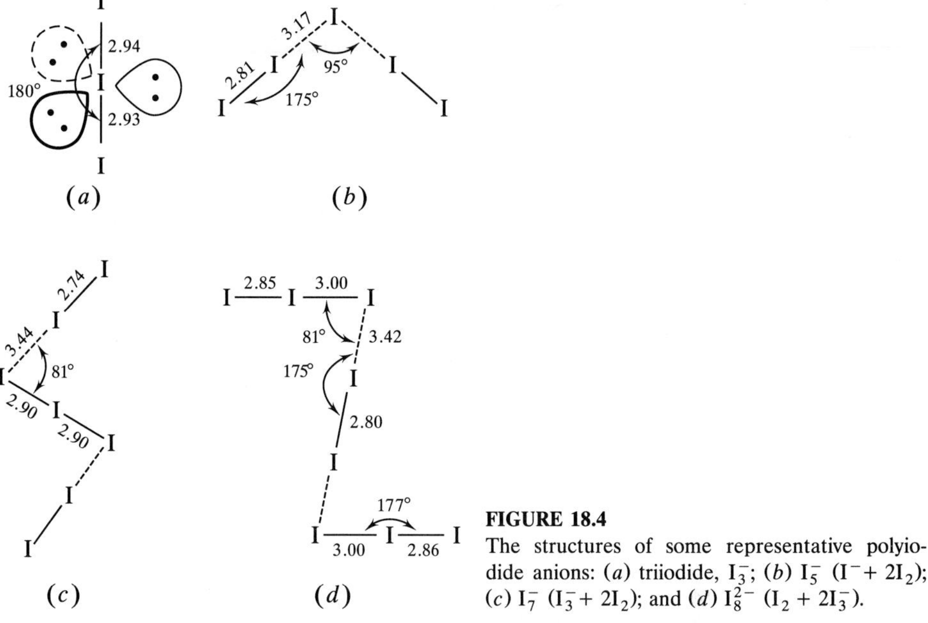

FIGURE 18.4
The structures of some representative polyiodide anions: (*a*) triiodide, I_3^-; (*b*) I_5^- ($I^- + 2I_2$); (*c*) I_7^- ($I_3^- + 2I_2$); and (*d*) I_8^{2-} ($I_2 + 2I_3^-$).

18.5 REACTIONS AND COMPOUNDS OF PRACTICAL IMPORTANCE

Fluoridation

The addition of fluoride to public water supplies has been controversial almost from the moment it was first initiated in 1945 in Grand Rapids, Michigan. Investigations of dental fluorosis (the mottling, or spotting and streaking, of tooth enamel caused by an excess of fluoride in drinking water) were conducted by the Public Health Service in the 1930s. These studies showed that at fluoride concentrations of 1 part per million, fluorosis was minimized and, somewhat unexpectedly, so also was the incidence of dental caries. The Grand Rapids project was meant to be a 10-year pilot study, but word of the effectiveness of fluoride in "reducing cavities" spread and, without the benefit of modern environmental and health effect studies, the movement to fluoridate community water supplies quickly became an intense political issue.

Fluoridation is carried out by adding sodium fluoride, (NaF), hydrogen hexafluorosilicate (H_2SiF_6), or sodium hexafluorosilicate (Na_2SiF_6) to a water supply to bring the concentration of ionic fluoride ion up to 1 part per million.

The mechanism or mechanisms by which a small amount of fluoride in drinking water may reduce the incidence of dental caries is not well understood. It is known, however, that as tooth enamel, largely calcium hydroxyapatite, $Ca_{10}(PO_4)_6(OH)_2$, is formed in children's teeth, the fluoride ion available in the water is incorporated into the tooth surfaces as calcium fluoroapatite, $Ca_{10}(PO_4)_6F_2$. The latter contains the less-basic fluoride ion and therefore is thought to be less susceptible to attack by the acids produced in the mouth when sugars are broken down by bacteria. Today, we know that the above explanation is misleadingly simple. For example, there may be competing mechanisms. It may be that the fluoride ions in saliva and dental plaque directly interfere with the bacterial conversion of sugars to acids rather than making the enamel less susceptible to attack by these acids. Or, it may be that fluoride interferes in a more complex manner with the equilibrium between mineralization and demineralization of tooth enamel surfaces.

Still another possibility is that fluoridation of public water supplies is not nearly as effective at reducing cavities as the early, rather incomplete studies had indicated. It may be that other factors such as changing patterns in diet, oral hygiene, dental care, or the use of fluoride in foods and toothpastes made it appear that water fluoridation decreased tooth decay. Consistent with such a conclusion are some studies which show that the reduction in tooth decay is as great in nonfluoridated as it is in fluoridated areas.

While the true effectiveness of fluoridation in reducing dental caries and the mechanisms that might be involved are now hotly debated, other concerns have risen. What risks to public health are involved? Various studies have linked skeletal fluorosis, kidney disease, hypersensitivity reactions, enzyme effects, genetic mutations, birth defects, and cancer with fluoridated water.

Alternatives to fluoridating public water supplies include over-the-counter fluoride tablets, fluoride rinses, and various toothpaste formulations. Stannous fluoride, SnF_2 (the "fluoristan" in Crest commercials), and sodium monofluorophosphate, $[Na^+]_2[PO_3F]^{2-}$ (the "MFP" in Colgate), were for a time the most common fluoride additives in toothpastes. As detailed in Chapter 15, p. 388, stannous fluoride is a source of ionic fluoride. Fluoristan has been replaced of late in Crest because there has been some evidence that it reacts with tooth enamel to form tin fluorophosphates with unknown physiological properties. Crest now contains "fluoristat," which is just ionic sodium fluoride. Monofluorophosphate is a salt of monofluorophosphoric acid, $OP(OH)_2F$, in which a fluoride ion has replaced an isoelectronic OH^- group in phosphoric acid. Monofluorophosphate releases fluoride ions by the hydrolysis shown in Equation (18.57):

$$PO_3F^-(aq) + H_2O \longrightarrow H_2PO_4^-(aq) + F^-(aq) \qquad (18.57)$$

Chlorination

The chlorination of public water supplies chiefly involves (1) wastewater treatment, (2) drinking water disinfection, and (3) chlorination of swimming facilities. One of the final steps in a municipal wastewater treatment plant is to bubble chlorine gas through the effluent to kill bacteria. The diatomic chlorine gas is quickly hydrolyzed to hypochlorous acid which, in turn, partially dissociates to hypochlorite ions. These reactions are represented in Equations (18.58) and (18.59):

$$Cl_2(g) + H_2O \longrightarrow HOCl(aq) + HCl(aq) \qquad (18.58)$$

$$HOCl(aq) \rightleftharpoons OCl^-(aq) + H^+(aq) \qquad (18.59)$$

The HOCl and OCl^- together are referred to as *free available chlorine*. It kills bacteria by oxidizing certain enzymes essential to bacterial metabolism. In small-scale operations, hypochlorite salts (sodium or calcium) are often substituted for the more-difficult-to-use chlorine gas.

Drinking water is similarly disinfected with chlorine gas. Extensive research has proven the benefits of this process in limiting various bacterial infections including typhoid fever. Knowing that there is "no such thing as a free lunch," we are not surprised to learn that there is a small but manageable danger associated with chlorination of drinking water. Chlorine can interact with organic materials in water to produce small amounts of mutagens. Mutagens cause genetic mutations and are a common measure of the ability of a substance to cause cancer. In Europe, ozonation is often used in place of chlorination (see Chapter 11, p. 288).

Swimming pool water is also chlorinated, but in this case solid chlorite or hypochlorite salts are used directly. Again, the resulting free available chlorine kills bacteria, but, on the smaller scale of a pool, these salts are more convenient and more economical than chlorine gas.

Bleaches

Recall that Scheele himself noted the ability of chlorine gas to decolorize, or *bleach* (a word derived from the Old English word for "pale"), flowers and green leaves. Household bleaches are commonly 5.25% solutions of sodium hypochlorite, NaOCl. The active ingredient of bleaching powders is calcium hypochlorite, $Ca(OCl)_2$. (Safe bleaches for colored materials commonly contain perborates. See Chapter 14, pp. 360–361.) These hypochlorite salts are commonly made by the reaction of chlorine with a base in aqueous solution [Equation (18.32)] or by the electrolysis of aqueous chloride solutions. On an industrial scale, hypochlorites are generated by the hydrolysis of dichlorine oxide [Equation (18.34)].

What is the mechanism of the bleaching action? Recall that what we perceive as color is the absorption of visible light. When visible frequencies are absorbed by an object, the remaining frequencies are reflected or transmitted to our eye and we perceive that the object is colored. Absorption of light energy, we recall, is due to the promotion of electrons from one energy level to a higher one. Bleaches are selective oxidizing agents which serve to remove these electrons from a variety of colored materials. Hypochlorite removes electrons and is itself reduced to chloride and hydroxide ions [see Equation (18.29)].

Bromides

Silver bromide has been known to be a light-sensitive compound for about 150 years. Black-and-white photography depends on the ability of AgBr, spread evenly in an undeveloped film, to absorb visible light thereby producing silver and bromine atoms as shown in Equation (18.60). The presence of silver atoms corresponds to a dark spot on the film. The greater the intensity of light which shines on a given area, the darker the spot. The result is a negative, or reversed, image of the pattern of light:

$$Ag^+ + Br^- \xrightarrow{h\nu} Ag + Br \tag{18.60}$$

In the mid-nineteenth century, potassium and sodium bromides were found to depress the central nervous system. They were widely used as mild sedatives in "headache powders" (and even in treatments for epilepsy) for a century. Largely replaced now by more modern pharmaceuticals, such products have left a legacy in the English language. A *bromide*, to this day, refers to a statement, notion, or even a person which is dull and boring to the point that the listener is put to sleep. It is hoped that this paragraph is not a bromide!

18.6 SELECTED TOPIC IN DEPTH: CHLOROFLUOROCARBONS (CFCS): A THREAT TO THE OZONE LAYER

In Chapter 11 (pp. 287ff) we discussed both the structure of ozone and the role it plays in absorbing dangerous uv radiation in the stratosphere. Now we turn to the threats to stratospheric ozone posed by chlorofluorocarbons (CFCs) and bromine-containing compounds called halons.

CFCs, or Freons, were originally developed by Thomas Midgley in the 1930s as stable, nontoxic, nonflammable replacements for ammonia gas used for refrigeration. (A refrigerant gas is compressed to a liquid usually by an electric motor and then circulated in small-diameter metal coils wrapped around the space to be cooled. When the liquid is allowed to expand to a gas again, it absorbs heat and cools the space.) Midgley demonstrated the desired properties of Freon-12, CCl_2F_2, in front of a meeting of the American Chemical Society in 1930. He inhaled a lungful of the gas and then used it to blow out a candle.

The role of CFCs has greatly expanded in the last 60 years. In addition to a continued use in refrigerators and air conditioners, they are employed as (1) solvents, (2) propellants for aerosol products such as hair sprays and deodorants, and (3) foaming agents to make insulation for the construction industry and packages for fast foods. The two most prevalent and widely used CFCs are Freon-11, $CFCl_3$, and Freon-12, both produced by the action of hydrogen fluoride on carbon tetrachloride as represented in Equation (18.61):

$$CCl_4 + HF \xrightarrow[100°C]{SbF_5} CFCl_3 + HCl \qquad (18.61)$$

How is it that these CFCs, highly valued for their stability, could be responsible for the widespread destruction of stratospheric ozone? In fact, as predicted by Mario J. Molina and F. Sherwood Rowland in 1974, it is their stability that makes them so dangerous. CFCs are virtually indestructible in the troposphere (near the ground) and so diffuse very slowly up into the stratosphere. Here, as shown for Freon-12 in Equation (18.62), they can be degraded by uv radiation into free chlorine atoms and various radicals. It is this free atomic chlorine that catalytically destroys ozone. The sequence of reactions is shown in Equations (18.63) and (18.64). Note that the chlorine atom is regenerated in Equation (18.64). In fact, the sum of Equations (18.63) and (18.64), shown in Equation (18.65), is just the conversion of ozone and atomic oxygen [a necessary reactant to make ozone, see Equation (11.16*b*)] to diatomic oxygen without the net consumption of chlorine:

$$CF_2Cl_2 \xrightarrow{h\nu} CF_2Cl + Cl \qquad (18.62)$$

$$Cl + O_3 \longrightarrow ClO + O_2 \qquad (18.63)$$

$$ClO + O \longrightarrow Cl + O_2 \qquad (18.64)$$

$$\text{Net: } O + O_3 \longrightarrow 2O_2 \qquad (18.65)$$

In this manner, it is estimated that one chlorine atom could destroy as many as 100,000 or 1 million ozone molecules before it is tied up in some inert form or moves out of the stratosphere. This threat of CFCs to the ozone layer was indeed a cause for great concern and, after considerable debate, the United States Congress in 1978 banned the use of CFCs as aerosol propellants.

In what manner could the above reaction cycle be interrupted and chlorine be rendered inert toward ozone? Equations (18.66) and (18.67) show two possibilities. In the first, the active chlorine monoxide reacts with nitrogen dioxide to produce inert chlorine nitrate, $ClONO_2$. In the second, a chlorine atom reacts with methane to produce hydrochloric acid, HCl, also an inactive or inert form of chlorine. HCl and $ClONO_2$, because they sequester chlorine away in an inert form, are called *chlorine reservoirs*:

$$ClO + NO_2 \longrightarrow ClONO_2 \quad (18.66)$$

$$CH_4 + Cl \longrightarrow HCl + CH_3 \quad (18.67)$$

In 1985, the British Antarctic Survey announced that they had found a "hole" in the ozone layer over Antarctica. Soon maps such as those shown in Figure 18.5 displaying the ozone hole started to show up in a variety of newspapers and news magazines. It seems that this large section of ozone-depleted stratospheric air had first appeared in the mid-1970s but had gone unnoticed for a decade. We now know that the hole appears late in each Antarctic winter, starting in late August. By October (or early in the Antarctic spring), the ozone levels stabilize and in November return to their previous levels. In more recent years, the ozone concentrations in the Antarctic have fallen by an astounding 50 percent or more (100 percent in some areas—now that's a hole). The decreases recorded in the Arctic are considerably less, but, even more worrisome, there is evidence of a small erosion of the entire ozone layer at all latitudes. These are indeed alarming developments, and the race has been on to answer the many questions they raise. What is the cause of such a huge decline in ozone concentrations in so short a time? Are CFCs or other chemical agents responsible for this effect? What role do dynamic meteorological processes play? If CFCs are responsible, do they act by the same catalytic cycle outlined above?

While the details of the mechanisms are still being debated, we can sketch some broad outlines of the answers to these above questions. First, as winter sets in, a polar vortex (a pattern of swirling polar winds that isolates a region of exceptionally still, very cold air) is set up. The temperatures in the vortex fall to

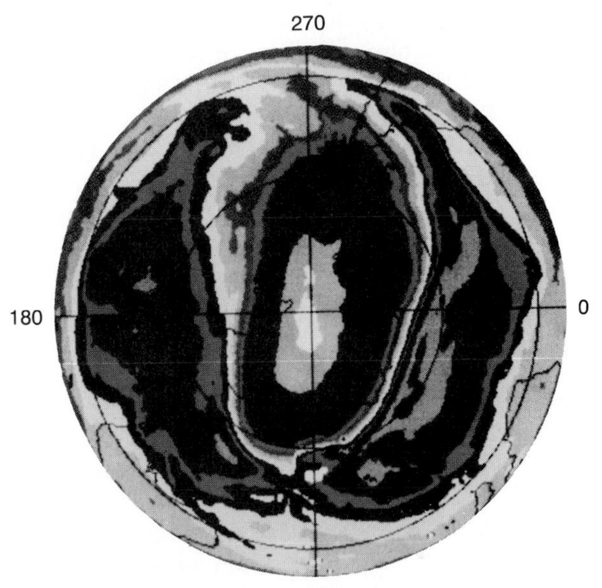

FIGURE 18.5
The Antarctic ozone hole. This map was produced from data taken on October 6, 1991, by the Total Ozone Mapping Satellite. The white and light gray areas in the very center (directly over Antarctica) represent the areas of greatest ozone depletion. (*NASA*.)

$-80°$ C at which polar stratospheric clouds (PSCs) start to form. These clouds, some of which possess a beautiful seashell-like iridescence, contain not only ice but, critically, small amounts of nitrogen compounds, including nitric acid trihydrate, $HNO_3 \cdot 3H_2O$. These nitrogen compounds are taken from chlorine reservoirs such as $ClONO_2$, and their formation releases more active forms of chlorine, such as molecular chlorine, Cl_2. When the sun returns to the region, it breaks apart the chlorine molecules to form ozone-eating chlorine atoms that react as given above in Equations (18.63) to (18.65). There is also evidence that the ice particles of PSCs provide a surface on which these chlorine-releasing reactions can readily occur.

So it appears that yes, CFCs are a major cause of the polar ozone holes because they ultimately supply the free atomic chlorine. In addition, special meteorological processes (polar vortex, PSC formation) also play a role because they provide a pathway whereby active chlorine and chlorine compounds can be released from previously inert chlorine reservoirs. The Arctic ozone hole is smaller, it appears, because its polar vortex is not as stable for as long a period of time and the Arctic temperatures are not as low as those in the Antarctic.

Bromine-containing fluorocarbons, called *halons*, are similarly implicated as posing a danger to ozone. (They may account for as much as 20 percent of the ozone depletion.) Bromine compounds are used as fumigants and in fire extinguishers. Again, free bromine and then bromine oxide, BrO, is photochemically produced and, working in conjunction with chlorine oxide, consumes ozone in a catalytic cycle similar to that shown above.

The possible effects of a decrease in stratospheric ozone concentrations include (1) an increase in skin cancers, (2) an increase in damage to crops and trees, (3) stratospheric cooling leading to large-scale climate changes, and (4) additional production of tropospheric ozone as more and more uv radiation is able to penetrate down into the atmosphere closer to the earth. In addition, CFCs themselves, like carbon dioxide and other compounds mentioned in Chapter 11, are greenhouse gases.

What are we doing to combat the destruction of the ozone layer? Since the scientific community has been able to pinpoint the cause of this problem so definitively, the obvious choice of action is to severely limit or stop altogether the release of CFCs to the atmosphere. Significant steps have been taken to do just that. In 1987, the United States and a number of other industrial nations signed the Montreal Protocol on Substances That Deplete the Ozone Layer. This initial agreement called for reducing CFC emissions up to 50 percent of the 1986 levels by the year 2000. In 1990, these same nations agreed to phase out CFCs completely by the turn of the century.

To carry out this complete elimination of CFCs, several important steps have been taken. First, the EPA has implemented a national recycling program to reduce CFC emissions. Second, and most importantly, alternatives to CFCs are being developed. These are principally the hydrochlorofluorocarbons (HCFCs) which are not as stable as CFCs and therefore significantly decompose before reaching the stratosphere. Unfortunately, these hydrogen derivatives still release some chlorine and therefore pose a threat, albeit considerably smaller, to the ozone layer. Consequently, HCFCs are somewhat controversial and considered only a stopgap measure on the way to producing compounds that can be used for air conditioning and refrigeration without posing any threat at all to the ozone layer. HCFCs will also be recycled as mandated by the EPA, and, ultimately, they too will be replaced by other compounds that release no chlorine whatsoever.

SUMMARY

Chlorine, the first halogen to be discovered, was isolated in the late eighteenth century by Scheele who believed it to be a compound of oxygen. Davy, representing the English school, asserted that chlorine was a new element and HCl a nonoxygen-containing acid. The French school, founded by Lavoisier and then led by Gay-Lussac, claimed that all acids contained oxygen. Iodine was isolated from seaweed by Courtois. Gay-Lussac and Davy quarreled over who was the first to claim it was an element. Iodine was quickly found to be a remedy for goiters. The properties of bromine, isolated by Balard, were immediately recognized to be intermediate between those of chlorine and iodine. This triad was among several to give impetus to the establishment of the periodic table. Fluorine proved more dangerous and more difficult to isolate than the first three halogens. A number of chemists suffered from hydrogen fluoride and fluorine poisoning. Moissan finally isolated this extremely reactive element toward the end of the nineteenth century. Astatine, a radioactive element, was not produced until 1940.

The halogens are all nonmetals with regularly varying periodic properties. Fluorine, however, like all the lightest elements in the main groups, is quite different from its congeners. Diatomic fluorine owes its extreme reactivity to (1) the weakness of the fluorine-fluorine bond, (2) the enhanced strength of covalent E—F bonds, and (3) the large lattice energies of ionic fluorides. The very high standard reduction potential of fluorine can be traced to the weak F—F bond combined with a large energy of hydration for the small fluoride ion. The ability of fluorine to oxidize water (therefore making it very difficult to isolate) can be demonstrated by analyzing standard reduction potentials. Chlorine, not as good an oxidizing agent, is more easily produced by the chlor-alkali process. The oxidizing ability of the second halogen can be used to produce bromine and iodine from their respective halides.

The hydrogen halides can be produced by direct combination of the elements. HF and HCl can also be prepared by treating the respective halides with strong acids. HCl is most commonly obtained as a by-product of the chlorination of hydrocarbons. The extremely reactive HF etches glass, approaches being a universal solvent, and is characterized by extremely strong hydrogen bonds.

The halides of the elements have been surveyed throughout the tour of the representative elements. Common methods to prepare them include (1) direct reaction of the elements, (2) reactions of oxides or hydroxides with hydrogen halides, (3) reactions of oxides or lower halides with covalent fluorides, and (4) halogen exchange reactions. Nonmetal halides usually hydrolyze to form the corresponding hydroxide, oxide, or oxoacid. Pseudohalides are anions which resemble halides in their chemical behavior.

The oxides of the halogens are generally highly reactive compounds. Chlorine oxide has been extensively studied due to its role in the destruction of stratospheric ozone. Chlorine dioxide is a powerful chlorinating and bleaching agent. Iodine pentoxide is used to quantitatively determine carbon monoxide concentrations.

The halogens (excluding fluorine) have the largest variety of oxoacids and corresponding oxoanions of any group yet encountered. Both the oxoacids and the oxoanions are good oxidizing agents, although the former are always the stronger of a given pair. The acid strengths generally increase with the oxidation state of the halogen. For a given oxidation state, acid strengths decrease down the group.

The hypohalous acids and hypohalites are generally prepared by the hydrolysis of the halogen, although this is complicated by the disproportionation of the hypohalite to the halide and halate. Hypochlorite is an excellent oxidizing agent and usually transfers one or more oxygen atoms to a given reactant. Chlorous acid is the only stable halous acid, and chlorate is a widely used industrial oxidizing agent.

Halates are readily prepared by the base hydrolysis of the halogen. Bromates and iodates can also be synthesized by the oxidation of the halide. Only iodic acid can be isolated outside of aqueous solution. Chlorates are excellent oxidizing agents. Iodate occurs naturally, is a source of elemental iodine, and is a primary standard for iodimetry. The perhalic acids of chlorine and iodine are much easier to prepare than that of bromine. Only iodine forms an expanded octet oxoacid, H_5IO_6, called orthoperiodic acid. Perchloric acid is a particularly powerful acid and oxidizing agent. Perchlorates are found in rocket fuels, fireworks, and other pyrotechnics.

The interhalogens, XX'_n (X = the larger, less electronegative halogen), are almost always prepared by direct combination of the elements. Their structures follow the rules of the VSEPR theory. Typical reactions include (1) oxidations resulting in X′ halogenation, (2) hydrolysis to HX′ and the oxoacid of X, and (3) halide donor-acceptor reactions. The latter is one route to a variety of interhalogen anions. The interhalogen cations are more difficult to prepare. There are a variety of polyatomic monohalogen anions, of which triiodide is the most common.

Reactions and compounds of practical importance include fluoridation, chlorination, bleaches, and bromides. The exact mechanism by which fluoride prevents dental caries is still unclear. Chlorination is widely accepted for wastewater treatment and the disinfection of drinking water and swimming facilities. Mild aqueous hypochlorite solutions are the most common bleaching agents. Silver bromide is the basis of black-and-white photography, while sodium and potassium bromides were once common sedatives.

Chlorofluorocarbons (CFCs) were originally developed as refrigerant gases but soon found a variety of uses. So stable that they slowly rise to the stratosphere unscathed, CFCs are broken apart only by uv solar radiation. The resulting chlorine atoms catalytically convert ozone and atomic oxygen to molecular oxygen. Some forms of chlorine can be converted into inert chlorine reservoirs which are not reactive with ozone.

The ozone hole over the Antarctic develops in the late winter and has been deepening since its discovery. Polar stratospheric clouds (PSCs) which are formed within the wintertime polar vortex seem to help break down chlorine reservoirs into active ozone-destroying chlorine compounds. Bromine compounds also play a role in ozone destruction. The effects of decreasing stratospheric ozone concentrations may include skin cancer, crop damage, climatic changes, and an increase in tropospheric ozone. CFCs are also greenhouse gases. The Montreal Protocol and subsequent agreements will eliminate CFC emissions by the turn of the century.

PROBLEMS

18.1. Write an equation for the reaction of sodium chloride with strong acids.

18.2. Identify the oxidizing and reducing agents in Scheele's method of preparing chlorine gas.

18.3. Of Arrhenius, Brønsted-Lowry, and Lewis, which would agree with Davy that hydrogen content is characteristic of an acid?

18.4. Write an equation to represent Courtois' preparation of iodine vapor. Would such a method be useful for preparing chlorine? Why or why not?

18.5. Iodine 131, a beta minus emitter, is useful in treating thyroid cancer. Briefly explain why this might be. Give an equation for the decay of this isotope of iodine.

18.6. J. W. Döbereiner, in 1829, one year after the discovery of bromine and about fifty years before Mendeleev's periodic tables, proposed his law of triads. He showed that chemically similar elements often appear in triads, with the middle member having nearly the same atomic weight as the mean of the lighter and heavier. Using this law, he predicted the atomic weight of the newly discovered bromine. How close to the actual value was he?

18.7. Suppose that Liebeg had really made iodine chloride rather than bromine. How much would they have differed in their molecular weights?

18.8. One of Moissan's most famous students was Alfred Stock. Trace the chemical genealogy (professor-student relationships) from Gay-Lussac to Stock.

18.9. Write an equation for the alpha decay of astatine 211.

18.10. Closely examine the variations in melting points, covalent radii, and ionization energies of the halogens. Comment on these trends.

18.11. Of what oxoacids are the following oxides the acid anhydrides:
(*a*) Cl_2O_7
(*b*) Br_2O
(*c*) I_2O_5
Write an equation for the relationship in each case.

18.12. Write equations to represent the reactions of diatomic fluorine with (*a*) methane gas and (*b*) gaseous ammonia.

18.13. Hydrofluoric acid is a weak acid, whereas all the other hydrogen halides are strong acids in aqueous solution. Briefly rationalize this observation.

18.14. The reaction between hydrogen and fluorine gases is extremely exothermic and results in the highest known flame temperature (in excess of 6000° C, or about the same as the surface of the sun). Briefly rationalize these observations.

18.15. The bond energies of the H—H, F—F, and H—F bonds are 436, 151, and 568 kJ/mol, respectively. Determine the standard energy of formation of hydrogen fluoride gas.

18.16. In Chapter 9, we briefly discussed the unusually low value of the electron affinity of fluorine. Summarize the rationalization of this vertical anomaly in electron affinities.

18.17. Oxygen, like fluorine, has a lower electron affinity than its congeners. Briefly rationalize this anomaly in the electron affinities of the chalcogens.

18.18. How would you compare the uniqueness of the F—F bond to that of the N—N bond in F_2 and N_2, respectively?

18.19. Write a Born-Haber cycle corresponding to the standard reduction potential of fluorine. What thermodynamic properties would be needed to estimate the value of the reduction potential? Using your cycle, discuss why the standard reduction potential of fluorine is so extremely positive.

* **18.20.** Analyze the reaction represented in Equation (18.1) and repeated below by using standard reduction potentials. Is this reaction spontaneous under standard state conditions? What changes in the concentrations of the reactants and products would increase the spontaneity of this reaction?

$$4NaCl(aq) + 2H_2SO_4(aq) + MnO_2(s) \xrightarrow{\text{heat}} 2Na_2SO_4(aq) + MnCl_2(aq) + 2H_2O(l) + Cl_2(g)$$

18.21. Suppose that the sodium halides, NaX (X = F, Cl, Br, I) were treated with sulfuric acid. Predict the likely products in each case and discuss why they differ.

* **18.22.** Use standard reduction potentials to show why sulfuric acid could not be used to oxidize fluorides to fluorine but was used by Courtois to produce iodine from iodides. Why did Courtois have to use concentrated sulfuric acid?

18.23. Given the standard reduction potentials of the halogens, briefly explain why it is difficult to keep solutions of hydrobromic and hydroiodic acids in contact with the air without having them become contaminated with the free halogens.

18.24. Hydrogen fluoride has much higher melting and boiling points compared to the other hydrogen halides. Briefly rationalize this observation.

18.25. Sketch the structure of the $(HF)_6$ hexamer characterized by H—F --- H hydrogen bonds.

18.26. Could hydrobromic and hydroiodic acids be prepared by the action of nitric acid on bromides and iodides? Why or why not? Using potassium iodide as an example, write an equation to represent what you think would happen if nitric acid were added.

18.27. Could chlorine be produced by the action of nitric acid on a chloride? Support your answer by referring to standard reduction potentials. Write an equation as part of your answer.

18.28. How might you prepare antimony trifluoride and antimony pentafluoride? Give equations as part of your answer.

18.29. How might you prepare aluminum trifluoride and aluminum trichloride? Give equations as part of your answer.

18.30. Propose two methods of preparing calcium fluoride. Give equations as part of your answer.

18.31. Complete and balance the following equations:
(*a*) $Co_3O_4 + ClF_3 \longrightarrow$
(*b*) $B_2O_3 + BrF_3 \longrightarrow$
(*c*) $SiO_2 + BrF_3 \longrightarrow$

18.32. Speculate whether silicon tetrachloride should react with hydrogen fluoride in a gaseous halogen exchange reaction. Write a balanced equation for the process and estimate the heat of reaction using the following bond energies. Bond energies: Si—F = 582, Si—Cl = 391, H—F = 566, H—Cl = 431 kJ/mol, respectively.

18.33. Complete and balance the following equations:
(*a*) $GeCl_2(s) + H_2O \longrightarrow$
(*b*) $PCl_3(s) + H_2O \longrightarrow$
(*c*) $PCl_5(s) + H_2O \longrightarrow$

18.34. Write an equation representing the reaction between cyanogen, $(CN)_2$, and sodium metal.

18.35. Write an equation representing the reaction between sodium thiocyanate and silver nitrate.

18.36. Draw diagrams showing the structures of oxygen difluoride and dioxygen difluoride. Estimate the bond angles in each.

18.37. Speculate on the structures of dichlorine oxide and chlorine dioxide.

18.38. What is the relationship between dichlorine oxide and hypochlorous acid? Write an equation showing how the latter is produced from the former.

18.39. Identify the oxidizing and reducing agents in the reaction between iodine pentoxide and carbon monoxide shown in Equation (18.27), repeated below.

$$5CO(g) + I_2O_5(s) \longrightarrow I_2(s) + 5CO_2(g)$$

18.40. The total concentration of hypochlorous acid when water is saturated with chlorine gas is about 0.030 *M*. Accordingly, hydrolysis is not a particularly good way to make this acid. How is the preparation made feasible by the addition of a finely divided suspension of mercuric oxide? Write equations to support your answer.

18.41. Analyze the hydrolysis of a halogen reaction given in Equation (18.30), repeated below. Is this an acid-base or a redox reaction? Identify the acid and base reactants if the former or the oxidizing and reducing agents if the latter.

$$X_2(g, l, s) + H_2O \rightleftharpoons HOX(aq) + H^+(aq) + X^-(aq)$$

18.42. Predict the products and write balanced equations representing the reactions between hypochlorite and:
(*a*) Iodate
(*b*) Chlorite
(*c*) Sulfite

18.43. How might you prepare an aqueous solution of bromic acid starting with barium bromate? Give an equation as part of your answer.

18.44. Iodic acid can be prepared from iodine by oxidation using hydrogen peroxide. Write an equation to represent this preparation.

18.45. In the iodine clock reaction, a colorless solution suddenly changes to a blue-black iodine-starch complex in a predetermined amount of time (dependent upon the temperature and the concentrations of the reactants). The iodine is produced by the reaction of iodic acid and hydrogen iodide as represented in the following unbalanced equation. Identify the oxidizing and reducing agents in this reaction and balance the equation.

$$HIO_3 + HI \longrightarrow I_2 + H_2O$$

18.46. Draw and carefully label a geometrically accurate diagram of a molecule of iodic acid.

18.47. Perbromic acid was actually first prepared by the beta decay of selenate (containing selenium 83) to perbromate. Write an equation for this process.

18.48. Use standard reduction potentials to analyze the thermodynamic possibility of using (*a*) peroxydisulfate ($S_2O_8^{2-} \longrightarrow SO_4^{2-}$, $E^0 = +2.01$ V) and (*b*) ozone ($O_3 \longrightarrow O_2$, $E° = +2.07$ V) to oxidize bromate to perbromate.

18.49. Use standard reduction potentials to analyze the thermodynamic possibility of using xenon difluoride to oxidize bromate to perbromate.

$$XeF_2(aq) + 2H^+ + 2e^- \longrightarrow Xe + 2HF(aq) \qquad E° = 2.64\text{ V}$$

18.50. When barium iodate is treated with chlorine in a strongly basic solution, barium orthoperiodate, $Ba_5(IO_6)_2$, is produced. Orthoperiodic acid can be isolated from this solution by adding strong acid. Write equations representing these reactions.

18.51. Two condensed orthoperiodic acids are diperiodic acid, $H_4I_2O_9$, and triperiodic acid, $H_7I_3O_{14}$. Draw diagrams showing the structures of these acids.

18.52. The decomposition of ammonium perchlorate in rocket fuel is given below. Why is so much energy released by this reaction?

$$2NH_4ClO_4 \longrightarrow N_2 + Cl_2 + 2O_2 + 4H_2O$$

18.53. Write equations representing the synthesis of BrF, BrF_3, and BrF_5.

18.54. Write equations representing the synthesis of IF_3, IF_5, and IF_7.

18.55. Write equations representing the production of:
(*a*) Selenium tetrafluoride from selenium using chlorine fluoride
(*b*) Nickel fluoride from the oxide using chlorine trifluoride
(*c*) Silicon tetrafluoride from silicon dioxide using bromine trifluoride
(*d*) Phosphorus pentachloride from elemental phosphorus using iodine chloride

18.56. Write equations representing the hydrolysis of bromine fluoride, trifluoride, and pentafluoride.

18.57. Write equations representing the hydrolysis of iodine trifluoride, pentafluoride, and heptafluoride.

18.58. Write equations for the self-ionization of
(*a*) Iodine trichloride
(*b*) Chlorine trifluoride

18.59. Draw Lewis diagrams and geometrically accurate diagrams showing the structural formulas (as determined from VSEPR theory) of (*a*) IF_4^+ and (*b*) I_3^+.

* **18.60.** Propose a reason why IF_3 should be trigonal planar rather than the expected distorted T-shape.

18.61. Fluorine 20, a beta minus emitter, is used to determine the age of hominid bones and teeth. Briefly outline why this is possible.

18.62. In one well-written paragraph, summarize the pros and cons of fluoridation.

18.63. Why is hydrogen peroxide a good bleach? Include an equation as part of your explanation.

18.64. Write equations showing the production of household bleaches and bleaching powders from chlorine gas.

18.65. Write an equation for the production of free chlorine atoms from Freon-11.

18.66. Briefly explain why the temperature must fall to about $-80°$ C in order to form ice crystals or PSCs in the stratosphere.

18.67. Briefly explain why ozone can be depleted so much more over the Antarctic than it apparently can be in other parts of the stratosphere.

CHAPTER 19

GROUP 8A: THE NOBLE GASES

The Group 8A elements (helium, neon, argon, krypton, xenon, and radon) have been referred to at various times as the "inert" or sometimes the "rare" gases. Given the abundance of helium and the dozens of known compounds of xenon, neither of these group names is particularly appropriate. On the other hand, since most of these elements are generally reluctant to react or otherwise become involved with any but the most reactive elements, this group has become known as the *noble gases*. Helium, neon, or argon are still regarded as chemically inert, but both krypton and especially xenon have been found to form an ever-growing list of compounds. Although radon also has been found to form a few compounds, it is mostly known as a heavy, dense, radioactive gas derived from the radioactive thorium, uranium, and actinium decay series.

The history of these elements and their few compounds dominates this chapter. We start with the usual treatment of the discovery and isolation of the elements followed by a much shorter than normal section on periodic properties and the network. Given that the actual chemistry of this group is dominated by xenon, the third section is devoted almost entirely to the preparation, structure, and reactions of its compounds. A short section on the practical importance of the elements is followed by the selected topic in depth, the carcinogenic threat of radon.

19.1 DISCOVERY AND ISOLATION OF THE ELEMENTS

As shown in Figure 19.1, all of the noble gases were discovered within the short span of just 6 years at the end of the nineteenth century. Prior to that time, there were various suspicions and speculations that such elements might exist. Henry

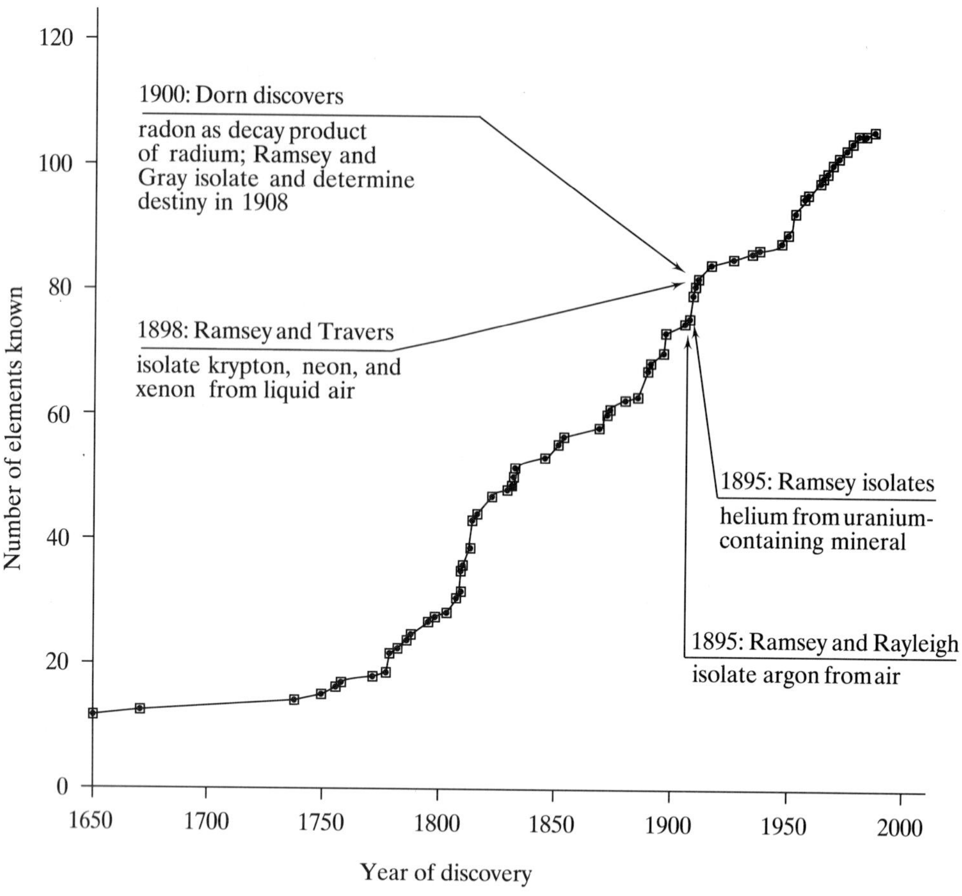

FIGURE 19.1
The discovery of the Group 8A elements superimposed on the plot of the number of known elements versus time.

Cavendish, for example, on the basis of experiments to be described shortly, had speculated that the atmosphere contained gases other than dephlogisticated air (oxygen), phlogisticated air (nitrogen), fixed air (carbon dioxide), and water vapor. Nevertheless, it took more than a century before the pivotal work of Lord Rayleigh and Sir William Ramsay firmly established the existence of the noble gases.

Argon

John William Strutt, the third Lord Rayleigh, was an English physicist primarily interested in the interactions between electromagnetic radiation and matter. In an early piece of work, he determined how the scattering of light is mathematically related to its wavelength. (Using this mathematics, John Tyndall explained how what is still known as the *Rayleigh scattering* of visible light is responsible for the blue color of the sky and the red colors of sunsets.) Rayleigh also investigated the mathematics of *blackbody radiation* that puzzled physicists for several decades and finally led Max Planck, in 1900, to reluctantly postulate that light comes in only certain allowed energies ($E = h\nu$) called *quanta*. In the 1880s, in a rather abrupt change in research interests, Rayleigh turned his attention to William Prout's hypothesis, first proposed in 1815, that all elements were composed of hydrogen atoms and that all atomic weights would be found to be exact multiples of that of hydrogen.

Like others before him, Rayleigh also disproved Prout's hypothesis. More importantly, he carefully investigated the densities of gases, particularly those which make up the atmosphere. He found, for example, that the nitrogen gas left over after all the oxygen, carbon dioxide, and water vapor had been removed from a sample of air was just a little denser than the nitrogen gas obtained by decomposing ammonia, ammonium nitrite, or other nitrogen-containing compounds. (Rayleigh, by the way, was not the first to note this discrepancy. Cavendish recorded in his notebook that about 1/120 of the volume of air could not be accounted for after the major gases had apparently been removed.) There was no easy explanation for this difference in nitrogen densities. Rayleigh was so puzzled that he even wrote to the journal *Nature* to ask "chemical readers" for suggestions.

There being no viable suggestions forthcoming from the *Nature* readership, Rayleigh was happy to honor the request of William Ramsay who wrote to ask if he might further investigate the nature of atmospheric nitrogen. Note from Figure 19.1 that this Scottish chemist was to play a prominent role in the discovery of each and every one of the noble gases. Ramsay repeated the experiments of Cavendish and Rayleigh and found that one-eightieth of his air samples remained unaccounted for. The spectrum of the remaining fraction revealed new red and green bands. Ramsay began to suspect that he and Rayleigh had found a new (or at least unrecognized) constituent of the atmosphere. But was it a mixture or a pure substance? If it was a pure substance, was it an element or a compound? Further investigation showed that this constituent could not be further separated and, furthermore, seemed to be completely unreactive. All the evidence seemed to point toward a new element, but there remained one major problem.

Assuming the new constituent of the atmosphere to be a pure, uncombined new element, the atomic weight (as calculated from the density) came out to be about 39.9 u. This would place such an element between potassium and calcium, but there was no room there! Yet there seemed to be, as Ramsay wrote to Rayleigh in 1894, "room for gaseous elements at the end... of the periodic table." Perhaps the atomic weight of the new element and that of potassium were inverted. (There were two other known atomic weight inversions, tellurium-iodine and cobalt-nickel, but many suspected these reversals were due to experimental difficulties.) All this presented a dilemma not easily resolved. Nevertheless, later in 1894, Ramsay and Rayleigh jointly announced the discovery of a "new Gaseous Constituent of the Atmosphere." They were careful not to claim that this was a new element.

In 1894, the periodic table was only about 25 years old and Mendeleev himself was only 60. The table was already recognized as one of the great empirical generalizations of science and Mendeleev as one of the greatest living chemists. The Russian was not prepared to accept atomic weight inversions, and he had several theories about this new *argon* (from the Greek *argos*, for "idle" or "lazy"). For example, he and others thought argon might be triatomic nitrogen (like ozone was to oxygen) or a diatomic molecule of a new element of atomic weight of 20 (which *would* fit between fluorine and sodium). Yet if argon was an unreactive element, its valence of 0 would certainly fit the spirit of Mendeleev's table. Perhaps time would solve this nagging dilemma. Lecoq de Boisbaudran, the discoverer of gallium, suggested, as did others, that a whole new family of these elements might exist. A detailed investigation of the nature of such a new group would then put argon into its proper place. The race to find these elements was on.

Helium

During a solar eclipse in 1868, the French astronomer Pierre Janssen observed a new spectral line in the chromosphere of the sun. He sent his results to Joseph Lockyer, an English astronomer and expert on the solar spectrum who, as it turned out, had found the same unexplained line in the spectra of solar prominences. They both speculated that this line could be that of a new element, but their proposal was greeted with a great deal of skepticism. After all, spectroscopy was a new technique, having been devised by Bunsen and Kirchhoff less than a decade before Janssen's and Lockyer's observations. (See Chapter 12, p. 299, for more about spectroscopy.) So although this new line was tentatively assigned to a new element called *helium* (from the Greek *hēlios*, for "sun"), many scientists remained unconvinced.

Nothing more was done on this idea of a new element found only on the sun until, in the late 1880s, an unknown, inert gas was found to be liberated from the mineral uraninite. The element had tentatively been identified as nitrogen, but when Ramsay read the report of the discovery, he had a different idea. This gas could very well be, he thought, another piece of the puzzle which, when put together, might resolve the dilemma of argon. Ramsay set to work, and when he isolated a gas from clevite, also a uranium-containing mineral, he gave some to Crookes who ran a spectrum. It showed the same line observed by Janssen and Lockyer. It was the sun's new element come to light on earth. Within a few years, helium was found to be present in significant quantities in natural gas deposits, particularly those found in the southwestern United States.

Krypton, Neon, and Xenon

The discoveries of these three elements were all announced in 1898 by Ramsay and his young co-worker, Morris Travers. Building on earlier attempts to isolate additional inert gases, they decided to practice some manipulative techniques on liquid air before risking their small supply of precious argon. Somewhat unexpectedly, the residue remaining after the liquid oxygen and nitrogen were allowed to boil away showed new green and orange spectral lines which were attributed to a new element they called *krypton* (from the Greek *kryptos*, for "hidden"). Working late into the night, Ramsay and Travers determined that its atomic weight placed their new element between bromine and rubidium in the periodic table. Travers was so excited he almost forgot his doctoral examination scheduled for the next day.

With their techniques perfected, Ramsay and Travers turned to repeated fractionations of argon. They soon discovered a lighter fraction with many spectral lines in the red, faint-green, and the violet. The discharge tube containing this fraction blazed with a bright-crimson light. Ramsay's 13-year-old son wanted to call this gas novum (from the Latin for "new"), but they settled on *neon* (from the Greek *neos*) having the same meaning. Ramsay and Travers were observing the first neon light.

By this time Ramsay and his colleague had a new "liquid air machine" from which they could get larger and larger quantities of krypton and neon. Repeated fractionations produced a heavier gas which produced a beautiful blue glow in their discharge tubes. They called it *xenon* (from the Greek *xenos*, for "strange").

Radon

Recall (Chapter 13, pp. 326–328; Chapter 17, p. 448) that the Curies also announced their discoveries of radium and polonium in 1898. (That makes five elements discovered in 1 year.) They had found that not only was radium itself intensely radioactive but so also was the air with which it came into contact. This was because the alpha decay of radium produces the final and heaviest inert gas [see Equation (13.6)] that we know today as radon. (Like radium, the name *radon* comes from the Latin *radius*, for "ray" or "beam.") For some years, there were various names applied to this element, with radon becoming the official name in 1923. Friedrich Dorn is generally credited with its actual discovery in 1900. He called it the "radium emanation." In 1908, Ramsay (again) and Gray isolated and determined the density of the last of the noble gases, which, it turns out, is the densest gaseous element known.

In 1904, a most unusual Noble Prize ceremony occurred. The winner in *physics* for his work on the densities of gases and the discovery of argon was Lord Rayleigh. The winner in *chemistry* for his work in discovering the inert gaseous elements in air and their place in the periodic table was William Ramsay. Sometimes the divisions between these two sciences are very small indeed.

19.2 FUNDAMENTAL PROPERTIES AND THE NETWORK

Figure 19.2 shows the noble gases superimposed on the network of interconnected ideas. Table 19.1 is a slightly amended version of the usual table of periodic properties. Note that these properties are exactly as expected on the basis of effective nuclear charge and the distance of the valence electrons from that charge. Consistent with the noble nature of these elements, the usual entries for atomic and ionic radii have been replaced by van der Waals radii. Only two entries, for xenon and krypton, have been made in the table under covalent radii. (Several radon compounds are known, but the covalent radius has not been well established.) As expected, these radii increase regularly down the group.

With very high effective nuclear charges operating on the valence electrons of the Group 8A elements, their ionization energies are exceedingly high. However, as the valence electrons are removed farther and farther from this effective nuclear charge, the ionization energies slowly decrease, and it gets easier to remove an electron from the heavier elements. As we will discuss in the next section, this trend in ionization energies played an important role in the thinking which lead to the synthesis of the very first compounds of xenon by Bartlett in 1962.

On the other hand, the effective nuclear charges operating on electrons coming into the range of Group 8A elements are very small. Accordingly, they have little ability to attract electrons to themselves. No electronegativities are listed for this reason. Electron affinities can be estimated, and, as shown in Table 19.1, they are slightly positive, an indication that the addition of electrons to these atoms is not a particularly favorable process. The electron affinities become more positive going down the group, a trend consistent with the idea that the electrons are being added to orbitals farther and farther removed from the effective nuclear charge.

As more electrons are associated with the heavier noble gases, the atoms are more polarizable and the van der Waals intermolecular forces become somewhat

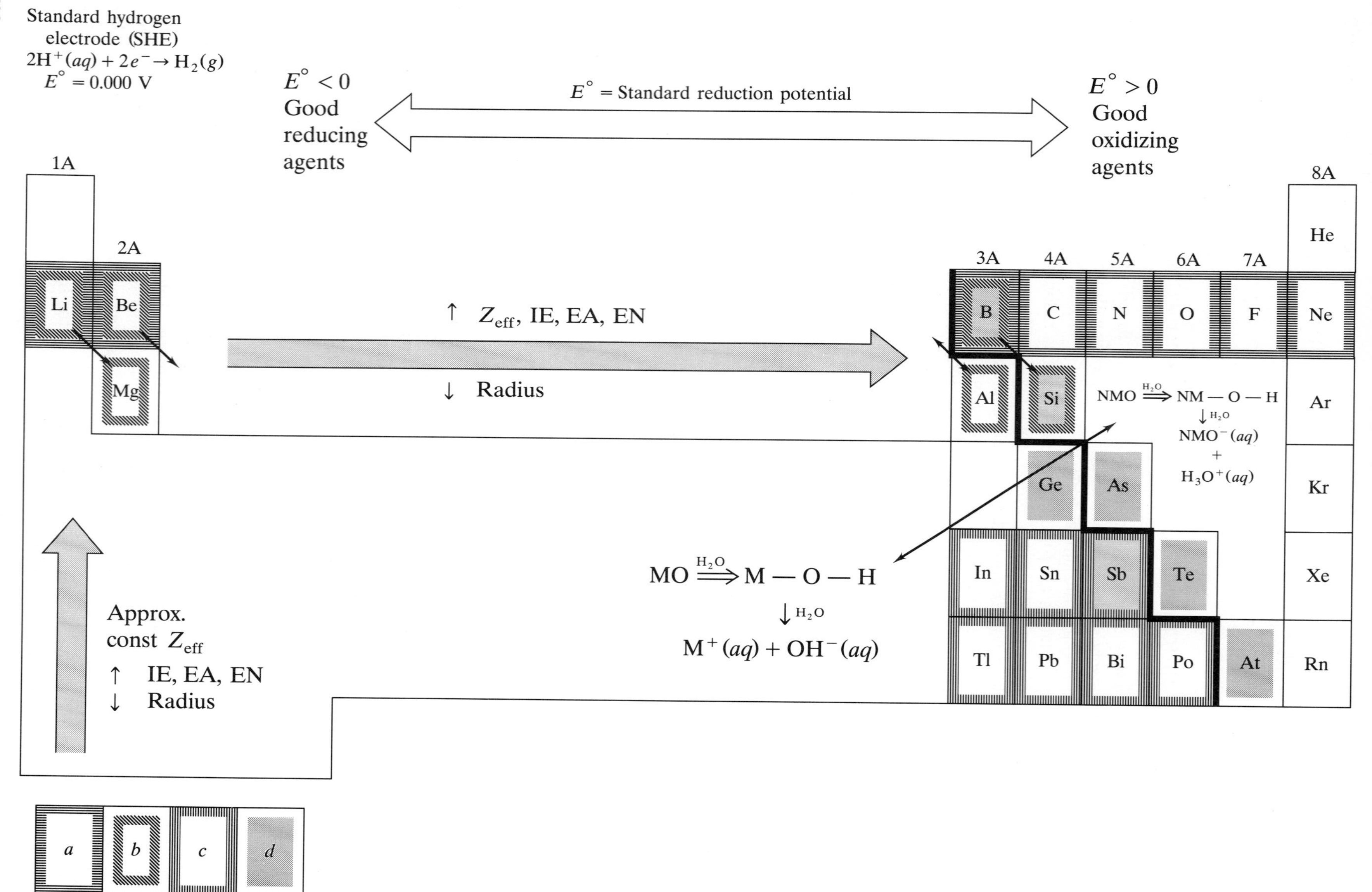

FIGURE 19.2
The noble gases (Group 8A) superimposed on the network of interconnected ideas including the trends in periodic properties, the acid-base character of metal and nonmetal oxides, trends in standard reduction potentials, and (*a*) the uniqueness principle, (*b*) the diagonal effect, (*c*) the inert-pair effect, and (*d*) the metal-nonmetal line.

TABLE 19.1
The fundamental properties of the Group 8A elements

	Helium	Neon	Argon	Krypton	Xenon	Radon
Symbol	He	Ne	Ar	Kr	Xe	Rn
Atomic number	2	10	18	36	54	86
Natural isotopes, A/% abundances	$^{4}He/100$ $^{3}He/0.00013$	$^{20}Ne/90.92$ $^{21}Ne/0.257$ $^{22}Ne/8.82$	$^{36}Ar/0.337$ $^{38}Ar/0.063$ $^{40}Ar/99.60$	$^{78}Kr/0.35$ $^{80}Kr/2.27$ $^{82}Kr/11.56$ $^{83}Kr/11.55$ $^{84}Kr/56.90$ $^{86}Kr/17.37$	$^{124}Xe/0.096$ $^{126}Xe/0.090$ $^{128}Xe/1.92$ $^{129}Xe/26.44$ $^{130}Xe/4.08$ $^{131}Xe/21.18$ $^{132}Xe/26.89$ $^{134}Xe/10.44$ $^{136}Xe/8.87$	^{222}Rn†
Total no. of isotopes	5	8	8	21	25	20
Atomic weight	4.003	20.18	39.95	83.80	131.3	(222)
Valence electrons	$1s^2$	$2s^22p^6$	$3s^23p^6$	$4s^24p^6$	$5s^25p^6$	$6s^26p^6$
mp/bp, °C	−272/− 269	−249/− 246	−189/− 186	−157/− 152	−112/− 107	−71/− 62
Density, g/liter	0.177	0.900	1.784	3.733	5.887	9.73
Covalent radius, Å	....	....	....	1.15	1.26	
Van der Waals radius, Å	1.40	1.54	1.88	2.02	2.16	
Oxidation states	0	0	0	0, +2	0, +2, +4, +6	0, +2
Ionization energy, kJ/mol	2373	2080	1521	1241	1167	
Electron affinity, kJ/mol (estimates)	21	29	35	39	40	
Isolated by/date	Ramsay 1895	Travers & Ramsay 1898	Rayleigh & Ramsay 1894	Travers & Ramsay 1898	Travers & Ramsay 1898	Dorn 1900
rpw‡ halogens	....	....	....	KrF_2	XeF_n (n = 2, 4, 6)	
Crystal structure	hcp	fcc	fcc	fcc	fcc	fcc

†Longest-lived isotope, $t_{1/2}$ = 3.82 days.
‡rpw = reaction product with.

stronger. Accordingly, the boiling points and melting points increase going down the group. Note, however, that the boiling and melting points, even of the heavier elements, are still very low. Helium has a melting point exceedingly close to absolute zero. At 2.2 K, incidentally, there is a transition from normal liquid helium, called helium I, to a superfluid form called helium II. Helium II has essentially zero surface tension and will spread out to cover all surfaces in a film only a few atoms thick. If helium II is placed in a prechilled beaker, it will climb up the inside walls and escape down the outside of the beaker.

The inertness of these elements provided a vital link in the empirical and theoretical understanding of the periodic table. Empirically, a 0 valence (atomic weight inversions notwithstanding) was immediately recognized as providing a bridge between the halogens (−1 valence) and the alkali metals (+1 valence). A theoretical interpretation of 0 valence, however, had to await the development of the nuclear atom in which the nucleus is surrounded by a field of electrons. These ideas were worked out by Rutherford, Thomson, and others in the early part of the twentieth century. Gilbert Newton (G. N.) Lewis, in 1916 or thereabouts, applied these new ideas to account for the high stability of the electronic configurations of the noble gases. The stability of the octet, or the *octet rule*, although recognized as a useful but limited generalization, is still a dominant organizing principle in the minds of all chemistry students.

Note that the "reaction products with" entries in Table 19.1 are much abbreviated compared to the analogous tables of earlier groups. Only xenon and krypton react with fluorine to produce fluorides. Therefore, instead of following the usual format of describing the hydrides, oxides, hydroxides, and halides of these elements (most of which do not exist), we will adopt a historical description of the synthesis of xenon compounds and then briefly expand the discussion to include the small number of examples drawn from krypton and radon chemistry.

19.3 COMPOUNDS OF NOBLE GASES

History

Even though Ramsay was unable to get his new elements to react with anything, he maintained from the beginning that it should be possible. Moissan, who had just succeeded in finally isolating fluorine (see Chapter 18, p. 476), could not get the most reactive element of all to combine with the limited amounts of these gases available to him. Linus Pauling, in 1933, predicted that XeF_6 and KrF_6 should be attainable, and various attempts were made to test his prediction. For example, an electric discharge was applied to mixtures of xenon and fluorine, and xenon and chlorine were subjected to various photochemical techniques. Nothing happened. As it turns out, xenon and fluorine will react photochemically, but this combination of reactants and technique was not tried. Had it been, the first noble gas compound would most likely have been synthesized some 30 years earlier.

Noble gas hydrates, in which the atoms are trapped in an icelike lattice when water is frozen under a high pressure of the gas, have been prepared. (See Chapter 11, pp. 270–271, for a description of the large hexagonal holes which form in ice.) A few organic compounds also form similar structures called *clathrates*. The hydrates are of general formula $E \cdot 6H_2O$, where E = Ar, Kr, Xe, but neither these nor the clathrates are compounds in the usual sense of that term. (They contain neither ionic nor covalent bonds involving noble gas atoms.) There has been, over the years, some spectroscopic evidence for a few fleetingly stable cations involving the noble gases (HeH^+, ArH^+, He_2^+, Kr_2^+, $NeXe^+$, and most recently Xe_2^+), but until 1962 no stable neutral compounds of the noble gases were known. Up to that year, the term *inert gases* was an appropriate group name for these elements.

Neil Bartlett, an English chemist working at the University of British Columbia, was interested in the fluorides of platinum and related metals. During World War II, uranium hexafluoride, UF_6, an easily vaporized compound, proved immensely useful in separating uranium isotopes by gaseous diffusion. The uranium 235 used to make the first atom bomb dropped over Hiroshima contained fissionable U 235 separated from the nonfissionable U 238 by this technique. The resulting interest in the synthesis and characterization of various metal fluorides had extended into the 1950s and was one reason Bartlett was pursuing his line of research.

Bartlett was investigating platinum hexafluoride, PtF_6, a deep-red gas which is an extremely strong oxidizing and fluorinating agent. It had to be manipulated in a completely air- and water-free environment, and even then it appeared to be reacting with the glass (silicon dioxide) of his apparatus to give an orange-brown solid. A closer investigation, however, showed that some small amounts of diatomic oxygen were present in his vacuum line and that the orange-brown solid was in fact a compound called oxygenyl hexafluoroplatinate(V). Equation (19.1) represents what was happening in Bartlett's vacuum line. Note that diatomic oxygen is being oxidized by the platinum hexafluoride. The latter must indeed be a strong oxidizing agent:

$$O_2(g) + PtF_6(g) \longrightarrow O_2^+PtF_6^- \tag{19.1}$$

It occurred to Bartlett that the ionization energies of the oxygen molecule (1180 kJ/mol) and the xenon atom (1167 kJ/mol) were remarkably similar. He decided to try the same reaction as above with xenon replacing the diatomic oxygen. He prepared known volumes of xenon (in slight excess) and platinum hexafluoride and carefully noted the pressure of each. When he allowed the two gases to mix, an orange-yellow solid was immediately formed and the pressure of the remaining xenon was consistent with the formation of a 1 : 1 compound. The reaction was initially represented as shown in Equation (19.2):

$$\underset{\text{Colorless}}{Xe(g)} + \underset{\text{Red}}{PtF_6(g)} \longrightarrow \underset{\text{Orange-yellow solid}}{Xe^+PtF_6^-(s)} \tag{19.2}$$

Further investigation demonstrated that the reaction was more complicated than Bartlett had originally thought. In fact, this reaction is still not well understood. Equation (19.3) is probably more representative of what happens when xenon and platinum hexafluoride come together:

$$Xe + 2PtF_6 \xrightarrow{25°C} [XeF^+][PtF_6^-] + PtF_5 \tag{19.3a}$$

$$[XeF^+][PtF_6^-] + PtF_5 \xrightarrow{60°C} [XeF^+][Pt_2F_{11}^-] \tag{19.3b}$$

The exact nature of this reaction, while important, was not as significant as the fact that it was no longer correct to regard xenon as an inert gas. "Noble" perhaps it was, but no longer "inert."

Fluorides

A flurry of activity followed Bartlett's announcement of the preparation of the first xenon compound. Only a few months later, a group at Argonne National Laboratory was able to prepare xenon tetrafluoride by *direct* reaction of the elements. They placed a 1 : 5 Xe/F_2 ratio of the gases in a nickel container, and, after an hour at 400°C and 6 atm, the xenon was completely consumed. They sublimed the mixture to produce brilliant colorless crystals of XeF_4. The reaction is summarized in Equation (19.4):

$$Xe(g) + 2F_2(g) \xrightarrow[6\text{ atm}]{400°C} XeF_4(s) \tag{19.4}$$

The tetrafluoride is the easiest to make, the most stable, and the best characterized xenon-fluoride compound. One must, however, be careful to keep it out of contact with moisture as it will react as shown in Equation (19.5) to produce the violently explosive (comparable to TNT) xenon trioxide. Xenon tetrafluoride is an excellent fluorinating agent as shown in Equations (19.6) and (19.7):

$$6XeF_4 + 12H_2O \longrightarrow 2XeO_3 + 4Xe + 3O_2 + 24HF \tag{19.5}$$

$$Pt(s) + XeF_4 \longrightarrow PtF_4 + Xe(g) \tag{19.6}$$

$$2SF_4 + XeF_4 \longrightarrow 2SF_6 + Xe(g) \tag{19.7}$$

Both the difluoride and the hexafluoride of xenon can also be made directly from the elements by varying the reaction conditions. An excess of xenon produces XeF_2. The reaction can even by carried out by exposing the mixture to ordinary sunlight. Large excesses of fluorine at high pressures yield the hexafluoride. These are both solids at room temperature and, like the tetrafluoride, vigorously react with water as shown in Equations (19.8) to (19.10):

$$2XeF_2 + 2H_2O \longrightarrow 2Xe + 4HF + O_2 \qquad (19.8)$$

$$XeF_6 + H_2O \longrightarrow XeOF_4 + 2HF \qquad (19.9)$$

$$\underset{\text{Excess}}{XeF_6 + 3H_2O} \longrightarrow XeO_3 + 6HF \qquad (19.10)$$

There are two characteristic reactions of xenon difluoride that need to be mentioned here. One is its action as an oxidizing agent. The standard reduction potential of XeF_2 to Xe is +2.64 V as shown in Equation (19.11):

$$XeF_2(aq) + 2H^+(aq) + 2e^- \longrightarrow Xe(g) + 2HF(aq) \qquad E° = +2.64 \text{ V} \qquad (19.11)$$

This makes XeF_2 so strong an oxidizing agent that it can even be used to oxidize bromate to perbromate (see Problem 18.51). It also will oxidize a variety of other substances, including chromium(III) to chromate and chloride to chlorine as shown in Equations (19.12) and (19.13):

$$XeF_2 + 2Cl^- + 2H^+ \longrightarrow Xe + Cl_2 + 2HF \qquad (19.12)$$

$$XeF_2 + 2Cr^{3+} + 7H_2O \longrightarrow 3Xe + Cr_2O_7^{2-} + 6HF + 8H^+ \qquad (19.13)$$

The second characteristic reaction of xenon difluoride is its combination with fluoride-ion acceptors such as various pentafluorides. Two typical reactions are represented in Equations (19.14) and (19.15):

$$MF_5 + XeF_2 \longrightarrow [XeF]^+[MF_6^-] \qquad (19.14)$$

$$MF_5 + 2XeF_2 \longrightarrow [Xe_2F_3^+][MF_6^-] \qquad (19.15)$$

where M = P, As, Sb, I, and some metals.

In addition to its characteristic hydrolysis discussed above, xenon hexafluoride also acts as both a fluoride acceptor and a donor as represented in Equations (19.16) and (19.17), respectively:

$$XeF_6 + RbF \longrightarrow RbXeF_7 \qquad (19.16a)$$

$$XeF_6 + 2CsF \longrightarrow Cs_2XeF_8 \qquad (19.16b)$$

$$XeF_6 + EF_5 \longrightarrow [XeF_5^+][EF_6^-] \qquad E = \text{As, Pt} \qquad (19.17)$$

Structures

Given their relative ease of preparation and high reactivity, it is not surprising that the binary fluorides constitute the major starting materials for other xenon compounds. The structures of some of these compounds are shown in Table 19.2. Although most of the molecular geometries follow readily from VSEPR considerations, three (the hexafluoride, XeF_6, the octafluoroxenate anion, XeF_8^{2-}, and the oxytetrafluoride, $XeOF_4$) require some brief additional comment.

Various viewpoints for XeF_6 are pictured in Figure 19.3. Note that the Lewis structure of Figure 19.3*a* shows seven electron pairs around the central xenon

TABLE 19.2
Structures of some representative noble gas molecules and ions

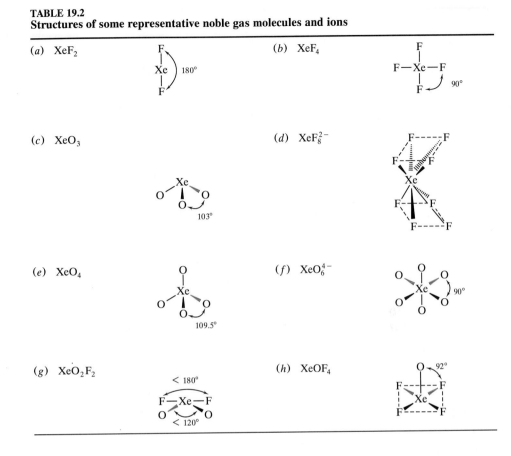

(*a*) XeF_2	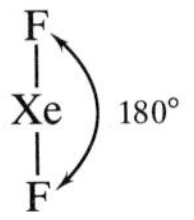 F–Xe–F, 180°	(*b*) XeF_4	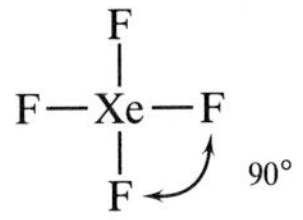 F–Xe–F, 90°
(*c*) XeO_3	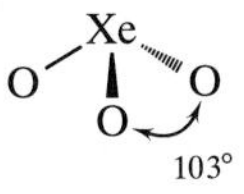 Xe, O, 103°	(*d*) XeF_8^{2-}	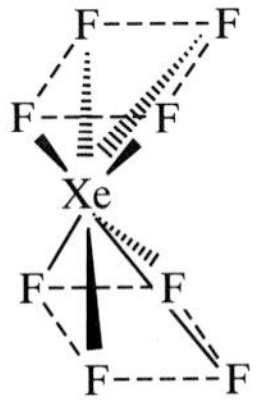 Xe, F
(*e*) XeO_4	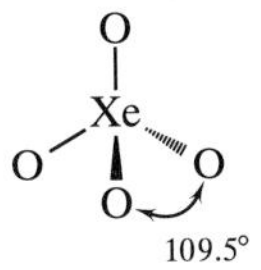 Xe, O, 109.5°	(*f*) XeO_6^{4-}	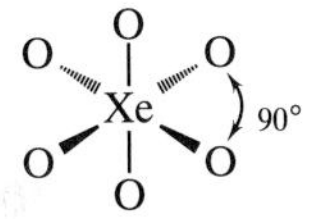 Xe, O, 90°
(*g*) XeO_2F_2	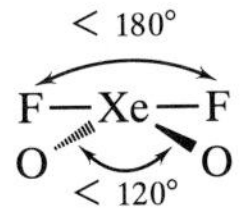 F–Xe–F, < 180°, O, < 120°	(*h*) $XeOF_4$	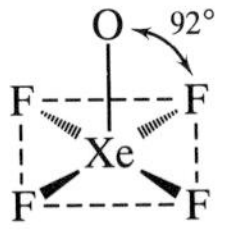 O, Xe, F, 92°

atom. One might suspect, then, that the molecule will assume a pentagonal pyramidal (Figure 19.3*b*) structure, in which a lone pair of electrons occupies one of the positions of the pentagonal bipyramidal configuration first introduced in Chapter 18 (p. 489). Or perhaps it might assume another geometry such as the monocapped octahedron shown in Figure 19.3*c*. It turns out, however, that the gas-phase molecular structure is an only slightly distorted, rather "soft" or non-rigid octahedron with a very small dipole moment. Apparently, the molecule slips rather easily from one structure to another. In the colorless, crystalline solid, structural studies show the existence of at least four different forms, all characterized by square pyramidal XeF_5^+ cations bridged by F^- ions. Three of the forms involve tetramers of formula $[(XeF_5^+)F^-]_4$, while the fourth is a hexamer, $[(XeF_5^+)F^-]_6$. These polymeric units are shown in Figure 19.3*d* and *e*.

The octafluoroxenate(VI) anion, XeF_8^{2-}, assumes a square antiprismatic geometry as shown in Table 19.2(*d*). Again, an ordinary Lewis structure would show nine pairs of electrons around the central xenon. In this case, however, it has been suggested that the lone pair occupies the spherical 5*s* orbital (another example of the inert-pair effect) and the remaining eight bonding pairs arrange themselves in the regular square antiprism. $XeOF_4$, shown in Table 19.2(*h*), assumes a rather regular square pyramid with the oxygen in the axial position. Apparently, the oxygen and the lone pair exert rather similar repulsive forces so that the O–Xe–F angle is very close to 90°.

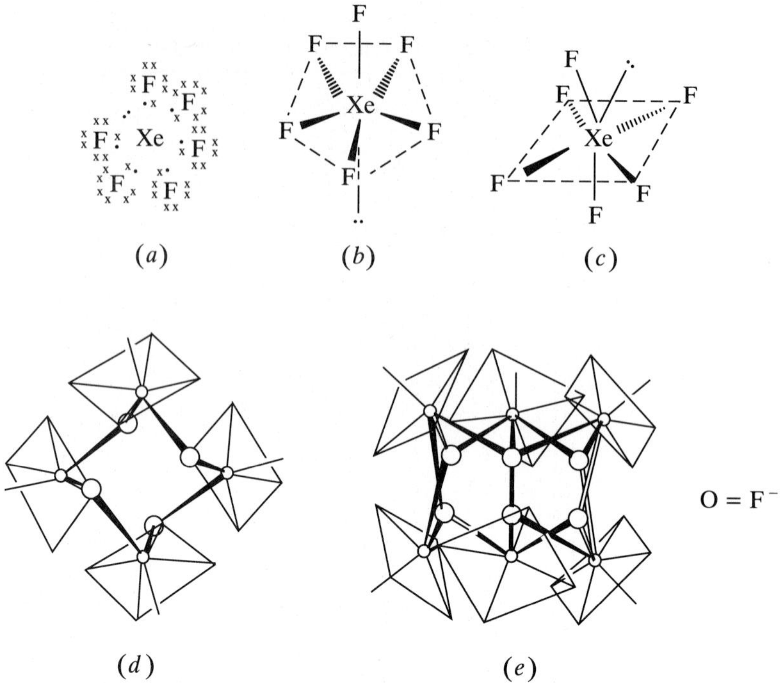

FIGURE 19.3
Various representations of the structure and bonding of xenon hexafluoride: (*a*) Lewis structure, (*b*) pentagonal bipyramid, (*c*) monocapped octahedron, (*d*) tetrameric form $[XeF_5^+)F^-]_4$, and (*e*) hexameric form $[(XeF_5)F^-]_6$.

Other Compounds

The fluorides are essentially the only stable halides of xenon. $XeCl_2$ has been obtained by condensing the products after a microwave discharge has been passed through a mixture of chlorine and an excess of xenon. Xenon tetrachloride and xenon dibromide have also been detected.

Xenon trioxide has been mentioned above as the product of the hydrolysis of several of the binary fluorides. It is an excellent oxidizing agent with a standard reduction potential of 2.10 V as shown in Equation (19.18):

$$XeO_3 + 6H^+ + 6e^- \rightleftharpoons Xe + 3H_2O \qquad E^\circ = 2.10\text{ V} \tag{19.18}$$

Aqueous solutions of XeO_3 are often referred to as *xenic acid*. Xenic acid can be oxidized by ozone, for example, to the perxenates as represented in Equation (19.19):

$$XeO_3 + 12NaOH(aq) + O_3 \longrightarrow Na_4XeO_6 + 6H_2O \tag{19.19}$$

A powerful and rapid oxidizing agent, the XeO_6^{4-} anion has, as shown in Table 19.2(*f*), an octahedral configuration.

Krypton difluoride can be synthesized by passing an electric discharge through the constituent elements at −196°C. It is a volatile white solid containing linear molecular KrF_2 units.

A few compounds with bonds to nitrogen are known. Both krypton and xenon have been reported to form a compound of general formula [HC≡N—

$EF^+][AsF_6^-]$, where E = Kr or Xe. The krypton compound detonates violently with a flash of white light.

Radon reacts spontaneously at room temperature with fluorine or chlorine trifluoride to form radon difluoride. This compound is not well characterized due to the difficulties of working with the extremely radioactive radon.

19.4 PHYSICAL PROPERTIES AND ELEMENTS OF PRACTICAL IMPORTANCE

Usually at this point we discuss *reactions and compounds* of practical importance. However, for Group 8A there are very few reactions and compounds of any kind, independent of their practical importance. On the other hand, there are, to use the analogous phrase, some *physical properties and elements* which bear upon how and why the world works as it does. These are briefly explored here. To start the discussion, the atmospheric abundances, sources, and the important uses of the noble gases are listed in Table 19.3.

Note that argon is in significantly greater supply than the others. Why should this be? The answer lies in the amount of argon 40 produced by the beta plus decay of potassium 40. Recall (from Chapter 12, p. 316) that it is this decay scheme

TABLE 19.3
Atmospheric abundances, sources, and uses of the noble gases

Element	Atmospheric abundance, % by volume	Sources	Uses
Helium	5×10^{-4}	Radioactive (α) decay Natural gas wells	Si/Ge crystal growth Ti/Zr production (inert blanket) Nuclear reactor coolant Diving artificial atmosphere Pressurizing liquid fuel rockets Inert atmosphere for welding Flow gas in chromatography Aircraft tire inflation Lifting gas
Neon	1.2×10^{-3}	Fractionation of liquid air	Neon lights Lasers High-voltage indicators Cryogenics
Argon	0.94	Radioactive decay of potassium 40 Fractionation of liquid aid	Si/Ge crystal growth Inert gas for incandescent bulbs and fluorescent tubes Inert atmosphere for welding Inert atmosphere for research Ti/Zr production (inert blanket)
Krypton	1.1×10^{-4}	Fractionation of liquid air	Standard meter Inert gas for fluorescent tubes High-speed photographic lamps Lasers
Xenon	9×10^{-6}	Fractionation of liquid air	Inert gas for fluorescent tubes Lasers
Radon	....	α decay of radium	Medicinal (obsolete)

which has been employed to indirectly determine the age of various hominids like the *Australopithecus afarensis* called "Lucy." While the argon gas remains trapped in some rock samples and is therefore useful for chronometric determinations, most of it escapes into the open atmosphere.

Why is it that helium is found in natural gas deposits? First, we should note that in 1907, Ernest Rutherford and his co-workers demonstrated that the alpha particles emitted by so many radioactive elements are, in fact, just helium nuclei. These nuclei go on to pick up electrons from the environment to become helium atoms. Eventually, enough helium gas is produced in the soil that it collects together in natural gas deposits. As we continue to use up our finite supplies of natural gas, so too will we deplete our major terrestrial source of helium.

Recall that helium, while being very scarce on earth, is the second most abundant element in the universe. In Chapter 10, p. 242, we noted that current thinking has the energy of the big bang condensing into matter, one-third of which is thought to have been helium. Helium is also produced in the proton-proton cycle of stars and is a major component in nucleosynthesis. Any helium which was present when our planet was formed from the original solar nebula would have quickly escaped because the earth's gravitational field was not strong enough to retain it.

Several of the uses of the noble gases noted in Table 19.3 involve the setting up of an inert atmosphere to carry out a process which would not work well in air. These include (1) passing an electric current through tungsten wire to produce incandescent light (the use of argon reduces the evaporation from the filament and so prolongs the lifetime of the bulb), (2) arc welding (helium is used in the United States because of its great abundance, but argon is used in most other places), and (3) preparations of reactive elements such as titanium and zirconium or elements which must be prepared in a very pure form such as silicon and germanium for use in semiconductors.

An 80 percent helium–20 percent oxygen mixture is used as an artificial atmosphere for drivers. Nitrogen cannot be used because it is more soluble in the blood. When a diver starts to ascend, the nitrogen escapes from the blood too slowly, causing the *bends* which can be fatal. Helium is less soluble in blood and so minimizes this problem.

Helium is also used in *cryogenics*, the study of the behavior of substances at very low temperatures. Helium has the lowest boiling point known and therefore is used to liquefy out other gases. Using liquid helium, temperatures can be reduced sufficiently that superconductivity becomes common. (*Superconductivity* is the property of certain metals, alloys, and compounds such that they conduct electricity with very little resistance. It used to be that superconductivity could only be achieved at liquid helium temperatures. More recently, however, compounds containing rare-earth metals, copper, and oxygen have been found to be "warm" superconductors at temperatures well over 100 K.)

In 1960, it was internationally agreed that the fundamental unit of length, the meter, would be defined in terms of the orange-red spectral line of krypton 86. In fact, 1 meter is exactly 1,650,763.73 wavelengths (in vacuo) of this line. This agreement replaced the standard platinum-iridium meter located in Paris.

Neon lights have become pervasive in today's cultures. Ramsay would be truly astonished to drive down a typical modern main street. To think that he had to work so hard to find this inert gas and here it is glowing red and (by appropriately mixing it with other gases) a variety of other colors in discharge tubes of a nearly inexhaustible and sometimes highly provocative supply of shapes.

Neon lighting certainly makes this gas the best known of all the Group 8A elements.

19.5 SELECTED TOPIC IN DEPTH: RADON AS A CARCINOGEN

In 1984, an engineer working at the Limerick nuclear power plant in eastern Pennsylvania had the alarming experience of setting off the warning signals on the plant's radiation counters. Even more unsettling was that these alarms sounded not after he had worked all day but when he first *arrived* at work from home. Eventually, it was discovered that there were very high concentrations of radon gas in the basement of his house. Other houses in the area were tested, with some showing similarly high radon levels. At first, it was hoped that the problem might be confined to special geographic areas, types of soil, or house construction, but this does not seem to be the case. The Environmental Protection Agency (EPA) now says that radon can be found throughout the country and that every home in the United States should be tested for its presence. In just a few short years, radon has become a national environmental health problem.

Most of the radon in homes is radon 222, a decay product of naturally occurring radium 226. Ra 226, in turn, is part of the naturally occurring decay series from uranium 238 to lead 206. (See Chapter 15, pp. 392–393, for more on these series.) So Rn 222 is released by rocks, soil, and groundwater and seems to enter homes from the soil, eventually seeping into homes through small cracks in basement floors and walls. It is now estimated that every square mile of soil 6 in deep contains about 1 g of radium which gradually decays to radon.

The health effects of radon had been studied prior to the 1980s. In fact, it now seems likely that a "miner's disease" of underground miners in central Europe in the 1500s was a radon-induced cancer. In the 1950s, the first detailed studies of the effects of radon on miners in the United States were conducted. These results, in conjunction with some animal studies, have been used to estimate the threats posed by various concentrations of radon.

It appears that it is the radioactive products, that is, the *daughters* of radon, which actually cause cancer. Radon 222 is an alpha emitter with a half-life of 3.82 days. Its decay, as shown in Equation (19.20), produces polonium 218, which itself is an alpha emitter with a half-life of 3.11 min:

$$^{222}_{86}\text{Rn} \xrightarrow{\alpha} {}^{218}_{84}\text{Po} + {}^{4}_{2}\text{He} \tag{19.20}$$

Because radon gas is continually breathed in and out, it does not stay in contact with the tissues of the respiratory tract long enough to cause significant damage. Polonium 218 and other radioactive daughters, however, are not gases and adhere to the surfaces of the lungs and airway passages. Here, they cause mutations in DNA and are carcinogenic. Studies have revealed that smokers are at a much higher risk due to radon than nonsmokers, probably because radon daughters adhere to smoke particles that linger in the air and eventually are likely to be deposited in the lungs or the upper respiratory tract.

How do you know if there is a dangerous amount of radon in a given home? Unfortunately, this needs to be done individually for each house because soil composition, type of construction, and the amount of ventilation varies considerably from house to house even within the same neighborhood. To determine radon levels, several different, effective, and usually inexpensive test kits are available. (Some of these directly collect radon decay products, while others provide a medium upon which alpha particles leave a track.) No matter which of these kits are used, however, it is important to realize that, even in a given house, radon levels are highly variable. For example, they are highly dependent upon such factors as (1) the amount of available ventilation that in turn depends upon weather and seasonal factors, (2) the floor level at which the measurements are carried out, (3) the amount of moisture in the soil, and (4) outdoor and indoor temperatures. Because of these variables, radon measurements must be carried out using proper sampling techniques.

Once a reliable level of radon is determined, radon concentrations may need to be reduced to a safe level. This can be done either by altering the amount of radon that is able to seep into a home through basement walls and/or by increasing the ventilation that carries radon out. The best method is the sealing of cracks and crevices in basement walls that allow the gas to diffuse into the home. In some cases this is easy to do, but in others, due to the varying porosity of foundation materials, it is more difficult. Increasing ventilation in the house also helps to reduce the amounts of the gas to acceptable levels. For example, replacing the air in a given section of a dwelling every 4 days allows radon levels to build up to only one-half of their maximum value for that area. Ventilating the area completely each day, on the other hand, cuts radon levels to roughly one-sixth of the maximum.

SUMMARY

The noble gases were all discovered during the last decade of the nineteenth century. Lord Rayleigh and William Ramsay isolated a new constituent of the air which they called argon in 1894. Helium was first observed in the solar spectrum but was isolated from a uranium mineral by Ramsay in 1895. Krypton, neon, and xenon were obtained by the fractionation of liquid air by Ramsay and Travers in 1898. Dorn discovered radon as a radioactive decay product of radium in 1900, but Ramsay isolated and determined the density of this highly radioactive gas in 1908.

The properties of the noble gases are exactly as expected on the basis of effective nuclear charge. They have high ionization energies, low electron affinities, and very low melting and boiling points. Helium II is a superfluid that exists below 2.2 K. The inertness of the elements provided a crucial link in the empirical and theoretical understanding of the still-young periodic table.

In the first six decades of this century, there were various unsuccessful attempts to get the noble gases to react. The hydrates, clathrates, and various unstable cations were prepared during that time, but only in 1962 did Bartlett prepare a genuinely stable compound of xenon. Having already characterized an oxygenyl compound made from the reaction of diatomic oxygen and platinum hexafluoride, he suspected that xenon might form an analogous compound. His experiments confirmed his hypothesis and demonstrated that Group 8A could no longer be classified as inert.

The xenon fluorides were synthesized by the direct reaction of the elements shortly after Bartlett's announcement. The tetrafluoride is the most stable and best characterized, although its hydrolysis yields the dangerously explosive xenon trioxide. Xenon difluoride is an excellent oxidizing agent and fluoride ion donor. Xenon hexafluoride is both a fluoride donor and acceptor.

For the most part, the structures of noble gas compounds follow readily from VSEPR theory. In the gas phase, xenon hexafluoride assumes a nonrigid, slightly distorted octahedral structure even though it has seven electron pairs around the central xenon atom. In the solid, it is characterized by square pyramidal XeF_5^+

cations bridged by fluoride ions. XeF_8^{2-} is a square antiprism, and $XeOF_4$ is nearly a perfect square pyramid.

Other xenon halides include the dichloride, the tetrachloride, and the dibromide, but these are not particularly stable. Solutions of xenon trioxide, called *xenic acid*, are excellent oxidizing agents as is the octahedral perxenate anion, XeO_6^{4-}. Krypton difluoride, a few nitrogen compounds of both xenon and krypton, and radon difluoride have also been prepared but are not well characterized.

The noble gases are in scarce supply on the earth. Argon is the most abundant because it is the product of the beta decay of potassium 40. Alpha particles are just helium nuclei and are the principal source of helium which concentrates in natural gas deposits. Although not retained by the gravitational field of the earth, helium is the second most abundant element in the cosmos and plays a central role in the proton-proton cycle of stars and nucleosynthesis.

Argon and helium are used to provide an inert atmosphere in incandescent light bulbs, arc welding, and in the preparation of various reactive elements. Helium is also used in the artificial atmosphere for divers. Because of its very low boiling point, helium is used in cryogenics and studies of superconductivity. Krypton now provides the standard for the meter, and lights filled with neon make the latter the best known of the noble gases.

Starting with discoveries made in the mid-1980s, radon gas is now recognized as a national environmental health problem. An alpha decay product of radium 226, radon 222 is released by rocks, soil, and groundwater and makes its way into the basements of homes. The radioactive products of radon, including polonium 218, are implicated as the actual carcinogens. They become lodged in the respiratory tract, particularly in smokers, where their alpha decay produces DNA mutations. Radon test kits must be used with proper sampling techniques as the concentrations of this gas are highly dependent on a number of factors. Methods of reducing radon concentrations include sealing basements and/or increasing the amount of ventilation in a house.

PROBLEMS

19.1. Gold and silver, which do not corrode or deteriorate easily, are often numbered among the noble metals. How is this designation related to the group name for helium, neon, argon, krypton, xenon, and radon?

19.2. Oxygen can be removed from air by reaction with white phosphorus. Write an equation for this process.

19.3. Water vapor can be removed from air by absorbing it with magnesium perchlorate. Write an equation for this process.

19.4. Carbon dioxide can be removed from air by absorbing it with sodium hydroxide. Write an equation for this process.

19.5. What is the ratio of the atomic weight of oxygen to hydrogen? How close is this to an integral ratio as required by Prout's hypothesis?

19.6. Calculate the average molecular weight of air and compare your result to that obtained by Ramsay for his unreacted one-eightieth part of air.

19.7. Ramsay eliminated nitrogen from his atmospheric samples by reacting it with hot magnesium. Write an equation to represent this reaction.

19.8. Ramsay calculated atomic weights from the densities he measured. Given the density of argon to be 1.78 g/liter at standard temperature and pressure, calculate its atomic weight.

19.9. Given a density of 3.70 g/liter for krypton at standard temperature and pressure, calculate the atomic weight of this element.

19.10. Explain why the ionization energies of the noble gases decrease going down the group.

19.11. Explain why the electron affinities of the noble gases increase in value going down the group.

19.12. If the effective nuclear charges in the noble gases are so great, why do they not have a great attraction for free electrons?

19.13. Using arguments based on effective nuclear charge, explain why the octet of valence electrons of neon ($2s^22p^6$) is such a stable configuration.

19.14. What type of forces would operate in the xenon hydrates? Would these forces include either ionic or covalent bonds involving xenon atoms? Briefly explain.

19.15. Uranium hexafluoride, UF_6, is not usually prepared by the direct reaction of the elements. Suggest a method for synthesizing this compound. (*Hint*: Recall the discussions of halide synthesis in Section 18.2.)

19.16. Which diffuses faster, $^{235}UF_6(g)$ or $^{238}UF_6(g)$? Calculate a ratio of rates of diffusion for these two gases. (*Hint*: The mass numbers of the uranium isotopes may be used as close approximations of their atomic weights.)

19.17. Analyze the change in oxidation numbers in the reaction between diatomic oxygen and platinum hexafluoride. Identify the reducing and oxidizing agents in this reaction.

19.18. Would you suspect that radon would be more or less reactive toward platinum hexafluoride than xenon is? Briefly justify your answer.

19.19. In Bartlett's original letter to the British journal *Nature*, he mentioned that he thought it probable that xenon would also react with ruthenium and rhodium hexafluorides. He was shortly proven correct. Write equations which might represent these reactions.

19.20. How might you prepare mercuric fluoride using xenon tetrafluoride? Write an equation as part of your answer.

19.21. Potassium fluoride can be prepared from potassium iodide and xenon tetrafluoride. Write an equation for this reaction.

19.22. Analyze the change in oxidation numbers for the hydrolysis of xenon tetrafluoride found in Equation (19.5), repeated below. Identify the oxidizing and reducing agents in this reaction.

$$6XeF_4 + 12H_2O \longrightarrow 2XeO_3 + 4Xe + 3O_2 + 24HF$$

19.23. Analyze the change in oxidation numbers for the hydrolysis of xenon difluoride found in Equation (19.8), repeated below. Identify the oxidizing and reducing agents in this reaction.

$$2XeF_2 + 2H_2O \longrightarrow 2Xe + O_2 + 4HF$$

19.24. Write balanced equations for the reactions in which xenon difluoride is used to oxidize (*a*) Ce(III) to Ce(IV) and (*b*) Ag(I) to Ag(II).

* **19.25.** When XeF_2 is added to antimony pentafluoride, a strong fluoride acceptor, a solution of $XeF^+Sb_2F_{11}^-$ is produced. When xenon gas is then added to this solution in SbF_5, it turns a bright-green color characteristic of the Xe_2^+ cation. Write equations corresponding to the above two reactions.

* **19.26.** Bartlett's original reaction between xenon and platinum hexafluoride is now known to also produce the bridged $[Pt_2F_{11}^-]$ species. Draw a diagram to represent the structure of this anion.

19.27. Determine whether or not the tenets of the VSEPR theory are consistent with the geometries of the noble gas molecules found in parts (*a*), (*c*), (*e*), and (*g*) of Table 19.2.

19.28. Determine whether or not the tenets of the VSEPR theory are consistent with the geometries of the noble gas molecules and anions found in parts (*b*), (*d*), (*f*), and (*h*) of Table 19.2.

***19.29.** Give a possible rationale for the fact that the F–Xe–O angle of $XeOF_4$ is not more substantially compressed from 90°.

19.30. Draw a Lewis structure and a diagram representing the geometry of the XeF_5^+ cation. Estimate the values of all bond angles.

19.31. Xenon hexafluoride cannot be handled in a glass apparatus because of the reactions represented in the following equations:

$$2XeF_6 + SiO_2 \longrightarrow 2XeOF_4 + SiF_4$$

$$2XeOF_4 + SiO_2 \longrightarrow 2XeO_2F_2 + SiF_4$$

$$2XeO_2F_2 + SiO_2 \longrightarrow 2XeO_3 + SiF_4$$

Draw diagrams showing the geometry of all the xenon-containing compounds in these reactions.

19.32. Draw a Lewis structure and a diagram representing the geometry of the XeO_3F_2 molecule.

19.33. Draw a Lewis structure and a diagram representing the geometry of the $XeOF_2$ molecule.

19.34. When the iodine 129 contained in ICl_2^- decays by beta minus emission, what compound would result? Write an equation as part of your answer. Draw a geometrically accurate diagram representing the molecular formula of the product.

19.35. When the iodine 129 contained in $KICl_4 \cdot 2H_2O$ decays by beta minus emission, what compound would result? Write an equation as part of your answer. Draw a geometrically accurate diagram representing the molecular formula of the product.

19.36. Identify the reducing and oxidizing agents in the oxidation of xenon trioxide to sodium perxenate by ozone in base solution as shown in Equation (19.19), repeated below.

$$XeO_3 + 12NaOH(aq) + O_3 \longrightarrow Na_4XeO_6 + 6H_2O$$

19.37. The perxenate-to-xenon standard reduction potential in acid solution is 2.18 V.
(*a*) Write a half-equation corresponding to this potential.
(*b*) Write a balanced equation for the oxidation of Mn(II) to permanganate by perxenate in acid solution.

19.38. The perxenate-to-xenon standard reduction potential in acid solution is 2.18 V. Write a balanced equation for the oxidation of Cr(III) to dichromate by perxenate in acid solution. Calculate the $E°$ of this reaction.

19.39. Write an equation corresponding to the beta decay of potassium 40 to an isotope of argon.

19.40. Write an equation that represents the nuclear process involved when clevite produces helium gas.

19.41. Why does radon gas tend to collect in the basement of a house? Put your answer into one well-constructed paragraph.

19.42. Write an equation for alpha decay of polonium 218.

19.43. Since the status of such environmental problems as the carcinogenic threat of radon changes rapidly, this text will not contain the latest story. Summarize, in one well-written paragraph, the current status of radon as an environmental health hazard.

19.44. In one well-written paragraph, summarize why smokers are at a greater risk to the hazards of radon gas than nonsmokers.

A WORD ABOUT SOURCES

In the course of writing any textbook, an author must necessarily draw upon the years of experience teaching the topic, in this case inorganic chemistry. As this author has developed a style of teaching over nearly a quarter of a century, he has (long before the opportunity came to write a textbook) always tried to keep current with the original literature and the many textbooks in the field. These articles and monographs have inevitably been the sources, direct and indirect, of the content of the pedagogy of both his courses and this book.

While this introduction to coordination, solid state, and descriptive inorganic chemistry often takes a different tact (with its emphasis in such areas as history, applications, and the network of interconnected ideas) in presenting inorganic chemistry to the post-introductory chemistry student audience, virtually all of the content of the book is more or less well-known inorganic chemistry or history of science. To specifically acknowledge the many sources of this content would be both cumbersome and difficult, and even then incomplete. Nevertheless, some general acknowledgment of these sources is in order. The following is an attempt to group them rather arbitrarily into four divisions: general inorganic chemistry, applications, history, and references by chapter.

General Inorganic Chemistry

Synthesis and Technique in Inorganic Chemistry, R. J. Angelici, Saunders, Philadelphia (1977).

Inorganic Chemistry, Principles and Applications, Ian S. Butler and John F. Harrod, Benjamin/Cummings, Redwood City, CA (1989).

Chemistry, Raymond Chang, 4th ed. McGraw-Hill, New York (1991).

Advanced Inorganic Chemistry, F. A. Cotton and G. Wilkinson, Wiley, New York; 4th Ed. (1980) and 5th ed. (1988).

Basic Inorganic Chemistry, F. A. Cotton, G. Wilkinson, and P. Gaus, 2nd ed. Wiley, New York (1987).

Concepts and Models of Inorganic Chemistry, Bodie Douglas, Darl H. McDaniel, and John J. Alexander, 2nd ed. Wiley, New York (1983).

Chemistry of the Elements, N. N. Greenwood and A. Earnshaw, Pergamon Press, Oxford (1984).

Inorganic Chemistry, Principles of Structure and Reactivity, James E. Huheey, 3rd ed. Harper Row, New York, (1983).

Structural and Comparative Inorganic Chemistry, Peter R. S. Murray and P. R. Dawson, Heinemann Educational Books, London (1976).

Introduction to Modern Inorganic Chemistry, K. M. Mackay and R. Ann Mackay, 2nd ed. Intext Educational Publishers, New York (1973).

Inorganic Chemistry, A Modern Introduction. Therald Moeller, Wiley, New York (1982).

Modern University Chemistry, Norbert T. Porile, Harcourt Brace Jovanovich, San Diego (1987).

Inorganic Chemistry, A Unified Approach, William M. Porterfield, Addison-Wesley, Reading, MA (1984).

An Introduction to Inorganic Chemistry, Keith A. Purcell and John C. Kotz, Saunders, Philadelphia (1980).

Inorganic Chemistry, Keith A. Purcell and John C. Kotz, Saunders, Philadelphia (1977).

Inorganic Chemistry, A. G. Sharpe, 2nd ed., Longman, London (1981).

Inorganic Chemistry, Duward Shriver, P. W. Atkins, Cooper, H. Langford, Freeman, New York (1990).

Principles of Descriptive Inorganic Chemistry, Gary Wulfsberg, Brooks/Cole, Monterey, CA (1987).

Applications

Handbook on Toxicity of Inorganic Compounds, edited by Hans G. Seiler and Helmut Sigel with Astrid Sigel, Marcel Dekker, New York (1988).

Descriptive Chemistry, Donald A. McQuarrie and Peter A. Rock, Freeman, New York (1985).

Extraordinary Origins of Everyday Things, Charles Panati, Harper Row, New York (1987).

Modern Descriptive Chemistry, Eugene G. Rochow, Saunders, Philadelphia (1977).

McGraw-Hill Encyclopedia of Science and Technology, 6th ed. McGraw-Hill, New York (1987).

Encyclopedia of Physical Science and Technology, Robert A. Meyers, ed., Academic Press, Harcourt, Brace Jovanovich, Orlando, FL (1987).

History

Humour and Humanism in Chemistry, John Read, G. Bell and Sons, London (1947).

The Development of Modern Chemistry, Aaron J. Ihde, Dover, New York (1984).

Crucibles: The Story of Chemistry, from Ancient Alchemy to Nuclear Fission, Bernard Jaffe, 4th ed., Dover, New York (1976).

Discovery of the Elements, Mary E. Weeks, edited, with a chapter on elements discovered by atomic bombardment, by Henry M. Leicester, 6th ed., *Journal of Chemical Education*, Easton, PA (1956).

Asimov's Biographical Encyclopedia of Science and Technology, Isaac Asimov, 2d ed., Doubleday, Garden City, NY (1982).

"The Elements," C. R. Hammond, *Handbook of Chemistry and Physics*, 58th ed., Robert C. Weast, ed., CRC Press, West Palm Beach, FL (1978).

References by Specific Chapters

Chapter 1

A History of the International Chemical Industry, Fred Aftalion (translated by Otto Theodor Benfey), University of Pennsylvania Press, Philadelphia (1991).

Chapters 2 to 4

Coordination Chemistry, Fred Basolo and Ronald C. Johnson, Science Reviews, Wilmington, DE (1986).

Introduction to Ligand Fields, B. N. Figgis, Interscience, New York (1966).

An Introduction to Transition-Metal Chemistry: Ligand-Field Theory, Leslie E. Orgel, Butler and Tanner, London (1960).

Chapter 5

Inorganic and Organometallic Reaction Mechanisms, Jim D. Atwood, Brooks/Cole, Monterey, CA (1985).

Mechanisms of Inorganic Reactions: A Study of Metal Complexes in Solution, Fred Basolo and Ralph G. Pearson, Wiley, New York (1958).

Chapter 6

"Search for More Metal-Containing Antitumor Agents Intensifies," *Chemical and Engineering News*, *64*(40), 21–22 (1986).

"Lead Poisoning," J. Julian Chisholm, *Scientific American*, *224*(8), 15–23 (1971).

"Lead Poisoning and the Fall of Rome," S. C. Gilfillan, *Journal of Occupational Medicine*, *7*(2), 53–60 (1965).

"Mercury and the Environment," Leonard J. Goldwater, *Scientific American*, *224*(5), 15–21, (1971).

Biochemistry, Donald Voet and Judith G. Voet, Wiley, New York (1990).

Chapter 7

"Predictions of Crystal Structure Based on Radius Ratio," Lawrence C. Nathan, *Journal of Chemical Education*, *62*(3), 215–218 (1985).

Structural Inorganic Chemistry, A. F. Wells, 4th ed. Clarendon Press, Oxford (1975).

Chapter 8

"Energy Cycles," G. P. Haight, Jr, *Journal of Chemical Education*, *45*(6), 420–422 (1968).

"Reappraisal of Thermochemical Radii for Complex Ions," H. D. B. Jenkins and K. P. Thakur, *Journal of Chemical Education*, *56*(9), 576–577 (1979).

"Enthalpy Cycles in Inorganic Chemistry," Jan Lutzow Holm, *Journal of Chemical Education*, *51*(7), 460–463 (1974).

Chapter 9

"Reexamining the Diagonal Relationships," Timothy P. Hanusa, *Journal of Chemical Education*, *64*(8), 686–786 (1987).

Chapter 10

"The Search for Tritium—the Hydrogen Isotope of Mass Three," M. L. Eidinoff, *Journal of Chemical Education*, *25*(1), 31–34 (1948).

"Heavy Water," P. W. Selwood, *Journal of Chemical Education*, *18*(11), 515–520 (1941).

Chapter 11

"Global Climatic Change," Richard A. Houghton and George M. Woodwell," *Scientific American*, *260*(4), 36–44 (1989).

"Joseph Priestley, Enlightened Chemist," David J. Rhees, Center for the History of Chemistry, Publication No. 1 (1983).

"Making Sense of the Nomenclature of the Oxyacids and Their Salts," Glen E. Rodgers, Harold M. State, and Richard L. Bivens, *Journal of Chemical Education*, *64*(5), 409–410 (1987).

Chapter 12

"Lithium Treatment of Manic-Depressive Illness: Past, Present, and Perspectives," Schou Mogens, *Journal of the American Medical Association*, *259*(12), 1834–1836 (1988).

Safe Storage of Laboratory Chemicals, edited by David A. Pipitone, Wiley-Interscience, New York (1984).

Chapter 13

"Chemistry of Fireworks," John A. Conkling, *Chemical and Engineering News*, *59*(26): 24–32 (1981).

"Beryllium and Berylliosis," Jack Schubert, *Scientific American*, *199*(2), 27–33 (1958).

"Marie Curie's Doctoral Thesis: Prelude to a Nobel Prize, Robert L. Wolke, *Journal of Chemical Education*, *65*(7), 561–573 (1983).

Chapter 14

Electron Deficient Compounds, K. Wade, Nelson, London (1971).

Chapter 15

"Lead Poisoning in Children," Robert L. Boeckx, *Analytical Chemistry*, *58*(2), 274A–287A (1986).
"Optimal Materials," A. M. Glass, *Science*, *235*, 1003–1009 (1987).
"Asbestos," W. J. Smither, *School Science Review*, *60*(210), 59–69 (1978).

Chapter 16

"A History of the Match Industry," M. F. Crass, Jr., *Journal of Chemical Education*, *18*(3), 116–120 (1941).
Phosphorus Chemistry in Everyday Living, Arthur D. F. Toy and Edward N. Walsh, 2d ed. American Chemical Society, Washington, D.C. (1987).

Chapter 17

"Ion Models Nitrogenase N_2-reduction Site," *Chemical and Engineering News*, *66*(32), 25 (1988).
"Removal of NO_x and NO_2 from Flue Gas Using Aqueous Emulsions of Yellow Phosphorus and Alkali," David K. Liu, Di-Xin Shen, and Shih-Ger Chang, *Environmental Science and Technology*, *25*(1), 55–60 (1991).
Sulfur, Energy and the Environment, Beat Meyer, Elsevier Scientific, Amsterdam (1977).
"The Challenge of Acid Rain," Volker A. Mohnen, *Scientific American*, *259*(22), 30–39 (1988).
"Extending the Bouncing Life of Tennis Balls," K. M. Reese, *Chemical and Engineering News*, *66*(9), 44 (1988).

Chapter 18

"Chlorofluorocarbons and Stratospheric Ozone," Scott Elliott and F. S. Rowland, *Journal of Chemical Education*, *64*(5), 387–391 (1987).
"The Blowing of the Ozone Shell," Joe Farman, *New Scientist*, 12 November 1987, pp. 50–54.
"Fluoridation of Water," Bette Hileman, *Chemical and Engineering News*, *66*(31), 26–42 (1988).
"In Search of the Safe CFCs," Meirion Jones, *New Scientists*, 26 May 1988, pp. 56–60.
"Winds, Pollutants Drive Ozone Hole," Richard A. Kerr, *Science*, *238*, 156–158 (1987).
"Fluorine," Richard H. Langley and Larry Welch, *Journal of Chemical Education*," *60*(9), 759–761 (1983).
"Clouds Without a Silver Lining," Richard Monastersky, *Science News*, *134*(16), 249–251 (1988).
"Arctic Air Primed to Destroy Ozone," R. Monastersky, *Science News*, *135*(8), 116 (1989).
"The Naming of Fluorine," William H. Waggoner, *Journal of Chemical Education*, *53*(1), 27 (1976).

Chapter 19

"Radon Tagged as Cancer Hazard by Most Studies, Researchers," David J. Hanson, *Chemical and Engineering News*, *67*(3), 7–13 (1989).
"A Decade of Xenon Chemistry," G. J. Moody, *Journal of Chemical Education*, *51*(10), 628–630 (1974).
"The Noble Gases and the Periodic Table," John H. Wolfenden, *Journal of Chemical Education*, *46*(9), 456–576 (1969).

CREDIT REFERENCES

1. From F. Basolo and R. C. Johnson, *Coordination Chemistry*, 2nd ed., p. 6 © 1986. Used by permission of Science Reviews Ltd., Northwood, Middlesex, England.
2. From Bodie Douglas, Darl H. McDaniel, and John J. Alexander, *Concepts and Models of Inorganic Chemistry*, 2nd ed., p. 305. Copyright © 1983. Reprinted by permission of John Wiley & Sons., Inc.
3. From Audrey Companion, *Chemical Bonding*. © 1979. Used by permission of Prof. Companion.
4. From F. A. Cotton and G. Wilkinson, *Advanced Inorganic Chemistry*, 4th ed., p. 643. Copyright © 1980. Reprinted by permission of John Wiley & Sons, Inc.
5. Figure from *Synthesis and Technique in Inorganic Chemistry*, Second Edition, by Robert J. Angelici, copyright © 1977 by Saunders College Publishing, reprinted by permission of the publisher.
6. From B. N. Figgis, *Introduction to Ligand Fields*. © 1966. Used by permission of the author.
7. Reprinted with permission from D. W. Margerone et al., *Coordination Chemistry*, vol. 2, A. E. Martell, ed., *ACS Monograph* #178. Copyright 1978 American Chemical Society.
8. Reprinted with permission from H. J. Buser, *Inorganic Chemistry*, vol. 11, p. 2704. Copyright 1977 American Chemical Society.
9. From M. F. Perutz et al., "Structure of Hemogolin," in *Nature*, vol. 185 (February 13, 1960), fig. 8, p. 420, as used in *Inorganic Chemistry*, *Principles of Structure and Reactivity*, 3rd ed., by James E. Huheey. Copyright © by James Huheey. Reprinted by permission of HarperCollins, Publishers, Inc.
10. From D. Voet and J. G. Voet, *Biochemistry*, p. 211. Copyright © 1990. Reprinted by permission of John Wiley & Sons, Inc.
11. Figure from *The Joy of Chemistry* by Stanley M. Cherim and Leo E. Kallan, copyright © 1976 by Saunders College Publishing, reprinted by permission of the publisher.
12. From R. E. Dickerson, Harry B. Gray, and Gilbert P. Haight, Jr., *Chemical Principles*, 3rd ed. © 1979. Used by permission of Benjamin/Cummings Publishing Company.
13. From Norbert T. Porile, *Modern University Chemistry*, in press. Used by permission of McGraw-Hill, Inc.
14. Reprinted with the permission of Macmillan Publishing Company from *General Chemistry: Principles of Modern Applications*, Fifth Edition, by Ralph H. Petrucci. Copyright © 1989 by Macmillan Publishing Company.
15. From Harry B. Gray, *Chemical Bond: An Introduction to Atomic and Molecular Structure*. © 1973. Used by permission of Benjamin/Cummings Publishing Company.

16. From Darrell D. Ebbing, *General Chemistry*, 1st ed. Copyright © 1984 by Houghton Mifflin Company. Used with permission.
17. From B. D. Culity, *Elements of X-Ray Diffraction*, 2nd ed. © 1978. Used by permission of Addison-Wesley Publishing Company.
18. From *Inorganic Chemistry, Principles of Structure and Reactivity*, 4rd ed., by James E. Huheey. Copyright © by James E. Huheey. Reprinted by permission of HarperCollins, Publishers, Inc.
19. From W. W. Porterfield, *Inorganic Chemistry: A Unified Approach*, Second Edition, Academic Press, San Diego, 1993. Copyright © 1993. Reprinted by permission of Academic Press.
20. K. M. Mackay and R. A. Mackay, *Introduction to Modern Inorganic Chemistry*, 4th ed., © 1989, p. 76. Reprinted by permission of Prentice Hall, Englewood Cliffs, New Jersey.
21. From Raymond Chang, *Chemistry*, 3rd ed. Copyright © 1988. Used with permission of McGraw-Hill, Inc.
22. From C. H. Langford and R. A. Beebe, *The Development of Chemical Principles*. © 1969. Used by permission of Addison-Wesley Publishing Company.
23. From Steven S. Zumdahl, *Chemistry*, 2nd ed. © 1989, Used by permission of D. C. Heath & Co.
24. Table from *Modern Descriptive Chemistry* by Eugene G. Rochow, copyright © 1977 by Saunders College Publishing, reprinted by permission of the publisher.
25. From Peter R. S. Murray and P. R. Dawson, *Structural and Comparative Inorganic Chemistry*, Heineman Educational Books, p. 139, Fig. 14.4 © 1976. Used by permission of Peter R. S. Murray.
26. Reprinted by N. N. Greenwood and A. Earnshaw, *Chemistry of the Elements*, Copyright 1984, with kind permission from Pergamon Press Ltd., Headington Hill Hall, Oxford OX3 OBW, UK.
27. From A. G. Sharpe, *Inorganic Chemistry*, 2nd ed. © 1981. Used by permission of Longman Group UK Ltd.
28. From Kenneth Wade, *Electron Deficient Compounds*, p. 51. © 1971. Used by permission of Thomas Nelson & Sons Ltd.
29. From E. Weiss and E. A. C. Lucken in *Journal of Organometallic Chemistry*, vol. 2 (1964), p. 197. © 1964. Used by permission of Elsevier Sequoia S. A.
30. From Robert F. Curl and Richard Smalley, "Fullerenes," in *Scientific American*, vol. 265, no. 4 (1991), p. 54. Used by permission of Prof. Richard Smalley.
31. Used by permission of Culligan International Co.
32. From John Read, *Humor and Humanism in Chemistry*, G. Bell & Sons, Ltd., London, 1947.
33. Illustration by Bob Conrad from Volker A. Mohnen, "The Challenge of Acid Rain." Copyright © 1993 by Scientific American, Inc. All rights reserved.

INDEX

P

S

Network Figures

Tables of Fundamental Properties of the Representative Elements

(For each element, each table includes the symbol, atomic number, natural isotopes, total number of isotopes, atomic mass, valence electrons, mp/bp, density, atomic and Shannon-Prewitt ionic radius, electronegativity, charge density, standard reduction potential, oxidation states, ionization energy, electron affinity, discoverer and date of discovery, acid/base character of oxide, crystal structure, and reaction products with oxygen, nitrogen, halogens, and hydrogen.)

Other Useful Constants

c, velocity of light $= 2.998 \times 10^{10}$ m/sec
h, Planck's constant $= 6.626 \times 10^{-34}$ J-sec
N, Avogadro's number $= 6.022 \times 10^{23}$ mol^{-1}
e, elementary charge $= 1.602 \times 10^{-19}$ C